Fundamental Theories of Physics

Volume 206

The international monograph series "Fundamental Theories of Physics" aims to stretch the boundaries of mainstream physics by clarifying and developing the theoretical and conceptual framework of physics and by applying it to a wide range of interdisciplinary scientific fields. Original contributions in well-established fields such as Quantum Physics, Relativity Theory, Cosmology, Quantum Field Theory, Statistical Mechanics and Nonlinear Dynamics are welcome. The series also provides a forum for non-conventional approaches to these fields. Publications should present new and promising ideas, with prospects for their further development, and carefully show how they connect to conventional views of the topic. Although the aim of this series is to go beyond established mainstream physics, a high profile and open-minded Editorial Board will evaluate all contributions carefully to ensure a high scientific standard.

More information about this series at https://link.springer.com/bookseries/6001

Teiji Kunihiro · Yuta Kikuchi · Kyosuke Tsumura

Geometrical Formulation of Renormalization-Group Method as an Asymptotic Analysis

With Applications to Derivation of Causal Fluid Dynamics

Springer

Teiji Kunihiro
Yukawa Institute for Theoretical Physics
Kyoto University
Sakyo-ku, Kyoto, Japan

Yuta Kikuchi
Brookhaven National Laboratory
Upton, NY, USA

Kyosuke Tsumura
Analysis Technology Center
Fujifilm Corporation
Minamiashigara, Kanagawa, Japan

ISSN 0168-1222 ISSN 2365-6425 (electronic)
Fundamental Theories of Physics
ISBN 978-981-16-8188-2 ISBN 978-981-16-8189-9 (eBook)
https://doi.org/10.1007/978-981-16-8189-9

This Springer imprint is published by the registered company Springer Nature Singapore Pte Ltd.
The registered company address is: 152 Beach Road, #21-01/04 Gateway East, Singapore 189721, Singapore

Preface

The purpose of this monograph is twofold: The one is to present a comprehensive account of the so-called Renormalization-Group (RG) method and its extension, which we named the doublet scheme, in a geometrical point of view with various examples. The second is the application of the doublet scheme in the RG method to the derivation of the so-called second-order (causal) fluid dynamics from the relativistic as well as non-relativistic Boltzmann equations with quantum statistics. All the contents of the monograph are virtually based on the authors original work except for some review parts.

The RG method is a global and asymptotic analysis of differential equations, which was developed by an Illinois group and others some thirty years ago. We introduce the method in a purely mathematical way on the basis of the classical theory of envelopes, a notion in elementary differential geometry.

Then a focus is put on the fact that the RG method provides us with a powerful and transparent method for the reduction of dynamics, which includes an elementary method of a construction of the invariant/attractive manifolds and reduced dynamical equations written in terms of the variables that constitute the coordinates of the manifolds. They are the key concepts in the reduction theory of dynamical systems, and thereby naturally lead to a foundation to existing theories in the specific physical systems such as Krylov-Bogoliubov-Mitropolsky theory for non-linear oscillators and Kuramoto's reduction theory for evolution equations. Examples treated include stochastic equations like Langevin and Fokker-Planck equations.

Although the RG method is applicable to discrete systems and thereby provides us with an optimized discretization scheme of differential equations, we have omitted the once prepared part on the discrete systems partly because of its irrelevance to the second part and partly for the sake of making the already lengthy monograph too voluminous. We hope that there will be a chance to publish the omitted part somewhere else in the future.

The usual reduction theory including the RG method based on the perturbation theory utilizes the zero modes of the linear operator in the unperturbed equation. However, in the derivation of the so-called causal second-order fluid dynamics, which has a nature of the mesoscopic dynamics of the given system, from the Boltzmann equation, one needs to extend the invariant manifold so as to incorporate appropriate excited modes as the additional coordinate variables. In this monograph, a general reduction theory is presented for constructing the mesoscopic dynamics with inclusion of appropriate excited modes from the given microscopic equation, which is formulated as an extension of the RG method and called the doublet scheme.

In the second part, we work out for the derivation of the second-order or causal fluid dynamics on the basis of the doublet scheme: Thus we obtain not only the fluid dynamic equations that are uniquely those in the energy frame but also the microscopic formulae of the transport coefficients and the relaxation times in Kubo-like formulae that admit natural physical interpretations. It is shown that the resultant fluid dynamics is not only causal but also stable dissipative (non-)relativistic fluid dynamics. The derivation of the mesoscopic dynamics beyond the fluid dynamics in the non-relativistic regime is also one of the hot topics in physics of cold atoms. The present monograph includes numerical analyses on these interesting physical systems. We also provide an accurate and efficient numerical method for computing the transport equations and relaxation times using the microscopic expressions. The numerical method utilizes the double exponential formulae for integrations and the direct matrix inversion method. The numerical calculations are fully worked out for typical model systems composed of classical particles, a fermion system with a Yukawa interaction, and boson system described by a chiral Lagrangian. The numerical calculations are also presented for the non-relativistic system and a critical comparison is made with the relaxation-time approximation, which is commonly used in the current literature.

The presentation of the monograph, at least in the first part, is intended to be as pedagogical as possible so that not only researchers who are not familiar with the RG theory in physics but also motivated undergraduate students with mathematical backgrounds such as introductory calculus including linear differential equations and linear algebra may appreciate and understand the method.

Conversely, the monograph is not intended to be a systematic review of the current status of the studies of the RG method and causal fluid dynamics either with respect to its foundations or applications. Since the literature on these subjects is huge now that making a systematic review on this subject is not the intention of the authors and beyond the authors' ability. Therefore we apologize to those whose important articles are not cited in this monograph in advance.

Sakyo-ku, Kyoto, Japan
Upton, NY, USA
Minamiashigara, Kanagawa, Japan
June 2021

Teiji Kunihiro
Yuta Kikuchi
Kyosuke Tsumura

Acknowledgements

One of the authors (T.K.) is grateful to Glenn Paquette who gave a nice seminar talk on the Illinois Renormalization-Group (RG) method at the Department of Applied Mathematics and Informatics, Ryukoku University, in the end of 1994, which was presented in a pedagogical way mostly with the use of elementary examples; however, honestly speaking, virtually all the audience of the seminar, the major part of which consisted of experts of differential equations/dynamical systems, did not understand what the lecturer was doing. This 'extraordinary' phenomenon may have reflected the fact that the content of the seminar, *i.e.*, the RG method, was so original and new. In any event, the seminar and the subsequent conversation with a couple of other attendants of the seminar strongly motivated T.K. to make an effort to understand what is being done in a purely mathematical way in the Illinois RG method.

He is indebted to Shoji Yotsutani for his question made to him, which prompted him to realize the significance of the 'initial' condition in the formulation of the RG method in a purely mathematical manner. He is also deeply indebted to late Professor Masaya Yamaguti for his interest in his work on the envelope formulation of the RG method and encouragement.

T.K. acknowledges Junta Matsukidaira and Shin-Ichiro Ei and Kazuyuki Fujii and Yoshitaka Hatta for their collaboration on the works on which this monograph is partly based.

We thank Steven Strogatz for his interest in this project to writing this monograph and elucidating questions/comments on a part of the manuscript in the final stage.

Finally, but not least, we are grateful to Yoshiki Kuramoto, Takao Ohta, and Alexander Gorban for their interest in our work on the RG method and encouragement.

Contents

Acronyms

BBGKY	Bogoliubov-Born-Green-Kirkwood-Yvon
CAM	Coherent anomaly method
CE	Chapman-Enskog
EOM	Equation of motion
F-P	Fokker-Planck
IS	Israel-Stewart
KBM	Krylov, Bogoliubov, and Mitropolski
KdV	Kortweg-de Vries
l.h.s.	left-hand side
LHC	Large Hadron Collider
L-P	Lindstedt-Poincaré
MVOC	Method of variation of constants
QCD	Quantum chromodynamics
QFT	Quantum field theory
QGP	QUARK-gluon plasma
r.h.s.	right-hand side
RBE	Relativistic Boltzmann equation
RG	Renormalization-group
RG/E	Renormalization-group/envelope
RHIC	Relativistic Heavy Ion Collider
RTA	Relaxation-time approximation
TDGL	Time-dependent Ginzburg-Landau equation

Chapter 1
Introduction: Reduction of Dynamics, Notion of Effective Theories, and Renormalization Groups

1.1 Reduction of Dynamics of a Simple Equation and the Notion of Effective Theory

Let us start with solving the following simple equation of a damped oscillator in classical mechanics:

$$\ddot{x} + 2\epsilon\dot{x} + x = 0, \tag{1.1}$$

where the positive parameter ϵ describes the strength of the friction, and is assumed to be small, say, less than 1. Actually, we know the exact solution to Eq. (1.1) as

$$x(t) = A(t)\sin\phi(t), \quad (A(t) := \bar{A}\mathrm{e}^{-\epsilon t},\ \phi(t) := \omega t + \bar{\theta}), \tag{1.2}$$

where $\omega := \sqrt{1-\epsilon^2}$ with $\bar{A}$ and $\bar{\theta}$ being constant. One sees that the angular velocity ω becomes small and the amplitude $A(t)$ shows a slow damping along with time owing to the friction ϵ. These are well known facts.

Now, let us dare to solve Eq. (1.1) by applying a simple perturbative expansion:

$$x(t) = x_0(t) + \epsilon x_1(t) + \epsilon^2 x_2(t) + \cdots, \tag{1.3}$$

substitution of which to (1.1) and the subsequent comparison of the coefficients of ϵ^n $(n = 0, 1, \ldots)$ leads to series of equations for $x_n(t)$, with $\ddot{x}_0 + x_0 = 0$. If we take

$$x_0(t) = A\sin(t+\theta), \quad (A,\ \theta;\ \text{constant}) \tag{1.4}$$

as the zeroth order solution, the first-order equation reads

$$Lx_1 = -A\cos(t+\theta), \quad \left(L := \frac{d^2}{dt^2} + 1\right), \tag{1.5}$$

which is formally the equation of motion (Newton equation) for a forced oscillation with the external force with the same frequency as that of the intrinsic one. Then

T. Kunihiro et al., *Geometrical Formulation of Renormalization-Group Method as an Asymptotic Analysis*, Fundamental Theories of Physics 206,
https://doi.org/10.1007/978-981-16-8189-9_1

there arises a resonance phenomenon, which leads to an ever increasing amplitude of the oscillation because no friction is present;

$$x_1 = -At\sin(t+\theta), \tag{1.6}$$

which is proportional to t and called a **secular term**. As will be worked out in Sect. 2.2, the second-order solution $x_2(t)$ contains similar terms, and the perturbative solution up to the second order is given by

$$x(t) = A\sin(t+\theta) - \epsilon At\sin(t+\theta) + \epsilon^2\frac{A}{2}\{t^2\sin(t+\theta) - t\cos(t+\theta)\}, \tag{1.7}$$

which shows that the amplitude of the oscillator increases with a polynomial of time owing to the secular terms. This behavior is quite different from that of the exact solution (1.2) of the damped oscillator, and hence a disaster.

We remark that the appearance of the secular terms are attributed to the fact that the zero modes of the linear operator L constitutes the inhomogeneous term in the higher-order equations.

Nevertheless, one might have recognized that the perturbative solution (1.7) is actually nothing but the first few terms of the expansion of the exact solution in terms of powers of ϵ: Indeed one can make the following manipulation,

$$\begin{aligned} x &\simeq A(1-\epsilon\cdot t+\epsilon^2/2\cdot t^2)\sin((1-\epsilon^2/2)t+\theta) \\ &\simeq A\exp(-\epsilon t)\sin(\sqrt{1-\epsilon^2}t+\theta), \end{aligned} \tag{1.8}$$

where one may have recognized that the secular terms are 'renormalized' into the slowly-varying amplitude and the shifted angular velocity.

Here, we see the typical problems in the (naïve) perturbative expansion of the solution of differential (or dynamical) equations; If the homogeneous equation is expressed as $Lx = 0$ and the linear operator L has zero modes, a naïve perturbation method of the solution gives rise to secular terms, which may give a valid description in a local domain but would lead to an inadequate or even disastrous result in a global domain.

Then the problem can be how to circumvent the appearance of secular terms and/or resum the seemingly divergent perturbation series of the solution. Furthermore, the secular terms appearing in the perturbation theory inherently contain small parameters, and are expected to be 'renormalized' into slow modes such as, say, some amplitudes or phases. Therefore it is most desirable to be able to extract not only the slow variables but also the dynamical equations that describe the slow motions explicitly. Indeed, the amplitude $A(t)$ and the phase $\phi(t)$ in the damped oscillator discussed above satisfy the following simple equations,

$$\frac{dA}{dt} = -\epsilon A, \quad \frac{d\phi}{dt} = 1 - \frac{1}{2}\epsilon^2. \tag{1.9}$$

This means that the intrinsic dynamics in the respective (time) scales is revealed, and also the reduction of the degrees of freedom is achieved. In other words, one may say that the respective **effective theories** in **different energy/time scales** are extracted.

One of the central aims of this monograph is to make an introductory account of the so-called renormalization-group method [1, 2] in a geometrical way [3–6], and thereby present elementary methods to achieve these tasks, *i.e.*, reduction of dynamics and construction of effective theories in a not only systematic but also elementary manner.

1.2 Notion of Effective Theories and Renormalization Group in Physical Sciences

Extracting low-energy slow dynamics with long wave lengths is of fundamental significance in physical sciences since the birth of physics; it was actually one of the essential ingredients of the method developed by Galileo Galiley [7], who invented the method of experimentation in which conjectures are tested by actively modifying Nature, effectively used mathematics, and utilized the notion of *idealization* as is seen in the discovery of the law of inertia that holds when resistance by the environment can be neglected. The last point, which is often overseen, is of essential importance for the success of the method of Galilei, and mostly based on a separation of (energy) scales in modern languages. The recognition of separation of scales constitutes the basis of the notion of (infrared) effective theories and the renormalization-group method in various fields of physical sciences.

The concept of the renormalization group (RG) was introduced by Stuckelberg and Petermann [8] as well as Gell-Mann and Low [9] in relation to an ambiguity in the renormalization procedure of the perturbation series in quantum field theory (QFT). The significance of the RG equation was greatly emphasized by Bogoliubov and Shirkov [10, 11]; see also [12]. However, the essential nature of the RG is non-perturbative [13–16]. Subsequently, as is well known, the machinery of the RG has been applied to various problems in QFT and statistical physics with a great success [17–19].

The essence of the RG in QFT and statistical physics may be stated as follows [13, 14, 18, 19]: Let $\Gamma(\phi, \boldsymbol{g}(\Lambda), \Lambda)$ be the effective action (or thermodynamic potential) obtained by integration of the field variable with the energy scale down to Λ from infinity or a very large cutoff Λ_0. Here $\boldsymbol{g}(\Lambda)$ is a collection of the coupling constants including the wave-function renormalization constant defined at the energy scale at Λ. Then the RG equation may be expressed as a simple fact that the effective action as a functional of the field variable ϕ should be the same, irrespective to how much the integration of the field variable is achieved, *i.e.*, $\Gamma(\phi, \boldsymbol{g}(\Lambda), \Lambda) = \Gamma(\phi, \boldsymbol{g}(\Lambda'), \Lambda')$. If we take the limit $\Lambda' \to \Lambda$, we have

$$d\Gamma(\phi, \boldsymbol{g}(\Lambda), \Lambda)/d\Lambda = 0, \tag{1.10}$$

which is the Wilson RG equation [13], or the flow equation in the Wegner's terminology [14]; notice that Eq. (1.10) is rewritten as

$$\frac{\partial \Gamma}{\partial g} \cdot \frac{dg}{d\Lambda} = -\frac{\partial \Gamma}{\partial \Lambda}. \tag{1.11}$$

If the number of the coupling constants is finite, the theory is called renormalizable. In this case, the functional space of the theory does not change in the flow given by the variation of Λ; one may say that the flow has an invariant manifold.

Owing to the very non-perturbative nature, the RG has at least two merits: (A) Construction of infrared effective actions, (B) Resummation of the perturbation series.

(A) Finding effective degrees of freedom and extract the reduced dynamics of the effective variables in fact have constituted and still constitute the core of various fields of theoretical physics. A notable aspect of the RG is that the RG equation gives a systematic tool for obtaining the infrared effective theories with fewer degrees of freedom than in the original Lagrangian relevant in the high-energy region [13]. This is a kind of reduction of the dynamics.

(B) Applying the RG equation of Gell-Mann-Low type [9–11] to perturbative calculations up to first lowest orders, a resummation in the infinite order of diagrams of some kind can be achieved. That is, the RG method gives a powerful resummation method [12].

An appearance of diverging series is a common phenomenon in every mathematical science not restricted in QFT, and some convenient resummation methods are needed and developed [20–22]. Deducing a slow and long-wave length motion is one of the basic problems in almost all the fields of physics. The problems may be collectively called the reduction problem of dynamics. The RG method might be a unified method for the reduction of dynamics.

1.3 The Renormalization Group Method in Global and Asymptotic Analysis

It was an Illinois group [1] and Bricmont and Kupiainen [23] who showed that the RG equations can be used for a global and asymptotic analysis of ordinary and partial differential equations, and thereby give a reduction theory of dynamical systems of some types.

Whereas the theory of Bricmont and Kupiainen [23] is based on a scaling transformation (block transformation) applied to nonlinear diffusion equations in a rigorous manner, a unique feature of the perturbative approach proposed by the Illinois group [1] is to allow secular terms to appear. Then introducing an intermediate time τ like a renormalization point in QFT, they rewrite the perturbative solution by 'renormalizing' the integral constants reminiscent of the procedure in QFT. Next declaring that the renormalized solution should not depend on τ, they write down a Gell-Mann-Low like equation with respect to τ, and finally mentioning that one is entitled to equate τ to the physical time t, they succeed in obtaining a global solution of differential equations.

Subsequently, one of the present authors [3–5] formulated the RG method in terms of the classical theory of envelopes [24], where a geometrical interpretation was given on the RG method[1] in terms of the classical theory of envelopes in elementary differential geometry. He also formulated a short-cut prescription for the renormalization-group (RG) method without introducing an intermediate time τ, which procedure was adopted in [1] but might have been somewhat mysterious to those who were not familiar with the renormalization group in physics. A detailed account of the geometrical formulation of the RG method is presented in Chap. 4. It will be also emphasized in Chaps. 4 and 5 that the RG method as formulated in [5, 6] and presented in this monograph may be viewed as a natural extension of the asymptotic method by Krylov, Bogoliubov, and Mitropolski for non-linear oscillators [25]. The conventional resummation methods and asymptotic analysis applied to differential equations are reviewed in Chap. 3. The formulation of the RG method in terms of envelope *surfaces* is given in [4], and an asymptotic analysis of partial differential equations such as Barlenblatt equation [26], Swift-Hohenberg equations [27], a damped Kuramoto-Sivashinsky equation [28–31].

It was also elucidated in [5, 6] that the RG method by the Illinois group can be nicely reformulated so that it provides us with a powerful systematic reduction theory of dynamics. In particular, it gives an elementary realization of the geometrical scenario of the reduction of dynamics proposed by Yoshiki Kuramoto [30, 32]: Some time ago, Kuramoto revealed the universal structure of all the existing perturbative methods for reduction of evolution equations [30, 32]; when a reduction of evolution equation is possible, the unperturbed equation admits neutrally stable solutions, and succeeded in describing the reduction of dynamics in a geometrical manner, without recourse to any particular mathematical theory as given in [33, 34].

In Chap. 5, the RG method is reformulated in a non-perturbative way, and then a comprehensive formulation [6] of the reduction theory of dynamics based on the perturbative RG method is given in terms of the notion of attractive/invariant manifold [33, 34]. Then, in this chapter, a fully systematic reduction theory is developed for generic system that contains zero modes in the homogeneous linear operator.[2] A notable point is that the formulation is developed for the case in which the linear operator is not semi-simple and has a Jordan cell structure as well as the semi-simple case [6]; examples of the non semi-simple case include a soliton-soliton interaction described by the KdV equation [6] and the extraction of the final speed of the Benney equation [36], which is treated in Chap. 5 for the first time in the RG method.

---

[1] We shall call the method by the Illinois group simply the renormalization-group method or RG method in short.

[2] We refer to the work by Gorban and Karlin [35], in which they present a unique reduction theory of dynamics with an emphasis on the notion of invariant manifold and show an extensive applications of it to physical and chemical kinetics.

The RG method has been applied to quite a wide class of problems by many authors; see review articles by some of the original authors [2] for some references on the subsequent works in some stage. To mention some; Graham [37] derived a rotationally invariant amplitude equation appearing in the problem of pattern formation. Sasa [38] derived a diffusion type phase equation, and Maruo, Nozaki, and Yoshimori [39] derived Kuramoto-Sivashinsky equation [28, 29]. The discrete RG method based on the notion of the discrete envelopes was developed by Kunihiro and Matsukidaira [40], where a global and asymptotic analysis of discrete systems was discussed and thereby an optimized discretization scheme of differential equations was proposed. The method was applied to analyze asymptotic behavior of the non-linear equations appearing in cosmology [41, 42]. Boyanovsky and de Vega also and their collaborators [43] apply the RG method to discuss anomalous transport and relaxation phenomena in the early universe and quark-gluon plasma (QGP). The RG method was also shown to be a powerful tool to resum divergent perturbation series appearing in problems of quantum mechanics [44–46]. Possible relation between renormalizability and integrability of Hamilton systems was discussed by Yamaguchi and Nambu [47].

As some effort for more rigorously formulate the Illinois RG method, we can refer to [48–52], although there should be more works of importance, in particular by others.

1.4 Derivation of Stochastic Equations and Fluid Dynamic Limit of Boltzmann Equation

Statistical physics, in particular, non-equilibrium statistical physics, is a collection of theories of how to reduce the dynamics of many-body systems to ones with fewer variables, since the work of Boltzmann [53]. The time-*irreversible* Boltzmann equation [55] , which is written solely in terms of the single-particle distribution function for dilute gas systems [54], can be derived from the Bogoliubov-Born-Green-Kirkwood-Yvon (BBGKY) hierarchy [55] , which is equivalent to Liouville equation hence time-*reversible*; Bogoliubov [54] showed that the dilute-gas dynamics as described by the hierarchy of the many-body distribution functions has an *attractive/invariant manifold* [34] spanned by the one-particle distribution function . In fact, a sketch for deriving Euler equation from the Boltzmann equation as the RG equation was given in [56]. In [57], the RG method was applied to derive the Boltzmann equation from the BBGKY hierarchy where the essential importance of the setting of the 'initial' condition was elucidated.

The Boltzmann equation in turn can be further reduced to the fluid dynamic equation (Navier-Stokes equation) by a perturbation theory like Chapman-Enskog method or Bogoliubov's method [25, 54]. In [57], an attempt of the derivation of the Euler equation from the classical non-relativistic Boltzmann equation with the use of the RG method was presented, which is, unfortunately, based on an inadequate inner product in the functional space composed of distribution functions. Subsequently a derivation of the dissipative fluid dynamics from the Boltzmann equation with the corrected inner product was reported in [58]. A comprehensive derivation of the Navier-Stokes equation based on [58] is presented in Chap. 8 on the basis of the geometrical formulation of the RG method [3, 5, 6].

Recently, there is a growing interest in the relativistic fluid dynamics [59, 60]. One might get surprised to find that there is no established relativistic dissipative fluid dynamic equations, as is detailed in Chap.10: For instance, there is an ambiguity of the choice of the rest frame of the fluid [61, 62] owing to the energy-mass equivalence inherent in the relativistic theory. Another problem is that the relativistic counter part of the Navier-Stokes equation, which is called the first-order (dissipative) equation, suffers from the loss of the causality because the equation becomes parabolic as the diffusion equation is. Furthermore, it is found that the thermal equilibrium state can be unstable if some relativistic dissipative fluid dynamic equations are applied [63].

A promising way of deriving the fluid dynamics that is free from these problems may be to start with the relativistic Boltzmann equation, which does not have such drawbacks, and derive the fluid dynamic limit of it by adopting some reduction theory of dynamics [64–68]. It should be naturally recognized, however, that the crucial point in such a project is how powerful and reliable is the reduction theory to be adopted. Popular methods among them are the multiple-scale method and/or use of a truncated functional space ; see [67, 68], for instance. In fact, the problem is rather involved because the derivation of a causal relativistic dissipative fluid dynamic equation must incorporate some excited modes as well as the usual zero modes originating from the conservation laws of energy-momenta and conserved charges. The excited modes reflect relatively short time dynamics, which is called the mesoscopic dynamics. One must say, however, there was no reliable reduction theory that properly identifies the appropriate excited modes as well as the zero modes.

In Chap. 9, a general theory for constructing mesoscopic dynamics is presented as an extension of the RG method utilizing the notion of envelopes, which is called the doublet scheme [69]. Then the doublet scheme in the RG method is applied to derive the relativistic second-order dissipative fluid dynamics in Chap. 12, which is extended to the reactive-multiple-component case in Chap. 15. An accurate and efficient numerical scheme is presented in Chap. 14, where numerical calculations are worked out for single component system.

The construction of mesoscopic dynamics in the non-relativistic case will be made in Chap. 16, where numerical results of the transport coefficients and the relaxation times are also presented using the derived microscopic expressions of them, and thereby some critical comparison is also made with the results obtained in the relaxation-time approximation that is commonly used in the current literature.

Langevin equation [70–76] is a kind of kinetic equations, and can be reduced to the time-irreversible Fokker-Planck equation [73–78], as is the fluid dynamics can be derived as an asymptotic dynamics of the Boltzmann equation. In Chap. 7, the RG method is applied to derive the Fokker-Planck equation from a generic multiplicative Langevin equation.

All the contents of the monograph are virtually based on the authors original works except for some review parts that include Chap. 2 in which a description is given on how secular terms appear ubiquitously in the perturbation theory, and Chap.3 in which an account of various conventional methods for the asymptotic analysis is made with a focus on their universal aspect that they all utilize the solvability condition of a linear inhomogeneous equation to make up some techniques with which the appearance of secular terms is circumvented. The other review part is Chap. 10, where a comprehensive account of the derivation of the relativistic dissipative fluid dynamics based on the Chapman-Enskog and Israel-Stewart methods is given with some comments.

The presentation of the monograph, at least in the first part, is intentionally made as pedagogical as possible so that not only researchers who are not familiar with the RG theory in physics but also undergraduate students with minimal mathematical backgrounds such as linear differential equations and linear algebra may appreciate and understand the method. Moreover we believe that simple but classical examples that are worked out will help the reader to understand what the RG method does in a geometrical way.

Conversely, the monograph is never intended to be a systematic review not only on the RG method and but also the relativistic dissipative fluid dynamics either with respect to its foundations or applications. In fact the literature on these subjects is so large that it is virtually impossible and beyond the authors' ability to make a systematic review on the current status on these subjects. Therefore we apologize those whose important articles are not cited in this monograph in advance.

Part I
Geometrical Formulation of Renormalization-Group Method and Its Extention for Global and Asymptotic Analysis with Examples

Chapter 2
Naïve Perturbation Method for Solving Ordinary Differential Equations and Notion of Secular Terms

2.1 Introduction

If a differential equation has a small parameter ϵ, one may be inclined to try to represent the solution by a power series of ϵ by applying the naïve perturbation theory to solve it. However, surprisingly enough, such a simple-minded method often leads to a disastrous result showing an unreasonable divergent behavior after a long time, or in a global domain of the independent variables. One of the typical causes of this undesired behavior is due to secular terms.

In this chapter, we analyze a few simple differential equations that have a small parameter ϵ, and introduce the notion of a secular term, and then demonstrate that secular terms are to appear in general in the naïve perturbation series of solutions of ordinary differential equations.

The presentation of this chapter is quite pedagogical and intended to be an elementary introduction to standard methods for solving linear inhomogeneous ordinary differential equations, say, in the undergraduate level. Indeed we present a detailed account of the Lagrange's method of variation of constants in the appendix of this chapter.

2.2 A Simple Example: Damped Oscillator

Let us first consider the following simple equation

$$m\frac{d^2x}{dt^2} = -kx - \kappa\frac{dx}{dt}. \tag{2.1}$$

This is a Newton equation for a particle with mass m in a harmonic potential

$$U(x) = \frac{1}{2}kx^2 \tag{2.2}$$

giving the mechanical force $-kx$ in accordance with the Hooke's law, and the second term in the right-hand side (r.h.s.) represents the air resistance (or friction) supposed

T. Kunihiro et al., *Geometrical Formulation of Renormalization-Group Method as an Asymptotic Analysis*, Fundamental Theories of Physics 206,
https://doi.org/10.1007/978-981-16-8189-9_2

to be proportional to the velocity

$$v = \frac{dx}{dt} =: \dot{x} \tag{2.3}$$

with a coefficient κ. Defining the angular velocity by

$$\omega_0 = \sqrt{\frac{k}{m}} \tag{2.4}$$

and making a replacement

$$\omega_0 t \;\rightarrow\; t, \tag{2.5}$$

Equation (2.1) is converted to a simple form as

$$\ddot{x} + 2\epsilon\dot{x} + x = 0, \quad \left(\epsilon := \frac{\kappa}{2m\omega_0} > 0\right) \tag{2.6}$$

with

$$\ddot{x} := \frac{d^2x}{dt^2}. \tag{2.7}$$

Since Eq. (2.6) is a second-order linear differential equation, there exist two independent solutions [21, 22], which may be obtained by inserting

$$x = \mathrm{e}^{\lambda t} \tag{2.8}$$

into (2.6); λ is found to be

$$\lambda = -\epsilon \pm i\sqrt{1-\epsilon^2} =: \lambda_\pm. \tag{2.9}$$

The general solution $x(t)$ of (2.6) is given by a linear combination of the independent solutions,

$$\mathrm{e}^{\lambda_\pm t} = \mathrm{e}^{-\epsilon t}\mathrm{e}^{\pm i\omega t}, \quad (\omega := \sqrt{1-\epsilon^2}), \tag{2.10}$$

as

$$x(t) = a\mathrm{e}^{-\epsilon t}\mathrm{e}^{i\omega t} + a^*\mathrm{e}^{-\epsilon t}\mathrm{e}^{-i\omega t}, \tag{2.11}$$

where the reality of $x(t)$ has been taken into account.

If we parametrize the coefficient as

$$a = -\frac{i}{2}\bar{A}e^{i\theta} \tag{2.12}$$

we have

$$x(t) = A(t)\sin\phi(t), \quad A(t) := \bar{A}e^{-\epsilon t}, \ \phi(t) := \omega t + \bar{\theta}, \tag{2.13}$$

where $\bar{A}$ and $\bar{\theta}$ are both constant real numbers.

If it were not for the resistance, *i.e.*, $\epsilon = 0$, then the amplitude $A(t)$ is time-independent and the angular velocity $\omega = 1$ with the period

$$T = \frac{2\pi}{\omega} = 2\pi. \tag{2.14}$$

In the following, we also use the word "frequency" in place of angular velocity if any misunderstanding will not be expected.

Owing to the resistance proportional to ϵ, the amplitude $A(t)$ decreases exponentially in time and the ω becomes smaller with a longer period.

Although we know the exact solution to (2.6), we shall dare to apply a naïve perturbative expansion [22] assuming that ϵ is small:

$$x = x_0 + \epsilon x_1 + \epsilon^2 x_2 + \cdots \tag{2.15}$$

Inserting (2.15) into (2.6), we have

$$\ddot{x}_0 + x_0 + \epsilon(\ddot{x}_1 + x_1) + \epsilon^2(\ddot{x}_2 + x_2) + \cdots = -2\epsilon(\dot{x}_0 + \epsilon\dot{x}_1 + \cdots). \tag{2.16}$$

Equating the coefficients of ϵ^n $(n = 0, 1, \ldots)$, we have a series of equations as follows,

$$\ddot{x}_0 + x_0 = 0, \quad \ddot{x}_1 + x_1 = -2\dot{x}_0, \quad \ddot{x}_2 + x_2 = -2\dot{x}_1, \tag{2.17}$$

and so on.

If we define the linear operator

$$L = \frac{d^2}{dt^2} + 1, \tag{2.18}$$

all the perturbative equations are expressed as

$$Lx_n = -2\dot{x}_{n-1}, \quad (n = 0, 1, \ldots) \tag{2.19}$$

with $x_{-1} = 0$.

The zeroth order solution may be given by

$$x_0 = \bar{A}\sin(t + \theta) \tag{2.20}$$

with $\bar{A}$ and θ being integral constants. Then the first-order equation has a form of an inhomogeneous equation as

$$\ddot{x}_1 + x_1 = -2\bar{A}\cos(t+\theta) \equiv F(t). \tag{2.21}$$

The inhomogeneous equation (2.21) may be solved using the Lagrange's method of variation of constants [21], an account of which is given below in Sect. 2.5. Using the two independent solutions

$$x^{(1)}(t) = \cos(t+\theta) \quad \text{and} \quad x^{(2)}(t) = \sin(t+\theta) \tag{2.22}$$

to the unperturbed equation,[1] we set

$$x_1 = C_1(t)x^{(1)}(t) + C_2(t)x^{(2)}(t) = C_1(t)\cos(t+\theta) + C_2(t)\sin(t+\theta) \tag{2.23}$$

with a constraint

$$\dot{C}_1 x^{(1)}(t) + \dot{C}_2 x^{(2)}(t) = 0, \tag{2.24}$$

where

$$\dot{C}_i := \frac{dC_i}{dt}, \qquad (i = 1,\ 2). \tag{2.25}$$

Inserting (2.23) into (2.21), we have

$$\dot{C}_1 \dot{x}^{(1)}(t) + \dot{C}_2 \dot{x}^{(2)}(t) = F(t), \tag{2.26}$$

where (2.24) has been utilized.

The set of Eqs. (2.24) and (2.26) for $\dot{C}_1$ and $\dot{C}_2$ yields

$$\dot{C}_1 = -F(t)x^{(2)}(t)/W(t), \quad \dot{C}_2 = F(t)x^{(1)}(t)/W(t), \tag{2.27}$$

where $W(t)$ is the Wronskian

$$W(t) = \begin{vmatrix} x^{(1)}(t) & x^{(2)}(t) \\ \dot{x}^{(1)}(t) & \dot{x}^{(2)}(t) \end{vmatrix} = \begin{vmatrix} \cos(t+\theta) & \sin(t+\theta) \\ -\sin(t+\theta) & \cos(t+\theta) \end{vmatrix} = 1. \tag{2.28}$$

Thus we have

$$\dot{C}_1 = \bar{A}\sin(2t+2\theta), \quad \dot{C}_2 = -\bar{A}(1+\cos(2t+2\theta)). \tag{2.29}$$

---

[1] Although the choice $x^{(1)}(t) = \cos t$ and $x^{(2)}(t) = \sin t$ also will do, θ has been introduced for later convenience.

A simple integration leads to

$$C_1 = -\frac{1}{2}\bar{A}\cos(2t + 2\theta) + \alpha, \quad C_2 = -\bar{A}t - \frac{1}{2}\bar{A}\sin(2t + 2\theta) + \beta \quad (2.30)$$

with α and β being constants. Thus we have for $x_1(t)$

$$x_1 = -\bar{A}t\sin(t + \theta) + \left(\alpha - \frac{\bar{A}}{2}\right)\cos(t + \theta) + \beta\sin(t + \theta), \quad (2.31)$$

where the last two terms satisfy the unperturbed equation and may be discarded because we are interested in deriving a general solution to (2.6); the prefactors of the two terms are interpreted to be renormalized into $\bar{A}$ and θ in $x_0(t)$. Thus we arrive at

$$x_1 = -\bar{A}t\sin(t + \theta), \quad (2.32)$$

which is proportional to time and called a **secular term** [21, 22].

This result can be understood in physical terms: Equation (2.21) is of the same form as that for a forced oscillator with no resistance but with an external force $f_{\rm ex}(t) = -2\bar{A}\cos(t + \theta)$ having the same frequency as the intrinsic one. Then it is expected that a resonance phenomenon occurs with an ever growing amplitude because of no resistance, which reflects in the appearance of the very secular term.

We can proceed to the second-order equation, which now reads

$$\ddot{x}_2 + x_2 = F_1(t) + F_2(t) \quad (2.33)$$

with

$$F_1(t) := 2\bar{A}\sin(t + \theta), \quad F_2(t) := 2\bar{A}t\cos(t + \theta). \quad (2.34)$$

Since (2.33) is a linear equation with $F_{1,2}(t)$ being the inhomogeneous terms, the solution is given as a sum $x_2(t) = x_2^{(1)} + x_2^{(2)}$ of the solutions $x_2^{(i)}$ to the same equations but with $F(t)$ being replaced by $F_i(t)$, respectively:

$$\ddot{x}_2^{(i)} + x_2^{(i)} = F_i(t), \quad (i = 1,\ 2). \quad (2.35)$$

Applying the method of variation of constants as before, we have

$$\begin{aligned} x_2^{(1)}(t) &= -\bar{A}t\cos(t + \theta) + \frac{\bar{A}}{2}\sin(t + \theta), \\ x_2^{(2)}(t) &= \frac{\bar{A}}{2}t^2\sin(t + \theta) + \frac{\bar{A}}{2}t\cos(t + \theta) - \frac{\bar{A}}{4}\sin(t + \theta), \end{aligned} \quad (2.36)$$

where the respective last term may be discarded because it is a solution to the homogeneous equation as was done before and implicitly for other similar terms even in the present case.

Thus summing all the terms obtained so far, we arrive at the perturbative solution to (2.6) in the second order as

$$\begin{aligned} x(t) &\simeq x_0(t) + \epsilon x_1(t) + \epsilon^2 x_2(t) \\ &= \bar{A}\sin(t+\theta) - \epsilon \bar{A} t \sin(t+\theta) \\ &\qquad + \epsilon^2 \frac{\bar{A}}{2}\{t^2 \sin(t+\theta) - t\cos(t+\theta)\} =: \bar{x}(t). \end{aligned} \tag{2.37}$$

Notice the appearance of the secular terms given by polynomials of t that cause the amplitude to ever increase with time contrary to the exact solution (2.13) in which the amplitude decreases in time exponentially. Furthermore we see that the powers of t in the secular terms in the perturbative (would-be) corrections becomes worse in higher orders.

The reason of the appearance of the secular terms can be traced back to the fact that the **zero modes** of the linear operator $L = d^2/dt^2 + 1$ in the unperturbed terms appear as the inhomogeneous term in the perturbative equations.

Nevertheless, it is noteworthy that the secular terms in (2.37) can be absorbed into the amplitude and phase with a weak time dependence of the unperturbed solution up to ϵ^3 as

$$\begin{aligned} \bar{x}(t) &\simeq \bar{A}\left(1 - \epsilon t + \frac{\epsilon^2}{2}t^2\right)\sin\left(\left(1 - \frac{\epsilon^2}{2}\right)t + \theta\right) \\ &\simeq \bar{A}\mathrm{e}^{-\epsilon t}\sin(\sqrt{1-\epsilon^2}\, t + \theta). \end{aligned} \tag{2.38}$$

Thus one sees that the perturbative solution (2.37) actually represents the first few terms of the expanded formula of the exact solution (2.13) with respect to ϵ. Furthermore the amplitude $A(t)$ of the damped oscillator just discussed satisfies the following equation

$$\dot{A} = -\epsilon A, \tag{2.39}$$

which shows that the time variation of A is proportional to ϵ and hence small.

What we have seen are typical phenomena and problems in naïve perturbation method for solving differential equations. When the inhomogeneous terms in the naïve perturbative equations contain zero modes of the linear operator L of the homogeneous equation, the appearance of secular terms are inevitable, which actually may be resummed into nonsingular expressions. Furthermore, it seems that the secular terms may be renormalized into the integral constants in the unperturbed solutions. Then it would be desirable to extract such slow motions and write down explicitly the equations to describe the slow motions.

2.3 Motion of a Particle in an Anharmonic Potential: Duffing Equation

In the previous section, we have dealt with a linear differential equation. In this section and the following one, we shall apply the perturbative expansion to nonlinear equations [21, 22].

Let us consider a particle with a mass m in a potential

$$U(x) = \frac{m\omega^2}{2}x^2 + \frac{k'}{4}x^4. \tag{2.40}$$

The equation of motion (EOM) reads

$$m\ddot{x} = -\frac{dU}{dx} = -m\omega^2 x - k'x^3, \tag{2.41}$$

which is called Duffing equation . It is found that the mechanical energy

$$E = \frac{m}{2}\dot{x}^2 + \frac{x^2}{2} + \frac{\epsilon}{4}x^4 \tag{2.42}$$

is conserved, *i.e.*, time-independent. In fact,

$$\frac{dE}{dt} = m\dot{x}\ddot{x} + \frac{dU}{dx}\dot{x} = \dot{x}\left[m\ddot{x} + \frac{dU}{dx}\right] = 0 \tag{2.43}$$

on account of (2.41). We can give a simple qualitative argument on the behavior of $x(t)$. Because the kinetic energy is semi-positive definite, we have an inequality $\frac{1}{2}m\dot{x}^2 = E - U(x) \geq 0$, leading to

$$\frac{m\omega^2}{2}x^2 + \frac{k'}{4}x^4 \leq E. \tag{2.44}$$

which implies that $|x(t)|$ is bounded for $E > 0$;

$$|x(t)|^2 \leq (m\omega^2/k')\left[\sqrt{1 + Ek'^2/(m\omega)^2} - 1\right]. \tag{2.45}$$

Now making a replacement $\omega t \to t$ and defining

$$\epsilon := \frac{k'}{m\omega^2}, \tag{2.46}$$

Equation (2.41) is cast into the following form,

$$\ddot{x} + x = -\epsilon x^3. \tag{2.47}$$

In the present analysis, we assume that the strength of the anharmonic potential is small, *i.e.*,

$$|\epsilon| < 1. \tag{2.48}$$

2.3.1 Exact Solution of Duffing Equation

Using the constancy of E, the time dependence of $x(t)$ can be exactly obtained in terms of elliptic functions[22]. Firstly, we note that Eq. (2.42) implies that

$$\left(\frac{dx}{dt}\right)^2 = 2E - x^2 - \frac{1}{2}\epsilon x^4, \tag{2.49}$$

which is reduced to a first-order equation as

$$\frac{dx}{dt} = \pm\sqrt{2E - x^2 - \epsilon x^4/2}. \tag{2.50}$$

Let x varies from x_0 to x as time does from t_0 to t. Then Eq. (2.50) is readily integrated out as

$$\int_{t_0}^{t} dt' = \pm \int_{x_0}^{x} \frac{dx'}{\sqrt{2E - x'^2 - \epsilon x'^4/2}}, \tag{2.51}$$

or

$$t - t_0 = \pm \int_{x_0}^{x} \frac{dx'}{\sqrt{2E - x'^2 - \epsilon x'^4/2}}. \tag{2.52}$$

To convert the integrand into a convenient form, we first determine the amplitude A, *i.e.*, the maximum value of x:

$$\frac{\epsilon}{2}A^4 + A^2 - 2E = 0, \tag{2.53}$$

which is solved for A^2 as

$$A^2 = \frac{1}{\epsilon}\left(-1 + \sqrt{1 + 4\epsilon E}\right) \simeq \frac{1}{\epsilon}\left[-1 + \left(1 + 2\epsilon E - \frac{1}{8}16\epsilon^2 E^2 + \cdots\right)\right]$$
$$\simeq 2E(1 - \epsilon E), \tag{2.54}$$

where the expansion formula $\sqrt{1+x} = 1 + x/2 - x^2/8 + \cdots$ has been used in the last equality. Using Eq. (2.53) for A^2, the function in the integrand is rewritten as

$$2E - x^2 - \frac{\epsilon}{2}x^4 = 2E - x^2 - \frac{\epsilon}{2}x^4 - \left(2E - A^2 - \frac{\epsilon}{2}A^4\right)$$
$$= \frac{\epsilon}{2}\left(A^2 - x^2)(x^2 + A^2 + \frac{2}{\epsilon}\right). \tag{2.55}$$

Then changing the integration variable by

$$x' = A\cos\phi', \tag{2.56}$$

Equation (2.52) is converted to a simple form as

$$t - t_0 = \mp\frac{1}{\sqrt{1+\epsilon A^2}}\int_{\phi_0}^{\phi}\frac{d\phi'}{\sqrt{1-k^2\sin^2\phi'}}, \qquad (x_0 = \sin\phi_0), \tag{2.57}$$

where

$$k^2 := \frac{\epsilon A^2/2}{1+\epsilon A^2}. \tag{2.58}$$

The integral in (2.57) is expressed in terms of the incomplete elliptic integral of the first kind

$$F(\phi, k) = \int_0^{\phi}\frac{d\phi'}{\sqrt{1-k^2\sin^2\phi'}} \tag{2.59}$$

with k being the modulus. Then from (2.57), the period T is given by the time needed for the phase to change from 0 to 2π;

$$T = \frac{1}{\sqrt{1+\epsilon A^2}}\int_0^{2\pi}\frac{d\phi'}{\sqrt{1-k^2\sin^2\phi'}} = \frac{1}{\sqrt{1+\epsilon A^2}}4\int_0^{\pi/2}\frac{d\phi}{\sqrt{1-k^2\sin^2\phi}}$$
$$= \frac{1}{\sqrt{1+\epsilon A^2}}4F(\pi/2, k) = \frac{4}{\sqrt{1+\epsilon A^2}}K(k), \tag{2.60}$$

where we have introduced the complete elliptic integral of the first kind

$$K(k) := F(\pi/2, k). \tag{2.61}$$

When ϵ is small, *i.e.*,

$$|\epsilon|A^2 < 1, \tag{2.62}$$

we have an approximate formula for $K(k)$ using the expansion $(1-x)^{-1/2} = 1 + x/2 + 3x^2/8 + \cdots$ as

$$K(k) \simeq \int_0^{\pi/2} d\phi' \left[1 + \frac{k^2}{2}\sin^2\phi' + \frac{3k^4}{8}\sin^4\phi' + \cdots\right]$$
$$= \frac{\pi}{2}[1 + \frac{1}{4}k^2 + \frac{9}{64}k^4 + \cdots], \tag{2.63}$$

where we have used the Wallis's formula with n being a positive integer

$$\int_0^{\pi/2} d\phi \sin^{2n}\phi = \frac{(2n-1)!!}{(2n)!!}\frac{\pi}{2}. \tag{2.64}$$

Thus an approximate formula for $K(k)$ is given as

$$K(k) \simeq \frac{\pi}{2}\left[1 + \frac{\epsilon}{8}A^2\right]. \tag{2.65}$$

Hence

$$T \simeq 2\pi\left(1 - \frac{\epsilon}{2}A^2\right)\left(1 + \frac{\epsilon}{8}A^2\right) \simeq 2\pi\left[1 - \frac{3\epsilon}{8}A^2\right], \tag{2.66}$$

accordingly the frequency is given by

$$\Omega \equiv \frac{2\pi}{T} \simeq \frac{2\pi}{2\pi(1 - \frac{3\epsilon}{8}A^2)} \simeq 1 + \frac{3\epsilon}{8}A^2. \tag{2.67}$$

We see that the anharmonic force makes the frequency larger and the period shorter. This is physically plausible results because for a large amplitude, the nonlinear term in the restoring force $-k'x^3$ would become significant, and a part of which effects may be absorbed into a *renormalization of the spring constant* $m\omega^2$ and hence the frequency.

2.3.2 Naïve Perturbation Theory Applied to Duffing Equation

Although the exact solution is known as has been shown above, we here dare to apply a naïve perturbative expansion

$$x = x_0 + \epsilon x_1 + \epsilon x_2 + \cdots, \tag{2.68}$$

assuming that ϵ is small. Inserting this expansion into (2.47), we have

$$L(x_0 + \epsilon x_1 + \epsilon^2 x_2 + \cdots) = -\epsilon(x_0 + \epsilon x_1 + \epsilon^2 x_2 + \cdots)^3, \tag{2.69}$$

with $L = \frac{d^2}{dt^2} + 1$ defined in (2.18). Then equating the coefficients of ϵ^n ($n = 0, 1, \ldots$), we have a series of equations as follows,

$$Lx_0 = 0, \quad Lx_1 = -x_0^3, \quad Lx_2 = -3x_0^2 x_1, \tag{2.70}$$

and so on. The zero-th order solution may be given by

$$x_0 = \bar{A}\cos(t + \theta) \tag{2.71}$$

with $\bar{A}$ and θ being integral constants.

The first-order equation has a form of an inhomogeneous equation which reads

$$\ddot{x}_1 + x_1 = -\bar{A}^3 \cos^3(t + \theta) = -\frac{\bar{A}^3}{4}\left(3\cos\phi(t) + \cos 3\phi(t)\right), \tag{2.72}$$

with

$$\phi(t) = t + \theta. \tag{2.73}$$

Since Eq. (2.72) is a linear equation, the particular solution to it is given as a linear combination $x_1 = x_1^{(1)}(t) + x_2^{(2)}(t)$ of those to the inhomogeneous equations with homogeneous terms, $\cos\phi(t)$ and $\cos 3\phi(t)$, respectively:

$$\ddot{x}_1^{(1)} + x_1^{(1)} = -\frac{3\bar{A}^3}{4}\cos\phi(t), \tag{2.74}$$

$$\ddot{x}_1^{(2)} + x_1^{(2)} = -\frac{\bar{A}^3}{4}\cos 3\phi(t). \tag{2.75}$$

These equations are of the same form as that of a forced oscillator without resistance with external forces proportional to $\cos\phi(t)$ and $\cos 3\phi(t)$, the former of which causes a resonance phenomenon described by a secular term as before. Although the particular solution can be obtained by the method of variation of constants , here we make a short cut by assuming that

$$x_1^{(1)} = at\cos\phi(t) + bt\sin\phi(t). \tag{2.76}$$

Inserting this ansatz to (2.74), we have

$$\ddot{x}_1^{(1)} + x_1^{(1)} = -2a\sin\phi(t) + 2b\cos\phi(t) = -\frac{3\bar{A}^3}{4}\cos\phi(t), \tag{2.77}$$

which gives

$$a = 0 \quad \text{and} \quad b = -3\bar{A}^3/8 \tag{2.78}$$

and hence

$$x_1^{(1)} = -\frac{3\bar{A}^3}{8} t \sin\phi(t). \tag{2.79}$$

On the other hand, a simple application of the method of variation of constants (2.75) gives

$$x_1^{(2)} = \frac{\bar{A}^3}{32} \cos 3\phi(t). \tag{2.80}$$

Thus we have for the particular solution $x_1(t)$

$$x_1(t) = -\frac{3\bar{A}^3}{8} t \sin\phi(t) + \frac{\bar{A}^3}{32} \cos 3\phi(t). \tag{2.81}$$

The approximate solution up to this order now reads

$$x(t) \simeq \bar{A}\cos\phi(t) - \frac{3\bar{A}^3}{8} \epsilon \left[t \sin\phi(t) - \frac{1}{12} \cos 3\phi(t) \right], \tag{2.82}$$

which contains a secular term causing the amplitude to grow infinitely when $t \to \infty$. This is a physically absurd result and a quite opposite behavior to the exact solution that is bounded because of the restoring (confining) force.

Nevertheless we can make the following argument which suggests a possible origin of the appearance of the secular term. Since

$$(3\bar{A}^2/8)\,\epsilon t \simeq \sin[(3\epsilon t \bar{A}^2/8] \tag{2.83}$$

for sufficiently small ϵt, the secular term and the unperturbed one in (2.82) can be combined into

$$\begin{aligned} \bar{A}\cos\phi(t) - \frac{3\bar{A}^3}{8} \epsilon t \sin\phi(t) &\simeq \bar{A}\cos\phi(t) - \sin(3\bar{A}^3 \epsilon t/8) \sin\phi(t) \\ &\simeq \bar{A}\cos[\phi(t) + 3\epsilon t \bar{A}^2/8] \\ &= \bar{A}\cos[(1 + 3\epsilon\bar{A}^2/8)t + \theta]. \end{aligned} \tag{2.84}$$

Thus we have

$$x(t) \simeq \bar{A}\cos[\Omega t + \theta] + \epsilon\frac{\bar{A}^3}{32} \cos 3\phi(t), \tag{2.85}$$

with

$$\Omega := 1 + \frac{3}{8}\epsilon\bar{A}^2, \tag{2.86}$$

which coincides with the shifted frequency (2.67) extracted from the exact solution for small ϵ.

It is also to be noted that even if a part of the effects of the nonlinear term could be absorbed through the *renormalization of the spring constant*, it is not the whole effects of it; the nonlinear term also gives rise to an additional term with *higher harmonics*.

2.4 van der Pol Equation

The final example is the van der Pol equation, which describes a self-sustained oscillation [22];

$$\ddot{x} + x = \epsilon\,(1 - x^2)\,\dot{x}, \tag{2.87}$$

where ϵ is supposed to be small so as to the perturbation analysis is valid. It is known that the van der Pol equation admits a limit cycle; see [79] for a pedagogical explanation on what a limit cycle is in non-linear oscillators.

Let us apply the perturbative expansion to (2.87) as before,

$$x(t) = x_0(t) + \epsilon\, x_1(t) + \epsilon^2\, x_2(t) + \cdots . \tag{2.88}$$

Inserting this expansion to (2.87), we have the following series of equations with $L = \frac{d^2}{dt^2} + 1$,

$$L x_0 = 0, \tag{2.89}$$
$$L x_1 = (1 - x_0^2)\,\dot{x}_0, \tag{2.90}$$
$$L x_2 = (1 - x_0^2)\,\dot{x}_1 - 2x_0 x_1 \dot{x}_0, \tag{2.91}$$

and so on.

We take for the zeroth solution

$$x_0(t) = A\cos(t + \theta). \tag{2.92}$$

Then Eq. (2.90) becomes

$$L x_1 = -A\left(1 - \frac{A^2}{4}\right)\sin\phi(t) + \frac{A^3}{4}\sin 3\phi(t), \quad (\phi(t) = t + \theta). \tag{2.93}$$

Again the first term of r.h.s. is a zero mode of L, and hence the particular solution to this equation will contain a secular term. As was done in the previous section, the application of the method of variation of constants leads to the solution as

$$x_1(t) = t\frac{A}{2}\left(1 - \frac{A^2}{4}\right)\cos\phi(t) - \frac{A^3}{32}\sin 3\phi(t). \tag{2.94}$$

Thus up to the second order, we have

$$x(t) \simeq A\cos\phi(t) + \epsilon\left[\frac{tA}{2}\left(1 - \frac{A^2}{4}\right)\cos\phi(t) - \frac{A^3}{32}\sin 3\phi(t)\right], \tag{2.95}$$

which contains a secular term proportional to t. Again the secular term can be absorbed into the unperturbed solution, and then (2.95) is rewritten as

$$x(t) \simeq \mathcal{A}(t)\cos\phi(t) - \epsilon\frac{A^3}{32}\sin 3\phi(t). \tag{2.96}$$

with

$$\mathcal{A}(t) := A\left[1 + \frac{\epsilon t}{2}\left(1 - \frac{A^2}{4}\right)\right], \tag{2.97}$$

which happens to satisfy the following equation up to ϵ^2

$$\frac{d\mathcal{A}}{dt} = \epsilon\frac{A}{2}\left(1 - \frac{A^2}{4}\right) \simeq \epsilon\frac{\mathcal{A}}{2}\left(1 - \frac{\mathcal{A}^2}{4}\right). \tag{2.98}$$

If one takes the last equality literally, it is readily verified that if $\mathcal{A}(0) \neq 0$, $\lim_{t\to\infty}\mathcal{A}(t) = 2$, which means that the van der Pol equation admits a limit cycle with a radius 2; this result is in accordance with the behavior of the solution obtained in numerical calculations and in resummation methods to be introduced later.

It is also noteworthy that the effects of the nonlinear term also create higher-harmonics terms that could not be absorbed or renormalized away to the zeroth order solution as might have been possible for the secular terms.

2.5 Concluding Remarks

We have seen typical phenomena and problems in the naïve perturbation method for solving differential expansions using a few examples. When the inhomogeneous terms in the naïve perturbative equations contain **zero modes** of the linear operator L of the homogeneous equation, the appearance of secular terms is inevitable, which actually do not appear in the exact solutions.

A rather generic phenomenon in such a resummation of the perturbation series is that the *resummed terms* may be *renormalized into* the integral constants in the unperturbed solution, like the amplitudes and/or phases, which should contain a small parameter and thus would describe slow motions (or long-wave length phenomena), *i.e.*, a kind of collective motion. Then it would be desirable to explicitly extract such slow variables and the dynamical equations to describe their slow motions.

In fact, many methods have been developed to avoid the appearance of secular terms and in the same time achieve a *resummation* of the seemingly divergent series with secular terms, which include Poincare-Lighthil-Luo, Krylov-Bogoliubov-Mitropolsky method, the reductive perturbation theory and so on [21, 22]. All of these existing methods are based on some techniques with which the appearance of secular terms or singular terms are circumvented.

The renormalization-group (RG) method that we are going to introduce is one of such methods: A unique feature of the method, however, lies in the fact that it allows the appearance of secular terms in contrast to the conventional resummation methods. Another merit of the RG method is that it provides us with an elementary but constructive method to extract the reduced dynamical equations explicitly for the slow motions embedded in the original dynamical equation.

Appendix: Method of Variation of Constants

Let us consider the following n-th order equation with the coefficients $a_i(t)$ $(i = 1, 2, \ldots, n)$ and the inhomogeneous term $b(t)$:

$$x^{(n)} + a_1(t)x^{(n-1)} + \cdots + a_{n-1}(t)\dot{x} + a_n(t)x = b(t), \tag{2.99}$$

where $x^{(k)} := d^k x/dt^k$.

The standard procedure of the method of variation of constants (MVOC) goes as follows [21]. Let $x_1(t), x_2(t), \ldots, x_n(t)$ be a set of independent solutions of the homogeneous equation

$$Lx_i := x^{(n)} + a_1(t)x^{(n-1)} + \cdots + a_{n-1}(t)\dot{x} + a_n(t)x = 0 \tag{2.100}$$

with $i = 1, 2, \ldots, n$. The general solution to (2.100) is given by

$$x(t) = \sum_{i=1}^{n} C_i x_i(t), \tag{2.101}$$

with C_i $(i = 1, 2, \ldots, n)$ being arbitrary constants. In the MVOC, we start with the following ansatz to a solution to (2.100),

$$x(t) = \sum_{i=1}^{n} C_i(t) x_i(t), \tag{2.102}$$

where the coefficients $C_i(t)$'s are now all functions of t in contrast to (2.101) where they are constants, hence the name of MVOC.

Differentiating Eq. (2.102) with respect to t, one has

$$\dot{x}(t) = \sum_{i=1}^{n} \dot{C}_i(t)x_i(t) + C_i(t)\dot{x}_i(t). \tag{2.103}$$

We demand the following condition to $\dot{C}_i(t)$'s

$$\sum_{i=1}^{n} \dot{C}_i(t)x_i(t) = 0. \tag{2.104}$$

Then we have

$$\dot{x}(t) = \sum_{i=1}^{n} C_i(t)\dot{x}_i(t). \tag{2.105}$$

For obtaining to convenient formulae, we demand the following conditions to the higher derivatives of $C_i(t)$'s in addition to (2.104),

$$\sum_{i=1}^{n} \frac{d^k C_i(t)}{dt^k} x_i(t) = 0, \quad (k = 2, \ldots, n). \tag{2.106}$$

Then the successive differentiations of (2.105) lead to

$$x^{(k)}(t) = \sum_{i=1}^{n} C_i(t)x_i^{(k)}(t) \tag{2.107}$$

for $k = 2, \ldots, n$. Inserting these formulae for $x^{(k)}(t)$ into (2.99), we arrive at

$$\sum_{i=1}^{n} \dot{C}_i(t)x_i^{(n-1)}(t) + \sum_{i=1}^{n} C_i(t)\,(Lx_i) = \sum_{i=1}^{n} \dot{C}_i(t)x_i^{(n-1)}(t) = b(t), \tag{2.108}$$

where (2.100) has been used.

Now Eqs. (2.104), (2.106) and (2.108) make a set of equations for $\dot{C}_i(t)$ $(i = 1, 2, \ldots, n)$. For notational convenience. let introduce the following vectors and matrix

$$\boldsymbol{C}(t) = \begin{pmatrix} C_1(t) \\ C_2(t) \\ \vdots \\ C_n(t) \end{pmatrix}, \quad \boldsymbol{b}(t) = \begin{pmatrix} 0 \\ 0 \\ \vdots \\ b(t) \end{pmatrix}, \quad \boldsymbol{x}(t) = \begin{pmatrix} x_1(t) \\ x_2(t) \\ \vdots \\ x_n(t) \end{pmatrix} \tag{2.109}$$

and

$$\hat{W}(t) := \begin{pmatrix} x_1 & x_2 & \dots & x_n \\ \dot{x}_1 & \dot{x}_2 & \dots & \dot{x}_n \\ \vdots & \vdots & \vdots & \vdots \\ x_1^{(n)} & x_2^{(n)} & \dots & x_n^{(n)} \end{pmatrix}. \tag{2.110}$$

Then the set of equations for $\dot{C}_i(t)$ $(i = 1, 2)$ is rewritten in a compact form of a vector equation as

$$\hat{W}(t)\frac{d\boldsymbol{C}(t)}{dt} = \boldsymbol{b}(t). \tag{2.111}$$

As is well known, when $x_1(t),\ x_2(t),\ \dots,\ x_n(t)$ are linearly independent, the Wronskian $W(t) := \det \hat{W}(t)$ does not vanish $W(t) \neq 0$, and hence the inverse $\hat{W}^{-1}(t)$ of $\hat{W}(t)$ exists [21]. Therefore the solution to (2.111) is readily obtained by multiplying the inverse $\hat{W}^{-1}(t)$ as

$$\frac{d\boldsymbol{C}(t)}{dt} = \hat{W}^{-1}(t)\,\boldsymbol{b}(t). \tag{2.112}$$

Since the r.h.s. is represented in terms of the known functions, (2.112) should be solved by a quadrature in principle.

As an example, let us give the case of a second-order equation:

$$\ddot{x} + a_1(t)\dot{x} + a_2(t)x = b(t), \tag{2.113}$$

for which (2.112) reads

$$\dot{C}_1(t) = -b(t)x_2(t)/W(t), \quad \dot{C}_2(t) = b(t)x_1(t)/W(t), \tag{2.114}$$

A simple integration gives

$$\begin{aligned} C_1(t) &= -\int_{t_0}^{t} dt'\, b(t')x_2(t')/W(t') + A \\ C_2(t) &= \int_{t_0}^{t} dt'\, b(t')x_1(t')/W(t') + B \end{aligned} \tag{2.115}$$

with A and B are integration constants, which are to be discarded when the particular solution is concerned. Thus we have for the particular solution to (2.113)

$$x(t) = -x_1(t) \int_{t_0}^{t} dt'\, b(t')x_2(t')/W(t') + x_2(t) \int_{t_0}^{t} dt'\, b(t')x_1(t')/W(t'). \tag{2.116}$$

This completes an elementary account of MVOC for the second-order inhomogeneous equation.

Finally we show some basic properties of the Wronskian:

$$\frac{dW}{dt} = \begin{vmatrix} \dot{x}_1 & \dot{x}_2 \\ \dot{x}_1 & \dot{x}_2 \end{vmatrix} + \begin{vmatrix} x_1 & x_2 \\ \ddot{x}_1 & \ddot{x}_2 \end{vmatrix} = \begin{vmatrix} x_1 & x_2 \\ -a_1(t)\dot{x}_1 - a_2(t)x_1 & -a_1(t)\dot{x}_2 - a_2(t)x_2 \end{vmatrix} = -a_1(t)W(t), \tag{2.117}$$

and hence

$$W(t) = \mathrm{e}^{-\int_{t_0}^{t} dt'\, a_1(t')} W(t_0). \tag{2.118}$$

which shows that

$$\text{if } W(t_0) \neq 0 \Rightarrow W(t) \neq 0 \quad \forall t. \tag{2.119}$$

Chapter 3
Conventional Resummation Methods for Differential Equations

3.1 Introduction

In the last chapter (Chap. 2), it was shown that the naïve perturbation method for solving differential equations does not necessarily work due to the ubiquitous appearance of secular terms in the perturbative solutions; the secular terms make the perturbative solutions become valid only in a local domain. Its appearance is inevitable when the inhomogeneous terms in the perturbative equations contain the **zero modes** of the linear operator L of the homogeneous equation.

In this chapter, we introduce several conventional methods [20–22, 30, 34] to resum secular terms in the perturbative expansions with use of the examples treated in the last chapter. It will be found that the essential notion involved in all the methods is the **solvability condition** of linear equations [80–83]. Therefore this chapter will start with an introductory account of it.

3.2 Solvability Condition of Linear Equations and Appearance of Secular Terms: Fredholm's Alternative

Since the conventional methods for circumventing secular terms to be introduced in this chapter are all based on the solvability condition of linear equations [80–83], we first make a basic account of it in general terms on the basis of the notion of **Fredholm's alternative** for linear operators [80–82]. Although some of the readers might have never heard of the name, Fredholm's alternative theorem [80–82] provides us with the solvability condition of the equation and also the structure of the solution when the solvability condition is not satisfied.

For an account of the theorem, let us start with a trivial equation for x

$$ax = b \tag{3.1}$$

T. Kunihiro et al., *Geometrical Formulation of Renormalization-Group Method as an Asymptotic Analysis*, Fundamental Theories of Physics 206,
https://doi.org/10.1007/978-981-16-8189-9_3

with a, x and b being complex numbers. A tricky point is that the solution may not be generally expressed as the quotient $x = b/a$, because a can be 0. When $a = 0$, we have two alternatives:

1. If $b = 0$, x can be any number, and the solution is undetermined.
2. If $b \neq 0$, no solution exists.

This is a well-known fact.

A similar alternatives exists in the solution of a system of linear equations. Let us consider a linear equation

$$A\boldsymbol{x} = \boldsymbol{b}, \quad \boldsymbol{x} = \begin{pmatrix} x_1 \\ x_2 \\ \vdots \\ x_n \end{pmatrix}, \tag{3.2}$$

where A is an $n \times n$ matrix and $\boldsymbol{b}$ a given n-dimensional vector.

(1) When the determinant $|A| \neq 0$, there exists the inverse A^{-1} and the solution is uniquely given as

$$\boldsymbol{x} = A^{-1}\boldsymbol{b}. \tag{3.3}$$

(2) Alternatively, if $|A| = 0$ and accordingly the inverse of A does not exist, the solution depends on the properties of $\boldsymbol{b}$. Note that $|A| = 0$ implies that A has zero eigenvalues with, say, m degeneracy. In this case, we have m independent left and right eigenvectors $\boldsymbol{U}_i(\neq \boldsymbol{0})$ and $\tilde{\boldsymbol{U}}_i$ $(\neq \boldsymbol{0})$ $(i = 1, 2, \ldots, m)$ belonging to the 0 eigenvalue as

$$A\,\boldsymbol{U}_i = 0, \qquad \tilde{\boldsymbol{U}}_i^{\dagger} A = 0, \quad (i = 1, 2, \ldots, m). \tag{3.4}$$

They are called the **zero modes**, and the subspace spanned by the zero modes is called the kernel and denoted by Ker$A$. Let us call the Ker$A$ the P space; the projection operator onto the P space is denoted by P. The compliment of the P space is called the Q space and the projection operator onto it is denoted by Q. Then the following alternative statements hold[1]:

(2-a) If $\boldsymbol{b}$ is orthogonal to all the zero modes, *i.e.*,

$$(\tilde{\boldsymbol{U}}_i^{(0)}, \boldsymbol{b}) = 0, \qquad (i = 1, 2, \ldots m), \tag{3.5}$$

then we have a series of the solution

[1] This is the simplest case of the theorem called Fredholm's alternative theorem in the theory of the linear operators in the functional analysis [80–82].

$$x = A_Q^{-1}\boldsymbol{b} + \sum_{i=1}^{m} c_i \boldsymbol{U}_i^{(0)} = A^{-1}Q\boldsymbol{b} + \sum_{i=1}^{m} c_i \boldsymbol{U}_i^{(0)}, \tag{3.6}$$

where

$$A_Q := QAQ = AQ = QA \tag{3.7}$$

and c_i's are arbitrary constants.

[A simple example] As a simple example, let us consider a linear equation (3.2) in two dimensions with

$$A = \begin{pmatrix} 1 & 1 \\ 2 & 2 \end{pmatrix} \quad \text{and} \quad \boldsymbol{x} = \begin{pmatrix} x_1 \\ x_2 \end{pmatrix}. \tag{3.8}$$

The inhomogeneous part $\boldsymbol{b}$ will be specified later. The eigenvalues of A reads

$$\lambda_1 = 0 \quad \text{and} \quad \lambda_2 = 3, \tag{3.9}$$

and the respective left eigenvectors are given by

$$\boldsymbol{U}_1 = \begin{pmatrix} 1 \\ -1 \end{pmatrix} \quad \text{and} \quad \boldsymbol{U}_2 = \begin{pmatrix} 1 \\ 2 \end{pmatrix}, \tag{3.10}$$

respectively, whereas the respective right eigenvectors are

$$\tilde{\boldsymbol{U}}_1 = \begin{pmatrix} 2/3 \\ -1/3 \end{pmatrix} \quad \text{and} \quad \tilde{\boldsymbol{U}}_2 = \begin{pmatrix} 1/3 \\ 1/3 \end{pmatrix}, \tag{3.11}$$

which satisfy the orthonormal condition

$$(\tilde{\boldsymbol{U}}_i, \boldsymbol{U}_j) = \delta_{ij}, \quad (i, j = 1, 2). \tag{3.12}$$

The projection operator onto the Q space is found to be

$$Q = \boldsymbol{U}_2 \tilde{\boldsymbol{U}}_2^{\dagger} = A/3. \tag{3.13}$$

Let us consider a case where the inhomogeneous term $\boldsymbol{b}$ satisfies the solvability condition. As an example, let

$$\boldsymbol{b} = \begin{pmatrix} 2 \\ 4 \end{pmatrix}, \tag{3.14}$$

which satisfies the condition

$$(\tilde{\boldsymbol{U}}_1,\, \boldsymbol{b}) = 0, \tag{3.15}$$

the solution to the linear equation $A\boldsymbol{x} = \boldsymbol{b}$ is readily found to be

$$x_1 + x_2 = 2. \tag{3.16}$$

If we apply the general formula (3.6), we find that this solution is also represented with a parameter c as

$$\begin{pmatrix} x_1 \\ x_2 \end{pmatrix} = A^{-1}Q\boldsymbol{b} + c\boldsymbol{U}_1 = \frac{1}{3}\begin{pmatrix} 2 \\ 4 \end{pmatrix} + \begin{pmatrix} c \\ -c \end{pmatrix}, \tag{3.17}$$

accordingly,

$$x_1 = 2/3 + c \quad \text{and} \quad x_2 = 4/3 - c, \tag{3.18}$$

which satisfy $x_1 + x_2 = 2$ with a special parametrization of the one-dimensional solution space.

(2-b) If $\boldsymbol{b}$ does not satisfy the solvability condition, implying that

$$(\tilde{\boldsymbol{U}}_k^{(0)},\, \boldsymbol{b}) \neq 0 \tag{3.19}$$

for some $(1 \leq) k (\leq m)$, then there exists no solution to Eq. (3.2), that is, Eq. (3.2) is not solvable.

For example, if $\boldsymbol{b} = \begin{pmatrix} 1 \\ -1 \end{pmatrix}$, for which $(\tilde{\boldsymbol{U}}_1,\, \boldsymbol{b}) = 1 \neq 0$, then one readily sees that the linear equation considered in (2-a) has no solution.

Thus (3.5) is called the **solvability condition** to the linear equation (3.2).

3.3 Solvability Condition of Linear Differential Equations with Hermitian Operator

Next let us take the differential operator that has appeared in the previous chapter as a linear operator;

$$L = \frac{d^2}{dt^2} + 1. \tag{3.20}$$

We define the inner product for periodic functions $u(t)$ and $v(t)$ as

$$(u,\, v) := \int_0^{2\pi} dt\, u^*(t)v(t). \tag{3.21}$$

Here the periodicity of $u(t)$ means that the following equalities hold

$$u(t+2\pi) = u(t), \quad \frac{du(t)}{dt} = \frac{du(t+2\pi)}{dt}. \tag{3.22}$$

Then applying the partial integrations twice, it is found that L is a symmetric operator with respect to this inner product for periodic functions,

$$(u,\ Lv) = (Lu,\ v). \tag{3.23}$$

L has two independent zero modes,

$$u_0^{(1)} = \frac{1}{\sqrt{\pi}} \sin t, \quad u_0^{(2)} = \frac{1}{\sqrt{\pi}} \cos t, \tag{3.24}$$

which are made orthonormal bases as

$$(u_0^{(i)},\ u_0^{(j)}) = \delta_{ij}. \tag{3.25}$$

The functional space spanned by the zero modes is called the P space and the projection operator P onto it is given in terms of the integral kernel

$$\mathcal{P}(t,\ t') = u_0^{(1)}(t)u_0^{(1)*}(t') + u_0^{(2)}(t)u_0^{(2)*}(t'), \tag{3.26}$$

as

$$\begin{aligned}(Px)(t) &:= \int_0^{2\pi} dt' \mathcal{P}(t,\ t')x(t') \\ &= u_0^{(1)}(t)(u_0^{(1)},\ x) + u_0^{(2)}(t)(u_0^{(2)},\ x).\end{aligned} \tag{3.27}$$

Let $u_\lambda(t)$ be a normalized eigenfunction belonging to a nonvanishing eigenvalue λ of L:

$$Lu_\lambda(t) = \lambda u_\lambda(t), \qquad (u_\lambda,\ u_\lambda) = 1. \tag{3.28}$$

As is well known, the eigenfunctions belong to different eigenvalues are orthogonal to each other. We assume that degenerate eigenfunctions are distinguished also by λ and made orthogonal to each other. Thus we have

$$(u_\lambda,\ u_{\lambda'}) = \delta_{\lambda\lambda'}. \tag{3.29}$$

The perturbative equations in the previous chapter take the following form

$$Lx = b(t). \tag{3.30}$$

The solvability condition for (3.30) reads

$$(u_0^{(i)},\, b) = 0, \quad (i = 1,\, 2). \tag{3.31}$$

When this condition is satisfied, the identity

$$b = Qb, \qquad (Q := 1 - P) \tag{3.32}$$

holds, and the solution to (3.30) may be formally given as

$$x = L^{-1}Qb = L_Q^{-1}b, \quad (L_Q := LQ). \tag{3.33}$$

However, if the inhomogeneous term b contains the zero modes, $\sin t$ and/or $\cos t$, as were the case in the examples treated in the previous chapter, b is expressed as

$$b = Pb + QB, \quad \text{with } Pb \neq 0, \tag{3.34}$$

and the solvability condition is not satisfied. Thus the *formal* solution has singular terms $L^{-1}Pb$, which turns out to be expressed as secular terms.

To give a more precise argument, it is convenient to introduce a resolvent operator [81]

$$R(z) := (z - L)^{-1}, \tag{3.35}$$

with z being a complex number $z \in \boldsymbol{C}$. Then let us consider the linear equation

$$(L - z)x = b, \tag{3.36}$$

the solution to which is given by

$$x = -R(z)b = -\frac{1}{z-L}Pb - \frac{1}{z-L}Qb = \frac{-1}{z}Pb + \frac{1}{L-z}Qb \tag{3.37}$$

when $z \neq \forall\lambda$. It is now clear that the solution is singular for the limit $z \to 0$ due to the existence of the zero mode $Pb \neq 0$ in the inhomogeneous term; the first term in the right-hand side diverges as $1/z \to \infty$. This singular behavior manifests itself in the form of secular terms.

All the conventional methods for resummation of the perturbative equations are based on some formal manipulations to modify the target equation so that the **solvability condition** is always satisfied. In these methods, some yet unknown constants and functions that specify the perturbative solutions are introduced from the outset, and the solvability conditions such as (3.31) for the perturbative equations are used to determine the constants and functions. The variety of the resummation methods is that of the ideas to keep the solvability condition and hence circumvent the appearance of the secular terms.

3.4 Lindstedt-Poincaré Method: Duffing Equation Revisited

We take the Duffing equation (2.47) for the demonstration of the Lindstedt-Poincaré method for resummation:

$$\ddot{x} + x = -\epsilon x^3. \tag{3.38}$$

In the Lindstedt-Poincaré method, one first takes into account the fact that the nonlinear term would modify the period or frequency by rewriting the equation as follows,

$$\ddot{x} + \Omega^2 x = (\Omega^2 - 1)x - \epsilon x^3, \tag{3.39}$$

which is reexpressed as

$$L_\Omega x = (\Omega^2 - 1)x - \epsilon x^3, \quad (L_\Omega := \frac{d^2}{dt^2} + \Omega^2). \tag{3.40}$$

The linear operator L_Ω has the following zero modes

$$X_0^{(1)} := \sqrt{\frac{\Omega}{\pi}} \cos \Omega t \quad \text{and} \quad X_0^{(2)} := \sqrt{\frac{\Omega}{\pi}} \sin \Omega t. \tag{3.41}$$

Here Ω^2 is assumed to be expanded as

$$\Omega^2 = 1 + \epsilon\omega_1^2 + \epsilon^2\omega_2^2 + \cdots \tag{3.42}$$

As was done in Sect. 2.3.2, we apply a perturbative expansion $x = x_0 + \epsilon x_1 + \epsilon^2 x_2 + \cdots$ to (3.40) with (3.42), then we have

$$L_\Omega x_0 = 0, \tag{3.43}$$
$$L_\Omega x_1 = \omega_1^2 x_0 - x_0^3, \tag{3.44}$$
$$L_\Omega x_2 = \omega_1^2 x_1 + \omega_2^2 x_0 - 3x_0^2 x_1, \tag{3.45}$$

and so on.

The zeroth-order solution may be given by

$$x_0 = \bar{A}\cos(\Omega t + \theta) \tag{3.46}$$

with $\bar{A}$ and θ being integral constants. Then the first-order equation takes the form

$$L_\Omega x_1 = \left(\bar{A}\omega_1^2 - \frac{3}{4}\bar{A}^3\right)\cos\phi(t) - \frac{\bar{A}^3}{4}\cos 3\phi(t) =: b_1(t) \tag{3.47}$$

with

$$\phi(t) := \Omega t + \theta. \tag{3.48}$$

Now we recognize the first term of the inhomogeneous term $b_1(t)$ is a zero mode of the unperturbed equation and would produce a *secular term*, which should be avoided. This condition can be formulated in terms of the solvability condition.

We define an inner product for two periodic functions u and v as

$$(u,\, v) := \int_0^{2\pi/\Omega} dt\, u^*(t) v(t). \tag{3.49}$$

Incidentally, the zero modes (3.41) satisfy the ortho-normal condition with this inner product as

$$(X_0^{(i)},\, X_0^{(j)}) = \delta_{ij}, \quad (i,\, j = 1,\, 2). \tag{3.50}$$

Then the solvability condition of Eq. (3.47) means that the inner product of the right-hand side of (3.47) with the zero modes of L_Ω should vanish: While one of them

$$(X_0^{(2)},\, b_1) = 0 \tag{3.51}$$

is satisfied automatically, the other yields

$$(X_0^{(1)},\, b_1) = \sqrt{\frac{\pi}{\Omega}} \left(\bar{A}\omega_1^2 - \frac{3}{4}\bar{A}^3 \right) = 0, \tag{3.52}$$

which leads to

$$\omega_1^2 = \frac{3}{4}\bar{A}^2. \tag{3.53}$$

Thus on account of (3.42), we have

$$\Omega^2 \simeq 1 + \epsilon\frac{3}{4}\bar{A}^2 \quad \text{or} \quad \Omega \simeq 1 + \epsilon\frac{3}{8}\bar{A}^2. \tag{3.54}$$

Then (3.47) now reads

$$L_\Omega x_1 = -\frac{\bar{A}^3}{4}\cos(3\Omega t + 3\theta), \tag{3.55}$$

which is readily solved as

$$x_1 = \frac{\bar{A}^3}{32\Omega^2}\cos(3\Omega t + 3\theta) \equiv A_1 \cos 3\phi(t), \quad \left(A_1 := \frac{\bar{A}^3}{32\Omega^2} \right). \tag{3.56}$$

We remark that the last result can be easily obtained by formally multiplying the inverse operator

$$L_{\Omega}^{-1} = \frac{1}{\frac{d^2}{dt^2} + \Omega^2} \tag{3.57}$$

to (3.55) from the left as

$$\begin{aligned} x_1 &= -\frac{\bar{A}^3}{4}\frac{1}{\frac{d^2}{dt^2} + \Omega^2}\cos(3\Omega t + 3\theta) = -\frac{\bar{A}^3}{4}\frac{1}{-(3\Omega)^2 + \Omega^2}\cos(3\Omega t + 3\theta) \\ &= \frac{\bar{A}^3}{32\Omega^2}\cos(3\Omega t + 3\theta), \end{aligned} \tag{3.58}$$

where the fact that $\cos(3\Omega t + 3\theta)$ is an eigen function of L_{Ω} with a non-vanishing eigenvalue has been utilized;

$$L_{\Omega}\cos(3\Omega t + \theta) = -8\Omega^2 \cos(3\Omega t + 3\theta). \tag{3.59}$$

Summing up the zeroth and first-order solutions, we get for $x(t)$ up to the second order of ϵ,

$$x(t) \simeq \bar{A}\cos(\Omega t + \theta) + \epsilon\frac{\bar{A}^3}{32\Omega^2}\cos(3\Omega t + 3\theta) \tag{3.60}$$

with Ω defined in (3.54), which is in accordance with the expanded formula of the exact solution (2.67).

3.5 Krylov-Bogoliubov-Mitropolsky Method

The perturbative solution (3.60) to Duffing equation suggests that the perturbative solution would be expressed as a sum of the unperturbed solution $x_0(t)$ with the amplitude and frequency modified by the perturbation and the terms with higher harmonics.

In the Krylov-Bogoliubov-Mitropolsky (KBM) method [20, 25] developed for weakly nonlinear oscillators, this is taken as the general ansatz of the solution, and transform the equations to those for the amplitude and phase. We shall see again that the solvability conditions for the converted equations play an essential role for the KBM method to work.

Because the KBM method involves a tedious calculation for converting the time derivatives to the differentiation with respect to the amplitude and phase, we first apply the method to a simplest example of a linear damped oscillator and then later to nonlinear oscillators such as Duffing and van der Pol equations.

3.5.1 Generalities

We treat weakly nonlinear oscillator which is written with use of the linear operator $L = \frac{d^2}{dt^2} + 1$ as

$$Lx = \epsilon F(x, \dot{x}, t). \tag{3.61}$$

Inserting the perturbative expansion $x = x_0 + \epsilon x_1 + \epsilon^2 x_2 + \cdots$ into (3.61) and equating the coefficients of ϵ^n $(n = 0, 1, 2, \ldots)$, we have

$$Lx_0 = 0, \tag{3.62}$$

$$Lx_1 = F(x_0, \dot{x}_0, t), \tag{3.63}$$

$$Lx_2 = \left.\frac{\partial F}{\partial x}\right|_{x_0, \dot{x}_0} x_1 + \left.\frac{\partial F}{\partial \dot{x}}\right|_{x_0, \dot{x}_1} \dot{x}_1, \tag{3.64}$$

and so on.

The zeroth order solution may be expressed as

$$x_0 = \bar{A}\,\sin(t + \theta) =: u_0(\bar{A}, \phi_0), \quad (\phi_0 := t + \theta). \tag{3.65}$$

In the KBM method, respecting the above form of the zeroth order solution, we assume that the exact solution takes the following form

$$x = u(A, \phi) = u_0(A, \phi) + \rho(A, \phi) = A \sin\phi + \rho(A, \phi) \tag{3.66}$$

with the expanded form of $\rho(A, \phi)$,

$$\rho(A, \phi) = \epsilon\rho_1(A, \phi) + \epsilon^2\rho_2(A, \phi) + \cdots. \tag{3.67}$$

To introduce another important condition for $\rho(A, \phi)$, let us introduce a linear operator L_ϕ defined by

$$L_\phi := \frac{d^2}{d\phi^2} + 1, \tag{3.68}$$

acting on function of $\phi \in [0, 2\pi)$. It has the zero modes given by $\cos\phi =: u_0^{(1)}(\phi)$ and $\sin\phi =: u_0^{(2)}$;

$$L_\phi u_0^{(i)} = 0, \quad (i = 1, 2). \tag{3.69}$$

The constraint to be imposed on $\rho_i(A, \phi)$ is that it does not contain the zero modes as expressed by

$$
\begin{aligned}
(u_0^{(1)}, \rho_i) &= \int_0^{2\pi} d\phi \cos\phi \, \rho_i(A, \phi) = 0, \\
(u_0^{(2)}, \rho_i) &= \int_0^{2\pi} d\phi \sin\phi \, \rho_i(A, \phi) = 0,
\end{aligned}
\tag{3.70}
$$

where (u, v) denotes an inner product for arbitrary functions $u(\phi)$ and $v(\phi)$ defined by

$$
(u, v) := \int_0^{2\pi} d\phi \, u^*(\phi) \, v(\phi). \tag{3.71}
$$

The whole time dependence of $x(t)$ is given through those of $A(t)$ and $\phi(t)$, and their time-dependence themselves are assumed to be given by the following differential equations

$$
\frac{dA}{dt} = F(A), \quad \frac{d\phi}{dt} = 1 + \Omega(A), \tag{3.72}
$$

respectively, where the following expansion formulae are assumed

$$
F(A) = \epsilon F_1(A) + \epsilon^2 F_2(A) + \cdots, \quad \Omega(A) = \epsilon\Omega_1(A) + \epsilon^2\Omega_2(A) + \cdots. \tag{3.73}
$$

The notable point is that their time-dependence is assumed to be governed solely by the amplitude A.

So far the rather long preparation to solve the equation in the KBM method. It should be noted that with the above set-up of the expansion ansatz, the naïve expansion scheme as given in (3.64) is abandoned and somehow reorganized.

Now inserting the form (3.66) into (3.61), the equation is converted into those with respect to $\frac{\partial}{\partial A}$ and $\frac{\partial}{\partial \phi}$ by the use of chain rule, as follows;

$$
\frac{dx}{dt} = \frac{du(A, \phi)}{dt} = \frac{dA}{dt}\frac{\partial u}{\partial A} + \frac{d\phi}{dt}\frac{\partial u}{\partial \phi} = F(A)\frac{\partial u}{\partial A} + \big(1 + \Omega(A)\big)\frac{\partial u}{\partial \phi}, \tag{3.74}
$$

i.e.,

$$
\frac{d}{dt} = F(A)\frac{\partial}{\partial A} + \big(1 + \Omega(A)\big)\frac{\partial}{\partial \phi}. \tag{3.75}
$$

Inserting (3.66) into the above formula, we have

$$
\begin{aligned}
\frac{dx}{dt} &= F(A)(\sin\phi + \frac{\partial \rho}{\partial A}) + \big(1 + \Omega(A)\big)\big(A\cos\phi + \frac{\partial \rho}{\partial \phi}\big) \\
&= A\cos\phi + \left(F(A)\sin\phi + \Omega(A)A\cos\phi + \frac{\partial \rho}{\partial \phi}\right) + \left[F(A)\frac{\partial \rho}{\partial A} + \Omega(A)\frac{\partial \rho}{\partial \phi}\right],
\end{aligned}
\tag{3.76}
$$

where the first, second and third terms are order ϵ^0, ϵ^1 and ϵ^2, respectively. Applying the formula (3.75), we get the second derivative $\frac{d^2x}{dt^2}$, for which we need the following formulae,

$$\begin{aligned}\frac{d}{dt}A\cos\phi &= \left[F(A)\frac{\partial}{\partial A} + (1+\Omega(A))\frac{\partial}{\partial\phi}\right]A\cos\phi = F(A)\cos\phi - (1+\Omega(A))A\sin\phi \\ &= -A\sin\phi + (F(A)\cos\phi - A\Omega(A)\sin\phi),\end{aligned} \tag{3.77}$$

$$\begin{aligned}&\frac{d}{dt}\left(F(A)\sin\phi + \Omega A\cos\phi + \frac{\partial\rho}{\partial\phi}\right) \\ &= F(A)\left(\frac{dF(A)}{dA}\sin\phi + \Omega\cos\phi + \frac{\partial^2\rho}{\partial A\partial\phi}\right) + (1+\Omega(A))\left(F(A)\cos\phi - \Omega A\sin\phi + \frac{\partial^2\rho}{\partial\phi^2}\right) \\ &= \left(F(A)\cos\phi - \Omega A\sin\phi + \frac{\partial^2\rho}{\partial\phi^2}\right) + \left[F(A)\left(\frac{dF(A)}{dA}\sin\phi + \Omega\cos\phi + \frac{\partial^2\rho}{\partial A\partial\phi}\right)\right. \\ &\qquad\left. + \Omega(A)\left(F(A)\cos\phi - \Omega A\sin\phi + \frac{\partial^2\rho}{\partial\phi^2}\right)\right],\end{aligned} \tag{3.78}$$

$$\frac{d}{dt}\left[F(A)\frac{\partial\rho}{\partial A} + \Omega(A)\frac{\partial\rho}{\partial\phi}\right] = \left[F(A)\frac{\partial^2\rho}{\partial\phi\partial A} + \Omega(A)\frac{\partial^2\rho}{\partial^2\phi}\right] + \mathrm{o}(\epsilon^3), \tag{3.79}$$

where $\mathrm{o}(\epsilon^3)$ denotes terms of the order ϵ^n with $n \geq 3$. Thus we finally arrive at

$$\begin{aligned}\ddot{x} + x &= \frac{\partial^2\rho}{\partial\phi^2} + \rho + (2F\cos\phi - 2A\Omega\sin\phi) + F\left(\frac{dF}{dA}\sin\phi + \Omega\cos\phi + \frac{\partial^2\rho}{\partial A\partial\phi}\right) \\ &\quad + \Omega(F\cos\phi - \Omega A\sin\phi + \frac{\partial^2\rho}{\partial\phi^2}) + F(A)\frac{\partial^2\rho}{\partial\phi\partial A} + \Omega(A)\frac{\partial^2\rho}{\partial\phi^2} + \mathrm{o}(\epsilon^3).\end{aligned} \tag{3.80}$$

3.5.2 *Damped Oscillator*

We first treat the simple linear equation of a damped oscillator (2.6), for which the inhomogeneous term in (3.61) reads

$$F(x, \dot{x}, t) = -2\dot{x}. \tag{3.81}$$

With use of (3.80) and (3.76), Eq. (3.61) takes the following form

$$
\begin{aligned}
\frac{\partial^2 \rho}{\partial \phi^2} + \rho = & -2(\epsilon A + F)\cos\phi + 2A\Omega\sin\phi \\
& - 2\epsilon\left(F\sin\phi + \Omega A\cos\phi + \frac{\partial \rho}{\partial \phi}\right) - F\left(\frac{dF}{dA}\sin\phi + \Omega\cos\phi + \frac{\partial^2 \rho}{\partial A \partial \phi}\right) \\
& - \Omega\left(F\cos\phi - \Omega A\sin\phi + \frac{\partial^2 \rho}{\partial \phi^2}\right) - F(A)\frac{\partial^2 \rho}{\partial \phi \partial A} - \Omega(A)\frac{\partial^2 \rho}{\partial \phi^2} \\
& + \mathrm{o}(\epsilon^3) \\
=: & \, b(\phi) + \mathrm{o}(\epsilon^3).
\end{aligned}
\tag{3.82}
$$

The first order in ϵ of Eq. (3.82) now takes the form

$$
L_\phi \rho_1 = -2(A + F_1)\cos\phi + 2A\Omega_1 \sin\phi =: b_1(\phi). \tag{3.83}
$$

The *solvability condition* of (3.83) is given by

$$
(u_0^{(1)}, b_1) = 0 \quad \text{and} \quad (u_0^{(2)}, b_1) = 0, \tag{3.84}
$$

which lead to

$$
F_1 = -A, \quad \Omega_1 = 0, \tag{3.85}
$$

and hence $b_1 = 0$ implying that

$$
\rho_1 = 0. \tag{3.86}
$$

Thus we have

$$
\frac{dA}{dt} = -\epsilon A, \quad \frac{d\phi}{dt} = 1 \tag{3.87}
$$

up to the second order of ϵ^2. The equations are readily solved to give

$$
A(t) = \bar{A}\mathrm{e}^{-\epsilon t}, \quad \phi(t) = t + \theta. \tag{3.88}
$$

Inserting Eqs. (3.85) and (3.86), the second-order terms of Eq. (3.82) reads

$$
\begin{aligned}
L_\phi \rho_2 = & -2F_2\cos\phi + 2A\Omega_2\sin\phi - 2(F_1\sin\phi + \Omega_1 A\cos\phi + \frac{\partial \rho_1}{\partial \phi}) \\
& - F_1\left(\frac{dF_1}{dA}\sin\phi + \Omega_1\cos\phi + \frac{\partial^2 \rho_1}{\partial A \partial \phi}\right) \\
= & -2F_2\cos\phi + 2\,A\Omega_2\sin\phi + 2A\sin\phi - A\sin\phi \\
= & -2F_2\cos\phi + A(2\Omega_2 + 1)\sin\phi \\
=: & \, b_2(\phi)
\end{aligned}
\tag{3.89}
$$

The solvability condition of the last equation reads

$$(u_0^{(1)},\ b_2) = 0, \quad (u_0^{(2)},\ b_2) = 0, \tag{3.90}$$

which lead to

$$F_2 = 0, \quad \Omega_2 = -\frac{1}{2} \tag{3.91}$$

and hence $b_2 = 0$ implying that

$$\rho_2 = 0. \tag{3.92}$$

Thus up to the third order we have

$$\frac{dA}{dt} = -\epsilon A, \quad \frac{d\phi}{dt} = 1 - \frac{1}{2}\epsilon^2, \tag{3.93}$$

the solutions to which read

$$A(t) = \bar{A}\mathrm{e}^{-\epsilon t}, \quad \phi(t) = (1 - \frac{1}{2}\epsilon^2)t + \theta, \tag{3.94}$$

respectively. Thus we have

$$x(t) = \bar{A}\mathrm{e}^{-\epsilon t} \sin\left((1 - \frac{1}{2}\epsilon^2)t + \theta\right). \tag{3.95}$$

We note that the frequency $\Omega = 1 - \frac{1}{2}\epsilon^2$ is correct up to the third order because the exact frequency is expanded as $\sqrt{1-\epsilon^2} \simeq 1 - \epsilon^2 - \frac{1}{8}\epsilon^4$.

3.5.3 *Duffing Equation*

Next we treat the Duffing equation in the KBM method: $\ddot{x} + x = -\epsilon x^3$.

On account of (3.80), the equation is converted to the following form;

$$\begin{aligned} L_\phi \rho = &-(2F\cos\phi - 2A\Omega\sin\phi) \ - F\left(\frac{dF}{dA}\sin\phi + \Omega\cos\phi + \frac{\partial^2\rho}{\partial A\partial\phi}\right) \\ &- \Omega\left(F\cos\phi - \Omega A\sin\phi + \frac{\partial^2\rho}{\partial\phi^2}\right) \ - F(A)\frac{\partial^2\rho}{\partial\phi\partial A} \ - \Omega(A)\frac{\partial^2\rho}{\partial\phi^2} \\ &- \epsilon(A\sin\phi + \rho)^3 \ + \mathrm{o}(\epsilon^3). \end{aligned} \tag{3.96}$$

Equating the terms of ϵ^n $(n = 0,\ 1,\ 2,\ \ldots)$, we have

$$\begin{aligned} L_\phi \rho_1 &= -2F_1 \cos\phi + 2A\Omega_1 \sin\phi - A^3 \sin^3\phi \\ &= -2F_1 \cos\phi + A\left(2\Omega_1 - \frac{3}{4}A^2\right)\sin\phi + \frac{A^3}{4}\sin 3\phi \\ &=: b_1(\phi). \end{aligned} \tag{3.97}$$

The solvability condition $(\cos\phi,\ b_1) = (\sin\phi,\ b_1) = 0$ leads to

$$F_1 = 0, \quad \Omega_1 = \frac{3}{8}A^2. \tag{3.98}$$

Then we have the equation for ρ_1

$$L_\phi \rho_1 = \frac{A^3}{4}\sin 3\phi, \tag{3.99}$$

which yields

$$\rho_1 = -\frac{A^3}{32}\sin 3\phi. \tag{3.100}$$

Thus we have up to the second order of ϵ

$$x(t) = A\sin(\Omega t + \theta) - \epsilon\frac{A^3}{32}\sin 3(\Omega t + \theta), \quad \text{with } \Omega = 1 + \epsilon\frac{3A^2}{8} \tag{3.101}$$

which agrees with the result given in the Lindstedt-Poincaré method (3.60) up to ϵ^2.

3.5.4 The van der Pol Equation

Next we consider the van der Pol equation given by (2.87), which is reproduced here,

$$\ddot{x} + x = \epsilon(1 - x^2)\dot{x}. \tag{3.102}$$

The first order-equation reads

$$\begin{aligned} L_\phi \rho_1 &= 2A\Omega_1 \cos\phi + 2F_1 \sin\phi - \left(A - \frac{A^2}{4}\right)\sin\phi + \frac{A^3}{4}\sin 3\phi \\ &=: b_1(\phi). \end{aligned} \tag{3.103}$$

The solvability condition $(\sin\phi,\ b_1) = (\cos\phi,\ b_1) = 0$ leads to

$$F_1 = \frac{1}{2}A\left(1 - \frac{A^2}{4}\right), \quad \Omega_1 = 0. \tag{3.104}$$

Thus the first-order equation is reduced to

$$\frac{\partial^2 \rho_1}{\partial \phi^2} + \rho_1 = \frac{A^3}{4} \sin 3\phi, \tag{3.105}$$

yielding the solution

$$\rho_1(\phi) = -\frac{A^3}{32} \sin 3\phi, \tag{3.106}$$

and the amplitude and phase equations are found to take the following forms, respectively,

$$\frac{dA}{dt} = \frac{\epsilon}{2} A \left(1 - \frac{A^2}{4}\right), \quad \frac{d\phi}{dt} = 1. \tag{3.107}$$

The phase equation is readily solved to give

$$\phi(t) = t + \theta_0. \tag{3.108}$$

The amplitude equation has fixed points $A = 0$ and $A = 2$, the latter of which suggests the existence of a limit cycle. Indeed, the solution to the amplitude equation expressed as

$$\frac{1}{2}\epsilon t = \int_{A_0}^{A(t)} \frac{dA}{A(1 - \frac{A^2}{4})} =: I(A(t),\, A_0). \tag{3.109}$$

The integral of the right-hand side can be done as follows;

$$I(A(t),\, A_0) = \frac{1}{2} \int_{A_0^2}^{A^2(t)} \frac{dB}{B(1 - \frac{B}{4})} = \ln \frac{A^2(t)}{|1 - A^2(t)/4|} + C \tag{3.110}$$

with $C = -\ln \frac{A_0^2}{|1 - A_0^2/4|}$. Thus we have

$$A(t) = \frac{2}{\sqrt{1 + (4/A_0^2 - 1)\mathrm{e}^{-\epsilon t}}}, \tag{3.111}$$

which approaches 2 when $t \to \infty$, showing a limit cycle with the radius 2. Then collecting all the results in this order, we have

$$x(t) = \frac{2 \sin(t + \theta_0)}{\sqrt{1 + (4/A_0^2 - 1)\mathrm{e}^{-\epsilon t}}} - \epsilon \frac{A^3(t)}{32} \sin(3t + 3\theta_0). \tag{3.112}$$

3.6 Multiple-Scale Method

In order to get motivated for the multiple-scale method, let us first look into the solution of Duffing equation given up to ϵ^2 in the previous sections; (3.60) or (3.101). The lowest harmonics part of the solution is rewritten as

$$A\sin(\Omega t+\theta)=A\cos\left(\epsilon\frac{3A^2}{8}t\right)\sin(t+\theta)+A\sin\left(\epsilon\frac{3A^2}{8}t\right)\cos(t+\theta). \quad (3.113)$$

One may recognize that it has two time scales, *i.e.*, the fast motion with frequency 1 and the slow motion with frequency $\epsilon\frac{3A^2}{8}$, which may be interpreted as a slowly varying amplitude.

Similarly, the solution of the van der Pol equation (3.112) up to ϵ^2 in the previous subsection also contains two time scales; a fast oscillation with frequency 1 and a slow variation of the amplitude with a time scale of $1/\epsilon$. It is naturally expected that more time scales will appear when higher orders of ϵ are incorporated.

With this observation kept in mind, one introduces *multiple times* [20–22] as

$$t,\ t_1:=\epsilon t,\ t_2:=\epsilon^2 t,\ \cdots, \quad (3.114)$$

and assumes that all the times t and t_n $(n=1,\ 2,\ \dots)$ are all mutually independent in the multiple-scale method. For instance, we consider a function $f(t_1)$ of $t_1=\epsilon t$. Differentiating it with respect to t, we have

$$\frac{df}{dt}=\frac{dt_1}{dt}\frac{df}{dt_1}=\epsilon\frac{df}{dt_1}, \quad (3.115)$$

which may be regarded to be vanishingly small when ϵ is sufficiently small, and hence $f(t_1)$ may regarded independent of t. In general, the solution to differential equations containing a small parameter ϵ is expressed as

$$x=x(t,t_1,\ t_2,\ \cdots). \quad (3.116)$$

Then

$$\begin{aligned}\frac{dx}{dt}&=\frac{dt}{dt}\frac{\partial x}{\partial t}+\frac{dt_1}{dt}\frac{\partial x}{\partial t_1}+\frac{dt_2}{dt}\frac{\partial x}{\partial t_2}+\cdots\\&=\frac{\partial x}{\partial t}+\epsilon\frac{\partial x}{\partial t_1}+\epsilon^2\frac{\partial x}{\partial t_2}+\cdots,\end{aligned} \quad (3.117)$$

or formally,

$$\frac{d}{dt}=\frac{\partial}{\partial t}+\epsilon\frac{\partial}{\partial t_1}+\epsilon^2\frac{\partial}{\partial t_2}+\cdots. \quad (3.118)$$

The second derivative is similarly expressed as

$$\begin{aligned}\frac{d^2}{dt^2} &= \frac{d}{dt}\frac{d}{dt} \\ &= \left(\frac{\partial}{\partial t} + \epsilon\frac{\partial}{\partial t_1} + \epsilon^2\frac{\partial}{\partial t_2} + \cdots\right)\left(\frac{\partial}{\partial t} + \epsilon\frac{\partial}{\partial t_1} + \epsilon^2\frac{\partial}{\partial t_2} + \cdots\right) \\ &= \frac{\partial^2}{\partial t^2} + 2\epsilon\frac{\partial^2}{\partial t\partial t_1} + \epsilon^2\left(\frac{\partial^2}{\partial t_1^2} + 2\frac{\partial^2}{\partial t\partial t_2}\right) + \cdots .\end{aligned} \tag{3.119}$$

In this subsection, we shall make an introductory account of this method using a few examples.

3.6.1 *Duffing Equation*

As the first example, let us take the Duffing equation $\frac{d^2x}{dt^2} + 1 = -\epsilon x^3$. We expand the solution x in terms of the powers of ϵ^n $(n = 0, 1, 2, \ldots)$;

$$x = x_0 + \epsilon x_1 + \epsilon^2 x_2 + \cdots . \tag{3.120}$$

Notice that all x_i's are functions of $t, t_1, t_2, \ldots$;

$$x_i = x_i(t, t_1, t_2, \ldots). \tag{3.121}$$

Then applying the formula (3.119) and equating the coefficients of the respective order of ϵ^n, we have a series of equations as follows

$$Lx_0 = 0, \tag{3.122}$$

$$Lx_1 = -\left(2\frac{\partial^2 x_0}{\partial t\partial t_1} + x_0^3\right), \tag{3.123}$$

$$Lx_2 = -\left(2\frac{\partial^2 x_1}{\partial t\partial t_1} + 3x_0^2 x_1 + \frac{\partial^2 x_0}{\partial t_1^2} + 2\frac{\partial^2 x_0}{\partial t\partial t_2}\right), \tag{3.124}$$

and so on, where $L := \frac{\partial^2}{\partial t^2} + 1$, which has two independent zero modes

$$x_0^{(1)}(t) := \sin(t + \theta). \quad x_0^{(2)} := \cos(t + \theta) \tag{3.125}$$

with θ being arbitrary constant. We define the inner product for two periodic functions u and v of t as follows

$$(u, v) = \int_0^{2\pi} dt\, u^*(t)v(t). \tag{3.126}$$

The zeroth-order solution reads

$$x_0 = A\sin(t+\theta), \tag{3.127}$$

where the would-be integral constants A and θ are actually functions of the remaining independent variables;

$$A = A(t_1, t_2, \ldots), \quad \theta = \theta(t_1, t_2, \ldots). \tag{3.128}$$

Thus, we have

$$\frac{\partial^2 x_0}{\partial t \partial t_1} = \frac{\partial}{\partial t_1} A\cos(t+\theta) = \frac{\partial A}{\partial t_1}\cos(t+\theta) - A\frac{\partial \theta}{\partial t_1}\sin(t+\theta). \tag{3.129}$$

The first-order equation now takes the following form

$$\begin{aligned}\frac{\partial^2 x_1}{\partial t^2} + x_1 &= -2\left(\frac{\partial A}{\partial t_1}\cos(t+\theta) - A\frac{\partial \theta}{\partial t_1}\sin(t+\theta)\right) - A^3\sin^3(t+\theta),\\ &= 2\left(A\frac{\partial \theta}{\partial t_1} - \frac{3A^3}{8}\right)\sin(t+\theta) - 2\frac{\partial A}{\partial t_1}\cos(t+\theta) + \frac{A^3}{4}\sin(3t+3\theta)\\ &=: b_1(t, t_1, t_2, \ldots).\end{aligned} \tag{3.130}$$

Here we have used the formula; $\sin^3\phi = \frac{1}{4}(3\sin\phi - \sin 3\phi)$.

The solvability condition of the last equation reads

$$(x_0^{(1)}, b_1) = (x_0^{(2)}, b_1) = 0, \tag{3.131}$$

which lead to the following equations

$$\frac{\partial \theta}{\partial t_1} = \frac{3A^2}{8}, \quad \frac{\partial A}{\partial t_1} = 0. \tag{3.132}$$

These equations are readily solved to give

$$\theta = \frac{3A^2}{8}t_1 + \theta_1, \quad A = A(t_2, t_3, \ldots), \tag{3.133}$$

where θ_1 is independent of t and t_1 but may depend on $(t_2, t_3, \ldots)$.

The first-order equation (3.130) is now solvable and yields

$$x_1 = -\frac{A^3}{32}\sin(3t + 3\theta_1). \tag{3.134}$$

Then we have the perturbative solution up to ϵ^2 as

$$x = A\sin(\Omega t + \theta_1) - \epsilon\frac{A^3}{32}\sin(3t + 3\theta_1), \quad \left(\Omega = 1 + \epsilon\frac{3A^2}{8}\right), \tag{3.135}$$

which agrees with the solution obtained repeatedly.

3.6.2 Bifurcation in the Lorenz Model

The Lorenz model or Lorenz equation [84] is a system consisting of three variables that is reduced from a simplified model for describing the atmospheric convection by B. Saltzman [85]; this was actually the model equation for a two-dimensional fluid layer heated from below and cooled from above uniformly. Interestingly enough, the Lorenz equation has also some relevance to other fields of science [86] such as laser physics, dynamos, Rikitake model of gyromagnetic reversal phenomenon and so on. The Lorenz model not only describes the onset of the convection with varied parameter but also shows a remarkable initial-value sensitivity of the solution for a range of the parameters and the initial conditions. This is a chaos, as we now call it.

Now the Lorenz equation is given by

$$\begin{aligned} \dot{\xi} &= \sigma(-\xi + \eta), \\ \dot{\eta} &= r\xi - \eta - \xi\zeta, \\ \dot{\zeta} &= \xi\eta - b\zeta. \end{aligned} \tag{3.136}$$

The equation is rewritten as follows

$$\frac{d\boldsymbol{X}}{dt} = \boldsymbol{F}(\boldsymbol{X}; r) \tag{3.137}$$

with

$$\boldsymbol{X} := \begin{pmatrix} \xi \\ \eta \\ \zeta \end{pmatrix}, \quad \boldsymbol{F}(\boldsymbol{X}; r) := \begin{pmatrix} \sigma(-\xi + \eta) \\ -\xi\zeta + r\xi - \eta \\ \xi\eta - b\zeta \end{pmatrix}. \tag{3.138}$$

When $\boldsymbol{F}(\boldsymbol{X}; r)$ vanishes at $\boldsymbol{X} = \boldsymbol{X}_0(r) = {}^t(\xi_0(r), \eta_0(r), \zeta_0(r))$, *i.e.*,

$$\boldsymbol{F}(\boldsymbol{X}_0(r); r) = 0, \tag{3.139}$$

$\dot{\boldsymbol{X}} = 0$ at this point. This means that $\boldsymbol{X} = \boldsymbol{X}_0(r) =$ const. is a special solution to (3.138). Such a solution is called a **steady solution** and the point $\boldsymbol{X} = \boldsymbol{X}_0(r)$ is called a **fixed point**.

The steady state is obtained from the following equation;

$$\sigma(-\xi_0 + \eta_0) = 0 \quad -\xi_0\zeta_0 + r\xi_0 - \eta_0 = 0 \quad \xi_0\eta_0 - b\zeta_0 = 0, \tag{3.140}$$

which is readily solved to give the following two solutions

$$X_0^{(A)} = \begin{pmatrix} \xi_0^{(A)} \\ \eta_0^{(A)} \\ \zeta_0^{(A)} \end{pmatrix} = \begin{pmatrix} 0 \\ 0 \\ 0 \end{pmatrix}, \quad X_0^{(B)} = \begin{pmatrix} \xi_0^{(B)} \\ \eta_0^{(B)} \\ \zeta_0^{(B)} \end{pmatrix} = \begin{pmatrix} \pm\sqrt{b(r-1)} \\ \pm\sqrt{b(r-1)} \\ r-1 \end{pmatrix}, \quad (r > 1), \tag{3.141}$$

the latter of which get to exist only when $r > 1$.

The linear stability analysis [34] shows that $X_0^{(A)}$ is stable for $0 < r < 1$ but unstable for $r > 1$, while $X_0^{(B)}$ is stable for $1 < r < \sigma(\sigma + b + 3)/(\sigma - b - 1) =: r_c$ but unstable for $r > r_c$. This is a typical bifurcation phenomenon known in dynamical systems [34]. In this subsection, we shall confine ourselves to examining the non-linear dynamics around the **origin** $X_0^{(A)}$ for $r \sim 1$, and derive a reduced dynamical equation and an invariant manifold of one dimension on which the variables (ξ, η, ζ) is confined by the multi-scale method. It should be mentioned that the existence of the reduced dynamics and the invariant manifold is supported by the center manifold theorem [34].

We parametrize r as

$$r = 1 + \mu \quad \text{and} \quad \mu = \chi\epsilon^2, \quad \chi = \text{sgn}\mu. \tag{3.142}$$

We expand the quantities as Taylor series of ϵ:

$$X = \epsilon X_1 + \epsilon^2 X_2 + \epsilon^3 X_3 + \cdots, \tag{3.143}$$

where

$$X_i = \begin{pmatrix} \xi_i \\ \eta_i \\ \zeta_i \end{pmatrix}, \quad (i = 1, 2, 3, \ldots). \tag{3.144}$$

To apply the multiple-scale method, we introduce the multi-scales of time as

$$t, \ t_1 := \epsilon t, \ t_2 := \epsilon^2 t, \ \cdots, \tag{3.145}$$

and assume that all the times t and t_n $(n = 1, 2, \ldots)$ are all mutually independent as before. Thus

$$X_i = X_i(t, t_1, t_2, \ldots). \tag{3.146}$$

The first-order equation reads

$$\left(\frac{\partial}{\partial t} - L_0\right) X_1 = \mathbf{0}, \tag{3.147}$$

where

$$L_0 = \begin{pmatrix} -\sigma & \sigma & 0 \\ 1 & -1 & 0 \\ 0 & 0 & -b \end{pmatrix}, \tag{3.148}$$

the eigenvalues of which are found to be

$$\lambda_1 = 0, \quad \lambda_2 = -\sigma - 1, \quad \lambda_3 = -b. \tag{3.149}$$

The respective eigenvectors are

$$\boldsymbol{U}_1 = \begin{pmatrix} 1 \\ 1 \\ 0 \end{pmatrix}, \quad \boldsymbol{U}_2 = \begin{pmatrix} \sigma \\ -1 \\ 0 \end{pmatrix}, \quad \boldsymbol{U}_3 = \begin{pmatrix} 0 \\ 0 \\ 1 \end{pmatrix}. \tag{3.150}$$

Since the linear operator L_0 is an asymmetric matrix, the left and the right eigenvectors are different[2]: The left eigenvector $\tilde{\boldsymbol{U}}_i$ satisfies the equation

$$L_0^{\dagger}\tilde{\boldsymbol{U}}_i = \lambda_i\tilde{\boldsymbol{U}}_i, \quad (i = 1,\, 2,\, 3). \tag{3.151}$$

We find the normalized left eigenvectors as follows,

$$\tilde{\boldsymbol{U}}_1 = \frac{1}{1+\sigma}\begin{pmatrix} 1 \\ \sigma \\ 0 \end{pmatrix}, \quad \tilde{\boldsymbol{U}}_2 = \frac{1}{1+\sigma}\begin{pmatrix} 1 \\ -1 \\ 0 \end{pmatrix}, \quad \tilde{\boldsymbol{U}}_3 = \begin{pmatrix} 0 \\ 0 \\ 1 \end{pmatrix}. \tag{3.152}$$

Indeed the following orthonormal conditions hold:

$$(\tilde{\boldsymbol{U}}_i,\, \boldsymbol{U}_j) = \tilde{\boldsymbol{U}}_i^{*} \cdot \boldsymbol{U}_j = \delta_{ij}. \tag{3.153}$$

The division of unity is represented as

$$I = \sum_{i=1}^{3} \boldsymbol{U}_i\,{}^t\tilde{\boldsymbol{U}}_i = \sum_{i=1}^{3} |\boldsymbol{U}_i\rangle\langle\tilde{\boldsymbol{U}}_i|, \tag{3.154}$$

where Dirac's bra-ket notation has been used. Then any three-dimensional vector $\boldsymbol{X}_0$ can be represented as a linear combination of $\boldsymbol{U}_i$'s ($i = 1,\, 2,\, 3$) as

[2] A somewhat detailed account of left and right eigenvectors of an asymmetric matrix is given in Sect. 3.6.2.

$$X_0 = IX_0 = \left(\sum_{i=1}^{3} U_i{}^t\tilde{U}_i\right) X_0 = \sum_{i=1}^{3} U_i\,(\tilde{U}_i,\, X_0)$$
$$= \sum_{i=1}^{3} C_i U_i, \qquad (C_i := (\tilde{U}_i,\, X_0)). \tag{3.155}$$

The general solution $X_1(t)$ of the first-order equation with the initial condition $X_1(0) = X_0 = \sum_{i=1}^{3} C_i U_i$ is expressed as

$$X_1(t) = \mathrm{e}^{L_0 t} X_0 = \mathrm{e}^{L_0 t} \sum_{i=1}^{3} C_i U_i = \sum_{i=1}^{3} C_i \mathrm{e}^{L_0 t} U_i$$
$$= AU_1 + \sum_{i=2}^{3} C_i e^{\lambda_i t} U_i, \tag{3.156}$$

where we renamed C_1 as A in the last equality.

Since both λ_2 and λ_3 are negative values, the terms with $\mathrm{e}^{\lambda_{2,3} t}$ die out asymptotically when $t \to \infty$. If we are interested in such an asymptotic state, we may only keep the neutrally stable solution

$$X_1(t, t_1, \ldots) \simeq A(t_1, \ldots) U_1, \qquad \text{for} \quad t \to \infty, \tag{3.157}$$

where we have made it explicit that the solution may depend on the slower times $t_i\,(i = 1, 2, \ldots)$:

$$\xi_1(t, t_1, \ldots) = A, \quad \eta_1(t, t_1, \ldots) = A, \quad \zeta_1(t, t_1, \ldots) = 0. \tag{3.158}$$

The second order equation now reads

$$\left(\frac{d}{dt} - L_0\right) X_2 = -\frac{\partial X_1}{\partial t_1} + \begin{pmatrix} 0 \\ -\xi_1\zeta_1 \\ \xi_1\eta_1 \end{pmatrix} = -\frac{\partial A}{\partial t_1} U_1 + A^2 U_3 =: b_2 \tag{3.159}$$

In contrast to all the previous examples, the unperturbed solutions are not of oscillatory nature but of exponential decay, and we need to construct the inner product properly to formulate the *solvability condition*. For this purpose, we recall what was the practical role of the solvability condition of the perturbed equations. That is to circumvent the appearance of the secular terms, which invalidates the perturbative expansion after a long time although they can be a remnant of hidden slow modes of the original equation.

Let us consider a particular solution to the equation given by

$$(d_t - L_0) X(t) = b, \tag{3.160}$$

where

$$d_t := \frac{d}{dt} \tag{3.161}$$

and $\boldsymbol{b}$ being a constant vector. The solution to the inhomogeneous equation (3.160) can be constructed by the method of variation of constants. First, the general solution to the unperturbed equation

$$(d_t - L_0)\boldsymbol{X}_0(t) = 0 \tag{3.162}$$

is given by

$$\boldsymbol{X}_0(t) = \mathrm{e}^{L_0 t}\boldsymbol{C}, \tag{3.163}$$

with $\boldsymbol{C}$ being a constant vector. Then we insert the ansatz

$$\boldsymbol{X}(t) = \mathrm{e}^{L_0 t}\boldsymbol{C}(t) \tag{3.164}$$

into (3.160), and we have an equation for $\boldsymbol{C}(t)$ as

$$\dot{\boldsymbol{C}}(t) = \mathrm{e}^{-L_0 t}\boldsymbol{b}, \tag{3.165}$$

which is solved to give

$$\boldsymbol{C}(t) = \boldsymbol{C}(t_0) + \int_{t_0}^{t} ds\, \mathrm{e}^{-L_0 s}\boldsymbol{b}. \tag{3.166}$$

Inserting this into the ansatz of the solution, we have

$$\begin{aligned}
\boldsymbol{X}(t) &= \mathrm{e}^{L_0(t-t_0)}\boldsymbol{C}(t_0) + \mathrm{e}^{L_0 t}\int_{t_0}^{t} ds\, \mathrm{e}^{-L_0 s}\boldsymbol{b} \\
&= \mathrm{e}^{L_0(t-t_0)}\boldsymbol{C}(t_0) + \mathrm{e}^{L_0 t}\int_{t_0}^{t} ds\, \mathrm{e}^{-L_0 s}\sum_{i=1}^{3}|\boldsymbol{U}_i\rangle\langle\tilde{\boldsymbol{U}}_i|\boldsymbol{b}\rangle \\
&= \mathrm{e}^{L_0(t-t_0)}\boldsymbol{C}(t_0) + \sum_{i=1}^{3}\mathrm{e}^{\lambda_i t}\int_{t_0}^{t} ds\, \mathrm{e}^{-\lambda_i s}\,|\boldsymbol{U}_i\rangle\langle\tilde{\boldsymbol{U}}_i|\boldsymbol{b}\rangle
\end{aligned} \tag{3.167}$$

where the division of unity (3.154) is inserted in the second equality.

Noting that $\lambda_1 = 0$, we find that if $\langle\tilde{\boldsymbol{U}}_1|\boldsymbol{b}\rangle \neq 0$, then there appears a secular term proportional to $t - t_0$. Thus, the solvability condition of (3.159) with which the appearance of a secular term is avoided is found to be

$$\langle\tilde{\boldsymbol{U}}_1|\boldsymbol{b}_2\rangle = 0, \tag{3.168}$$

which gives

$$\frac{\partial A}{\partial t_1} = 0, \tag{3.169}$$

implying that A has no dependence on t_1;

$$A = A(t_2, \ldots). \tag{3.170}$$

Then, (3.159) is reduced to

$$(d_t - L_0)\boldsymbol{X}_2 = A^2\boldsymbol{U}_3, \tag{3.171}$$

the special solution to which formally reads

$$\boldsymbol{X}_2 = \frac{1}{d_t - L_0}A^2\boldsymbol{U}_3 = \frac{1}{b}A^2\boldsymbol{U}_3 = \begin{pmatrix} 0 \\ 0 \\ \zeta_2 \end{pmatrix}, \quad (\zeta_2 = \frac{A^2}{b}). \tag{3.172}$$

The third-order equation reads

$$\begin{aligned}\left(\frac{d}{dt} - L_0\right)\boldsymbol{X}_3 &= -\frac{\partial \boldsymbol{X}_2}{\partial t_1} - \frac{\partial \boldsymbol{X}_1}{\partial t_2} + \begin{pmatrix} 0 \\ \chi\xi_1 - \xi_1\zeta_2 \\ 0 \end{pmatrix} \\ &= -\frac{\partial A}{\partial t_2}\boldsymbol{U}_1 + \left(\chi A - \frac{1}{b}A^3\right)\begin{pmatrix} 0 \\ 1 \\ 0 \end{pmatrix} =: \boldsymbol{b}_3,\end{aligned} \tag{3.173}$$

where we have used the fact that

$$\frac{\partial \boldsymbol{X}_2}{\partial t_1} = 0 \tag{3.174}$$

because $\partial A/\partial t_1 = 0$.

The solvability condition of the third-order equation reads

$$\langle \tilde{\boldsymbol{U}}_1 | \boldsymbol{b}_3 \rangle = 0, \tag{3.175}$$

which leads to

$$\frac{\partial A}{\partial t_2} = \frac{\sigma}{\sigma + 1}\left(\chi A - \frac{1}{b}A^3\right). \tag{3.176}$$

Here we have used the formula

$$\begin{pmatrix} 0 \\ 1 \\ 0 \end{pmatrix} = \frac{\sigma}{\sigma+1} \boldsymbol{U}_1 - \frac{1}{\sigma+1} \boldsymbol{U}_2. \tag{3.177}$$

Recalling that $t_2 = \epsilon^2 t$, we arrive at

$$\frac{dA}{dt} = \epsilon^2 \frac{\sigma}{\sigma+1} \left(\chi A - \frac{1}{b} A^3 \right). \tag{3.178}$$

The third-order equation now reads

$$(d_t - L_0) \boldsymbol{X}_3 = -\frac{\chi A - \frac{1}{b} A^3}{\sigma+1} \boldsymbol{U}_2 \tag{3.179}$$

which yields

$$\boldsymbol{X}_3 = \frac{\chi A - \frac{1}{b} A^3}{(\sigma+1)^2} \boldsymbol{U}_2. \tag{3.180}$$

Thus collecting all the terms thus obtained, we have

$$\boldsymbol{X}(t) = \epsilon A \boldsymbol{U}_1 + \frac{\epsilon^2}{b} A^2 \boldsymbol{U}_3 + \frac{\epsilon^3}{(1+\sigma)^2} \left(\chi A - \frac{1}{b} A^3 \right) \boldsymbol{U}_2 \tag{3.181}$$

up to $O(\epsilon^3)$. The respective components take the following forms

$$\begin{aligned} \xi(t) &= \epsilon A + \frac{\epsilon^3 \sigma}{(1+\sigma)^2} \left(\chi A - \frac{1}{b} A^3 \right), \\ \eta(t) &= \epsilon A - \frac{\epsilon^3}{(1+\sigma)^2} \left(\chi A - \frac{1}{b} A^3 \right), \\ \zeta(t) &= \epsilon^2 \frac{A^2}{b}. \end{aligned} \tag{3.182}$$

The time-dependence of the trajectory is governed through that of the amplitude as given by (3.178). One sees that

(i) $\chi = -1$:
The right-hand side of (3.178) is negative definite (the equality holds only when $A = 0$), and thus the origin $A = 0$ is stable.

(ii) $\chi = 1$:

Equation (3.178) has two fixed points, $A = 0$ and $A = \sqrt{b}$. The former (latter) is an unstable (stable) fixed point. The amplitude equation can be solved by quadrature as

$$\int_{A_0}^{A} \frac{dA}{A - \frac{1}{b}A^3} = \lambda \int_0^t dt = \lambda t \quad \left(\lambda := \epsilon^2 \frac{\sigma}{\sigma + 1}\right). \tag{3.183}$$

The left-hand side is evaluated as follows:

$$\int_{A_0}^{A} \frac{dA}{A - \frac{1}{b}A^3} = \frac{1}{2} \int_{A_0^2}^{A^2} \frac{dB}{B(1 - \frac{B}{b})} = \frac{1}{2} \ln \left| \frac{B}{1 - \frac{B}{b}} \right| \Bigg|_{A_0^2}^{A^2} \tag{3.184}$$

which implies that

$$\frac{A^2(t)}{1 - \frac{A^2(t)}{b}} = C\mathrm{e}^{2\lambda t}, \tag{3.185}$$

with C being a constant ensuring that $A(0) = A_0$. After some manipulation, we have

$$A(t) = \frac{A_0}{\sqrt{\frac{A_0^2}{b} + (1 - \frac{A_0^2}{b})\mathrm{e}^{-2\lambda t}}}, \quad \left(\lambda = \epsilon^2 \frac{\sigma}{1 + \sigma}\right). \tag{3.186}$$

This expression shows that $A(0) = A_0$ and $A(t)$ approaches b as $t \to \infty$.

Appendix 1: Asymmetric Matrices, Left and Right Eigenvectors, and Symmetrized Inner Product

In this appendix, we make an account of asymmetric matrices and their left and right eigenvectors. We also construct an inner product with which the asymmetric matrix is made symmetric. One will see that projection operators play significant roles in the formulations. The formulae below are given when the eigenvalues are discrete, but should be valid for general linear operators with due modifications caring continuous spectra.

1.1 Eigen Values and Vectors of an Asymmetric Matrix

Definition
For a given $n \times n$ matrix A, the right and left eigenvalues, λ_i and μ_i, are defined by the following equations, respectively,

$$A\,\boldsymbol{U}_i = \lambda_i\,\boldsymbol{U}_i, \tag{3.187}$$

$$\tilde{\boldsymbol{U}}_i^{\dagger} A = \mu_i \tilde{\boldsymbol{U}}_i^{\dagger}, \tag{3.188}$$

with non-vanishing vectors $\boldsymbol{U}_i$ and $\tilde{\boldsymbol{U}}_i^{\dagger}$. The Hermite conjugate of the last equation reads

$$A^{\dagger}\tilde{\boldsymbol{U}}_i = \mu_i^* \tilde{\boldsymbol{U}}_i \tag{3.189}$$

with μ^* denoting the complex conjugate of μ. The characteristic equation for the left eigenvalue μ reads with I being the n-dimensional unit matrix

$$0 = \det(A^{\dagger} - \mu^* I) = [\det({}^{\mathrm{t}}A - \mu I)]^* = [\det(A - \mu I)]^*, \tag{3.190}$$

which means that $\det(A - \mu I) = 0$. This is the same equation as the characteristic equation for the right eigenvalue λ. Thus we see that

$$\mu_i = \lambda_i. \tag{3.191}$$

From now on, the left eigenvalues are also denoted by λ_i.

We define a naïve inner product $\langle \boldsymbol{U}, \boldsymbol{V} \rangle$ for arbitrary vectors $\boldsymbol{U}$ and $\boldsymbol{V}$ by

$$\langle \boldsymbol{U}, \boldsymbol{V} \rangle := \boldsymbol{U}^{\dagger}\boldsymbol{V} = \sum_{k=1}^{n} U_k^* V_k. \tag{3.192}$$

We also use the notation for the inner product with the comma ‘,’ being replaced by a vertical line ‘|’ as

$$\langle \boldsymbol{U}, \boldsymbol{V} \rangle = \langle \boldsymbol{U} \mid \boldsymbol{V} \rangle. \tag{3.193}$$

It is easy to see that the right and left eigenvectors belonging to different eigenvalues are orthogonal to each other. In fact, since

$$\langle \tilde{\boldsymbol{U}}_j, A\boldsymbol{U}_i \rangle = \lambda_i \langle \tilde{\boldsymbol{U}}_j, \boldsymbol{U}_i \rangle = \langle A^{\dagger}\tilde{\boldsymbol{U}}_j, \boldsymbol{U}_i \rangle = \lambda_j \langle \tilde{\boldsymbol{U}}_j, \boldsymbol{U}_i \rangle, \tag{3.194}$$

we have

$$(\lambda_i - \lambda_j) \langle \tilde{\boldsymbol{U}}_j, \boldsymbol{U}_i \rangle = 0. \tag{3.195}$$

Thus

$$\lambda_i \neq \lambda_i \implies \langle \tilde{\boldsymbol{U}}_j, \boldsymbol{U}_i \rangle = 0. \tag{3.196}$$

When λ_i is not degenerate, the eigenvectors are normalized as

$$\langle \tilde{\boldsymbol{U}}_i, \boldsymbol{U}_i \rangle = 1. \tag{3.197}$$

Thus we have the ortho-normal conditions as

$$\langle \tilde{\boldsymbol{U}}_j, \boldsymbol{U}_i \rangle = \delta_{ij}. \tag{3.198}$$

Let us rewrite the above facts using the **bra-ket notaion** for conciseness:

$$A\boldsymbol{U}_i =: A|\boldsymbol{U}_i\rangle = \lambda_i|\boldsymbol{U}_i\rangle, \quad \tilde{\boldsymbol{U}}_i^\dagger A =: \langle \tilde{\boldsymbol{U}}_i|A = \lambda_i \langle \tilde{\boldsymbol{U}}_i|, \tag{3.199}$$

$$\langle \tilde{\boldsymbol{U}}_j, A\boldsymbol{U}_i \rangle =: \langle \tilde{\boldsymbol{U}}_j|A|\boldsymbol{U}_i\rangle, \quad \langle \tilde{\boldsymbol{U}}_i|\boldsymbol{U}_j\rangle = \delta_{ij}. \tag{3.200}$$

An arbitrary ket vector $|X\rangle$ is represented as a linear combination of the right eigenvectors as

$$|X\rangle = \sum_{i=1}^{n} c_i|\boldsymbol{U}_i\rangle, \quad \text{with } c_i = \langle \tilde{\boldsymbol{U}}_i|X\rangle. \tag{3.201}$$

Similarly a bra vector $\langle Y|$ is expressed as

$$\langle Y| = \sum_{i=1}^{n} d_i \langle \tilde{\boldsymbol{U}}_i|, \quad \text{with } d_i = \langle Y|\boldsymbol{U}_i\rangle. \tag{3.202}$$

Projection Operators

First we discuss the simple case where the eigenvalue λ_i is not degenerate. In this case, the projection operator to the eigenspace is given by

$$P_i = \boldsymbol{U}_i \tilde{\boldsymbol{U}}_i^\dagger = |\boldsymbol{U}_i\rangle\langle \tilde{\boldsymbol{U}}_i|. \tag{3.203}$$

In fact, since

$$P_i|\boldsymbol{U}_j\rangle = |\boldsymbol{U}_i\rangle\langle \tilde{\boldsymbol{U}}_i|\boldsymbol{U}_j\rangle = |\boldsymbol{U}_i\rangle\delta_{\rm ij}, \tag{3.204}$$

we have for an arbitrary vector $|W\rangle = \sum_j C_j|\boldsymbol{U}_j\rangle$,

$$P_i|W\rangle = \sum_j C_j|\boldsymbol{U}_i\rangle\langle \tilde{\boldsymbol{U}}_i|\boldsymbol{U}_j\rangle = \sum_j C_j\delta_{ij}|\boldsymbol{U}_i\rangle = C_i|\boldsymbol{U}_i\rangle, \tag{3.205}$$

which shows that P_i is the projection operator onto the eigenspace belonging to the eigenvalue λ_i. Furthermore P_i satisfies the idempotency as

$$P_i P_j = |\boldsymbol{U}_i\rangle\langle \tilde{\boldsymbol{U}}_i|\boldsymbol{U}_j\rangle\langle \tilde{\boldsymbol{U}}_j| = |\boldsymbol{U}_i\rangle\delta_{ij}\langle \tilde{\boldsymbol{U}}_j| = \delta_{ij}P_i. \tag{3.206}$$

The completeness of the eigenvectors is expressed as

$$I = \sum_{i=1}^{n} P_i \tag{3.207}$$

Indeed multiplying the last formula to a ket vector $|X\rangle$, we have

$$I|X\rangle = \sum_{i=1}^{n} P_i|X\rangle = \sum_{i=1}^{n} |\boldsymbol{U}_i\rangle\langle\tilde{\boldsymbol{U}}_i|X\rangle = \sum_{i=1}^{n} C_i|\boldsymbol{U}_i\rangle = |X\rangle, \tag{3.208}$$

where (3.201) has been used. Similarly,

$$\langle Y|I = \sum_{i=1}^{n} \langle Y|P_i = \sum_{i=1}^{n} \langle Y|\boldsymbol{U}_i\rangle\langle\tilde{\boldsymbol{U}}_i| = \sum_{i=1}^{n} d_i\langle\tilde{\boldsymbol{U}}_i| = \langle Y|, \tag{3.209}$$

where (3.202) has been used.

1.2 Spectral Representation

With the use of the projection operators P_i, we have a spectral representation of A as

$$A = \sum_{i=1}^{n} \lambda_i P_i = \sum_{i=1}^{n} \lambda_i|\boldsymbol{U}_i\rangle\langle\tilde{\boldsymbol{U}}_i|. \tag{3.210}$$

When A Has Zero Modes

When $\lambda_1 = 0$ and $\lambda_j \neq 0$ ($j \neq 1$), let us define the projection operator Q onto the compliment of the zero mode as

$$Q := I - P_1 = \sum_{i\neq 1}^{n} P_i = \sum_{j\neq 1}^{n} |\boldsymbol{U}_j\rangle\langle\tilde{\boldsymbol{U}}_j|. \tag{3.211}$$

Then

$$\begin{aligned} AQ &= \sum_{i=1, j\neq 1}^{n} \lambda_i|\boldsymbol{U}_i\rangle\langle\tilde{\boldsymbol{U}}_i|\boldsymbol{U}_j\rangle\langle\tilde{\boldsymbol{U}}_j| = \sum_{i=1, j\neq 1}^{n} \delta_{ij}\lambda_i|\boldsymbol{U}_i\rangle\langle\tilde{\boldsymbol{U}}_j| \\ &= \sum_{j\neq 1}^{n} \lambda_j|\boldsymbol{U}_j\rangle\langle\tilde{\boldsymbol{U}}_j|. \end{aligned} \tag{3.212}$$

Similary,

$$\begin{aligned} QA &= \sum_{i=1, j\neq 1}^{n} \lambda_i|\boldsymbol{U}_j\rangle\langle\tilde{\boldsymbol{U}}_j|\boldsymbol{U}_i\rangle\langle\tilde{\boldsymbol{U}}_i| = \sum_{i=1, j\neq 1}^{n} \delta_{ij}\lambda_i|\boldsymbol{U}_j\rangle\langle\tilde{\boldsymbol{U}}_i| = \sum_{j\neq 1}^{n} \lambda_j|\boldsymbol{U}_j\rangle\langle\tilde{\boldsymbol{U}}_j| \\ &= AQ. \end{aligned} \tag{3.213}$$

We have also

$$A_Q := QAQ = AQ = QA \tag{3.214}$$

because $Q^2 = Q$.

Using the above formulae, we can define the inverse of A applying only to the Q space as follows

$$A_Q^{-1} = (AQ)^{-1} = (QA)^{-1} = \sum_{j\neq 1}^{n} \lambda_j^{-1} |\boldsymbol{U}_j\rangle\langle\tilde{\boldsymbol{U}}_j|. \tag{3.215}$$

1.3 The Case Where λ_i Is Degenerate

Let the eigenvalue λ_i is m_i-tiply degenerate as

$$A\,|\boldsymbol{U}_{i\alpha}\rangle = \lambda_i|\boldsymbol{U}_{i\alpha}\rangle, \quad \langle\tilde{\boldsymbol{U}}_{i\alpha}|\,A = \lambda_i\langle\tilde{\boldsymbol{U}}_{i\alpha}|. \quad (\alpha = 1, 2, \ldots, m_i) \tag{3.216}$$

We define the metric tensor $\hat{\eta}_i$ by

$$\langle\tilde{\boldsymbol{U}}_{i\alpha}|\boldsymbol{U}_{i\beta}\rangle =: \eta_{i\,\alpha\beta}. \quad (\alpha,\ \beta = 1, 2, \ldots, m_i), \tag{3.217}$$

with $(\hat{\eta})_{i\,\alpha\beta} = \eta_{i\,\alpha\beta}$. The inverse of $\hat{\eta}_i$ is denoted by $\hat{\eta}_i^{-1}$ which satisfies

$$(\eta_i^{-1})^{\alpha\gamma}\eta_{i\,\gamma\beta} = \delta_{\alpha\beta}. \tag{3.218}$$

Then projection operator onto the eigenspace belonging to the eigenvalue λ_i is given by

$$P_i = \sum_{\alpha\beta} |\boldsymbol{U}_{i\alpha}\rangle(\eta_i^{-1})^{\alpha\beta}\langle\tilde{\boldsymbol{U}}_{i\beta}|. \tag{3.219}$$

In fact, the following relations hold for P_i:

$$P_i\,|\boldsymbol{U}_{j\alpha}\rangle = \delta_{ij}|\boldsymbol{U}_{i\alpha}\rangle, \quad P_iP_j = \delta_{ij}P_i, \tag{3.220}$$

the proof for which is given as follows:

$$
\begin{aligned}
P_i\,|U_{j\alpha}\rangle &= \sum_{\alpha'\beta} |U_{i\alpha'}\rangle(\eta_i^{-1})^{\alpha'\beta}\langle\tilde{U}_{i\beta}|U_{j\alpha}\rangle = \sum_{\alpha'\beta} |U_{i\alpha'}\rangle(\eta_i^{-1})^{\alpha'\beta}\eta_{i\,\beta\alpha}\delta_{ij} \\
&= \delta_{ij}\sum_{\alpha'} |U_{i\alpha'}\rangle\delta_{\alpha}^{\alpha'} \\
&= \delta_{ij}\,|U_{i\alpha}\rangle,
\end{aligned}
\tag{3.221}
$$

$$
\begin{aligned}
P_iP_j &= \sum_{\alpha\beta} |U_{i\alpha}\rangle(\eta_i^{-1})^{\alpha\beta}\langle\tilde{U}_{i\beta}|\sum_{\alpha'\beta'} |U_{j\alpha'}\rangle(\eta_j^{-1})^{\alpha'\beta'}\langle\tilde{U}_{j\beta'}| \\
&= \sum_{\alpha\beta}\sum_{\alpha'\beta'} |U_{i\alpha}\rangle(\eta_i^{-1})^{\alpha\beta}\langle\tilde{U}_{i\beta}|U_{j\alpha'}\rangle(\eta_j^{-1})^{\alpha'\beta'}\langle\tilde{U}_{j\beta'}| \\
&= \sum_{\alpha\beta}\sum_{\alpha'\beta'} |U_{i\alpha}\rangle(\eta_i^{-1})^{\alpha\beta}\eta_{i\,\beta\alpha'}\delta_{ij}(\eta_j^{-1})^{\alpha'\beta'}\langle\tilde{U}_{j\beta'}| \\
&= \delta_{ij}\sum_{\alpha\beta}\sum_{\beta'} |U_{i\alpha}\rangle(\eta_i^{-1})^{\alpha\beta}\delta_{\beta}^{\beta'}\langle\tilde{U}_{i\beta'}| \\
&= \delta_{ij}\sum_{\alpha\beta} |U_{i\alpha}\rangle(\eta_i^{-1})^{\alpha\beta}\langle\tilde{U}_{i\beta}| \\
&= \delta_{ij}P_i.
\end{aligned}
\tag{3.222}
$$

If the number of different eigenvalues of A is $\bar{n}$ and the degeneracy of the eigenvalue λ_i $(i = 1, 2, \ldots, \bar{n})$ is m_i, then the spectral representation of A is given by

$$
A = \sum_{i=1}^{\bar{n}} \lambda_i \left(\sum_{\alpha_i=1}^{m_i} |U_{i\alpha_i}\rangle\langle\tilde{U}_{i\alpha_i}| \right) = \sum_{i=1}^{\bar{n}}\sum_{\alpha_i=1}^{m_i} \lambda_i P_{i\alpha_i}
\tag{3.223}
$$

with

$$
P_{i\alpha_i} := |U_{i\alpha_i}\rangle\langle\tilde{U}_{i\alpha_i}|.
\tag{3.224}
$$

1.4 Symmetrized Inner Product

For notational simplicity, the indices $i\alpha$ are written as i when no confusion is expected. We assume that the set of the right eigenvectors $\{U_i\}$ constitute a complete set. Arbitrary vectors u_1 and u_2 are represented as linear combinations of the right eigenvectors as

$$
u_i = \sum_j c_j^{(i)} U_j, \quad (i = 1, 2).
\tag{3.225}
$$

We define the metric tensor g in terms of the left eigenvectors $\langle\tilde{U}_i|$ as follows:

$$g = \sum_i \tilde{U}_i \tilde{U}_i^\dagger = \sum_i |\tilde{U}_i\rangle\langle\tilde{U}_i|. \tag{3.226}$$

Now we define the novel inner product $(\boldsymbol{u}_1, \boldsymbol{u}_2)$ for arbitrary two vectors $\boldsymbol{u}_1, \boldsymbol{u}_2$ as follows:

$$(\boldsymbol{u}_1, \boldsymbol{u}_2) := \langle \boldsymbol{u}_1, \, g\boldsymbol{u}_2\rangle = \sum_i \langle \boldsymbol{u}_1, \, \tilde{U}_i\rangle\langle\tilde{U}_i, \, \boldsymbol{u}_2\rangle. \tag{3.227}$$

Then

$$\begin{aligned}(\boldsymbol{u}_1, A\boldsymbol{u}_2) &= \sum_i \langle \boldsymbol{u}_1, \tilde{U}_i\rangle\langle\tilde{U}_i \, A\boldsymbol{u}_2\rangle = \sum_i \langle \boldsymbol{u}_1, \, \tilde{U}_i\rangle\langle A^\dagger\tilde{U}_i, \, \boldsymbol{u}_2\rangle \\ &= \sum_i \lambda_i \langle \boldsymbol{u}_1, \, \tilde{U}_i\rangle\langle\tilde{U}_i, \, \boldsymbol{u}_2\rangle.\end{aligned} \tag{3.228}$$

On the other hand,

$$\begin{aligned}(A\boldsymbol{u}_1, \boldsymbol{u}_2) &= \sum_i \langle A\boldsymbol{u}_1, \, \tilde{U}_i\rangle\langle\tilde{U}_i, \, \boldsymbol{u}_2\rangle = \sum_i \langle \boldsymbol{u}_1, \, A^\dagger\tilde{U}_i\rangle\langle\tilde{U}_i, \, \boldsymbol{u}_2\rangle \\ &= \sum_i \lambda_i^* \langle \boldsymbol{u}_1, \, \tilde{U}_i\rangle\langle\tilde{U}_i, \, \boldsymbol{u}_2\rangle,\end{aligned} \tag{3.229}$$

which coincides with (3.228) when all the eigenvalues are real numbers, and hence we have

$$(\boldsymbol{u}_1, A\boldsymbol{u}_2) = (A\boldsymbol{u}_1, \boldsymbol{u}_2) \tag{3.230}$$

which means that A that has only real eigenvalues is manifestly Hermitian (symmetric) under the new inner product.

Appendix 2: Solvability Condition of Linear Equations: Fredholm's Alternative Theorem

In this section, we summarize the condition for a linear equation can have solutions, that is, the solvability condition, which is to be generalized to Fredholm's alternative theorem and further Riesz-Schauder's theorem in functional analysis theory [80–82].

In the text, we started the discussion from a trivial equation $ax = b$ for x, the solution to which becomes somewhat complicated when $a = 0$. It was also mentioned that we also encounter a quite similar situation when solving a system of linear equations; $A\boldsymbol{x} = \boldsymbol{b}$, where $\boldsymbol{x} = {}^t(x_1, \, x_2, \, \ldots, \, x_n)$ and A is an $n \times n$ matrix. In this appendix, we shall make a more detailed account of the solvability condition of linear equations though without a mathematical rigor.

For notational convenience, we here adopt the bra-ket notation of Dirac and write the linear equation above as

$$A|\boldsymbol{x}\rangle = |\boldsymbol{b}\rangle. \tag{3.231}$$

In this section, we denote the eigenvector belonging to the eigenvalue λ_i of A as $|\lambda_i\rangle$ instead of $|\boldsymbol{U}_i\rangle$;

$$A|\lambda_i\rangle = \lambda_i|\lambda_i\rangle, \qquad A^\dagger|\tilde{\lambda}_i\rangle = \lambda_i|\tilde{\lambda}_i\rangle. \tag{3.232}$$

When A is regular, all the eigenvalues are non-vanishing, and

$$|\boldsymbol{x}\rangle = A^{-1}|\boldsymbol{b}\rangle = \sum_{i=1}^{n} \lambda_i^{-1}|\lambda_i\rangle\langle\tilde{\lambda}_i|\boldsymbol{b}\rangle, \tag{3.233}$$

where i takes into account the degeneracy.

On the other hand, When $|A| = 0$, and accordingly A is singular, A has zero eigenvalues. Let us suppose that the degeneracy of the zero eigenvalue is m, and the other eigenvalues are non-vanishing;

$$\lambda_1 = \cdots = \lambda_m = 0, \quad \lambda_i \neq 0 \quad (i = m+1, \ldots, n). \tag{3.234}$$

We denote the m independent eigenvectors with zero eigenvalues as $|0;\, \alpha\rangle$ $(\alpha = 1, 2, \ldots, m)$;

$$A|0;\, \alpha\rangle = 0, \quad A^\dagger|0;\, \tilde{\alpha}\rangle = 0, \tag{3.235}$$

which are called the zero modes. In mathematical terms,

$$|0;\, \alpha\rangle \in \ker A, \quad |0;\, \tilde{\alpha}\rangle \in \ker A^\dagger, \tag{3.236}$$

where $\ker A$ $(\ker A)$ denotes the subspace spanned by the zero modes of A $(A^\dagger)$, and is called the kernel of A $(A^\dagger)$. We also call the kernel the P space, and the projection operator onto the P space is denoted by P

$$P = \sum_{\alpha=1}^{m} |0;\, \alpha\rangle\langle 0;\, \tilde{\alpha}|. \tag{3.237}$$

The compliment of the P space is called the Q space and the projection operator onto it is denoted by $Q = I - P$:

$$Q = \sum_{i=m+1}^{n} |\lambda_i\rangle\langle\tilde{\lambda}_i|. \tag{3.238}$$

The following alternative statements hold.

1. If $|b\rangle$ is orthogonal to all the zero modes,

$$\langle 0;\ \tilde{\alpha}|b\rangle = 0, \qquad (i = 1, 2, ..m), \tag{3.239}$$

then $|b\rangle$ is expressed as

$$|b\rangle = \sum_{i=m+1}^{n} b_i|\lambda_i\rangle = \sum_{i=m+1}^{n} |\lambda_i\rangle\langle\tilde{\lambda}_i|b\rangle = Q|b\rangle. \tag{3.240}$$

Then we have a series of the solution

$$|x\rangle = A_Q^{-1}|b\rangle + \sum_{\alpha=1}^{m} c_\alpha|0;\ \alpha\rangle, \tag{3.241}$$

with c_i's being arbitrary constants and

$$A_Q := AQ = \sum_{i=m+1}^{n} \lambda_i|\lambda_i\rangle\langle\tilde{\lambda}_i|, \quad A_Q^{-1} = \sum_{i=m+1}^{n} \lambda_i^{-1}|\lambda_i\rangle\langle\tilde{\lambda}_i|. \tag{3.242}$$

2. If

$$P|b\rangle \neq 0, \tag{3.243}$$

i.e., $\langle 0;\ \tilde{\alpha}|b\rangle \neq 0$ $(1 \leq \exists\alpha \leq m)$, then there exists no solution to Eq. (3.231), that is, Eq. (3.231) is not solvable. Indeed if $|x\rangle$ is expressed by the complete set spanned by the eigenvectors as

$$|\boldsymbol{x}\rangle = \sum_{\alpha=1}^{m} c_\alpha|0;\ \alpha\rangle + \sum_{i=m+1}^{n} d_i|\lambda_i\rangle \tag{3.244}$$

the left-hand side of (3.231) becomes

$$A|\boldsymbol{x}\rangle = \sum_{i=m+1}^{n} d_i\lambda_i|\lambda_i\rangle, \tag{3.245}$$

which lacks in components belonging to the P space in contrast to $|b\rangle$.

Thus (3.239) certainly gives the solvability condition to the linear equation (3.231).

Chapter 4
Renormalization Group Method for Global Analysis: A Geometrical Formulation and Simple Examples

4.1 Introduction

We have seen that various methods have been developed to avoid the appearance of secular terms in the perturbative solutions and thereby obtain a sensible solution of differential equations in a global domain; it is to be noted that all of them are based on the notion of the solvability condition of linear equations.

In this chapter, we introduce a quite different method [1] based on the *renormalization group (RG) equation* [8, 9, 14, 18, 87, 88] for obtaining perturbative solutions of differential equations that are valid in a global domain through a resummation of the perturbative expansions; the method is also found useful to get the asymptotic behavior of solutions of differential equations. The method is relatively simple in manipulation and has a wide variety of applications including singular and reductive perturbation problems in a unified manner. It might be, however, difficult to understand the method even for those who are familiar with the renormalization groups in physics in understanding the reason why the very RG equation can be relevant to and useful for global analysis of differential equations in the way presented in [1] because the RG equation in physics is usually related with the scale invariance of the system under consideration; for those who are not familiar with the notion of the RG in physics, the method may have just looked mysterious.

Actually, what was done in [1] is a construction an approximate but globally valid solution from a local solution of the given differential equation through a seemingly mysterious equation that they call the RG equation. This fact suggests that the RG method or rather RG equation in [1] may be formulated solely in terms of purely mathematical notions without recourse to any physical intuitions. In the following sections, it will be elucidated that this is the case, and clarified that the RG equation may be better characterized as the *envelope equation* known in elementary differential geometry [24], as was first shown in [3–5]; see [6] for a later development.

The theory of the envelopes of a family of curves is long known in elementary differential geometry [24], and one might have recognized that it has an improved global nature in comparison with the member curves in the family. Therefore it could have been expected that the theory of envelopes may have some usefulness for global

T. Kunihiro et al., *Geometrical Formulation of Renormalization-Group Method as an Asymptotic Analysis*, Fundamental Theories of Physics 206,
https://doi.org/10.1007/978-981-16-8189-9_4

analysis of dynamical equations, although such a recognition had not been fully made until the work [3–5].[1] Interestingly enough, the notion of envelopes, in turn, can make an account of the very RG equation in the quantum field theory, as is shown in [3]. Therefore one may call the RG equation as used in the global and asymptotic analysis in [1] may be called the renormalization group/envelope equation or RG/E equation in short.

We note that the present geometrical formulation [3–6] of the RG method for a global analysis of differential equations naturally leads to some elaborated prescriptions of the method: It starts with solving the original equation faithfully as an 'initial' condition problem at an arbitrary time, say, t_0, in the perturbation theory without any modifications or rearrangements of the equation as are common in the traditional methods as explained in Chap. 3; then it accordingly allows appearance of secular terms in the perturbative solutions [1] in contrast to the traditional methods. An important set up of the novel prescription of the method [3–6] is to suppose that the 'initial value' at $t = t_0$ is on an exact solution that is yet to be determined, and the integral constants contained in the perturbative solutions are made dependent on the 'initial time' t_0 [3–6], whereby the renormalization procedure through an introduction of an intermediate time scale τ [1] is totally unnecessary. The would-be integral constants are then lifted to the dynamical variables by the renormalization group/envelope equation . We shall also give a proof that the envelope function thus constructed through the RG/E equation satisfy the original differential equation in a global domain .

In this chapter, we start with giving an elementary account of the classical theory of envelopes of a family of curves [3–5, 24], which will be also adapted so that it can be applicable to the perturbative solutions of differential equations [3–5] as a family of curves or trajectories. Then some simple examples that have been treated in the previous chapters will be analyzed in the RG method.

4.2 Classical Theory of Envelopes and Its Adaptation for Global Analysis of Differential Equations

In this section, we give an elementary account of the theory of envelopes of curves/trajectories.

4.2.1 Envelope Curve in Two-Dimensional Space

Taking a family of curves in two-dimensional space, we first describes the very basic mathematics of the classical theory of envelopes [3–5, 24]. We remark that when the

[1] It should be remarked, however, that M. Suzuki [89] had shown that the notion of envelopes can play a significant role in describing the properties of critical points and developed the coherent anomaly method (CAM) for computing the critical exponents of the critical points.

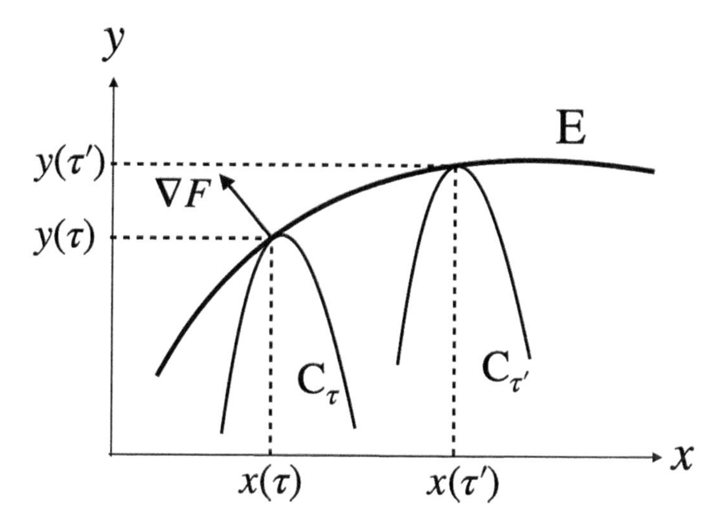

Fig. 4.1 The curve E is the envelope of the family of curves $\{C_\tau\}_\tau$. ∇F denote the normal vector of C_τ to which the tangent vector of E should be perpendicular

theory is applied to the curves obtained as the solutions of a differential equation, the curves or trajectories in the x-y plane are given in a parameter representation with t being the parameter as

$$\boldsymbol{u}(t;\,\tau) = (x(t;\,\tau),\ y(t;\,\tau)), \tag{4.1}$$

where τ is another parameter with which the curves in the family are distinguished.

Let $\{C_\tau\}_\tau$ be a family of curves parametrized by a parameter τ in x-y plane and given by the equation

$$F(x,\ y;\ \tau) = 0. \tag{4.2}$$

An envelope E of $\{C_\tau\}_\tau$ is a curve in the x-y plane that share the tangent at a contact point with every curve C_τ; see Fig. 4.1.

Our problem is to obtain the equation

$$G(x,\ y) = 0 \tag{4.3}$$

that represents the envelope curve E of from the function $F(x,\ y;\ \tau)$ that gives the family of curves $\{C_\tau\}_\tau$.

This problem can be solved as follows. Let an arbitrary point $(x,\ y)$ on E is a point of tangency with

$$C_\tau\ :\ F(x,\ y;\ \tau(x,\ y)) = 0. \tag{4.4}$$

Conversely, the coordinate of the contact point are functions of τ;

$$(x,\ y) = (x(\tau),\ y(\tau)) =: (\phi(\tau),\ \psi(\tau)). \tag{4.5}$$

Along with a variation of τ, the point $(\phi(\tau),\ \psi(\tau))$ moves on E, and hence the tangent vector at that point is parallel to

$$\left(\frac{d\phi}{d\tau},\ \frac{d\psi}{d\tau}\right) =: \boldsymbol{t}. \tag{4.6}$$

Thus, for an infinitesimal variation $\tau \rightarrow \tau + \Delta\tau$, we have

$$(d\phi,\ d\psi) = \boldsymbol{t}\,\Delta\tau. \tag{4.7}$$

The condition that E and C_τ share a tangent implies that $\boldsymbol{t}(x,\ y)$ is perpendicular to the normal vector of C_τ at this point, which is proportional to

$$\nabla F = \left(\frac{\partial F}{\partial x},\ \frac{\partial F}{\partial y}\right) =: \boldsymbol{n}(x,\ y). \tag{4.8}$$

Thus the condition of sharing a tangent at the contact point can be expressed by the following two equations

$$F(\phi(\tau),\ \psi(\tau);\ \tau) = 0, \tag{4.9}$$

$$\boldsymbol{n}(x,\ y)\cdot\boldsymbol{t}(x,\ y) = F_x(x,\ y;\ \tau)\frac{\partial\phi}{d\tau} + F_y(x,\ y;\ \tau)\frac{\partial\psi}{d\tau} = 0, \tag{4.10}$$

where

$$F_x = \frac{\partial F}{\partial x},\quad F_y = \frac{\partial F}{\partial y}. \tag{4.11}$$

We remark that the condition of a common tangent (4.10) may be expressed as

$$\boldsymbol{n}(x,\ y)\cdot\boldsymbol{t}(x,\ y)\,\Delta\tau = F_x(x,\ y;\ \tau)d\phi + F_y(x,\ y;\ \tau)d\psi = 0. \tag{4.12}$$

For $\tau' \neq \tau$ corresponding to another point on E,

$$F(\phi(\tau'),\ \psi(\tau');\ \tau') = 0 = F(\phi(\tau),\ \psi(\tau);\ \tau). \tag{4.13}$$

Accordingly, for an infinitesimal $\Delta\tau$,

$$\begin{aligned}0 &= F(\phi(\tau+\Delta\tau),\ \psi(\tau+\Delta\tau);\ \tau+\Delta\tau) - F(\phi(\tau),\ \psi(\tau);\ \tau)\\ &= F_x\frac{\partial\phi}{d\tau}\Delta\tau + F_y\frac{\partial\psi}{d\tau}\Delta\tau + \frac{\partial F}{\partial\tau}\Delta\tau.\end{aligned} \tag{4.14}$$

Dividing the both sides by $\Delta\tau$ and then taking the limit $\Delta\tau \rightarrow 0+$ we have

$$F_x \frac{\partial \phi}{d\tau} + F_y \frac{\partial \psi}{d\tau} + \frac{\partial F}{\partial \tau} = 0, \tag{4.15}$$

which is combined with (4.10) to give

$$\frac{\partial F(x, y; \tau)}{\partial \tau} = 0. \tag{4.16}$$

This is the basic equation of the classical theory of envelopes.

From (4.16), we can solve τ as

$$\tau = \tau(x, y), \tag{4.17}$$

which is in turn inserted into F to give a function of (x, y) only. The resultant function would be the desired envelope function of $G(x, y)$,

$$G(x, y) \equiv F(x, y; \tau(x, y)) = 0. \tag{4.18}$$

An important caution is in order here: The equation $G(x, y) = 0$ may give not only the envelope E but also a set of *singular points* of the family of curves $\{C_\tau\}_\tau$, where

$$\frac{\partial F}{\partial x} = \frac{\partial F}{\partial y} = 0, \tag{4.19}$$

because the last condition is compatible with Eq. (4.16).

When the equation of each curve is expressed simply as

$$y = f(x, \tau), \tag{4.20}$$

the condition (4.16) is reduced to a simpler one

$$\frac{\partial f}{\partial \tau} = 0, \tag{4.21}$$

and the envelope function is given by

$$y = f(x, \tau(x)). \tag{4.22}$$

As an example, let us take the following family of curves $\{\mathrm{C}_\tau\}_\tau$ parametrized by τ,

$$y = f(x, \tau) = e^{-\epsilon\tau}(1 - \epsilon \cdot (x - \tau)), \tag{4.23}$$

which goes minus-infinity when $x - \tau \to \infty$. We try to find out the function $y = g(x)$ describing the possible envelope E of the family of curves $\{C_\tau\}_\tau$. From the envelope equation

$$0 = \frac{\partial f}{\partial \tau} = -\epsilon e^{-\epsilon\tau}(1 - \epsilon \cdot (x - \tau)) + \epsilon e^{-\epsilon\tau} = \epsilon^2 (x - \tau), \tag{4.24}$$

one has

$$\tau = x, \tag{4.25}$$

meaning that the parameter τ is identical with the x-coordinate of the point of the tangency of E and C_τ. Thus the envelope function is found to be

$$y = g(x) = f(x, x) = e^{-\epsilon x}, \tag{4.26}$$

which converges to 0 when $x \to \infty$ in contrast to all members of the family of curves.

Thus, we see that the envelope function has a better (moderate) properties in a global domain even though all members of the family of curves are bounded only locally and divergent asymptotically in a global domain. This property of the envelope function has a great significance when the notion of envelopes is applied to a global analysis of differential equations [3–5] and difference ones [40].

4.2.2 *Envelope Curves/Trajectories in n-Dimensional Space*

The notion of envelopes can be readily extended to the curves or trajectories in the n-dimensional space [5].

A curve C in the n-dimensional space may be represented as

$$\boldsymbol{X}(t) = {}^t(X_1(t), X_2(t), \ldots, X_n(t)) \tag{4.27}$$

with one parameter (time) $t \in [a, b]$ where a and b are real numbers ($a < b$). From now on, the parameter t will be called 'time' and the word 'trajectory' will be also used occasionally in place of 'curve'.

Now let us consider a family of curves in the n-dimensional space where each curve is represented by

$$C_\tau : \boldsymbol{X}(t; \tau) = {}^t(X_1(t; \tau), X_2(t; \tau), \ldots, X_n(t; \tau)), \tag{4.28}$$

where each member of curves is distinguished by a parameter τ. We suppose that the family of curves $\{C_\tau\}_\tau$ has an envelope trajectory E:

$$\mathrm{E}: \boldsymbol{X}_{\mathrm{E}}(t) = {}^t(X_{\mathrm{E}1}(t),\ X_{\mathrm{E}2}(t),\ \ldots,\ X_{\mathrm{E}n}(t)). \tag{4.29}$$

The function $\boldsymbol{X}_{\mathrm{E}}(t)$ may be constructed from $\boldsymbol{X}(t;\tau)$ as follows. Let C_τ and E share the tangent line at time $t = t(\tau)$

$$\boldsymbol{X}_{\mathrm{E}}(t(\tau)) = \boldsymbol{X}(t(\tau);\tau). \tag{4.30}$$

We note that the parameter τ can be reparametrized so as to be the contact time $t(\tau)$, *i.e.*,

$$t(\tau) = \tau, \quad \text{and} \quad \boldsymbol{X}_{\mathrm{E}}(\tau) = \boldsymbol{X}(\tau;\tau). \tag{4.31}$$

The condition that E and C_τ has the common tangent at a time $t = \tau$ may imply that the 'velocities' at this point are the same, *i.e.*,

$$\left.\frac{d\boldsymbol{X}_{\mathrm{E}}(t)}{dt}\right|_{t=t(\tau)} = \left.\frac{\partial \boldsymbol{X}(t,\tau)}{\partial t}\right|_{t=t(\tau)}. \tag{4.32}$$

On the other hand, differentiating Eq. (4.30) with respect to τ, one has

$$\left.\frac{d\boldsymbol{X}_{\mathrm{E}}(t)}{dt}\right|_{t=t(\tau)} \frac{dt}{d\tau} = \left.\frac{\partial \boldsymbol{X}(t,\tau)}{\partial t}\right|_{t=t(\tau)} \frac{dt}{d\tau} + \left.\frac{\partial \boldsymbol{X}(t,\tau)}{\partial \tau}\right|_{t=t(\tau)}. \tag{4.33}$$

Thus combining (4.32) and (4.33), we arrive at

$$\left.\frac{\partial \boldsymbol{X}(t,\tau)}{\partial \tau}\right|_{t=t(\tau)} = \boldsymbol{0}. \tag{4.34}$$

This is the basic equation in the theory of envelope trajectories : From this equation, the parameter value $\tau = \tau(t)$ at the contact point can be extracted, and the envelope function is given by

$$\boldsymbol{X}_{\mathrm{E}}(t) = \boldsymbol{X}(t;\tau(t)). \tag{4.35}$$

A caution is that (4.34) only gives a necessary condition for constructing the envelope, because of the possible existence of singular points , as has been indicated in the preceding subsection.

4.2.3 Adaptation of the Envelope Theory in a Form Applicable to Dynamical Equations

In the application of the envelope theory for constructing a solution of differential equations valid in a global domain [3–5], the parameter is the 'observation or probe time' t_0, *i.e.*, $\tau \equiv t_0$, at which the solution is to be 'observed' or 'probed'. In addition to t_0, we have a set of integral constants $\boldsymbol{C}$ which determine the initial value and hence conversely depend on t_0;

$$\boldsymbol{C} = \boldsymbol{C}(t_0), \tag{4.36}$$

the functional form of which is yet totally unknown. Then the family of the curves and trajectories are given by

$$\{\boldsymbol{X}(t;\ t_0,\ \boldsymbol{C}(t_0))\}_{t_0}. \tag{4.37}$$

Furthermore, we impose that

$$t_0 = \tau = t, \tag{4.38}$$

i.e., each curve in the family is parametrized at the *contact point* with the envelope that is supposed to be the exact solution of the differential equation. Then the envelope equation (4.34) now reads

$$\left.\frac{\partial \boldsymbol{X}(t,\ t_0,\ \boldsymbol{C}(t_0))}{\partial t_0}\right|_{t=t_0} + \left.\frac{\partial \boldsymbol{X}(t,;\ t_0,\ \boldsymbol{C}(t_0))}{\partial \boldsymbol{C}} \cdot \frac{d\boldsymbol{C}}{dt_0}\right|_{t=t_0} = \boldsymbol{0}. \tag{4.39}$$

It should be noted here that the meaning of the present Eq. (4.39) is totally different from that of (4.34) because Eq. (4.39) gives the equation of the unknown function $\boldsymbol{C}(t)$. In other words, the unknown function $\boldsymbol{C}(t)$ is determined so that $\boldsymbol{X}(t;\ t,\ \boldsymbol{C}(t))$ becomes the envelope of the family of curves given by $\boldsymbol{X}(t;\ t_0,\ \boldsymbol{C}(t_0))$. We shall show that the resultant envelope function $\boldsymbol{X}_{\mathrm{E}}(t) = \boldsymbol{X}(t;\ t,\ \boldsymbol{C}(t))$ thus obtained becomes an approximate but uniformly valid solution in a global domain.

A couple of comments are in order here:

(1) Equation (4.39) is essentially of the same form as the RG equation postulated in [1], and those used in quantum field theory and statistical physics [8, 9, 14, 18, 87, 88], although it has been derived as the envelope equation adapted to the analysis of dynamical equations solely with geometrical terms. Thus we shall call Eq. (4.39) and its variants the renormalization group/envelope equation or RG/E equation as well as the RG equation.

(2) The arbitrary time t_0 corresponds to the renormalization point in the renormalization program in quantum field theory [18, 88], and we shall call it 'initial' time, because the perturbed solutions will be constructed around $t \sim t_0$, although it may be more adequate to call it the 'probe time' or 'observation time'.

4.3 Damped Oscillator in RG Method

In this section, we take up the simple linear equation of a damped oscillator and demonstrate how the RG method works in detail. We shall see that the RG/E equation leads to the dynamical equations for the would-be integral constants, the amplitude $A(t)$ and the phase $\theta(t)$, separately, and a resummation of infinite series of the perturbation series is achieved through the RG/E equation.

4.3.1 *Treatment as a Second-Order Differential Equation for Single Dependent Variable x*

The equation of the damped oscillator is given in (2.6), which we reproduce here

$$Lx = -2\epsilon\dot{x}, \quad \text{with} \quad L := \frac{d^2}{dt^2} + 1, \tag{4.40}$$

where ϵ (> 0) is supposed to be small. The exact solution to (4.40) is given by (2.13), which reads

$$x(t) = \bar{A}\mathrm{e}^{-\epsilon t}\sin(\sqrt{1-\epsilon^2}t + \bar{\theta}), \tag{4.41}$$

with $\bar{A}$ and $\bar{\theta}$ being constants to be determined by some initial condition, say, at $t = 0$.

We try to obtain the solution around the 'initial time' $t = t_0$ in a perturbative way, expanding x as

$$x(t;\, t_0) = x_0(t;\, t_0) + \epsilon x_1(t;\, t_0) + \epsilon^2 x_2(t;\, t_0) + \ldots, \tag{4.42}$$

where x_n's ($n = 0, 1, 2, \ldots$) satisfy

$$Lx_0 = 0, \quad Lx_1 = -2\dot{x}_0, \quad Lx_2 = -2\dot{x}_1, \tag{4.43}$$

and so on.

In the RG method, we make the following special set up, as shown in Fig. 4.2: We suppose that the 'initial' value $W(t_0)$ of the solution at an arbitrary time $t = t_0$ is always on an exact solution of (4.40) as

$$x(t_0;\, t_0) = W(t_0). \tag{4.44}$$

Conversely speaking, we suppose that the exact solution $x(t)$ is given by

$$x(t) = W(t) = x(t;\, t). \tag{4.45}$$

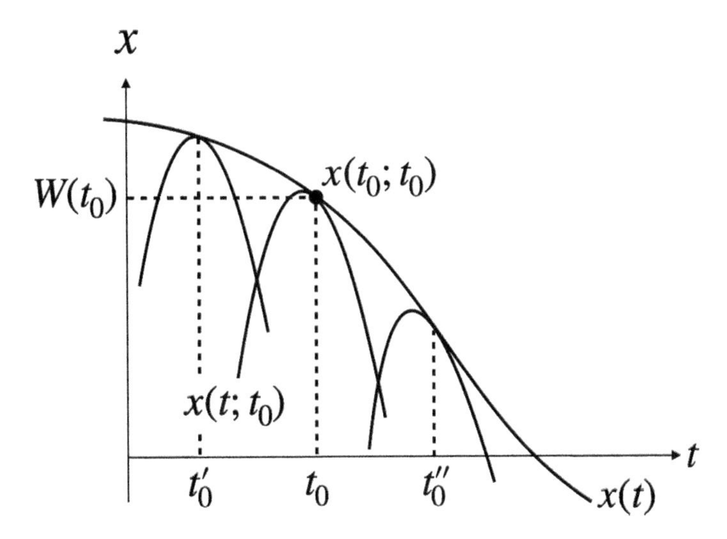

Fig. 4.2 The unperturbed solution $x(t;\, t_0)$ which has the 'initial' value $W(t_0)$ that coincides with the exact solution $x(t)$ at $t = t_0$. The 'initial' time t_0 is arbitrary and hence may be t_0', t_0'' and so on

Then the exact solution as the 'initial' value $W(t_0)$ at $t = t_0$ should be also expanded as

$$W(t_0) = W_0(t_0) \ + \ \epsilon W_1(t_0) \ + \ \epsilon^2 W_2(t_0) \ + \ \dots. \tag{4.46}$$

The solution to the lowest-order equation (4.43) is given by the zero modes of the linear operator L and may be written as

$$x_0(t;\ t_0) = A(t_0)\sin(t + \theta(t_0)), \tag{4.47}$$

where we have made it explicit that the *integral constants* A and θ may depend on the 'initial' time t_0. The 'initial' value $W_0(t_0)$ in the zeroth order is naturally given by

$$W_0(t_0) = x_0(t_0;\ t_0) = A(t_0)\sin(t_0 + \theta(t_0)). \tag{4.48}$$

We remark that the zeroth-order solution is a neutrally stable solution in the sense that the absolute value of it will not diverge when t becomes large.

The first-order Eq. (4.43) now takes the form

$$Lx_1 \ = \ -2A\cos(t + \theta). \tag{4.49}$$

In fact, the solution to this equation was worked out by the method of variation of constants in Chap. 2, and the general solution is given by (2.31), which can be cast into the following form[2]

$$x_1 = -At\,\sin(t+\theta) + C_1^{(1)}\sin(t+\theta) + C_1^{(2)}\cos(t+\theta), \tag{4.50}$$

with $C_1^{(i)}$ $(i = 1,\ 2)$ being arbitrary constants.[3] Here the last two terms are the general solution to the unperturbed equation, that is , the zero modes of L.

Utilizing the ambiguities owing to the arbitrariness of the addition of the zero modes (the independent solutions of L), the form of the particular solutions to the inhomogeneous equations as the higher-order equations in the RG method are determined according to the following principle. First of all since we suppose that the 'initial' value at $t = t_0$ is on an exact solution, the corrections from the zeroth-order solution should be made as small as possible. Then it is an important observation that possible secular terms contained in the general solution (4.50) are made to cancel out with the zero modes at $t = t_0$. Thus we arrive at the following prescription for the construction of the particular solution to the higher-order equations:

(I) Only the functions independent of the zero modes are retained except when the secular terms proportional to the zero modes are present, for which the zero modes are added so that the secular terms cancel out with them at $t = t_0$.

By this way, the possible contributions from the higher-orders to the zero modes are *renormalized away* in the integral constants in the zeroth solution. The added zero modes play a kind of '*counter terms*' in the renormalization program in quantum field theory[83, 88]. It should be emphasized here that the prescription given above has no ambiguities for constructing the perturbative solutions.

In accordance with the above prescription (I), the constants $C_1^{(i)}$ $(i = 1,\ 2)$ are chosen as

$$C_1^{(1)} = At_0, \quad C_1^{(2)} = 0, \tag{4.51}$$

which leads to

$$x_1(t;\ t_0) = -A(t-t_0)\sin(t+\theta), \tag{4.52}$$

and accordingly,

$$x_1(t = t_0;\ t_0) = W_1(t_0) = 0. \tag{4.53}$$

[2] The following elementary way of construction of the perturbative solutions is aimed at those who are not familiar with inhomogeneous differential equations.

[3] Since we are dealing with a second-order differential equation with respect to time, the number of integral constants are two, and we have already introduced two integral constants A and θ. Thus the seemingly new constants $C_1^{(i)}$ $(i = 1,\ 2)$ should be functions of A and θ.

With this choice, one see that the secular term vanishes at the arbitrary 'initial' time t_0 and in the same time the possible difference between the exact value $W(t_0)$ and the lowest one $W_0(t_0)$ is made of higher order of ϵ.

The second-order equation now takes the form

$$Lx_2 = 2A(t - t_0)\cos(t + \theta) + 2A\sin(t + \theta). \tag{4.54}$$

The general solution to the equation is given in (2.36) by the method of variation of constants. Then adding a linear combination of the unpertubed solutions so as to satisfy the prescription (I), we have

$$x_2(t;\ t_0) = \frac{A}{2}(t - t_0)^2 \sin(t + \theta) - \frac{A}{2}(t - t_0)\cos(t + \theta), \tag{4.55}$$

so that

$$W_2(t_0) = 0. \tag{4.56}$$

We remark that there appear no higher-order corrections to the 'initial' value $W(t_0)$ because of the linearity of the equation. As has been noticed already, the choice of the coefficients of the unperturbed solution or the zero modes corresponds to a *renormalization of the integral constants* $A(t_0)$ and $\theta(t_0)$ in the unperturbed solution.

Thus the perturbative solution up to $O(\epsilon^2)$ reads

$$\begin{aligned} x(t;\ t_0;\ A(t_0), \theta(t_0)) = {} & A(t_0)\sin(t + \theta(t_0)) - \epsilon A(t_0)(t - t_0)\sin(t + \theta(t_0)) \\ & + \epsilon^2 \frac{A(t_0)}{2}\{(t - t_0)^2 \sin(t + \theta(t_0)) \\ & - (t - t_0)\cos(t + \theta(t_0))\}. \end{aligned} \tag{4.57}$$

Here we have made explicit the fact that the solution depends on the integral constants $A(t_0)$ and $\theta(t_0)$. Accordingly, the 'initial' value that is supposed to be the exact solution at $t = t_0$ is given by

$$W(t_0) = W_0(t_0) = A(t_0)\sin(t_0 + \theta(t_0)), \tag{4.58}$$

up to $O(\epsilon^2)$.

Admittedly, the solution (4.57) contains secular terms, which vanish at $t = t_0$ in accordance with the prescription (I) in contrast to the perturbative solution (2.37) given in the naïve perturbation method. As emphasized above, by way of making the secular terms vanish at $t = t_0$, the higher-order terms are renormalized into the t_0-dependent integral constants $A(t_0)$ and $\theta(t_0)$, although the definite t_0 dependence of them are yet to be determined.

Here let us take a *geometrical point of view*: We have a family of curves $\{C_{t_0}\}_{t_0}$ given by functions $\{x(t;\ t_0)\}_{t_0}$ parametrized with t_0. They are all solutions of Eq. (4.40) up to $O(\epsilon^2)$ which are supposed to coincide with the exact solution around $t = t_0$ locally up to the same order. The idea is that the envelope curve of the family of the curves $\{C_{t_0}\}_{t_0}$ will give a solution that coincides with the exact solution up to to $O(\epsilon^2)$ *uniformly in a global domain* of t, which will be shown to be the case. From the way of construction of the perturbative solution, the point $P_{t_0}(t_0, x(t_0, ; t_0, A(t_0), \theta(t_0)))$ should be on the exact solution, which means that P_{t_0} is the contact point of each curve with the envelope E. These conditions are neatly summarized in the following envelope equation :

$$\left.\frac{dx(t\,; t_0, A(t_0), \theta(t_0))}{dt_0}\right|_{t_0=t} = \left.\frac{\partial x}{\partial t_0}\right|_{t_0=t} + \left.\frac{\partial x}{\partial A}\frac{dA}{dt_0}\right|_{t_0=t} + \left.\frac{\partial x}{\partial \theta}\frac{d\theta}{dt_0}\right|_{t_0=t} = 0. \quad (4.59)$$

Here we repeat the important observations made in Sect. 4.2.3: The envelope Eq. (4.59) has the same form as the renormalization-group equation in quantum field theory [8, 9] to express the natural requirement that the values of physical quantities should be independent of the renormalization point . Hence the authors in [1] called essentially the same equation as (4.59) the renormalization group (RG) equation.

Inserting the perturbative solution (4.57) into (4.59) we have up to $O(\epsilon^2)$,

$$\left.\left(\frac{dA(t_0)}{dt_0} + \epsilon A(t_0)\right)\right|_{t_0=t} \sin\phi(t) + A(t_0)\left.\left(\frac{d\theta(t_0)}{dt_0} + \frac{1}{2}\epsilon^2\right)\right|_{t_0=t} \cos\phi(t) = 0, \quad \forall t \quad (4.60)$$

with $\phi(t) \equiv t + \theta(t)$. A natural condition for the equation to hold $\forall t$ is that the coefficients of the independent functions $\sin\phi(t)$ and $\cos\phi(t)$ should vanish, and thus we have the following amplitude and phase equations,

$$\frac{dA}{dt} = -\epsilon A, \qquad \frac{d\theta}{dt} = -\frac{\epsilon^2}{2}. \quad (4.61)$$

The readers should be familiar with these equations now because they appeared in the conventional resummation methods such as Krylov-Bogoliubov-Mitropolsky (KBM) and multiple-scale method described in Sect. 3.5 and Sect. 3.6, respectively; The solutions to these equations read

$$A(t) = \bar{A}e^{-\epsilon t}, \quad \theta(t) = -\frac{\epsilon^2}{2}t + \bar{\theta}, \quad (4.62)$$

respectively.

Thus the 'initial' value $W(t)$ in this approximation is given as the envelope

$$x_E(t) = W(t) = x(t\,; t_0 = t, A(t), \theta(t)) = \bar{A}e^{-\epsilon t}\sin\left((1 - \epsilon^2/2)t + \bar{\theta}\right), \quad (4.63)$$

which is certainly a uniformly valid in a global domain. Recall that the 'initial value' $W(t)$ has been supposed to be the exact solution.

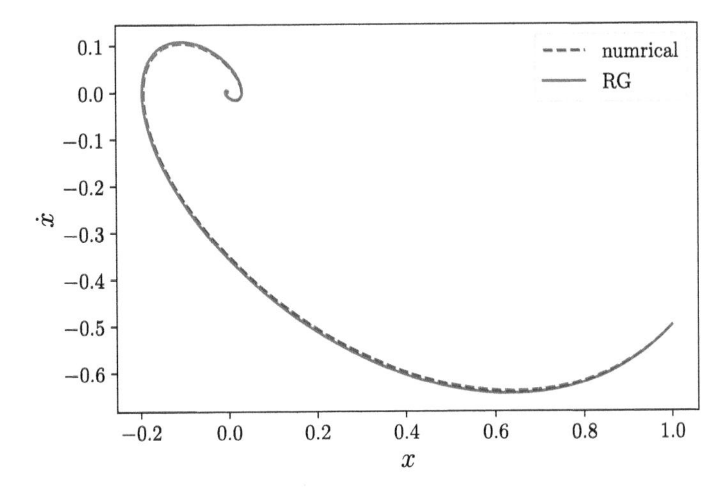

Fig. 4.3 The solutions of the damped oscillator based on numerics and RG method (4.63) with $\epsilon = 0.5$ and initial values $x(0) = 1$ and $\dot{x} = -0.5$

Figure 4.3 shows a comparison of the behavior in the phase space $(x, \dot{x})$ of the RG solution $x_{\mathrm{E}}(t)$ give by (4.63) and the exact solution (obtained by the numerical calculation) for $\epsilon = 0.5$. One sees that the agreement is excellent for a global domain of time until the solution shows the complete damping, although $x_{\mathrm{E}}(t)$ has been constructed through the perturbative expansion only up to the second order.

Here we remark that the solution (4.63) is written as

$$x_E(t) = A\cos\phi + \rho(A, \phi), \tag{4.64}$$

where $\rho(A, \phi) = 0$, and the amplitude A and phase ϕ satisfy the following equations, respectively,

$$\frac{dA}{dt} = F(A), \qquad \text{with} \quad F(A) = 0, \tag{4.65}$$

$$\frac{d\phi}{dt} = 1 + \Omega(A), \qquad \text{with} \quad \Omega(A) = -\frac{\epsilon^2}{2}. \tag{4.66}$$

This structure of the presentation of the solution exactly in accord with that of the ansatz for the solution of nonlinear oscillators adopted in the KBM method, which might look mysterious in the first glance, but has been naturally *derived* in the present RG method. It will be shown that this is generally true when the RG method is applied, at least, to weakly nonlinear oscillators in Sects. 4.5 and 4.6.

The recipe of the RG method is summarized as follows:

1. Starting from the zero mode solution around arbitrary $t = t_0$,
2. the higher-order solutions are constructed according to the prescription (I).

3. Then the RG Eq. (4.59) is applied to construct the envelope of the family of the curves, which is reduced to the dynamical equations of the would-be integral constants of the zero modes, and the solution of the RG/E equation is inserted to the initial value $x(t\,;t,\,A(t),\theta(t))$, which gives the approximate but uniformly valid solution in a global domain by construction.

Finally a couple of comments are in order on the significance and 'uniqueness' of Eq. (4.61) as extracted from Eq. (4.60):

(1) Strictly speaking, Eq. (4.61) is a sufficient condition for Eq. (4.60) to holds, because the argumentation based on the functional independence of $\sin\phi(t)$ and $\cos\phi(t)$ is sloppy, because the coefficients of them are also functions of t. It should be emphasized, however, a rigorous reduction of Eq. (4.61) from the RG/E equation is possible by converting the equation to a *system of first-order equations* , as is shown in Sect. 3.2 of Ref. [6] and below in Sect. 4.3.2.
(2) What we are doing is a kind of the **reduction of dynamics**, *i.e.*, extracting simpler equations from the original equation. As Yoshiki Kuramoto emphasizes [32], one of the principles of the reduction of dynamics is to obtain reduced equations that are as simple as possible, even when the functional form of the solution could become complicated. In the present case, we choose Eq. (4.61) because it not only satisfies the required conditions but also simple although a more complicated set of equations for $A(t)$ and $\theta(t)$ may satisfy Eq. (4.60).

4.3.2 *Treatment of Damped Oscillator as a System of First-Order Equations*

By defining a vector by

$$\boldsymbol{u} = \begin{pmatrix} x \\ y \end{pmatrix}, \quad (y := \dot{x}), \tag{4.67}$$

Equation (4.40) for the damped oscillator is converted to a system of first-order equations as

$$\left(\frac{d}{dt} - L\right)\boldsymbol{u} = -2\epsilon y \begin{pmatrix} 0 \\ 1 \end{pmatrix}, \tag{4.68}$$

with

$$L = \begin{pmatrix} 0 & 1 \\ -1 & 0 \end{pmatrix}. \tag{4.69}$$

We try to obtain a local solution to Eq. (4.68) around $t = \forall t_0$ in the perturbation theory. We expand the dependent variable and the initial value in Taylor series as

$$\boldsymbol{u}(t; t_0) = \boldsymbol{u}_0 + \epsilon \boldsymbol{u}_1 + \epsilon^2 \boldsymbol{u}_2 + \cdots \tag{4.70}$$

with $\boldsymbol{u}_i = {}^t(x_i, y_i)$, and the 'initial' value $\boldsymbol{W}(t_0)$ at $t = t_0$ is chosen to be an exact solution, which is also expanded as

$$\boldsymbol{W}(t_0) = \boldsymbol{W}_0 + \epsilon \boldsymbol{W}_1 + \epsilon^2 \boldsymbol{W}_2 + \cdots, \tag{4.71}$$

The lowest-order equation reads

$$\left(\frac{d}{dt} - L\right) \boldsymbol{u}_0 = 0, \tag{4.72}$$

the solution to which is given by

$$\boldsymbol{u}_0(t; t_0) = C(t_0)\mathrm{e}^{it}\boldsymbol{U}_+ + C^*(t_0)\mathrm{e}^{-it}\boldsymbol{U}_- =: \begin{pmatrix} x_0(t; t_0) \\ y_0(t; t_0) \end{pmatrix}, \tag{4.73}$$

where

$$\boldsymbol{U}_\pm = \begin{pmatrix} 1 \\ \pm i \end{pmatrix} \tag{4.74}$$

are the eigenfunctions of L belonging to the eigenvalues $\pm i$, respectively;

$$L\boldsymbol{U}_\pm = \pm i \boldsymbol{U}_\pm. \tag{4.75}$$

This is a *neutrally stable solution* . We have made it explicit that the integral constants $C(t_0)$ and $C^*(t_0)$ may dependent on t_0. The 'initial' value at $t = t_0$ in this order reads,

$$\boldsymbol{W}_0(t_0) = z(t_0)\boldsymbol{U}_+ + \text{c.c.}, \tag{4.76}$$

with

$$z(t_0) = C(t_0)\mathrm{e}^{it_0}. \tag{4.77}$$

Here c.c. denotes the complex conjugate. We remark that the lowest-order solution can be expressed in terms of the initial value as

$$\boldsymbol{u}_0(t; t_0) = \mathrm{e}^{(t-t_0)L}\boldsymbol{W}_0(t_0). \tag{4.78}$$

We remark that

$$y_0(t; t_0) = iC(t_0)\mathrm{e}^{it} + \text{c.c.}. \tag{4.79}$$

Owing to the equality

$$\begin{pmatrix} 0 \\ 1 \end{pmatrix} = (\boldsymbol{U}_+ - \boldsymbol{U}_-)/2i, \tag{4.80}$$

the first-order equation reads

$$\begin{aligned} \left(\frac{d}{dt} - L\right)\boldsymbol{u}_1 &= iy_0(t; t_0)(\boldsymbol{U}_+ - \boldsymbol{U}_-) \\ &= (-C\mathrm{e}^{it} - C^*\mathrm{e}^{-it})(\boldsymbol{U}_+ - \boldsymbol{U}_-). \end{aligned} \tag{4.81}$$

Applying the general formula (4.225) for the solution of the inhomogeneous equation given in Sect. 4.7, Eq. (4.81) is formally solved as

$$\begin{aligned} \boldsymbol{u}_1(t; t_0) &= \mathrm{e}^{(t-t_0)L}\boldsymbol{W}_1(t_0) + i\int_{t_0}^{t} ds\mathrm{e}^{(t-s)L}y_0(s)(\boldsymbol{U}_+ - \boldsymbol{U}_-), \\ &= \mathrm{e}^{(t-t_0)L}\left[\boldsymbol{W}_1(t_0) + \frac{1}{2}\{iC\mathrm{e}^{it_0}\boldsymbol{U}_- + \text{c.c.}\}\right] \\ &\quad - \left[\{(t - t_0)C\mathrm{e}^{it}\boldsymbol{U}_+ + i\frac{1}{2}C\mathrm{e}^{it}\boldsymbol{U}_-\} + \text{c.c.}\right]. \end{aligned} \tag{4.82}$$

One observes that the first line of (4.82) would gives rise to terms that could be included in the unperturbed solution, and hence should be avoided according to the prescription (I). This requirement is fulfilled by simply choosing the 'initial' value $\boldsymbol{W}_1(t_0)$ yet to be determined as

$$\boldsymbol{W}_1(t_0) = -\frac{1}{2}\{iz(t_0)\boldsymbol{U}_- + \text{c.c.}\} =: \boldsymbol{\rho}_1[z, z^*]. \tag{4.83}$$

One may say that this is a kind of a renormalization of the integral constants in the unperturbed solution. Thus we have

$$\begin{aligned} \boldsymbol{u}_1(t; t_0) &= -C(t_0)\mathrm{e}^{it}\{(t - t_0)\boldsymbol{U}_+ + \frac{i}{2}\boldsymbol{U}_-\} + \text{c.c.}, \\ &=: \begin{pmatrix} x_1(t; t_0) \\ y_1(t; t_0) \end{pmatrix}. \end{aligned} \tag{4.84}$$

A comment is in order: The above solution could be more efficiently obtained by using the operator method [3, 5, 6] presented in Appendix 2 Sect. 4.7 in this chapter. In fact, the solution to (4.81) is formally expressed in the operator method as

$$\boldsymbol{u}_1(t;t_0) = -\frac{1}{d_t - L}(Ce^{it} - C^*e^{-it})(\boldsymbol{U}_+ - \boldsymbol{U}_-), \tag{4.85}$$

with $d_t := \frac{d}{dt}$. However, using the formulae (4.238) and (4.237), we have

$$\frac{1}{d_t - L}e^{it}\boldsymbol{U}_+ = (t - t_0)e^{it}\boldsymbol{U}_+, \tag{4.86}$$

$$\frac{1}{d_t - L}e^{it}\boldsymbol{U}_- = \frac{1}{i - (-i)}e^{it}\boldsymbol{U}_- = -\frac{i}{2}e^{it}\boldsymbol{U}_-, \tag{4.87}$$

respectively. Thus

$$\boldsymbol{u}_1(t;t_0) = -C(t_0)\left[(t - t_0)e^{it}\boldsymbol{U}_+ + \frac{i}{2}e^{it}\boldsymbol{U}_-\right] + \text{c.c.}, \tag{4.88}$$

which exactly coincides with (4.84), which implies that

$$y_1(t;t_0) = C(t_0)e^{it}\{-i(t - t_0) - 1/2\} + \text{c.c.}. \tag{4.89}$$

The second-order equation reads

$$\begin{aligned}\left(\frac{d}{dt} - L\right)\boldsymbol{u}_2 &= iy_1(t;t_0)(\boldsymbol{U}_+ - \boldsymbol{U}_-)\\ &= \left[C(t_0)e^{it}\{(t - t_0) - i/2\} + \text{c.c.}\right](\boldsymbol{U}_+ - \boldsymbol{U}_-)\\ &= C(t_0)\Big(\left[(t - t_0)e^{it}\boldsymbol{U}_+ - (i/2)e^{it}\boldsymbol{U}_+\right]\\ &\quad - \left[(t - t_0)e^{it}\boldsymbol{U}_- - (i/2)e^{it}\boldsymbol{U}_-\right]\Big) + \text{c.c.}.\end{aligned} \tag{4.90}$$

with the 'initial' condition $\boldsymbol{u}_2(t_0;t_0) = \boldsymbol{W}_2(t_0)$. By using the operator method [3, 5, 6] presented in Appendix 2 of this chapter, we have

$$\frac{1}{d_t - L}(t - t_0)e^{it}\boldsymbol{U}_+ = \frac{1}{2}(t - t_0)^2e^{it}\boldsymbol{U}_+ \tag{4.91}$$

$$\begin{aligned}\frac{1}{d_t - L}(t - t_0)e^{it}\boldsymbol{U}_- &= \frac{1}{2i}\left\{(t - t_0) - \frac{1}{2i}\right\}e^{it}\boldsymbol{U}_-\\ &= \left\{-\frac{i}{2}(t - t_0) + \frac{1}{4}\right\}e^{it}\boldsymbol{U}_-,\end{aligned} \tag{4.92}$$

respectively. Thus we have

$$\begin{aligned} u_2(t;t_0) &= C(t_0)\Big(\Big[\frac{1}{2}(t-t_0)^2 e^{it} U_+ - (i/2)(t-t_0)e^{it} U_+\Big] \\ &\quad - \Big[\{-\frac{i}{2}(t-t_0) + \frac{1}{4}\} e^{it} U_- - (1/4)e^{it} U_-\Big]\Big) + \text{c.c.} \\ &= \frac{1}{2}C(t_0)\Big(\big[(t-t_0)^2 - i(t-t_0)\big] e^{it} U_+ + i(t-t_0)e^{it} U_-\Big) + \text{c.c.}. \end{aligned} \tag{4.93}$$

Collecting all the terms up to the second order, we have

$$\begin{aligned} u(t;t_0) &= Ce^{it}\Big[U_+ - \epsilon\{(t-t_0)U_+ + \frac{i}{2}U_-\} \\ &\quad + \frac{\epsilon^2}{2}[\{(t-t_0)^2 - i(t-t_0)\}U_+ + i(t-t_0)U_-]\Big] + \text{c.c.}, \end{aligned} \tag{4.94}$$

with the 'initial' value

$$W(t_0) = C(t_0)e^{it_0}\{U_+ - i\frac{\epsilon}{2}U_-\} + \text{c.c.}. \tag{4.95}$$

The RG/E equation

$$\left.\frac{du(t;t_0)}{dt_0}\right|_{t_0=t} = 0 \tag{4.96}$$

gives

$$\dot{C}e^{it}[U_+ - (i/2)U_-] + Ce^{it}[\epsilon U_+ + (i\epsilon^2/2)(U_+ - U_-)] + \text{c.c.} = 0, \tag{4.97}$$

$\forall t$ up to the second order of ϵ. From the equality of the coefficient of U_+, we have

$$\dot{C} = -(\epsilon + i\epsilon^2/2)C, \tag{4.98}$$

which implies that the coefficient of U_- vanishes up to the second order of ϵ because

$$(i\epsilon/2)(\dot{C} + \epsilon C) = O(\epsilon^3). \tag{4.99}$$

With the parameterization

$$C(t) = \frac{1}{2}A(t)e^{i\theta(t)}, \tag{4.100}$$

(4.98) leads to the amplitude and the phase, respectively, as

$$\dot{A} = -\epsilon A, \quad \dot{\theta} = -\epsilon^2/2, \tag{4.101}$$

which yield

$$A(t) = A_0 e^{-\epsilon t} \quad \text{and} \quad \theta(t) = -\epsilon^2 t/2 + \theta_0 \tag{4.102}$$

with A_0 and θ_0 being constants. The approximate solution as the envelope function is given by

$$\begin{aligned} \boldsymbol{u}_{\mathrm{E}}(t) &= \boldsymbol{u}(t;t) = \boldsymbol{W}(t) \\ &= C(t)e^{it}\{\boldsymbol{U}_+ - i\frac{\epsilon}{2}\boldsymbol{U}_-\} + \text{c.c.} \\ &= \frac{1}{2}A_0 e^{-\epsilon t} e^{i\{(1-\epsilon^2/2)t+\theta_0\}}\{\boldsymbol{U}_+ - i\frac{\epsilon}{2}\boldsymbol{U}_-\} + \text{c.c.}. \end{aligned} \tag{4.103}$$

Thus we have the final solution to the damped oscillator that is valid in a global domain as

$$x(t) = \bar{A}\sin(\omega t + \bar{\theta}) \tag{4.104}$$

where $\omega = 1 - \epsilon^2/2$ and $\bar{A}$ and $\bar{\theta}$ are constants. The above expression coincides with (4.63), and thus confirms the somewhat heuristic derivation of the amplitude and phase qaution (4.61) from (4.60).

4.4 RG/E Analysis of a Boundary-Layer Problem Without Matching; A System Treatment

Next we take up a linear but a typical boundary-layer problem [21, Chap. 9]:

$$\epsilon\frac{d^2y}{dx^2} + (1+\epsilon)\frac{dy}{dx} + y = 0, \tag{4.105}$$

with the boundary condition

$$y(0) = 0, \quad y(1) = 1. \tag{4.106}$$

Because of the linearity, the exact solution is readily found to be

$$y(x) = \frac{e^{-x} - e^{-x/\epsilon}}{e^{-1} - e^{-1/\epsilon}}. \tag{4.107}$$

To convert the equation in a form for which the regular perturbation theory is applicable, we introduce the so called *inner variable* X by

$$\epsilon X := x, \tag{4.108}$$

and define the function of X by

$$Y(X) := y(x). \tag{4.109}$$

Then the equation takes the following form,

$$L_B Y = -\epsilon \left(\frac{d}{dX} + 1 \right) Y, \quad L_B := \frac{d}{dX} \left(\frac{d}{dX} + 1 \right). \tag{4.110}$$

We try to obtain the solution $Y(X; X_0)$ around an arbitrary point $X = X_0$, where the boundary value $Y(X = X_0; X_0)$ is supposed to takes an exact solution, which we denote as[4]

$$Y(X_0; X_0) = W(X_0). \tag{4.111}$$

From now on, we treat the problem as a system of first-order equations . One will see tha the RG/E equation will lead to a couple of reduced equations in a rigorous way.[5]

Equation (4.110) written inner variable $X = x/\epsilon$ is converted to a system of first-order equations for

$$\mathbf{Z}(X) = \begin{pmatrix} Y \\ dY/dX \end{pmatrix} \equiv \begin{pmatrix} Z_1(X) \\ Z_2(X) \end{pmatrix}. \tag{4.112}$$

Defining the coefficient matrix as

$$L := \begin{pmatrix} 0 & 1 \\ 0 & -1 \end{pmatrix}, \tag{4.113}$$

we have

$$\frac{d\mathbf{Z}}{dX} = L\mathbf{Z} + \epsilon(Z_1 + Z_2) \begin{pmatrix} 0 \\ -1 \end{pmatrix}. \tag{4.114}$$

We try to solve Eq. (4.114) by the perturbation theory around $X = X_0$ with the expansion $\mathbf{Z} = \mathbf{Z}^{(0)} + \mathbf{Z}^{(1)} + \cdots$.

The zeroth-order equation reads

---

[4] The prescription for treating the higher order terms given in Ref. [3] was based on Ref. [1] which is different from that given in Sect. 3 of [3] nor that presented in this monograph. As a result, the formulae given in [3], unfortunately, suffere from some errors. Equations (4.5) through (4.10) in Ref. [3] should be substituted by the corresponding expressions given below and Appendix 1 of this chapter.

[5] An RG analysis in the scalar representation is given in Appendix 1 of this chapter, for an instructive purpose.

$$\frac{d}{dX}\boldsymbol{Z}^{(0)} = L\boldsymbol{Z}^{(0)}. \tag{4.115}$$

The eigenvalues and the respective eigenvectors of L are found to be

$$\lambda_1 = 0, \quad \boldsymbol{U}_1 := \begin{pmatrix} 1 \\ 0 \end{pmatrix}; \ \lambda_2 = -1, \quad \boldsymbol{U}_2 := \begin{pmatrix} -1 \\ 1 \end{pmatrix}. \tag{4.116}$$

We remark the equality

$$\begin{pmatrix} 0 \\ -1 \end{pmatrix} = -(\boldsymbol{U}_1 + \boldsymbol{U}_2). \tag{4.117}$$

Thus the zeroth-order solution is given by

$$\boldsymbol{Z}^{(0)}(X) = A(X_0)\boldsymbol{U}_1 + B(X_0)\mathrm{e}^{-X}\boldsymbol{U}_2 =: \begin{pmatrix} Z_1^{(0)} \\ Z_2^{(0)} \end{pmatrix}, \tag{4.118}$$

which gives

$$Z_1^{(0)} + Z_2^{(0)} = A(X_0). \tag{4.119}$$

Then the first-order equation is given by

$$\left(\frac{d}{dX} - L\right)\boldsymbol{Z}_1 = -A(X_0)(\boldsymbol{U}_1 + \boldsymbol{U}_2). \tag{4.120}$$

The particular solution to (4.120) in the form suitable for the RG method is given in the operator method[6] as

$$\begin{aligned} \boldsymbol{Z}_1 &= -A(X_0)\left[\frac{1}{\frac{d}{dX} - L}\boldsymbol{U}_1 + \frac{1}{\frac{d}{dX} - L}\boldsymbol{U}_2\right] \\ &= -A(X_0)(X - X_0)\boldsymbol{U}_1 - A(X_0)\boldsymbol{U}_2, \end{aligned} \tag{4.121}$$

where the formulae (4.238) and (4.237) have been used; we remark that (4.121) satisfies the principle (I).

Thus we have for the perturbative solution up to the first order

$$\begin{aligned} \boldsymbol{Z}(X; X_0) &= \boldsymbol{Z}^{(0)} + \epsilon\boldsymbol{Z}^{(1)} \\ &= A(X_0)\boldsymbol{U}_1 + B(X_0)\mathrm{e}^{-X}\boldsymbol{U}_2 \\ &\quad + \epsilon\left[-A(X_0)(X - X_0)\boldsymbol{U}_1 - A(X_0)\boldsymbol{U}_2\right] + \mathrm{O}(\epsilon^2). \end{aligned} \tag{4.122}$$

[6] See Appendix 2 in this chapter.

The RG/E equation up to the first order of ϵ

$$\frac{d}{dX_0}\boldsymbol{Z}\Big|_{X_0=X} = 0 \tag{4.123}$$

is reduced to

$$\begin{aligned}\boldsymbol{0} &= \frac{dA}{dX}\boldsymbol{U}_1 + \frac{dB}{dX}\mathrm{e}^{-X}\boldsymbol{U}_2 + \epsilon\left[A(X)\boldsymbol{U}_1 - \frac{dA}{dX}\boldsymbol{U}_2\right] \\ &= \left(\frac{dA}{dX} + \epsilon A(X)\right)\boldsymbol{U}_1 + \left(\frac{dB}{dX}\mathrm{e}^{-X} - \epsilon\frac{dA}{dX}\right)\boldsymbol{U}_2.\end{aligned} \tag{4.124}$$

Then, because of the linear independence of $\boldsymbol{U}_1$ and $\boldsymbol{U}_2$, we have

$$\frac{dA}{dX} = -\epsilon A(X) \quad \text{and} \quad \frac{dB}{dX}\mathrm{e}^{-X} - \epsilon\frac{dA}{dX} = 0. \tag{4.125}$$

We remark, however, that the first equation tells us that $dA/dX = \mathrm{O}(\epsilon)$, and hence the second equation actually means $\frac{dB}{dX}\mathrm{e}^{-X} = 0$ or

$$\frac{dB}{dX} = 0, \tag{4.126}$$

up to the first order of ϵ. Thus solving the simple equations, we have

$$A(X) = A_0\mathrm{e}^{-\epsilon X} \quad \text{and} \quad B(X) = \bar{B}. \tag{4.127}$$

with A_0 and $\bar{B}$ being constant.

The envelope function is now given by

$$\begin{aligned}\boldsymbol{Z}_{\mathrm{E}}(X) &= \boldsymbol{Z}(Z; X) = A(X)\boldsymbol{U}_1 + B(X)\mathrm{e}^{-X}\boldsymbol{U}_2 - \epsilon A(X)\boldsymbol{U}_2 \\ &= \begin{pmatrix}(1+\epsilon)A_0\mathrm{e}^{-\epsilon X} - \bar{B}\mathrm{e}^{-X} \\ -\epsilon A_0\mathrm{e}^{-\epsilon X} + \bar{B}\mathrm{e}^{-X}\end{pmatrix} \\ &= \begin{pmatrix}\bar{A}\mathrm{e}^{-\epsilon X} - \bar{B}\mathrm{e}^{-X} \\ -\epsilon\bar{A}\mathrm{e}^{-\epsilon X} + \bar{B}\mathrm{e}^{-X}\end{pmatrix}\end{aligned} \tag{4.128}$$

up to the first order of ϵ with $\bar{A} := (1+\epsilon)A_0$. Thus

$$Y_{\mathrm{E}}(X) = Z_1(X; X) = \bar{A}\mathrm{e}^{-\epsilon X} - \bar{B}\mathrm{e}^{-X}. \tag{4.129}$$

In terms of the original variable x,

$$y_{\mathrm{E}}(x) = \bar{A}\mathrm{e}^{-x} - \bar{B}\mathrm{e}^{-x/\epsilon}. \tag{4.130}$$

It is noteworthy that the resultant $y_E(x)$ can admit both the inner and outer boundary conditions (4.106), simultaneously;

$$y_E(0) = \bar{A} - \bar{B} = 0, \quad y(1) = \bar{A}\mathrm{e}^{-1} - \bar{B}\mathrm{e}^{-1/\epsilon} = 1, \tag{4.131}$$

which yield

$$\bar{A} = \bar{B} = \frac{1}{\mathrm{e}^{-1} - \mathrm{e}^{-1/\epsilon}} \tag{4.132}$$

and hence $y_E(x)$ coincides with the exact solution $y(x)$ given in Eq. (4.107).

4.5 The van der Pol Equation in RG Method

The van der Pol equation was treated already in the previous chapters,[7] which is reproduced here;

$$\ddot{x} + x = \epsilon\,(1 - x^2)\,\dot{x}, \tag{4.133}$$

with ϵ being small so as to the perturbation analysis is valid.

We first convert the Eq. (4.133) to a system of first-order differential equations as

$$\left(\frac{d}{dt} - L\right)\boldsymbol{u} = \epsilon\boldsymbol{b}(\boldsymbol{u}) \tag{4.134}$$

with

$$\boldsymbol{u} = \begin{pmatrix} x \\ y = \dot{x} \end{pmatrix} \quad \text{and} \quad \boldsymbol{b} = (1 - x^2)\,y\begin{pmatrix} 0 \\ 1 \end{pmatrix}, \tag{4.135}$$

and

$$L = \begin{pmatrix} 0 & 1 \\ -1 & 0 \end{pmatrix} \tag{4.136}$$

is the linear operator defined in Sect. 4.3.2 for describing the harmonic oscillator with the angular velocity 1: Their eigenvalues are $\pm i$ and the respective eigenvectors are given in (4.74), which we reproduce here; $\boldsymbol{U}_{\pm} = {}^t(1, \pm i)$.

We try to obtain a local solution to Eq. (4.134) around $t = \forall t_0$ in the perturbation theory. We expand the dependent variable in Taylor series as

$$\boldsymbol{u}(t; t_0) = \boldsymbol{u}_0 + \epsilon\boldsymbol{u}_1 + \epsilon^2\boldsymbol{u}_2 + \cdots \tag{4.137}$$

[7] See Sect. 3.5.4, for instance.

with $\boldsymbol{u}_i = {}^t(x_i, y_i)$, and the 'initial' value $W(t_0)$ at $t = t_0$ is chosen to be an exact solution, which is also expanded as

$$\boldsymbol{W}(t_0) = \boldsymbol{W}_0 + \epsilon \boldsymbol{W}_1 + \epsilon^2 \boldsymbol{W}_2 + \cdots , \tag{4.138}$$

The lowest-order equation is the same as that in the case of the damped oscillator and the solution to it has been also given in (4.73) as

$$\boldsymbol{u}_0(t; t_0) = C(t_0)\mathrm{e}^{it}\boldsymbol{U}_+ + C^*(t_0)\mathrm{e}^{-it}\boldsymbol{U}_- =: \begin{pmatrix} x_0(t; t_0) \\ y_0(t; t_0) \end{pmatrix}, \tag{4.139}$$

where

$$x_0(t; t_0) = z(t; t_0) + z^*(t; t_0), \quad y_0(t; t_0) = i(z(t; t_0) - z^*(t; t_0)) \tag{4.140}$$

with

$$z(t; t_0) := C(t_0)\mathrm{e}^{it}. \tag{4.141}$$

The first-order equation is given by

$$\left(\frac{d}{dt} - L\right)\boldsymbol{u}_1 = \frac{(1 - x_0^2)y_0}{2i}(\boldsymbol{U}_+ - \boldsymbol{U}_-). \tag{4.142}$$

However,

$$x_0^2 y_0 = i(z^2 + 2|z|^2 + z^{*2})(z - z^*) = i\left[(z^3 + |z|^2 z) - \text{c.c.}\right]. \tag{4.143}$$

Thus the right-hand side of (4.142) reads

$$\begin{aligned}
\frac{1}{2}\left[\{z - (z^3 + |z|^2 z)\} - \text{c.c.}\right](\boldsymbol{U}_+ - \boldsymbol{U}_-) &= \frac{1}{2}\left[\{(1 - |C|^2)C\mathrm{e}^{it} - C^3\mathrm{e}^{3it}\} - \text{c.c.}\right] \\
&\quad \times (\boldsymbol{U}_+ - \boldsymbol{U}_-) \\
&= \frac{1}{2}\Big[\{(1 - |C|^2)C\mathrm{e}^{it} - C^3\mathrm{e}^{3it}\}\boldsymbol{U}_+ \\
&\quad - \{(1 - |C|^2)C\mathrm{e}^{it} - C^3\mathrm{e}^{3it}\}\boldsymbol{U}_-\Big] \\
&\quad + \text{c.c.}.
\end{aligned} \tag{4.144}$$

However, on account of the formulae (4.238) and (4.237), we have

$$\frac{1}{\frac{d}{dt} - L} e^{it} \boldsymbol{U}_+ = (t - t_0) e^{it} \boldsymbol{U}_+, \tag{4.145}$$

$$\frac{1}{\frac{d}{dt} - L} e^{3it} \boldsymbol{U}_+ = \frac{1}{3i - i} e^{3it} \boldsymbol{U}_+ = -\frac{i}{2} e^{3it} \boldsymbol{U}_+, \tag{4.146}$$

$$\frac{1}{\frac{d}{dt} - L} e^{it} \boldsymbol{U}_- = \frac{1}{i + i} e^{it} \boldsymbol{U}_+ = -\frac{i}{2} e^{it} \boldsymbol{U}_-, \tag{4.147}$$

$$\frac{1}{\frac{d}{dt} - L} e^{3it} \boldsymbol{U}_- = \frac{1}{3i + i} e^{3it} \boldsymbol{U}_- = -\frac{i}{4} e^{3it} \boldsymbol{U}_-. \tag{4.148}$$

Thus the solution to the first-order equation is given by

$$\begin{aligned} \boldsymbol{u}_1(t; t_0) = \frac{1}{2} \Big[& \{(1 - |C|^2) C (t - t_0) e^{it} - C^3 (-i/2) e^{3it}\} \boldsymbol{U}_+ \\ & - \{(1 - |C|^2) C (-i/2) e^{it} - C^3 (-i/4) e^{3it}\} \boldsymbol{U}_- \Big] + \text{c.c.}. \end{aligned} \tag{4.149}$$

Summing the perturbative solutions up to the first order, we have

$$\begin{aligned} \boldsymbol{u}(t; t_0) &= \boldsymbol{u}_0(t; t_0) + \epsilon \boldsymbol{u}_1(t; t_0) \\ &= C(t_0) e^{it} \boldsymbol{U}_+ + \frac{\epsilon}{2} \Big[\{(1 - |C|^2) C (t - t_0) e^{it} - C^3 (-i/2) e^{3it}\} \boldsymbol{U}_+ \\ &\quad - \{(1 - |C|^2) C (-i/2) e^{it} - C^3 (-i/4) e^{3it}\} \boldsymbol{U}_- \Big] + \text{c.c.}. \end{aligned} \tag{4.150}$$

Now the RG/E equation

$$\frac{d}{dt_0} \boldsymbol{u}(t; t_0) \Big|_{t_0 = t} = 0 \tag{4.151}$$

is found to lead to

$$\frac{dC}{dt} = \frac{\epsilon}{2} (1 - |C|^2) C \tag{4.152}$$

up to the first order of ϵ. We remark that the remaining terms are of higher order because $\dot{C} = \mathrm{O}(\epsilon)$. The solution as the envelope trajectory is given by

$$\begin{aligned} \boldsymbol{u}_{\mathrm{E}}(t) &= \boldsymbol{u}(t; t) \\ &= \left[C(t) e^{it} + i \frac{\epsilon}{4} C^3(t) e^{3it} \right] \boldsymbol{U}_+ \\ &\quad + i \frac{\epsilon}{4} \left[(1 - |C|^2) C e^{it} - \frac{1}{2} C^3 e^{3it} \right] \boldsymbol{U}_- + \text{c.c.}, \end{aligned} \tag{4.153}$$

up to the first order of ϵ.

With the parametrization

$$C(t) = \frac{A(t)}{2} \mathrm{e}^{i\theta(t)} \tag{4.154}$$

with $A\,(\geq 0)$ and θ being real numbers, the reduced Eq. (4.152) leads to the amplitude and phase equations, separately;

$$\dot{A} = \frac{\epsilon}{2} A \left(1 - \frac{A^2}{4}\right), \quad \dot{\theta} = 0. \tag{4.155}$$

These equations are the same as those derived in Sect. 3.5.4 by the KBM method, and the solution to the amplitude equation is given in (3.111), which admits a limit cycle with a radius 2.

One can proceed to the second-order analysis [90], which leads to a non-trivial phase equation but no change in the amplitude equation as was given, for instance, in [22].

4.6 Jump Phenomenon in Forced Duffing Equation

Here we consider an extension of the Duffing Eq. (2.47) by adding a weak dissipative effect proportional to the velocity and external harmonic force [22, 25] as given by

$$\ddot{x} + 2\kappa\dot{x} + \omega^2 x + \alpha x^3 = F \cos \Omega t, \tag{4.156}$$

which is called the forced Duffing equation. In the following, we are going to discuss the case where external force is weak with the angular velocity Ω^2 being close to the intrinsic one ω^2 and the resistivity 2κ and the anharmonicity of the potential α are all small;

$$\omega^2 = \Omega^2 + \epsilon\delta\omega^2, \quad (0 <)\kappa = \epsilon\gamma, \quad \alpha = \epsilon h, \tag{4.157}$$

where $|\epsilon|$ is small but the sign of ϵ is not specified. Thus we have

$$\ddot{x} + 2\epsilon\gamma\dot{x} + (\Omega^2 + \epsilon\delta\omega^2)x + \epsilon h x^3 = \epsilon f \cos \Omega t. \tag{4.158}$$

A comment is in order here: In practice, people are interested in analyzing the jump phenomenon [22, 25, 91] in which the amplitude of the excited mode with the frequency close to the external one Ω or its multiples with some rational numbers shows a drastic change along with a slow variation of Ω. For this purpose, it may be more convenient to use the parametrization

$$\omega = \Omega + \epsilon\sigma, \quad \text{or} \quad \Omega = \omega - \epsilon\sigma. \tag{4.159}$$

One can translate it to our parametrization through the formula

$$\epsilon\delta\omega^2 = 2\epsilon\Omega\sigma, \quad \text{or} \quad \epsilon\sigma = \epsilon\frac{\delta\omega^2}{2\Omega}. \tag{4.160}$$

For calculational convenience, we also consider a similar equation with a different external force as

$$\ddot{y} + 2\epsilon\gamma\dot{y} + (\Omega^2 + \epsilon\delta\omega^2)y + \epsilon h y^3 = \epsilon f \sin\Omega t. \tag{4.161}$$

Then defining a complex variable $z = x + iy$, (4.158) and (4.161) are combined to a single equation as

$$\ddot{z} + \Omega^2 z = -\epsilon\left[\delta\omega^2 z + 2\gamma\dot{z} + \frac{h}{4}(3|z|^2 z + z^{*3}) - f\mathrm{e}^{i\Omega t}\right]. \tag{4.162}$$

Expanding z as

$$z = z_0 + \epsilon z_1 + \epsilon^2 z_2 + \cdots, \tag{4.163}$$

we try to solve Eq. (4.162) in the perturbation theory around an arbitrary time $t = t_0$ with the 'initial' value $W(t_o)$, which is supposed to be an exact solution as a function of t_0 and expanded as; $W(t_0) = W_0(t_0) + \epsilon W_1(t_0) + \epsilon^2 W_2(t_0) + \cdots$.

The zeroth-order equation reads

$$\ddot{z}_0 + \Omega^2 z_0 = 0, \tag{4.164}$$

the solution to which may be written as

$$z_0(t; t_0) = A(t_0)\mathrm{e}^{i\Omega t + \theta(t_0)}, \tag{4.165}$$

where A and θ are assumed to be real numbers without loss of generality and may depend on the initial time t_0. Defining the complex amplitude by

$$\mathcal{A}(t_0) := A\mathrm{e}^{i\theta(t_0)}, \tag{4.166}$$

we can write the zeroth order solution as

$$z_0(t; t_0) = \mathcal{A}(t_0)\mathrm{e}^{i\Omega t}. \tag{4.167}$$

The first-order equation now takes the form

$$\ddot{z}_1 + \Omega^2 z_1 = -2\gamma \dot{z}_0 - \delta\omega^2 z_0 - \frac{h}{4}(3|z_0|^2 z_0 + z_0^{*\,3}) + f e^{i\Omega t}$$
$$= \left(-2i\gamma\Omega\mathcal{A} - \delta\omega^2\mathcal{A} - \frac{3hA^2}{4}\mathcal{A} + f\right) e^{i\Omega t} - \frac{h}{4}\mathcal{A}^{*3} e^{-3i\Omega t}, \tag{4.168}$$

the solution to which with a suitable form to apply the RG method is given by

$$z_1(t; t_0) = -\frac{i}{2\Omega}(t - t_0)\left[-2i\gamma\Omega\mathcal{A} - \delta\omega^2\mathcal{A} - \frac{3h}{4}|\mathcal{A}|^2\mathcal{A} + f\right] e^{i\Omega t}$$
$$+ \frac{h}{32}\mathcal{A}^{*3} e^{-3i\Omega t}. \tag{4.169}$$

Thus up to the first order, we have

$$z(t; t_0) = \mathcal{A}(t_0) e^{i\Omega t} - \epsilon\frac{i}{2\Omega}(t - t_0)\left[-2i\gamma\Omega\mathcal{A} - \delta\omega^2\mathcal{A} - \frac{3h}{4}|\mathcal{A}|^2\mathcal{A} + f\right] e^{i\Omega t}$$
$$+ \frac{\epsilon h}{32}\mathcal{A}^{*3} e^{-3i\Omega t}. \tag{4.170}$$

Note that there exists a secular term in the first-order term.

Then one finds that the RG/E equation $\frac{dz}{dt_0}|_{t_0=t} = 0$ leads to

$$\dot{\mathcal{A}} = \epsilon\frac{i}{2\Omega}\left[2i\gamma\Omega\mathcal{A} + \delta\omega^2\mathcal{A} + \frac{3h}{4}|\mathcal{A}|^2\mathcal{A} - f\right]. \tag{4.171}$$

Here we have neglected terms such as $\epsilon d\mathcal{A}/dt$, which is $O(\epsilon^2)$ because $d\mathcal{A}/dt = O(\epsilon)$.

The resultant reduced Eq. (4.171) is the complex time-dependent Ginzburg-Landau equation (complex TDGL). From the real and imaginary part of (4.171), the coupled equation for the amplitude A and the phase θ is obtained as

$$\dot{A} = -\epsilon\gamma A - \epsilon\frac{\epsilon f}{2\Omega}\sin\theta, \tag{4.172}$$
$$\dot{\theta} = \frac{\epsilon}{2\Omega}\delta\omega^2 + \frac{3\epsilon h}{8\Omega}A^2 - \frac{\epsilon f}{2A\Omega}\cos\theta. \tag{4.173}$$

The approximate but globally valid solution is given by the envelope function as

$$z_E(t) = x_E(t) + iy_E(t) = z(t; t_0 = t) = \mathcal{A}(t) e^{i\Omega t} + \frac{\epsilon h}{32\Omega^2}\mathcal{A}^{*3}(t) e^{-3i\Omega t}$$
$$= A(t) e^{i(\Omega t + \theta(t))} + \frac{\epsilon h}{32\Omega^2} A^3(t) e^{-3i(\Omega t + \theta(t))}, \tag{4.174}$$

where the time dependence of $\mathcal{A}(t)$ is given by (4.171).

Notice that we have the mode with the one-third period of the normal mode with the amplitude proportional to $A^3(t)$, which would be absent when some other resummation methods like the averaging [22, 34] were used.

For completeness, we write down the expressions of $x_E(t)$ and $y_E(t)$:

$$x_E(t) = A(t)\cos(\Omega t + \theta(t)) + \frac{\epsilon h}{32\Omega^2} A^3(t)\cos\{3(\Omega t + \theta(t))\}, \quad (4.175)$$

$$y_E(t) = A(t)\sin(\Omega t + \theta(t)) - \frac{\epsilon h}{32\Omega^2} A^3(t)\sin\{3(\Omega t + \theta(t))\}. \quad (4.176)$$

Now it would be of interest to examine the basic properties of the complex TDGL given by (4.171). First let us see various limits of the equation.

1. $f = 0,\ \gamma \neq 0,\ h \neq 0$: No external force.In this case, (4.172) and (4.173) are reduced to

$$\dot{A} = -\epsilon\gamma A, \quad \dot{\theta} = \frac{\epsilon}{2\Omega}\delta\omega^2 + \frac{3\epsilon h}{8\Omega} A^2, \quad (4.177)$$

respectively. The amplitude equation is readily solved by the quadrature to give

$$A(t) = \bar{A}\mathrm{e}^{-\epsilon\gamma t}, \quad (4.178)$$

with $\bar{A}$ being a constant. Then the phase equation is also integrated out as follows:

$$\begin{aligned}\theta(t) &= \theta(0) + \frac{\epsilon}{2\Omega}\delta\omega^2 t + \frac{3\epsilon h}{8\Omega}\int_0^t ds\, \bar{A}^2 \mathrm{e}^{-2\gamma s} \\ &= \theta(0) + \frac{\omega^2 - \Omega^2}{2\Omega} t + \frac{3\epsilon h \bar{A}^2}{16\gamma\Omega}(1 - \mathrm{e}^{-2\gamma t}). \end{aligned} \quad (4.179)$$

Thus inserting these results into (4.174), we have

$$z_E(t) = \bar{A}\mathrm{e}^{-\epsilon\gamma t}\mathrm{e}^{i(\omega' t + \phi(t))} + \frac{\epsilon h \bar{A}^2}{32\Omega^2}\mathrm{e}^{-3\epsilon\gamma t}\mathrm{e}^{-3i(\omega' t + \phi(t))} \quad (4.180)$$

with

$$\omega' = \Omega + \frac{\omega^2 - \Omega^2}{2\Omega} = \Omega[1 + \frac{\omega^2 - \Omega^2}{\Omega^2}] \simeq (\Omega^2 + \omega^2 - \Omega^2)^{1/2} = \omega, \quad (4.181)$$

$$\phi(t) = \frac{3\epsilon h \bar{A}^2}{16\gamma\Omega}(1 - \mathrm{e}^{-2\gamma t}) + \theta(0) \xrightarrow[t\to\infty]{} \frac{3\epsilon h \bar{A}^2}{16\gamma\Omega} + \bar{\theta}, \quad (\bar{\theta} := \theta(0)). \quad (4.182)$$

2. $f = 0,\ \gamma = 0,\ h \neq 0$: No external force without resistance. This is the pure Duffing equation given in (2.47). In this case, (4.172) and (4.173) are reduced to

$$\dot{A} = 0, \quad \dot{\theta} = \frac{\epsilon}{2\Omega}\delta\omega^2 + \frac{3\epsilon h}{8\Omega}A^2, \tag{4.183}$$

respectively, which are readily integrated out to yield

$$A = \text{constant}, \quad \theta(t) = \left(\frac{\omega^2 - \Omega^2}{2\Omega} + \frac{3\epsilon h}{8\Omega}A^2\right)t + \bar{\theta}. \tag{4.184}$$

Inserting these results into (4.174), we reproduce the result given in the previous chapters; see (3.60), for instance.

3. $f \neq 0,\ \gamma \neq 0,\ h = 0$: No nonlinear potential. In this case, we start from (4.171), which is now reduced to

$$\dot{\mathcal{A}} = -\left(\gamma - \frac{i\delta\omega^2}{2\Omega}\right)\epsilon\mathcal{A} - \epsilon\frac{i}{2\Omega}f. \tag{4.185}$$

This inhomogeneous linear equation is readily solved by the method of variation of constant, as follows. From the formal solution to the homogeneous equation, we have

$$\mathcal{A}(t) = C(t)\mathrm{e}^{-\epsilon\Gamma t}, \quad (\Gamma := \gamma - \frac{i\delta\omega^2}{2\Omega}) \tag{4.186}$$

where $C(t)$ satisfies the equation

$$\dot{C} = -\epsilon\frac{i}{2\Omega}f\mathrm{e}^{\epsilon\Gamma t}, \tag{4.187}$$

which is readily integrated out to yield

$$C(t) = \bar{C} - i\frac{f}{2\Omega\Gamma}\left(\mathrm{e}^{\epsilon\Gamma t} - 1\right), \quad (\bar{C} : \text{constant}). \tag{4.188}$$

Thus we have for $\mathcal{A}(t)$

$$\begin{aligned}\mathcal{A}(t) &= \bar{C}\mathrm{e}^{-\epsilon\Gamma t} - i\frac{f}{2\Omega\Gamma}\left(1 - \mathrm{e}^{-\epsilon\Gamma t}\right)\\ &= -i\frac{f}{2\Omega\Gamma} + \left(\bar{C} + i\frac{f}{2\Omega\Gamma}\right)\mathrm{e}^{-\epsilon\gamma t + i\frac{\delta\omega^2}{2\Omega}t}, \end{aligned}\tag{4.189}$$

$$\xrightarrow[t\to\infty]{} -i\frac{f}{2\Omega\Gamma}. \tag{4.190}$$

Inserting this result into (4.174), we have the solution

$$z_E(t) = \mathcal{A}(t)\mathrm{e}^{i\Omega t} = \bar{A}\mathrm{e}^{i(\Omega t-\delta)} + \left(\bar{C} + i\frac{f}{2\Omega\Gamma}\right)\mathrm{e}^{-\epsilon\gamma t + i\omega' t}, \tag{4.191}$$

$$\xrightarrow[t\to\infty]{} \bar{A}\mathrm{e}^{i(\Omega t-\delta)}, \tag{4.192}$$

where

$$\bar{A} = \frac{f}{\sqrt{(\omega^2-\Omega^2)^2+4\Omega^2\gamma^2}}, \quad \tan\delta = \frac{2\Omega\gamma}{\omega^2-\Omega^2}. \tag{4.193}$$

This is the familiar resonance formula. Figure 4.4 shows the trajectories in the phase space with $\epsilon = 0.1$ for the cases of 1$\sim$ 3; the solid lines denote the results obtained in the RG method in the first order, while the dashed lines the exact ones obtained by numerics. One sees that the RG results well reproduce the exact solutions in the global domain in the asymptotic region.

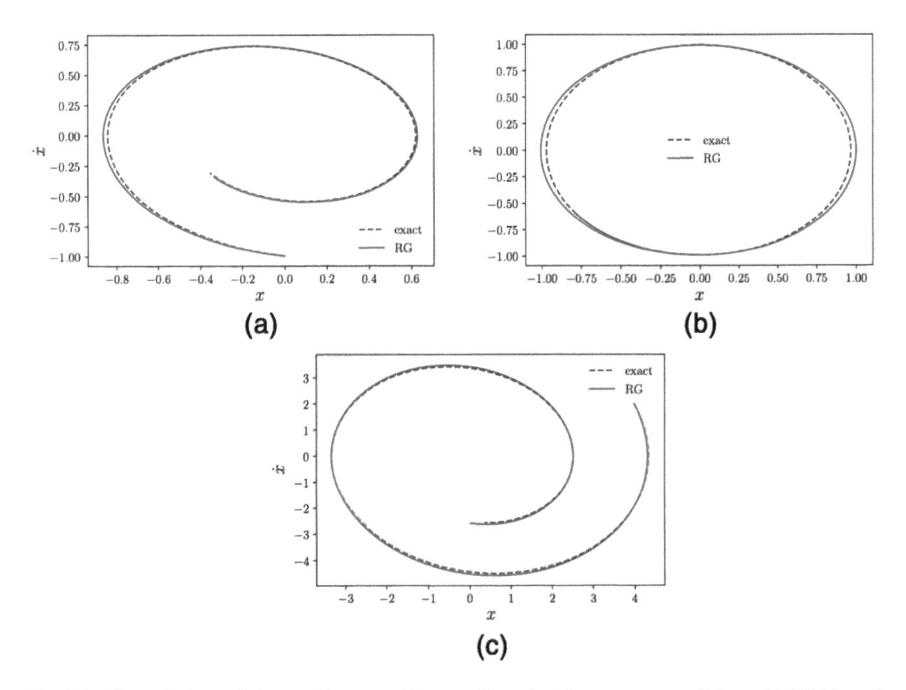

Fig. 4.4 The solutions of the nonlinear oscillator with and without an external force (4.158) based on numerics and RG method (4.171) with $\epsilon = 0.1$. (1) $f = 0$, $\gamma = 1$, $h = 1$, $\Omega = 1$, $\delta\omega = 0$, and initial values $x(0) = 0$ and $\dot{x}(0) = -0.953$. (2) $f = 0$, $\gamma = 0$, $h = 1$, and initial values $x(0) = 0$ and $\dot{x}(0) = -0.953$. (3) $f = 1$, $\gamma = 1$, $h = 0$, $\Omega = 1$, $\delta\omega = 2$, and initial values $x(0) = 4$ and $\dot{x}(0) = 2$

4. Stationary state. The stationary state that satisfies $\dot{\mathcal{A}} = 0$ or $\dot{A} = \dot{\theta} = 0$ is given by

$$0 = -\epsilon\gamma A - \epsilon\frac{\epsilon f}{2\Omega}\sin\theta, \tag{4.194}$$

$$0 = \frac{\epsilon}{2\Omega}\delta\omega^2 + \frac{3\epsilon h}{8\Omega}A^2 - \frac{\epsilon f}{2A\Omega}\cos\theta. \tag{4.195}$$

Making $\cos^2\theta + \sin^2\theta = 1$, we can eliminate θ and end up with

$$A^2\left[\left(\frac{3h}{4}A^2 + (\omega^2 - \Omega^2)\right)^2 + 4\gamma^2\Omega^2\right] = f^2, \tag{4.196}$$

which is known to describe the jump phenomenon in the nonlinear oscillator under the external periodic force [22, 25, 91]. Because the analysis of the jump phenomenon is beyond the scope of the present monograph, and there are many good literatures [22, 25, 91] on this subject, we shall not enter this interesting problem here.

4.7 Proof of a Global Validness of the Envelope Function as The approximate Solution to the Differential Equation

In this section, we shall show that the envelope function constructed through the RG/E equation indeed satisfies the original differential equation in the given order of approximation in a global domain using a generic equation [3–6].

First we notice that the differential equations considered so far can be converted to a system of differential equations with single time derivative:

$$\frac{d\boldsymbol{X}(t)}{dt} = \boldsymbol{F}(\boldsymbol{X}(t); \epsilon), \tag{4.197}$$

where $\boldsymbol{F}$ may be a non-linear function of $\boldsymbol{X}$ and t. For example, van der Pol equation is converted to the above form with the replacement of

$$\boldsymbol{X} = \begin{pmatrix} X_1 = x \\ X_2 = \dot{x} \end{pmatrix}, \quad \boldsymbol{F} = \begin{pmatrix} X_2 \\ -X_1 + \epsilon\,(1 - X_1^2)\,X_2 \end{pmatrix}. \tag{4.198}$$

Let us suppose that we have an approximate solution $\boldsymbol{X}(t; t_0)$ to Eq. (4.197) that is locally valid around $t = \forall t_0$ up to $O(\epsilon^n)$; namely, the following equation is satisfied **for all** t_0 *belonging to a global domain,*

$$\frac{d\boldsymbol{X}}{dt} - \boldsymbol{F}(\boldsymbol{X}(t; t_0); \epsilon) + O(\epsilon^{n+1}), \quad \forall t_0. \tag{4.199}$$

Now, the RG/E equation reads

$$\left.\frac{d\boldsymbol{X}(t;t_0)}{dt_0}\right|_{t=t_0} = 0, \qquad \forall\, t_0, \tag{4.200}$$

and the envelope function $\boldsymbol{X}_{\mathrm{E}}(t)$ is given by

$$\boldsymbol{X}_{\mathrm{E}}(t) = \boldsymbol{X}(t;t). \tag{4.201}$$

It is now easy to show that $\boldsymbol{X}_{\mathrm{E}}(t)$ satisfies Eq. (4.197) for arbitrary t in the global domain in which the 'initial' time t_0 can be located, up to the same order as $\boldsymbol{X}(t;t_0)$ does locally at $t \sim t_0$. In fact, $\forall t = t_0$ in the global domain,

$$\begin{aligned}\left.\frac{d\boldsymbol{X}_{\mathrm{E}}(t)}{dt}\right|_{t=t_0} &= \left.\frac{d\boldsymbol{X}(t;t_0)}{dt}\right|_{t=t_0} + \left.\frac{d\boldsymbol{X}(t;t_0)}{dt_0}\right|_{t=t_0} = \left.\frac{d\boldsymbol{X}(t;t_0)}{dt}\right|_{t=t_0} \\ &= \boldsymbol{F}(\boldsymbol{X}(t;t);\epsilon) + \mathrm{O}(\epsilon^{n+1}) \\ &= \boldsymbol{F}(\boldsymbol{X}_{\mathrm{E}}(t);\epsilon) + \mathrm{O}(\epsilon^{n+1}).\end{aligned} \tag{4.202}$$

This completes the proof. Here Eqs. (4.200) and (4.199) have been used together with the definition of $\boldsymbol{X}_{\mathrm{E}}(t)$, Eq. (4.201).

It should be stressed that Eq. (4.202) is valid uniformly $\forall t$ in the global domain of t, in contrast to Eq. (4.199) which is in a local domain around $t = t_0$.

Appendix 1: An RG Analysis of The Boundary-Layer Problem in the Scalar Representation

In this Appendix, we take up again the boundary-layer problem analysed in Sect. 4.4, and make an RG analysis of the Eq. (4.105) with the boundary condition (4.106) without converting it into a system.

With the use of the inner variable $X := x/\epsilon$, Eq. (4.105) takes the form given in (4.110), which we reproduce here;

$$L_B Y = -\epsilon\left(\frac{d}{dX} + 1\right)Y, \quad L_B := \frac{d}{dX}\left(\frac{d}{dX} + 1\right). \tag{4.203}$$

We first obtain the solution $Y(X;X_0)$ around an arbitrary point $X = X_0$, at which $Y(X;X_0)$ is supposed to coincide with an exact solution $W(X)$; $Y(X_0;X_0) = W(X_0)$. As in the text Sect. 4.4, $Y(X;X_0)$ and the exact solution $W(X_0)$ at $X = X_0$ are expanded as

$$Y(X;X_0) = Y_0(X;X_0) + \epsilon Y_1(X;X_0) + \epsilon^2 Y_2(X;X_0) + \cdots, \tag{4.204}$$

and

$$W(X_0) = W_0(X_0) + \epsilon W_1(X_0) + \epsilon^2 W_2(X_0) + \cdots, \tag{4.205}$$

respectively.

Substituting the above expansion into (4.203), we have for the first few orders of equations as

$$L_B Y_0 = 0, \quad L_B Y_1 = -\left(\frac{d}{dX} + 1\right) Y_0 \tag{4.206}$$

and so on.

Inserting the ansatz $Y_0(X) = e^{\lambda X}$ into the zeroth order equation, we have $\lambda(\lambda + 1) = 0$, accordingly, $\lambda = 0 \quad \text{or} \quad \lambda = -1$. Hence we have for the zeroth order solution

$$Y_0(X;\, X_0) = A(X_0) + B(X_0)e^{-X}, \tag{4.207}$$

with $A(X_0)$ and $B(X_0)$ being the (X_0-dependent) integral constants. We note that the initial value at $X = X_0$ in the zeroth order is given by $W_0(X_0) = A(X_0) + B(X_0)e^{-X_0}$.

The first-order equation now takes the form

$$L_B Y_1(X;\, X_0) = -A(X_0). \tag{4.208}$$

Since a constant and hence $A(X_0)$ is a zero mode of the linear operator L_B, the general solution to (4.208) admits a secular term as

$$Y_1(X;\, X_0) = -A(X_0)X + C_1^{(1)} + C_1^{(2)}e^{-X}. \tag{4.209}$$

However, following the prescription (I) for the RG method given in Sect. 4.3, the integral constants are chosen as

$$C_1^{(1)} = A(X_0)X_0 \quad \text{and} \quad C_1^{(2)} = 0, \tag{4.210}$$

which leads to

$$Y_1(X;\, X_0) = -A(X_0)(X - X_0). \tag{4.211}$$

Thus up to the first order, we have

$$Y(X;\, X_0) = A(X_0) + B(X_0)e^{-X} - \epsilon A(X_0)(X - X_0). \tag{4.212}$$

Now let us obtain the envelope $Y_E(X)$ of the family of functions $\{Y(X;\, X_0)\}_{X_0}$ each of which has the common tangent with $Y_E(X)$ at $X = X_0$. As we now know, the condition for the construction of the envelope function with this property is obtained by solving the envelope equation or RG/E equation as given by

$$\left.\frac{dY(X;X_0)}{dX_0}\right|_{X_0=X} = 0, \tag{4.213}$$

which is reduced to

$$\left.\left(\frac{dA(X_0)}{dX_0} + \epsilon A(X_0)\right)\right|_{X_0=X} - \left.\frac{dB(X_0)}{dX_0}\right|_{X_0=X} \mathrm{e}^{-X} = 0, \quad \forall X. \tag{4.214}$$

For this equation to satisfy for arbitrary X, one may demand that the coefficients of the (independent) unperturbed solutions should vanish;

$$\frac{dA}{dX} = -\epsilon A, \quad \frac{dB}{dX} = 0, \tag{4.215}$$

the solutions to which read

$$A(X) = \bar{A}\mathrm{e}^{-\epsilon X}, \quad B = \bar{B} = \text{ const.} \tag{4.216}$$

Then the envelope function is obtained as

$$\begin{aligned} Y_E(X) &= Y(X;X) = W(X) \\ &= A(X) + \bar{B}\mathrm{e}^{-X} = \bar{A}\mathrm{e}^{-\epsilon X} + \bar{B}\mathrm{e}^{-X}, \end{aligned} \tag{4.217}$$

which coincides with (4.129) that is derived rigorously in the text Sect. 4.4. Thus although Eq. (4.215) is seemingly mere a sufficient condition for Eq. (4.214) to hold, the consequence is rigorously correct, which fact may suggest that there may be more persuasive mathematical arguments for deducing the vanishing of the coefficients of the unperturbed solutions.

Appendix 2: Useful Formulae of Solutions for Inhomogeneous Differential Equations for the RG Method

In the present formulation of the RG method, the particular solutions in the higher orders are set up so that (1) possible fast modes should disappear and (2) the contributions of possible secular terms vanish at an arbitrary 'initial' time $t = t_0$. We shall show that the simple rules to write down the particular solutions with such conditions are expressed with the use of an operator method.

Appendix 2.1: General Formulae of the Particular Solution of Inhomogeneous Equations Suitable For RG Method

W are interested in finding the particular solution to the equation given by

$$\left(\frac{d}{dt} - A\right) \boldsymbol{u}(t; t_0) = \boldsymbol{F}(t), \tag{4.218}$$

with the 'initial' condition at $t = t_0$,

$$\boldsymbol{u}(t_0, t_0) = \boldsymbol{W}(t_0), \tag{4.219}$$

where A is a time-independent linear operator. The solution to the inhomogeneous Eq. (4.218) can be constructed by the method of variation of constants. First, the general solution to the unperturbed equation

$$\left(\frac{d}{dt} - A\right) \boldsymbol{u}(t; t_0) = 0 \tag{4.220}$$

is readily solved to be

$$\boldsymbol{u}_0(t) = \mathrm{e}^{At} \boldsymbol{C} \tag{4.221}$$

with $\boldsymbol{C}$ being a constant vector at this stage. Then we set the solution to (4.218) as

$$\boldsymbol{u}(t; t_0) = \mathrm{e}^{At} \boldsymbol{C}(t; t_0) \tag{4.222}$$

where $\boldsymbol{C}(t; t_0)$ is now a time-dependent vector, which may depend on t_0. Inserting this expression into (4.218), we have $\mathrm{e}^{At} \dot{\boldsymbol{C}}(t) = \boldsymbol{F}(t)$ or

$$\dot{\boldsymbol{C}}(t : t_0) = \mathrm{e}^{-At} \boldsymbol{F}(t). \tag{4.223}$$

Integrating the last equation from $t = t_0$ to $t = t$, we have

$$\boldsymbol{C}(t; t_0) = \boldsymbol{C}(t_0; t_0) + \int_{t_0}^{t} ds\, \mathrm{e}^{-As} \boldsymbol{F}(s). \tag{4.224}$$

Inserting this into (4.222), we have the solution to (4.218) with the initial condition (4.219) as follows;

$$\boldsymbol{u}(t; t_0) = \mathrm{e}^{A(t-t_0)} \boldsymbol{W}(t_0) + \mathrm{e}^{At} \int_{t_0}^{t} ds \mathrm{e}^{-As} \boldsymbol{F}(s), \tag{4.225}$$

where $\boldsymbol{C}(t_0;\, t_0)$ has been determined to be $\boldsymbol{C}(t_0;\, t_0) = \mathrm{e}^{-At_0}\boldsymbol{W}(t_0)$ to satisfy (4.219). This is the basic formula of the solution to the inhomogeneous Eq. (4.218), which is utilized in several ways through this monograph.

Appendix 2.2: Typical Examples with A Being Semi-simple

For definiteness, let us take the case where A is semi-simple[8] having eigenvalues λ_α ($\alpha = 1, 2, ...$) with the corresponding eigenvectors $\boldsymbol{U}_\alpha$ as

$$A\boldsymbol{U}_\alpha = \lambda_\alpha \boldsymbol{U}_\alpha. \tag{4.226}$$

We shall consider some typical cases where the inhomogeneous term $\boldsymbol{F}(t)$ in (4.218) takes the following forms in order: (1) $\boldsymbol{F}(t) = \mathrm{e}^{\lambda t}\boldsymbol{U}_\alpha$, $(\lambda \neq \lambda_\alpha)$ and (2) $\boldsymbol{F}(t) = \mathrm{e}^{\lambda_\alpha t}\boldsymbol{U}_\alpha$. Notice that when $\boldsymbol{F}(t)$ is given as a sum of them, the particular solution is given by a sum of those for the respective inhomogeneous term because the equation is linear.

(1) When $\boldsymbol{F}(t) = \mathrm{e}^{\lambda t}\boldsymbol{U}_\alpha$, $(\lambda \neq \lambda_\alpha)$, Eq. (4.218) reads

$$\left(\frac{d}{dt} - A\right)\boldsymbol{u}(t;\, t_0) = \mathrm{e}^{\lambda t}\boldsymbol{U}_\alpha. \tag{4.227}$$

Then the integral appearing in (4.225) is evaluated as follows:

$$\begin{aligned}
\mathrm{e}^{At}\int_{t_0}^{t} ds\, \mathrm{e}^{-As}\boldsymbol{F}(s) &= \mathrm{e}^{At}\int_{t_0}^{t} ds\, \mathrm{e}^{-As}\mathrm{e}^{\lambda s}\boldsymbol{U}_\alpha = \mathrm{e}^{\lambda_\alpha t}\int_{t_0}^{t} ds\, \mathrm{e}^{-(\lambda_\alpha - \lambda)s}\boldsymbol{U}_\alpha \\
&= \frac{\mathrm{e}^{\lambda_\alpha t}}{\lambda - \lambda_\alpha}[\mathrm{e}^{-(\lambda_\alpha - \lambda)t} - \mathrm{e}^{-(\lambda_\alpha - \lambda)t_0}]\boldsymbol{U}_\alpha \\
&= \frac{\mathrm{e}^{\lambda t_0}}{\lambda - \lambda_\alpha}[\mathrm{e}^{\lambda(t-t_0)} - \mathrm{e}^{\lambda_\alpha(t-t_0)}]\boldsymbol{U}_\alpha.
\end{aligned} \tag{4.228}$$

We remark that the second term is a solution of the unperturbed equation, which implies that an unperturbed solution is produced from the second term in (4.225) as well as the first term. Inserting (4.228) into (4.225), and noting that λ_α in the denominator may be replaced by A, we have

$$\boldsymbol{u}(t;\, t_0) = \mathrm{e}^{A(t-t_0)}\left[\boldsymbol{W}(t_0) - \frac{1}{\lambda - A}\mathrm{e}^{\lambda t_0}\boldsymbol{U}_\alpha\right] + \frac{1}{\lambda - A}\mathrm{e}^{\lambda t}\boldsymbol{U}_\alpha. \tag{4.229}$$

In the present formulation of the RG method based on the perturbation theory, the solution should be well approximated by the lowest perturbation to make

[8] The case when A is not diagonalizable and has a Jordan cell will be considered later.

the perturbative expansion as valid as possible. In the present case, $\boldsymbol{u}(t; t_0)$ is supposed to be an perturbative correction to the solution and hence should be made as small as possible. Now the first term of (4.229) is an unperturbed solution and hence should be discarded in the perturbative expansion. Thus we end up with the particular solution given by

$$\boldsymbol{u}(t; t_0) = \frac{1}{\lambda - A} \mathrm{e}^{\lambda t} \boldsymbol{U}_\alpha. \tag{4.230}$$

We note that this final form is simply obtained by a formal manipulation: Multiplying the inverse operator $\left(\frac{d}{dt} - A\right)^{-1}$ to the both side of (4.227) as

$$\boldsymbol{u}(t; t_0) = \frac{1}{\frac{d}{dt} - A} \mathrm{e}^{\lambda t} \boldsymbol{U}_\alpha = \frac{1}{\lambda - A} \mathrm{e}^{\lambda t} \boldsymbol{U}_\alpha \tag{4.231}$$

where the relation $\frac{d}{dt} \mathrm{e}^{\lambda t} = \lambda \mathrm{e}^{\lambda t} \boldsymbol{U}_\alpha$ has been used.

(2) When $\boldsymbol{F}(t) = \mathrm{e}^{\lambda_\alpha t} \boldsymbol{U}_\alpha$, Eq. (4.218) reads

$$\left(\frac{d}{dt} - A\right) \boldsymbol{u}(t; t_0) = \mathrm{e}^{\lambda_\alpha t} \boldsymbol{U}_\alpha. \tag{4.232}$$

Then the integral appearing in (4.225) is evaluated as follows:

$$\begin{aligned} \mathrm{e}^{At} \int_{t_0}^{t} ds\, \mathrm{e}^{-As} \boldsymbol{F}(s) &= \mathrm{e}^{At} \int_{t_0}^{t} ds\, \mathrm{e}^{-As} \mathrm{e}^{\lambda_\alpha s} \boldsymbol{U}_\alpha = \mathrm{e}^{\lambda_\alpha t} \int_{t_0}^{t} ds\, \mathrm{e}^{-(\lambda_\alpha - \lambda_\alpha)s} \boldsymbol{U}_\alpha \\ &= \mathrm{e}^{\lambda_\alpha t} \int_{t_0}^{t} ds\, \boldsymbol{U}_\alpha \\ &= \mathrm{e}^{\lambda_\alpha t} (t - t_0)\, \boldsymbol{U}_\alpha. \end{aligned} \tag{4.233}$$

We note the appearance of the secular term. Then inserting this formula into (4.225), we have for the particular solution

$$\boldsymbol{u}(t; t_0) = \mathrm{e}^{A(t-t_0)} \boldsymbol{W}(t_0) + (t - t_0) \mathrm{e}^{\lambda_\alpha t} \boldsymbol{U}_\alpha. \tag{4.234}$$

Since the first term is an unperturbed solution in the perturbative expansion, we may discard it for the perturbed solution. Thus we have

$$\boldsymbol{u}(t; t_0) = (t - t_0) \mathrm{e}^{\lambda_\alpha t} \boldsymbol{U}_\alpha, \tag{4.235}$$

for the particular solution, which is composed so as to vanish at $t = t_0$ in accordance with the construction prescription of the perturbative solution in the RG method.

This form of solution can be given by a formal prescription as

$$\boldsymbol{u}(t; t_0) = \frac{1}{\frac{d}{dt} - A} \mathrm{e}^{\lambda_\alpha t} \boldsymbol{U}_\alpha = (t - t_0) \mathrm{e}^{\lambda_\alpha t} \boldsymbol{U}_\alpha. \tag{4.236}$$

Appendix 2.3: Formal Prescription for Obtaining Particular Solutions of Inhomogeneous Equations Suitable for RG Method

On the basis of the results and suggested formal prescription, we here generalize and summarize the the rules of an operator method for obtaining the particular solutions suitable for the application of the RG method:

$$\frac{1}{\frac{d}{dt} - A} \mathrm{e}^{\lambda t} \boldsymbol{U}_\alpha = \frac{1}{\frac{d}{dt} - \lambda_\alpha} \mathrm{e}^{\lambda t} \boldsymbol{U}_\alpha = \frac{1}{\lambda - \lambda_\alpha} \mathrm{e}^{\lambda t} \boldsymbol{U}_\alpha, \quad (\lambda \neq \lambda_\alpha), \tag{4.237}$$

$$\frac{1}{\frac{d}{dt} - A} \mathrm{e}^{\lambda_\alpha t} \boldsymbol{U}_\alpha = \frac{1}{\frac{d}{dt} - \lambda_\alpha} \mathrm{e}^{\lambda_\alpha t} \boldsymbol{U}_\alpha = (t - t_0) \mathrm{e}^{\lambda_\alpha t} \boldsymbol{U}_\alpha. \tag{4.238}$$

Similary, one can verify that

$$\begin{aligned} \frac{1}{\frac{d}{dt} - A} (t - t_0)^n \mathrm{e}^{\lambda_\alpha t} \boldsymbol{U}_\alpha &= \frac{1}{\frac{d}{dt} - \lambda_\alpha} (t - t_0)^n \mathrm{e}^{\lambda_\alpha t} \boldsymbol{U}_\alpha \\ &= \frac{1}{n+1} (t - t_0)^{n+1} \mathrm{e}^{\lambda_\alpha t} \boldsymbol{U}_\alpha. \end{aligned} \tag{4.239}$$

Furthermore, when $\lambda \neq \lambda_\alpha$,

$$\begin{aligned} \frac{1}{\frac{d}{dt} - A} (t - t_0)^n \mathrm{e}^{\lambda t} \boldsymbol{U}_\alpha &= \mathrm{e}^{\lambda t_0} \frac{1}{d_\tau - A} \tau^n \mathrm{e}^{\lambda \tau} \boldsymbol{U}_\alpha \Big|_{\tau = t - t_0} \\ &= \mathrm{e}^{\lambda t_0} \partial_\lambda^n \frac{1}{d_\tau - A} \mathrm{e}^{\lambda \tau} \boldsymbol{U}_\alpha \Big|_{\tau = t - t_0}, \end{aligned} \tag{4.240}$$

where $d_\tau := \frac{d}{d\tau}$. Hence, for example,

$$\frac{1}{\frac{d}{dt} - A} (t - t_0) \mathrm{e}^{\lambda t} \boldsymbol{U}_\alpha = \frac{1}{\lambda - A} \{(t - t_0) - \frac{1}{\lambda - A}\} \mathrm{e}^{\lambda t} \boldsymbol{U}_\alpha, \tag{4.241}$$

$$\frac{1}{\frac{d}{dt} - A} (t - t_0)^2 \mathrm{e}^{\lambda t} \boldsymbol{U}_\alpha = \frac{1}{\lambda - A} \{(t - t_0)^2 - \frac{2}{\lambda - A}(t - t_0) + \frac{2}{(\lambda - A)^2}\} \mathrm{e}^{\lambda t} \boldsymbol{U}_\alpha, \tag{4.242}$$

where A may be replaced with λ_α.

Chapter 5
RG Method for Asymptotic Analysis with Reduction of Dynamics: An Elementary Construction of Attractive/Invariant Manifold

5.1 Introduction

In Chap. 4 (the last chapter), we have introduced the RG method [1, 38, 39, 48–52, 92–96] for obtaining perturbative solutions of differential equations that are valid in a global domain through a resummation of the perturbative expansions, on the basis of the formulation given in [3–5]. In this chapter, we reformulate the method in a non-perturbative way [6] by adapting the Wilsonian renormalization-group (RG) or Wegner-Houghton's flow equations developed in statistical physics and quantum field theory [14, 87]; The formulation will be given also in a way that it is clear that the RG method provides us with a powerful reduction theory of dynamical systems [6]. Then confining ourselves to cases where a perturbative treatment is possible, we shall show that the RG method gives a powerful but also mechanical method to construct the attractive/invariant manifold and the reduced dynamics on it of the dynamical systems [6], which are the essential ingredients of the center manifold theory in the conventional reduction theory [33, 34, 97].

One of the merits of the RG method lies in the fact that the natural coordinates for describing the invariant or attractive manifold are explicitly provided by the integral constants of the unperturbed solution [3–5]. It is to be noted that the reduction scheme provided by the RG method is also in accordance with that elucidated by Yoshiki Kuramoto three decades ago [32]: Kuramoto noticed that when a reduction of evolution equation is possible, the unperturbed equation admits neutrally stable solutions, and elucidated that the universal structure of the perturbative reduction of dynamics can be formulated as an extended scheme of the Krylov-Bogoliubov-Mitropolsky (KBM) method [25] for weakly nonlinear oscillators, on the basis of the reductive perturbation theory [98] with some ansatz on the form of the solutions as is the case in KBM method. In this respect, we remark that the KBM scheme naturally emerges in the RG method as was shown in the last chapter (Chap. 4).

T. Kunihiro et al., *Geometrical Formulation of Renormalization-Group Method as an Asymptotic Analysis*, Fundamental Theories of Physics 206,
https://doi.org/10.1007/978-981-16-8189-9_5

5.2 Non-perturbative RG/E Equation for Reduction of Dynamics

We begin with the following generic dynamical equation with n degrees of freedom;

$$\frac{d\boldsymbol{X}}{dt} = \boldsymbol{F}(\boldsymbol{X}, t), \tag{5.1}$$

with $\boldsymbol{X} = {}^t(X_1, X_2, \ldots, X_n)$ and $\boldsymbol{F} = {}^t(F_1, F_2, \ldots, F_n)$, where n may be infinity.

We assume that there exists at least a solution $\boldsymbol{X}(t) = \boldsymbol{W}(t)$ to Eq. (5.1). with some initial condition, say, at $t = 0$. Then we try to solve the equation with an 'initial' condition at an arbitrary time $t = t_0 > 0$.

$$\boldsymbol{X}(t = t_0) = \boldsymbol{W}(t_0). \tag{5.2}$$

As is indicated in Sect. 4.2.3, the arbitrary time t_0 corresponds to the renormalization point in the renormalization theory in quantum field theory [18, 88]. The reason why we call it 'initial' time lies in the fact that the perturbed solutions are to be constructed around $t \sim t_0$, although it may be more adequate to call it the 'probe time' or 'observation time'.

We write the solution with the 'initial' condition (5.2) as

$$\boldsymbol{X}(t) = \boldsymbol{X}(t; t_0, \boldsymbol{W}(t_0)), \quad \text{with} \quad \boldsymbol{X}(t_0; t_0, \boldsymbol{W}(t_0)) = \boldsymbol{W}(t_0). \tag{5.3}$$

This is a rather trivial equation.

In some situation where the initial values at $t = 0$ are confined in a some domain, the solution after a long time $t \to \infty$ is confined in a well defined manifold with m degrees of freedom, where m may be less than or equal to n. Such a manifold is called the invariant/attractive manifold of the evolution equation [33, 34].

If t_0 is such a time for which such an asymptotic behavior is realized, then the 'initial' value $\boldsymbol{W}(t_0)$ should be parametrized by a vector $\boldsymbol{C}(t_0)$ with m dimensions, which may be given by the collection of the integral constants at $t = t_0$:

$$\boldsymbol{W}(t_0) = \boldsymbol{W}[\boldsymbol{C}(t_0)], \quad \text{with} \quad \dim \boldsymbol{C} = m. \tag{5.4}$$

As is seen in Fig. 5.1, even when the 'initial' point $(t_0, \boldsymbol{W}(t_0))$ may be shifted to $(t_0', \boldsymbol{W}(t_0'))$, the values on the solution at t does not change; $\boldsymbol{X}(t; t_0, \boldsymbol{W}(t_0)) = \boldsymbol{X}(t; t_0', \boldsymbol{W}(t_0'))$ or

$$\boldsymbol{X}(t; t_0', \boldsymbol{W}(t_0')) - \boldsymbol{X}(t; t_0, \boldsymbol{W}(t_0)) = \boldsymbol{0}. \tag{5.5}$$

Dividing it by $t_0' - t_0$ and then taking the limit

$$t_0' - t_0 =: \Delta t_0 \to 0+, \tag{5.6}$$

one has

$$\frac{d\boldsymbol{X}}{dt_0} = \frac{\partial \boldsymbol{X}}{\partial t_0} + \frac{d\boldsymbol{X}}{d\boldsymbol{W}} \cdot \frac{d\boldsymbol{W}}{dt_0} = \boldsymbol{0}. \tag{5.7}$$

Note that this is an exact equation from which the time evolution of $\boldsymbol{W}(t)$ can be extracted. This gives a non-perturbative foundation of the RG/E equation to be applied to a global analysis, say, in the asymptotic region.

A remark is in order here: The Eq. (5.7) may be compared to the non-perturbative RG equations by Wilson [87] or flow equations by Wegner-Houghton [14] and so on in quantum field theory (QFT) and statistical physics [17–19]. The arbitrary time t_0 in the present case corresponds to (the logarithm of) the renormalization point with the energy scale, and the integral constants $\boldsymbol{C}(t_0)$ to physical values such as the coupling constants and the masses to be renormalized in QFT. One will also recognize that the existence of an invariant manifold of the dynamical system may correspond to the renormalizability of the theory in QFT.

5.3 Perturbative RG/E Equation

When the perturbative theory is used to obtain $\boldsymbol{X}(t; t_0, \boldsymbol{W}(t_0))$, the equality of $\boldsymbol{X}(t; t_0, \boldsymbol{W}(t_0))$ and $\boldsymbol{X}(t; t_0', \boldsymbol{W}(t_0'))$ in the perturbation theory may be valid only for

$$t \sim t_0 \quad \text{and} \quad t \sim t_0', \tag{5.8}$$

as shown in Fig. 5.1. Then, in view that the limit

$$t_0' \to t_0 + \tag{5.9}$$

is taken in deriving the RG/E Eq. (5.7), we can demand the equality $t_0 = t+$ for obtaining as good as possible approximation. In practice, the infinitesimal difference between t_0 and t will be discarded and simply use the equality[1]

$$t_0 = t. \tag{5.10}$$

Thus we arrive at a restrictive RG/E equation as

$$\left.\frac{d\boldsymbol{X}}{dt_0}\right|_{t_0=t} = \left.\frac{\partial \boldsymbol{X}}{\partial t_0}\right|_{t_0=t} + \frac{d\boldsymbol{X}}{d\boldsymbol{W}} \cdot \left.\frac{d\boldsymbol{W}}{dt_0}\right|_{t_0=t} = \boldsymbol{0}. \tag{5.11}$$

[1] This equality $t_0 = t$ will be understood with a caution in Chap. 7 where stochastic equations are treated: The times t_0 and t may represent the coarse-grained and microscopic ones, respectively, and hence may have quite different time scales from each other.

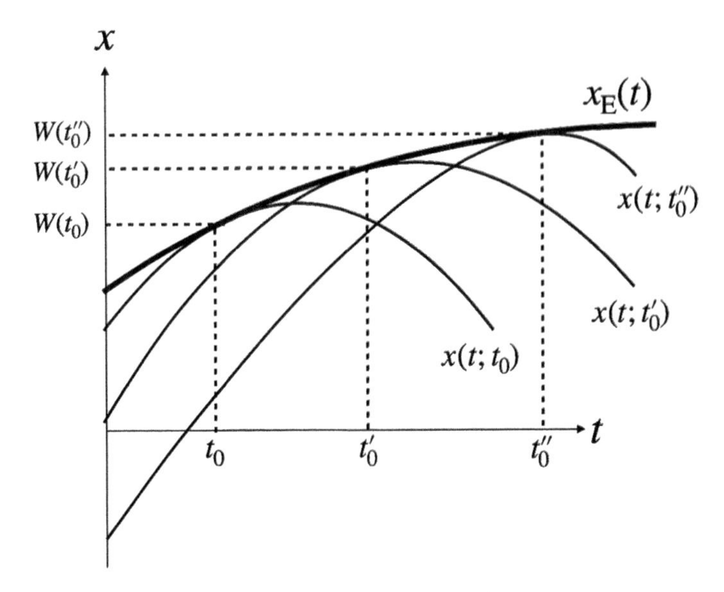

Fig. 5.1 The perturbative solutions $x(t;\ t_0)$, $x(t.;\ t_0')$, $x(t;\ t_0'')$ and so on have the exact values $W(t_0)$, $W(t_0')$ and $W(t_0'')$ of the solution $x_{\mathrm{E}}(t)$ at the respective 'initial' times t_0, t_0' and t_0'' and so on. Conversely, the exact solution $x_{\mathrm{E}}(t)$ can be obtained as the envelope curve of the family of curves $x(t;\ t_0)$, $x(t;\ t_0')$ and $x(t;\ t_0'')$

In Sect. 4.2.3, we have given a geometrical interpretation of Eq. (5.11) that it is the envelope equation for constructing an envelope curve from the family of curves as the perturbative solutions with different 'initial' time t_0.[2] Accordingly the solution to be valid in a global domain is given by the would-be 'initial' value

$$\boldsymbol{W}(t) = \boldsymbol{X}(t;\ t) = \boldsymbol{X}_{\mathrm{E}}(t) \tag{5.12}$$

which is interpreted as the envelope of the family of curves given by the perturbative solutions.

As was done in Sect. 4.7, It can be proved that $\boldsymbol{W}(t)$ satisfies the original equation (5.1) in a global domain up to the order with which $\boldsymbol{X}(t;\ t_0)$ satisfies around $t \sim t_0$ [3, 5, 6] because of the following equality

$$\frac{d\boldsymbol{W}(t)}{dt} = \left.\frac{d\boldsymbol{X}(t;\ t_0)}{dt}\right|_{t_0=t} + \left.\frac{d\boldsymbol{X}(t;\ t_0)}{dt_0}\right|_{t_0=t} = \left.\frac{d\boldsymbol{X}(t;\ t_0)}{dt}\right|_{t_0=t}, \tag{5.13}$$

due to (5.11).

[2] It is worth emphasizing that a reasoning to set t_0 equal to t (or $t+$) has been naturally derived here, which setting was done in a rather ad hoc way in [1].

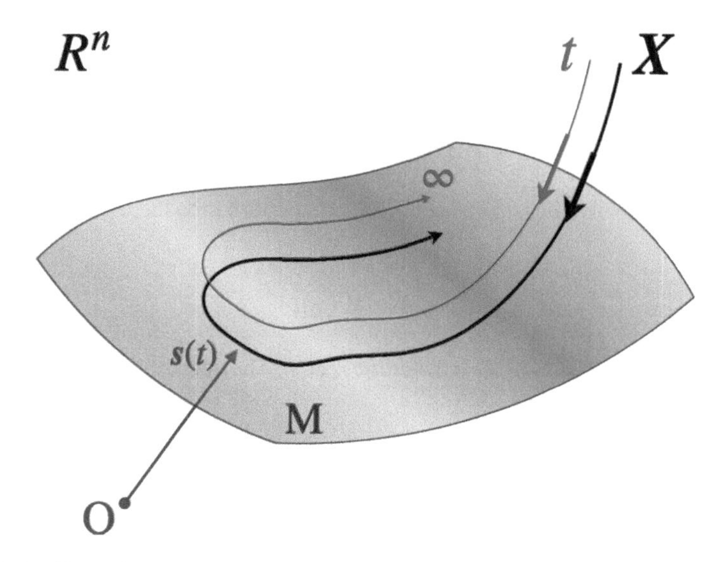

Fig. 5.2 An illustration of the invariant/attractive manifold with m dimensions of n-dimensional dynamical system. After a long time, the state vector $\boldsymbol{X}$ is attracted to a well-defined manifold M with m dimensions and to be confined there

5.4 Invariant/Attractive Manifold and Renormalizability

As we will see in the generic example to be treated in the subsequent section, there is a universal structure of the reduction of dynamics. In this section, we shall give some conceptual exposition of the general structure of the reduction of dynamics in the RG method based on the perturbation theory with an emphasis of a *trade-off* relation of the simplicity of the reduced equation and that of the representation of the invariant/attractive manifold.

Let the invariant manifold M is represented by the coordinate $\boldsymbol{s}$ with $\dim[\boldsymbol{s}] = m \leq n$: See Fig. 5.2. The reduced dynamics of Eq. (5.1) on M may be given in terms of a vector field $\boldsymbol{G}$ by

$$\frac{d\boldsymbol{s}}{dt} = \boldsymbol{G}(\boldsymbol{s}), \tag{5.14}$$

and the manifold M is represented by

$$\boldsymbol{X} = \boldsymbol{R}(\boldsymbol{s}). \tag{5.15}$$

Our task is to obtain the vector field $\boldsymbol{G}$ and the representation of the manifold $\boldsymbol{R}$ in a perturbation method. We consider a situation where the vector field $\boldsymbol{F}$ is composed of an unperturbed part $\boldsymbol{F}_0$ and the perturbative one $\boldsymbol{P}$, *i.e.*,

$$\boldsymbol{F} = \boldsymbol{F}_0(\boldsymbol{X}) + \epsilon \cdot \boldsymbol{P}(\boldsymbol{X}, t). \tag{5.16}$$

Here notice that $\boldsymbol{F}_0$ has no explicit t-dependence, while $\boldsymbol{P}(\boldsymbol{X}, t)$ does. We assume that the unperturbed problem is solved and an attractive/invariant manifold M_0 is easily found.

Now we try to solve Eqs. (5.1) and (5.16) by a perturbation theory with the 'initial' condition

$$\boldsymbol{X}(t_0) = \boldsymbol{W}(t_0), \tag{5.17}$$

at *an arbitrary time* $t = t_0$ *in the asymptotic region*. The decisive point of our method is to assume that

$$\boldsymbol{W}(t_0) = \boldsymbol{R}(\boldsymbol{s}(t_0)), \tag{5.18}$$

that is, the 'initial' point is supposed to be on the invariant manifold M to be determined. Now we apply the perturbation theory with an expansion

$$\boldsymbol{X}(t; t_0, \boldsymbol{W}(t_0)) = \boldsymbol{X}_0 + \sum_{n=1}^{\infty} \epsilon^n \boldsymbol{X}_n(t; t_0, \boldsymbol{W}(t_0)). \tag{5.19}$$

Here we have made it explicit that $\boldsymbol{X}$ is dependent on the 'initial' condition. We also expand the 'initial' value as

$$\boldsymbol{W}(t_0) = \boldsymbol{W}_0(t_0) + \boldsymbol{\rho}(t_0), \tag{5.20}$$

with

$$\boldsymbol{\rho}(t_0) = \sum_{n=1}^{\infty} \epsilon^n \boldsymbol{W}_n(t_0). \tag{5.21}$$

Now the unperturbed equation reads

$$\frac{d\boldsymbol{X}_0}{dt} = \boldsymbol{F}_0(\boldsymbol{X}_0). \tag{5.22}$$

As promised, we suppose that an attractive manifold is found for this equation as

$$\boldsymbol{X}_0(t) = \boldsymbol{R}(\boldsymbol{s}(t; \boldsymbol{C}(t_0))), \tag{5.23}$$

where $\boldsymbol{C}(t_0)$ is the integral constant (vector) with

$$\dim \boldsymbol{C}(t_0) = m \leq n \tag{5.24}$$

and may depend on t_0.

Here comes an important point of our method; we identify that

$$s(t_0) = C(t_0), \tag{5.25}$$

which gives a natural parameterization of the manifold M_0. This is a simple but a significant observation: It is not necessary to give any ansatz for the representation of the manifold because we only have to solve the unperturbed equation and the integral constants are trivially obtained.

The deformation of the manifold $\rho(t_0)$ at $t = \forall t_0$ should be, by definition, determined so as to be independent of the unperturbed solution at $t = t_0$, *i.e.*, $W_0(t_0)$, which implies that possible perturbative corrections proportional to the unperturbed solution is renormalized away into $W_0(t_0)$ [3, 5].

In fact, it is a general rule that there exists a *trade-off* relation between the simplicity of the reduced dynamical equation and that of the representation of the invariant/attractive manifold. In the present case, the choice of the ρ is intimately related to the resulting reduced differential equation. The notion of the reduction of the dynamics may mean that the resulting reduced equation should be as simple as possible because we are interested to reduce the dynamics to a simpler one. We shall see that the above choice of the deformation leads to the simpler form of the reduced differential equation without terms coming from the redundant unperturbed solutions.

Let us show more details of the above prescription by proceeding to the higher-order equations. The first-order equation reads

$$\frac{dX_1}{dt} = F_0'(X_0)X_1 + P(X_0). \tag{5.26}$$

The solution to this inhomogeneous equation is composed of a sum of the general solution of the homogeneous equation and the particular solution of the inhomogeneous equation. If the unperturbed solution $X_0(t)$ has a part of neutrally stable solution, there appear *secular terms* in the special particular as well as genuinely independent functions. It is a simple fact that the secular terms can be arranged so as to vanish at $t = t_0$, which may be interpreted as a renormalization condition to the unperturbed solution. Then the correction of the 'initial' value is now determined as

$$W_1(t_0) = X_1(t = t_0). \tag{5.27}$$

Notice that the 'initial' value is determined after solving the equation, hence the functional form of it as a function of t_0 is explicitly given without recourse to any ansatz because we are *solving the equation faithfully* in principle. If it were that $W_1(t_0) = 0$ contains additional unperturbed solutions, it would lead to a more complicated dynamics as noticed above.

We can proceed to any higher orders of the perturbative expansion keeping the above basic prescription. Then the deformation of the 'initial' value $\rho(t_0)$ and hence the total 'initial' value $W(t_0)$ are given solely in terms of $C(t_0)$,

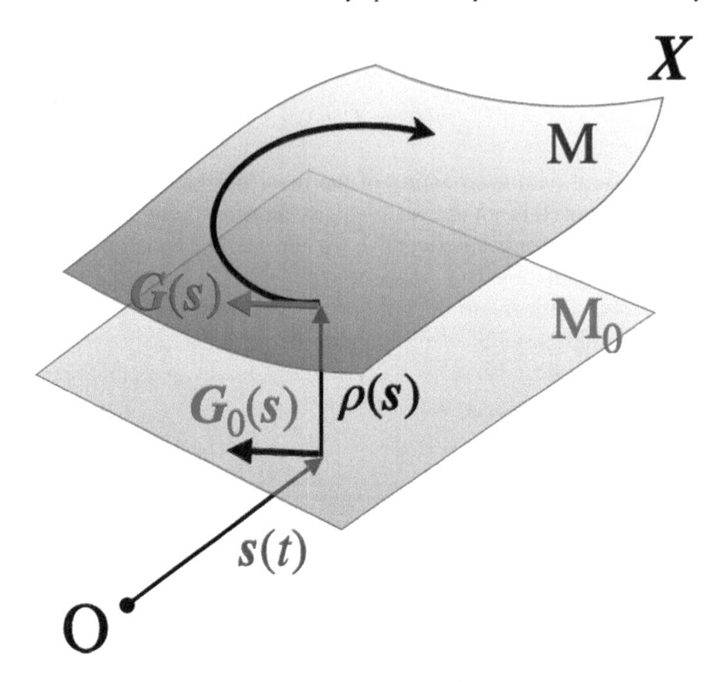

Fig. 5.3 An illustration of the perturbative construction of the invariant/attractive manifold with m dimensions of n-dimensional dynamical system. The zeroth-order manifold with dimension m is denoted as M_0; the coordinates of M_0 is given by a vector $\boldsymbol{s}$, which is nicely identified with the integral constants $\boldsymbol{C}$ in the zeroth-order solution. The nonperturbative manifold is denoted by M, which is deformed from M_0 by the perturbation. The rate of the deformation $\boldsymbol{\rho}$ is, however, given as a function of $\boldsymbol{s} = \boldsymbol{C}$, the coordinates of the unperturbed manifold M_0

$$\boldsymbol{W}(t_0) = \boldsymbol{W}_0[\boldsymbol{C}(t_0)] + \boldsymbol{\rho}[\boldsymbol{C}(t_0)]. \tag{5.28}$$

Now the dynamics of $\boldsymbol{C}(t)$ is determined by the **RG/E equation** Eq.(5.11);

$$\left.\frac{d\boldsymbol{X}}{dt_0}\right|_{t_0=t} = \left.\frac{\partial \boldsymbol{X}}{\partial t_0}\right|_{t_0=t} + \left.\frac{\partial \boldsymbol{X}}{\partial \boldsymbol{C}} \cdot \frac{d\boldsymbol{C}}{dt_0}\right|_{t_0=t} = 0, \tag{5.29}$$

which leads to the dynamical equation of $\boldsymbol{C}(t)$. Then the manifold M (more precisely, the trajectory on it) is represented as

$$\boldsymbol{X}(t) = \boldsymbol{W}(t) = \boldsymbol{W}_0[\boldsymbol{C}(t)] + \boldsymbol{\rho}[\boldsymbol{C}(t)], \qquad (\boldsymbol{s} = \boldsymbol{C}). \tag{5.30}$$

Equations (5.29) and (5.30) constitute our basic equations in the reduction theory of dynamics based on the RG method. The geometrical picture of the perturbative construction of the attractive/invariant manifold is illustrated in Fig. 5.3. We remark that these equations correspond to the basic postulates in Kuramoto's theory of reduction of evolution equations [32].

Equation (5.30) shows that the whole dynamics is given solely in terms of $\boldsymbol{C}(t)$ with dimension $m \leq n$. One may have recalled a correspondence between the renormalizability in quantum field theory [19] and the existence of a finite dimensional invariant manifold in the theory of dynamical systems where the would-be integral constants $\boldsymbol{C} = (C_1, C_2, \ldots, C_m)$ in the unperturbed solution play a quite similar role to the renormalized coupling constants including the renormalized masses, and $\boldsymbol{\rho}[\boldsymbol{C}]$ to the unrenormalizable operators that are represented by the renormalized constants. The last aspect that $\boldsymbol{\rho}[\boldsymbol{C}]$ is represented solely in terms of $\boldsymbol{C}$ is analogous to the **Polchinski theorem** [15, 19] in quantum field theory. We also remark that Eq.(5.30) justifies the slaving principle by Haken [99].

A comment is in order; When the unperturbed system is given by a linear operator is not semi-simple and has a Jordan cell, there will be a slight modification of the above scenario due to a technical complexity; see Sect. 5.6.

5.5 Example I: A Generic System with the Linear Operator Having Semi-simple Zero Eigenvalues

In the Lorenz model analyzed in the previous chapters, the asymptotic state around the fixed point $(\xi, \eta, \zeta) = \boldsymbol{0}$ for $r \sim 1$ is confined in a one-dimensional manifold embedded in the three-dimensional phase space as given by (3.182):

$$\left.\begin{aligned} \xi(t) &= \epsilon A + \tfrac{\epsilon^3 \sigma}{(1+\sigma)^2}(\chi A - \tfrac{1}{b}A^3), \\ \eta(t) &= \epsilon A - \tfrac{\epsilon^3}{(1+\sigma)^2}(\chi A - \tfrac{1}{b}A^3), \\ \zeta(t) &= \epsilon^2 \tfrac{A^2}{b}, \end{aligned}\right\} \tag{5.31}$$

and the original dynamics given in terms of the non-linear system consisting of three variables are reduced to single equation given by (3.178);

$$\frac{dA}{dt} = \epsilon^2 \frac{\sigma}{\sigma + 1}\left(\chi A - \frac{1}{b}A^3\right). \tag{5.32}$$

The manifold (with dimensions less than the original one of the phase space) in which the asymptotic state is confined is called the invariant manifold or center manifold [33, 34]. In the present case, the state variables with any initial values tend to be attracted to this manifold, hence the manifold is also an attractive manifold.

In this section, we shall show how the RG method gives a natural way for constructing the invariant manifold and deducing the reduced dynamics on it, taking up a generic system in which the linear operator A has semi-simple zero eigenvalues and those with negative real parts: In this case, the invariant manifold may be also called the center manifold [33, 34].

5.5.1 Generic Model that Admits an Attractive/Invariant Manifold

We treat the following rather generic n-dimensional system of first-order equations:

$$\left(\frac{d}{dt} - A\right)\boldsymbol{u} = \epsilon \boldsymbol{F}(\boldsymbol{u}), \tag{5.33}$$

where A is a linear operator (a matrix and/or differential/integral operator) and $\boldsymbol{F}$ a nonlinear function of $\boldsymbol{u}$ with ϵ being a small parameter, say, satisfying the inequality $|\epsilon| < 1$. We assume that A is semi-simple and has multiply degenerated zero eigenvalues and other eigenvalues of A have a negative real part.

We are interested in constructing the attractive/invariant manifold M and the reduced dynamics on it to be realized asymptotically when $t \to \infty$ in the perturbation theory. For this purpose, we try to solve the equation in the perturbation theory by expanding $\boldsymbol{u}$ as

$$\boldsymbol{u}(t; t_0) = \boldsymbol{u}_0(t; t_0) + \epsilon \boldsymbol{u}_1(t; t_0) + \epsilon^2 \boldsymbol{u}_2(t; t_0) + \cdots, \tag{5.34}$$

with the 'initial' value $\boldsymbol{W}(t_0)$ at an arbitrary time t_0 in the asymptotic regime. We suppose that the equation has been solved up to $t = t_0$ and the solution has the value $\boldsymbol{W}(t_0)$ at t_0 on M. In other words, $\boldsymbol{u}(t) = \boldsymbol{W}(t)$ gives a global solution to (5.33) confined on M. We assume that the exact solution can be expanded as follows:

$$\boldsymbol{W}(t_0) = \boldsymbol{W}_0(t_0) + \epsilon \boldsymbol{W}_1(t_0) + \epsilon^2 \boldsymbol{W}_2(t_0) + \cdots. \tag{5.35}$$

They are yet to be determined in this stage of the procedure.

The lowest-order equation reads

$$\left(\frac{d}{dt} - A\right)\boldsymbol{u}_0 = 0. \tag{5.36}$$

We assume that the multiplicity of the zero eigenvalues is m, which implies that the dimension of $\ker A$ is m;

$$A\boldsymbol{U}_i = 0, \quad (i = 1, 2, \ldots, m). \tag{5.37}$$

It is also assumed that the other eigenvalues have negative real parts;

$$A\boldsymbol{U}_\alpha = \lambda_\alpha \boldsymbol{U}_\alpha, \quad \mathrm{Re}\lambda_\alpha < 0, \quad (\alpha = m+1, m+2, \ldots, n). \tag{5.38}$$

It is taken for granted that $\boldsymbol{U}_i$'s and $\boldsymbol{U}_\alpha$'s are linearly independent. Since the linear operator A may not be symmetric (self-adjoint), we need to specify the properties of the adjoint operator $A^\dagger$, which has the same eigenvalues as A has;

$$A^{\dagger}\tilde{\boldsymbol{U}}_i = 0, \quad (i = 1, 2, \ldots, m), \tag{5.39}$$

$$A^{\dagger}\tilde{\boldsymbol{U}}_{\alpha} = \lambda^*_{\alpha}\tilde{\boldsymbol{U}}_{\alpha}, \quad (\alpha = m+1, m+2, \ldots, n). \tag{5.40}$$

Here we suppose that $\tilde{\boldsymbol{U}}_i$' and $\tilde{\boldsymbol{U}}_{\alpha}$'s are linearly independent. We have an orthogonality condition

$$(\tilde{\boldsymbol{U}}_a, \boldsymbol{U}_b) = \langle\tilde{\boldsymbol{U}}_a|\boldsymbol{U}_b\rangle = \delta_{ab}, \quad (a,\ b = 1,\ 2,\ \ldots,\ n) \tag{5.41}$$

where we have demanded the normalization condition for $a = b$. In the second equality, we have used Dirac's bra-ket notation for an inner product.

The projection operator P_a onto the eigenstate $\boldsymbol{U}_a$ is given by

$$P_a = |\boldsymbol{U}_a\rangle\langle\tilde{\boldsymbol{U}}_a| \tag{5.42}$$

in the bra-ket notation. Then the projection operator onto the kernel ker A reads

$$P = \sum_{i=1}^{m} P_i. \tag{5.43}$$

The projection operator Q onto the compliment to the kernel is given by

$$Q = I - P = \sum_{\alpha=m+1}^{n} P_{\alpha}, \tag{5.44}$$

and accordingly

$$P + Q = I. \tag{5.45}$$

We have also the idempotency of the projection operators as

$$P^2 = P, \quad Q^2 = Q. \tag{5.46}$$

Now the general solution to the homogeneous Eq. (5.36) is readily given by

$$\boldsymbol{u}_0(t;\ t_0) = \mathrm{e}^{At}\bar{\boldsymbol{C}}(t_0) \tag{5.47}$$

with $\bar{\boldsymbol{C}}$ being a constant vector, which can be expressed as a linear combination of the eigenvectors of A as

$$\bar{\boldsymbol{C}} = \sum_{a=1}^{n} C_a(t_0)\boldsymbol{U}_a. \tag{5.48}$$

Then (5.47) is reduced to

$$\begin{aligned} \boldsymbol{u}_0(t;\ t_0) &= \sum_{a=1}^{n} C_a(t_0)\mathrm{e}^{At}\boldsymbol{U}_a = \sum_{a=1}^{n} \mathrm{e}^{\lambda_a t} C_a(t_0)\boldsymbol{U}_a \\ &= \sum_{i=1}^{m} C_i(t_0)\boldsymbol{U}_i + \sum_{\alpha=m+1}^{n} \mathrm{e}^{\lambda_\alpha t} C_\alpha(t_0)\boldsymbol{U}_\alpha. \end{aligned} \tag{5.49}$$

Since we are interested in the asymptotic state in the limit $t \to \infty$, we may discard the second part in (5.49) composed of $n - m$ terms which would die out in this asymptotic state, and hence we take for the zero-th solution

$$\boldsymbol{u}_0(t;\ t_0) = \sum_{i=1}^{m} C_i(t_0)\boldsymbol{U}_i \tag{5.50}$$

with the 'initial' value

$$\boldsymbol{u}_0(t_0;\ t_0) = \boldsymbol{W}_0(t_0) = \sum_{i=1}^{m} C_i(t_0)\boldsymbol{U}_i =: \boldsymbol{W}_0[\boldsymbol{C}(t_0)], \tag{5.51}$$

where $\boldsymbol{C}(t_0)$ is an m-dimensional vector defined by

$$\boldsymbol{C}(t_0) = {}^t(C_1(t_0),\ C_2(t_0),\ \dots,\ C_m(t_0)). \tag{5.52}$$

5.5.2 First-Order Analysis

The first-order equation reads

$$\left(\frac{d}{dt} - A\right)\boldsymbol{u}_1 = \boldsymbol{F}(\boldsymbol{u}_0(t;\ t_0)). \tag{5.53}$$

This is a linear inhomogeneous equation, which may be solved using the method of variation of constants for a system of first-order equation, as has been already utilized in Chap. 3. For completeness, we repeat it here. We start with the formal solution to the homogeneous equation

$$\boldsymbol{u}_1(t;\, t_0) = \mathrm{e}^{At}\tilde{\boldsymbol{u}}, \tag{5.54}$$

which is inserted into (5.53) but now with $\tilde{\boldsymbol{u}}$ being supposed to be time-dependent. Then we have an equation for $\tilde{\boldsymbol{u}}$ as $\dot{\tilde{\boldsymbol{u}}} = \mathrm{e}^{-At}\boldsymbol{F}(\boldsymbol{u}_0)$, which is readily integrated out to give

$$\tilde{\boldsymbol{u}}(t;t_0) = \int_{t_0}^{t} ds\, \mathrm{e}^{-As} \boldsymbol{F}(\boldsymbol{u}_0) + \tilde{\boldsymbol{u}}(t_0;t_0). \tag{5.55}$$

Thus the solution of (5.53) with the 'initial' condition $\boldsymbol{u}_1(t_0;\ t_0) = \boldsymbol{W}_1(t_0)$ is found to be

$$\begin{aligned}\boldsymbol{u}_1(t;t_0) &= \mathrm{e}^{At}\left[\int_{t_0}^{t} ds\, \mathrm{e}^{-As} \boldsymbol{F}(\boldsymbol{u}_0) + \tilde{\boldsymbol{u}}(t_0;t_0)\right] \\ &= \mathrm{e}^{(t-t_0)A}\boldsymbol{W}_1(t_0) + \int_{t_0}^{t} ds\, \mathrm{e}^{(t-s)A}\boldsymbol{F}(\boldsymbol{u}_0(s;t_0)),\end{aligned} \tag{5.56}$$

where $\tilde{\boldsymbol{u}}(t_0;t_0) = \mathrm{e}^{-At_0}\boldsymbol{W}_1(t_0)$.

Here a couple of remarks are in order:

(1) The first-order correction to the 'initial' value $\boldsymbol{W}_1(t_0)$ is and can be taken so as to belong to the Q-space, because if $\boldsymbol{W}_1(t_0)$ had a component belonging to $\ker A$ or P space, such a component could be "renormalized away" into $\boldsymbol{W}_0$.
(2) Although $\boldsymbol{F}(\boldsymbol{u}_0(s;t_0))$ is seemingly time-dependent, it is a constant vector $\boldsymbol{F}(\boldsymbol{u}_0(s;t_0)) = \boldsymbol{F}(\boldsymbol{W}_0(t_0))$.

Inserting the identity $I = P + Q$ in front of $\boldsymbol{F}(\boldsymbol{u}_0)$ in the integral, the integral in (5.56) is evaluated as follows,

$$\begin{aligned}\int_{t_0}^{t} ds\, \mathrm{e}^{(t-s)A}\boldsymbol{F}(\boldsymbol{u}_0(s;t_0)) &= \int_{t_0}^{t} ds\, \mathrm{e}^{(t-s)A}(P+Q)\boldsymbol{F}(\boldsymbol{W}_0(t_0)) \\ &= \int_{t_0}^{t} ds\, \mathrm{e}^{(t-s)A}\, P\boldsymbol{F}(\boldsymbol{W}_0(t_0)) \\ &\quad + \int_{t_0}^{t} ds\, \mathrm{e}^{(t-s)A}\, Q\boldsymbol{F}(\boldsymbol{W}_0(t_0)).\end{aligned} \tag{5.57}$$

However,

$$\begin{aligned}\text{the 1st term} &= \int_{t_0}^{t} ds\, \mathrm{e}^{(t-s)A}\, P\boldsymbol{F}(\boldsymbol{W}_0(t_0)) = \int_{t_0}^{t} ds\, P\boldsymbol{F}(\boldsymbol{W}_0(t_0)) \\ &= (t-t_0)\, P\boldsymbol{F}(\boldsymbol{W}_0(t_0)).\end{aligned} \tag{5.58}$$

Since

$$AQ = QA = QAQ = \sum_{\alpha=m+1}^{n} |\boldsymbol{U}_\alpha\rangle\langle\tilde{\boldsymbol{U}}_\alpha|A|\boldsymbol{U}_\alpha\rangle\langle\tilde{\boldsymbol{U}}_\alpha| =: A_Q, \tag{5.59}$$

has no zero eigenvalues, the second integral is evaluated as follows,

$$
\begin{aligned}
\text{the 2nd integral} &= \int_{t_0}^{t} ds\, \mathrm{e}^{(t-s)A_Q}\, Q\boldsymbol{F}(\boldsymbol{W}_0(t_0)) \\
&= \mathrm{e}^{A_Q t}(-A_Q^{-1})\left[\mathrm{e}^{-A_Q s}\right]_{t_0}^{t}\, Q\boldsymbol{F}(\boldsymbol{W}_0(t_0)) \\
&= \mathrm{e}^{A_Q(t-t_0)} A_Q^{-1} Q\boldsymbol{F}(\boldsymbol{W}_0(t_0)) - A_Q^{-1} Q\boldsymbol{F}(\boldsymbol{W}_0(t_0)). \qquad (5.60)
\end{aligned}
$$

Thus we have for the first-order solution

$$
\begin{aligned}
\boldsymbol{u}_1(t;t_0) &= \mathrm{e}^{(t-t_0)A_Q}[\boldsymbol{W}_1(t_0) + A_Q^{-1} Q\boldsymbol{F}(\boldsymbol{W}_0(t_0))] \\
&\quad + (t-t_0)P\boldsymbol{F}(\boldsymbol{W}_0(t_0)) - A_Q^{-1} Q\boldsymbol{F}(\boldsymbol{W}_0(t_0)). \qquad (5.61)
\end{aligned}
$$

The first term containing the 'initial' value correction $\boldsymbol{W}_1(t_0)$ formally belongs to the Q space and is composed of fast modes which would die out asymptotically as $t \to \infty$. However, we have supposed that we are dealing with an asymptotic analysis in which such fast modes have already died out. Our point is that such a construction of the solution is possible by choosing the yet-unknown 'initial' value $\boldsymbol{W}_1(t_0)$ so that such undesired terms disappear. Indeed the first term disappear by choosing $\boldsymbol{W}_1(t_0)$ as

$$
\boldsymbol{W}_1(t_0) = -A_Q^{-1} Q\boldsymbol{F}(\boldsymbol{W}_0(t_0)) = -A^{-1} Q\boldsymbol{F}(\boldsymbol{W}_0(t_0)). \qquad (5.62)
$$

Here and from now on, we shall use a simplified notation for

$$
f(A_Q)Q = f(A)Q \qquad (5.63)
$$

for any function $f(x)$ even for the inverse as $f(A) = A^{-1}$. We note that $\boldsymbol{W}_1(t_0)$ is orthogonal to the P space, and depends on t_0 only through $\boldsymbol{C}(t_0)$. With this choice, we have for the first-order solution

$$
\boldsymbol{u}_1(t;t_0) = (t-t_0)P\boldsymbol{F} - A^{-1} Q\boldsymbol{F}. \qquad (5.64)
$$

It is to be noted that the argument of $\boldsymbol{F}$ is $\boldsymbol{W}_0(t_0)$ and hence a function of $\boldsymbol{C}(t_0)$. In a later section, we shall show a short-cut prescription to reach the final form (5.64).

Since the 'initial' value $\boldsymbol{W}(t_0)$ is supposed to be the exact solution, the above construction of the first-order solution implies that the invariant manifold is modified to M_1, which is represented by

$$
\mathrm{M}_1 = \{\boldsymbol{u} | \boldsymbol{u} = \boldsymbol{W}_0 - \epsilon A^{-1} Q\boldsymbol{F}(\boldsymbol{W}_0)\}. \qquad (5.65)
$$

If one stops in this order, the approximate solution reads

$$
\boldsymbol{u}(t;t_0) = \boldsymbol{W}_0 + \epsilon\{(t-t_0)P\boldsymbol{F} - A^{-1} Q\boldsymbol{F}\}, \qquad (5.66)
$$

which contains a secular term proportional to a function belonging to the P space, and should be valid locally around $t \sim t_0$.

To obtain the solution that is uniformly valid in a global domain, we apply the RG/E equation to (5.66),

$$du/dt_0|_{t_0=t} = 0, \tag{5.67}$$

which leads to

$$0 = \left.\frac{d\boldsymbol{W}_0(t_0)}{dt_0}\right|_{t_0=t} - \epsilon\left[P\boldsymbol{F}(\boldsymbol{W}_0(t)) + A^{-1}Q\frac{d\boldsymbol{F}}{d\boldsymbol{W}_0}\left.\frac{d\boldsymbol{W}_0}{dt_0}\right|_{t_0=t}\right]. \tag{5.68}$$

However, since $d\boldsymbol{W}_0(t)/dt$ is already of order ϵ^1, the second term in the big bracket is of higher order and may be neglected. Thus we arrive at

$$\frac{d\boldsymbol{W}_0(t)}{dt} = \epsilon P\boldsymbol{F}(\boldsymbol{W}_0(t)), \tag{5.69}$$

which is reduced to an m-dimensional coupled equation by making an inner product with $\tilde{\boldsymbol{U}}_i$ as

$$\frac{dC_i(t)}{dt} = \epsilon\langle\tilde{\boldsymbol{U}}_i, \boldsymbol{F}(\boldsymbol{W}_0[\boldsymbol{C}])\rangle, \quad (i = 1, 2, \cdots, m). \tag{5.70}$$

The global solution representing a trajectory on the invariant manifold up to this order is given by

$$\boldsymbol{u}(t) = \boldsymbol{u}(t; t_0 = t) = \sum_{i=1}^{m} C_i(t)\boldsymbol{U}_i - \epsilon A^{-1}Q\boldsymbol{F}(\boldsymbol{W}_0[\boldsymbol{C}]), \tag{5.71}$$

with $\boldsymbol{C}(t)$ being the solution to (5.70).

In short, we have derived the invariant manifold as the 'initial' value represented by (5.71) and the reduced dynamics (5.70) on it in the RG method in the first-order approximation.

5.5.3 Second-Order Analysis

We can proceed to the second-order analysis naturally. The second-order equation reads

$$\left(\frac{d}{dt} - A\right)\boldsymbol{u}_2 = \boldsymbol{F}'(\boldsymbol{u}_0)\boldsymbol{u}_1, \tag{5.72}$$

where

$$(\boldsymbol{F}'(\boldsymbol{u}_0)\boldsymbol{u}_1)_i = \sum_{j=1}^{n} \left\{\partial(F'(\boldsymbol{u}_0))_i/\partial(u_0)_j\right\}(u_1)_j. \tag{5.73}$$

Much the same way as in the first-order case, the second-order solution is formally obtained to give

$$\boldsymbol{u}_2(t;t_0) = \mathrm{e}^{(t-t_0)A}\boldsymbol{W}_2(t_0) + \int_{t_0}^{t} ds \mathrm{e}^{(t-s)A}\boldsymbol{F}'(\boldsymbol{u}_0(s;t_0))\boldsymbol{u}_1(s;t_0). \tag{5.74}$$

After inserting $\boldsymbol{u}_1(t;t_0)$ given in (5.64) into the integrand, a straightforward but somewhat tedius evaluation of the integral yields

$$\begin{aligned}\boldsymbol{u}_2(t;t_0) = &\mathrm{e}^{(t-t_0)A}\left[\boldsymbol{W}_2(t_0) - \left\{A^{-1}\boldsymbol{Q}\boldsymbol{F}'A^{-1}\boldsymbol{Q}\boldsymbol{F} - A^{-2}\boldsymbol{Q}\boldsymbol{F}'\boldsymbol{P}\boldsymbol{F}\right\}\right]\\ &+A^{-1}\boldsymbol{Q}\boldsymbol{F}'A^{-1}\boldsymbol{Q}\boldsymbol{F} - A^{-2}\boldsymbol{Q}\boldsymbol{F}'\boldsymbol{P}\boldsymbol{F}\\ &-(t-t_0)\left\{\boldsymbol{P}\boldsymbol{F}'A^{-1}\boldsymbol{Q}\boldsymbol{F} + A^{-1}\boldsymbol{Q}\boldsymbol{F}'\boldsymbol{P}\boldsymbol{F}\right\}\\ &+\frac{1}{2}(t-t_0)^2\boldsymbol{P}\boldsymbol{F}'\boldsymbol{P}\boldsymbol{F},\end{aligned} \tag{5.75}$$

where the argument of $\boldsymbol{F}$ and $\boldsymbol{F}'$ is $\boldsymbol{W}_0[\boldsymbol{C}]$. The 'initial' value $\boldsymbol{W}_2(t_0)$ that actually gives the second correction of the global solution is determined so as to cancel out the fast modes in the first line in (5.75) as before;

$$\boldsymbol{W}_2(t_0) = A^{-1}\boldsymbol{Q}\boldsymbol{F}'(\boldsymbol{W}_0)A^{-1}\boldsymbol{Q}\boldsymbol{F}(\boldsymbol{W}_0) - A^{-2}\boldsymbol{Q}\boldsymbol{F}'\boldsymbol{P}\boldsymbol{F}, \tag{5.76}$$

which implies that the invariant manifold is modified to M_2 as represented by

$$\mathrm{M}_2 = \{\boldsymbol{u}|\boldsymbol{u} = \boldsymbol{W}_0[\boldsymbol{C}] + \boldsymbol{\rho}[\boldsymbol{C}]\}; \quad \boldsymbol{\rho} = \epsilon\boldsymbol{W}_1 + \epsilon^2\boldsymbol{W}_2, \tag{5.77}$$

in the second-order approximation.

Now the second-order solution reads

$$\begin{aligned}\boldsymbol{u}_2(t;t_0) = &A^{-1}\boldsymbol{Q}\boldsymbol{F}'A^{-1}\boldsymbol{Q}\boldsymbol{F} - A^{-2}\boldsymbol{Q}\boldsymbol{F}'\boldsymbol{P}\boldsymbol{F}\\ &-(t-t_0)\left\{\boldsymbol{P}\boldsymbol{F}'A^{-1}\boldsymbol{Q}\boldsymbol{F} + A^{-1}\boldsymbol{Q}\boldsymbol{F}'\boldsymbol{P}\boldsymbol{F}\right\}\\ &+\frac{1}{2}(t-t_0)^2\boldsymbol{P}\boldsymbol{F}'\boldsymbol{P}\boldsymbol{F}.\end{aligned} \tag{5.78}$$

We note that the 'initial' value $\boldsymbol{u}_2(t = t_0; t_0)$ of which at $t = t_0$ coincides with $\boldsymbol{W}_2(t_0)$ given by (5.76). Thus the full expression of the solution up to the second order is given by

$$\begin{aligned}
\boldsymbol{u}(t;t_0) &= \boldsymbol{W}_0(t_0) + \epsilon\{(t-t_0)P\boldsymbol{F} - A^{-1}Q\boldsymbol{F}\} \\
&+\epsilon^2\Big[A^{-1}Q\boldsymbol{F}'A^{-1}Q\boldsymbol{F} - A^{-2}Q\boldsymbol{F}'P\boldsymbol{F} - (t-t_0)\left\{P\boldsymbol{F}'A^{-1}Q\boldsymbol{F} + A^{-1}Q\boldsymbol{F}'P\boldsymbol{F}\right\} \\
&+\frac{1}{2}(t-t_0)^2 P\boldsymbol{F}'P\boldsymbol{F}\Big].
\end{aligned} \tag{5.79}$$

This is a locally valid solution around $t = \forall t_0$. To construct a globally valid solution from it, we apply the RG/E equation $d\boldsymbol{u}/dt_0|_{t_0=t} = 0$ to (5.79), which is reduced to

$$\dot{\boldsymbol{W}}_0(t) - \epsilon P\boldsymbol{F} - \epsilon A^{-1}Q\boldsymbol{F}'\dot{\boldsymbol{W}}_0 + \epsilon^2\left\{P\boldsymbol{F}'A^{-1}Q\boldsymbol{F} + A^{-1}Q\boldsymbol{F}'P\boldsymbol{F}\right\} = 0. \tag{5.80}$$

Operating the projections P and Q to the both sides of (5.80), respectively, we have

$$\dot{\boldsymbol{W}}_0(t) - \epsilon P\boldsymbol{F} + \epsilon^2 P\boldsymbol{F}'A^{-1}Q\boldsymbol{F} = 0, \tag{5.81}$$
$$-\epsilon A^{-1}Q\boldsymbol{F}'\dot{\boldsymbol{W}}_0 + \epsilon^2 A^{-1}Q\boldsymbol{F}'P\boldsymbol{F} = 0. \tag{5.82}$$

Firstly, we notice that the last equation (5.82) is reduced to

$$\epsilon A^{-1}Q\boldsymbol{F}'(-\dot{\boldsymbol{W}}_0 + \epsilon P\boldsymbol{F}) = 0, \tag{5.83}$$

which is identically satisfied on account of (5.81) up to this order. Thus, we end up with the reduced equation given by

$$\dot{\boldsymbol{W}}_0(t) = \epsilon P\boldsymbol{F} - \epsilon^2 P\boldsymbol{F}'A^{-1}Q\boldsymbol{F}, \tag{5.84}$$

which is further reduced by taking an inner product with $\tilde{\boldsymbol{U}}_i$ to

$$\dot{C}_i = \epsilon\langle\tilde{\boldsymbol{U}}_i, \boldsymbol{F} - \epsilon\boldsymbol{F}'A^{-1}Q\boldsymbol{F}\rangle, \quad (i = 1, 2, \ldots, m), \tag{5.85}$$

i.e., m-dimensional coupled equations. Note that the time derivative is an order of ϵ, implying that the equation describes a slow dynamics with a fewer degrees of freedom than that of the microscopic Eq. (5.33).

The global solution giving the trajectory on the invariant manifold is given by the 'initial' value as

$$\begin{aligned}
\boldsymbol{u}(t) &= \boldsymbol{W}(t) = \boldsymbol{W}_0[\boldsymbol{C}] + \boldsymbol{\rho}[\boldsymbol{C}] \\
&= \boldsymbol{W}_0[\boldsymbol{C}] - \epsilon A^{-1}Q\boldsymbol{F} + \epsilon^2\{A^{-1}Q\boldsymbol{F}'A^{-1}Q\boldsymbol{F} - A^{-2}Q\boldsymbol{F}'P\boldsymbol{F}\},
\end{aligned} \tag{5.86}$$

with $\boldsymbol{C}(t)$ being the solution to (5.85). Notice that the argument of $\boldsymbol{F}$ and $\boldsymbol{F}'$ in the above expression is all $\boldsymbol{W}_0$, hence the right-hand side is a function of $\boldsymbol{C}$, *i.e.*,

$$\boldsymbol{u}(t) = \boldsymbol{u}[\boldsymbol{C}]. \tag{5.87}$$

Recall that $\boldsymbol{C}$ was introduced as the integral constants of the unperturbed solution.

A remark is in order here: The present formulation using the projection operators and the resultant RG/E Eq. (5.85) governing the slow motion resemble those by Mori theory [100] for deriving the Langevin equation from a generic dynamical equation.

5.6 Example II: The Case With the Generic System with the Linear Operator Having a Jordan Cell Structure

We start with the same n-dimensional system as Eq. (5.33);

$$\left(\frac{d}{dt} - A\right)\boldsymbol{u} = \epsilon \boldsymbol{F}(\boldsymbol{u}), \qquad (5.33).$$

In this section, however, we shall deals with the case where the linear operator A has 0 eigenvalues with degeneracy of m but not digonalizable and accordingly A has an m-dimensional **Jordan cell** structure. The other eigenvalues of A are assumed to have negative real parts as in the last section.

5.6.1 *Preliminaries for a Linear Operator with Two-Dimensional Jordan Cell*

We shall specifically deal with a simple case where the Jordan cell is two-dimensional; $m = 2$. In this case, we have the normalized vectors $\boldsymbol{U}_1$ and $\boldsymbol{U}_2$ that satisfy

$$A\boldsymbol{U}_1 = 0, \quad A\boldsymbol{U}_2 = \boldsymbol{U}_1, \tag{5.88}$$

while other eigenvalues have negative real parts;

$$A\boldsymbol{U}_\alpha = \lambda_\alpha \boldsymbol{U}_\alpha, \quad \mathrm{Re}\lambda_\alpha < 0, \quad (\alpha = 3,\ 4,\ \cdots,\ n). \tag{5.89}$$

We assume that $\boldsymbol{U}_i$ $(i = 1,\ 2)$ and $\boldsymbol{U}_\alpha$'s are linearly independent. Since the linear operator A may not be symmetric (self-adjoint), we need to specify the properties of the *adjoint operator* $A^\dagger$. The adjoint operator $A^\dagger$ is defined by

$$\langle \boldsymbol{V}, A\boldsymbol{U}\rangle = \langle A^\dagger \boldsymbol{V}, \boldsymbol{U}\rangle, \tag{5.90}$$

where $\langle \boldsymbol{V},\ \boldsymbol{U}\rangle$ is the Hermitian inner product. The adjoint $A^\dagger$ has also the same Jordan cell structure as A has;

$$A^\dagger \tilde{\boldsymbol{U}}_1 = 0, \quad A^\dagger \tilde{\boldsymbol{U}}_2 = \tilde{\boldsymbol{U}}_1, \tag{5.91}$$

and

$$A^{\dagger}\tilde{\boldsymbol{U}}_{\alpha} = \lambda_{\alpha}^{*}\tilde{\boldsymbol{U}}_{\alpha}, \quad \mathrm{Re}\lambda_{\alpha}^{*} < 0, \quad (\alpha = 3, 4, \cdots, n). \tag{5.92}$$

We assume that $\tilde{\boldsymbol{U}}_i$' and $\tilde{\boldsymbol{U}}_{\alpha}$'s are linearly independent.

We take the following normalization condition for the Jordan cell vectors as[3]

$$\langle \tilde{\boldsymbol{U}}_2, \boldsymbol{U}_1 \rangle = 1, \quad \langle \tilde{\boldsymbol{U}}_1, \boldsymbol{U}_2 \rangle = 1, \tag{5.93}$$

where $\langle \ , \ \rangle$ denotes the inner product. In the Dirac's bra-ket notation for an inner product, they are expressed as

$$\langle \tilde{\boldsymbol{U}}_2 | \boldsymbol{U}_1 \rangle = 1, \quad \langle \tilde{\boldsymbol{U}}_1 | \boldsymbol{U}_2 \rangle = 1. \tag{5.94}$$

We note also the following orthogonality

$$\langle \tilde{\boldsymbol{U}}_1, \boldsymbol{U}_1 \rangle = \langle A^{\dagger}\tilde{\boldsymbol{U}}_2, \boldsymbol{U}_1 \rangle = \langle \tilde{\boldsymbol{U}}_2, A\boldsymbol{U}_1 \rangle = \langle \tilde{\boldsymbol{U}}_2, 0 \rangle = 0. \tag{5.95}$$

For the rest of the eigenvectors, we have the orthogonal condition

$$\langle \tilde{\boldsymbol{U}}_{\alpha} | \boldsymbol{U}_{\beta} \rangle = \delta_{\alpha\beta}, \quad (\alpha, \ \beta = 3, \ 4, \ \ldots, \ n) \tag{5.96}$$

where we have demand the normalization condition for $\alpha = \beta$.

Let P and Q be the projection operators onto the subspace spanned by $\{\boldsymbol{U}_1, \boldsymbol{U}_2\}$ and its orthogonal compliment spanned by $\boldsymbol{U}_{\alpha}$'s, respectively:

$$P + Q = \mathbf{1}. \tag{5.97}$$

The projection operator to the P space is expressed as

$$P = |\boldsymbol{U}_1\rangle\langle\tilde{\boldsymbol{U}}_2| + |\boldsymbol{U}_2\rangle\langle\tilde{\boldsymbol{U}}_1|. \tag{5.98}$$

which implies that for an arbitrary vector $\boldsymbol{u}$ belonging to the P space is expressed as

$$P|\boldsymbol{u}\rangle = \langle\tilde{\boldsymbol{U}}_2|\boldsymbol{u}\rangle|\boldsymbol{U}_1\rangle + \langle\tilde{\boldsymbol{U}}_1|\boldsymbol{u}\rangle|\boldsymbol{U}_2\rangle = C_1\boldsymbol{U}_1 + C_2\boldsymbol{U}_2, \tag{5.99}$$

with

$$C_1 := \langle\tilde{\boldsymbol{U}}_2|\boldsymbol{u}\rangle, \quad C_2 := \langle\tilde{\boldsymbol{U}}_1|\boldsymbol{u}\rangle. \tag{5.100}$$

[3] The present ortho-normalization conditions are different from those adopted in [6], but the standard ones.

If we confine ourselves in the P space and choose $\boldsymbol{U}_i$ $(i = 1, 2)$ as the bases of the P space, A and $\boldsymbol{U}_i$'s are represented as

$$A \sim \begin{pmatrix} 0 & 1 \\ 0 & 0 \end{pmatrix}, \quad \boldsymbol{U}_1 = \begin{pmatrix} 1 \\ 0 \end{pmatrix} = \tilde{\boldsymbol{U}}_2, \quad \boldsymbol{U}_2 = \begin{pmatrix} 0 \\ 1 \end{pmatrix} = \tilde{\boldsymbol{U}}_1, \tag{5.101}$$

which has the orthogonality condition $\langle \tilde{\boldsymbol{U}}_2 | \boldsymbol{U}_2 \rangle = 0$. The projection operator P reads

$$P = \begin{pmatrix} 1 \\ 0 \end{pmatrix} (1\ 0) + \begin{pmatrix} 0 \\ 1 \end{pmatrix} (0\ 1) = \begin{pmatrix} 1 & 0 \\ 0 & 1 \end{pmatrix}. \tag{5.102}$$

The projection operator Q_a onto the eigenstate $\boldsymbol{U}_\alpha$ $(\alpha = 3, 4, \ldots, n)$ is expressed by

$$Q_\alpha = |\boldsymbol{U}_\alpha\rangle\langle\tilde{\boldsymbol{U}}_\alpha| \tag{5.103}$$

in the bra-ket notation. Then the projection operator onto the Q space reads

$$Q = \sum_{\alpha=3}^{n} Q_\alpha = \sum_{\alpha=3}^{n} |\boldsymbol{U}_\alpha\rangle\langle\tilde{\boldsymbol{U}}_\alpha| = \mathbf{1} - P. \tag{5.104}$$

We also note the idempotency of the projection operators as

$$P^2 = P, \quad Q^2 = Q. \tag{5.105}$$

Then we have the following useful formula for the operation of the time-evolution operator e^{tA},

$$\begin{aligned} \mathrm{e}^{tA}\boldsymbol{u} &= \mathrm{e}^{tA}(P + Q)\boldsymbol{u} = \mathrm{e}^{tA}(C_1\boldsymbol{U}_1 + C_2\boldsymbol{U}_2) + \mathrm{e}^{tA}Q\boldsymbol{u} \\ &= C_1\mathrm{e}^{tA}\boldsymbol{U}_1 + C_2\mathrm{e}^{tA}\boldsymbol{U}_2 + \mathrm{e}^{tA}Q\boldsymbol{u} \\ &= C_1\boldsymbol{U}_1 + C_2(1 + tA)\boldsymbol{U}_2 + \mathrm{e}^{tA}Q\boldsymbol{u} \\ &= (C_1 + C_2 t)\boldsymbol{U}_1 + C_2\boldsymbol{U}_2 + \mathrm{e}^{tA}Q\boldsymbol{u}, \end{aligned} \tag{5.106}$$

with $C_1 := \langle\tilde{\boldsymbol{U}}_2|\boldsymbol{u}\rangle$ and $C_2 := \langle\tilde{\boldsymbol{U}}_1|\boldsymbol{u}\rangle$. Here the last term in (5.106) is further reduced on account of (5.104) as

$$\mathrm{e}^{tA}Q\boldsymbol{u} = \sum_{\alpha=3}^{n} \mathrm{e}^{tA}|\boldsymbol{U}_\alpha\rangle\langle\tilde{\boldsymbol{U}}_\alpha|\boldsymbol{u}\rangle = \sum_{\alpha=3}^{n} \mathrm{e}^{\lambda_\alpha t} C_\alpha |\boldsymbol{U}_\alpha\rangle \tag{5.107}$$

with $C_\alpha := \langle\tilde{\boldsymbol{U}}_\alpha|\boldsymbol{u}\rangle$, which tends to vanish when $t \to \infty$ because

$$|\mathrm{e}^{\lambda_\alpha t}| = \mathrm{e}^{-t|\mathrm{Re}\lambda_\alpha|} \to 0. \tag{5.108}$$

5.6.2 *Perturbative Construction of the Attractive/Invariant Manifold*

We are in a position to construct the attractive/invariant manifold M and the reduced dynamics on it for the system (5.33). Since these notions are to become valid asymptotically as $t \to \infty$, we are going to make an asymptotic analysis in the perturbation theory utilizing the RG method.

As before, we first expand the solution $\boldsymbol{u}(t)$ at an arbitrary but large time t_0 with respect to ϵ as

$$\boldsymbol{u}(t; t_0) = \boldsymbol{u}_0(t; t_0) + \epsilon \boldsymbol{u}_1(t; t_0) + \epsilon^2 \boldsymbol{u}_2(t; t_0) + \cdots, \tag{5.109}$$

with the 'initial' value $\boldsymbol{u}(t_0; t_0) = \boldsymbol{W}(t_0)$ being supposed to be an asymptotic form of an exact solution: More specifically, $t = t_0$ is large enough so that the asymptotic solution $\boldsymbol{u}(t; t_0)$ has a value on the invariant manifold M. In other words, $\boldsymbol{u}(t) = \boldsymbol{W}(t)$ gives an asymptotic but global solution to (5.33) confined on M. Therefore we should also expand the "initial' value' $\boldsymbol{W}(t_0)$ as

$$\begin{aligned} \boldsymbol{W}(t_0) &= \boldsymbol{W}_0(t_0) + \epsilon \boldsymbol{W}_1(t_0) + \epsilon^2 \boldsymbol{W}_2(t_0) + \cdots . \\ &= \boldsymbol{W}_0(t_0) + \boldsymbol{\rho}(t_0), \end{aligned} \tag{5.110}$$

where $\boldsymbol{\rho}(t_0)$ is supposed to be independent of $\boldsymbol{W}_0$. They are yet-unknown functions to be determined through faithfully solving the equation in the perturbation theory.

The lowest-order equation takes the same form as that given in (5.36), which reads $(d/dt - A)\boldsymbol{u}_0 = 0$ with the 'initial' condition

$$\boldsymbol{u}_0(t_0; t_0) = \boldsymbol{W}_0(t_0). \tag{5.111}$$

Utilizing the formula (5.106), the solution to the lowest-order equation is readily given by

$$\begin{aligned} \boldsymbol{u}_0(t; t_0) &= \mathrm{e}^{(t-t_0)A} \boldsymbol{u}_0(t_0; t_0) \\ &= [C_1^{(0)} + C_2^{(0)}(t - t_0)]\boldsymbol{U}_1 + C_2^{(0)}\boldsymbol{U}_2 + \mathrm{e}^{(t-t_0)A} Q \boldsymbol{W}_0 \end{aligned} \tag{5.112}$$

where

$$C_1^{(0)} := \langle \tilde{\boldsymbol{U}}_2 | \boldsymbol{u}_0 \rangle = \langle \tilde{\boldsymbol{U}}_2 | \boldsymbol{W}_0 \rangle, \quad C_2^{(0)} := \langle \tilde{\boldsymbol{U}}_1 | \boldsymbol{W}_0 \rangle. \tag{5.113}$$

Since we are interested in constructing the invariant manifold to be realized asymptotically for $t \to \infty$, the lowest-order solution should be as stationary as possible. The formula (5.112) tells us that such a solution is provided with the choice of the 'initial' condition

$$C_2^{(0)} = \langle \tilde{\boldsymbol{U}}_1 | \boldsymbol{W}_0 \rangle = 0 \quad \text{and} \quad Q \boldsymbol{W}_0 = 0, \tag{5.114}$$

and we have

$$u_0(t; t_0) = C_1^{(0)}(t_0)U_1. \tag{5.115}$$

Accordingly, the 'initial' value reads

$$W_0(t_0) = C_1^{(0)}(t_0)U_1, \tag{5.116}$$

and the lowest order manifold is represented as

$$\mathrm{M}_0 = \{u_0 | u_0 = C_1^{(0)}U_1\}. \tag{5.117}$$

We remark that the second condition of Eq. (5.114) implies that

$$\langle \tilde{U}_\alpha | W_0 \rangle = 0, \quad (\alpha = 3, 4, \ldots, n). \tag{5.118}$$

A comment is in order here:
The zero-th order solution is constituted only by the genuine zero mode U_1, but not U_2 that is the other constituent of the Jordan cell in contrast to the semi-simple case. This is a point for the construction of the invariant manifold using the RG method when A has a Jordan cell. We shall see that the following perturbative calculations are performed with ease by this set up where the zeroth-order solution has no time dependence: If we had included the U_2 in the zeroth-order solution, it necessarily becomes time dependent. The component in the U_2 direction will be incorporated by the perturbation together with those in the Q space. However, we will see that the amplitude of U_2 in turn constitutes the reduced dynamics together with $C_1^{(0)}$. We shall give more words on the treatment of the reduced dynamics and the coordinates of the invariant manifold later in the final stage.

The first-order equation has the same form as that given by (5.53), which reads $(d/dt - A)u_1 = F(u_0(t; t_0))$, the formal solution to which is also given by (5.56),

$$u_1(t; t_0) = \mathrm{e}^{(t-t_0)A}W_1(t_0) + \int_{t_0}^{t} ds\, \mathrm{e}^{(t-s)A}F(u_0(s; t_0)). \tag{5.119}$$

In accordance with the prescription of the RG method that the perturbed solutions are chosen so that the part of unperturbed solutions vanish at $t = t_0$, the form of the 'initial' value $W_1(t_0)$ is chosen so as to be independent of W_0;

$$W_1(t_0) = C_2^{(1)}(t_0)U_2 + QW_1(t_0), \tag{5.120}$$

with the component QW_1 yet to be specified later in the process of the solution. Then the first term of the solution (5.119) is evaluated as follows:

$$
\begin{aligned}
\mathrm{e}^{(t-t_0)A}\boldsymbol{W}_1(t_0) &= \mathrm{e}^{(t-t_0)A}C_2^{(1)}(t_0)\boldsymbol{U}_2 + \mathrm{e}^{(t-t_0)A}Q\boldsymbol{W}_1(t_0) \\
&= C_2^{(1)}(t_0)[1+(t-t_0)A]\boldsymbol{U}_2 + \mathrm{e}^{(t-t_0)A}Q\boldsymbol{W}_1(t_0) \\
&= C_2^{(1)}(t_0)[(t-t_0)\boldsymbol{U}_1 + \boldsymbol{U}_2] + \mathrm{e}^{(t-t_0)A}Q\boldsymbol{W}_1(t_0). \quad (5.121)
\end{aligned}
$$

Next, we evaluate the second term of (5.119). Noting that $\boldsymbol{u}_0(s;t_0)$ is actually independent of s, the integrand in (5.119) is reduced as follows:

$$
\begin{aligned}
\mathrm{e}^{(t-s)A}\boldsymbol{F}(\boldsymbol{u}_0(s;t_0)) &= \mathrm{e}^{(t-s)A}(P+Q)\boldsymbol{F}(\boldsymbol{u}_0(s;t_0)) \\
&= \mathrm{e}^{(t-s)A}P\boldsymbol{F}(\boldsymbol{u}_0(s;t_0)) + e^{(t-s)A}Q\boldsymbol{F}(\boldsymbol{u}_0(s;t_0)) \\
&= (C_1^{(F)}(t_0) + (t-s)C_2^{(F)}(t_0))\boldsymbol{U}_1 \\
&\quad + C_2^{(F)}(t_0)\boldsymbol{U}_2 + e^{(t-s)A}Q\boldsymbol{F}, \quad (5.122)
\end{aligned}
$$

with

$$
C_1^F(t_0) := \langle \tilde{\boldsymbol{U}}_2 | \boldsymbol{F} \rangle, \quad C_2^F(t_0) := \langle \tilde{\boldsymbol{U}}_1 | \boldsymbol{F} \rangle. \quad (5.123)
$$

Then the integral is evaluated as

$$
\begin{aligned}
\int_{t_0}^{t} ds\, \mathrm{e}^{(t-s)A}\boldsymbol{F}(\boldsymbol{u}_0(s;t_0)) &= ((t-t_0)C_1^{(F)} + \frac{1}{2}(t-t_0)^2C_2^{(F)})\boldsymbol{U}_1 + (t-t_0)C_2^{(F)}\boldsymbol{U}_2 \\
&\quad - A^{-1}Q\boldsymbol{F}(\boldsymbol{u}_0(t;t_0)) + e^{(t-t_0)A}A^{-1}Q\boldsymbol{F}(\boldsymbol{u}_0(t;t_0)).
\end{aligned}
\quad (5.124)
$$

Collecting all the terms, we have the first-order solution as

$$
\begin{aligned}
\boldsymbol{u}_1(t;t_0) &= \mathrm{e}^{(t-t_0)A}[Q\boldsymbol{W}_1(t_0) + A^{-1}Q\boldsymbol{F}] \\
&\quad + \{C_2^{(1)}(t_0)(t-t_0) + C_1^{(F)}(t_0)(t-t_0) + C_2^{(F)}(t_0)\frac{1}{2}(t-t_0)^2\}\boldsymbol{U}_1 \\
&\quad + \{C_2^{(1)}(t_0) + C_2^{(F)}(t_0)(t-t_0)\}\boldsymbol{U}_2 - A^{-1}Q\boldsymbol{F}, \quad (5.125)
\end{aligned}
$$

where we have suppressed the terms proportional to $(t-t_0)^{n\geq 2}$ which do not contribute to the RG/E equation nor to the invariant manifold. We remark that the argument of $\boldsymbol{F}$ is $\boldsymbol{W}_0[C_1^{(0)}]$.

The yet unspecified Q-component of the 'initial' value $Q\boldsymbol{W}_1(t_0)$ can be now determined so that the possible fast mode in the first line of (5.125) vanishes. Then we have

$$
Q\boldsymbol{W}_1(t_0) = -A^{-1}Q\boldsymbol{F}, \quad (5.126)
$$

which implies that the first-order 'initial' value $\boldsymbol{W}_1(t_0)$ is now written as

$$\boldsymbol{W}_1(t_0) = \boldsymbol{u}_1(t_0; t_0) = C_2^{(1)}(t_0)\boldsymbol{U}_2 \; - \; A^{-1}Q\boldsymbol{F}. \tag{5.127}$$

Thus the first-order solution now takes a somewhat simpler form

$$\boldsymbol{u}_1(t; t_0) = \{C_2^{(1)}(t_0)(t - t_0) + C_1^{(F)}(t_0)(t - t_0) + C_2^{(F)}(t_0)\frac{1}{2}(t - t_0)^2\}\boldsymbol{U}_1 \\ +\{C_2^{(1)}(t_0) + C_2^{(F)}(t_0)(t - t_0)\}\boldsymbol{U}_2 - A^{-1}Q\boldsymbol{F}. \tag{5.128}$$

Then the solution to (5.33) in the first-order approximation is given by

$$\boldsymbol{u}(t; t_0) = C_1^{(0)}(t_0)\boldsymbol{U}_1 + \epsilon\Big[\{C_1^{(F)}(t_0)(t - t_0) + C_2^{(1)}(t_0)(t - t_0) \\ +\frac{1}{2}C_2^{(F)}(t_0)(t - t_0)^2\}\boldsymbol{U}_1 + \{C_2^{(F)}(t_0)(t - t_0) + C_2^{(1)}(t_0)\}\boldsymbol{U}_2 - A^{-1}Q\boldsymbol{F}\Big]. \tag{5.129}$$

The RG/E equation

$$\left.\frac{d\boldsymbol{u}(t; t_0)}{dt_0}\right|_{t_0=t} = 0 \tag{5.130}$$

in this order now leads to

$$0 = \dot{C}_1^{(0)}\boldsymbol{U}_1 - \epsilon\{(C_1^{(F)} + C_2^{(1)})\boldsymbol{U}_1 + (C_2^{(F)} - \dot{C}_2^{(1)})\boldsymbol{U}_2 + \frac{1}{A}Q\boldsymbol{F}'\dot{C}_1^{(0)}\boldsymbol{U}_1 \tag{5.131}$$

Equating the coefficients of the independent vectors, we arrive at the reduced dynamical equation with two degrees of freedom as

$$\dot{C}_1^{(0)} = \epsilon\left(\langle\tilde{\boldsymbol{U}}_2|\boldsymbol{F}\rangle + C_2^{(1)}\right), \quad \dot{C}_2^{(1)} = \langle\tilde{\boldsymbol{U}}_1|\boldsymbol{F}\rangle, \tag{5.132}$$

where we have inserted the definition of $C_i^{(F)}$ $(i = 1, 2)$ given in (5.123). We now see that the attractive/invariant manifold M_1 is given by

$$\boldsymbol{u}(t) = \boldsymbol{W}(t) \simeq \boldsymbol{W}_0[C_1^{(0)}] + \boldsymbol{W}_1[C_1^{(0)}], \\ = C_1^{(0)}(t)\boldsymbol{U}_1 + \epsilon C_2^{(1)}(t)\boldsymbol{U}_2 - \epsilon A^{-1}Q\boldsymbol{F}, \tag{5.133}$$

with $C_1^{(0)}(t)$ and $C_2^{(1)}(t)$ being governed by (5.132). Notice that $\boldsymbol{u}(t)$ is a functional of $C_1^{(0)}(t)$ and $C_2^{(1)}(t)$. One finds that the invariant manifold is deformed by the perturbation to the Q space as well as the remaining direction in the P space. The deformation in the Q direction is a function solely of $C_1^{(0)}$.

An important remark is in order here: There is an imbalance of the treatment of the coordinate $C_2^{(0)}$, which gives a deformation of the manifold in the P space with the amplitude of order $O(\epsilon)$, whereas the time dependence is not small without the small parameter as seen in the right-hand side of (5.132). Such an imbalance can be naturally made disappear by the redefinition of

$$\epsilon C_2^{(1)} = C_2^{(0)}. \tag{5.134}$$

Then the reduced evolution equation (5.132) and the expression of the invariant manifold (5.133) rewritten, respectively, as

$$\dot{C}_1^{(0)} = C_2^{(0)} + \epsilon\langle \tilde{\boldsymbol{U}}_2 | \boldsymbol{F}\rangle, \quad \dot{C}_2^{(0)} = \epsilon\langle \tilde{\boldsymbol{U}}_1 | \boldsymbol{F}\rangle, \tag{5.135}$$

and

$$\begin{aligned} \boldsymbol{u}(t) &= \boldsymbol{W}(t) \simeq \boldsymbol{W}_0[C_1^{(0)}] + \boldsymbol{W}_1[C_1^{(0)}], \\ &= C_1^{(0)}(t)\boldsymbol{U}_1 + C_1^{(0)}(t)\boldsymbol{U}_2 - \epsilon A^{-1} Q\boldsymbol{F}. \end{aligned} \tag{5.136}$$

We now see that both $C_{1,2}^{(0)}(t)$ become the slow variables and give the zero-th order coordinates of the invariant manifold, although the deformation

$$-\epsilon A^{-1} Q\boldsymbol{F} =: \rho(C_1^0(t)) \tag{5.137}$$

of it in the Q direction is solely given by the amplitude of the zero eigenvector $C_1^{(0)}$.

In the second-order analysis given below, we shall also use the notation $C_2^{(0)}$ instead of $C_2^{(1)}$ in the final stage. We shall see that the invariant manifold is expressed in the form of

$$\boldsymbol{u}(t) = C_1^{(0)}(t)\boldsymbol{U}_1 + C_2^{(0)}(t)\boldsymbol{U}_2 + \rho(C_1^0(t), C_2^{(0)}). \tag{5.138}$$

This form has been taken as the ansatz by Y. Kuramoto [32] for the functional form of the invariant manifold when the linear operator of the unperturbed equation has a two-dimensional Jordan cell.

Second-Order Analysis

We can proceed to incorporate the second-order corrections: Using the first-order solution $\boldsymbol{u}_1(t; t_0)$ given in (5.128), we solve the second-order Eq. (5.72), which reads $(d/dt - A)\boldsymbol{u}_2 = \boldsymbol{F}'(\boldsymbol{u}_0)\boldsymbol{u}_1$. The formal solution to (5.72) is already presented in (5.74), which reads

$$\boldsymbol{u}_2(t; t_0) = \mathrm{e}^{(t-t_0)A}\boldsymbol{W}_2(t_0) + \int_{t_0}^{t} ds \mathrm{e}^{(t-s)A}\boldsymbol{F}'(\boldsymbol{u}_0(s; t_0))\boldsymbol{u}_1(s; t_0), \tag{5.139}$$

where $\boldsymbol{u}_1(t; t_0)$ given in (5.128) is to be inserted.

For a further reduction of (5.139), it is found convenient to first define the following quantities;

$$\boldsymbol{g} = C_2^{(1)}(t_0)\boldsymbol{U}_1 + C_1^{(F)}\boldsymbol{U}_1 + C_2^{(F)}\boldsymbol{U}_2, \quad \boldsymbol{h} = -A^{-1}Q\boldsymbol{F} + C_2^{(1)}\boldsymbol{U}_2. \tag{5.140}$$

Then making a straightforward but rather tedious manipulations, we arrive at

$$\begin{aligned}\boldsymbol{u}_2(t;t_0) &= \mathrm{e}^{(t-t_0)A}[\boldsymbol{W}_2(t_0) + \{A^{-1}Q\boldsymbol{F}'\boldsymbol{h} + A^{-2}Q\boldsymbol{F}'(\boldsymbol{g}\}] \\ &+(t-t_0)\left\{-A^{-1}Q\boldsymbol{F}'\boldsymbol{g} + P\boldsymbol{F}'\boldsymbol{h}\right\} - \{A^{-1}Q\boldsymbol{F}'\boldsymbol{h} + A^{-2}Q\boldsymbol{F}'\boldsymbol{g}\} \\ &+O((t-t_0)^2)\end{aligned} \tag{5.141}$$

where the last term denote those that vanish when the RG/E equation is applied. As was done before, we demand that the fast motion as described by the exponential functions as $\mathrm{e}^{\lambda_\alpha t}$ disappear, which is achieved by the choice of the 'initial' value as

$$\boldsymbol{W}_2(t_0) = -\left(A^{-1}Q\boldsymbol{F}'\boldsymbol{h} + A^{-2}Q\boldsymbol{F}'\boldsymbol{g}\right), \tag{5.142}$$

which belongs to the Q-space and give rise to a deformation of the invariant manifold. With this choice of the 'initial' value, we have for the second-order solution

$$\begin{aligned}\boldsymbol{u}_2(t;t_0) &= (t-t_0)\left\{-A^{-1}Q\boldsymbol{F}'\boldsymbol{g} + P\boldsymbol{F}'\boldsymbol{h}\right\} - \{A^{-1}Q\boldsymbol{F}'\boldsymbol{h} + A^{-2}Q\boldsymbol{F}'\boldsymbol{g}\} \\ &+O((t-t_0)^2),\end{aligned} \tag{5.143}$$

where the last term denotes the term that vanishes when the RG/E equation is applied. Collecting all the terms, the solution to (5.33) in the second-order approximation is given by

$$\begin{aligned}\boldsymbol{u}(t;t_0) &= C_1^{(0)}(t_0)\boldsymbol{U}_1 + \epsilon\Big[\{C_1^{(F)}(t-t_0) + C_2^{(1)}(t_0)(t-t_0) + \frac{1}{2}C_2^{(F)}(t-t_0)^2\}\boldsymbol{U}_1 \\ &+\{C_2^{(F)}(t-t_0) + C_2^{(1)}(t_0)\}\boldsymbol{U}_2 - A^{-1}Q\boldsymbol{F}\Big] \\ &+\epsilon^2\Big[(t-t_0)\left\{-A^{-1}Q\boldsymbol{F}'\boldsymbol{g} + P\boldsymbol{F}'\boldsymbol{h}\right\} - \{A^{-1}Q\boldsymbol{F}'\boldsymbol{h} + A^{-2}Q\boldsymbol{F}'\boldsymbol{g}\}\Big] \\ &+O((t-t_0)^2,\end{aligned} \tag{5.144}$$

where the terms which vanish when the RG/E equation is applied are suppressed. Here $\boldsymbol{g}$ and $\boldsymbol{h}$ are defined in (5.140).

The RG/E equation $d\boldsymbol{u}/dt_0|_{t_0=t} = 0$ leads to

$$\begin{aligned}&\dot{C}_1^{(0)}\boldsymbol{U}_1 - \epsilon(C_1^{(F)} + C_2^{(1)})\boldsymbol{U}_1 + \epsilon(\dot{C}_2^{(1)} - C_2^{(F)})\boldsymbol{U}_2 - \epsilon A^{-1}Q\boldsymbol{F}'\dot{C}_1^{(0)}\boldsymbol{U}_1 \\ &-\epsilon^2\left\{-A^{-1}Q\boldsymbol{F}'\boldsymbol{g} + P\boldsymbol{F}'\boldsymbol{h}\right\} - \epsilon^2\left\{A^{-1}Q\boldsymbol{F}'\dot{C}_2^{(1)}\boldsymbol{U}_2 + A^{-2}Q\boldsymbol{F}'\dot{C}_2^{(1)}\boldsymbol{U}_1\right\} = 0,\end{aligned} \tag{5.145}$$

because of $\dot{C}_1^{(0)} = \mathrm{O}(\epsilon)$. Operating P and Q onto (5.145), we obtain

$$0 = \{\dot{C}_1^{(0)} - \epsilon C_2^{(1)} - \epsilon C_1^{(F)})\}\boldsymbol{U}_1 + (\epsilon \dot{C}_2^{(1)} - \epsilon C_2^{(F)})\boldsymbol{U}_2 - \epsilon^2 P\boldsymbol{F}'\boldsymbol{h}, \quad (5.146)$$

$$0 = -\epsilon^2 A^{-2} Q\boldsymbol{F}'\dot{C}_2^{(1)}\boldsymbol{U}_1 - \epsilon A^{-1} Q\boldsymbol{F}' \left\{\dot{C}_1^{(0)}\boldsymbol{U}_1 + \epsilon \dot{C}_2^{(1)}\boldsymbol{U}_2 - \epsilon \boldsymbol{g}\right\}. \quad (5.147)$$

Taking the inner product with $\tilde{\boldsymbol{U}}_2$ and $\tilde{\boldsymbol{U}}_1$ with (5.146), respectively, we have

$$\begin{cases} \dot{C}_1^{(0)} = \epsilon C_2^{(1)} + \epsilon\langle\tilde{\boldsymbol{U}}_2|\boldsymbol{F} + \epsilon \boldsymbol{F}'\boldsymbol{h}\rangle, \\ \dot{C}_2^{(1)} = \langle\tilde{\boldsymbol{U}}_1|\boldsymbol{F} + \epsilon \boldsymbol{F}'\boldsymbol{h}\rangle, \end{cases} \quad (5.148)$$

where we have inserted the definition of $C_i^{(F)}$ $(i = 1,\ 2)$ given in (5.123). We remark that (5.147) is a higher order quantity of $\mathrm{O}(\epsilon^3)$ because of (5.148), provided that

$$Q\boldsymbol{F}'\boldsymbol{U}_1 = 0, \quad (5.149)$$

which is a kind of the necessary condition for the reduction of the dynamics is realized. Equating the components in the $\boldsymbol{U}_1, \boldsymbol{U}_2$ in (5.146), we have the reduced dynamics with the replacement $\epsilon C_2^{(1)} = C_2^{(0)}$ as follows,

$$\begin{cases} \dot{C}_1^{(0)} = C_2^{(0)} + \epsilon\langle\tilde{\boldsymbol{U}}_1, \boldsymbol{F} + \epsilon \boldsymbol{F}'\boldsymbol{h}\rangle, \\ \dot{C}_2^{(0)} = \epsilon\langle\tilde{\boldsymbol{U}}_2, \boldsymbol{F} + \epsilon \boldsymbol{F}'\boldsymbol{h}\rangle, \end{cases} \quad (5.150)$$

where $\boldsymbol{F}$ and $\boldsymbol{F}'$ are functions solely of $\boldsymbol{W}_0(t) = C_!^{(0)}(t)\boldsymbol{U}_1$. The trajectory on the manifold M_2 is given by

$$\begin{aligned} \boldsymbol{u}(t) &= \boldsymbol{W}(t) = \boldsymbol{W}_0(t) + \epsilon \boldsymbol{W}_1(t) + \epsilon^2 \boldsymbol{W}_2(t), \\ &= C_1^{(0)}(t)\boldsymbol{U}_1 + C_2^{(0)}(t)\boldsymbol{U}_2 - \epsilon A^{-1}Q\boldsymbol{F} \\ &\quad -\epsilon^2\{A^{-1}Q\boldsymbol{F}'\boldsymbol{h} + \langle\tilde{\boldsymbol{U}}_2, \boldsymbol{F}\rangle A^{-2}Q\boldsymbol{F}'\boldsymbol{U}_2\}, \\ &= C_1^{(0)}(t)\boldsymbol{U}_1 + C_2^{(0)}(t)\boldsymbol{U}_2 + \boldsymbol{\rho}(C_1^0(t), C_2^{(0)}), \end{aligned} \quad (5.151)$$

with

$$\boldsymbol{\rho}(C_1^0(t), C_2^{(0)}(t)) = -\epsilon A^{-1}Q\boldsymbol{F} - \epsilon^2\{A^{-1}Q\boldsymbol{F}'\boldsymbol{h} + \langle\tilde{\boldsymbol{U}}_2, \boldsymbol{F}\rangle A^{-2}Q\boldsymbol{F}'\boldsymbol{U}_2\}. \quad (5.152)$$

Comments are in order: The solution is solely described by the coordinates $C_1^{(0)}$ and $C_2^{(0)}$ representing the P space as in the previous subsection. Conversely, the dynamics can not be described only by the coordinate in the zero-th manifold in this case; the dimension of the invariant manifold is increased from that of the unperturbed invariant manifold. The deformation of the zero-th order invariant manifold $C_1^{(0)}\boldsymbol{U}_1$ is given by not only the counter part of the Jordan doublet $C_2^{(0)}\boldsymbol{U}_2$ but also the

Q-space vector which is, however, a function of the coefficients C_1^0 and $C_2^{(0)}$, as was taken for granted by Kuramoto [32] for the form of the invariant manifold when the linear operator in the unperturbed equation has a two-dimensional Jordan cell.

5.7 Concluding Remarks

The reduction theory presented in this chapter has applicabilities to various problems in various sciences, in particular, non-equilibrium physics. We shall see such examples in Chap. 7 where stochastic equations are treated and extensive developments in Part II where derivation of fluid dynamical equations in the quantum (non)relativistic systems are derived and their properties are extensively examined.

Appendix: Useful Formulae of Solutions for Inhomogeneous Differential Equations for the RG Method II: When *A* Is Non Semi-Simple

In the RG method as a reduction theory of dynamics [5, 6], the particular solutions in the higher orders are set up with the rules that (1) the contributions of possible secular terms disappear at an arbitrary 'initial' time $t = t_0$, and (2) possible fast modes should be absent. Some simple formulae for particular solutions that satisfy the rule (1) are presented in Sect. 4.7 in the case where the linear operator appearing in the inhomogeneous differential equation is semi-simple. It is to be noted, however, the formulae presented in in Sect. 4.7 do satisfy the rule (2), too, for the semi-simple operator.

In the present Appendix, we shall give some useful formulae [6] fo the solutions of the inhomogeneous differential equations where the linear operator A is not semi-simple but has a Jordan cell structure.

1. two-dimensional Jordan cell

   Let us consider the case where A has a two dimensional Jordan cell structure as

$$AU_1 = 0, \quad AU_2 = U_1. \tag{5.153}$$

   The adjoint $A^\dagger$ has also a Jordan cell structure[4];

$$A^\dagger \tilde{U}_1 = 0, \quad A^\dagger \tilde{U}_2 = \tilde{U}_1. \tag{5.154}$$

[4] The adjoint operator $A^\dagger$ is defined by $\langle V, AU\rangle = \langle A^\dagger V, U\rangle$, where $\langle V, U\rangle$ is the Hermitian inner product.

We define the projection operators P and Q onto the subspace $\{\boldsymbol{U}_1, \boldsymbol{U}_2\}$ and its orthogonal compliment, respectively: $P + Q = \mathbf{1}$. We take the following normalization condition,

$$\langle \tilde{\boldsymbol{U}}_2, \boldsymbol{U}_1 \rangle = 1, \quad \langle \tilde{\boldsymbol{U}}_1, \boldsymbol{U}_2 \rangle = 1. \tag{5.155}$$

If we confine ourselves in the P space and choose $\boldsymbol{U}_i$ $(i = 1, 2)$ as the bases of the P space,

$$A = \begin{pmatrix} 0 & 1 \\ 0 & 0 \end{pmatrix}, \tag{5.156}$$

$$\boldsymbol{U}_1 = \begin{pmatrix} 1 \\ 0 \end{pmatrix} = \tilde{\boldsymbol{U}}_2, \quad \boldsymbol{U}_2 = \begin{pmatrix} 0 \\ 1 \end{pmatrix} = \tilde{\boldsymbol{U}}_1. \tag{5.157}$$

The projection operator onto the P space is expressed as $P = \boldsymbol{U}_1 \tilde{\boldsymbol{U}}_2^{\dagger} + \boldsymbol{U}_2 \tilde{\boldsymbol{U}}_1^{\dagger}$, and accordingly, an arbitrary vector $\boldsymbol{U}$ can be expressed as

$$P\boldsymbol{U} = \langle \tilde{\boldsymbol{U}}_2, \boldsymbol{U} \rangle \boldsymbol{U}_1 + \langle \tilde{\boldsymbol{U}}_1, \boldsymbol{U} \rangle \boldsymbol{U}_2. \tag{5.158}$$

Let $\boldsymbol{F} = f(t)\boldsymbol{G}$ with $\boldsymbol{G}$ being a constant vector, then the contribution of the P space to the integral in (4.225) can be performed as follows:

$$\begin{aligned}
\mathrm{e}^{At} \int_{t_0}^{t} ds\, \mathrm{e}^{-As} f(s) P\boldsymbol{G} &= \int_{t_0}^{t} ds\, \mathrm{e}^{A(t-s)} f(s) \left[\langle \tilde{\boldsymbol{U}}_2, \boldsymbol{G} \rangle \boldsymbol{U}_1 + \langle \tilde{\boldsymbol{U}}_1, \boldsymbol{G} \rangle \boldsymbol{U}_2 \right] \\
&= \int_{t_0}^{t} ds\, f(s) \left[\langle \tilde{\boldsymbol{U}}_2, \boldsymbol{G} \rangle \boldsymbol{U}_1 + \langle \tilde{\boldsymbol{U}}_1, \boldsymbol{G} \rangle \boldsymbol{U}_2 + (t-s) \langle \tilde{\boldsymbol{U}}_1, \boldsymbol{G} \rangle \boldsymbol{U}_1 \right] \\
&= \int_{t_0}^{t} ds \left[f(s) P\boldsymbol{G} + f(s)(t-s) \langle \tilde{\boldsymbol{U}}_1, \boldsymbol{G} \rangle \boldsymbol{U}_1 \} \right], \qquad (5.159) \\
&= \int_{t_0}^{t} ds f(s) P\boldsymbol{G} + \int_{t_0}^{t} ds \int_{t_0}^{s} ds' f(s') \langle \tilde{\boldsymbol{U}}_1, \boldsymbol{G} \rangle \boldsymbol{U}_1. \qquad (5.160)
\end{aligned}$$

In the last equality, a partial integration has been made. Thus, making the operator identity

$$\mathrm{e}^{At} \int_{t_0}^{t} ds\, \mathrm{e}^{-As} \equiv \frac{1}{\frac{d}{dt} - A} \tag{5.161}$$

as before, we have, for example,

$$\frac{1}{\frac{d}{dt} - A} PG = (t - t_0)PG + \frac{1}{2}(t - t_0)^2 \langle \tilde{U}_1, G \rangle U_1, \tag{5.162}$$

$$\frac{1}{\frac{d}{dt} - A}(t - t_0)^n PG = \frac{1}{(n+1)}(t - t_0)^{n+1} PG, \\ + \frac{1}{(n+2)(n+1)}(t - t_0)^{n+2} \langle \tilde{U}_1, G \rangle U_1. \tag{5.163}$$

The formulae involving QG are the same as those in the semi-simple case.

2. Three-dimensional Jordan cell

Next let us take the case where A has a three-dimensional Jordan cell [6] such as

$$AU_1 = 0, \quad AU_2 = U_1, \quad AU_3 = U_2. \tag{5.164}$$

The adjoint equation reads

$$A^\dagger \tilde{U}_1 = 0, \quad A^\dagger \tilde{U}_2 = \tilde{U}_1, \quad A^\dagger \tilde{U}_3 = \tilde{U}_2. \tag{5.165}$$

The normalization condition reads

$$\langle \tilde{U}_3, U_1 \rangle = \langle \tilde{U}_2, U_2 \rangle = \langle \tilde{U}_1, U_3 \rangle = 1. \tag{5.166}$$

Let us call the sub-vector space spanned by $\{U_1, U_2, U_3\}$ the P space and the compliment of it Q space. The projection operator onto the P space is given by

$$P = U_1 \tilde{U}_3^\dagger + U_2 \tilde{U}_2^\dagger + U_3 \tilde{U}_1^\dagger, \tag{5.167}$$

and the projection operator onto Q space is given by $Q = \mathbf{1} - P$. If we confine ourselves in the P space and choose U_i $(i = 1, 2, 3)$ as the bases of the P space,

$$A = \begin{pmatrix} 0 & 1 & 0 \\ 0 & 0 & 1 \\ 0 & 0 & 0 \end{pmatrix}, \tag{5.168}$$

$$U_1 = \begin{pmatrix} 1 \\ 0 \\ 0 \end{pmatrix} = \tilde{U}_3, \quad U_2 = \begin{pmatrix} 0 \\ 1 \\ 0 \end{pmatrix} = \tilde{U}_2, \quad U_3 = \begin{pmatrix} 0 \\ 0 \\ 1 \end{pmatrix} = \tilde{U}_1. \tag{5.169}$$

Let $F = f(t)G$ with G being a constant vector, then the contribution of the P space to the integral in (4.225) is calculated as follows:

$$
\begin{aligned}
\mathrm{e}^{At}\int_{t_0}^{t} ds\, \mathrm{e}^{-As} f(s) P\boldsymbol{G} &= \int_{t_0}^{t} ds\, \mathrm{e}^{A(t-s)} f(s) \left[\langle \tilde{\boldsymbol{U}}_3, \boldsymbol{G}\rangle \boldsymbol{U}_1 + \langle \tilde{\boldsymbol{U}}_2, \boldsymbol{G}\rangle \boldsymbol{U}_2 + \langle \tilde{\boldsymbol{U}}_1, \boldsymbol{G}\rangle \boldsymbol{U}_3 \right] \\
&= \int_{t_0}^{t} ds\, f(s) \big[\langle \tilde{\boldsymbol{U}}_3, \boldsymbol{G}\rangle \boldsymbol{U}_1 + \langle \tilde{\boldsymbol{U}}_2, \boldsymbol{G}\rangle \boldsymbol{U}_2 + (t-s)\langle \tilde{\boldsymbol{U}}_2, \boldsymbol{G}\rangle \boldsymbol{U}_1 \\
&\quad + \langle \tilde{\boldsymbol{U}}_1, \boldsymbol{G}\rangle \boldsymbol{U}_3 + (t-s)\langle \tilde{\boldsymbol{U}}_1, \boldsymbol{G}\rangle \boldsymbol{U}_2 + \frac{1}{2}(t-s)^2 \langle \tilde{\boldsymbol{U}}_1, \boldsymbol{G}\rangle \boldsymbol{U}_1 \big] \\
&= \int_{t_0}^{t} ds \big[f(s) P\boldsymbol{G} + f(s)(t-s) \{ \langle \tilde{\boldsymbol{U}}_2, \boldsymbol{G}\rangle \boldsymbol{U}_1 + \langle \tilde{\boldsymbol{U}}_1, \boldsymbol{G}\rangle \boldsymbol{U}_2 \} \\
&\quad + \frac{1}{2} f(s)(t-s)^2 \langle \tilde{\boldsymbol{U}}_1, \boldsymbol{G}\rangle \boldsymbol{U}_1 \big]. \qquad (5.170)
\end{aligned}
$$

Using the formal identification (5.161), we have

$$
\begin{aligned}
\frac{1}{\frac{d}{dt} - A} f(t) P\boldsymbol{G} &= \int_{t_0}^{t} ds f(s) P\boldsymbol{G} + \int_{t_0}^{t} ds \int_{t_0}^{s} ds' f(s') \{ \langle \tilde{\boldsymbol{U}}_2, \boldsymbol{G}\rangle \boldsymbol{U}_1 + \langle \tilde{\boldsymbol{U}}_1, \boldsymbol{G}\rangle \boldsymbol{U}_2 \} \\
&\quad + \int_{t_0}^{t} ds \int_{t_0}^{s} ds_1 \int_{t_0}^{s_1} ds_2 f(s_2) \langle \tilde{\boldsymbol{U}}_1, \boldsymbol{G}\rangle \boldsymbol{U}_1, \qquad (5.171)
\end{aligned}
$$

where partial integrations have been made as in (5.160).

Chapter 6
Miscellaneous Examples of Reduction of Dynamics

6.1 Introduction

In this chapter, we take up several examples of a system of differential equations and apply the renormalization-group (RG) method [1, 38, 39, 48–52, 92–96] as formulated in [3–6] for obtaining the reduced dynamics of them with different characteristics. The examples include a bifurcation in the Lorenz model [84, 86], and the Hopf bifurcation [34, 86] in Brusselator [30, 83] with the diffusion term. Furthermore we shall analyze an extended Takens equation [101] and Benney equation [36], both of which have a Jordan cell structure in the linear operator appearing in the zeroth-order equation.

6.2 RG/E Analysis of a Bifurcation in The Lorenz Model

In this section, we shall show how a bifurcation phenomenon of the Lorenz model [84, 86] (3.136) (or (3.138) in a vector notation) is described in the RG method, which was also performed with the use of the multiple-scale method in Chap. 3: As was mentioned there, what we are doing is to construct a center manifold and the reduced dynamics [33, 34, 86] on it in a perturbative way.

We reproduce the Lorenz equation (3.138) [84, 86] here;

$$\frac{d\boldsymbol{X}}{dt} = \boldsymbol{F}(\boldsymbol{X}; r), \tag{6.1}$$

with

$$\boldsymbol{X} := \begin{pmatrix} \xi \\ \eta \\ \zeta \end{pmatrix}, \quad \boldsymbol{F}(\boldsymbol{X}; r) := \begin{pmatrix} \sigma(-\xi + \eta) \\ -\xi\zeta + r\xi - \eta \\ \xi\eta - b\zeta \end{pmatrix}. \tag{6.2}$$

We shall dare to recapitulate the formulae written in Chap. 3 for convenience and completeness.

T. Kunihiro et al., *Geometrical Formulation of Renormalization-Group Method as an Asymptotic Analysis*, Fundamental Theories of Physics 206,
https://doi.org/10.1007/978-981-16-8189-9_6

There are two steady states of the equation; the one is the origin

$$\mathbf{0} =: \boldsymbol{X}_0^{(A)} \tag{6.3}$$

and the others are given by

$$\boldsymbol{X}_0^{(B)} = {}^t(\pm\sqrt{b(r-1)},\ \pm\sqrt{b(r-1)},\ r-1). \tag{6.4}$$

As in Chap. 3, we shall try to obtain the slow dynamics and the invariant manifold or center manifold [33, 34, 86] around the origin $\boldsymbol{X}_0^{(A)}$ for $r \sim 1$. For describing r around 1, we parametrize it as $r = 1 + \mu$ with $\mu = \chi\epsilon^2$ and $\chi = \mathrm{sgn}\mu$.

Let $\boldsymbol{X}(t;\ t_0)$ be a local solution around $t \sim \forall t_0$ in the *asymptotic regime*, and represent it as a perturbation series;

$$\boldsymbol{X}(t;\ t_0) = \epsilon\boldsymbol{X}_1(t;\ t_0) + \epsilon^2\boldsymbol{X}_2(t;\ t_0) + \epsilon^3\boldsymbol{X}_3(t;\ t_0) + \cdots, \tag{6.5}$$

with

$$\boldsymbol{X}_i = \begin{pmatrix} \xi_i \\ \eta_i \\ \zeta_i \end{pmatrix}, \quad (i = 1, 2, 3, \ldots). \tag{6.6}$$

We suppose that the 'initial' value $\boldsymbol{W}(t_0)$ of $\boldsymbol{X}(t;\ t_0)$ at $t = t_0$ is made equal to that of an exact solution, which we denote as $\boldsymbol{X}_{\mathrm{E}}(t)$,

$$\boldsymbol{W}(t_0) := \boldsymbol{X}(t_0;\ t_0) = \boldsymbol{X}_{\mathrm{E}}(t_0). \tag{6.7}$$

The 'initial' value as the exact solution should also be expanded as

$$\boldsymbol{W}(t_0) = \boldsymbol{W}_0(t_0) + \epsilon\,\boldsymbol{W}_1(t_0) + \epsilon^2\,\boldsymbol{W}_2(t_0) + \cdots. \tag{6.8}$$

The first order equation reads

$$\left(\frac{d}{dt} - L_0\right)\boldsymbol{X}_1 = \mathbf{0}, \quad \text{with} \quad L_0 = \begin{pmatrix} -\sigma & \sigma & 0 \\ 1 & -1 & 0 \\ 0 & 0 & -b \end{pmatrix}. \tag{6.9}$$

The linear operator L_0 has one zero eigenvalue $\lambda_1 = 0$ and two negative ones $\lambda_2 = -\sigma - 1$ and $\lambda_3 = -b$, and the respective eigenvectors are given as follows,

$$\boldsymbol{U}_1 = \begin{pmatrix} 1 \\ 1 \\ 0 \end{pmatrix}, \quad \boldsymbol{U}_2 = \begin{pmatrix} \sigma \\ -1 \\ 0 \end{pmatrix} \quad \text{and} \quad \boldsymbol{U}_3 = \begin{pmatrix} 0 \\ 0 \\ 1 \end{pmatrix}. \tag{6.10}$$

The general solution of the first-order equation is expressed as a linear combination of the normal modes given by

$$AU_1 + Be^{\lambda_2 t}U_2 + Ce^{\lambda_3 t}U_3. \tag{6.11}$$

However, since we are interested in the asymptotic state after a long time, we take the neutrally stable solution

$$X_1(t;\, t_0) = A(t_0)U_1 = {}^t(A(t_0),\, A(t_0),\, 0) = {}^t(\xi_1,\, \eta_1,\, \zeta_1), \tag{6.12}$$

which implies that

$$W_1(t_0;\, t_0) = A(t_0)U_1. \tag{6.13}$$

The second-order equation is now expressed as

$$\left(\frac{d}{dt} - L_0\right)X_2 = \begin{pmatrix} 0 \\ -\xi_1\zeta_1 \\ \xi_1\eta_1 \end{pmatrix} = A^2U_3. \tag{6.14}$$

Using the formula (4.237), we find that the particular solution to which is given by

$$X_2 = \frac{1}{\frac{d}{dt} - L_0}A^2U_3 = \frac{1}{0-(-b)}A^2U_3 = \begin{pmatrix} 0 \\ 0 \\ \zeta_2 \end{pmatrix} \quad \text{with} \quad \zeta_2 = \frac{A^2}{b}. \tag{6.15}$$

Accordingly, the second-order correction of the 'initial' value reads

$$W_2(t_0) = {}^t(0,\, 0,\, \frac{1}{b}A^2(t_0)). \tag{6.16}$$

The third-order equation now takes the form of

$$\begin{aligned}\left(\frac{d}{dt} - L_0\right)X_3 &= \begin{pmatrix} 0 \\ \chi\xi_1 - \xi_1\zeta_2 \\ 0 \end{pmatrix} = \left(\chi A - \frac{1}{b}A^3\right)\begin{pmatrix} 0 \\ 1 \\ 0 \end{pmatrix} \\ &= \left(\chi A - \frac{1}{b}A^3\right)\left[\frac{\sigma}{\sigma+1}U_1 - \frac{1}{\sigma+1}U_2\right].\end{aligned} \tag{6.17}$$

According to the formulae (4.238) and (4.237), we have

$$\frac{1}{\frac{d}{dt} - L_0}U_1 = (t - t_0)U_1, \tag{6.18}$$

$$\frac{1}{\frac{d}{dt} - L_0}U_2 = \frac{1}{0-(-\sigma-1)}U_2 = \frac{1}{\sigma+1}U_2, \tag{6.19}$$

respectively. Thus we have

$$X_3 = (\chi A - \frac{1}{b}A^3)\left\{\frac{\sigma}{\sigma+1}(t-t_0)U_1 + \frac{1}{(\sigma+1)^2}U_2\right\}, \tag{6.20}$$

implying that the third-order correction to the 'initial' value is

$$W_3(t_0) = \frac{\chi A - \frac{1}{b}A^3}{(\sigma+1)^2}U_2. \tag{6.21}$$

Note the appearance of a secular term in (6.20).

Now collecting the all the terms obtained so far, we have

$$X(t;\, t_0, A(t_0)) = \epsilon A(t_0)U_1 + \epsilon^2\frac{1}{b}A^2U_3 + \epsilon^3(\chi A - \frac{1}{b}A^3) \times\left\{\frac{\sigma}{\sigma+1}(t-t_0)U_1 + \frac{1}{(\sigma+1)^2}U_2\right\}, \tag{6.22}$$

where we have remade explicit the t_0-dependence of the integral constant $A(t_0)$.

Now we take a geometrical point of view. We have a family of trajectories $\{X(t;\, t_0, A(t_0))\}_{t_0}$ parametrized by t_0. Let us obtain the envelope trajectory E of this family with t_0 being the contact point with the envelope. Thus we postulate the envelope equation

$$\left.\frac{dX}{dt_0}\right|_{t_0=t} = 0, \tag{6.23}$$

which leads to

$$\epsilon\frac{dA}{dt}U_1 - \epsilon^3\frac{\sigma}{\sigma+1}(\chi A - \frac{1}{b}A^3)U_1 = 0, \tag{6.24}$$

up to (ϵ^3). Thus we arrive at

$$\frac{dA}{dt} = \epsilon^2\frac{\sigma}{\sigma+1}\left(\chi A - \frac{1}{b}A^3\right), \tag{6.25}$$

on account of the linear independence of $U_i\,(i = 1, 2, 3)$. This is the same as the reduced equation (3.178) obtained in the multiple-scale method in Sect. 3.6.2.

The envelope trajectory as the (approximate) solution to be valid in a global domain is now given by

$$\begin{aligned} X_{\mathrm{E}}(t) &= \epsilon X_1(t,\, t_0 = t, A(t)) + \epsilon^2 X_2(t,\, t_0 = t, A(t)) + \epsilon^3 X_3(t,\, t_0 = t, A(t)) \\ &= = \epsilon W_1(t) + \epsilon^2 W_2(t) + \epsilon^3 X_3(t) \\ &= \epsilon A(t)U_1 + \frac{\epsilon^2}{b}A^2(t)U_3 + \frac{\epsilon^3}{(1+\sigma)^2}(\chi A(t) - \frac{1}{b}A^3(t))U_2, \end{aligned} \tag{6.26}$$

up to $O(\epsilon^3)$, which again coincides with the results (3.181) given in the multiple-scale method in Sect. 3.6.2. Further detailed account of the results are discussed there.

6.3 RG/E Analysis of the Brusselator with a Diffusion Term: Extraction of Slow Dynamics Around The Hopf Bifurcation Point

In Chap.3 and Sect. 6.2, we saw that the qualitative behavior in the Lorenz model changes along with that of the parameter contained in the model. This phenomenon is called bifurcation [34, 86].

In this section, we take a celebrated model system which admits a Hopf bifurcation [34, 86], the Brusselator [30, 83] with and without a diffusion term.

6.3.1 *The Model Equation*

The Brusselator [30, 83] is an example of systems showing a Hopf bifurcation [34, 86]. Here we treat this interesting example in the RG/E method, and thereby demonstrating its simple and transparent manipulation for getting the reduction of the dynamics as well as the invariant manifold.

The Brusselator is given by[1]

$$\frac{\partial X}{\partial t} = A - (B+1)X + X^2 Y + D_X \frac{\partial^2 X}{\partial x^2}, \tag{6.27}$$

$$\frac{\partial Y}{\partial t} = BX - X^2 Y + D_Y \frac{\partial^2 Y}{\partial x^2}, \tag{6.28}$$

where A, B, D_X and D_y are constant; it is further assumed that A and B are positive numbers. In the present analysis, we fix the value of A, and vary B as the control parameter of the system.

In this model, there is a homogeneous steady state, which satisfies the following equation,

$$A - (B+1)X + X^2 Y = 0, \quad BX - X^2 Y = 0. \tag{6.29}$$

We find that the steady state is given by the fixed point

$$(X_0, Y_0) = (A, B/A). \tag{6.30}$$

[1] We follow [30] for the presentation and the notations.

We are interested in the dynamical behavior around the steady state. So we define the new variables as

$$\xi := X - X_0, \quad \eta := Y - Y_0. \tag{6.31}$$

Then we have a vector equation for

$$\boldsymbol{u}(t, \boldsymbol{r}) := \begin{pmatrix} \xi(t, \boldsymbol{r}) \\ \eta(t, \boldsymbol{r}) \end{pmatrix} \tag{6.32}$$

as

$$\frac{\partial \boldsymbol{u}}{\partial t} = (L + \hat{D}\nabla^2)\boldsymbol{u} + F(\xi, \eta)\begin{pmatrix} 1 \\ -1 \end{pmatrix}, \tag{6.33}$$

where

$$L = \begin{pmatrix} (B-1) & A^2 \\ -B & -A^2 \end{pmatrix}, \quad \hat{D} = \begin{pmatrix} D_x & 0 \\ 0 & D_y \end{pmatrix} \tag{6.34}$$

and

$$F(\xi, \eta) = \frac{B}{A}\xi^2 + 2A\xi\eta + \xi^2\eta. \tag{6.35}$$

6.3.2 *Linear Stability Analysis*

Let us first make a linear stability analysis of the fixed point (6.30) with $F(\xi, \eta)$ being neglected. Inserting the ansatz $\boldsymbol{u} = \boldsymbol{u}_0 \exp[i\boldsymbol{k} \cdot \boldsymbol{r} + \Omega t]$ with $\boldsymbol{u}_0$ being non-vanishing into the linearized equation $\dot{\boldsymbol{u}} = L\boldsymbol{u}$, we have

$$\begin{pmatrix} (B-1) - D_x k^2 & A^2 \\ -B & -A^2 - D_y k^2 \end{pmatrix} \boldsymbol{u}_0 = \Omega \boldsymbol{u}_0, \tag{6.36}$$

the characteristic equation to which reads

$$\Omega^2 + b(k)\Omega + c(k) = 0, \tag{6.37}$$

with

$$\begin{cases} b(k) = 1 - B + A^2 + (D_x + D_y)k^2, \\ c(k) = A^2 + \{A^2 D_x + (1-B)D_y\}k^2 + D_x D_y k^4. \end{cases} \tag{6.38}$$

The fixed point is stable (unstable) when the real part(s) of the roots $\Omega_{\pm}$ of (6.37) are negative (positive), for which $\Omega_{\pm}$ may be either real or complex. In the present analysis, we assume that $D_x > D_y$, which implies that

$$\sqrt{\frac{D_x}{D_y}} > \frac{\sqrt{A^2+1}-1}{A}. \tag{6.39}$$

In this case, a detailed analysis shows that the fixed point is stable when $b(k) > 0$, and $c(k) > 0$, $\forall k$, *i.e.*,

$$1 + A^2 > B. \tag{6.40}$$

Thus the critical condition for that is given by

$$k_c = 0 \quad \text{and} \quad B \nearrow 1 + A^2 \equiv B_C, \tag{6.41}$$

at which the fixed point becomes unstable.[2]

6.3.3 *Perturbative Expansion with the Diffusion Term*

Now let us analyze the slow motion and the slow manifold around the critical point $B = B_c$ by defining the following variables with a small parameter ϵ

$$B = B_C(1+\mu), \ \epsilon = \sqrt{|\mu|}, \text{ and } \ \chi = \text{sgn}(\mu), \tag{6.42}$$

which implies that

$$\mu = \chi\epsilon^2 = (B - B_c)/B_c. \tag{6.43}$$

In this case,

$$b(k) = 1 - B + A^2 + (D_x + D_y)k^2 = B_c\mu + (D_x + D_y)k^2. \tag{6.44}$$

Thus we assume that the order of the spatial derivative

$$\hat{D}\nabla^2 \boldsymbol{u} \sim \epsilon^2. \tag{6.45}$$

Therefore, we attach the factor ϵ^2 to this term so as to make explicit that this term is of $O(\epsilon^2)$.

[2] If B is decreased from a large value to $B_c' \equiv \left(1 + A\sqrt{\frac{D_x}{D_y}}\right)^2 > B_C$, another instability (Turing instability) occurs [30]. But we shall not discuss this instability in the present analysis.

Then we first try to obtain an approximate solution $\boldsymbol{u}(t, \boldsymbol{r}; t_0)$ around arbitrary $t = t_0$ by expanding it w.r.t. ϵ:

$$\boldsymbol{u}(t, \boldsymbol{r}; t_0) = \epsilon \boldsymbol{u}_1(t, \boldsymbol{r}; t_0) + \epsilon^2 \boldsymbol{u}_2(t, \boldsymbol{r}; t_0) + \epsilon^3 \boldsymbol{u}_3(t, \boldsymbol{r}; t_0) + \cdots; \tag{6.46}$$

$$\boldsymbol{u}_i(t, \boldsymbol{r}; t_0) = \begin{pmatrix} \xi_i(t, \boldsymbol{r}; t_0) \\ \eta_i(t, \boldsymbol{r}; t_0) \end{pmatrix}, \quad (i = 1, 2, \ldots). \tag{6.47}$$

Correspondingly the initial value or space profile $\boldsymbol{u}(t = t_0, \boldsymbol{r}; t_0) =: \boldsymbol{W}(t_0, \boldsymbol{r})$ at $t = t_0$ is also expanded as

$$\boldsymbol{W}(t_0, \boldsymbol{r}) = \epsilon \boldsymbol{W}_1(t_0, \boldsymbol{r}) + \epsilon^2 \boldsymbol{W}_2(t_0, \boldsymbol{r}) + \epsilon^3 \boldsymbol{W}_3(t_0, \boldsymbol{r}) + \cdots. \tag{6.48}$$

The Eq. (6.33) is rewritten as

$$(\partial_t - L_0)\boldsymbol{u} = \epsilon^2 \hat{D} \nabla^2 \boldsymbol{u} + \Big[\epsilon^2 (B_c \xi^2 / A + 2A\xi_1 \eta_1) + \epsilon^3 \{\chi B_c \xi_1 + 2B_c \xi_1 \xi_2 / A + 2A(\xi_1 \eta_2 + \xi_2 \eta_1) + \xi_1^2 \eta_1\}\Big] \boldsymbol{U}_0 + \cdots, \tag{6.49}$$

with

$$L_0 = \begin{pmatrix} A^2 & A^2 \\ -(A^2 + 1) & -A^2 \end{pmatrix}, \quad \boldsymbol{U}_0 = \begin{pmatrix} 1 \\ -1 \end{pmatrix}. \tag{6.50}$$

Then the first few-order equations read

$$(\partial_t - L_0)\boldsymbol{u}_1 = \boldsymbol{0}, \tag{6.51}$$

$$(\partial_t - L_0)\boldsymbol{u}_2 = \left(\frac{B_c}{A}\xi_1^2 + 2A\xi_1\eta_1\right)\boldsymbol{U}_0, \tag{6.52}$$

$$(\partial_t - L_0)\boldsymbol{u}_3 = \left(\chi B_c \xi_1 + \frac{2B_c}{A}\xi_1\xi_2 + 2A(\xi_1\eta_2 + \xi_2\eta_1) + \xi_1^2\eta_1\right)\boldsymbol{U}_0 + \hat{D}\nabla^2 \boldsymbol{u}_1, \tag{6.53}$$

and so on. The eigenvalues of the asymmetric matrix L_0 are readily found to be

$$\pm iA =: \pm i\omega, \tag{6.54}$$

and the respective eigenvectors are given by

$$\boldsymbol{U}_\omega = \begin{pmatrix} 1 \\ -1 + \frac{i}{A} \end{pmatrix}, \quad \boldsymbol{U}_{-\omega} = \bar{\boldsymbol{U}}_\omega = \begin{pmatrix} 1 \\ -1 - \frac{i}{A} \end{pmatrix}, \tag{6.55}$$

where $\bar{z}$ denotes the complex conjugate of z:

$$L_0 \boldsymbol{U}_\omega = i\omega \boldsymbol{U}_\omega, \quad L_0 \boldsymbol{U}_{-\omega} = -i\omega \boldsymbol{U}_{-\omega}. \tag{6.56}$$

The adjoint eigenvector $\boldsymbol{U}_{\omega}^{*}$ satisfying the eigenvalue equation

$$ {}^{t}L_0 \boldsymbol{U}_{\omega}^{*} = i\omega \boldsymbol{U}_{\omega}^{*} \tag{6.57} $$

is given by

$$ \boldsymbol{U}_{\omega}^{*} = \frac{1}{2}\begin{pmatrix} 1 - iA \\ -iA \end{pmatrix}, \tag{6.58} $$

which satisfies the orthonormal relation

$$ (\boldsymbol{U}_{\omega}^{*}, \boldsymbol{U}_{\omega}) \equiv {}^{t}\boldsymbol{U}_{\omega}^{*}\boldsymbol{U}_{\omega} = 1. \tag{6.59} $$

Note that the following orthogonal property is automatically satisfied,

$$ (\boldsymbol{U}_{\omega}^{*}, \bar{\boldsymbol{U}}_{\omega}) = {}^{t}\boldsymbol{U}_{\omega}^{*}\bar{\boldsymbol{U}}_{\omega} = 0. \tag{6.60} $$

Now the solution to the first-order equation (6.51) is readily found to be

$$ \boldsymbol{u}_1(t, \boldsymbol{r}; t_0) = C(t_0, \boldsymbol{r})\boldsymbol{U}_{\omega}\mathrm{e}^{i\omega t} + \text{c.c.}, \quad (\omega \equiv A), \tag{6.61} $$

which implies that

$$ \xi_1(t, \boldsymbol{r} : t_0) = C(t_0, \boldsymbol{r})\mathrm{e}^{i\omega t} + \text{c.c.}, \quad \eta_1(t, \boldsymbol{r} : t_0) = C(t_0, \boldsymbol{r})\left(-1 + \frac{i}{A}\right)\mathrm{e}^{i\omega t} + \text{c.c.} \tag{6.62} $$

Here $C(t_0, \boldsymbol{r})$ is the (complex) integral constant, which may depend on $\boldsymbol{r}$ as well as t_0. Accordingly, the initial value or space profile at $t = t_0$ in this order is given by

$$ \boldsymbol{W}_1(t_0, \boldsymbol{r}) = C(t_0, \boldsymbol{r})\boldsymbol{U}_{\omega}\mathrm{e}^{i\omega t_0} + \text{c.c.} \tag{6.63} $$

A simple manipulation shows that the second-order equation (6.52) now takes the form

$$ (\partial_t - L_0)\boldsymbol{u}_2 = \left[C^2\frac{(1 + iA)^2}{A}\boldsymbol{U}_0\mathrm{e}^{2i\omega t} + \text{c.c.}\right] + 2|C|\frac{(1 - A)^2}{A}\boldsymbol{U}_0. \tag{6.64} $$

Noting that $\boldsymbol{U}_0$ defined in (6.50) is expressed in terms of the eigenvectors as

$$ \boldsymbol{U}_0 = \frac{1}{2}(\boldsymbol{U}_{\omega} + \boldsymbol{U}_{-\omega}), \tag{6.65} $$

the solution to Eq. (6.64) is calculated as follows: Using the formula (4.237) given in Sect. 4.7, we have

$$\frac{1}{\partial_t - L_0} \boldsymbol{U}_{\pm\omega} e^{2i\omega t} = \frac{1}{2i\omega \mp i\omega} \boldsymbol{U}_{\pm\omega} e^{2i\omega t}, \tag{6.66}$$

which leads to

$$\frac{1}{\partial_t - L_0} \boldsymbol{U}_0 = \frac{1}{2} \left[\frac{1}{i\omega} \boldsymbol{U}_\omega + \frac{1}{32i\omega} \boldsymbol{U}_{-\omega} \right] = \frac{1}{2A} \begin{pmatrix} -2iA \\ 1 + 2iA \end{pmatrix}. \tag{6.67}$$

Similarly,

$$\begin{aligned} \frac{1}{\partial_t - L_0} \boldsymbol{U}_0 &= \frac{1}{2} \frac{1}{-L_0} (\boldsymbol{U}_\omega + \boldsymbol{U}_{-\omega}) = \frac{1}{2} \left(\frac{1}{-i\omega} \boldsymbol{U}_\omega + \frac{1}{i\omega} \boldsymbol{U}_{-\omega} \right) \\ &= \frac{-1}{A^2} \begin{pmatrix} 0 \\ 1 \end{pmatrix}. \end{aligned} \tag{6.68}$$

Thus we have for the solution to the second-order equation (6.64)

$$\boldsymbol{u}_2(t, \boldsymbol{r}; t_0) = \{C^2(t_0, \boldsymbol{r}) \boldsymbol{V}_+ e^{2i\omega t} + \text{c.c.}\} + |C(t_0, \boldsymbol{r})|^2 \boldsymbol{V}_0, \tag{6.69}$$

where

$$\boldsymbol{V}_+ = \frac{1 + iA}{3A^3} \begin{pmatrix} -2iA \\ 1 + 2iA \end{pmatrix}, \quad \boldsymbol{V}_0 = 2 \frac{A^2 - 1}{A^3} \begin{pmatrix} 0 \\ 1 \end{pmatrix}. \tag{6.70}$$

Accordingly,

$$\xi_2(t, \boldsymbol{r}; t_0) = -\frac{2}{3} i \frac{(1 + iA)^2}{A^2} C^2(t_0, \boldsymbol{r}) e^{2i\omega t} + \text{c.c.}, \tag{6.71}$$

$$\eta_2(t, \boldsymbol{r}; t_0) = \left(\frac{(1 + iA)^2}{3A^3} (1 + 2iA) C^2(t_0, \boldsymbol{r}) e^{2i\omega t} + \text{c.c.} \right) + \frac{2(A^2 - 1)}{A^3} |C(t_0, \boldsymbol{r})|^2. \tag{6.72}$$

We note that the initial profile at $t = t_0$ in this order is given by

$$\boldsymbol{W}_2(t_0, \boldsymbol{r}) = \{C^2(t_0, \boldsymbol{r}) \boldsymbol{V}_+ e^{2i\omega t_0} + \text{c.c.}\} + |C(t_0, \boldsymbol{r})|^2 \boldsymbol{V}_0. \tag{6.73}$$

After straightforward but tedious manipulations with use of the above results, we find that the third order equation (6.53) takes the following form,

$$\begin{aligned} (\partial_t - L_0) \boldsymbol{u}_3 &= \left[\left(\mathcal{W}_1(t_0, \boldsymbol{r}) e^{i\omega t} + \mathcal{W}_3(t_0, \boldsymbol{r}) e^{3i\omega t} \right) + \text{c.c.} \right] \frac{1}{2} (\boldsymbol{U}_\omega + \boldsymbol{U}_{-\omega}) \\ &\quad + \hat{D} \nabla^2 \left(C(t_0, \boldsymbol{r}) \boldsymbol{U}_\omega e^{i\omega t} + \text{c.c.} \right). \end{aligned} \tag{6.74}$$

Here

$$\mathcal{W}_1(t_0, \boldsymbol{r}) = \alpha C(t_0, \boldsymbol{r}) + \beta |C|^2 C(t_0, \boldsymbol{r}), \ \ \mathcal{W}_3(t_0, \boldsymbol{r}) = \gamma C^3(t_0, \boldsymbol{r}), \tag{6.75}$$

where

$$\alpha = \chi B_c = \chi(1 + A^2), \quad \beta = -\frac{2 + A^2}{A^2} + i\frac{-4A^4 + 7A^2 - 4}{3A^3}, \tag{6.76}$$

$$\gamma = 2\frac{B_c}{A} V_{+\xi} + 2A(V_{+\eta} + V_{+\xi}\bar{U}_\eta + U_\eta), \tag{6.77}$$

with

$$V_{+\xi} = -2i\frac{(1 + iA)^2}{3A^2}, \quad V_{+\eta} = (1 + 2iA)\frac{(1 + iA)^2}{3A^3}, \quad U_\eta = -\frac{A - i}{A}. \tag{6.78}$$

According to the formula (4.237), the solution to the equation

$$(\partial_t - L_0)\boldsymbol{v} = f(\boldsymbol{r})\mathrm{e}^{\lambda t}\boldsymbol{U}_\omega \text{ with } \lambda \neq i\omega \tag{6.79}$$

is given by

$$\boldsymbol{v} = \frac{1}{\partial_t - L_0} f(\boldsymbol{r})\mathrm{e}^{\lambda t}\boldsymbol{U}_\omega = \frac{1}{\lambda - i\omega} f(\boldsymbol{r})\mathrm{e}^{\lambda t}\boldsymbol{U}_\omega. \tag{6.80}$$

On the other hand, when $\lambda = i\omega$, the solution to

$$(\partial_t - L_0)\boldsymbol{v} = \mathrm{e}^{i\omega t} f(\boldsymbol{r})\boldsymbol{U}_\omega \tag{6.81}$$

is given by

$$\boldsymbol{v} = (t - t_0)\mathrm{e}^{i\omega t} f(\boldsymbol{r})\boldsymbol{U}_\omega, \tag{6.82}$$

as shown in (4.238).

Thus, we have the solution to the third order equation (6.74) as

$$\begin{aligned}\boldsymbol{u}_3(t; t_0) &= \left[\frac{\mathcal{W}_1}{2}\{(t - t_0)\boldsymbol{U}_\omega + \frac{1}{2i\omega}\boldsymbol{U}_{-\omega}\}\mathrm{e}^{i\omega t} + \frac{\mathcal{W}_3}{4i\omega}(\boldsymbol{U}_\omega + \frac{1}{2}\boldsymbol{U}_{-\omega})\mathrm{e}^{3i\omega t}\right] + \text{c.c.} \\ &+ (t - t_0)\hat{D}\nabla^2\left(C\boldsymbol{U}_\omega \mathrm{e}^{i\omega t} + \text{c.c.}\right).\end{aligned} \tag{6.83}$$

Here the initial values have been chosen to be

$$\boldsymbol{W}_3(t_0, \boldsymbol{r}) = \left[\frac{\mathcal{W}_1(t_0, \boldsymbol{r})}{4i\omega}\boldsymbol{U}_{-\omega}\mathrm{e}^{i\omega t_0} + \frac{\mathcal{W}_3(t_0, \boldsymbol{r})}{4i\omega}(\boldsymbol{U}_\omega + \frac{1}{2}\boldsymbol{U}_{-\omega})\mathrm{e}^{3i\omega t_0}\right] + \text{c.c.} \tag{6.84}$$

Here notice that $\boldsymbol{U}_{-\omega} = \bar{\boldsymbol{U}}_\omega$ is the complex conjugate of $\boldsymbol{U}_\omega$.

Collecting all the terms thus obtained to have the approximate $\boldsymbol{u}(t, \boldsymbol{r}; t_0)$:

$$\begin{aligned}
\boldsymbol{u}(t, \boldsymbol{r}; t_0) &= \epsilon \left(C(t_0, \boldsymbol{r}) \boldsymbol{U}_\omega \mathrm{e}^{i\omega t} + \text{c.c.} \right) \\
&+ \epsilon^2 \left(\{ C^2 \boldsymbol{V}_+ \mathrm{e}^{2i\omega t} + \text{c.c.} \} + |C|^2 \boldsymbol{V}_0 \right) \\
&+ \epsilon^3 \Bigg[\Bigg[\frac{\mathcal{W}_1}{2} \{ (t - t_0) \boldsymbol{U}_\omega + \frac{1}{2i\omega} \bar{\boldsymbol{U}}_\omega \} \mathrm{e}^{i\omega t} + \frac{\mathcal{W}_3}{4i\omega} (\boldsymbol{U}_\omega + \frac{1}{2} \bar{\boldsymbol{U}}_\omega) \mathrm{e}^{3i\omega t} \Bigg] + \text{c.c.} \\
&+ (t - t_0) \hat{D} \nabla^2 \left(C \boldsymbol{U}_\omega \mathrm{e}^{i\omega t} + \text{c.c.} \right) \Bigg]. \qquad (6.85)
\end{aligned}$$

6.3.4 *The Reduced Dynamics and Invariant Manifold*

Now applying the RG/E equation

$$\left. \frac{\partial \boldsymbol{u}}{\partial t} \right|_{t_0 = t} = 0 \qquad (6.86)$$

to (6.85), we have

$$\epsilon \frac{\partial C(t, \boldsymbol{r})}{\partial t} \boldsymbol{U}_\omega - \epsilon^3 \left\{ \frac{1}{2} \mathcal{W}_1(t, \boldsymbol{r}) + \hat{D} \nabla^2 C(t, \boldsymbol{r}) \right\} \boldsymbol{U}_\omega = 0, \qquad (6.87)$$

where terms of the order $O(\epsilon^4)$ or higher are neglected. Taking the inner product with $\boldsymbol{U}_\omega^*$, we have the following complex Ginzburg-Landau equation

$$\frac{\partial C(t, \boldsymbol{r})}{\partial t} - \bar{D} \nabla^2 C(t, \boldsymbol{r}) = \frac{1}{2} \alpha C(t, \boldsymbol{r}) + \frac{1}{2} \beta |C(t, \boldsymbol{r})|^2 C(t, \boldsymbol{r}), \qquad (6.88)$$

where

$$\bar{D} = (\boldsymbol{U}_\omega^*, \hat{D} \boldsymbol{U}_\omega) = \frac{1}{2} (D_X + D_Y) - i \frac{A}{2} (D_X - D_Y), \qquad (6.89)$$

and α and β are defined by (6.76). Note that β is a complex number, and so is $\bar{D}$ when $D_X \neq D_Y$.

The attractive manifold is given by the initial value as

$$
\begin{aligned}
\boldsymbol{u}(t, \boldsymbol{r}) = \boldsymbol{W}(t, \boldsymbol{r}) &= \epsilon \boldsymbol{W}_1(t, \boldsymbol{r}) + \epsilon^2 \boldsymbol{W}_2(t, \boldsymbol{r}) + \epsilon^3 \boldsymbol{W}(t, \boldsymbol{r}) \\
&= \epsilon \left[C(t, \boldsymbol{r}) \boldsymbol{U}_\omega \mathrm{e}^{i\omega t} + \text{c.c.} \right] + \epsilon^2 \left[(C^2(t, \boldsymbol{r}) \boldsymbol{V}_+ \mathrm{e}^{2i\omega t} + \text{c.c.}) + |C(t, \boldsymbol{r})|^2 \boldsymbol{V}_0 \right] \\
&+ \epsilon^3 \left[\left\{ \frac{\mathcal{W}_1(t, \boldsymbol{r})}{4i\omega} \bar{\boldsymbol{U}}_\omega \mathrm{e}^{i\omega t} + \frac{\mathcal{W}_3(t, \boldsymbol{r})}{4i\omega} (\boldsymbol{U}_\omega + \frac{1}{2} \bar{\boldsymbol{U}}_\omega) \mathrm{e}^{3i\omega t} \right\} + \text{c.c.} \right].
\end{aligned}
\tag{6.90}
$$

These results coincide with those obtained in the reductive perturbation method [30]. In contrast to the previous methods in which identification of the solvability condition for avoiding the appearance of secular terms is essential, the present method has no such adhoc procedure because secular terms are allowed to appear, which are to be renormalized away by means of the RG/E equation mechanically.

6.4 Example with a Jordan Cell I: Extended Takens Equation

As an example for the dynamical system with a linear operator of a Jordan cell structure, let us take an extended Takens equation [101] with three-degrees of freedom, which reads

$$
\dot{x} = y + \epsilon a x^2, \quad \dot{y} = \epsilon b x^2, \quad \dot{z} = -z + \epsilon f(x, y, z), \tag{6.91}
$$

where $f(x, y, z)$ is analytic function of (x, y, z). It will be shown that the seemingly ad hoc condition (5.149), which becomes relevant only when the system has more than two-degrees of freedom, gives a restriction to the form of $f(x, y, z)$.

Let us convert Eq. (6.91) to a system with the definition $\boldsymbol{u} = {}^t(x, y, z)$;

$$
\left(\frac{d}{dt} - A \right) \boldsymbol{u} = \epsilon \boldsymbol{F}(\boldsymbol{u}), \tag{6.92}
$$

where

$$
A = \begin{pmatrix} 0 & 1 & 0 \\ 0 & 0 & 0 \\ 0 & 0 & -1 \end{pmatrix}, \quad \boldsymbol{F}(\boldsymbol{u}) = \begin{pmatrix} a x^2 \\ b x^2 \\ f(x, y, z) \end{pmatrix}. \tag{6.93}
$$

Notice that A has a two-dimensional Jordan cell;

$$
A \boldsymbol{U}_1 = 0, \quad A \boldsymbol{U}_2 = \boldsymbol{U}_1, \quad A \boldsymbol{U}_3 = -\boldsymbol{U}_3, \tag{6.94}
$$

where

$$\boldsymbol{U}_1 = \begin{pmatrix} 1 \\ 0 \\ 0 \end{pmatrix} = \tilde{\boldsymbol{U}}_2, \quad \boldsymbol{U}_2 = \begin{pmatrix} 0 \\ 1 \\ 0 \end{pmatrix} = \tilde{\boldsymbol{U}}_1, \quad \boldsymbol{U}_3 = \begin{pmatrix} 0 \\ 0 \\ 1 \end{pmatrix} = \tilde{\boldsymbol{U}}_3. \tag{6.95}$$

Here $\tilde{\boldsymbol{U}}_i$ $(i = 1, 2, 3)$ are the adjoint eigenvectors and a pair of a cell;

$$A^\dagger \tilde{\boldsymbol{U}}_1 = 0, \quad A^\dagger \tilde{\boldsymbol{U}}_2 = \tilde{\boldsymbol{U}}_1, \quad A^\dagger \tilde{\boldsymbol{U}}_3 = -\tilde{\boldsymbol{U}}_3. \tag{6.96}$$

The projection operators onto the P and Q subspaces are given by

$$P = |\boldsymbol{U}_1\rangle\langle\tilde{\boldsymbol{U}}_2| + |\boldsymbol{U}_2\rangle\langle\tilde{\boldsymbol{U}}_1| = \mathrm{diag}(1, 1, 0), \tag{6.97}$$

$$Q = 1 - P = \mathrm{diag}(0, 0, 1), \tag{6.98}$$

respectively.

We are interested in the asymptotic behavior of the solution at $t \to \infty$. The solution to (6.92) around $t \sim t_0$ with the 'initial' value $\boldsymbol{W}(t_0)$ at $t = t_0$ is denoted by $\boldsymbol{u}(t; t_0)$, which is expanded as

$$\boldsymbol{u} = \boldsymbol{u}_0 + \epsilon \boldsymbol{u}_1 + \epsilon^2 \boldsymbol{u}_2 + \cdots, \tag{6.99}$$

together with the 'initial' value $\boldsymbol{W} = \boldsymbol{W}_0 + \epsilon \boldsymbol{W}_1 + \epsilon^2 \boldsymbol{W}_2 + \cdots$.

The lowest order equation reads

$$(d_t - A)\boldsymbol{u}_0 = 0, \tag{6.100}$$

the solution to which is expressed in the asymptotic region as

$$\boldsymbol{u}_0(t; t_0) = C_1^{(0)}(t_0)\boldsymbol{U}_1 = {}^t(C_1^{(0)}(t_0), 0, 0). \tag{6.101}$$

According to the general formulation given in the last section, the quantities which we only have to evaluate are as follows,

$$C_1^{(F)} = \langle\tilde{\boldsymbol{U}}_2, \boldsymbol{F}(\boldsymbol{u}_0)\rangle = a(C_1^{(0)})^2, \tag{6.102}$$

$$C_2^{(F)} = \langle\tilde{\boldsymbol{U}}_1, \boldsymbol{F}(\boldsymbol{u}_0)\rangle = b(C_1^{(0)})^2, \tag{6.103}$$

$$A^{-1} Q \boldsymbol{F}(\boldsymbol{u}_0) = -f(\boldsymbol{u}_0)\boldsymbol{U}_3. \tag{6.104}$$

Then the solution to the first order equation $(d_t - A)\boldsymbol{u}_1 = \boldsymbol{F}(\boldsymbol{u}_0)$ reads

$$\begin{aligned} \boldsymbol{u}_1(t; t_0) = &\{a(C_1^{(0)})^2(t - t_0) + C_2^{(1)}(t - t_0) + \frac{1}{2} b(C_1^{(0)})^2(t - t_0)^2\}\boldsymbol{U}_1 \\ &+ \{b(C_1^{(0)})^2(t - t_0) + C_2^{(1)}\}\boldsymbol{U}_2 + f(\boldsymbol{u}_0)\boldsymbol{U}_3. \end{aligned} \tag{6.105}$$

If we stop at this order, the full solution is given by $\boldsymbol{u} = \boldsymbol{u}_0 + \epsilon \boldsymbol{u}_1$. Applying the RG/E equation, we have

$$\dot{C}_1^{(0)} = \epsilon(a(C_1^{(0)})^2 + C_2^{(1)}), \quad \dot{C}_2^{(1)} = b(C_1^{(0)})^2. \tag{6.106}$$

And the trajectory on the manifold M$\simeq$M$_1$ is given by

$$\boldsymbol{u}(t) = C_1^{(0)}(t)\boldsymbol{U}_1 + \epsilon C_2^{(1)}(t)\boldsymbol{U}_2 + \epsilon f(C_1^{(0)}(t), 0, 0)\boldsymbol{U}_3. \tag{6.107}$$

Let us proceed to the second-order analysis. We first examine the necessary condition (5.149) for the reduction, which reads in the present case

$$0 = Q\boldsymbol{F}'(\boldsymbol{u}_0)\boldsymbol{U}_1 = Q\begin{pmatrix} 2aC_1^{(0)} & 0 & 0 \\ 2bC_1^{(0)} & 0 & 0 \\ \frac{\partial f}{\partial x} & \frac{\partial f}{\partial y} & \frac{\partial f}{\partial z} \end{pmatrix}\boldsymbol{U}_1 = Q\begin{pmatrix} 2aC_1^{(0)} \\ 2bC_1^{(0)} \\ \frac{\partial f}{\partial x} \end{pmatrix} = \left.\frac{\partial f(x, 0, 0)}{\partial x}\right|_{x=C_1^{(0)}}\boldsymbol{U}_3.$$

This demands that $f(x, y, z)$ should not depend on x when $y = z = 0$. For instance, the function $f(x, y, z) = x^3y + xz^2 + g(y, z)$ will do.

With this condition taken for granted, let us analyze the RG/E equation in the second order. We first notice the following relations,

$$\boldsymbol{h} = -A^{-1}Q\boldsymbol{F}(\boldsymbol{u}_0) + C_2^{(1)}\boldsymbol{U}_2 = f(\boldsymbol{u}_0)\boldsymbol{U}_3 + C_2^{(1)}\boldsymbol{U}_2 = \begin{pmatrix} 0 \\ C_2^{(1)} \\ f(\boldsymbol{u}_0) \end{pmatrix} \tag{6.108}$$

and hence

$$\boldsymbol{F}'(\boldsymbol{u}_0)\boldsymbol{h} = \left.\begin{pmatrix} 2aC_1^{(0)} & 0 & 0 \\ 2bC_1^{(0)} & 0 & 0 \\ 0 & \frac{\partial f}{\partial y} & \frac{\partial f}{\partial z} \end{pmatrix}\right|_{\boldsymbol{u}=\boldsymbol{u}_0}\begin{pmatrix} 0 \\ C_2^{(1)} \\ f(\boldsymbol{u}_0) \end{pmatrix} = \begin{pmatrix} 0 \\ 0 \\ \frac{\partial f}{\partial y}C_2^{(1)} + \frac{\partial f}{\partial z}f(\boldsymbol{u}_0) \end{pmatrix}, \tag{6.109}$$

which implies that

$$\langle\tilde{\boldsymbol{U}}_2|\boldsymbol{F}'(\boldsymbol{u}_0)\boldsymbol{h}\rangle = 0, \quad \langle\tilde{\boldsymbol{U}}_1|\boldsymbol{F}'(\boldsymbol{u}_0)\boldsymbol{h}\rangle = 0. \tag{6.110}$$

Thus we see that the RG/E equation has no corrections in this order.

To obtain the second-order correction to the trajectory, we need to evaluate the following;

$$-A^{-1}Q\boldsymbol{F}'\boldsymbol{h} = (C_2^{(1)}\frac{\partial f}{\partial y} + f(\boldsymbol{u}_0)\frac{\partial f}{\partial z})\boldsymbol{U}_3, \quad -A^{-2}Q\boldsymbol{F}'\boldsymbol{U}_2 = -\frac{\partial f}{\partial y}\boldsymbol{U}_3, \tag{6.111}$$

where the derivatives are evaluated at $\boldsymbol{u} = \boldsymbol{u}_0 = {}^t(C_1^{(0)}(t), 0, 0)$ as before. Thus the second-order correction of the 'initial' value reads

$$W_2(t_0) = (f(u_0)\frac{\partial f}{\partial z} + C_2^{(1)}\frac{\partial f}{\partial y} - \langle \tilde{U}_1|F\rangle\frac{\partial f}{\partial y})U_3. \tag{6.112}$$

Hence the trajectory in the second-order approximation is given by

$$u(t) = W(t) = C_1^{(0)}(t)U_1 + \epsilon C_2^{(1)}(t)U_2 + \epsilon f(u_0)U_3 \\ +\epsilon^2\left[f(u_0)\frac{\partial f}{\partial z} + C_2^{(1)}\frac{\partial f}{\partial y} - \langle \tilde{U}_1|F\rangle\frac{\partial f}{\partial y}\right]_{u=u_0} U_3. \tag{6.113}$$

Here $C_1^{(0)}(t)$ and $C_2^{(1)}(t)$ are governed by the Takens equation (6.106). We see that the higher order terms does not affect the dynamics but modifies the trajectory only in the U_2 and the Q-direction.

6.5 Example with a Jordan Cell II: The Asymptotic Speed of a Pulse Given in the Benney Equation

The Kortweg-de Vries (KdV) equation [102] itself can be derived as an asymptotic equation for describing the weakly nonlinear shallow water waves by Kortweg and de Vries and as the long-wave length limit of the one-dimensional non-linear coupled oscillators as investigated by Fermi, Pasta, Ulam and Tsingou [102, 103]. It reads

$$\frac{\partial u}{\partial t} + 6u\frac{\partial u}{\partial x} + \frac{\partial^3 u}{\partial x^3} = 0. \tag{6.114}$$

The KdV equation is integrable and admits one-pulse solution given by

$$u(x,t) = \frac{c}{2}\mathrm{sech}^2\left[\frac{\sqrt{c}(x-ct)}{2}\right] =: \varphi(x-ct;c), \tag{6.115}$$

where the velocity c is an arbitrary constant. We are interested in what occurs when the remaining few parts G of the derivative expansion are added to it;

$$\partial_t u + 6u\partial_x u + \partial_x^3 u + \epsilon F(\partial_x u) = 0, \quad F(u) := \partial_x^2 u + \partial_x^4 u, \tag{6.116}$$

where

$$\partial_t u := \frac{\partial u}{\partial t}, \quad \partial_x u := \frac{\partial u}{\partial x} \tag{6.117}$$

and so on. Equation (6.116) was first derived and studied by Benney [36] and bears his name.

To study the problem, it is found convenient to change the coordinate to that moving with the pulse. With the change of the independent variables to

$$t = t, \quad z = x - ct, \tag{6.118}$$

the partial derivatives are converted as

$$\partial_t = \partial_t, \quad \partial_x = \partial_z \tag{6.119}$$

and hence the KdV and Benney equations get to have the following forms

$$\partial_t u + I[u] = 0, \quad I[u] = -c\partial_z u + 6u\partial_z u + \partial_z^3 u, \tag{6.120}$$

and

$$\partial_t u + I[u] + \epsilon F(u) = 0, \tag{6.121}$$

respectively. We remark that the following relation holds,

$$I[\varphi(z - b; c)] = 0, \tag{6.122}$$

for an arbitrary constant b.

Making the perturbative expansion $u = u_0 + \epsilon u + \ldots$, we try to solve the Benney equation (6.121) around $t \sim t_0$. The zeroth order equation is given by

$$\partial_t u_0 + I[u_0] = 0, \tag{6.123}$$

for which we take one-soliton solution with the velocity c located around $z \sim b(t_0)$ as given by

$$u_0(z, t; t_0, b(t_0), c(t_0)) = \varphi(z - b(t_0); c(t_0)) \tag{6.124}$$

where it is made explicit that the velocity as well as the position of the soliton may be t_0 dependent.

The first-order equation reads

$$\partial_t u_1 = -\frac{\partial I}{\partial u}\bigg|_{u=u_0} u_1 + F(u_0) \equiv Au_1 + F(u_0). \tag{6.125}$$

Here we have defined the linear operator

$$A \equiv c\partial_z - \partial_z^3 - 6(\partial_z u_0 + u_0\partial_z). \tag{6.126}$$

It is noteworthy that A has a two-dimensional Jordan cell with a zero eigenvalue;

$$AU_1 = 0, \quad AU_2 = U_1, \tag{6.127}$$

where

$$U_1 = \partial_z u_0 = \partial_z \varphi(z - b(t_0); c(t_0)), \quad U_2 = -\partial_c u_0 = -\partial_c \varphi(z - b(t_0); c)|_{c=c(t_0)}. \tag{6.128}$$

The adjoint operator $A^\dagger$ is given by

$$A^\dagger = -c\partial_z + \partial_z^3 + 6u_0\partial_z, \tag{6.129}$$

which has also a Jordan cell structure

$$A^\dagger \tilde{U}_1 = 0, \quad A^\dagger \tilde{U}_2 = \tilde{U}_1, \tag{6.130}$$

where $\tilde{U}_1$ is identified with u_0

$$\tilde{U}_1 = u_0(z, t; t_0, b(t_0), c(t_0)). \tag{6.131}$$

The following orthogonality relations hold:

$$\langle \tilde{U}_1 | U_1 \rangle = \langle \tilde{U}_2 | U_2 \rangle = 0. \tag{6.132}$$

We call the subspace spanned by U_1 and U_2 the P space, while the complement of the P space called the Q space, and their respective projection operators are given by

$$P = \frac{|U_1\rangle\langle\tilde{U}_2|}{\langle\tilde{U}_2|U_1\rangle} + \frac{|U_2\rangle\langle\tilde{U}_1|}{\langle\tilde{U}_1|U_2\rangle}, \quad Q = \mathbf{1} - P. \tag{6.133}$$

The solution to (6.125) is given by

$$u_1(t; t_0, b(t_0), c(t_0)) = \mathrm{e}^{(t-t_0)A} W_1(t_0) + \int_{t_0}^{t} ds\, \mathrm{e}^{(t-s)A} F(u_0). \tag{6.134}$$

Here the 'initial' value may be set to be independent to U_1 and hence given by

$$W_1(t_0) = Q W_1(t_0). \tag{6.135}$$

A comment is in order here: The possible term $C_2^{(1)}(t_0)U_2$ in $W_1(t_0)$ is intentionally omitted here because its effect may be renormalized into the t_0 dependence of the velocity $c(t_0)$ in the zero-th solution.

The second term of (6.134) can be evaluated as follows: First we note that $F(u_0)$ in the integrand is independent of time s, and hence we can apply the formulae (5.106) to get

$$\begin{aligned}
e^{(t-s)A}F &= e^{tA}(P+Q)F = e^{(t-s)A}(C_1^{(F)}(t_0)U_1 + C_2^{(F)}(t_0)U_2) + e^{(t-s)A}QF \\
&= \left\{C_1^{(F)}(t_0) + C_2^{(F)}(t_0)(t-s)\right\} U_1 + C_2^{(F)}(t_0)U_2 + e^{(t-s)A}QF,
\end{aligned} \tag{6.136}$$

where

$$C_1^{(F)}(t_0) = \frac{\langle\tilde{U}_2|F\rangle}{\langle\tilde{U}_2|U_1\rangle}, \quad C_2^{(F)}(t_0) = \frac{\langle\tilde{U}_1|F\rangle}{\langle\tilde{U}_1|U_2\rangle}. \tag{6.137}$$

Then using the formulae (5.125) we see (6.134) is evaluated to be

$$\begin{aligned}
u_1(t;t_0) &= e^{(t-t_0)A}[QW_1(t_0) + A^{-1}QF] + \{C_1^{(F)}(t-t_0) + C_2^{(F)}\frac{1}{2}(t-t_0)^2\}U_1 \\
&\quad + C_2^{(F)}(t_0)(t-t_0)U_2 - A^{-1}QF,
\end{aligned} \tag{6.138}$$

where we have suppressed the terms proportional to $(t-t_0)^{n\geq 2}$ which do not contribute the RG/E equation nor the invariant manifold, as before. To avoid the appearance of the rapid mode, the Q space part of the 'initial' value is determined as

$$QW_1(t_0) = -A^{-1}QF. \tag{6.139}$$

Thus we arrive at the perturbative solution to the Benney equation up to the second order as

$$\begin{aligned}
u(z,t;t_0) &= u_0(z,t;t_0) + \epsilon\{C_1^{(F)}(t-t_0) + C_2^{(F)}\frac{1}{2}(t-t_0)^2\}U_1 \\
&\quad + \epsilon\{C_2^{(F)}(t_0)(t-t_0)U_2 - A^{-1}QF\}, \\
&= \varphi(z-b(t_0);c) + \epsilon\{C_1^{(F)}(t-t_0) + C_2^{(F)}\frac{1}{2}(t-t_0)^2\}U_1 \\
&\quad + \epsilon\{C_2^{(F)}(t_0)(t-t_0)U_2 - A^{-1}QF\}.
\end{aligned} \tag{6.140}$$

Then applying the RG/E equation, we have

$$\begin{aligned}
0 = \left.\frac{\partial u(z,t;t_0)}{\partial t_0}\right|_{t_0=t} &= -\dot{b}\partial_z\varphi(z-b(t);c) + \dot{c}\partial_c\varphi(z-b(t);c) \\
&\quad - \epsilon\left.\left\{C_1^{(F)}(t))U_1 + C_2^{(F)}(t)U_2\right\}\right|_{t_0=t}.
\end{aligned} \tag{6.141}$$

Since $\partial_z\varphi = U_1$ and $\partial_c\varphi = -U_2$ with $t_0 = t$ on account of (6.128), we have the following coupled equation

$$\dot{b} = -\epsilon C_1^{(F)}(t), \quad \dot{c}(t) = -C_2^{(F)}(t). \tag{6.142}$$

Here the higher-order terms of ϵ^n $(n \geq 2)$ are neglected. The asymptotically global solution is given as the envelope function

$$u_E(z,t) = u(z,t;t\_0 = t) = \varphi(z - b(t); c(t)) \; - \; \epsilon A^{-1} Q F. \qquad (6.143)$$

As an example, let us calculate the asymptotic velocity the pulse $c(t \to \infty) \equiv c^*$ for which $\dot{c} = 0$, or

$$\begin{aligned} 0 = C_2^{(F)} \propto \langle \tilde{U}_1 | F \rangle &= \int_{-\infty}^{\infty} dz \, u_0 \left(\frac{\partial^2 u_0}{\partial z^2} + \frac{\partial^4 u_0}{\partial z^4} \right) \\ &= -\int_{-\infty}^{\infty} dz \left[\left(\frac{\partial u_0}{\partial z} \right)^2 - \left(\frac{\partial^2 u_0}{\partial z^2} \right)^2 \right] \\ &= -\frac{c^2\sqrt{c}}{8} \left[\int_{-\infty}^{\infty} d\zeta \left(\frac{d}{d\zeta} \text{sech}^2 \zeta \right)^2 - \frac{c}{4} \int_{-\infty}^{\infty} d\zeta \left(\frac{d^2}{d\zeta^2} \text{sech}^2 \zeta \right)^2 \right]. \end{aligned} \qquad (6.144)$$

Then using the following formulae

$$\left(\frac{d}{d\zeta} \text{sech}^2 \zeta \right)^2 = 4(\text{sech}^4 \zeta \; - \; \text{sech}^6 \zeta), \qquad (6.145)$$

$$\left(\frac{d^2}{d\zeta^2} \text{sech}^2 \zeta \right)^2 = 4(2\text{sech}^2 \zeta \; - \; 3\text{sech}^4 \zeta)^2, \qquad (6.146)$$

$$\int_{-\infty}^{\infty} d\zeta \; \text{sech}^{2n} \zeta = 2 \int_0^1 du \, (1 - u^2)^{n-1} = 2 \frac{(2n-2)!!}{(2n-1)!!}, \qquad (6.147)$$

we get

$$\langle \tilde{U}_1 | F \rangle = -\frac{c^2\sqrt{c}}{8} \left(\frac{16}{15} - \frac{c}{4} \frac{64}{21} \right) = \frac{2c^2\sqrt{c}}{21} \left(c - \frac{7}{5} \right). \qquad (6.148)$$

Thus the stationary condition $C_2^{(F)} = 0$ determine the asymptotic velocity as

$$c^* = \frac{7}{5}, \qquad (6.149)$$

which agrees with the result given in other methods [104].

We note that the RG method is also applied to the pulse interaction described by the KdV equation in [6], where the same linear operator with a Jordan cell appears. The present analysis of the Benney equation based on the RG method has been first shown in this monograph [105].

Chapter 7
RG Method Applied to Stochastic Equations

7.1 Introduction

In this chapter, after a brief historical and introductory account of the celebrated Brownian motion [70, 106–109] and the Langevin equation [70–72], the Fokker-Planck (F-P) equation [77, 78] is derived as a reduced equation in a coarse-grained invariant manifold [34, 110] spanned by the averaged distribution function from a generic Langevin equation with time-reversal invariance on the basis of the RG method [1, 3–6, 38, 39, 48–52, 92–96] in the way as given in [57].

In derivation of kinetic or transport equations for describing the nonequilibrium properties of a physical system the following two basic ingredients are commonly seen [32, 54, 111–113]:

(a) The reduced dynamics is characterized with a longer time scale than that appearing in the original (microscopic) evolution equation, and
(b) the reduced dynamics is described by a time-irreversible equation even when the original microscopic equation is time-reversible,

which are interrelated with but relatively independent of each other. For instance, the derivation of the Boltzmann equation [53] by Bogoliubov [54] shows that the dilute-gas dynamics as a *dynamical system* with many-degrees of freedom has an *attractive/invariant manifold* [34, 110] spanned by the one-particle distribution function in the asymptotic regime after a long time; the Boltzmann equation in turn can be further reduced to the fluid dynamic equation (Euler/Navier-Stokes equations) by a perturbation theory like Chapman-Enskog method [113] or Bogoliubov's method [25, 54] and the RG method as is extensively discussed in the subsequent chapters of this monograph.

Similarly, the Langevin equation which may be time-reversible can be reduced to the time-irreversible F-P equation with a longer time scale than the scale in Langevin equation [74–76].

In the derivation of the F-P equation based on the RG method, the averaged distribution functions [114, 115] will be prepared as the integral constants of the solution of microscopic evolution equations. Then, as was the case with other examples treated in the previous chapters, the RG/E equation will lift the integral constants

T. Kunihiro et al., *Geometrical Formulation of Renormalization-Group Method as an Asymptotic Analysis*, Fundamental Theories of Physics 206,
https://doi.org/10.1007/978-981-16-8189-9_7

in the unperturbed solution to the dynamical variables, which are slow variables and the reduced dynamical equation is expressed solely in terms of the lifted integral constants.

It often occurs that a hierarchy in the time scales still remains within the dynamical variables in the kinetic/transport equations thus obtained [32, 74–76, 112]. For instance, the F-P equation may be further reduced to, say, a slow Fokker-Planck equation asymptotically after a long time [74–76]. In that case, an adiabatic elimination of (relatively) fast variables in the Langevin or the F-P equations is possible. In the second half of this chapter, we shall show that the RG method as developed in Chap. 5 provides us with a powerful method for an adiabatic elimination of fast variables from the F-P equation [57].

7.2 Langevin Equation: Simple Examples

As an example of a stochastic motion, we first consider Brownian motion of a tiny particle popping out of a pollen, with a mass m immersed and moving in water at room temperature. A biologist Robert Brown [106] observed in 1826–1827 that the tiny particles show rapid and random motions as if it were alive, which was later disproved by himself. In 1905, Einstein [107] proposed a systematic theory for describing such particles, Brownian particles, although Einstein himself was not so enthusiastic in directly relating his particles to Brownian particles because experimental results were still not so clear as those presented by Perrin later in 1910 [109]: Einstein argued in [107] that such a motion could be an evidence of the existence of molecules or atoms. Incidentally, it is amazing, as is emphasized by Gardiner [74], that the paper by Einstein contains almost all the pivotal theories that now constitute the essential ingredients of the nonequilibrium statistical mechanics which includes Chapman-Kolmogorov equation, a simple Fokker-Planck equation, a typical fluctuation-dissipation theorem and so on. Later the motion of the Brownian particle was nicely given a mathematical formulation by P. Langevin [70], who showed that the correct application of the theory by Smoluchowski [108] gives the same result as that given by Einstein: In his paper, Langevin introduced a novel differential equation, a stochastic equation , which is now called the Langevin equation , although the standard formulation of the Langevin equation with an explicit incorporation of the correlators of the noises, which we shall follow here, seems to have been first given by Uhlenbeck and Ornstein [71].

For simplicity, let us consider the one-dimensional case and assume that the particle moves along the x axis:

$$\begin{cases} \frac{dx}{dt} = v, \\ m\frac{dv}{dt} = -\mu v + R(t), \end{cases} \tag{7.1}$$

where μ denotes the friction coefficient which comes from the averaging effect of the collisions of the water molecules with the Brownian particle, and $R(t)$ represents the remaining fluctuations that can not be taken into account by the averaging which is nicely represented by the frictional force. If the Brownian particle is a sphere with radius a, then Stokes's law tells us that $\mu = 6\pi a\eta$ with η being the shear viscosity of the water. Einstein used the value $\eta = 1.35 \times 10^{-3}\,\mathrm{Pa}\cdot\mathrm{s}\,(\mathrm{Pa} = \mathrm{kg}/(\mathrm{m}\cdot\mathrm{s}^2))$, while it takes the value $1.002 \times 10^{-3}\,\mathrm{kg}/(\mathrm{m}\cdot\mathrm{s})$ for water at $T = 20\,^\circ\mathrm{C}$. The number density of the water in which the Brownian particles are suspended may be identified with the number of the molecules contained in the volume of $1\,\mathrm{m}^3$, the weight of which roughly amounts to 10^6 g at $T = 300$ K, while 1 mol of water is composed of $N_0 \simeq 6 \times 10^{23}$ molecules and weighs 18 g. Thus the number density of the water is estimated to be

$$n = N_0 \times \frac{10^6}{18} = 3.3 \times 10^{28}\,(/\mathrm{m}^3). \tag{7.2}$$

The average velocity of the water molecule in one direction is estimated to be $\bar{u} \simeq 300\,\mathrm{m/s}$. Let the radius a of the Brownian particle be 1×10^{-6} m. Then the number of the collisions of the water molecule with a Brownian particle is roughly[1]

$$N_{\mathrm{coll}} \simeq \frac{1}{2}\bar{u}\pi a^2 n \simeq 1.6 \times 10^{19}/\mathrm{s}, \tag{7.3}$$

which implies that the time duration between subsequent collisions of water molecules and a Brownian particle is as tiny as 10^{-19} s.

It will be found that the relaxation time is given by $\tau_r = 1/\gamma$, which may be as short as 10^{-8} s for a Brownian particle with a mass $\sim 10^{-15}$ kg but is also long enough so that the number of collisions of the water molecules with the Brownian particle within the relaxation time is as huge as 10^{11}. Thus the fluctuations of the force, $R(t)$, may be treated as a stochastic variable with, say, 10^{11} samples, and affects the Brownian particle as background noise. Then it should be noted that the dynamical variables x and v governed by $R(t)$ also become stochastic variables.

The average of $R(t)$ over such samples should vanish by its definition as

$$\langle R(t)\rangle_R = 0. \tag{7.4}$$

This treatment of the residual force $R(t)$ may implicitly imply that the infinitesimal time step Δt assumed in (7.1) is actually much larger than the each interval of the collisions of the water molecules with the Brownian particle. We further assume that the noises in different times are independent, which means that the time-correlation function of the noises is proportional to the delta function;

$$\langle R(t_1)R(t_2)\rangle_R = 2D\delta(t_1 - t_2). \tag{7.5}$$

[1] The number is claimed to be as larger as 10^{21} in [72] with no reasoning provided.

Since the spectral function of such a correlation function is flat and does not peaks, such a noise is called 'white'. We remark also that the noise is **stationary**, *i.e.*, invariant under the shift of time

$$t_i \to t_i + t' \quad (\forall t'), \tag{7.6}$$

in accordance with the Brownian particles are immersed in the 'water' in thermal equilibrium.

Since the second equation of (7.1) is an inhomogeneous but linear equation, it can be formally solved as follows. The solution to the homogeneous equation $\dot{v} = -(\mu/m) \cdot v$ reads

$$v(t) = C\mathrm{e}^{-\gamma t} \quad \text{with} \quad \gamma = \frac{\mu}{m}. \tag{7.7}$$

Assuming that C is now time dependent and inserting the homogeneous solution to (7.1), we have $\dot{C} = \frac{1}{m}\mathrm{e}^{\gamma t} R(t)$, the solution to which is given by

$$C(t) = \frac{1}{m} \int_0^t dt' \, \mathrm{e}^{\gamma t'} R(t') + v_0 \tag{7.8}$$

with v_0 being the integral constant. Thus the solution to (7.1) is found to be

$$v(t) = v_0 \mathrm{e}^{-\gamma t} + \frac{1}{m} \int_0^t dt' \, \mathrm{e}^{-\gamma (t-t')} R(t'), \tag{7.9}$$

with the 'initial' condition $v(0) = v_0$.

With the 'initial' velocity v_0 fixed, let us take the average over $R(t')$. Noting that $R(t')$ for $t' > 0$ is independent of v_0 and accordingly

$$\langle R(t') \rangle_R = 0 \quad \text{for } t' > 0, \tag{7.10}$$

we have

$$\langle v(t) \rangle_R = v_0 \mathrm{e}^{-\gamma t} + \frac{1}{m} \int_0^t dt' \, \mathrm{e}^{-\gamma (t-t')} \langle R(t') \rangle_R = v_0 \mathrm{e}^{-\gamma t}. \tag{7.11}$$

On the other hand, the time-correlation function for $v(t_1)$ and $v(t_2)$ with $t_i > 0$ ($i = 1, 2$) is evaluated as follows,

$$\begin{aligned} \langle v(t_1) v(t_2) \rangle_R &= \mathrm{e}^{-\gamma (t_1 + t_2)} \left[v_0^2 + \frac{1}{m^2} \int_0^{t_1} ds_1 \int_0^{t_2} ds_2 \, \mathrm{e}^{\gamma (s_1 + s_2)} \langle R(s_1) R(s_2) \rangle_R \right] \\ &= \mathrm{e}^{-\gamma (t_1 + t_2)} \left[v_0^2 + \frac{2D}{m^2} \int_0^{t_1} ds_1 \int_0^{t_2} ds_2 \, \mathrm{e}^{\gamma (s_1 + s_2)} \delta(s_1 - s_2) \right]. \end{aligned} \tag{7.12}$$

By considering the cases for $t_1 > t_2$ and $t_2 > t_1$, separately, the second term is found to take the form

$$\frac{D}{\gamma m^2}\left(e^{2\gamma t_<} - 1\right), \tag{7.13}$$

where $t_<$ denotes the smaller one of t_1 and t_2. Thus after some simple manipulations, we arrive at

$$\langle v(t_1)v(t_2)\rangle_R = v_0^2 e^{-\gamma(t_1+t_2)} + \frac{D}{\gamma m^2}\left(e^{-\gamma|t_1-t_2|} - e^{-\gamma(t_1+t_2)}\right), \tag{7.14}$$

which tends to

$$\langle v(t_1)v(t_2)\rangle_R \xrightarrow{1/\gamma \ll t,t'} \frac{D}{\gamma m^2} e^{-\gamma|t_1-t_2|}. \tag{7.15}$$

For

$$\gamma^{-1} \ll t_1 = t_2 =: t, \tag{7.16}$$

the average of the kinetic energy of the Brownian particle is found to be

$$\left\langle \frac{1}{2} m v^2(t) \right\rangle_R = \frac{D}{2\gamma m} = \frac{1}{2} k_B T, \tag{7.17}$$

where the *equipartition law* has been assumed in the last equality. Thus we have

$$\frac{D}{k_B T} = m\gamma = \mu, \tag{7.18}$$

which is the fluctuation-dissipation relation first derived by A. Einstein [107].

Now let us represent the dynamical variables $x(t) =: u_1(t)$ and $v =: u_2(t)$ collectively by

$$\boldsymbol{u}(t) := \begin{pmatrix} x(t) \\ v(t) \end{pmatrix} = \begin{pmatrix} u_1(t) \\ u_2(t) \end{pmatrix}, \tag{7.19}$$

and introduce the stochastic distribution function [114, 115]

$$f(\boldsymbol{x}, t) := \delta(\boldsymbol{u}(t) - \boldsymbol{x}), \tag{7.20}$$

where $\boldsymbol{u}(t)$ satisfies the stochastic differential equation

$$\frac{d\boldsymbol{u}}{dt} = \boldsymbol{h}(\boldsymbol{u}) + \hat{\boldsymbol{g}}\boldsymbol{R}(t), \tag{7.21}$$

with

$$h(u) := \begin{pmatrix} u_2(t) \\ -\gamma u_2 \end{pmatrix}, \quad \hat{g} := \begin{pmatrix} 0 & 0 \\ 0 & m^{-1} \end{pmatrix}, \quad \boldsymbol{R}(t) := \begin{pmatrix} 0 \\ R(t) \end{pmatrix}. \tag{7.22}$$

Then we have the following continuity equation as will be shown below:

$$\frac{\partial f(\boldsymbol{x}, t)}{\partial t} = -\nabla_x \cdot [(\boldsymbol{h}(\boldsymbol{x}) + \hat{g}(\boldsymbol{x})\boldsymbol{R}) f(\boldsymbol{x}, t)], \tag{7.23}$$

where we have written down when $\hat{g}$ can be a function of the dynamical variables $\boldsymbol{u}(t)$, *i.e.*, the case of the multiplicative noise [74–76]. This continuity equation (7.23) for the stochastic distribution function is called Kubo's stochastic Liouville equation. Equation (7.23) can be derived with use of the definition (7.20), as follows,

$$\begin{aligned}
\frac{\partial f(\boldsymbol{x}, t)}{\partial t} &= \frac{\partial}{\partial t}\delta(\boldsymbol{u}(t) - \boldsymbol{x}) = \dot{\boldsymbol{u}}(t) \cdot \frac{\partial}{\partial \boldsymbol{u}(t)}\delta(\boldsymbol{u}(t) - \boldsymbol{x}) \\
&= -(\boldsymbol{h}(\boldsymbol{u}(t)) + \hat{g}(\boldsymbol{u}(t))\boldsymbol{R}(t)) \cdot \frac{\partial}{\partial \boldsymbol{x}}\delta(\boldsymbol{u}(t) - \boldsymbol{x}) \\
&= -\nabla_x \left[(\boldsymbol{h}(\boldsymbol{u}(t)) + \hat{g}(\boldsymbol{u}(t))\boldsymbol{R}(t))\delta(\boldsymbol{u}(t) - \boldsymbol{x})\right] \\
&= -\nabla_x \left[(\boldsymbol{h}(\boldsymbol{x}) + \hat{g}(\boldsymbol{x})\boldsymbol{R}(t))\delta(\boldsymbol{u}(t) - \boldsymbol{x})\right] \\
&= -\nabla_x \left[(\boldsymbol{h}(\boldsymbol{x}) + \hat{g}(\boldsymbol{x})\boldsymbol{R}(t)) f(\boldsymbol{x}, t)\right].
\end{aligned} \tag{7.24}$$

We emphasize again that this formal derivation shows that (7.23) is valid even when $\hat{g}$ is a function of the dynamical variables $\boldsymbol{u}(t)$, *i.e.*, the case of the multiplicative noise [74–76].

In the subsequent section, we shall take a generic Langevin equation with a more general structure and derive the F-P equation governing the averaged distribution function.

7.3 RG/E Derivation of Fokker-Planck Equation from a Generic Langevin Equation

In this section, the RG method is applied for the derivation of the Fokker-Planck (F-P) equation [74–76, 112, 115] from Kubo's stochastic Liouville equation (7.23), as advertised in the last section. One of the important ingredients in the derivation is that the 'initial' value of the stochastic distribution function at arbitrary time t_0 is chosen to be on the averaged distribution function for the RG/E equation, which leads to the F-P equation governing the averaged distribution function. It will be found the time t_0 appearing in the RG/E equation is a coarse-grained macroscopic time.

7.3.1 A Generic Langevin Equation with a Multiplicative Noise

As a generalization of (7.21), let us take the following system of dynamical equations for the n-dimensional vector[2] $\boldsymbol{u} = {}^t(u_1, u_2, \ldots, u_n)$

$$\frac{du_i}{dt} = h_i(\boldsymbol{u}) + \sum_{j=1}^{n} g_{ij}(\boldsymbol{u})R_j(t), \tag{7.25}$$

where $R_i(t)$ $(i = 1, 2, ..., n)$ denotes stochastic variables to be realized with the probability $P(\boldsymbol{R}) = \mathcal{P}(\boldsymbol{R})d\boldsymbol{R}$ with the vanishing average

$$\langle R_i(t)\rangle_{\boldsymbol{R}} = 0, \quad (i = 1, 2, \ldots, n). \tag{7.26}$$

Here $\langle O(t)\rangle_{\boldsymbol{R}}$ denotes the average of $O(t)$ with respect to the noise $\boldsymbol{R}$:

$$\langle O(t)\rangle_{\boldsymbol{R}} := \int d\boldsymbol{R}\,\mathcal{P}(\boldsymbol{R})O(t). \tag{7.27}$$

We assume that the noise are stationary though the noise may be or not may be Gaussian.

In the vector notation, (7.25) is written in a compact form as

$$\frac{d\boldsymbol{u}}{dt} = \boldsymbol{h}(\boldsymbol{u}) + \hat{g}(\boldsymbol{u})\boldsymbol{R}, \tag{7.28}$$

with $\boldsymbol{h}(\boldsymbol{u}) = {}^t(h_1(\boldsymbol{u}), h_2(\boldsymbol{u}), \ldots, h_n(\boldsymbol{u}))$ and $\hat{g}(\boldsymbol{u})$ denotes an n times n matrix, the (i, j) component is given by $g_{ij}(\boldsymbol{u})$. Although the similarity of (7.28) with (7.21) is apparent, the coefficient matrix $\hat{g}$ of the noise is not a constant but now allowed to depend on the dynamical variables $\boldsymbol{u}$: Such a noise term is said to be *multiplicative noise*, where as the way of the noise entering (7.21) is said to be *additive noise* [74–76]. From now on, we shall suppress the subscript $\boldsymbol{R}$ for the averaging.

As in the simple case, let

$$f(\boldsymbol{u}, t) = \delta(\boldsymbol{u}(t) - \boldsymbol{u}) \tag{7.29}$$

be the distribution function of $\boldsymbol{u}$ at time t when the stochastic time evolution $\boldsymbol{u}$ is governed by Eq. (7.28). From the way of the derivation given in (7.24), it is clear that the distribution function satisfies the continuity equation or the Kubo's stochastic Liouville equation (7.23)

[2] We have changed an argument $\boldsymbol{x}$ of the distribution function to $\boldsymbol{u}$.

$$\frac{\partial f}{\partial t} + \nabla_u \cdot [(\boldsymbol{h} + \hat{g}\boldsymbol{R})f] = 0, \tag{7.30}$$

with $\nabla_u = (\frac{\partial}{\partial u_1}, \frac{\partial}{\partial u_2}, \ldots, \frac{\partial}{\partial u_n})$. We remark that ∇_u acts on f as well as $\boldsymbol{h}$ and $\hat{g}$.

7.3.2 The RG/E Derivation of the Fokker-Planck Equation

In the following derivation, we shall take the perturbative RG method for the derivation, which is of approximate nature but will be found to be applicable even when the noise is non-Gaussian.

The method which we shall take for derivation of the F-P equation for the multiplicative noise is an adaptation [57] of that proposed in [56] for a simpler equation[3]; we shall make clear the importance of the identification of the 'initial' condition for obtaining the averaged equation, thereby keep the conformity in the presentation of the RG method with those applied to other problems in the previous chapters.

The method begins with a change of independent variables [56] for (7.30)

$$(t, \boldsymbol{u}) \rightarrow (\tau, \boldsymbol{X}) \tag{7.31}$$

by

$$\tau = t, \quad \boldsymbol{X} = \boldsymbol{u} - \int^t ds\, \boldsymbol{h}(\boldsymbol{u}(s)). \tag{7.32}$$

Then the derivatives with respect to t and $\boldsymbol{u}$ become

$$\frac{\partial}{\partial t} = \frac{\partial \tau}{\partial t}\frac{\partial}{\partial \tau} + \frac{\partial \boldsymbol{X}}{\partial t} \cdot \frac{\partial}{\partial \boldsymbol{X}} = \frac{\partial}{\partial \tau} - \boldsymbol{h} \cdot \frac{\partial}{\partial \boldsymbol{X}}, \tag{7.33}$$

$$\frac{\partial}{\partial u_i} = \frac{\partial \tau}{\partial u_i}\frac{\partial}{\partial \tau} + \frac{\partial \boldsymbol{X}}{\partial u_i} \cdot \frac{\partial}{\partial \boldsymbol{X}} = \frac{\partial}{\partial X_i}, \tag{7.34}$$

respectively, where

$$\frac{\partial}{\partial \boldsymbol{X}} = (\frac{\partial}{\partial X_1}, \frac{\partial}{\partial X_2}, \ldots, \frac{\partial}{\partial X_n}) =: \nabla_X. \tag{7.35}$$

With this new independent variables, the Langevin equation (7.30) becomes

[3] In [57], a different method that utilizes a kind of the 'interaction picture' familiar in the time-dependent perturbation theory in quantum mechanics [116–118] was also presented.

$$\frac{\partial f}{\partial \tau} - \boldsymbol{h} \cdot \frac{\partial f}{\partial \boldsymbol{X}} + (\nabla_x \cdot \boldsymbol{h}) f + \boldsymbol{h} \cdot \nabla_x f + \nabla_x (\hat{g} \boldsymbol{R} f)$$
$$= \frac{\partial f}{\partial \tau} + (\nabla_x \cdot \boldsymbol{h}) f + \nabla_x (\hat{g} \boldsymbol{R} f) = 0. \tag{7.36}$$

Next, we introduce a small parameter ϵ and define a new variable $\boldsymbol{x}$ by

$$\epsilon \boldsymbol{X} = \boldsymbol{x}. \tag{7.37}$$

Then we have

$$\frac{\partial}{\partial X_i} = \frac{\partial \boldsymbol{x}}{\partial X_i} \cdot \frac{\partial}{\partial \boldsymbol{x}} = \epsilon \frac{\partial}{\partial x_i}. \tag{7.38}$$

One finds that Eq. (7.30) now takes the form

$$\frac{\partial f}{\partial \tau} = -\epsilon [\nabla_{\boldsymbol{x}} (\hat{g} \boldsymbol{R} f) + (\nabla_{\boldsymbol{x}} \cdot \boldsymbol{h}) f], \tag{7.39}$$

with

$$\nabla_x = (\frac{\partial}{\partial x_1}, \frac{\partial}{\partial x_2}, \ldots, \frac{\partial}{\partial x_n}). \tag{7.40}$$

To apply the RG method, we first try to solve Eq. (7.39) around an arbitrary time $\tau = \forall \tau_0$ by the perturbation theory. The solution is denoted by $\tilde{f}(\boldsymbol{u}, \tau; \tau_0)$ and expanded as

$$\tilde{f}(\boldsymbol{u}, \tau; \tau_0) = \tilde{f}_0(\boldsymbol{u}, \tau; \tau_0) + \epsilon \tilde{f}_1(\boldsymbol{u}, \tau; \tau_0) + \epsilon^2 \tilde{f}_2(\boldsymbol{u}, \tau; \tau_0) + \ldots, \tag{7.41}$$

with the 'initial' condition

$$\tilde{f}(\boldsymbol{u}, \tau_0; \tau_0) = \bar{f}(\boldsymbol{u}, \tau_0). \tag{7.42}$$

The 'initial' distribution function $\bar{f}(\boldsymbol{u}, \tau_0)$ is also expanded as

$$\bar{f}(\boldsymbol{u}, \tau_0) = \bar{f}_0(\boldsymbol{u}, \tau_0) + \epsilon \bar{f}_1(\boldsymbol{u}, \tau_0) + \epsilon^2 \bar{f}_2(\boldsymbol{u}, \tau_0) + \ldots, \tag{7.43}$$

where

$$\tilde{f}_i(\boldsymbol{u}, \tau_0; \tau_0) = \bar{f}_i(\boldsymbol{u}, \tau_0), \quad (i = 1, 2, \ldots). \tag{7.44}$$

As an important ansatz, we demand that the 'initial' distribution function $\bar{f}(\boldsymbol{u}, \tau_0)$ takes the value of the *averaged* distribution function $P(\boldsymbol{u}, \tau_0)$ at $\tau = \tau_0$;

$$\bar{f}(\boldsymbol{u}, \tau_0) = \tilde{f}(\boldsymbol{u}, \tau = \tau_0; \tau_0) = P(\boldsymbol{u}, \tau_0). \tag{7.45}$$

One will recognize that this choice of the 'initial' condition naturally makes the RG/E equation identified with the F-P equation.

Inserting (7.41) into (7.39) and equating the coefficient of ϵ^n $(n = 0, 1, \ldots)$, we have the following series of equations,

$$\frac{\partial \tilde{f}_0}{\partial \tau} = 0, \tag{7.46}$$

$$\frac{\partial \tilde{f}_1}{\partial \tau} = -[\boldsymbol{\nabla}_x(\hat{g}\boldsymbol{R}\tilde{f}_0) + \boldsymbol{\nabla}_x \cdot \boldsymbol{h}\tilde{f}_0], \tag{7.47}$$

$$\frac{\partial \tilde{f}_2}{\partial \tau} = -[\boldsymbol{\nabla}_x(\hat{g}\boldsymbol{R}\tilde{f}_1) + \boldsymbol{\nabla}_x \cdot \boldsymbol{h}\tilde{f}_1], \tag{7.48}$$

and so on.

The zeroth-order equation (7.46) shows that the solution $\tilde{f}_0(\boldsymbol{x}, \tau; \tau_0)$ is time independent;

$$\tilde{f}_0(\boldsymbol{x}, \tau; \tau_0) = \tilde{f}_0(\boldsymbol{x}, \tau_0; \tau_0) = \bar{f}_0(\boldsymbol{u}, \tau_0), \tag{7.49}$$

where we have made explicit that $\tilde{f}_0(\boldsymbol{x}, \tau; \tau_0)$ may depend on $\boldsymbol{x}$ as well as the 'initial' time τ_0.

Inserting (7.49) into (7.47), we see that the r.h.s. of (7.47) has no time dependence. Then the first-order solution is readily obtained by a quadrature as

$$\tilde{f}_1(\boldsymbol{x}, \tau; \tau_0) = -\int_{\tau_0}^{\tau} ds\, \boldsymbol{\nabla}_x \cdot (\hat{g}\boldsymbol{R}(s)\bar{f}_0) - (\tau - \tau_0)(\boldsymbol{\nabla}_x \cdot \boldsymbol{h})\bar{f}_0(\boldsymbol{u}, \tau_0). \tag{7.50}$$

Notice the appearance of the secular term.

Inserting $\tilde{f}_1(\boldsymbol{x}, \tau; \tau_0)$ into (7.48), we find that the second-order solution is given by

$$\tilde{f}_2(\boldsymbol{x}, \tau; \tau_0) = \int_{\tau_0}^{\tau} ds_1 \int_{\tau_0}^{s_1} ds_2\, L_1(s_1)L_1(s_2)\bar{f}_0(\boldsymbol{u}, \tau_0) + \frac{(\tau - \tau_0)^2}{2}(\boldsymbol{\nabla}_x \cdot \boldsymbol{h})(\boldsymbol{\nabla}_x \cdot \boldsymbol{h})\bar{f}_0(\boldsymbol{u}, \tau_0) \tag{7.51}$$

with

$$L_1(s) := -\boldsymbol{\nabla}_x \hat{g}\boldsymbol{R}(s), \tag{7.52}$$

up to terms linear in $\boldsymbol{R}$, which vanish when averaged over the noise.

The distribution function $\tilde{P}(\boldsymbol{u}, \tau; \tau_0)$ averaged over the noise is given by

$$\begin{aligned}\tilde{P}(\boldsymbol{u},\tau;\tau_0) &= \langle \tilde{f}(\boldsymbol{u},\tau;\tau_0)\rangle \\ &= \bar{f}_0(\boldsymbol{u},\tau_0) - \epsilon(\tau-\tau_0)(\nabla_{\boldsymbol{x}}\cdot\boldsymbol{h})\bar{f}_0(\boldsymbol{u},\tau_0) \\ &\quad +\epsilon^2\int_{\tau_0}^{\tau} ds_1 \int_{\tau_0}^{s_1} ds_2 \langle L_1(s_1)L_1(s_2)\rangle \bar{f}_0(\boldsymbol{u},\tau_0) \\ &\quad +\epsilon^2\frac{1}{2}(\tau-\tau_0)^2\nabla_{\boldsymbol{x}}\cdot\boldsymbol{h}\nabla_{\boldsymbol{x}}\cdot\boldsymbol{h}\bar{f}_0(\boldsymbol{u},\tau_0). \end{aligned} \tag{7.53}$$

For steady noises, their time correlation function

$$\langle R_i(s_1)R_j(s_2)\rangle =: \Gamma_{ij}(s_1.s_2) \tag{7.54}$$

is a function of the time difference $s_1 - s_2$ and may be written as

$$\Gamma_{ij}(s_1.s_2) = \Gamma_{ij}(s_1 - s_2). \tag{7.55}$$

For definiteness, we furthermore assume that the noise is a white noise as given by

$$\Gamma_{ij}(s_1 - s_2) = 2\delta_{ij}D_i\delta(s_1 - s_2). \tag{7.56}$$

Then the integral of the correlation function $\int_{t_0}^{t} ds_1 \int_{t_0}^{s_1} ds_2\langle L_1(s_1)L_1(s_2)\rangle$ can be further reduced as follows:

$$\begin{aligned}\int_{\tau_0}^{\tau} ds_1 \int_{\tau_0}^{s_1} ds_2 \,\langle L_1(s_1)L_1(s_2)\rangle &= \int_{\tau_0}^{\tau} ds_1 \int_{\tau_0}^{s_1} ds_2\, \frac{\partial}{\partial x_i}g_{ij}\frac{\partial}{\partial x_k}g_{kl}\langle R_j(s_1)R_k(s_2)\rangle \\ &= \frac{\partial}{\partial x_i}g_{ij}\frac{\partial}{\partial x_k}g_{kl}\int_{\tau_0}^{\tau} ds_1 \int_{\tau_0}^{s_1} ds_2\, 2\delta_{jk}D_j\delta(s_1-s_2).\end{aligned}$$

Here we make the change of the integral variables as follows

$$s_1 = s, \quad s_1 - s_2 = \sigma. \tag{7.57}$$

Then

$$\begin{aligned}\int_{\tau_0}^{\tau} ds_1 \int_{\tau_0}^{s_1} ds_2 \,\langle L_1(s_1)L_1(s_2)\rangle &= 2\frac{\partial}{\partial x_i}g_{ij}D_j\frac{\partial}{\partial x_k}g_{kj}\int_{\tau_0}^{\tau} ds \int_{0}^{\tau-\tau_0} d\sigma\,\delta(\sigma) \\ &= 2(\tau-\tau_0)\frac{\partial}{\partial x_i}g_{ij}D_j\frac{\partial}{\partial x_k}g_{kj}\theta(\tau-\tau_0).\end{aligned} \tag{7.58}$$

Thus collecting all the terms up to this order and recovering the variable $\boldsymbol{u}$ instead of $\boldsymbol{x}$, we have[4]

[4] $\partial/\partial u_i = \partial/\partial X_i = (\partial x_i/\partial X_i)(\partial/\partial x_i) = \epsilon\partial/\partial x_i$.

$$\begin{aligned}\tilde{P}(\boldsymbol{u}, \tau; \tau_0) = &\ \bar{f}_0(\boldsymbol{u}, \tau_0) - (\tau - \tau_0)(\boldsymbol{\nabla}_{\boldsymbol{u}} \cdot \boldsymbol{h})\bar{f}_0(\boldsymbol{u}, \tau_0) \\ &+ 2(\tau - \tau_0)\frac{\partial}{\partial u_i} g_{ij} D_j \frac{\partial}{\partial u_k} g_{kj} \theta(\tau - \tau_0)\bar{f}_0(\boldsymbol{u}, \tau_0) \\ &+ \frac{1}{2}(\tau - \tau_0)^2 \boldsymbol{\nabla}_{\boldsymbol{u}} \cdot \boldsymbol{h} \boldsymbol{\nabla}_{\boldsymbol{u}} \cdot \boldsymbol{h} \bar{f}_0(\boldsymbol{u}, \tau_0). \end{aligned} \tag{7.59}$$

The RG/E equation

$$\left.\frac{\partial \tilde{P}}{\partial \tau_0}\right|_{\tau_0 = \tau} = 0 \tag{7.60}$$

gives

$$\begin{aligned} 0 &= \frac{\partial \bar{f}_0(\boldsymbol{u}, \tau)}{\partial \tau} + (\boldsymbol{\nabla}_{\boldsymbol{u}} \cdot \boldsymbol{h})\bar{f}_0(\boldsymbol{u}, \tau) - 2\frac{\partial}{\partial u_i} g_{ij} D_j \frac{\partial}{\partial u_k} g_{kj} \theta(0)\bar{f}_0(\boldsymbol{u}, \tau) \\ &= \frac{\partial \bar{f}_0(\boldsymbol{u}, t)}{\partial t} + \left(\boldsymbol{\nabla}_{\boldsymbol{u}} \cdot \boldsymbol{h} \bar{f}_0(\boldsymbol{u}, t)\right) - \frac{\partial}{\partial u_i} g_{ij} D_j \frac{\partial}{\partial u_k} g_{kj} \bar{f}_0(\boldsymbol{u}, t). \end{aligned} \tag{7.61}$$

In the last equality, the τ derivative is changed to the t derivative with the use of the formula (7.33); *i.e.*, the operator $\boldsymbol{\nabla}$ hits not only $\boldsymbol{h}$ but also $f_0(\boldsymbol{u}, t)$. Here we have used the identity

$$\theta(0) = 1/2. \tag{7.62}$$

Then noting the choice of the 'initial' value (7.45), we arrive at the familiar form of the F-P equation for the multiplicative Gaussian noise [74–76],

$$\frac{\partial P(\boldsymbol{u}, t)}{\partial t} = -\boldsymbol{\nabla}_{\boldsymbol{u}} \cdot (\boldsymbol{h} P(\boldsymbol{u}, t)) + D_j \frac{\partial}{\partial u_i}\left[g_{ij}\frac{\partial}{\partial u_k}\left(g_{kj} P(\boldsymbol{u}, t)\right)\right]. \tag{7.63}$$

This means that the 'initial' distribution function at an arbitrary time $t = t_0$ before averaging must coincide with the averaged distribution to be determined. The 'initial' value may be considered as the integral constant in the unperturbed equation, which would move slowly being governed by the RG/E equation. In other words, the averaging is automatically made by the RG method.

Now it is often the case that the matrix $\hat{g}$ is a diagonal one as

$$g_{ij} = \delta_{ij} g_j, \tag{7.64}$$

for which, (7.63) is further reduced to a simpler form as

$$\frac{\partial P(\boldsymbol{u}, t)}{\partial t} = -\boldsymbol{\nabla}_{\boldsymbol{u}} \cdot (\boldsymbol{h} P(\boldsymbol{u}, t)) + \sum_{j=1}^{n} D_j \frac{\partial}{\partial u_j}\left[g_j \frac{\partial}{\partial u_j}\left(g_j P(\boldsymbol{u}, t)\right)\right]. \tag{7.65}$$

We note the following trivial identity,

$$\frac{\partial^2}{\partial u^2}\left(g^2 P\right) = \frac{\partial}{\partial u}\left(g\frac{\partial gP}{\partial u}\right) + \frac{\partial}{\partial u}\left(gP\frac{\partial g}{\partial u}\right), \tag{7.66}$$

which implies that

$$\frac{\partial}{\partial u_j}\left[g_j\frac{\partial}{\partial u_j}\left(g_j P(\boldsymbol{u},t)\right)\right] = \frac{\partial^2}{\partial u_j^2}\left(g_j^2 P\right) - \frac{\partial}{\partial u_j}\left[g_j\left(\frac{\partial g_j}{\partial u_j}\right)P\right]. \tag{7.67}$$

Then (7.65) is rewritten as

$$\frac{\partial P(\boldsymbol{u},t)}{\partial t} = -\nabla_u\cdot\left(\tilde{\boldsymbol{h}}P(\boldsymbol{u},t)\right) + \sum_{j=1}^{n}\frac{\partial^2}{\partial u_j^2}\left(D_j g_j^2 P(\boldsymbol{u},t)\right), \tag{7.68}$$

with

$$(\tilde{\boldsymbol{h}})_i = h_i + D_i g_i\left(\frac{\partial g_i}{\partial u_i}\right). \tag{7.69}$$

For the Brownian motion (7.21), (7.63) takes the familiar form [74–76],

$$\frac{\partial P(x,v,t)}{\partial t} = -v\frac{\partial}{\partial x}P(x,v,t) + \gamma\frac{\partial}{\partial v}vP(x,v,t) + \frac{D}{m^2}\frac{\partial^2}{\partial v^2}P(x,v,t), \tag{7.70}$$

which is called the Kramers equation [119].

In conclusion, we emphasize the fact that the coarse graining of time is inevitably incorporated in the derivation of the F-P equation from the Langevin equation through the stochastic Liouville equation. Indeed, Eq. (7.45) shows that the 'initial' value of the stochastic distribution function $f(\boldsymbol{u},t,t_0)$ at $t = t_0$ is set to that of the macroscopic distribution function $P(\boldsymbol{u},t_0)$ for which effects from all the fluctuating 'forces' $\boldsymbol{R}(t)$ due to the 'water molecules' acting on the 'Brownian particle' during the order of the relaxation time are averaged up. If we recall that the time scale of fluctuating forces is much less than the relaxation time, we see that the 'initial' time t_0 entering $P(\boldsymbol{u},t_0)$ in the RG method is a *coarse-grained time*, different from that of the time t in the Langevin equation, and hence the time derivative $\partial/\partial t_0$ in the RG/E equation is a *macroscopic-time derivative* [57]. Moreover, the solution $P(\boldsymbol{u},t)$ of the RG/E as the F-P equation may be interpreted as an '*averaged invariant manifold*' onto which the stochastic distribution function is to get relaxed within the relaxation time [57].

7.4 Adiabatic Elimination of Fast Variables in Fokker-Planck Equation

In this section, we shall show that the reduction theory [6] based on the RG method [1, 3, 5, 6], which is presented in Chap. 5, nicely provides us with a simple but systematic method of the adiabatic elimination of fast variables appearing in the F-P equations [57]. In fact, because of the linearity of the F-P equation, it is a rather easy task once the slow and fast variables are identified as is often the case. Incidentally, the techniques developed for it [73, 120–125] has a noticeable affinity with the perturbation theory in quantum mechanics [117, 118, 126, 127], as emphasized in [123, 125].

A typical problem in this category [73, 120–122, 124] is to derive the Smoluchowski equation [128], hopefully with corrections, describing the time evolution of the *space distribution* $\tilde{P}(x, t)$ of the Brownian particles in the large friction limit $\gamma = \mu/m \to \infty$ from the Kramers equation (7.70) for describing the time evolution of the *space-velocity distribution* $P(x, v, t)$ of the particles.

It seems that Brinkman [120] was the first who considered the problem seriously, though more reliable derivations were provided later [73, 121, 122, 124]; see [73, 125] for a review. It should be also remarked that Mastuo and Sasa [129] was the first who utilized the RG method given in [1] as a tool for eliminating the velocity in a F-P equation for describing molecular engines [130–132].

We shall show that our reduction theory based on the RG method as given in [3, 5, 6] applied for the elimination of fast variables in the F-P equation [57] is free from any ansatz and has a clear correspondence with and hence provide a foundation to those given in other methods.

For definiteness, we take the following 2-dimensional Langevin equation [73] with γ being a positive large number,

$$\begin{cases} \dot{x} = h_x(x, y) + g_x(x, y)\Gamma_x(t), \\ \dot{y} = \gamma h_y(x, y) + f(x, y) + \sqrt{\gamma} g_y(x, y)\Gamma_y(t), \end{cases} \tag{7.71}$$

where $\Gamma_i(t)$ $(i = x, y)$ are Gaussian noises satisfying

$$\langle \Gamma_i(t)\Gamma_j(t') \rangle = 2D_i \delta_{ij} \delta(t - t'). \tag{7.72}$$

We follow Ref. [73] for the notations. Because of the large friction given by the term containing γ, the variable y is a fast variable. Our task is to eliminate the fast variable y adiabatically, and this obtain a reduced evolution equation given solely in terms of x.

Before doing this task, we first eliminate the most rapid variables, *i.e.*, $\Gamma_i(t)$ $(i = x, y)$ to obtain the dynamics written only in terms of x and y. This is tantamount to transforming the Langevin equation (7.71) to a F-P equation as was done in the previous section.

In terms of the notation used in the previous section, we find that

$$\tilde{h}_x = h_x + D_x g_x \frac{\partial g_x}{\partial x}, \tag{7.73}$$

$$\tilde{h}_y = \gamma h_y + f + \gamma D_y g_y \frac{\partial g_y}{\partial y}, \tag{7.74}$$

$$\hat{g} = \begin{pmatrix} g_x & 0 \\ 0 & \sqrt{\gamma} g_y \end{pmatrix}. \tag{7.75}$$

The corresponding F-P equation for the probability $W(x, y, t)$ reads[5]

$$\begin{aligned} \frac{\partial W}{\partial t} &= -\frac{\partial}{\partial x}\left(h_x + D_x g_x \frac{\partial g_x}{\partial x}\right) - \frac{\partial}{\partial y}\left(\gamma h_y + f + \gamma D_y g_y \frac{\partial g_y}{\partial y}\right) \\ &\quad + \frac{\partial^2}{\partial x^2}\left(D_x g_x^2 W\right) + \gamma \frac{\partial^2}{\partial x^2}\left(D_y g_y^2 W\right) \\ &=: [\hat{L}_1 + \gamma \hat{L}_0] W, \end{aligned} \tag{7.76}$$

where

$$\hat{L}_1 = -\frac{\partial}{\partial x}\left(h_x + D_x g_x \frac{\partial g_x}{\partial x}\right) - \frac{\partial}{\partial y} f(x, y) + D_x \frac{\partial^2}{\partial x^2} g_x^2, \tag{7.77}$$

$$\hat{L}_0 = -\frac{\partial}{\partial y}\left(h_y + D_y g_y \frac{\partial g_y}{\partial y}\right) + D_y \frac{\partial^2}{\partial y^2} g_y^2. \tag{7.78}$$

7.4.1 *Perturbative Expansion in the Case of a Strong Friction*

When γ is large, this equation clearly shows that the effects of $\hat{L}_0$ overwhelms that of $\hat{L}_1$, which may be thus treated as a perturbation. To implement this, we introduce a scaled time τ by

$$\epsilon \tau = t \quad \text{with } \epsilon := \frac{1}{\gamma} < 1. \tag{7.79}$$

Our equation now takes the form

$$\left(\frac{\partial}{\partial \tau} - \hat{L}_0\right) W = \epsilon \hat{L}_1 W, \tag{7.80}$$

[5] We use the notation W instead of P for the probability density.

We shall apply the perturbation theory to (7.80) with $\hat{L}_0$ and $\hat{L}_1$ being the unperturbed and perturbative part, respectively. Let us suppose that a solution is given in the perturbation series;

$$W(\tau, x, y) = W_0(\tau, x, y) + \epsilon W_1(\tau, x, y) + \epsilon^2 W_2(\tau, x, y) + \dots. \tag{7.81}$$

Following the general scheme of the RG method [3–6] presented in the previous chapters, we first try to construct the perturbed solution around an arbitrary time $\tau \sim \tau_0$ in the *asymptotic region* as

$$\tilde{W}(\tau, x, y; \tau_0) = \tilde{W}_0(\tau, x, y; \tau_0) + \epsilon \tilde{W}_1(\tau, x, y) + \epsilon^2 \tilde{W}_2(\tau, x, y; \tau_0) + \dots. \tag{7.82}$$

with the 'initial' conditions given at $\tau = \tau_0$ belonging to the asymptotic region;

$$\tilde{W}_i(\tau_0, x, y; \tau_0) = W_i(\tau_0, x, y), \tag{7.83}$$

which implies that

$$W(\tau, x, y) = \tilde{W}(\tau, x, y; \tau). \tag{7.84}$$

It is readily found that $\tilde{W}_n$'s ($n = 0, 1, 2, \dots$) satisfy the following equations, respectively,

$$\left(\frac{\partial}{\partial \tau} - \hat{L}_0\right) \tilde{W}_0 = 0, \tag{7.85}$$

$$\left(\frac{\partial}{\partial \tau} - \hat{L}_0\right) \tilde{W}_n = \hat{L}_1 \tilde{W}_{n-1}, \quad (n = 1, 2, \dots). \tag{7.86}$$

7.4.2 The Eigenvalue Problem of $\hat{L}_0$

To proceed, we must first solve the eigenvalue problem of the unperturbed operator $\hat{L}_0(y, x)$ acting on functions of y with x being a mere parameter:

$$\hat{L}_0 \varphi_n(y; x) = -\lambda_n(x) \varphi_n(y; x). \tag{7.87}$$

Here we assume the eigenvalues are all discrete ones, mainly for notational simplicity. As is the case with the Kramers equation, we are interested in the case where $\hat{L}_0(y, x)$ has unique zero eigenvalue; we denote the corresponding eigenfunction as $\varphi_0(y; x) =: \varphi_{st}(y; x)$,

$$\hat{L}_0(y, x)\varphi_0(y; x) = 0, \quad \text{with} \quad \lambda_0(x) = 0. \tag{7.88}$$

We call $\varphi_0(y; x) = \varphi_{st}(y; x)$ the *stationary distribution function* because it satisfies the zeroth order evolution equation (7.85).

Since $\hat{L}_0(y, x)$ is not a symmetric (Hermitian) operator,[6] we introduce the corresponding *adjoint operator* $\hat{L}_0^\dagger$, which satisfies

$$\hat{L}_0^\dagger \tilde{\varphi}_n(y; x) = -\lambda_n(x)\tilde{\varphi}_n(y; x). \tag{7.89}$$

The explicit form of $\hat{L}_0^\dagger$ will be given later.

Now it is readily verified[7] that Eq. (7.88) is rewritten as [75]

$$\hat{L}_0\varphi_0(y; x) = -\frac{\partial}{\partial y}\left[D^{(1)} - \frac{\partial}{\partial y}D^{(2)}\right]\varphi_0(y; x) = 0, \tag{7.90}$$

with

$$D^{(1)} := h_y(x; y) + D_y g_y(x, y)\frac{\partial g_y(x, y)}{\partial y}, \quad D^{(2)} := D_y g_y^2(x; y). \tag{7.91}$$

Equation (7.90) implies that

$$\frac{\partial(D^{(2)}\varphi_0)}{\partial y} = D^{(1)}\varphi_0 = \frac{D^{(1)}}{D^{(2)}}(D^{(2)}\varphi_0), \tag{7.92}$$

which is readily solved to yield

$$\varphi_0(y; x) = \frac{N}{D^{(2)}}\exp\left[\int^y dy' \frac{D^{(1)}}{D^{(2)}}\right] = N\mathrm{e}^{-\Phi(y;x)}, \tag{7.93}$$

where

$$\Phi(y; x) := \ln D^{(2)} - \int^y dy' \frac{D^{(1)}}{D^{(2)}}. \tag{7.94}$$

Then it is a remarkable fact that $\hat{L}_0$ can be cast into a form of a *Sturm-Liouville operator* [80] in terms of Φ as follows

$$\hat{L}_0 = \frac{\partial}{\partial y}\left(D^{(2)}(y; x)\mathrm{e}^{-\Phi(y;x)}\frac{\partial}{\partial y}\mathrm{e}^{\Phi(y;x)}\right). \tag{7.95}$$

---

[6] See Sect. 3.6.2 for an introductory account of asymmetric matrices and related subjects.

[7] In the following, the presentation is somewhat based on [75].

In fact,

$$
\begin{aligned}
\frac{\partial}{\partial y}\left(D^{(2)}(y;x)\mathrm{e}^{-\Phi(y;x)}\frac{\partial}{\partial y}\mathrm{e}^{\Phi(y;x)}\right) &= \frac{\partial}{\partial y}\left[D^{(2)}(y;x)\left(\frac{\partial \Phi}{\partial y}+\frac{\partial}{\partial y}\right)\right] \\
&= \frac{\partial}{\partial y}\left(-D^{(1)}+\frac{\partial D^{(2)}}{\partial y}+D^{(2)}\frac{\partial}{\partial y}\right) \\
&= -\frac{\partial}{\partial y}D^{(1)}+\frac{\partial^2}{\partial y^2}D^{(2)}=\hat{L}_0. \qquad (7.96)
\end{aligned}
$$

Then we find that the following operator is an Hermite operator,[8]

$$
\mathrm{e}^{\Phi(y;x)}\hat{L}_0 =: \hat{\mathcal{H}}_0. \qquad (7.97)
$$

To show that an operator is Hermitian or not, one must specify the functional space on which the operator acts. We note that the unperturbed equation (7.85) is expressed in a form of the continuity equation as

$$
\frac{\partial \tilde{W}_0}{\partial \tau}+\frac{\partial S[\tilde{W}_0]}{\partial y}=0, \qquad (7.98)
$$

with

$$
S[\tilde{W}_0]=\left(D^{(1)}-\frac{\partial}{\partial y}\right)\tilde{W}_0=\left(D^{(2)}(y;x)\mathrm{e}^{-\Phi(y;x)}\frac{\partial}{\partial y}\mathrm{e}^{\Phi(y;x)}\right)\tilde{W}_0 \qquad (7.99)
$$

which is the probability current or *probability flux*. For simplicity, we demand that any solution $\tilde{W}_0$ to (7.85) satisfies the (natural) boundary condition;

$$
\lim_{|y|\to\infty} S[\tilde{W}_0]=0. \qquad (7.100)
$$

Now that the functional space has been specified, let us examine the properties of the operator $\hat{\mathcal{H}}_0$ by calculating the following 'matrix element' given in terms of two arbitrary functions $Y_1(y)$ and $Y_2(y)$ satisfying the natural boundary condition

[8] The fact that $\hat{L}_0$ is written in a form of a Sturm-Liouville operator [80] suggests that the operator can be made a symmetric or Hermitian operator with a weighted inner product.

$$
\begin{aligned}
\langle Y_1|\hat{\mathcal{H}}_0|Y_2\rangle &:= \int_{\infty}^{\infty} dy\, Y_1(y)\mathrm{e}^{\Phi}\hat{L}_0 Y_2(y) \\
&= \int_{\infty}^{\infty} dy\, Y_1(y)\mathrm{e}^{\Phi}\frac{\partial}{\partial y}\left(D^{(2)}(y;x)\mathrm{e}^{-\Phi(y;x)}\frac{\partial}{\partial y}\mathrm{e}^{\Phi(y;x)}\right) Y_2(y) \\
&= Y_1(y)\mathrm{e}^{\Phi}\left(D^{(2)}(y;x)\mathrm{e}^{-\Phi(y;x)}\frac{\partial}{\partial y}\mathrm{e}^{\Phi(y;x)}\right) Y_2(y)\Big|_{-\infty}^{\infty} \\
&\quad - \int_{\infty}^{\infty} dy\, \frac{\partial}{\partial y}\left(Y_1(y)\mathrm{e}^{\Phi}\right) D^{(2)}(y;x)\mathrm{e}^{-\Phi(y;x)}\frac{\partial}{\partial y}\left(\mathrm{e}^{\Phi(y;x)}Y_2(y)\right).
\end{aligned}
\tag{7.101}
$$

The first term is rewritten in terms of the flux as

$$
Y_1(y)\mathrm{e}^{\Phi}\left(-S[Y_2]\right)\Big|_{-\infty}^{\infty} = 0, \tag{7.102}
$$

where the last equality is due to the boundary condition posed on $Y_2(y)$. Then performing a partial integration to the second term, we have

$$
\begin{aligned}
\langle Y_1|\hat{\mathcal{H}}_0|Y_2\rangle &= -\left(D^{(2)}(y;x)\mathrm{e}^{-\Phi(y;x)}\frac{\partial}{\partial y}Y_1(y)\mathrm{e}^{\Phi}\right)\left(\mathrm{e}^{\Phi(y;x)}Y_2(y)\right)\Big|_{-\infty}^{\infty} \\
&\quad + \int_{\infty}^{\infty} dy\, \frac{\partial}{\partial y}\left(D^{(2)}(y;x)\mathrm{e}^{-\Phi(y;x)}\frac{\partial}{\partial y}Y_1(y)\mathrm{e}^{\Phi}\right)\mathrm{e}^{\Phi(y;x)}Y_2(y) \\
&= \int_{\infty}^{\infty} dy\, \left(\mathrm{e}^{\Phi(y;x)}\hat{L}_0 Y_1(y)\right) Y_2(y) = \langle\hat{\mathcal{H}}_0 Y_1|Y_2\rangle,
\end{aligned}
\tag{7.103}
$$

where the boundary term vanishes because of the boundary condition posed on Y_1;

$$
\lim_{|y|\to\infty} S[Y_1] = 0. \tag{7.104}
$$

Thus we have established that $\hat{\mathcal{H}}_0$ is a *Hermitian operator*:

$$
\hat{\mathcal{H}}_0^{\dagger} = \hat{\mathcal{H}}_0. \tag{7.105}
$$

Comments are in order here.

1. The following operator is also a Hermitian operator

$$
\hat{H}_0 = \mathrm{e}^{-\Phi/2}\hat{\mathcal{H}}_0\mathrm{e}^{-\Phi/2}, \tag{7.106}
$$

since

$$
\hat{H}_0^{\dagger} = \mathrm{e}^{-\Phi/2}\hat{\mathcal{H}}_0^{\dagger}\mathrm{e}^{-\Phi/2} = \hat{H}_0. \tag{7.107}
$$

2. If we use the stationary distribution function $\varphi_0(y; x)$ in place of Φ, $\hat{H}_0$ is expressed as

$$\hat{H}_0 = \frac{1}{\sqrt{\varphi_0(y; x)}} \hat{L}_0 \sqrt{\varphi_0(y; x)}. \tag{7.108}$$

Then what are eigenvalues and eigenfunctions of $\hat{H}_0$? In fact, it is readily found that $\psi_n(y; x)$ defined below is the eigenfunctions of $\hat{H}_0$ with eigenvalue $-\lambda_n$:

$$\hat{H}_0 \psi_n(y; x) = -\lambda_n \psi_(y; x), \quad \psi_n(y; x) := \frac{1}{\sqrt{\varphi_0(y; x)}} \varphi_n(y; x). \tag{7.109}$$

In fact,

$$\begin{aligned}\hat{H}_0 \psi_n(y; x) &= \left(\frac{1}{\sqrt{\varphi_0(y; x)}} \hat{L}_0 \sqrt{\varphi_0(y; x)} \right) \left(\frac{1}{\sqrt{\varphi_0(y; x)}} \varphi_n(y; x) \right) \\ &= \frac{1}{\sqrt{\varphi_0(y; x)}} \hat{L}_0 \varphi_n(y; x) = -\lambda_n \frac{1}{\sqrt{\varphi_0(y; x)}} \varphi_n(y; x) = -\lambda_n \psi(y; x).\end{aligned} \tag{7.110}$$

Similarly, the *adjoint eigenfunctions*[9] are formally given by

$$\tilde{\psi}_n(y; x) = \sqrt{\varphi_0(y; x)} \tilde{\varphi}_n(y; x). \tag{7.111}$$

Indeed, with the use of the adjoint form of (7.108), we have

$$\begin{aligned}\hat{H}_0^\dagger \tilde{\psi}_n(y; x) &= \left(\sqrt{\varphi_0(y; x)} \hat{L}_0^\dagger \frac{1}{\sqrt{\varphi_0(y; x)}} \right) \sqrt{\varphi_0(y; x)} \tilde{\varphi}_n(y; x) \\ &= \sqrt{\varphi_0(y; x)} \hat{L}_0^\dagger \tilde{\varphi}_n(y; x) = -\lambda_n \sqrt{\varphi_0(y; x)} \tilde{\varphi}_n(y; x) = -\lambda_n \tilde{\psi}_n(y; x).\end{aligned} \tag{7.112}$$

As is well known in quantum mechanics [117, 118], however, if the number of degrees of freedom is one, then there is no degeneracy in the eigenvalues and the eigenfunctions can be made real functions. In the case of $\hat{H}_0$, it implies that

$$\tilde{\psi}_n(y; x) = \psi_n(y; x). \tag{7.113}$$

Then combining (7.109) and (7.111), we have

$$\tilde{\varphi}_n(y; x) = \frac{1}{\varphi_0(y; x)} \varphi_n(y; x), \tag{7.114}$$

---

[9] An adjoint eigenvector is an adjoint vector of the left eigenvector in the terminology given in Sect. 3.6.2.

which implies that

$$\tilde{\varphi}_0(y; x) = 1. \tag{7.115}$$

Owing to the Hermiticity of $\hat{H}_0$, the *orthonormality* holds;

$$\int \psi_n(y; x)\psi_m(y; x)dy = \int \tilde{\varphi}_n(y; x)\varphi_m(y; x)dy = \delta_{nm}. \tag{7.116}$$

We also assume the *completeness* of the eigenfunctions

$$\sum_{n=0}^{\infty} \psi_n(y; x)\psi_n(y'; x) = \sum_{n=0}^{\infty} \tilde{\varphi}_n(y; x)\varphi_n(y'; x) = \delta(y - y'). \tag{7.117}$$

Next let us show that all the eigenvalues λ_n for $n \neq 0$ is *positive definite*. Utilizing that Hermitness of $\hat{\mathcal{H}} = \mathrm{e}^{\Phi}\hat{L}_0$, we have

$$\begin{aligned}
&\int_{\infty}^{\infty} dy\, \psi_n(y; x)\hat{H}_0\psi_n(y; x) = -\lambda_n \\
&= \int_{\infty}^{\infty} dy\, \psi_n(y; x)\mathrm{e}^{\Phi(y;x)/2}\frac{\partial}{\partial y}\left(D^{(2)}(y; x)\mathrm{e}^{-\Phi(y;x)}\frac{\partial}{\partial y}\mathrm{e}^{\Phi(y;x)}\mathrm{e}^{-\Phi(y;x)/2}\psi_n(y; x)\right) \\
&= -\int_{\infty}^{\infty} dy\, \frac{\partial}{\partial y}\left(\psi_n(y; x)\mathrm{e}^{\Phi(y;x)/2}\right) D^{(2)}(y; x)\mathrm{e}^{-\Phi(y;x)}\frac{\partial}{\partial y}\left(\mathrm{e}^{\Phi(y;x)/2}\psi_n(y; x)\right) \\
&= -\int_{\infty}^{\infty} dy\, \left(\frac{\partial}{\partial y}\psi_n(y; x)\mathrm{e}^{\Phi(y;x)/2}\right)^2 D^{(2)}(y; x)\mathrm{e}^{-\Phi(y;x)} \leq 0,
\end{aligned} \tag{7.118}$$

where the equality holds only for $n = 0$, which shows that

$$\lambda_n > 0, \qquad (n \neq 0). \tag{7.119}$$

Now we define the projection operator $\hat{P}$ to the kernel of $\hat{L}_0$ by

$$[\hat{P}\varphi](y; x) = \varphi_0(y; x)\int \tilde{\varphi}_0(y'; x)\varphi(y'; x)dy'. \tag{7.120}$$

From (7.95), one finds that the adjoint of $\hat{L}_0$ is given by

$$\hat{L}_0^{\dagger} = \mathrm{e}^{\Phi(y;x)}\frac{\partial}{\partial y}\left(D^{(2)}(y; x)\mathrm{e}^{-\Phi(y;x)}\right)\frac{\partial}{\partial y}. \tag{7.121}$$

Then we see that the zero mode for the adjoint operator is given by $\tilde{\varphi}_0(y; x) = 1$ as has been already deduced in (7.115). The projection operator to the complement to the kernel is denoted by $\hat{Q}$,

$$[\hat{Q}\varphi](y;x) = [(1-\hat{P})\varphi](y;x)$$
$$= \sum_{n=1}^{\infty} \varphi_n(y;x) \int dy' \tilde{\varphi}_n(y';x)\varphi(y';x). \tag{7.122}$$

In the following, we assume that

$$\hat{P}\hat{L}_1\hat{P} = 0, \tag{7.123}$$

which can be always satisfied by redefinition of $\hat{L}_0$ and $\hat{L}_1$. In terms of the notions in quantum field theory [133], $\hat{L}_1$ is *normal-ordered* with respect to the vacuum $|0\rangle := \varphi_0$; $\langle 0|\hat{L}_1|0\rangle = 0$.

7.4.3 The Solution to the Perturbative Equations

The general solution to the lowest-order equation (7.85) is given by

$$\tilde{W}_0(\tau, x, y; \tau_0) = \sum_{n=0}^{\infty} \mathcal{P}_n(x; \tau_0)\varphi_n(y; x)\mathrm{e}^{-\lambda_n \tau}, \tag{7.124}$$

where we have made explicit that the integral constants $\mathcal{P}_n(x;\tau_0)$ may depend on the 'initial' time τ_0 as well as the other variable x. We remark that (7.124) has the 'initial' value

$$W(\tau_0, x, y) = \tilde{W}_0(\tau_0, x, y; \tau_0) = \sum_{n=0}^{\infty} \mathcal{P}_n(x; \tau_0)\varphi_n(y; x)\mathrm{e}^{-\lambda_n \tau_0}. \tag{7.125}$$

Since we are interested in the *asymptotic (long time) behavior* at $t \to \infty$, we may keep only the *stationary solution* in the sum as

$$\tilde{W}_0(\tau, x, y; \tau_0) \simeq \mathcal{P}_0(x; \tau_0)\varphi_0(y; x) = W_0(\tau_0, x, y). \tag{7.126}$$

Notice the last equality holds because of the stationarity of the solution. Accordingly,

$$\hat{P}\tilde{W}_0(\tau, x, y; \tau_0) = \tilde{W}_0(\tau, x, y; \tau_0). \tag{7.127}$$

Let us proceed to the higher-order equations (7.86). The solution to Eq. (7.86) is formally given by

$$\tilde{W}_n(\tau, x, y; \tau_0) = \frac{1}{\partial_\tau - \hat{L}_0}\hat{L}_1\tilde{W}_{n-1} = \frac{1}{\partial_\tau - \hat{L}_0}(\hat{P} + \hat{Q})\hat{L}_1\tilde{W}_{n-1}$$
$$= X_n(\tau, x, y; \tau_0) + Y_n(\tau, x, y; \tau_0), \tag{7.128}$$

where

$$X_n := \hat{P}\tilde{W}_n(\tau, x, y; \tau_0) = \frac{1}{\partial_\tau - \hat{L}_0}\hat{P}\hat{L}_1\tilde{W}_{n-1}, \tag{7.129}$$

$$Y_n := \hat{Q}\tilde{W}_n(\tau, x, y; \tau_0) = \frac{1}{\partial_\tau - \hat{L}_0}\hat{Q}\hat{L}_1\tilde{W}_{n-1}, \tag{7.130}$$

for $n = 1, 2, \ldots$. Here we have utilized the commutability of $\hat{L}_0$ and the projection operators,

$$[\hat{L}_0, \hat{P}] = 0, \quad [\hat{L}_0, \hat{Q}] = 0. \tag{7.131}$$

It is well known that the solution to an inhomogeneous equation has an ambiguity due to the freedom to add a function proportional to the unperturbed solution. We determine the form of the particular solution to the inhomogeneous equations (7.86) so that the possible P-space component in the perturbed solutions for $n \geq 1$ vanishes at $\tau = \tau_0$;

$$\hat{P}\tilde{W}_n(\tau_0, x, y; \tau_0) = X_n(\tau_0, x, y; \tau_0) = 0. \tag{7.132}$$

In fact, it will be found that this condition is always made satisfied owing to the appearance of secular terms proportional to $(\tau - \tau_0)^n$ multiplied to the P-space component.

The particular solutions to the perturbative equations (7.86) with the 'initial' condition (7.132) can be obtained in a mechanical way, as is presented in Sect. 4.7. We first note that for any constant vector U the following formulae hold, as given by (4.237) and (4.239), respectively,

$$\frac{1}{\partial_\tau - \hat{L}_0}\hat{Q}U = \frac{1}{-\hat{L}_0}\hat{Q}U, \tag{7.133}$$

$$\frac{1}{\partial_\tau - \hat{L}_0}(\tau - \tau_0)^n\hat{P}U = \frac{1}{(n+1)}(\tau - \tau_0)^{n+1}\hat{P}U, \tag{7.134}$$

for $n = 0, 1, 2\ldots$. In particular, the formula (4.238) gives

$$\frac{1}{\partial_\tau - \hat{L}_0}\hat{P}U = (\tau - \tau_0)\hat{P}U. \tag{7.135}$$

We also note the following formula, as given by (4.241)

$$\frac{1}{\partial_\tau - \hat{L}_0}(\tau - \tau_0)\hat{Q}\hat{L}_1U = (\tau - \tau_0)\frac{1}{-\hat{L}_0}\hat{Q}\hat{L}_1U - \frac{1}{(-\hat{L}_0)^2}\hat{Q}\hat{L}_1U. \tag{7.136}$$

Here, notice that $[\hat{L}_0, \hat{Q}] = 0$, so $\hat{Q}/\hat{L}_0 = \hat{L}_0^{-1}\hat{Q} - \hat{Q}\hat{L}_0^{-1}\hat{Q}$. We remark that these particular solutions are all compatible with the 'initial' condition (7.132).

It is readily seen that X_1 identically vanishes because

$$\hat{P}\hat{L}_1\tilde{W}_0 = \hat{P}\hat{L}_1\hat{P}\tilde{W}_0 = 0, \tag{7.137}$$

on account of (7.123) and (7.127). Thus we have for the first-order solution

$$\tilde{W}_1(\tau, x, y; \tau_0) = Y_1 = \frac{1}{-\hat{L}_0}\hat{Q}\hat{L}_1\tilde{W}_0. \tag{7.138}$$

A couple of remarks are in order:

1. One may replace the operator $\hat{Q}$ with $\hat{P} + \hat{Q} = 1$ in this expression owing to Eq. (7.137). Although such simplification in the expressions may be done also for higher-order solutions, we shall retain the $\hat{Q}$ operator for definiteness, since the expression $(-\hat{L}_0)^{-1}\hat{L}_1$ itself is ill-defined because of the zero eigenvalue of $\hat{L}_0$.
2. The solution (7.138) satisfies the 'initial' condition (7.132) as

$$\tilde{W}_1(\tau_0, x, y, \tau_0) = W_1(\tau_0, x, y) = \frac{1}{-\hat{L}_0}\hat{Q}\hat{L}_1 W_0(\tau_0, x, y). \tag{7.139}$$

[The 2nd order]
Using the formula (7.135), we have

$$X_2 = \frac{1}{\partial_\tau - \hat{L}_0}\hat{P}\hat{L}_1\frac{1}{-\hat{L}_0}\hat{Q}\hat{L}_1\tilde{W}_0 = (\tau - \tau_0)\hat{P}\hat{L}_1\frac{\hat{Q}}{-\hat{L}_0}\hat{L}_1\tilde{W}_0. \tag{7.140}$$

Similarly with use of the formula (7.133), we have for the Q-space component

$$Y_2 = \frac{1}{\partial_\tau - \hat{L}_0}\hat{Q}\hat{L}_1\frac{1}{-\hat{L}_0}\hat{Q}\hat{L}_1\tilde{W}_0 = \left(\frac{\hat{Q}}{-\hat{L}_0}\hat{L}_1\right)^2\tilde{W}_0. \tag{7.141}$$

$\tilde{W}_2(\tau, x, y, \tau_0)$ is given by a sum of these terms;

$$\tilde{W}_2(\tau, x, y, \tau_0) = (\tau - \tau_0)\hat{P}\hat{L}_1\frac{\hat{Q}}{-\hat{L}_0}\hat{L}_1\tilde{W}_0 + \left(\frac{\hat{Q}}{-\hat{L}_0}\hat{L}_1\right)^2\tilde{W}_0, \tag{7.142}$$

which gives the 'initial' value at $\tau = \tau_0$ as

$$\tilde{W}_2(\tau_0, x, y, \tau_0) = W_2(\tau_0, x, y) = \frac{1}{\hat{L}_0}\hat{Q}\hat{L}_1\frac{1}{\hat{L}_0}\hat{Q}\hat{L}_1 W_0(\tau_0, x, y), \tag{7.143}$$

in accordance with (7.132).

[The 3rd order]
Since $\hat{P}\hat{L}_1 X_2 = 0$ because of (7.123),

$$X_3 = \frac{1}{\partial_\tau - \hat{L}_0}\hat{P}\hat{L}_1 Y_2 = (\tau - \tau_0)\hat{P}\hat{L}_1\left(\frac{\hat{Q}}{-\hat{L}_0}\hat{L}_1\right)^2 \tilde{W}_0, \tag{7.144}$$

where (7.135) has been used in the last equality. With use of (7.133) and (7.136), the Q-space component becomes

$$Y_3 = [(\tau - \tau_0)\frac{\hat{Q}}{-\hat{L}_0}\hat{L}_1\hat{P}\hat{L}_1\frac{\hat{Q}}{-\hat{L}_0}\hat{L}_1 - \frac{\hat{Q}}{(-\hat{L}_0)^2}\hat{L}_1\hat{P}\hat{L}_1\frac{\hat{Q}}{-\hat{L}_0}\hat{L}_1 + \left(\frac{\hat{Q}}{-\hat{L}_0}\hat{L}_1\right)^3]\tilde{W}_0. \tag{7.145}$$

Thus we have for the third-order solution

$$\begin{aligned}\tilde{W}_3(\tau, x, y; \tau_0) = \Big[&(\tau - \tau_0)\{\hat{P}\hat{L}_1\left(\frac{\hat{Q}}{-\hat{L}_0}\hat{L}_1\right)^2 + \frac{\hat{Q}}{-\hat{L}_0}\hat{L}_1\hat{P}\hat{L}_1\frac{\hat{Q}}{-\hat{L}_0}\hat{L}_1\} \\ &- \frac{\hat{Q}}{(-\hat{L}_0)^2}\hat{L}_1\hat{P}\hat{L}_1\frac{\hat{Q}}{-\hat{L}_0}\hat{L}_1 + \left(\frac{\hat{Q}}{-\hat{L}_0}\hat{L}_1\right)^3\Big]\tilde{W}_0,\end{aligned} \tag{7.146}$$

which satisfies the 'initial' condition (7.132) as

$$\tilde{W}_3(\tau_0, x, y; \tau_0) = \Big[-\frac{\hat{Q}}{(-\hat{L}_0)^2}\hat{L}_1\hat{P}\hat{L}_1\frac{\hat{Q}}{-\hat{L}_0}\hat{L}_1 + \left(\frac{\hat{Q}}{-\hat{L}_0}\hat{L}_1\right)^3\Big]\tilde{W}_0. \tag{7.147}$$

[The 4th order]
With use of the formulae (7.134), one obtains

$$\begin{aligned}X_4 = \Big[&\frac{1}{2}(\tau - \tau_0)^2\hat{P}\hat{L}_1\frac{\hat{Q}}{-\hat{L}_0}\hat{L}_1\hat{P}\hat{L}_1\frac{\hat{Q}}{-\hat{L}_0}\hat{L}_1 + (\tau - \tau_0)\{-\hat{P}\hat{L}_1\frac{\hat{Q}}{(-\hat{L}_0)^2}\hat{L}_1\hat{P}\hat{L}_1\frac{\hat{Q}}{-\hat{L}_0}\hat{L}_1 \\ &+\hat{P}\hat{L}_1(\frac{\hat{Q}}{-\hat{L}_0}\hat{L}_1)^3\}\Big]\tilde{W}_0.\end{aligned} \tag{7.148}$$

Similarly,

$$\begin{aligned}Y_4 = [&(\tau - \tau_0)\left\{\frac{\hat{Q}}{-\hat{L}_0}\hat{L}_1\hat{P}\hat{L}_1\left(\frac{\hat{Q}}{-\hat{L}_0}\hat{L}_1\right)^2 + \left(\frac{\hat{Q}}{-\hat{L}_0}\hat{L}_1\right)^2\hat{P}\hat{L}_1\frac{\hat{Q}}{-\hat{L}_0}\hat{L}_1\right\} \\ &- \frac{\hat{Q}}{(-\hat{L}_0)^2}\hat{L}_1\hat{P}\hat{L}_1\left(\frac{\hat{Q}}{-\hat{L}_0}\hat{L}_1\right)^2 - \frac{\hat{Q}}{-\hat{L}_0}\left(\frac{\hat{Q}}{-\hat{L}_0}\hat{L}_1\right)^2\hat{P}\hat{L}_1\frac{\hat{Q}}{-\hat{L}_0}\hat{L}_1 \\ &- \frac{\hat{Q}}{-\hat{L}_0}\hat{L}_1\frac{\hat{Q}}{(-\hat{L}_0)^2}\hat{L}_1\hat{P}\hat{L}_1\frac{\hat{Q}}{-\hat{L}_0}\hat{L}_1 + \left(\frac{\hat{Q}}{-\hat{L}_0}\hat{L}_1\right)^4]\tilde{W}_0.\end{aligned} \tag{7.149}$$

Adding the two terms, we have

$$\begin{aligned}\tilde{W}_4(\tau, x, y; \tau_0) = \Big[&\frac{1}{2}(\tau-\tau_0)^2 \hat{P}\hat{L}_1 \frac{\hat{Q}}{-\hat{L}_0}\hat{L}_1\hat{P}\hat{L}_1\frac{\hat{Q}}{-\hat{L}_0}\hat{L}_1 \\ &+(\tau-\tau_0)\{-\hat{P}\hat{L}_1\frac{\hat{Q}}{\left(-\hat{L}_0\right)^2}\hat{L}_1\hat{P}\hat{L}_1\frac{\hat{Q}}{-\hat{L}_0}\hat{L}_1 + \hat{P}\hat{L}_1\left(\frac{\hat{Q}}{-\hat{L}_0}\hat{L}_1\right)^3 \\ &+\frac{\hat{Q}}{-\hat{L}_0}\hat{L}_1\hat{P}\hat{L}_1\left(\frac{\hat{Q}}{-\hat{L}_0}\hat{L}_1\right)^2 + (\frac{\hat{Q}}{-\hat{L}_0}\hat{L}_1)^2\hat{P}\hat{L}_1\frac{\hat{Q}}{-\hat{L}_0}\hat{L}_1\} \\ &-\frac{\hat{Q}}{(-\hat{L}_0)^2}\hat{L}_1\hat{P}\hat{L}_1\left(\frac{\hat{Q}}{-\hat{L}_0}\hat{L}_1\right)^2 - \frac{\hat{Q}}{-\hat{L}_0}\left(\frac{\hat{Q}}{-\hat{L}_0}\hat{L}_1\right)^2\hat{L}_1\hat{P}\hat{L}_1\frac{\hat{Q}}{-\hat{L}_0}\hat{L}_1 \\ &-\frac{\hat{Q}}{(-\hat{L}_0)^2}\hat{L}_1\frac{\hat{Q}}{-\hat{L}_0}\hat{L}_1\hat{P}\hat{L}_1\frac{\hat{Q}}{-\hat{L}_0}\hat{L}_1 + \left(\frac{\hat{Q}}{-\hat{L}_0}\hat{L}_1\right)^4\Big]\tilde{W}_0. \end{aligned} \tag{7.150}$$

We remark that the 'initial' value at $\tau = \tau_0$ has no P-space component;

$$\hat{P}\tilde{W}_4(\tau_0, x, y; \tau_0) = 0. \tag{7.151}$$

[The perturbative result in the 4th order]
Summing up the results up to the fourth order, we have the approximate solution with *secular terms*

$$\tilde{W}(\tau, x, y; \tau_0) \simeq \sum_{i=0}^{4} \epsilon^i \tilde{W}_i(\tau, x, y; \tau_0). \tag{7.152}$$

Since it is quite lengthy, we do not write down the explicit expression of the sum. Because of the secular terms, this solution given in the naive perturbation theory becomes invalid as $\tau - \tau_0 \to \infty$.

7.4.4 Application of the RG/E Equation

Now we apply the RG/E equation to the perturbative solution with secular terms

$$\left.\frac{d\tilde{W}}{d\tau_0}\right|_{\tau_0=\tau} = 0. \tag{7.153}$$

Applying the projection operators and recovering the original time variable $\tau = t/\epsilon$, we have for the P- and Q-space components

$$\frac{\partial}{\partial t}\hat{P}W_0(\tau, x, y) = \epsilon\hat{P}\hat{L}_1\left[\sum_{n=1}^{3}\epsilon^{n-1}\left(\frac{\hat{Q}}{-\hat{L}_0}\hat{L}_1\right)^n - \epsilon^2\frac{\hat{Q}}{-\hat{L}_0}\frac{\hat{Q}}{-\hat{L}_0}\hat{L}_1\hat{P}\hat{L}_1\frac{\hat{Q}}{-\hat{L}_0}\hat{L}_1\right]W_0(\tau, x, y), \tag{7.154}$$

and

$$\left[\frac{\hat{Q}}{-\hat{L}_0}\hat{L}_1+\epsilon\left(\frac{\hat{Q}}{-\hat{L}_0}\hat{L}_1\right)^2\right]\frac{\partial}{\partial t}W_0=\epsilon\Big\{\frac{\hat{Q}}{-\hat{L}_0}\hat{L}_1\hat{P}\hat{L}_1\frac{\hat{Q}}{-\hat{L}_0}\hat{L}_1+\epsilon^2\frac{\hat{Q}}{-\hat{L}_0}\hat{L}_1\hat{P}\hat{L}_1\Big(\frac{\hat{Q}}{-\hat{L}_0}\hat{L}_1\Big)^2$$
$$+\epsilon^2\left(\frac{\hat{Q}}{-\hat{L}_0}\hat{L}_1\right)^2\hat{P}\hat{L}_1\frac{\hat{Q}}{-\hat{L}_0}\hat{L}_1\Big\}W_0, \tag{7.155}$$

respectively. Here we have utilized the fact that $\tilde{W}_0(\tau,x,y;\tau_0=\tau)=W_0(\tau,x,y)$ and Eq. (7.127). Here it is found that the second equation (7.155) turns out redundant because it follows from the first one (7.154).

We note that Eq. (7.154) is exactly the same as the one given in [74, 122, 125] in quite different methods. Indeed, our $\hat{L}_0$ and $\hat{L}_1$ correspond to $-L_1$ and L_2 in Sect. 6.4 of Ref. [75], respectively, and then Eq. (7.154) is recognized to be the same as Eq. (6.4.101) in Ref. [75] up to the 'initial' value due to the Laplace transformation. We note that the fourth-order term in (7.154) may be rearranged as a multiplicative operator in the left hand side as if a wave-function renormalization in quantum field theory [57], as

$$\left(1+\epsilon^2\hat{P}\hat{L}_1\frac{\hat{Q}}{(-\hat{L}_0)^2}\hat{L}_1\right)\frac{\partial}{\partial t}W_0(\tau,x,y)=\epsilon\hat{P}\hat{L}_1\left[\sum_{n=1}^{3}\epsilon^{n-1}\Big(\frac{\hat{Q}}{-\hat{L}_0}\hat{L}_1\Big)^n\right]W_0(\tau,x,y). \tag{7.156}$$

In fact, multiplying Eq. (7.155) by $-\epsilon\hat{P}\hat{L}_1\hat{L}_0^{-1}$ and summing the result with Eq. (7.154), one arrives at Eq. (7.156). Equation (7.156) corresponds to Eq. (6.4.100) in [74] with identification

$$\begin{cases}\hat{A}=\hat{P}\hat{L}_1\frac{\hat{Q}}{-\hat{L}_0}\hat{L}_1,\\ \hat{B}=\hat{P}\hat{L}_1\frac{\hat{Q}}{-\hat{L}_0}\hat{L}_1\frac{\hat{Q}}{-\hat{L}_0}\hat{L}_1,\\ \hat{C}=\hat{P}\hat{L}_1\frac{\hat{Q}}{-\hat{L}_0}\hat{L}_1\frac{\hat{Q}}{-\hat{L}_0}\hat{L}_1\frac{\hat{Q}}{-\hat{L}_0}\hat{L}_1,\\ \hat{D}=-\hat{P}\hat{L}_1\frac{\hat{Q}}{(-\hat{L}_0)^2}\hat{L}_1,\end{cases} \tag{7.157}$$

where we have attached the hat ^ for indicating the operator nature. In terms of these symbols, Eq. (7.154) is expressed as

$$\frac{\partial}{\partial t}\hat{P}W_0(\tau,x,y)=\epsilon\left[\hat{A}+\epsilon\hat{B}+\epsilon^2(\hat{C}+\hat{D}\hat{A})\right]W_0(\tau,x,y) \tag{7.158}$$

with $\epsilon=\gamma^{-1}$. Equation (7.158) is to be compared with (6.4.101) in [74].

We shall not use the form (7.156) in the following, as is usual in the literature [74, 122, 125].

7.4.5 Smoluchowski Equation with Corrections

Next let us calculate the action of the projection operator $\hat{P}$ in (7.154), which leads to the equation of the space-time dependence of $\mathcal{P}_0(x;\tau)$. To show that this is the case, let us operating the projection $\hat{P}$ on the both sides of (7.156).[10] We shall show a couple of typical integrations. Firstly, noting that the $\hat{P}$-projection and τ-derivative are commutable, we have

$$\hat{P}\frac{\partial}{\partial\tau}W_0(\tau,x,y)=\frac{\partial}{\partial\tau}\int_{\infty}^{\infty}dy\,\tilde{\varphi}_0(y;x)W_0(\tau,x,y)=\frac{\partial\mathcal{P}_0(x;\tau)}{\partial\tau}\int_{\infty}^{\infty}dy\,\tilde{\varphi}_0(y;x)\varphi_0(y;x)$$
$$=\frac{\partial\mathcal{P}_0(x;\tau)}{\partial\tau}, \tag{7.159}$$

where we have used the expression

$$W_0(\tau,x,y)=\mathcal{P}_0(x;\tau)\varphi_0(y;x) \tag{7.160}$$

owing to (7.126).

Inserting the completeness condition (7.117) between the operators and using the bra-ket notation as in Sect. 3.6.2

$$|n\rangle\langle\tilde{n}|=\varphi_n(y;x)\int_{-\infty}^{\infty}dy'\,\tilde{\varphi}_n(y';x) \tag{7.161}$$

we have

$$\begin{aligned}\hat{A}W_0(\tau,x,y) &= \hat{P}\hat{L}_1\frac{\hat{Q}}{-\hat{L}_0}\hat{L}_1W_0(\tau,x,y)=\hat{P}\hat{L}_1\frac{\hat{Q}}{-\hat{L}_0}\hat{L}_1|0\rangle\mathcal{P}_0(x;\tau)\\ &= \sum_{n,m,l}\langle\tilde{\varphi}_0|\hat{L}_1|n\rangle\langle\tilde{n}|\frac{\hat{Q}}{-\hat{L}_0}|m\rangle\langle\tilde{m}|\hat{L}_1|l\rangle\langle\tilde{l}|0\rangle\mathcal{P}_0(x;\tau)\\ &= \sum_{n\neq 0}\langle\tilde{0}|\hat{L}_1|n\rangle\frac{1}{\lambda_n(x)}\langle\tilde{n}|\hat{L}_1|0\rangle\mathcal{P}_0(x;\tau)\\ &= \sum_{n=1}^{\infty}\hat{\mathcal{L}}_{0,n}\frac{1}{\lambda_n(x)}\hat{\mathcal{L}}_{n,0}\mathcal{P}_0(x;\tau)\\ &=: \hat{\mathcal{L}}_A\mathcal{P}_0(x;\tau),\end{aligned} \tag{7.162}$$

where $\hat{L}_{n,m}(x)$ denotes the y-averaged operator as given by

$$\hat{\mathcal{L}}_{n,m}(x):=\int dy\tilde{\varphi}_n(y;x)\hat{L}_1(x,y)\varphi_m(y;x). \tag{7.163}$$

[10] Note that this operation is nothing but performing the integration $\int dy$ with respect to the fast variable y because $\tilde{\varphi}(y;x)=1$.

Quite similarly,

$$\begin{aligned}
\hat{B}W_0(\tau,x,y) &= \hat{P}\hat{L}_1\frac{\hat{Q}}{-\hat{L}_0}\hat{L}_1\frac{\hat{Q}}{-\hat{L}_0}\hat{L}_1W_0(\tau,x,y)\\
&= \sum_{n,m=1}^{\infty}\hat{\mathcal{L}}_{0,n}\frac{1}{\lambda_n(x)}\hat{\mathcal{L}}_{n,m}\frac{1}{\lambda_m(x)}\hat{\mathcal{L}}_{m,0}\mathcal{P}_0(x;\tau)\\
&=: \hat{\mathcal{L}}_B\mathcal{P}_0(x;\tau), \qquad (7.164)\\
\hat{C}W_0(\tau,x,y) &= \hat{P}\hat{L}_1\frac{\hat{Q}}{-\hat{L}_0}\hat{L}_1\frac{\hat{Q}}{-\hat{L}_0}\hat{L}_1\frac{\hat{Q}}{-\hat{L}_0}\hat{L}_1W_0(\tau,x,y)\\
&= \sum_{n,m,l=1}^{\infty}\hat{\mathcal{L}}_{0,n}\frac{1}{\lambda_n(x)}\hat{\mathcal{L}}_{n,m}\frac{1}{\lambda_m(x)}\hat{\mathcal{L}}_{m,l}\frac{1}{\lambda_l(x)}\hat{\mathcal{L}}_{l,0}\mathcal{P}_0(x;\tau)\\
&=: \hat{\mathcal{L}}_C\mathcal{P}_0(x;\tau). \qquad (7.165)
\end{aligned}$$

We have also

$$\begin{aligned}
\hat{D}|0\rangle &= -\hat{P}\hat{L}_1\frac{\hat{Q}}{(-\hat{L}_0)^2}\hat{L}_1|0\rangle = -\sum_{n,m,l}\langle\tilde{0}|\hat{L}_1|n\rangle\langle\tilde{n}|\frac{\hat{Q}}{(-\hat{L}_0)^2}|m\rangle\langle\tilde{m}|\hat{L}_1|0\rangle\\
&= -\sum_{n\neq 0}\langle\tilde{0}|\hat{L}_1|n\rangle\frac{1}{\lambda_n^2}\langle\tilde{n}|\hat{L}_1|0\rangle = -\sum_{n=1}^{\infty}\hat{\mathcal{L}}_{0,n}\frac{1}{\lambda_n^2(x)}\hat{\mathcal{L}}_{n,0} =: \hat{\mathcal{L}}_D.
\end{aligned} \tag{7.166}$$

Thus (7.158) takes the form

$$\frac{\partial}{\partial t}\mathcal{P}_0(x,t) = \gamma^{-1}\left[\hat{\mathcal{L}}_A + \gamma^{-1}\hat{\mathcal{L}}_B + \gamma^{-2}(\hat{\mathcal{L}}_C + \hat{\mathcal{L}}_D\hat{\mathcal{L}}_A)\right]\mathcal{P}_0(x,t). \tag{7.167}$$

This is a generalized corrected Smoluchowski equation corresponding to the Langevin equation with a multiplicative noise.

7.4.6 *Simple Examples*

As examples, we shall give a Smoluchowski equation for a multiplicative noise and the corrected Smoluchowski equation.

Let us take the following model equation with a multiplicative noise [129];

$$\begin{cases}\dot{x} = y,\\ \dot{y} = -\gamma y - U'(x) + \sqrt{\gamma T(x)}\Gamma_y(t),\end{cases} \tag{7.168}$$

with $U'(x) = dU(x)/dx$. This equation is obtained from (7.71) with the replacement

$$h_x(x, y) = y, \quad g_x(x, y) = 0, \quad h_y(x, y) = -y,$$
$$f(x, y) = -U'(x), \quad g_y(x, y) = \sqrt{T(x)}. \tag{7.169}$$

We assume for the time correlation of the noise as

$$\langle \Gamma_y(t_1)\Gamma_y(t_2)\rangle = 2\delta(t_1 - t_2), \tag{7.170}$$

where we have set $D_y = 1$ in (7.72) for simplicity. Then $D^{(i)}$ $(i = 1, 2)$ defined in (7.91) now read

$$D^{(1)} = -y, \quad D^{(2)} = T(x). \tag{7.171}$$

The corresponding F-P equation reads

$$\frac{\partial W(x, y, \tau)}{\partial \tau} = (\hat{L_0} + \gamma^{-1}\hat{L}_1)W(x, y, \tau) \tag{7.172}$$

with

$$\hat{L}_0 = \frac{\partial}{\partial y}y + T(x)\frac{\partial^2}{\partial y^2}, \quad \hat{L}_1 = -y\frac{\partial}{\partial x} + U'(x)\frac{\partial}{\partial y}. \tag{7.173}$$

The normalized stationary state is given from the general formula of the zero mode (7.93) as

$$\begin{aligned}\varphi_0(y; x) &= \frac{N}{D^{(2)}}\exp\left[\int_0^y dy'\frac{D^{(1)}}{D^{(2)}}\right] = \frac{N}{T(x)}\exp\left[-\frac{1}{T(x)}\int_0^y dy'\, y'\right] \\ &= \frac{N}{T(x)}\mathrm{e}^{-y^2/(2T(x))} = \frac{1}{\sqrt{2T(x)}}\mathrm{e}^{-y^2/(2T(x))},\end{aligned} \tag{7.174}$$

where the normalization constant N has been chosen to be

$$N = \sqrt{T(x)/2}. \tag{7.175}$$

Then the operation of the hermitian operator $\hat{H}_0$ on $\psi(y; x)$ is evaluated as follows,

$$\begin{aligned}\hat{H}_0\psi(y; x) &= \mathrm{e}^{y^2/4T(x)}\left(\frac{\partial}{\partial y}y + T(x)\frac{\partial^2}{\partial y^2}\right)\mathrm{e}^{-y^2/4Tx}\psi(y; x) \\ &= \left(T(x)\frac{\partial^2}{\partial y^2} - \frac{y^2}{4T(x)} + \frac{1}{2}\right)\psi(y; x) \\ &= -\left(-\frac{1}{2}\frac{\partial^2}{\partial \eta^2} + \frac{1}{2}\eta^2\right)\psi(y; x) + \frac{1}{2}\psi(y; x)\end{aligned} \tag{7.176}$$

with

$$\eta := y/\sqrt{2T(x)}. \tag{7.177}$$

Then the eigenvalue equation for $\hat{H}_0$ (7.109) now reads

$$\left(-\frac{1}{2}\frac{\partial^2}{\partial\eta^2} + \frac{1}{2}\eta^2\right)\psi_n(\eta; x) = \left(\lambda_n(x) + \frac{1}{2}\right)\psi_n(\eta; x). \tag{7.178}$$

This is exactly the eigenvalue equation for a harmonic oscillator with the potential $\eta^2/2$ in elementary quantum mechanics [117, 118]. Thus we have

$$\lambda_n = n, \quad (n = 0, 1, 2\ldots) \tag{7.179}$$

$$\varphi_{n\geq 1}(y; x) = \frac{1}{2^n n!}H_n(y/\sqrt{2T(x)})\varphi_0(y; x), \tag{7.180}$$

where $H_n(x)$ is the Hermite polynomial in the n-th order. The conjugate eigenfunctions are

$$\tilde{\varphi}_n(y; x) = H_n(y/\sqrt{2T(x)}), \quad \tilde{\varphi}_0(y; x) = 1. \tag{7.181}$$

In accordance with the general argument presented before, the projection operator $\hat{P}$ is given by a simple integration

$$\hat{P} = \varphi_0(y; x)\int dy. \tag{7.182}$$

To obtain the reduced dynamics, we only have to calculate the y-averaged operators $\hat{\mathcal{L}}_{n,0}(x)$ and $\hat{\mathcal{L}}_{0,n}(x)$.

Since the eigenfunctions are given by those of the harmonic oscillator in elementary quantum mechanics [117, 118], the necessary calculations to be done below are straightforward even when somewhat tedious. Therefore we shall omit the details of the computational processes in the following.

First we notice that since $\hat{P}\hat{L}_1\hat{P} = 0$,

$$\hat{\mathcal{L}}_1 = \hat{\mathcal{L}}_{0,0}(x) = 0. \tag{7.183}$$

The non-vanishing term is evaluated to be

$$\hat{\mathcal{L}}_{0,n}(x) = \int dy\hat{L}_1\varphi_n(y; x) = -\int dy(y\partial_x - U'(x)\partial_y)\varphi_n(y; x) = -\delta_{1,n}\frac{\partial}{\partial x}\sqrt{\frac{T}{2}}. \tag{7.184}$$

So we only have to calculate $\hat{\mathcal{L}}_{1,0}$: A straightforward calculation gives

$$\begin{aligned}\hat{L}_1\varphi_0(y;x) &= -(y\partial_x - U'(x)\partial_y)\varphi_0(y;x)\\ &= -\varphi_1(y;x)\sqrt{\frac{2}{T}}\left(T\frac{\partial}{\partial x}+U'(x)+\frac{\partial T}{\partial x}\right) - \varphi_3(y;x)\sqrt{\frac{72}{T}}\frac{\partial T}{\partial x},\end{aligned} \tag{7.185}$$

which implies that

$$\hat{\mathcal{L}}_{m,0} = \begin{cases} -\sqrt{\frac{2}{T}}\left(T\frac{\partial}{\partial x}+U'(x)+\frac{\partial T}{\partial x}\right); & m=1,\\ -\sqrt{\frac{72}{T}}\frac{\partial T}{\partial x}; & m=3,\\ 0; & \text{others}.\end{cases} \tag{7.186}$$

Using the above results, one obtains $\hat{\mathcal{L}}_A$ defined in (7.162) as

$$\hat{\mathcal{L}}_A = \hat{L}_{0,1}\hat{L}_{1,0} = \frac{\partial}{\partial x}\left(T\frac{\partial}{\partial x}+U'(x)+\frac{\partial T}{\partial x}\right). \tag{7.187}$$

Because of (7.167), we now find that the generalized Smoluchowski equation for multiplicative noise without corrections takes the form [129]

$$\begin{aligned}\frac{\partial \mathcal{P}_0(x,t)}{\partial t} &= \gamma^{-1}\hat{\mathcal{L}}_A\mathcal{P}_0(x,t)\\ &= \gamma^{-1}\frac{\partial}{\partial x}\left(T\frac{\partial}{\partial x}+U'(x)+\frac{\partial T}{\partial x}\right)\mathcal{P}_0(x,t).\end{aligned} \tag{7.188}$$

If $T(x)$ is independent of x, (7.188) reduces exactly to the Smoluchowski equation [128].

Next let us investigate corrections to the Smoluchowski equation (7.188). According to (7.167), we need to calculate the operator $\hat{\mathcal{L}}_B$ for the γ^{-2} correction and $\hat{\mathcal{L}}_C + \hat{\mathcal{L}}_D\hat{\mathcal{L}}_A$ for the γ^{-3} one.

On account of (7.184) and (7.186),

$$\hat{\mathcal{L}}_B = \sum_{m=1,3}\hat{\mathcal{L}}_{0,1}\frac{1}{\lambda_1(x)}\hat{\mathcal{L}}_{1,m}\frac{1}{\lambda_m(x)}\hat{\mathcal{L}}_{m,0}. \tag{7.189}$$

However, owing to the parity conservation, it is found that

$$\hat{\mathcal{L}}_{1,1} = \hat{\mathcal{L}}_{1,3} = 0, \tag{7.190}$$

and hence

$$\hat{\mathcal{L}}_B = 0. \tag{7.191}$$

Again on account of (7.184) and (7.186),

$$\hat{\mathcal{L}}_C = \sum_{m=1}^{\infty} \sum_{l=1,3} \hat{\mathcal{L}}_{0,1} \hat{\mathcal{L}}_{1,m} \frac{1}{m} \hat{\mathcal{L}}_{m,l} \frac{1}{l} \hat{\mathcal{L}}_{l,0}. \tag{7.192}$$

In the following, we shall deal with the case where T has no x dependence.Then, a straightforward computation gives

$$\hat{\mathcal{L}}_{1,m} = -\delta_{m,2}\sqrt{T/2}\frac{\partial}{\partial x}, \tag{7.193}$$

$$\hat{\mathcal{L}}_{2,l} = \sqrt{T/2}\left\{\delta_{l,3}\frac{\partial}{\partial x} - \delta_{l,1}4\left(\frac{\partial}{\partial x} + U'(x)/T\right)\right\}, \tag{7.194}$$

$$\hat{\mathcal{L}}_{3,0} = 0. \tag{7.195}$$

Thus we have

$$\hat{\mathcal{L}}_C = \frac{\partial^2}{\partial x^2}\left(T\frac{\partial}{\partial x} + U'(x)\right)^2. \tag{7.196}$$

Because of (7.184), $\mathcal{L}_{0,n} \propto \delta_{1,n}$, then (7.166) tells us that

$$\hat{\mathcal{L}}_D = -\hat{\mathcal{L}}_A, \tag{7.197}$$

which implies that

$$\hat{\mathcal{L}}_D\hat{\mathcal{L}}_A = -\frac{\partial}{\partial x}\left(T\frac{\partial}{\partial x} + U'(x)\right)\frac{\partial}{\partial x}\left(T\frac{\partial}{\partial x} + U'(x)\right). \tag{7.198}$$

Then putting

$$T\frac{\partial}{\partial x} + U'(x) =: \hat{F}, \tag{7.199}$$

we have

$$\hat{\mathcal{L}}_C + \hat{\mathcal{L}}_D\hat{\mathcal{L}}_A = \partial_X^2 \hat{F}\hat{F} - \partial_x\hat{F}\partial_x\hat{F} = \partial_X\left[\partial_x\hat{F} - \hat{F}\partial_x\right]\hat{F}. \tag{7.200}$$

However,

$$\partial_x\hat{F} - \hat{F}\partial_x = (T\partial_x^2 + U''(x) + U'(x)\partial_x) - (T\partial_x + U'(x))\partial_x = U''(x), \tag{7.201}$$

which leads to

$$\hat{\mathcal{L}}_C + \hat{\mathcal{L}}_D \hat{\mathcal{L}}_A = \frac{\partial}{\partial x} U''(x) \left(T \frac{\partial}{\partial x} + U'(x) \right). \tag{7.202}$$

Thus we arrive at the following corrected Smoluchowski equation [74, 122, 125]

$$\frac{\partial \mathcal{P}_0(x,t)}{\partial t} = \gamma^{-1} \frac{\partial}{\partial x} \left[1 + \gamma^{-2} U''(x) \right] \left(T \frac{\partial}{\partial x} + U'(x) \right) \mathcal{P}_0(x,t). \tag{7.203}$$

7.5 Concluding Remarks

The RG method has been used for deriving the Fokker-Planck (F-P) equation from the stochastic Liouville equation (Kubo's equation) in such a way that the coarse graining of time [134–136] is naturally achieved through the construction of the asymptotic invariant manifold [34, 110]. Technically speaking, the essential role the choice of the 'initial' value on the invariant manifold has been elucidated, which is reminiscent of the work by Bogoliubov [54], Lebowitz [111], Kubo [115] and Kawasaki [112].

We have also shown that the reduction theory [6] based on the RG method as presented in Chap. 5 makes a systematic method for obtaining the reduced evolution equation of the distribution function given solely in terms of *slow variables*, that is, the *adiabatic elimination of fast variables* appearing in the F-P equations.

The F-P equation is a linear equation with a form similar to Schrödinger equation, and the perturbation theory developed here has also quite similar structure as the usual time-(in)dependent perturbation theory in quantum mechanics, which includes the Born-Oppenheimer approximation [123, 125]. Thus the techniques developed in the present section may have useful implications to problems in quantum mechanics [137].

Chapter 8
RG/E Derivation of Dissipative Fluid Dynamics from Classical Non-relativistic Boltzmann Equation

8.1 Introduction: Fluid Dynamics as Asymptotic Slow Dynamics of Boltzmann Equation

The fluid dynamics may be identified with the slow dynamics of the kinetic equation of a many-body system in the asymptotic regime after a long time. For example, the time evolution of a rarefied many-body system can be described as a sequential relaxation process with different time scales, as follows [54, 112, 115].

(I) In the beginning of the time evolution of a prepared state, the whole dynamical evolution of the system will be governed by Hamiltonian dynamics that is time-reversal invariant, and the Liouville equation for the distribution functions holds.

(II) As the system gets old, the dynamics is relaxed into the kinetic regime, where the time evolution of the system is well described by a partial truncation of the BBGKY (Bogoliubov-Born-Green-Kirkwood-Yvon) hierarchy [54, 55, 115, 138] for the one-, two-, and s-body distribution functions with $s = 3, 4, \ldots$, and the Boltzmann equation for one-body distribution function describes a coarse grained slower dynamics, in which the time-reversal invariance is lost by a loss of information by the truncation.

(III) Then, as the system is further relaxed, the time evolution will be described in terms of the fluid dynamical variables, *i.e.*, the flow velocity, the particle-number density, and the local temperature. In this sense, the Navier-Stokes equation together with the energy equation with the heat current is the asymptotic slow dynamics of the kinetic equation. In the rest of the chapter, we shall collectively use the term of the Navier-Stokes equation including the energy equation with the heat current.

Thus one sees that the derivation of the Navier-Stokes equation provides us with a nice 'playground' to apply some reduction theories of dynamics as emphasized by Kuramoto [32]. Indeed the celebrated Chapman-Enskog method [113] for deriving the fluid dynamics from the Boltzmann equation can be viewed as an application of

T. Kunihiro et al., *Geometrical Formulation of Renormalization-Group Method as an Asymptotic Analysis*, Fundamental Theories of Physics 206,
https://doi.org/10.1007/978-981-16-8189-9_8

the multiple-scale method [20, 21]. Thus it is natural and intriguing to apply the RG method [1] as introduced in the previous chapters [3, 5, 6] as a powerful reduction theory to such an interesting task.[1] In this chapter, we deals with a non-relativistic rarefied system with a classical statistics and apply the RG method based on the perturbation theory [57, 58] to derive the Navier-Stokes equation from the Boltzmann equation [53, 113, 139–141] with the small inhomogeneity in the asymptotic regime being identified as a small parameter for the perturbative expansion.[2] The derivation consists of a two-fold construction, *i.e.*, the explicit construction of deformed distribution function as the attractive/invariant manifold and the reduced differential equation for the slow variables to be identified with the fluid dynamic equation where the microscopic expressions of the transport coefficients are explicitly given in terms of the distribution function in a form of the Kubo formula [142].

8.2 Basics of Non-relativistic Classical Boltzmann Equation

Let us consider a non-relativistic many body system composed of single species with a mass m. The Boltzmann equation [53, 113, 139–141] is a transport equation describing the time evolution of one-particle distribution function defined in the phase space $(\boldsymbol{r},\ \boldsymbol{p} = m\boldsymbol{v})$:

$$\frac{\partial}{\partial t} f(\boldsymbol{r},\ \boldsymbol{v},\ t) + \boldsymbol{v} \cdot \nabla f(\boldsymbol{r},\ \boldsymbol{v},\ t) = C[f](\boldsymbol{r},\ \boldsymbol{v},\ t), \tag{8.1}$$

with $\nabla = \partial/\partial \boldsymbol{r}$. The right-hand side of the above equation is called the collision integral,

$$\begin{aligned} C[f](\boldsymbol{r},\ \boldsymbol{v},\ t) = \frac{1}{2!} \int \mathrm{d}^3 \boldsymbol{v}_1 \int \mathrm{d}^3 \boldsymbol{v}_2 \int \mathrm{d}^3 \boldsymbol{v}_3\, \omega(\boldsymbol{v},\ \boldsymbol{v}_1 | \boldsymbol{v}_2,\ \boldsymbol{v}_3) \\ \times \Big(f(\boldsymbol{r},\ \boldsymbol{v}_2,\ t)\, f(\boldsymbol{r},\ \boldsymbol{v}_3,\ t) - f(\boldsymbol{r},\ \boldsymbol{v},\ t)\, f(\boldsymbol{r},\ \boldsymbol{v}_1,\ t) \Big), \end{aligned} \tag{8.2}$$

where $\omega(\boldsymbol{v},\ \boldsymbol{v}_1 | \boldsymbol{v}_2,\ \boldsymbol{v}_3)$ denotes the transition probability which comes from the microscopic two-particle interaction. We remark that $\omega(\boldsymbol{v},\ \boldsymbol{v}_1 | \boldsymbol{v}_2,\ \boldsymbol{v}_3)$ includes the delta functions reflecting the energy-momentum conservation laws as

$$\begin{aligned} \omega(\boldsymbol{v},\ \boldsymbol{v}_1 | \boldsymbol{v}_2,\ \boldsymbol{v}_3) \propto \delta(m\,|\boldsymbol{v}|^2/2 + m\,|\boldsymbol{v}_1|^2/2 - m\,|\boldsymbol{v}_2|^2/2 - m\,|\boldsymbol{v}_3|^2/2) \\ \times\, \delta^3(m\,\boldsymbol{v} + m\,\boldsymbol{v}_1 - m\,\boldsymbol{v}_2 - m\,\boldsymbol{v}_3), \end{aligned} \tag{8.3}$$

---

[1] As for other references of the RG method with a comment on its historical development, see Sect. 1.3.

[2] See also [56] where a sketch is given for deriving the Euler equation from the Boltzmann equation on the basis of the RG method as given in [1].

and satisfies the following relations based on the indistinguishability of identical particles and the time reversal symmetry in the scattering process:

$$\omega(\boldsymbol{v},\,\boldsymbol{v}_1|\boldsymbol{v}_2,\,\boldsymbol{v}_3)=\omega(\boldsymbol{v}_1,\,\boldsymbol{v}|\boldsymbol{v}_3,\,\boldsymbol{v}_2)=\omega(\boldsymbol{v}_2,\,\boldsymbol{v}_3|\boldsymbol{v},\,\boldsymbol{v}_1)=\omega(\boldsymbol{v}_3,\,\boldsymbol{v}_2|\boldsymbol{v}_1,\,\boldsymbol{v}). \tag{8.4}$$

On account of the symmetry property (8.4), the convolution of the collision integral and an arbitrary function $\Phi(\boldsymbol{v})$ of $\boldsymbol{v}$ can be into the following symmetric form,

$$\begin{aligned}&\int \mathrm{d}^3\boldsymbol{v}\,\Phi(\boldsymbol{v})\,C[f](\boldsymbol{r},\,\boldsymbol{v},\,t)\\&=\frac{1}{2!}\frac{1}{4}\int \mathrm{d}^3\boldsymbol{v}\int \mathrm{d}^3\boldsymbol{v}_1\int \mathrm{d}^3\boldsymbol{v}_2\int \mathrm{d}^3\boldsymbol{v}_3\,\omega(\boldsymbol{v},\,\boldsymbol{v}_1|\boldsymbol{v}_2,\,\boldsymbol{v}_3)\\&\quad\times(\Phi(\boldsymbol{v})+\Phi(\boldsymbol{v}_1)-\Phi(\boldsymbol{v}_2)-\Phi(\boldsymbol{v}_3))\\&\quad\times\Big(f(\boldsymbol{r},\,\boldsymbol{v}_2,\,t)\,f(\boldsymbol{r},\,\boldsymbol{v}_3,\,t)-f(\boldsymbol{r},\,\boldsymbol{v},\,t)\,f(\boldsymbol{r},\,\boldsymbol{v}_1,\,t)\Big).\end{aligned}\tag{8.5}$$

We call a function $\Phi_c(\boldsymbol{v})$ a **collision invariant** if the convolution vanishes as

$$\int \mathrm{d}^3\boldsymbol{v}\,\Phi_c(\boldsymbol{v})\,C[f](\boldsymbol{r},\,\boldsymbol{v},\,t)=0. \tag{8.6}$$

Now, from the expression (8.5), it is clear that if $\Phi_c(\boldsymbol{v})$ is a conserved quantity in the collision process satisfying

$$\Phi_c(\boldsymbol{v}_2)\,+\,\Phi_c(\boldsymbol{v}_3)=\Phi_c(\boldsymbol{v}_1)\,+\,\Phi_c(\boldsymbol{v}), \tag{8.7}$$

$\Phi_c(\boldsymbol{v})$ is a collision invariant. Then one sees that the energy $m\,|\boldsymbol{v}|^2/2$ and the momentum $m\boldsymbol{v}$ are collision invariants due to the energy and momentum conservation, and a constant, *i.e.*, 1, is also a collision invariant due to the particle number conservation law. Thus we have

$$\int \mathrm{d}^3\boldsymbol{v}\,1\,C[f](\boldsymbol{r},\,\boldsymbol{v},\,t)=0, \tag{8.8}$$

$$\int \mathrm{d}^3\boldsymbol{v}\,mv^i\,C[f](\boldsymbol{r},\,\boldsymbol{v},\,t)=0, \qquad (i=1,\,2,\,3) \tag{8.9}$$

$$\int \mathrm{d}^3\boldsymbol{v}\,m|\boldsymbol{v}|^2/2\,C[f](\boldsymbol{r},\,\boldsymbol{v},\,t)=0. \tag{8.10}$$

In the present chapter, we assume that there is no other conserved quantities than the above five quantities. Then it is concluded that any collision invariant $\Phi_c(\boldsymbol{v})$ can be expressed as a linear combination of the five collision invariants given above

$$\Phi_c(\boldsymbol{v})=\alpha(\boldsymbol{r},\,t)+\boldsymbol{\beta}(\boldsymbol{r},\,t)\cdot m\boldsymbol{v}+\gamma(\boldsymbol{r},\,t)\,m\,|\boldsymbol{v}|^2/2. \tag{8.11}$$

An important notice is in order here: The coefficients $\alpha(\boldsymbol{r},\, t)$, $\boldsymbol{\beta}(\boldsymbol{r},\, t)$, and $\gamma(\boldsymbol{r},\, t)$ may depend on $(\boldsymbol{r},\, t)$.

Corresponding to the five collision invariants $(1,\, m\boldsymbol{v},\, m\,|\boldsymbol{v}|^2/2)$, the Boltzmann equation (8.1) leads to the following balance equations

$$\frac{\partial}{\partial t}\rho + \nabla \cdot (\rho\, \boldsymbol{U}) = 0, \quad (8.12)$$

$$\frac{\partial}{\partial t}(\rho\, U^i) + \nabla^j (\rho\, U^j\, U^i + P^{ji}) = 0, \quad (8.13)$$

$$\frac{\partial}{\partial t}(\rho\,|\boldsymbol{U}|^2/2 + E) + \nabla^j \big((\rho\,|\boldsymbol{U}|^2/2 + E)\, U^j + Q^j + P^{ji}\, U^i\big) = 0, \quad (8.14)$$

where we have introduced the *mass* density ρ, the flow velocity U^i of the whole system, the internal energy E, the pressure tensor P^{ij}, and the heat flux Q^i, which are defined by

$$\rho(\boldsymbol{r},\, t) := m \int \mathrm{d}^3\boldsymbol{v}\, f(\boldsymbol{r},\, \boldsymbol{v},\, t), \quad (8.15)$$

$$U^i(\boldsymbol{r},\, t) := \frac{m}{\rho} \int \mathrm{d}^3\boldsymbol{v}\, v^i\, f(\boldsymbol{r},\, \boldsymbol{v},\, t), \quad (8.16)$$

$$E(\boldsymbol{r},\, t) := \int \mathrm{d}^3\boldsymbol{v}\, \frac{m\,|\boldsymbol{v} - \boldsymbol{U}|^2}{2}\, f(\boldsymbol{r},\, \boldsymbol{v},\, t), \quad (8.17)$$

$$P^{ij}(\boldsymbol{r},\, t) := \int \mathrm{d}^3\boldsymbol{v}\, m\, (v^i - U^i)\, (v^j - U^j)\, f(\boldsymbol{r},\, \boldsymbol{v},\, t), \quad (8.18)$$

$$Q^i(\boldsymbol{r},\, t) := \int \mathrm{d}^3\boldsymbol{v}\, \frac{m\,|\boldsymbol{v} - \boldsymbol{U}|^2}{2}\, (v^i - U^i)\, f(\boldsymbol{r},\, \boldsymbol{v},\, t), \quad (8.19)$$

respectively. With the use of Eqs. (8.12) and (8.13), Eqs. (8.13) and (8.14) can be reorganized to make the following forms

$$\rho\, \frac{\partial}{\partial t} U^i + \rho\, \boldsymbol{U} \cdot \nabla U^i + \nabla^j P^{ji} = 0, \quad (8.20)$$

$$\rho\, \frac{\partial}{\partial t}(E/\rho) + \rho\, \boldsymbol{U} \cdot \nabla (E/\rho) + P^{ij}\, \nabla^i U^j + \nabla \cdot \boldsymbol{Q} = 0. \quad (8.21)$$

Here the first and the second equations may be viewed as the Newton equation and the energy equation, respectively.

We note that although the set of the balance equations (8.12), (8.20), and (8.21) takes seemingly the same form as the set of the fluid dynamic equations, it is not closed and has no dynamical information unless an explicit time dependence of $f(\boldsymbol{r},\, \boldsymbol{v},\, t)$ is obtained as a solution to the Boltzmann equation (8.1).

In the Boltzmann theory, an entropy density and an entropy current are defined by

$$s(\boldsymbol{r},\, t) := -\int \mathrm{d}^3\boldsymbol{v}\, f(\boldsymbol{r},\, \boldsymbol{v},\, t)\left(\ln f(\boldsymbol{r},\, \boldsymbol{v},\, t) - 1\right), \tag{8.22}$$

$$J_s^i(\boldsymbol{r},\, t) := -\int \mathrm{d}^3\boldsymbol{v}\, v^i\, f(\boldsymbol{r},\, \boldsymbol{v},\, t)\left(\ln f(\boldsymbol{r},\, \boldsymbol{v},\, t) - 1\right), \tag{8.23}$$

respectively. Then the Boltzmann equation (8.1) leads to the balance equation for the entropy density as

$$\frac{\partial}{\partial t} s(\boldsymbol{r},\, t) + \nabla \cdot \boldsymbol{J}_s(\boldsymbol{r},\, t) = \int \mathrm{d}^3\boldsymbol{v}\, C[f](\boldsymbol{r},\, \boldsymbol{v},\, t)\, \ln f(\boldsymbol{r},\, \boldsymbol{v},\, t). \tag{8.24}$$

If we define the entropy of the system by

$$S(t) := \int \mathrm{d}^3\boldsymbol{r}\, s(\boldsymbol{r},\, t), \tag{8.25}$$

it is found that the entropy $S(t)$ is conserved when $\ln f(\boldsymbol{r},\, \boldsymbol{v},\, t)$ is a collision invariant, and hence $f(\boldsymbol{r},\, \boldsymbol{v},\, t)$ is expressed as

$$f(\boldsymbol{r},\, \boldsymbol{v},\, t) = \exp\left[\alpha(\boldsymbol{r},\, t) + \boldsymbol{\beta}(\boldsymbol{r},\, t) \cdot m\boldsymbol{v} + \gamma(\boldsymbol{r},\, t)\, m\, |\boldsymbol{v}|^2/2\right], \tag{8.26}$$

which can be cast into the form of the Maxwellian by a slight rearrangement of the variables as

$$f^{\mathrm{eq}}(\boldsymbol{r},\, \boldsymbol{v},\, t) = n(\boldsymbol{r},\, t)\left[\frac{m}{2\pi\, T(\boldsymbol{r},\, t)}\right]^{\frac{3}{2}} \exp\left[\frac{m}{2T(\boldsymbol{r},\, t)}\, |\boldsymbol{v} - \boldsymbol{u}(\boldsymbol{r},\, t)|^2\right], \tag{8.27}$$

where the five fields n, T, and u^i are called fluid dynamical variables.

We note that the collision integral of $f^{\mathrm{eq}}(\boldsymbol{r},\, \boldsymbol{v},\, t)$ vanishes:

$$C[f^{\mathrm{eq}}](\boldsymbol{r},\, \boldsymbol{v},\, t) = 0, \tag{8.28}$$

due to the detailed balance condition for the equilibrium distribution,

$$\omega(\boldsymbol{v},\, \boldsymbol{v}_1|\boldsymbol{v}_2,\, \boldsymbol{v}_3)\left(f^{\mathrm{eq}}(\boldsymbol{r},\, \boldsymbol{v}_2,\, t)\, f^{\mathrm{eq}}(\boldsymbol{r},\, \boldsymbol{v}_3,\, t) - f^{\mathrm{eq}}(\boldsymbol{r},\, \boldsymbol{v},\, t)\, f^{\mathrm{eq}}(\boldsymbol{r},\, \boldsymbol{v}_1,\, t)\right) = 0, \tag{8.29}$$

which itself is owing to the energy-momentum conservation law (8.3).

Let us consider the simple case of the equilibrium state putting $f(\boldsymbol{r},\, \boldsymbol{v},\, t) = f^{\mathrm{eq}}(\boldsymbol{r},\, \boldsymbol{v},\, t)$. Then we find that the quantities defined by Eqs. (8.15)–(8.19) take the following forms in the equilibrium state

$$\rho = m\,n, \quad U^i = u^i, \quad E = \frac{3}{2}\,n\,T, \tag{8.30}$$

$$P^{ij} = n\,T\,\delta^{ij}, \quad Q^i = 0, \tag{8.31}$$

respectively. Substituting these expressions, we see that the balance equations (8.12), (8.20), and (8.21) take the following forms

$$\frac{\partial}{\partial t}n + \nabla \cdot (n\,\boldsymbol{u}) = 0, \tag{8.32}$$

$$m\,n\,\frac{\partial}{\partial t}u^i + m\,n\,\boldsymbol{u} \cdot \nabla u^i + \nabla^i(n\,T) = 0, \tag{8.33}$$

$$n\,\frac{\partial}{\partial t}(3\,T/2) + n\,\boldsymbol{u} \cdot \nabla(3\,T/2) + n\,T\,\nabla \cdot \boldsymbol{u} = 0, \tag{8.34}$$

respectively. These are nothing but the Euler equation for an ideal fluid without dissipative effects. One thus finds that the dissipative effects such as viscosities that are present in the Navier-Stokes equation come from the deviation or deformation of $f(\boldsymbol{r},\ \boldsymbol{v},\ t)$ from $f^{\rm eq}(\boldsymbol{r},\ \boldsymbol{v},\ t)$. In the subsequent sections, we shall obtain the very deformation by faithfully solving the Boltzmann equation in the asymptotic regime in the RG method.

8.3 Asymptotic Analysis and Dynamical Reduction of Boltzmann Equation in RG Method

In this section, we shall apply the RG method [57, 58] as formulated in the previous chapters [3, 5, 6] to make an asymptotic analysis of the Boltzmann equation in the asymptotic regime with a small spatial inhomogeneity, and deduce not only the attractive/invariant manifold spanned by the fluid dynamical variables but also the reduced differential equation for the fluid dynamical variables, which is to take the form of the Navier-Stokes equation.

8.3.1 Preliminaries and Set Up

To circumvent possible complexities inherent in continuous variables, we treat the arguments $\boldsymbol{v}$ as discrete variables, following [32]: With this treatment, one will readily recognize the apparent correspondence of the present treatment to the general formulation of the RG method [6] presented in Chap. 5.

Now let us consider a system confined in a finite volume V. Then using the relation

$$\int d^3 \boldsymbol{v} = \int d^3 \boldsymbol{p} \, \frac{1}{m^3} = \sum_{\boldsymbol{v}} \frac{(2\pi)^3}{V} \frac{1}{m^3} = \sum_{\boldsymbol{v}} h^3, \quad h := \frac{1}{m} \frac{2\pi}{V^{1/3}}, \tag{8.35}$$

we define a discrete form of the distribution function $f(\boldsymbol{r}, \boldsymbol{v}, t)$ by

$$f_{\boldsymbol{v}}(\boldsymbol{r}, t) := h^3 \, f(\boldsymbol{r}, \boldsymbol{v}, t), \tag{8.36}$$

which satisfies the relation

$$\int d^3 \boldsymbol{v} \, f(\boldsymbol{r}, \boldsymbol{v}, t) = \sum_{\boldsymbol{v}} f_{\boldsymbol{v}}(\boldsymbol{r}, t). \tag{8.37}$$

Then the Boltzmann equation (8.1) now reads

$$\frac{\partial}{\partial t} f_{\boldsymbol{v}}(\boldsymbol{r}, t) + \boldsymbol{v} \cdot \boldsymbol{\nabla} f_{\boldsymbol{v}}(\boldsymbol{r}, t) = C[f]_{\boldsymbol{v}}(\boldsymbol{r}, t), \tag{8.38}$$

where

$$\begin{aligned} C[f]_{\boldsymbol{v}}(\boldsymbol{r}, t) &:= h^3 \, C[f](\boldsymbol{r}, \boldsymbol{v}, t) \\ &= \frac{1}{2!} \sum_{\boldsymbol{v}_1} \sum_{\boldsymbol{v}_2} \sum_{\boldsymbol{v}_3} \omega(\boldsymbol{v}, \boldsymbol{v}_1 | \boldsymbol{v}_2, \boldsymbol{v}_3) \\ &\quad \times \big(f_{\boldsymbol{v}_2}(\boldsymbol{r}, t) f_{\boldsymbol{v}_3}(\boldsymbol{r}, t) - f_{\boldsymbol{v}}(\boldsymbol{r}, t) f_{\boldsymbol{v}_1}(\boldsymbol{r}, t) \big). \end{aligned} \tag{8.39}$$

As promised, we apply the RG method to extract the low-frequency and long-wavelength dynamics from the Boltzmann equation in the asymptotic regime, where we suppose that the spatial variation governed by Eq. (8.38) is small, that is,

$$\boldsymbol{\nabla} f_{\boldsymbol{v}}(\boldsymbol{r}, t) = \mathrm{O}(\epsilon). \tag{8.40}$$

For making this smallness explicit, we use a scaled coordinate $\bar{\boldsymbol{r}}$ defined by

$$\bar{\boldsymbol{r}} := \epsilon \, \boldsymbol{r}, \tag{8.41}$$

accordingly,

$$\boldsymbol{\nabla} = \epsilon \, \frac{\partial}{\partial \bar{\boldsymbol{r}}} =: \epsilon \, \bar{\boldsymbol{\nabla}}. \tag{8.42}$$

Then Eq. (8.38) is converted to a form to which the perturbation theory is readily applicable as

$$\frac{\partial}{\partial t} f_{\boldsymbol{v}}(\bar{\boldsymbol{r}}, t) = C[f]_{\boldsymbol{v}}(\bar{\boldsymbol{r}}, t) - \epsilon \, \boldsymbol{v} \cdot \bar{\boldsymbol{\nabla}} f_{\boldsymbol{v}}(\bar{\boldsymbol{r}}, t). \tag{8.43}$$

We note that this ϵ may be identified with the Knudsen number.

From now on, we write $\bar{\boldsymbol{r}}$ and $\bar{\nabla}$ as $\boldsymbol{r}$ and ∇, respectively, for simplicity.

Let $\tilde{f}_v(\boldsymbol{r},\, t\,;\, t_0)$ be an approximate solution to Eq. (8.43) around $t = t_0$ in the asymptotic region where the spatial inhomogeneity has become small. We make a perturbative expansion for $\tilde{f}_v(\boldsymbol{r},\, t\,;\, t_0)$ as

$$\tilde{f}_v(\boldsymbol{r},\, t\,;\, t_0) = \tilde{f}_v^{(0)}(\boldsymbol{r},\, t\,;\, t_0) + \epsilon\, \tilde{f}_v^{(1)}(\boldsymbol{r},\, t\,;\, t_0) + \epsilon^2\, \tilde{f}_v^{(2)}(\boldsymbol{r},\, t\,;\, t_0) + \cdots . \tag{8.44}$$

We impose that $\tilde{f}_v(\boldsymbol{r},\, t\,;\, t_0)$ satisfies the ‘initial’ condition at $t = t_0$ as

$$\tilde{f}_v(\boldsymbol{r},\, t = t_0\,;\, t_0) = f_v(\boldsymbol{r}\,;\, t_0), \tag{8.45}$$

where $f_v(\boldsymbol{r},\, t_0)$ is supposed to be the value of the exact solution $f_v(\boldsymbol{r},\, t)$ yet to be determined at $t = t_0$, and is also expanded as follows,

$$f_v(\boldsymbol{r}\,;\, t_0) = f_v^{(0)}(\boldsymbol{r}\,;\, t_0) + \epsilon\, f_v^{(1)}(\boldsymbol{r}\,;\, t_0) + \epsilon^2\, f_v^{(2)}(\boldsymbol{r}\,;\, t_0) + \cdots . \tag{8.46}$$

We set the ‘initial’ conditions order by order as

$$\tilde{f}_v^{(\ell)}(\boldsymbol{r},\, t = t_0\,;\, t_0) = f_v^{(\ell)}(\boldsymbol{r}\,;\, t_0), \quad \ell = 0,\, 1,\, 2,\, \ldots . \tag{8.47}$$

Substituting the expansion (8.44) into Eq. (8.43), we obtain the series of the perturbative equations. We will suppress the arguments $(\boldsymbol{r},\, t\,;\, t_0)$ and $(\boldsymbol{r}\,;\, t_0)$ when misunderstanding is not expected.

8.3.2 *Analysis of Unperturbed Solution*

The zeroth-order equation reads

$$\frac{\partial}{\partial t}\tilde{f}_v^{(0)} = C[\tilde{f}^{(0)}]_v. \tag{8.48}$$

Since we are interested in the slow motion which is to be realized asymptotically as $t \to \infty$, we look for the stationary solution

$$\frac{\partial}{\partial t}\tilde{f}_v^{(0)} = 0, \tag{8.49}$$

which is satisfied when

$$C[\tilde{f}^{(0)}]_v = 0. \tag{8.50}$$

This equation is identical to Eq. (8.28), and hence the solution is given by the local Maxwellian (8.27):

$$\tilde{f}_v^{(0)}(\boldsymbol{r}, t; t_0) = f_v^{\mathrm{eq}}(\boldsymbol{r}; t_0)$$
$$= h^3 n(\boldsymbol{r}; t_0) \left[\frac{m}{2\pi T(\boldsymbol{r}; t_0)} \right]^{\frac{3}{2}} \exp\left[-\frac{m\,|\boldsymbol{v} - \boldsymbol{u}(\boldsymbol{r}; t_0)|^2}{2T(\boldsymbol{r}; t_0)} \right]. \tag{8.51}$$

Here, the particle-number density $n(\boldsymbol{r}; t_0)$, local temperature $T(\boldsymbol{r}; t_0)$, and flow velocity $\boldsymbol{u}(\boldsymbol{r}; t_0)$ are integration constants which are independent of t but may depend on t_0 and $\boldsymbol{r}$.

The zeroth-order 'initial' value is given by

$$f_v^{(0)}(\boldsymbol{r}; t_0) = f_v^{\mathrm{eq}}(\boldsymbol{r}; t_0), \tag{8.52}$$

or in the vector notation,

$$\tilde{\boldsymbol{f}}^{(0)}(\boldsymbol{r}, t; t_0) = \boldsymbol{f}^{\mathrm{eq}}(\boldsymbol{r}; t_0), \tag{8.53}$$
$$\boldsymbol{f}^{(0)}(\boldsymbol{r}; t_0) = \boldsymbol{f}^{\mathrm{eq}}(\boldsymbol{r}; t_0), \tag{8.54}$$

where the components of the respective components are given by

$$\left[\tilde{\boldsymbol{f}}^{(0)}(\boldsymbol{r}, t; t_0)\right]_v := \tilde{f}_v^{(0)}(\boldsymbol{r}, t; t_0), \tag{8.55}$$

and so on.

8.3.3 First-Order Equation

The first-order equation now reads

$$\frac{\partial}{\partial t}\tilde{\boldsymbol{f}}^{(1)} = A\,\tilde{\boldsymbol{f}}^{(1)} + f^{\mathrm{eq}}\,\boldsymbol{F}, \tag{8.56}$$

where we have introduced an operator A and a vector $\boldsymbol{F}$, components of which are given by

$$A_{vk} := \left.\frac{\partial}{\partial f_k} C[f]_v \right|_{f=f^{\mathrm{eq}}}, \tag{8.57}$$

$$F_v := -\frac{1}{f_v^{\mathrm{eq}}}\,\boldsymbol{v}\cdot\nabla f_v^{\mathrm{eq}}, \tag{8.58}$$

respectively.

We shall solve Eq. (8.56) to construct a solution which describes a slow motion realized in the asymptotic regime. To this end, we clarify the properties of the linear operator A to proceed further.

First, we introduce a new operator L defined in terms of A as

$$\begin{aligned} L_{vk} &:= f_v^{\mathrm{eq}-1}\, A_{vk}\, f_k^{\mathrm{eq}} \\ &= -\frac{1}{2!}\sum_{v_1}\sum_{v_2}\sum_{v_3}\omega(v,\, v_1|v_2,\, v_3)\, f_{v_1}^{\mathrm{eq}}\,(\delta_{vk}+\delta_{v_1k}-\delta_{v_2k}-\delta_{v_3k}), \end{aligned} \tag{8.59}$$

which is called the collision operator [112, 113].

Let us define the inner product for arbitrary two vectors φ and ψ by

$$\langle\, \varphi\, ,\, \psi\, \rangle := \sum_v f_v^{\mathrm{eq}}\, \varphi_v\, \psi_v = \langle\, \psi\, ,\, \varphi\, \rangle. \tag{8.60}$$

Then it is found that the collision operator L is self-adjoint (Hermitian) with respect to this inner product.[3] In fact, we have for arbitrary vectors φ and ψ,

$$\langle\, \varphi\, ,\, L\, \psi\, \rangle = \langle\, L\, \varphi\, ,\, \psi\, \rangle. \tag{8.61}$$

A proof of Eq. (8.61) is given as follows: With the use of the explicit form of L in Eq. (8.59), the left-hand side of Eq. (8.61) is calculated to be

$$\begin{aligned} \langle\, \varphi\, ,\, L\, \psi\, \rangle &= \sum_v\sum_k f_v^{\mathrm{eq}}\, \varphi_v\, L_{vk}\, \psi_k \\ &= -\frac{1}{2!}\sum_v\sum_{v_1}\sum_{v_2}\sum_{v_3}\omega(v,\, v_1|v_2,\, v_3)\, f_v^{\mathrm{eq}}\, f_{v_1}^{\mathrm{eq}} \\ &\quad \times \varphi_v\,(\psi_v+\psi_{v_1}-\psi_{v_2}-\psi_{v_3}). \end{aligned} \tag{8.62}$$

Then bearing the symmetry property (8.4) of $\omega(v,\, v_1|v_2,\, v_3)$, we make the change of the dummy variables $(v,\, v_1,\, v_2,\, v_3)$ in the following three ways,

$$(v,\, v_1,\, v_2,\, v_3) \to (v_1,\, v,\, v_3,\, v_2),\ (v_2,\, v_3,\, v,\, v_1),\ \text{and}\ (v_3,\, v_2,\, v_1,\, v), \tag{8.63}$$

which give the following three equalities

$$\begin{aligned} \langle\, \varphi\, ,\, L\, \psi\, \rangle &= -\frac{1}{2!}\sum_v\sum_{v_1}\sum_{v_2}\sum_{v_3}\omega(v,\, v_1|v_2,\, v_3)\, f_v^{\mathrm{eq}}\, f_{v_1}^{\mathrm{eq}} \\ &\quad \times \varphi_{v_1}\,(\psi_v+\psi_{v_1}-\psi_{v_2}-\psi_{v_3}), \end{aligned} \tag{8.64}$$

[3] A proof is given in Appendix A (Sect. 8.5) that the inner product that makes the self-adjointness of L apparent is uniquely given as (8.60).

$$\langle \boldsymbol{\varphi}, L\boldsymbol{\psi} \rangle = \frac{1}{2!}\sum_{v}\sum_{v_1}\sum_{v_2}\sum_{v_3} \omega(\boldsymbol{v}, \boldsymbol{v}_1|\boldsymbol{v}_2, \boldsymbol{v}_3)\, f^{\mathrm{eq}}_{v}\, f^{\mathrm{eq}}_{v_1} \times \varphi_{v_2}\,(\psi_v + \psi_{v_1} - \psi_{v_2} - \psi_{v_3}), \tag{8.65}$$

$$\langle \boldsymbol{\varphi}, L\boldsymbol{\psi} \rangle = \frac{1}{2!}\sum_{v}\sum_{v_1}\sum_{v_2}\sum_{v_3} \omega(\boldsymbol{v}, \boldsymbol{v}_1|\boldsymbol{v}_2, \boldsymbol{v}_3)\, f^{\mathrm{eq}}_{v}\, f^{\mathrm{eq}}_{v_1} \times \varphi_{v_3}\,(\psi_v + \psi_{v_1} - \psi_{v_2} - \psi_{v_3}), \tag{8.66}$$

respectively. Adding the above four expressions of $\langle \boldsymbol{\varphi}, L\boldsymbol{\psi} \rangle$ and dividing by four, we have

$$\begin{aligned}\langle \boldsymbol{\varphi}, L\boldsymbol{\psi} \rangle &= -\frac{1}{2!}\frac{1}{4}\sum_{v}\sum_{v_1}\sum_{v_2}\sum_{v_3} \omega(\boldsymbol{v}, \boldsymbol{v}_1|\boldsymbol{v}_2, \boldsymbol{v}_3)\, f^{\mathrm{eq}}_{v}\, f^{\mathrm{eq}}_{v_1} \\ &\quad \times (\varphi_{v_2} + \varphi_{v_3} - \varphi_v - \varphi_{v_1})\,(\psi_{v_2} + \psi_{v_3} - \psi_v - \psi_{v_1}) \\ &= \langle \boldsymbol{\psi}, L\boldsymbol{\varphi} \rangle,\end{aligned} \tag{8.67}$$

where we have used the invariance under the exchange of $\boldsymbol{\varphi}$ and $\boldsymbol{\psi}$ in the last equality. However, on account of the symmetry property of the inner product (8.60), we have the equality

$$\langle \boldsymbol{\psi}, L\boldsymbol{\varphi} \rangle = \langle L\boldsymbol{\varphi}, \boldsymbol{\psi} \rangle, \tag{8.68}$$

which, combined with (8.67), completes the proof of the self-adjointness of L.

8.3.4 Spectral Analysis of Collision Operator L

Next, we investigate the spectral properties of the linear operator L. It is found that L has five zero modes, that is, eigenvectors whose eigenvalues are zero, and the dimension of the kernel of L is five, *i.e.*, $\dim[\mathrm{Ker}L] = 5$ [139]:

$$L\boldsymbol{\varphi}_0^{\alpha} = 0, \quad \alpha = 0, 1, 2, 3, 4, \tag{8.69}$$

where the five zero modes $\boldsymbol{\varphi}_{\alpha}^{0}$ are given by

$$\varphi^{0}_{0\,v} := 1, \tag{8.70}$$

$$\varphi^{i}_{0\,v} := m\,\delta v^{i}, \quad i = 1, 2, 3, \tag{8.71}$$

$$\varphi^{4}_{0\,v} := \frac{m}{2}\,|\boldsymbol{\delta v}|^2 - \frac{3}{2}\,T, \tag{8.72}$$

with $\boldsymbol{\delta v} := \boldsymbol{v} - \boldsymbol{u}$. We note that $\boldsymbol{\varphi}_0^{\alpha}$ are nothing but the collision invariants, which are orthogonal to each other as

$$\langle \boldsymbol{\varphi}_0^\alpha , \boldsymbol{\varphi}_0^\beta \rangle = c^\alpha \, \delta^{\alpha\beta}, \tag{8.73}$$

with

$$c^0 := n, \tag{8.74}$$
$$c^i := m\, n\, T, \quad i = 1,\, 2,\, 3, \tag{8.75}$$
$$c^4 := \frac{3}{2}\, n\, T^2. \tag{8.76}$$

The other eigenvalues than the zero modes are found to be negative because the following inequality holds for an arbitrary vector $\boldsymbol{\varphi}$

$$\begin{aligned}\langle \boldsymbol{\varphi} , L\, \boldsymbol{\varphi} \rangle &= -\frac{1}{2!}\,\frac{1}{4} \sum_{\boldsymbol{v}} \sum_{\boldsymbol{v}_1} \sum_{\boldsymbol{v}_2} \sum_{\boldsymbol{v}_3} \omega(\boldsymbol{v},\, \boldsymbol{v}_1 | \boldsymbol{v}_2,\, \boldsymbol{v}_3)\, f^{\mathrm{eq}}_{\boldsymbol{v}}\, f^{\mathrm{eq}}_{\boldsymbol{v}_1} \\ &\quad \times (\varphi_{\boldsymbol{v}_2} + \varphi_{\boldsymbol{v}_3} - \varphi_{\boldsymbol{v}} - \varphi_{\boldsymbol{v}_1})^2 \\ &\le 0, \end{aligned} \tag{8.77}$$

where the first line has been derived from Eq. (8.67) with $\boldsymbol{\psi} = \boldsymbol{\varphi}$. The equality is satisfied when

$$\varphi_{\boldsymbol{v}_2} + \varphi_{\boldsymbol{v}_3} = \varphi_{\boldsymbol{v}} + \varphi_{\boldsymbol{v}_1}, \tag{8.78}$$

which implies that $\varphi_{\boldsymbol{v}}$ must be a collision invariant. Thus we see that any eigenvector other than the zero modes must belong to a negative (definite) eigenvalue.

Finally, we define a projection operator P onto Ker$L$ spanned by the zero modes $\boldsymbol{\varphi}_0^\alpha$, by

$$\big[P\, \boldsymbol{\psi}\big]_{\boldsymbol{v}} := \sum_{\alpha=0}^{4} \varphi_{0\,\boldsymbol{v}}^\alpha \, \frac{1}{c^\alpha} \langle \boldsymbol{\varphi}_0^\alpha , \boldsymbol{\psi} \rangle, \tag{8.79}$$

and denote the projection operator to the space complement to Ker$L$ as

$$Q := 1 - P. \tag{8.80}$$

We call Ker$L$ the P space and the space complement to the P space the Q space.

8.3.5 *Solution to First-Order Equation*

Multiplying the first-order equation (8.56) by $f_{\boldsymbol{v}}^{\mathrm{eq}-1}$, we have

$$\frac{\partial}{\partial t}\big(f^{\mathrm{eq}-1}\, \tilde{\boldsymbol{f}}^{(1)}\big) = L\,\big(f^{\mathrm{eq}-1}\, \tilde{\boldsymbol{f}}^{(1)}\big) + \boldsymbol{F}. \tag{8.81}$$

The first-order solution is readily obtained as

$$f^{\mathrm{eq}-1}\,\tilde{\boldsymbol{f}}^{(1)}(t) = \mathrm{e}^{(t-t_0)L}\left[f^{\mathrm{eq}-1}\,\boldsymbol{f}^{(1)} + L^{-1}\,Q\,\boldsymbol{F}\right] + (t-t_0)\,P\,\boldsymbol{F} - L^{-1}\,Q\,\boldsymbol{F}. \tag{8.82}$$

The first-order 'initial' value vector $\boldsymbol{f}^{(1)}(\boldsymbol{r};t_0)$ is now determined so that the would-be fast motion coming from the Q space disappears. Thus, we can set

$$\boldsymbol{f}^{(1)}(\boldsymbol{r};t_0) = -f^{\mathrm{eq}}\,L^{-1}\,Q\,\boldsymbol{F}, \tag{8.83}$$

with which the first-order solution (8.82) is reduced to

$$\tilde{\boldsymbol{f}}^{(1)}(\boldsymbol{r},t;t_0) = f^{\mathrm{eq}}\left\{(t-t_0)\,P\,\boldsymbol{F} - L^{-1}\,Q\,\boldsymbol{F}\right\}. \tag{8.84}$$

We note the appearance of the secular term proportional to $t - t_0$.

8.3.6 Second-Order Solution

Then the second-order equation reads

$$\frac{\partial}{\partial t}(f^{\mathrm{eq}-1}\,\tilde{\boldsymbol{f}}^{(2)}) = L\,(f^{\mathrm{eq}-1}\,\tilde{\boldsymbol{f}}^{(2)}) + (t-t_0)^2\,\boldsymbol{G} + (t-t_0)\,\boldsymbol{H} + \boldsymbol{I}, \tag{8.85}$$

where $\boldsymbol{G}$, $\boldsymbol{H}$, and $\boldsymbol{I}$ denote vectors, the respective components of which are defined by

$$G_{\boldsymbol{v}} := \frac{1}{2}\left[B[P\,\boldsymbol{F}\,,\,P\,\boldsymbol{F}]\right]_{\boldsymbol{v}}, \tag{8.86}$$

$$H_{\boldsymbol{v}} := -\left[B[P\,\boldsymbol{F}\,,\,L^{-1}\,Q\,\boldsymbol{F}]\right]_{\boldsymbol{v}} - \frac{1}{f^{\mathrm{eq}}_{\boldsymbol{v}}}\,\boldsymbol{v}\cdot\nabla\left[f^{\mathrm{eq}}\,P\,\boldsymbol{F}\right]_{\boldsymbol{v}}, \tag{8.87}$$

$$I_{\boldsymbol{v}} := \frac{1}{2}\left[B[L^{-1}\,Q\,\boldsymbol{F}\,,\,L^{-1}\,Q\,\boldsymbol{F}]\right]_{\boldsymbol{v}} + \frac{1}{f^{\mathrm{eq}}_{\boldsymbol{v}}}\,\boldsymbol{v}\cdot\nabla\left[f^{\mathrm{eq}}\,L^{-1}\,Q\,\boldsymbol{F}\right]_{\boldsymbol{v}}, \tag{8.88}$$

respectively. Here, we have introduced a vector $\left[B[\boldsymbol{\varphi}\,,\,\boldsymbol{\psi}]\right]_{\boldsymbol{v}}$ defined for arbitrary vectors $\boldsymbol{\varphi}$ and $\boldsymbol{\psi}$ as

$$\left[B[\boldsymbol{\varphi}\,,\,\boldsymbol{\psi}]\right]_{\boldsymbol{v}} := \sum_{\boldsymbol{k}}\sum_{\boldsymbol{l}} B_{\boldsymbol{v}\boldsymbol{k}\boldsymbol{l}}\,\varphi_{\boldsymbol{k}}\,\psi_{\boldsymbol{l}}, \tag{8.89}$$

with

$$
\begin{aligned}
B_{\boldsymbol{v}kl} &:= \frac{1}{f^{\mathrm{eq}}_{\boldsymbol{v}}} \frac{\partial^2}{\partial f_k \, \partial f_l} C[f]_{\boldsymbol{v}} \bigg|_{f=f^{\mathrm{eq}}} f^{\mathrm{eq}}_k \, f^{\mathrm{eq}}_l \\
&= \frac{1}{2!} \sum_{\boldsymbol{v}_1} \sum_{\boldsymbol{v}_2} \sum_{\boldsymbol{v}_3} \omega(\boldsymbol{v},\, \boldsymbol{v}_1 | \boldsymbol{v}_2,\, \boldsymbol{v}_3) \\
&\quad \times \, f^{\mathrm{eq}}_{\boldsymbol{v}_1} (\delta_{\boldsymbol{v}_2 k} \, \delta_{\boldsymbol{v}_3 l} + \delta_{\boldsymbol{v}_3 k} \, \delta_{\boldsymbol{v}_2 l} - \delta_{\boldsymbol{v} k} \, \delta_{\boldsymbol{v}_1 l} - \delta_{\boldsymbol{v}_1 k} \, \delta_{\boldsymbol{v} l}).
\end{aligned} \tag{8.90}
$$

The solution to this equation is found to be

$$
\begin{aligned}
f^{\mathrm{eq}-1} \, \tilde{\boldsymbol{f}}^{(2)}(\boldsymbol{r}, t; t_0) &= \mathrm{e}^{(t-t_0)L} \left[f^{\mathrm{eq}-1} \, \boldsymbol{f}^{(2)} + 2\, L^3 \, Q \, \boldsymbol{G} + L^{-2} \, Q \, \boldsymbol{H} + L^{-1} \, Q \, \boldsymbol{I} \right] \\
&\quad + \frac{1}{3} \, (t-t_0)^3 \, P \, \boldsymbol{G} + \frac{1}{2} \, (t-t_0)^2 \, (P \, \boldsymbol{H} - 2\, L^{-1} \, Q \, \boldsymbol{G}) \\
&\quad + (t-t_0) \, (P \, \boldsymbol{I} - 2\, L^{-2} \, Q \, \boldsymbol{G} - L^{-1} \, Q \, \boldsymbol{H}) \\
&\quad - 2\, L^{-3} \, Q \, \boldsymbol{G} - L^{-2} \, Q \, \boldsymbol{H} - L^{-1} \, Q \, \boldsymbol{I}.
\end{aligned} \tag{8.91}
$$

By setting the 'initial' value in the second order as

$$
\boldsymbol{f}^{(2)}(\boldsymbol{r}; t_0) = f^{\mathrm{eq}} \, (-2\, L^3 \, Q \, \boldsymbol{G} - L^{-2} \, Q \, \boldsymbol{H} - L^{-1} \, Q \, \boldsymbol{I}), \tag{8.92}
$$

we can eliminate the fast motion due to the Q-space component and obtain the second-order solution as

$$
\begin{aligned}
\tilde{\boldsymbol{f}}^{(2)}(\boldsymbol{r}, t; t_0) = f^{\mathrm{eq}} \bigg\{ &\frac{1}{3} \, (t-t_0)^3 \, P \, \boldsymbol{G} + \frac{1}{2} \, (t-t_0)^2 \, (P \, \boldsymbol{H} - 2\, L^{-1} \, Q \, \boldsymbol{G}) \\
&+ (t-t_0) \, (P \, \boldsymbol{I} - 2\, L^{-2} \, Q \, \boldsymbol{G} - L^{-1} \, Q \, \boldsymbol{H}) \\
&- 2\, L^{-3} \, Q \, \boldsymbol{G} - L^{-2} \, Q \, \boldsymbol{H} - L^{-1} \, Q \, \boldsymbol{I} \bigg\}.
\end{aligned} \tag{8.93}
$$

Notice again the appearance of the secular terms in Eq. (8.93).

8.3.7 *Application of RG/E Equation and Construction of a Global Solution*

Collecting all the terms obtained in the perturbative analysis so far, the 'initial' value and the solution in the second-order approximation are given by

$$
\begin{aligned}
\boldsymbol{f}(\boldsymbol{r}; t_0) &= \boldsymbol{f}^{\mathrm{eq}} - \epsilon \, f^{\mathrm{eq}} \, L^{-1} \, Q \, \boldsymbol{F} \\
&\quad - \epsilon^2 \, f^{\mathrm{eq}} \, (2\, L^{-3} \, Q \, \boldsymbol{G} + L^{-2} \, Q \, \boldsymbol{H} + L^{-1} \, Q \, \boldsymbol{I}) + \mathrm{O}(\epsilon^3),
\end{aligned} \tag{8.94}
$$

$$\begin{aligned}\tilde{\boldsymbol{f}}(\boldsymbol{r},t;t_0) &= \boldsymbol{f}^{\mathrm{eq}} + \epsilon\, \boldsymbol{f}^{\mathrm{eq}} \left\{ (t-t_0)\, P\, \boldsymbol{F} - L^{-1}\, Q\, \boldsymbol{F} \right\} \\ &+ \epsilon^2\, \boldsymbol{f}^{\mathrm{eq}} \left\{ \frac{1}{3}\,(t-t_0)^3\, P\, \boldsymbol{G} + \frac{1}{2}\,(t-t_0)^2\,(P\, \boldsymbol{H} - 2\, L^{-1}\, Q\, \boldsymbol{G}) \right. \\ &+ (t-t_0)\,(P\, \boldsymbol{I} - 2\, L^{-2}\, Q\, \boldsymbol{G} - L^{-1}\, Q\, \boldsymbol{H}) \\ &\left. - 2\, L^{-3}\, Q\, \boldsymbol{G} - L^{-2}\, Q\, \boldsymbol{H} - L^{-1}\, Q\, \boldsymbol{I} \right\} + \mathrm{O}(\epsilon^3), \end{aligned} \tag{8.95}$$

respectively. Needless to say, we have

$$\tilde{\boldsymbol{f}}(\boldsymbol{r}, t_0; t_0) = \boldsymbol{f}(\boldsymbol{r}; t_0), \tag{8.96}$$

by construction.

We note the approximate solution (8.95) is valid only for $t \simeq t_0$ due to the secular terms. As has been done through this monograph, we can, however, obtain the solution that is valid in a global domain through constructing the envelope of these diverging local solutions parametrized by t_0. A key to construct the envelope is a set of the envelope equation or RG/E equation and an envelope function.

Now recalling that $\left[\tilde{\boldsymbol{f}}(\boldsymbol{r},\, t\,;\, t_0)\right]_{\boldsymbol{v}} := \tilde{f}_{\boldsymbol{v}}(\boldsymbol{r},\, t\,;\, t_0)$, the RG/E equation reads

$$\left.\frac{\mathrm{d}}{\mathrm{d}t_0}\tilde{f}_{\boldsymbol{v}}(\boldsymbol{r},\, t\,;\, t_0)\right|_{t_0=t} = 0, \tag{8.97}$$

which is reduced to

$$\begin{aligned}&\frac{\partial}{\partial t}\left\{ \boldsymbol{f}^{\mathrm{eq}} - \epsilon\, \boldsymbol{f}^{\mathrm{eq}}\, L^{-1}\, Q\, \boldsymbol{F} - \epsilon^2\, \boldsymbol{f}^{\mathrm{eq}}\,(2\, L^{-3}\, Q\, \boldsymbol{G} + L^{-2}\, Q\, \boldsymbol{H} + L^{-1}\, Q\, \boldsymbol{I}) \right\} \\ &- \epsilon\, \boldsymbol{f}^{\mathrm{eq}}\, P\, \boldsymbol{F} - \epsilon^2\, \boldsymbol{f}^{\mathrm{eq}}\,(P\, \boldsymbol{I} - 2\, L^{-2}\, Q\, \boldsymbol{G} - L^{-1}\, Q\, \boldsymbol{H}) + \mathrm{O}(\epsilon^3) = 0. \end{aligned} \tag{8.98}$$

It is noteworthy that since the derivative with respect to t_0 only hits the fluid dynamical variables as the integral constants in the unperturbed solution, Eq. (8.98) actually gives the equation governing the time evolution of the five variables $n(\boldsymbol{r}\,;\, t)$, $T(\boldsymbol{r}\,;\, t)$ and $\boldsymbol{u}(\boldsymbol{r}\,;\, t)$ in $\boldsymbol{f}^{\mathrm{eq}}$.

The envelope function $f_{\boldsymbol{v}}^{\mathrm{E}}(\boldsymbol{r},\, t)$ to give the solution valid in a global domain of t is given by the 'initial' value (8.94) with the replacement of $t_0 = t$ as

$$f_{\boldsymbol{v}}^{\mathrm{E}}(\boldsymbol{r},\, t) = \tilde{f}_{\boldsymbol{v}}(\boldsymbol{r},\, t\,;\, t_0 = t) = f_{\boldsymbol{v}}(\boldsymbol{r}\,;\, t), \tag{8.99}$$

$$\begin{aligned}&= \left[\boldsymbol{f}^{\mathrm{eq}} - \epsilon\, \boldsymbol{f}^{\mathrm{eq}}\, L^{-1}\, Q\, \boldsymbol{F} - \epsilon^2\, \boldsymbol{f}^{\mathrm{eq}}\,(2\, L^{-3}\, Q\, \boldsymbol{G} \right. \\ &\left. + L^{-2}\, Q\, \boldsymbol{H} + L^{-1}\, Q\, \boldsymbol{I}) \right]_{\boldsymbol{v}}(\boldsymbol{r}\,;\, t_0 = t) + \mathrm{O}(\epsilon^3). \end{aligned} \tag{8.100}$$

where the functions $n(\boldsymbol{r}\,;\, t)$, $T(\boldsymbol{r}\,;\, t)$ and $\boldsymbol{u}(\boldsymbol{r}\,;\, t)$ therein are the solutions to Eq. (8.98). In accordance with the proof given for a generic case in Sect. 4.7, it is proved that the envelope function (8.100) satisfies the Boltzmann equation (8.43) up to $\mathrm{O}(\epsilon^2)$ in a global domain of t, as follows: For arbitrary t, the time derivative of the envelope function reads

$$\begin{aligned}\frac{\partial}{\partial t} f_{\boldsymbol{v}}^{\mathrm{E}}(\boldsymbol{r},\, t) &= \frac{\partial}{\partial t} \tilde{f}_{\boldsymbol{v}}(\boldsymbol{r},\, t\,;\, t) \\ &= \frac{\partial}{\partial t} \tilde{f}_{\boldsymbol{v}}(\boldsymbol{r},\, t\,;\, t_0)\bigg|_{t_0=t} + \frac{\mathrm{d}}{\mathrm{d}t_0} \tilde{f}_{\boldsymbol{v}}(\boldsymbol{r},\, t\,;\, t_0)\bigg|_{t_0=t} \\ &= \frac{\partial}{\partial t} \tilde{f}_{\boldsymbol{v}}(\boldsymbol{r},\, t\,;\, t_0)\bigg|_{t_0=t}, \end{aligned} \tag{8.101}$$

where the relation (8.99) and the RG/E equation (8.97) have been used. Since $\tilde{f}_{\boldsymbol{v}}(\boldsymbol{r},\, t\,;\, t_0)$ satisfies the Boltzmann equation (8.43) up to $\mathrm{O}(\epsilon^2)$ at $t \sim t_0$, the following equality holds at $t \sim t_0$:

$$\frac{\partial}{\partial t} \tilde{f}_{\boldsymbol{v}}(\boldsymbol{r},\, t\,;\, t_0) = C[\tilde{f}]_{\boldsymbol{v}}(\boldsymbol{r},\, t\,;\, t_0) - \epsilon\, \boldsymbol{v} \cdot \boldsymbol{\nabla} \tilde{f}(\boldsymbol{r},\, t\,;\, t_0) + \mathrm{O}(\epsilon^3). \tag{8.102}$$

Substituting Eq. (8.102) into Eq. (8.101) and using the definition of the envelope function (8.99), we have

$$\frac{\partial}{\partial t} f_{\boldsymbol{v}}^{\mathrm{E}}(\boldsymbol{r},\, t) = C[f^{\mathrm{E}}]_{\boldsymbol{v}}(\boldsymbol{r},\, t) - \epsilon\, \boldsymbol{v} \cdot \boldsymbol{\nabla} f^{\mathrm{E}}(\boldsymbol{r},\, t) + \mathrm{O}(\epsilon^3), \tag{8.103}$$

valid for arbitrary t. This concludes that $f_{\boldsymbol{v}}^{\mathrm{E}}(\boldsymbol{r},\, t)$ is actually the global solution to the Boltzmann equation (8.43) up to $\mathrm{O}(\epsilon^2)$.

Equation (8.98) tells us that all the fluid dynamic variables are slow variables because

$$\frac{\partial}{\partial t} \boldsymbol{f}^{\mathrm{eq}}(\boldsymbol{r};\, t) = \mathrm{O}(\epsilon). \tag{8.104}$$

Thus the RG/E equation (8.98) describes the macroscopic dynamics as expected.

Finally, we note that on account of Eq. (8.104), Eq. (8.98) is further simplified as

$$\begin{aligned}&\frac{\partial}{\partial t} \Big\{ \boldsymbol{f}^{\mathrm{eq}}(\boldsymbol{r};\, t) - \epsilon\, f^{\mathrm{eq}}\, L^{-1}\, Q\, \boldsymbol{F} \Big\} \\ &- \epsilon\, f^{\mathrm{eq}}\, P\, \boldsymbol{F} - \epsilon^2\, f^{\mathrm{eq}}\, (P\, \boldsymbol{I} - 2\, L^{-2}\, Q\, \boldsymbol{G} - L^{-1}\, Q\, \boldsymbol{H}) + \mathrm{O}(\epsilon^3) = 0, \end{aligned} \tag{8.105}$$

because

$$\frac{\partial}{\partial t} \Big\{ - \epsilon^2\, f^{\mathrm{eq}}\, (2\, L^{-3}\, Q\, \boldsymbol{G} + L^{-2}\, Q\, \boldsymbol{H} + L^{-1}\, Q\, \boldsymbol{I}) \Big\} = \mathrm{O}(\epsilon^3), \tag{8.106}$$

which follows from the fact that L, Q, $\boldsymbol{G}$, $\boldsymbol{H}$, and $\boldsymbol{I}$ are functions of f_v^{eq} and the time derivative hits f_v^{eq} in these quantities.

8.4 Reduction of RG/E Equation To fluid Dynamic Equation: Derivation Of Navier-Stokes Equation

We are now in a position to derive fluid dynamic equation from (8.105).[4] First we note that the RG/E equation (8.105) contains fast modes belonging to the Q space as well as the fluid dynamic modes or slow modes belonging to the P space. Since the fluid dynamic equations should be closed within the the latter variables, the former components should be projected out from the equation. This task can be simply accomplished by applying a projection operator $P\, f^{\mathrm{eq}-1}$ from the left of Eq. (8.105) as

$$P\, f^{\mathrm{eq}-1}\, \frac{\partial}{\partial t} \left\{ f^{\mathrm{eq}} - \epsilon\, f^{\mathrm{eq}}\, L^{-1}\, Q\, \boldsymbol{F} \right\} - \epsilon\, P\, \boldsymbol{F} - \epsilon^2\, P\, \boldsymbol{I} = 0, \tag{8.107}$$

where we have omitted $\mathrm{O}(\epsilon^3)$.

Next, in order to extract a more concrete form of the slow dynamics, which is to be identified with the fluid dynamic equations, we take the inner product of (8.107) with the zero modes $\varphi_{0\,v}^{\alpha}$, which may be interpreted as a kind of averaging. The result takes the form

$$\langle\, \varphi_0^{\alpha}\,,\, f^{\mathrm{eq}-1}\, \frac{\partial}{\partial t} f^{\mathrm{eq}}\, \rangle - \epsilon\, \langle\, \varphi_0^{\alpha}\,,\, \boldsymbol{F}\, \rangle - \epsilon^2\, \langle\, \varphi_0^{\alpha}\,,\, \boldsymbol{I}\, \rangle = 0, \tag{8.108}$$

where we have used the relations

$$\langle\, \varphi_0^{\alpha}\,,\, P\, \boldsymbol{F}\, \rangle = \langle\, \varphi_0^{\alpha}\,,\, \boldsymbol{F}\, \rangle, \quad \langle\, \varphi_0^{\alpha}\,,\, P\, \boldsymbol{I}\, \rangle = \langle\, \varphi_0^{\alpha}\,,\, \boldsymbol{I}\, \rangle, \tag{8.109}$$

which follow from the definitions (8.60) and (8.79), and the equality

$$\langle\, \varphi_0^{\alpha}\,,\, f^{\mathrm{eq}-1}\, \frac{\partial}{\partial t} \left\{ - \epsilon\, f^{\mathrm{eq}}\, L^{-1}\, Q\, \boldsymbol{F} \right\} \rangle = 0, \tag{8.110}$$

which follows from the expansion

---

[4] It is to be remarked that the substitution of the one-particle distribution function (8.100) into the **exact** balance equations (8.12), (8.20), and (8.21), which are already in the forms of fluid dynamic equations, leads to the Navier-Stokes equation. An explicit and detailed derivation is given in Sect. 8.5.

$$\langle \boldsymbol{\varphi}_0^\alpha \,,\, f^{\rm eq-1} \frac{\partial}{\partial t} \Big\{ f^{\rm eq} \, L^{-1} \, Q \, \boldsymbol{F} \Big\} \rangle = \frac{\partial}{\partial t} \Big\{ \langle \boldsymbol{\varphi}_0^\alpha \,,\, L^{-1} \, Q \, \boldsymbol{F} \rangle \Big\} - \langle \frac{\partial}{\partial t} \boldsymbol{\varphi}_0^\alpha \,,\, L^{-1} \, Q \, \boldsymbol{F} \rangle, \tag{8.111}$$

and the fact that not only $\boldsymbol{\varphi}_0^\alpha$ but also $\partial \boldsymbol{\varphi}_0^\alpha / \partial t$ belong to the P space as follows:

$$\frac{\partial}{\partial t} \varphi_{0v}^0 = 0 = 0 \times \varphi_{0v}^0, \tag{8.112}$$

$$\frac{\partial}{\partial t} \varphi_{0v}^i = -m \, \frac{\partial}{\partial t} u^i = -m \, \frac{\partial}{\partial t} u^i \times \varphi_{0v}^0, \tag{8.113}$$

$$\frac{\partial}{\partial t} \varphi_{0v}^4 = -m \, \delta v^i \, \frac{\partial}{\partial t} u^i - \frac{3}{2} \frac{\partial}{\partial t} T = -\varphi_{0v}^i \, \frac{\partial}{\partial t} u^i - \frac{3}{2} \frac{\partial}{\partial t} T \times \varphi_{0v}^0, \tag{8.114}$$

which have been calculated from Eqs. (8.70)–(8.72).

From now on, we shall show that the reduced RG/E equation (8.108) is nothing but the fluid dynamic equation, *i.e.*, the Navier-Stokes equation [139].

First, we try to express the first term of Eq. (8.108) in terms of the fluid dynamic variables. From the definition of the equilibrium distribution function (8.51), we have

$$\big[f^{\rm eq-1} \, \frac{\partial}{\partial t} f^{\rm eq} \big]_{\boldsymbol{v}} = \varphi_{0\,\boldsymbol{v}}^0 \, \frac{1}{n} \, \frac{\partial}{\partial t} n + \varphi_{0\,\boldsymbol{v}}^j \, \frac{1}{T} \, \frac{\partial}{\partial t} u^j + \varphi_{0\,\boldsymbol{v}}^4 \, \frac{1}{T^2} \, \frac{\partial}{\partial t} T. \tag{8.115}$$

Then using the orthogonality condition (8.73) of the zero modes together with the normalization constants (8.74)–(8.76), we find the first term of Eq. (8.108) is reduced to

$$\langle \boldsymbol{\varphi}_0^0 \,,\, f^{\rm eq-1} \, \frac{\partial}{\partial t} f^{\rm eq} \rangle = \frac{\partial}{\partial t} n, \tag{8.116}$$

$$\langle \boldsymbol{\varphi}_0^i \,,\, f^{\rm eq-1} \, \frac{\partial}{\partial t} f^{\rm eq} \rangle = m \, n \, \frac{\partial}{\partial t} u^i, \quad i = 1,\, 2,\, 3, \tag{8.117}$$

$$\langle \boldsymbol{\varphi}_0^4 \,,\, f^{\rm eq-1} \, \frac{\partial}{\partial t} f^{\rm eq} \rangle = n \, \frac{\partial}{\partial t} (3\,T/2). \tag{8.118}$$

Next, let us also try to express the second term of Eq. (8.108) in terms of the fluid dynamic variables. First, using the definition (8.58), $\boldsymbol{F}$ is found to be expressed in terms of the fluid dynamic variables as

$$\begin{aligned} F_{\boldsymbol{v}} = & -\varphi_{0\,\boldsymbol{v}}^0 \, \frac{1}{n} \, \boldsymbol{u} \cdot \nabla n - \varphi_{0\,\boldsymbol{v}}^j \, \frac{1}{T} \, \boldsymbol{u} \cdot \nabla u^j - \varphi_{0\,\boldsymbol{v}}^4 \, \frac{1}{T^2} \, \boldsymbol{u} \cdot \nabla T \\ & - \tilde{\varphi}_{1\,\boldsymbol{v}}^{i0} \, \frac{1}{n} \, \nabla^i n - \tilde{\varphi}_{1\,\boldsymbol{v}}^{ij} \, \frac{1}{T} \, \nabla^i u^j - \tilde{\varphi}_{1\,\boldsymbol{v}}^{i4} \, \frac{1}{T^2} \, \nabla^i T, \end{aligned} \tag{8.119}$$

where

$$\tilde{\varphi}_{1\,\boldsymbol{v}}^{i\alpha} := \delta v^i \, \varphi_{0\,\boldsymbol{v}}^\alpha. \tag{8.120}$$

Then taking the inner product of Eq. (8.119) with $\boldsymbol{\varphi}_0^{\alpha}$, we find that the second term of Eq. (8.108) is reduced to

$$\langle \boldsymbol{\varphi}_0^0 , \boldsymbol{F} \rangle = -\boldsymbol{\nabla} \cdot (n\, \boldsymbol{u}), \tag{8.121}$$
$$\langle \boldsymbol{\varphi}_0^i , \boldsymbol{F} \rangle = -m\, n\, \boldsymbol{u} \cdot \boldsymbol{\nabla} u^i - \nabla^i (n\, T), \quad i = 1,\, 2,\, 3, \tag{8.122}$$
$$\langle \boldsymbol{\varphi}_0^4 , \boldsymbol{F} \rangle = -n\, \boldsymbol{u} \cdot \boldsymbol{\nabla}(3\, T/2) - n\, T\, \boldsymbol{\nabla} \cdot \boldsymbol{u}. \tag{8.123}$$

In the derivation, we have used the orthogonality relation (8.73) and the following formulae

$$\langle \boldsymbol{\varphi}_0^0 , \tilde{\boldsymbol{\varphi}}_1^{j0} \rangle = 0, \quad \langle \boldsymbol{\varphi}_0^0 , \tilde{\boldsymbol{\varphi}}_1^{jk} \rangle = n\, T\, \delta^{jk}, \quad \langle \boldsymbol{\varphi}_0^0 , \tilde{\boldsymbol{\varphi}}_1^{j4} \rangle = 0, \tag{8.124}$$
$$\langle \boldsymbol{\varphi}_0^i , \tilde{\boldsymbol{\varphi}}_1^{j0} \rangle = n\, T\, \delta^{ij}, \quad \langle \boldsymbol{\varphi}_0^i , \tilde{\boldsymbol{\varphi}}_1^{jk} \rangle = 0, \quad \langle \boldsymbol{\varphi}_0^i , \tilde{\boldsymbol{\varphi}}_1^{j4} \rangle = n\, T^2\, \delta^{ij}, \tag{8.125}$$
$$\langle \boldsymbol{\varphi}_0^4 , \tilde{\boldsymbol{\varphi}}_1^{j0} \rangle = 0, \quad \langle \boldsymbol{\varphi}_0^4 , \tilde{\boldsymbol{\varphi}}_1^{jk} \rangle = n\, T^2\, \delta^{jk}, \quad \langle \boldsymbol{\varphi}_0^4 , \tilde{\boldsymbol{\varphi}}_1^{j4} \rangle = 0, \tag{8.126}$$

for $i,\ j,\ k = 1,\, 2,\, 3$, which are results of a direct calculation.

Finally, we also express the third term of Eq. (8.108) in terms of the fluid dynamic variables. Using the formula (8.88) of $\boldsymbol{I}$, we have

$$\langle \boldsymbol{\varphi}_0^{\alpha} , \boldsymbol{I} \rangle = \frac{1}{2} \langle \boldsymbol{\varphi}_0^{\alpha} , B[L^{-1}\, Q\, \boldsymbol{F} , L^{-1}\, Q\, \boldsymbol{F}] \rangle + \langle \tilde{\boldsymbol{\varphi}}_1^{i\alpha} , f^{\mathrm{eq}-1}\, \nabla^i \Big[f^{\mathrm{eq}}\, L^{-1}\, Q\, \boldsymbol{F} \Big] \rangle. \tag{8.127}$$

We remark that the first term containing B turns out to vanish nicely in the present formulation:

$$\langle \boldsymbol{\varphi}_0^{\alpha} , B[L^{-1}\, Q\, \boldsymbol{F} , L^{-1}\, Q\, \boldsymbol{F}] \rangle = 0, \tag{8.128}$$

which is due to the fact that $\boldsymbol{\varphi}_0^{\alpha}$ are the collision invariants as shown in Eq. (8.6). A detailed proof is presented in Sect. 8.5.

On account of Eq. (8.128) and a chain rule with ∇^i, Eq. (8.127) is cast into the form

$$\langle \boldsymbol{\varphi}_0^{\alpha} , \boldsymbol{I} \rangle = \nabla^i \Big[\langle \tilde{\boldsymbol{\varphi}}_1^{i\alpha} , L^{-1}\, Q\, \boldsymbol{F} \rangle \Big] - \langle \nabla^i \tilde{\boldsymbol{\varphi}}_1^{i\alpha} , L^{-1}\, Q\, \boldsymbol{F} \rangle. \tag{8.129}$$

Since $L^{-1}\, Q\, \boldsymbol{F}$ belong to the Q space, an insertion of Q in front of $\tilde{\boldsymbol{\varphi}}_1^{i\alpha}$ in the first term and $\nabla^i \tilde{\boldsymbol{\varphi}}_1^{i\alpha}$ in the second term does not change the values of the inner products as,

$$\langle \tilde{\boldsymbol{\varphi}}_1^{i\alpha} , L^{-1}\, Q\, \boldsymbol{F} \rangle = \langle Q\, \tilde{\boldsymbol{\varphi}}_1^{i\alpha} , L^{-1}\, Q\, \boldsymbol{F} \rangle, \tag{8.130}$$
$$\langle \nabla^i \tilde{\boldsymbol{\varphi}}_1^{i\alpha} , L^{-1}\, Q\, \boldsymbol{F} \rangle = \langle Q\, \nabla^i \tilde{\boldsymbol{\varphi}}_1^{i\alpha} , L^{-1}\, Q\, \boldsymbol{F} \rangle. \tag{8.131}$$

Thus, we have

$$\langle \boldsymbol{\varphi}_0^{\alpha} , \boldsymbol{I} \rangle = \nabla^i \Big[\langle Q \tilde{\boldsymbol{\varphi}}_1^{i\alpha} , L^{-1} Q \boldsymbol{F} \rangle \Big] - \langle Q \nabla^i \tilde{\boldsymbol{\varphi}}_1^{i\alpha} , L^{-1} Q \boldsymbol{F} \rangle. \tag{8.132}$$

For the further reduction of Eq. (8.132), we shall calculate $Q \tilde{\boldsymbol{\varphi}}_1^{i\alpha}$, $Q \nabla^i \tilde{\boldsymbol{\varphi}}_1^{i\alpha}$, and $L^{-1} Q \boldsymbol{F}$ one by one. A straightforward calculation based on Eqs. (8.124)–(8.126) reduces $Q \tilde{\boldsymbol{\varphi}}_1^{i\alpha}$ to

$$\big[Q \tilde{\boldsymbol{\varphi}}_1^{i0} \big]_{\boldsymbol{v}} = 0, \tag{8.133}$$
$$\big[Q \tilde{\boldsymbol{\varphi}}_1^{ij} \big]_{\boldsymbol{v}} = \pi_{\boldsymbol{v}}^{ij}, \quad j = 1, 2, 3, \tag{8.134}$$
$$\big[Q \tilde{\boldsymbol{\varphi}}_1^{i4} \big]_{\boldsymbol{v}} = J_{\boldsymbol{v}}^{i}, \tag{8.135}$$

where

$$\pi_{\boldsymbol{v}}^{ij} := m \, \Delta^{ijkl} \, \delta v^k \, \delta v^l, \tag{8.136}$$
$$J_{\boldsymbol{v}}^{i} := \Big(\frac{m}{2} \, |\delta \boldsymbol{v}|^2 - \frac{5}{2} T \Big) \, \delta v^i, \tag{8.137}$$

with

$$\Delta^{ijkl} := \frac{1}{2} \Big(\delta^{ik} \, \delta^{jl} + \delta^{il} \, \delta^{jk} - \frac{2}{3} \, \delta^{ij} \, \delta^{kl} \Big). \tag{8.138}$$

In a similar way, $Q \nabla^i \tilde{\boldsymbol{\varphi}}_1^{i\alpha}$ is calculated to be

$$\big[Q \nabla^i \tilde{\boldsymbol{\varphi}}_1^{i0} \big]_{\boldsymbol{v}} = 0, \tag{8.139}$$
$$\big[Q \nabla^i \tilde{\boldsymbol{\varphi}}_1^{ij} \big]_{\boldsymbol{v}} = 0, \quad j = 1, 2, 3, \tag{8.140}$$
$$\big[Q \nabla^i \tilde{\boldsymbol{\varphi}}_1^{i4} \big]_{\boldsymbol{v}} = -\pi_{\boldsymbol{v}}^{ij} \, \sigma^{ij}, \tag{8.141}$$

with

$$\sigma^{ij} := \Delta^{ijkl} \, \nabla^k u^l. \tag{8.142}$$

By combining Eq. (8.119) with the equalities (8.133)–(8.135), we have

$$Q \boldsymbol{F} = -\frac{1}{T} \, \boldsymbol{\pi}^{ij} \, \sigma^{ij} - \frac{1}{T^2} \, \boldsymbol{J}^i \, \nabla^i T. \tag{8.143}$$

We note that $\boldsymbol{\pi}^{ij}$ and $\boldsymbol{J}^i$ in Eqs. (8.136) and (8.137) are the microscopic representations of the dissipative currents.

Collecting the results obtained so far, we arrive at the expression of the third term of Eq. (8.108) that is nicely written in terms of the transport coefficients:

$$\langle \varphi_0^0 , \boldsymbol{I} \rangle = 0, \tag{8.144}$$
$$\langle \varphi_0^i , \boldsymbol{I} \rangle = \nabla^j (2\,\eta\,\sigma^{ij}), \quad i = 1,\, 2,\, 3, \tag{8.145}$$
$$\langle \varphi_0^4 , \boldsymbol{I} \rangle = \nabla^i (\lambda\,\nabla^i T) + 2\,\eta\,\sigma^{ij}\,\sigma^{ij}. \tag{8.146}$$

Here we have defined

$$\eta := -\frac{1}{10\,T}\,\langle \boldsymbol{\pi}^{ij} , L^{-1}\,\boldsymbol{\pi}^{ij} \rangle, \tag{8.147}$$
$$\lambda := -\frac{1}{3\,T^2}\,\langle \boldsymbol{J}^i , L^{-1}\,\boldsymbol{J}^i \rangle, \tag{8.148}$$

and utilized the space rotational symmetry [112, 113],

$$\langle \boldsymbol{\pi}^{ij} , L^{-1}\,\boldsymbol{\pi}^{kl} \rangle = \frac{1}{5}\,\Delta^{ijkl}\,\langle \boldsymbol{\pi}^{mn} , L^{-1}\,\boldsymbol{\pi}^{mn} \rangle, \tag{8.149}$$
$$\langle \boldsymbol{J}^i , L^{-1}\,\boldsymbol{J}^j \rangle = \frac{1}{3}\,\delta^{ij}\,\langle \boldsymbol{J}^k , L^{-1}\,\boldsymbol{J}^k \rangle, \tag{8.150}$$
$$\langle \boldsymbol{\pi}^{ij} , L^{-1}\,\boldsymbol{J}^k \rangle = 0, \tag{8.151}$$

We stress that η and λ are transport coefficients, called shear viscosity and heat conductivity, respectively, and Eqs. (8.147) and (8.148) provide with their microscopic representations, which are the same as those by the Chapman-Enskog expansion method [112, 113].

Now let us define the following "time-evolved" microscopic currents,

$$\boldsymbol{\pi}^{ij}(\tau) := \mathrm{e}^{L\tau}\,\boldsymbol{\pi}^{ij}, \quad \boldsymbol{J}^i(\tau) := \mathrm{e}^{L\tau}\,\boldsymbol{J}^i. \tag{8.152}$$

Then it is remarkable that (8.147) and (8.148) can be written as the time-correlation functions as

$$\eta = \frac{1}{10\,T}\int_0^\infty \mathrm{d}\tau\,\langle \boldsymbol{\pi}^{ij}(0) , \boldsymbol{\pi}^{ij}(\tau) \rangle, \tag{8.153}$$
$$\lambda = \frac{1}{3\,T^2}\int_0^\infty \mathrm{d}\tau\,\langle \boldsymbol{J}^i(0) , \boldsymbol{J}^i(\tau) \rangle, \tag{8.154}$$

which are in the same form as the Kubo formula of the same quantities [115].

Collecting Eqs. (8.116)–(8.118), (8.121)–(8.123), and (8.144)–(8.146) and putting back $\epsilon = 1$, the reduced RG/E equation (8.108) is found to be

$$\frac{\partial}{\partial t} n = -\boldsymbol{\nabla}\cdot(n\,\boldsymbol{u}), \tag{8.155}$$
$$m\,n\,\frac{\partial}{\partial t} u^i = -m\,n\,\boldsymbol{u}\cdot\boldsymbol{\nabla} u^i - \nabla^i p + \nabla^j (2\,\eta\,\sigma^{ji}), \quad i = 1,\, 2,\, 3, \tag{8.156}$$
$$n\,\frac{\partial}{\partial t} e = -n\,\boldsymbol{u}\cdot\boldsymbol{\nabla} e - p\,\boldsymbol{\nabla}\cdot\boldsymbol{u} + 2\,\eta\,\sigma^{ij}\,\sigma^{ij} + \boldsymbol{\nabla}\cdot(\lambda\,\boldsymbol{\nabla} T), \tag{8.157}$$

with $e = 3T/2$ and $p = nT$. We emphasize that the set of Eqs. (8.155)–(8.157) perfectly agrees with the Navier-Stokes equation [139].

8.5 Summary

We have shown that the Navier-Stokes equation is extracted from the Boltzmann equation as the macroscopic slow dynamics with use of the RG method. In this method, one begins with solving the Boltzmann equation faithfully by the perturbation method in the asymptotic regime where the spatial inhomogeneity is expected small; the expansion parameter is the Knudsen number. The zeroth-order solution is chosen to be the local equilibrium distribution function, which is identical to the Maxwellian expressed in terms of the five fluid dynamic variables corresponding to the collision invariants. The dissipative effects are taken into account as a deformation of the distribution function caused by the spatial inhomogeneity as the perturbation. After defining the inner product in the function space spanned by the distribution function, the deviation from the Maxwellian that gives rise to the dissipative effects is constructed so that it is precisely orthogonal to the zero modes with respect to the inner product.

The present scheme based on the RG method for deriving the dissipative fluid dynamics from the underlying kinetic theory is of a generic nature and should be applicable in the relativistic and quantum statistical cases, which will be shown to be the case in Chap. 11.

Appendix 1: Foundation of the Symmetrized Inner Product (8.60)

In this Appendix, on the basis of the general theory for asymmetric linear operators given in Sect. 3.6.2, we shall show the inner product defined by (8.60) is the unique choice for making the self-adjointness of the collision operator L apparent.

In general, L is an asymmetric matrix and hence L has right eigenvectors $\boldsymbol{U}_i$ and left eigenvectors $\tilde{\boldsymbol{U}}_i^{\dagger}$ with $i = 1, 2, \ldots$, which satisfy

$$L\,\boldsymbol{U}_i = \lambda_i\,\boldsymbol{U}_i, \tag{8.158}$$

$$\tilde{\boldsymbol{U}}_i^{\dagger}\,L = \lambda_i\,\tilde{\boldsymbol{U}}_i^{\dagger}, \tag{8.159}$$

respectively. Here, λ_i with $i = 1, 2, \ldots$ are corresponding eigenvalues. A Hermitian conjugate of Eq. (8.159) is

$$L^{\dagger}\,\tilde{\boldsymbol{U}}_i = \lambda_i^{*}\,\tilde{\boldsymbol{U}}_i, \tag{8.160}$$

with λ_i^* being a complex conjugate of λ_i. Without loss of generality, we can impose the orthogonality and completeness as follows:

$$\tilde{\boldsymbol{U}}_i^\dagger \boldsymbol{U}_j = \delta_{ij}, \tag{8.161}$$

$$\sum_i \boldsymbol{U}_i \tilde{\boldsymbol{U}}_i^\dagger = 1, \tag{8.162}$$

respectively. An inner product with respect to the right eigenvectors $\boldsymbol{U}_i$ with $i = 1, 2, \ldots$ is defined as

$$\langle \boldsymbol{\varphi}, \boldsymbol{\psi} \rangle := \boldsymbol{\varphi}^\dagger g \boldsymbol{\psi}, \tag{8.163}$$

where g is a metric tensor

$$g = \sum_i \tilde{\boldsymbol{U}}_i \tilde{\boldsymbol{U}}_i^\dagger. \tag{8.164}$$

We note that the following symmetry property with respect to $\boldsymbol{U}_i$ with $i = 1, 2, \ldots$ is realized:

$$\langle \boldsymbol{U}_i, \boldsymbol{U}_j \rangle = \delta_{ij}, \tag{8.165}$$

which can be derived from Eq. (8.161). This inner product leads to

$$\begin{aligned} \langle \boldsymbol{\varphi}, L\boldsymbol{\psi} \rangle = \boldsymbol{\varphi}^\dagger g L \boldsymbol{\psi} &= \sum_i \boldsymbol{\varphi}^\dagger \tilde{\boldsymbol{U}}_i \tilde{\boldsymbol{U}}_i^\dagger L \boldsymbol{\psi} \\ &= \sum_i \lambda_i \boldsymbol{\varphi}^\dagger \tilde{\boldsymbol{U}}_i \tilde{\boldsymbol{U}}_i^\dagger \boldsymbol{\psi}, \end{aligned} \tag{8.166}$$

and

$$\begin{aligned} \langle L\boldsymbol{\varphi}, \boldsymbol{\psi} \rangle = (L\boldsymbol{\varphi})^\dagger g \boldsymbol{\psi} = \boldsymbol{\varphi}^\dagger L^\dagger g \boldsymbol{\psi} &= \sum_i \boldsymbol{\varphi}^\dagger L^\dagger \tilde{\boldsymbol{U}}_i \tilde{\boldsymbol{U}}_i^\dagger \boldsymbol{\psi} \\ &= \sum_i \lambda_i^* \boldsymbol{\varphi}^\dagger \tilde{\boldsymbol{U}}_i \tilde{\boldsymbol{U}}_i^\dagger \boldsymbol{\psi}. \end{aligned} \tag{8.167}$$

Accordingly, when all the eigenvalues are real number as $\lambda_i^* = \lambda_i$ with $i = 1, 2, \ldots$, the self-adjoint nature of L is automatically respected,

$$\langle \boldsymbol{\varphi}, L\boldsymbol{\psi} \rangle = \langle L\boldsymbol{\varphi}, \boldsymbol{\psi} \rangle. \tag{8.168}$$

From now on, we show that the eigenvalues λ_i with $i = 1, 2, \ldots$ are actually real number and then derive an explicit form of the metric g. We will see that the resultant g reproduces the natural inner product (8.60).

Using the definition of L in Eq. (8.59), we find that $f_{\boldsymbol{v}}^{\text{eq}}\, L_{\boldsymbol{v}\boldsymbol{k}}$ is a symmetric matrix, that is,

$$f_{\boldsymbol{v}}^{\text{eq}}\, L_{\boldsymbol{v}\boldsymbol{k}} = f_{\boldsymbol{k}}^{\text{eq}}\, L_{\boldsymbol{k}\boldsymbol{v}}. \tag{8.169}$$

In fact, we calculate $f_{\boldsymbol{v}}^{\text{eq}}\, L_{\boldsymbol{v}\boldsymbol{k}}$ to be

$$\begin{aligned} f_{\boldsymbol{v}}^{\text{eq}}\, L_{\boldsymbol{v}\boldsymbol{k}} &= -\frac{1}{2!} \sum_{\boldsymbol{v}_1} \sum_{\boldsymbol{v}_2} \sum_{\boldsymbol{v}_3} \omega(\boldsymbol{v},\, \boldsymbol{v}_1 | \boldsymbol{v}_2,\, \boldsymbol{v}_3)\, f_{\boldsymbol{v}}^{\text{eq}}\, f_{\boldsymbol{v}_1}^{\text{eq}}\, (\delta_{\boldsymbol{v}\boldsymbol{k}} + \delta_{\boldsymbol{v}_1\boldsymbol{k}} - \delta_{\boldsymbol{v}_2\boldsymbol{k}} - \delta_{\boldsymbol{v}_3\boldsymbol{k}}) \\ &= -\frac{1}{2!}\, (\delta_{\boldsymbol{v}\boldsymbol{k}}\, a_{\boldsymbol{v}} + b_{\boldsymbol{v}\boldsymbol{k}} - c_{\boldsymbol{v}\boldsymbol{k}} - d_{\boldsymbol{v}\boldsymbol{k}}), \end{aligned} \tag{8.170}$$

with

$$a_{\boldsymbol{v}} := \sum_{\boldsymbol{v}_1} \sum_{\boldsymbol{v}_2} \sum_{\boldsymbol{v}_3} \omega(\boldsymbol{v},\, \boldsymbol{v}_1 | \boldsymbol{v}_2,\, \boldsymbol{v}_3)\, f_{\boldsymbol{v}}^{\text{eq}}\, f_{\boldsymbol{v}_1}^{\text{eq}}, \tag{8.171}$$

$$b_{\boldsymbol{v}\boldsymbol{k}} := \sum_{\boldsymbol{v}_2} \sum_{\boldsymbol{v}_3} \omega(\boldsymbol{v},\, \boldsymbol{k} | \boldsymbol{v}_2,\, \boldsymbol{v}_3)\, f_{\boldsymbol{v}}^{\text{eq}}\, f_{\boldsymbol{k}}^{\text{eq}}, \tag{8.172}$$

$$c_{\boldsymbol{v}\boldsymbol{k}} := \sum_{\boldsymbol{v}_1} \sum_{\boldsymbol{v}_3} \omega(\boldsymbol{v},\, \boldsymbol{v}_1 | \boldsymbol{k},\, \boldsymbol{v}_3)\, f_{\boldsymbol{v}}^{\text{eq}}\, f_{\boldsymbol{v}_1}^{\text{eq}}, \tag{8.173}$$

$$d_{\boldsymbol{v}\boldsymbol{k}} := \sum_{\boldsymbol{v}_1} \sum_{\boldsymbol{v}_2} \omega(\boldsymbol{v},\, \boldsymbol{v}_1 | \boldsymbol{v}_2,\, \boldsymbol{k})\, f_{\boldsymbol{v}}^{\text{eq}}\, f_{\boldsymbol{v}_1}^{\text{eq}}. \tag{8.174}$$

We note that

$$b_{\boldsymbol{v}\boldsymbol{k}} = b_{\boldsymbol{k}\boldsymbol{v}}, \quad c_{\boldsymbol{v}\boldsymbol{k}} = c_{\boldsymbol{k}\boldsymbol{v}}, \quad d_{\boldsymbol{v}\boldsymbol{k}} = d_{\boldsymbol{k}\boldsymbol{v}}, \tag{8.175}$$

which lead to Eq. (8.169). As an instance, we present a proof in the case of $d_{\boldsymbol{v}\boldsymbol{k}}$:

$$\begin{aligned} d_{\boldsymbol{v}\boldsymbol{k}} &= \sum_{\boldsymbol{v}_1} \sum_{\boldsymbol{v}_2} \omega(\boldsymbol{v},\, \boldsymbol{v}_1 | \boldsymbol{v}_2,\, \boldsymbol{k})\, f_{\boldsymbol{v}}^{\text{eq}}\, f_{\boldsymbol{v}_1}^{\text{eq}} \\ &= \sum_{\boldsymbol{v}_1} \sum_{\boldsymbol{v}_2} \omega(\boldsymbol{k},\, \boldsymbol{v}_2 | \boldsymbol{v}_1,\, \boldsymbol{v})\, f_{\boldsymbol{v}}^{\text{eq}}\, f_{\boldsymbol{v}_1}^{\text{eq}} \\ &= \sum_{\boldsymbol{v}_1} \sum_{\boldsymbol{v}_2} \omega(\boldsymbol{k},\, \boldsymbol{v}_2 | \boldsymbol{v}_1,\, \boldsymbol{v})\, f_{\boldsymbol{k}}^{\text{eq}}\, f_{\boldsymbol{v}_2}^{\text{eq}} \\ &= \sum_{\boldsymbol{v}_1} \sum_{\boldsymbol{v}_2} \omega(\boldsymbol{k},\, \boldsymbol{v}_1 | \boldsymbol{v}_2,\, \boldsymbol{v})\, f_{\boldsymbol{k}}^{\text{eq}}\, f_{\boldsymbol{v}_1}^{\text{eq}} = d_{\boldsymbol{k}\boldsymbol{v}}, \end{aligned} \tag{8.176}$$

where the symmetry property of the transition probability (8.4) and the detailed balance (8.29) have been used in the second and third lines, respectively, and the dummy variables $\boldsymbol{v}_1$ and $\boldsymbol{v}_2$ have been interchanged in the final line.

Since L_{vk} is real as

$$L_{vk} = L^*_{vk} = L^\dagger_{kv}, \tag{8.177}$$

we rewrite Eq. (8.169) as

$$L^\dagger = f^{\text{eq}}\, L\, f^{\text{eq}-1}, \tag{8.178}$$

with

$$f^{\text{eq}}_{vk} := f^{\text{eq}}_v\, \delta_{vk}. \tag{8.179}$$

We note that $f^{\text{eq}-1}$ is the inverse of f^{eq}.

Substituting Eq. (8.178) into Eq. (8.160), we have

$$f^{\text{eq}}\, L\, f^{\text{eq}-1}\, \tilde{\boldsymbol{U}}_i = \lambda^*_i\, \tilde{\boldsymbol{U}}_i, \tag{8.180}$$

which can be converted to

$$L\, f^{\text{eq}-1}\, \tilde{\boldsymbol{U}}_i = \lambda^*_i\, f^{\text{eq}-1}\, \tilde{\boldsymbol{U}}_i. \tag{8.181}$$

By comparing this equation with Eq. (8.158), we can read

$$\lambda^*_i - \lambda_i, \tag{8.182}$$

$$f^{\text{eq}-1}\, \tilde{\boldsymbol{U}}_i = \boldsymbol{U}_i. \tag{8.183}$$

From Eq. (8.182), we conclude that all eigenvalues of L are real number and hence the self-adjoint nature of L is respected under the inner product (8.163). Furthermore, using Eq. (8.183) and the completeness (8.162), we find that the metric tensor g is nothing but f^{eq}:

$$g = \sum_i f^{\text{eq}}\, \boldsymbol{U}_i\, \tilde{\boldsymbol{U}}^\dagger_i = f^{\text{eq}}. \tag{8.184}$$

Thus, an explicit form of the inner product (8.163) reads

$$\begin{aligned} \langle \boldsymbol{\varphi}\,,\, \boldsymbol{\psi} \rangle &= \sum_v \sum_k \varphi_v\, f^{\text{eq}}_{vk}\, \psi_k \\ &= \sum_v \sum_k \varphi_v\, f^{\text{eq}}_v\, \delta_{vk}\, \psi_k = \sum_v f^{\text{eq}}_v\, \varphi_v\, \psi_k. \end{aligned} \tag{8.185}$$

We stress that the derived inner product (8.185) is identical to the inner product (8.60). The proof for the uniqueness of the inner product (8.60) is completed.

Appendix 2: Alternative Derivation of Navier-Stokes Equation from RG/E Equation

In this Appendix, we present an alternative but somewhat short-cut derivation of the Navier-Stokes equation, utilizing the fact that the RG/E equation (8.97) implies that the envelope function $f^{\mathrm{E}}_{\boldsymbol{v}}$ given by Eq. (8.100) satisfies the Boltzmann equation up to $\mathrm{O}(\epsilon^2)$, as was shown in Sect. 8.3.7;

$$\frac{\partial}{\partial t} f^{\mathrm{E}}_{\boldsymbol{v}} + \epsilon\, \boldsymbol{v} \cdot \nabla f^{\mathrm{E}}_{\boldsymbol{v}} = C[f^{\mathrm{E}}]_{\boldsymbol{v}}. \tag{8.186}$$

In other words, Eq. (8.186) contains the same dynamical information as the RG/E equation (8.97) does. Therefore it is expected that Eq. (8.186) leads to the Navier-Stokes equation, as the RG/E equation (8.97) did.

First we notice the following identities hold for the envelope function,

$$\sum_{\boldsymbol{v}} C[f^{\mathrm{E}}]_{\boldsymbol{v}} = 0, \tag{8.187}$$

$$\sum_{\boldsymbol{v}} m\, v^i\, C[f^{\mathrm{E}}]_{\boldsymbol{v}} = 0, \tag{8.188}$$

$$\sum_{\boldsymbol{v}} m\, |\boldsymbol{v}|^2/2\, C[f^{\mathrm{E}}]_{\boldsymbol{v}} = 0. \tag{8.189}$$

Then it is apparent that Eq. (8.186) can be reduced to the following balance equations up to $\mathrm{O}(\epsilon^2)$ in the same manner as was done in Sect. 8.2 for the exact distribution function:

$$\frac{\partial}{\partial t}\rho + \epsilon\, \nabla \cdot (\rho\, \boldsymbol{U}) = 0, \tag{8.190}$$

$$\rho\, \frac{\partial}{\partial t} U^i + \epsilon \left[\rho\, \boldsymbol{U} \cdot \nabla U^i + \nabla^j P^{ji}\right] = 0, \tag{8.191}$$

$$\rho\, \frac{\partial}{\partial t}(E/\rho) + \epsilon \left[\rho\, \boldsymbol{U} \cdot \nabla (E/\rho) + P^{ij}\, \nabla^i U^j + \nabla \cdot \boldsymbol{Q}\right] = 0, \tag{8.192}$$

where ρ, U^i, E, P^{ij}, and Q^i are given in terms of the envelop function as

$$\rho = m \sum_{\boldsymbol{v}} f^{\mathrm{E}}_{\boldsymbol{v}}, \tag{8.193}$$

$$U^i = \frac{m}{\rho} \sum_{\boldsymbol{v}} v^i\, f^{\mathrm{E}}_{\boldsymbol{v}}, \tag{8.194}$$

$$E = \sum_{\boldsymbol{v}} \frac{m\, |\boldsymbol{v} - \boldsymbol{U}|^2}{2}\, f^{\mathrm{E}}_{\boldsymbol{v}}, \tag{8.195}$$

$$P^{ij} = \sum_{v} m\,(v^i - U^i)\,(v^j - U^j)\, f_v^{\mathrm{E}}, \tag{8.196}$$

$$Q^i = \sum_{v} \frac{m\,|\boldsymbol{v} - \boldsymbol{U}|^2}{2}\,(v^i - U^i)\, f_v^{\mathrm{E}}, \tag{8.197}$$

respectively.

Now for calculational convenience, we express f_v^{E} in the following form,

$$f_v^{\mathrm{E}} = f_v^{\mathrm{eq}}\,(1 + \phi_v) = f_v^{\mathrm{eq}}\,(1 + [\boldsymbol{\phi}]_v), \tag{8.198}$$

with $\boldsymbol{\phi}$ being the Q-space vector defined by

$$\boldsymbol{\phi} := -\epsilon\, L^{-1}\, Q\, \boldsymbol{F} - \epsilon^2\,(2\, L^{-3}\, Q\, \boldsymbol{G} + L^{-2}\, Q\, \boldsymbol{H} + L^{-1}\, Q\, \boldsymbol{I}) + \mathrm{O}(\epsilon^3), \tag{8.199}$$

which are to be inserted into Eqs. (8.193)–(8.197) in order.

First the mass density ρ (8.193) is calculated as

$$\begin{aligned}\rho &= m \sum_{v} f_v^{\mathrm{eq}}\,(1 + \phi_v) \\ &= m\,\langle\, \boldsymbol{\varphi}_0^0\,,\, \boldsymbol{\varphi}_0^0\,\rangle + m\,\langle\, \boldsymbol{\varphi}_0^0\,,\, \boldsymbol{\phi}\,\rangle \\ &= m\, n,\end{aligned} \tag{8.200}$$

where the formula $\varphi_{0v}^0 = 1$, the definition of the inner product (8.60), the orthogonality relation (8.73), and the orthogonality between the zero modes $\boldsymbol{\varphi}_0^\alpha$ and the Q-space vector $\boldsymbol{\phi}$ have been used. Much the same way, the flow velocity U^i and the total energy E given in (8.194) and (8.195), respectively, are calculated to give

$$\begin{aligned}U^i &= \frac{1}{n} \sum_{v} v^i\, f_v^{\mathrm{eq}}\,(1 + \phi_v) = \frac{1}{m\, n} \sum_{v} (\varphi_{0v}^i + m\, u^i)\, f_v^{\mathrm{eq}}\,(1 + \phi_v) \\ &= \frac{1}{m\, n} \Big[\langle\, \boldsymbol{\varphi}_0^i\,,\, \boldsymbol{\varphi}_0^0\,\rangle + m\, u^i\, \langle\, \boldsymbol{\varphi}_0^0\,,\, \boldsymbol{\varphi}_0^0\,\rangle + \langle\, \boldsymbol{\varphi}_0^i\,,\, \boldsymbol{\phi}\,\rangle + m\, u^i\, \langle\, \boldsymbol{\varphi}_0^0\,,\, \boldsymbol{\phi}\,\rangle \Big] \\ &= u^i,\end{aligned} \tag{8.201}$$

$$\begin{aligned}E &= \sum_{v} \frac{m\,|\boldsymbol{v} - \boldsymbol{u}|^2}{2}\, f_v^{\mathrm{eq}}\,(1 + \phi_v) = \sum_{v} (\varphi_{0v}^4 + \frac{3}{2}\, T)\, f_v^{\mathrm{eq}}\,(1 + \phi_v) \\ &= \langle\, \boldsymbol{\varphi}_0^4\,,\, \boldsymbol{\varphi}_0^0\,\rangle + \frac{3}{2}\, T\, \langle\, \boldsymbol{\varphi}_0^0\,,\, \boldsymbol{\varphi}_0^0\,\rangle + \langle\, \boldsymbol{\varphi}_0^4\,,\, \boldsymbol{\phi}\,\rangle + \frac{3}{2}\, T\, \langle\, \boldsymbol{\varphi}_0^0\,,\, \boldsymbol{\phi}\,\rangle \\ &= \frac{3}{2}\, n\, T,\end{aligned} \tag{8.202}$$

respectively.

The calculation of the pressure tensor P^{ij} given in (8.196) is slightly involved:

$$\begin{aligned} P^{ij} &= \sum_{v} m\,(v^i - u^i)\,(v^j - u^j)\, f_v^{\rm eq}\,(1+\phi_v) = \sum_{v} \tilde{\varphi}_{1v}^{ij}\, f_v^{\rm eq}\,(1+\phi_v) \\ &= \langle\, \tilde{\boldsymbol{\varphi}}_1^{ij}\,,\, \boldsymbol{\varphi}_0^0\,\rangle + \langle\, \tilde{\boldsymbol{\varphi}}_1^{ij}\,,\, \boldsymbol{\phi}\,\rangle \\ &= n\,T\,\delta^{ij} + \langle\, Q\,\tilde{\boldsymbol{\varphi}}_1^{ij}\,,\, \boldsymbol{\phi}\,\rangle, \end{aligned} \tag{8.203}$$

where the formulae $\varphi_{0v}^i = m\,\delta v^i = m\,(v^i - u^i)$ and $\tilde{\varphi}_{1v}^{\alpha i} = \varphi_{0v}^{\alpha}\,\delta v^i$ have been inserted, and the inner products between $\boldsymbol{\varphi}_0^{\alpha}$ and $\tilde{\boldsymbol{\varphi}}_1^{\alpha i}$ given by Eqs. (8.124)–(8.126) have been used. Now it is noteworthy here that we only have to evaluate P^{ij} up to $\mathrm{O}(\epsilon)$, because P^{ij} enters as the form of $\epsilon\, P^{ij}$ into the balance equations (8.190)–(8.192), which are valid up to $\mathrm{O}(\epsilon^2)$. Noticing that

$$\begin{aligned} \langle\, Q\,\tilde{\boldsymbol{\varphi}}_1^{ij}\,,\, \boldsymbol{\phi}\,\rangle &= -\epsilon\,\langle\, \boldsymbol{\pi}^{ij}\,,\, L^{-1}\,Q\,\boldsymbol{F}\,\rangle + \mathrm{O}(\epsilon^2) \\ &= -\epsilon\,2\,\eta\,\sigma^{ij} + \mathrm{O}(\epsilon^2), \end{aligned} \tag{8.204}$$

where Eqs. (8.134), (8.143), and (8.147) have been used, we end up with

$$P^{ij} = n\,T\,\delta^{ij} - \epsilon\,2\,\eta\,\sigma^{ij} + \mathrm{O}(\epsilon^2). \tag{8.205}$$

In a quite similar way, the heat flow Q^i given in (8.197) is calculated to give

$$\begin{aligned} Q^i &= \sum_{v} \frac{m\,|\boldsymbol{v}-\boldsymbol{u}|^2}{2}\,(v^i - u^i)\, f_v^{\rm eq}\,(1+\phi_v) = \sum_{v} (\tilde{\varphi}_{1v}^{4i} + \frac{3}{2}\,T\,\varphi_{0v}^i)\, f_v^{\rm eq}\,(1+\phi_v) \\ &= \langle\, \tilde{\boldsymbol{\varphi}}_1^{4i}\,,\, \boldsymbol{\varphi}_0^0\,\rangle + \frac{3}{2}\,T\,\langle\, \boldsymbol{\varphi}_0^i\,,\, \boldsymbol{\varphi}_0^0\,\rangle + \langle\, \tilde{\boldsymbol{\varphi}}_1^{4i}\,,\, \boldsymbol{\phi}\,\rangle + \frac{3}{2}\,T\,\langle\, \boldsymbol{\varphi}_0^i\,,\, \boldsymbol{\phi}\,\rangle = \langle\, \tilde{\boldsymbol{\varphi}}_1^{4i}\,,\, \boldsymbol{\phi}\,\rangle \\ &= \langle\, Q\,\tilde{\boldsymbol{\varphi}}_1^{4i}\,,\, \boldsymbol{\phi}\,\rangle \\ &= -\epsilon\,\langle\, \boldsymbol{J}^i\,,\, L^{-1}\,Q\,\boldsymbol{F}\,\rangle + \mathrm{O}(\epsilon^2) \\ &= -\epsilon\,\lambda\,\nabla^i T + \mathrm{O}(\epsilon^2). \end{aligned} \tag{8.206}$$

In summary, we have obtained the explicit forms of ρ, U^i, E, P^{ij}, and Q^i, as follows,

$$\rho = m\,n, \quad U^i = u^i, \quad E = e\,n, \tag{8.207}$$

$$P^{ij} = p\,\delta^{ij} - \epsilon\,2\,\eta\,\sigma^{ij} + \mathrm{O}(\epsilon^2), \quad Q^i = -\epsilon\,\lambda\,\nabla^i T + \mathrm{O}(\epsilon^2), \tag{8.208}$$

respectively, where $e = 3\,T/2$ and $p = n\,T$.

Substituting these expressions into the balance equations (8.190)–(8.192), we have the equations which describe the slow dynamics of T, n, and u^i up to $\mathrm{O}(\epsilon^2)$, as follows,

$$\frac{\partial}{\partial t} n = -\epsilon\,\boldsymbol{\nabla}\cdot(n\,\boldsymbol{u}), \tag{8.209}$$

$$m\,n\,\frac{\partial}{\partial t}u^i = -\epsilon\, m\,n\,\boldsymbol{u}\cdot\nabla u^i - \epsilon\,\nabla^i p + \epsilon^2\,\nabla^j(2\,\eta\,\sigma^{ji}), \tag{8.210}$$

$$n\,\frac{\partial}{\partial t}e = -\epsilon\,n\,\boldsymbol{u}\cdot\nabla e - \epsilon\,p\,\nabla\cdot\boldsymbol{u} + \epsilon^2\,2\,\eta\,\sigma^{ij}\,\sigma^{ij} + \epsilon^2\,\nabla\cdot(\lambda\,\nabla T), \tag{8.211}$$

where the following formula has been utilized

$$\sigma^{ij}\,\nabla^i u^j = \sigma^{kl}\,\Delta^{klij}\,\nabla^i u^j = \sigma^{kl}\,\sigma^{kl}. \tag{8.212}$$

Putting back $\epsilon = 1$, we find that the set of these equations is nothing but the Navier-Stokes equation given in Eqs. (8.155)–(8.157).

In conclusion of this Appendix, we give a comment on the next-order equation, the Burnett equation. In the present derivation of the Navier-Stokes equation given by Eqs. (8.209)–(8.211), the expressions of P^{ij} and Q^i up to $\mathrm{O}(\epsilon)$ were needed and computed using only the first-order terms of $\boldsymbol{\phi}$, although the expression of $\boldsymbol{\phi}$ up to $\mathrm{O}(\epsilon^2)$ is available. It is noteworthy that it can be utilized to obtain the next-order equation of the Navier-Stokes equation. Indeed P^{ij} and Q^i are calculated up to $\mathrm{O}(\epsilon^2)$ to give

$$\begin{aligned} P^{ij} &= n\,T\,\delta^{ij} - \epsilon\,2\,\eta\,\sigma^{ij} - \epsilon^2\Big[2\,\langle\,\boldsymbol{\pi}^{ij}\,,\,L^{-3}\,Q\,\boldsymbol{G}\,\rangle + \langle\,\boldsymbol{\pi}^{ij}\,,\,L^{-2}\,Q\,\boldsymbol{H}\,\rangle \\ &\quad + \langle\,\boldsymbol{\pi}^{ij}\,,\,L^{-1}\,Q\,\boldsymbol{I}\,\rangle\Big], \end{aligned} \tag{8.213}$$

$$Q^i = -\epsilon\,\lambda\,\nabla^i T - \epsilon^2\Big[2\,\langle\,\boldsymbol{J}^i\,,\,L^{-3}\,Q\,\boldsymbol{G}\,\rangle + \langle\,\boldsymbol{J}^i\,,\,L^{-2}\,Q\,\boldsymbol{H}\,\rangle + \langle\,\boldsymbol{J}^i\,,\,L^{-1}\,Q\,\boldsymbol{I}\,\rangle\Big]. \tag{8.214}$$

Here the second-order terms in the above equations can be converted into

$$\begin{aligned} &2\,\langle\,\boldsymbol{\pi}^{ij}\,,\,L^{-3}\,Q\,\boldsymbol{G}\,\rangle + \langle\,\boldsymbol{\pi}^{ij}\,,\,L^{-2}\,Q\,\boldsymbol{H}\,\rangle + \langle\,\boldsymbol{\pi}^{ij}\,,\,L^{-1}\,Q\,\boldsymbol{I}\,\rangle \\ &= 2\,\langle\,L^{-3}\,\boldsymbol{\pi}^{ij}\,,\,\boldsymbol{G}\,\rangle + \langle\,L^{-2}\,\boldsymbol{\pi}^{ij}\,,\,\boldsymbol{H}\,\rangle + \langle\,L^{-1}\,\boldsymbol{\pi}^{ij}\,,\,\boldsymbol{I}\,\rangle, \end{aligned} \tag{8.215}$$

$$\begin{aligned} &2\,\langle\,\boldsymbol{J}^i\,,\,L^{-3}\,Q\,\boldsymbol{G}\,\rangle + \langle\,\boldsymbol{J}^i\,,\,L^{-2}\,Q\,\boldsymbol{H}\,\rangle + \langle\,\boldsymbol{J}^i\,,\,L^{-1}\,Q\,\boldsymbol{I}\,\rangle \\ &= 2\,\langle\,L^{-3}\,\boldsymbol{J}^i\,,\,\boldsymbol{G}\,\rangle + \langle\,L^{-2}\,\boldsymbol{J}^i\,,\,\boldsymbol{H}\,\rangle + \langle\,L^{-1}\,\boldsymbol{J}^i\,,\,\boldsymbol{I}\,\rangle, \end{aligned} \tag{8.216}$$

where we have used the self-adjointness of L and the fact that $L^{-1}\,\pi^{ij}$, $L^{-2}\,\pi^{ij}$, $L^{-3}\,\pi^{ij}$, $L^{-1}\,J^i$, $L^{-2}\,J^i$, and $L^{-3}\,J^i$ belong to the Q space. With the use of $\boldsymbol{G}$, $\boldsymbol{H}$, and $\boldsymbol{I}$ given by Eqs. (8.86)–(8.88), we have

$$\begin{aligned} P^{ij} &= n\,T\,\delta^{ij} - \epsilon\,2\,\eta\,\sigma^{ij} \\ &\quad - \epsilon^2\Big\{\langle\,L^{-3}\,\boldsymbol{\pi}^{ij}\,,\,B[P\,\boldsymbol{F}\,,\,P\,\boldsymbol{F}]\,\rangle - \langle\,L^{-2}\,\boldsymbol{\pi}^{ij}\,,\,B[P\,\boldsymbol{F}\,,\,L^{-1}\,Q\,\boldsymbol{F}]\,\rangle \\ &\quad + \frac{1}{2}\,\langle\,L^{-1}\,\boldsymbol{\pi}^{ij}\,,\,B[L^{-1}\,Q\,\boldsymbol{F}\,,\,L^{-1}\,Q\,\boldsymbol{F}]\,\rangle \\ &\quad - \langle\,L^{-2}\,\boldsymbol{\pi}^{ij}\,,\,f^{\mathrm{eq}-1}\,\boldsymbol{v}\cdot\nabla(f^{\mathrm{eq}}\,P\,\boldsymbol{F})\,\rangle \\ &\quad + \langle\,L^{-1}\,\boldsymbol{\pi}^{ij}\,,\,f^{\mathrm{eq}-1}\,\boldsymbol{v}\cdot\nabla(f^{\mathrm{eq}}\,L^{-1}\,Q\,\boldsymbol{F})\,\rangle\Big\} + \mathrm{O}(\epsilon^3), \end{aligned} \tag{8.217}$$

$$
\begin{aligned}
Q^i &= -\epsilon\,\lambda\,\nabla^i T \\
&\quad - \epsilon^2 \Big\{ \langle L^{-3}\,\boldsymbol{J}^i\,,\,B[P\,\boldsymbol{F}\,,\,P\,\boldsymbol{F}] \rangle - \langle L^{-2}\,\boldsymbol{J}^i\,,\,B[P\,\boldsymbol{F}\,,\,L^{-1}\,Q\,\boldsymbol{F}] \rangle \\
&\quad + \frac{1}{2}\,\langle L^{-1}\,\boldsymbol{J}^i\,,\,B[L^{-1}\,Q\,\boldsymbol{F}\,,\,L^{-1}\,Q\,\boldsymbol{F}] \rangle \\
&\quad - \langle L^{-2}\,\boldsymbol{J}^i\,,\,f^{\mathrm{eq}-1}\,\boldsymbol{v}\cdot\nabla(f^{\mathrm{eq}}\,P\,\boldsymbol{F}) \rangle \\
&\quad + \langle L^{-1}\,\boldsymbol{J}^i\,,\,f^{\mathrm{eq}-1}\,\boldsymbol{v}\cdot\nabla(f^{\mathrm{eq}}\,L^{-1}\,Q\,\boldsymbol{F}) \rangle \Big\} + \mathrm{O}(\epsilon^3).
\end{aligned}
\tag{8.218}
$$

Although the further reduction is beyond the scope of this chapter, we point out that P^{ij} and Q^i contain bilinear terms with respect to σ^{ij} and $\nabla^i T$ owing to the terms with B. For instance,

$$
\begin{aligned}
&\frac{1}{2}\,\langle L^{-1}\,\boldsymbol{\pi}^{ij}\,,\,B[L^{-1}\,Q\,\boldsymbol{F}\,,\,L^{-1}\,Q\,\boldsymbol{F}] \rangle \\
&= b_{\pi\pi\pi}\,\Delta^{ijkl}\,\sigma^{km}\,\sigma^{lm} + b_{\pi JJ}\,\Delta^{ijkl}\,(\nabla^k T)\,(\nabla^l T),
\end{aligned}
\tag{8.219}
$$

with

$$
b_{\pi\pi\pi} := \frac{6}{35\,T^2}\,\langle L^{-1}\,\boldsymbol{\pi}^{ij}\,,\,B[L^{-1}\,\boldsymbol{\pi}^{ik}\,,\,L^{-1}\,\boldsymbol{\pi}^{jk}] \rangle, \tag{8.220}
$$

$$
b_{\pi JJ} := \frac{1}{10\,T^4}\,\langle L^{-1}\,\boldsymbol{\pi}^{ij}\,,\,B[L^{-1}\,\boldsymbol{J}^i\,,\,L^{-1}\,\boldsymbol{J}^j] \rangle, \tag{8.221}
$$

where Eq. (8.143) and the following formulae have been used,

$$
\Delta^{ijlm}\,\Delta^{ikln}\,\Delta^{jkmn} = 35/12, \quad \Delta^{ijkl}\,\Delta^{ijkl} = 5. \tag{8.222}
$$

These bilinear terms are called the Burnett term [113, 141], which is known to be derived also by the Chapman-Enskog expansion method as a next order equation of the Navier-Stokes equation, *i.e.*, the Burnett equation [113].

Appendix 3: Proof of Vanishing of Inner Product Between Collision Invariants and B

In this Appendix, we present the proof for the identity given by Eq. (8.128).

We start with the definition of the collision invariants:

$$
\sum_{\boldsymbol{v}} \varphi^{\alpha}_{0\boldsymbol{v}}\,C[f]_{\boldsymbol{v}} = 0. \tag{8.223}
$$

Since the above equation is valid for an arbitrary f_v, we can take the derivatives with respect to f_v. Then the second derivative reads

$$\sum_v \varphi_{0v}^{\alpha} \frac{\partial^2}{\partial f_k \partial f_l} C[f]_v = 0. \tag{8.224}$$

Taking the value of Eq. (8.224) at $f_v = f_v^{\mathrm{eq}}$, we have an identity

$$\sum_v \varphi_{0v}^{\alpha}\, f_v^{\mathrm{eq}}\, B_{vkl}\, f_k^{\mathrm{eq}-1}\, f_l^{\mathrm{eq}-1} = 0 \tag{8.225}$$

with B being defined in Eq. (8.89).

By multiplying arbitrary vectors $f_k^{\mathrm{eq}}\, \psi_k$ and $f_l^{\mathrm{eq}}\, \chi_l$ to the identity (8.225), and then summing up with respect to $\boldsymbol{k}$ and $\boldsymbol{l}$, we have

$$\begin{aligned} 0 &= \sum_v \varphi_{0v}^{\alpha}\, f_v^{\mathrm{eq}} \sum_k \sum_l B_{vkl}\, \psi_k\, \chi_l \\ &= \sum_v \varphi_{0v}^{\alpha}\, f_v^{\mathrm{eq}}\, B[\boldsymbol{\psi}\,,\, \boldsymbol{\chi}]_v \\ &= \langle \varphi_0^{\alpha}\,,\, B[\boldsymbol{\psi}\,,\, \boldsymbol{\chi}] \rangle, \end{aligned} \tag{8.226}$$

where the notation (8.90) has been used. Finally, putting $\boldsymbol{\psi} = \boldsymbol{\chi} = L^{-1}\, Q\, \boldsymbol{F}$, we find that Eq. (8.226) becomes Eq. (8.128). This completes the proof of Eq. (8.128).

Chapter 9
A General Theory for Constructing Mesoscopic Dynamics: Doublet Scheme in RG Method

9.1 Introduction

We have seen that the RG method [1, 38, 39, 48–52, 92–96] as described in Chaps. 4 and 5 based on [3, 5, 6, 40, 46, 57, 58] has a wide applicability for deriving the slow dynamics with fewer degrees of freedom from the underlying microscopic equations in the perturbation theory.[1] The slow dynamics can be described by coarse-grained collective variables or macroscopic variables such as the amplitudes and phases of oscillators. The key ingredient of the method is the zero modes of the unperturbed evolution operator, which are to be promoted by the nonlinear terms to the dynamical variables forming the invariant manifold [33, 34] on which the dynamical motion is confined: The reduced evolution equation of the would-be zero modes on the manifold is provided through the RG/E equation.[2]

In some interesting problems in physics, however, it becomes necessary to construct a mesoscopic dynamics which would be given, say, by partial coarse-graining of the microscopic equation [144–151], [62, pp. 78–81], [152–154]. A natural way to achieve this task is to incorporate appropriate excited or fast modes as well as the zero modes to construct the invariant manifold on which the mesoscopic variables are defined: Fig. 9.1 gives a schematic picture of the invariant/attractive manifold composed of the zero and excited modes.

Then how can one identify the appropriate excited modes describing the invariant manifold of the mesoscopic scale and derive the mesoscopic dynamics on it. In this chapter, we shall present a general scheme [69] to overcome these problems by extending the RG method presented in the previous chapters. The general scheme is called the *doublet scheme* in the RG method because the excited modes necessary to

[1] As for some historical development of the RG method, see Sect. 1.3.

[2] We remark that one can see in [35, 143] extensive applications of a reduction theory to physical and chemical kinetics based on the notion of invariant manifolds with quite different formulation from the RG method.

T. Kunihiro et al., *Geometrical Formulation of Renormalization-Group Method as an Asymptotic Analysis*, Fundamental Theories of Physics 206,
https://doi.org/10.1007/978-981-16-8189-9_9

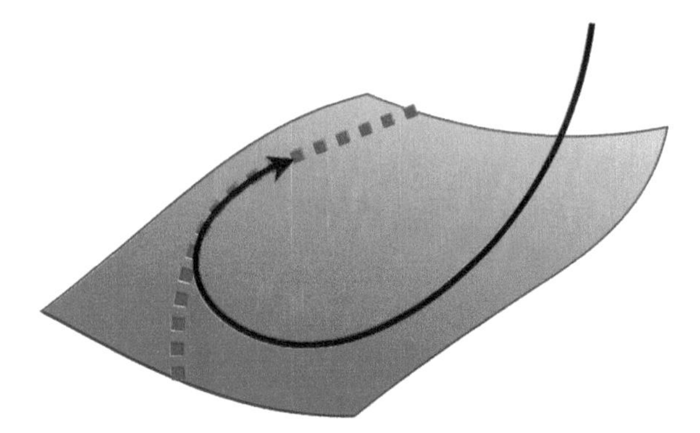

Fig. 9.1 A schematic picture of the invariant/attractive manifold spanned by the zero and excited modes: The zero-mode manifold is represented by the (blue) dashed line, while the solid line denotes the orbit of an exact solution to the microscopic equation. The surface shows the invariant/attractive manifold extended by incorporating excited modes. Under the time evolution of the system, the exact solution starting from a point away from the surface is rapidly attracted along the solid line to the surface, then after performing a fast motion on it, the solution approaches the dashed line, and eventually shows a slow motion confined on it

describe the mesoscopic dynamics turn out to always appear pairwise and the RG/E equation still plays the essential role to give the reduced evolution equation [69]. It is found that the doublet scheme can be applicable to derive a mesoscopic dynamics from a wide class of evolution equations, although only one example is treated in this chapter: In Part II of the present monograph, the scheme will be applied to derive the relativistic and nonrelativistic fluid dynamics in the mesoscopic scales [155–157].

This chapter is based on [69] with some elucidations, and organized as follows: In Sect. 9.2, we describe the doublet scheme in the RG method. In Sect. 9.3, we analyze the Lorenz model [84] in the doublet scheme and demonstrate the validity of the doublet scheme as a method for constructing the invariant/attractive manifold that incorporates the excited modes as well as the zero modes.

9.2 General Formulation

In this section, using a generic evolution equation we extract its mesoscopic dynamics by constructing the invariant/attractive manifold incorporating the appropriate excited modes as well as the zero modes of its linearized evolution operator: The appropriate excited modes are identified on the basis of the natural principle inherent in the reduction theory [32] that the resultant dynamics should be as simple as possible with as few as possible variables. In fact, there is still a trade-off between the simplicity/complexity of the reduced equation and the representation of the invariant/attractive manifold. We shall choose the variables so that the reduced equation becomes simpler. These principles lead to a unique extraction of the excited modes to be incorporated for describing the mesoscopic dynamics, and it turns out that these additional modes always appear pairwise, and hence the name of the doublet scheme.

9.2.1 Preliminaries

As a generic evolution equation, we treat the following system of nonlinear differential equations for $\boldsymbol{X}$ with $\dim\boldsymbol{X} = N\,(1 < N \leq \infty)$,

$$\frac{\partial}{\partial t}\boldsymbol{X} = \boldsymbol{G}(\boldsymbol{X}) + \epsilon\,\boldsymbol{F}(\boldsymbol{X}), \tag{9.1}$$

or in terms of components

$$\frac{\partial}{\partial t}X_i = G_i(X_1,\,\ldots,\,X_N) + \epsilon\,F_i(X_1,\,\ldots,\,X_N), \quad i = 1,\,\ldots,\,N. \tag{9.2}$$

Here $\boldsymbol{G}(\boldsymbol{X})$ and $\boldsymbol{F}(\boldsymbol{X})$ are non-linear functions of $\boldsymbol{X}$, and ϵ is introduced as an indicator of the smallness of $\boldsymbol{F}(\boldsymbol{X})$ that is to be set to 1 in the final stage. When $\boldsymbol{F}(\boldsymbol{X})$ is absent, the vector $\boldsymbol{X}(t)$ governed by Eq. (9.1) is supposed to be relaxed to a stationary solution $\boldsymbol{X}^{\text{eq}}$ after a long time as

$$\boldsymbol{X}(t \to \infty) \to \boldsymbol{X}^{\text{eq}}, \tag{9.3}$$

which satisfies the stationary condition

$$\boldsymbol{G}(\boldsymbol{X}^{\text{eq}}) = 0. \tag{9.4}$$

Here, we assume that when the initial conditions are varied, the resultant stationary solution $\boldsymbol{X}^{\text{eq}}$ forms a well-defined M_0-dimensional manifold where $1 \leq M_0 \leq N$, and accordingly the whole manifold $\boldsymbol{X}^{\text{eq}}$ is parametrized by M_0 integral constants

$$(C_1,\,\ldots,\,C_{M_0}) =: \boldsymbol{C}, \tag{9.5}$$

belonging to some M_0-dimensional domain;

$$\boldsymbol{X}^{\text{eq}} = \boldsymbol{X}^{\text{eq}}(\boldsymbol{C}). \tag{9.6}$$

For a given set of $\{C_\alpha\}_{\alpha=1,\,\ldots,\,M_0}$, let us define the linearized evolution operator A by

$$A_{ij} := \frac{\partial}{\partial X_j} G_i(X_1,\,\ldots,\,X_N)\bigg|_{\boldsymbol{X}=\boldsymbol{X}^{\text{eq}}}. \tag{9.7}$$

By differentiating the i-th component of Eq. (9.4) with respect to C_α with $\alpha = 1, \ldots, M_0$, we have

$$0 = \frac{\partial G_i}{\partial C_\alpha} = \sum_{j=1}^{N} \frac{\partial G_i}{\partial X_j^{\mathrm{eq}}} \frac{\partial X_j^{\mathrm{eq}}}{\partial C_\alpha} = \sum_{j=1}^{N} A_{ij} \frac{\partial X_j^{\mathrm{eq}}}{\partial C_\alpha}, \tag{9.8}$$

or

$$A \cdot (\partial \boldsymbol{X}^{\mathrm{eq}} / \partial C_\alpha) = 0, \tag{9.9}$$

in the matrix notation. Equation (9.9) tells us that the linear operator A has M_0 independent (right) eigenvectors belonging to the zero eigenvalue, *i.e.*, zero modes

$$\boldsymbol{\varphi}_0^\alpha := \partial \boldsymbol{X}^{\mathrm{eq}} / \partial C_\alpha, \quad \varphi_{0i}^\alpha = \partial X_i^{\mathrm{eq}} / \partial C_\alpha, \tag{9.10}$$

and accordingly

$$\dim [\ker A] = M_0. \tag{9.11}$$

It should be noted that the matrix A is not necessarily symmetric but generally asymmetric one.[3] Then, let us denote the left eigenvectors belonging to zero eigenvalue by, $\tilde{\boldsymbol{\varphi}}_0^\alpha$, which satisfy

$$A^\dagger \tilde{\boldsymbol{\varphi}}_0^\alpha = 0. \tag{9.12}$$

We assume that the other than zero eigenvalues λ_a of A are discrete real negative values:

$$A \boldsymbol{\varphi}_{\lambda_a} = \lambda_a \boldsymbol{\varphi}_{\lambda_a}, \quad \lambda_a < 0 \quad (a = M_0 + 1, \ldots, N). \tag{9.13}$$

We assume that they are not degenerate: $\lambda_a \neq \lambda_b$ for $a \neq b$. The right eigenvector belonging to λ_a is denoted by $\tilde{\boldsymbol{\varphi}}_{\lambda_a}$:

$$A^\dagger \tilde{\boldsymbol{\varphi}}_{\lambda_a} = \lambda_a \tilde{\boldsymbol{\varphi}}_{\lambda_a}. \tag{9.14}$$

We note that the tangent space of the invariant manifold at $\boldsymbol{X}^{\mathrm{eq}} = \boldsymbol{X}(\boldsymbol{C})$ is spanned locally by the zero modes $\boldsymbol{\varphi}_0^\alpha$ with $\alpha = 1, \ldots, M_0$.

Let us introduce an inner product between arbitrary vectors $\boldsymbol{\psi} = (\psi_1, \ldots, \psi_N)$ and $\boldsymbol{\chi} = (\chi_1, \ldots, \chi_N)$ as

$$\langle \boldsymbol{\psi} | \boldsymbol{\chi} \rangle := \sum_{i=1}^{N} \psi_i^* \chi_i, \tag{9.15}$$

[3] See Sect. 3.6.2 for basic mathematical facts about an asymmetric matrix.

which has the positive definiteness of the norm as

$$\langle \boldsymbol{\psi} | \boldsymbol{\psi} \rangle > 0, \quad \text{for } \boldsymbol{\psi} \neq 0. \tag{9.16}$$

Since the basis vectors $\boldsymbol{\varphi}_0^\alpha$ with $\alpha = 1, \ldots, M_0$ are not necessarily orthogonal, we define the 'metric' tensor in the kernel space of A by

$$\eta_0^{\alpha\beta} := \langle \tilde{\boldsymbol{\varphi}}_0^\alpha | \boldsymbol{\varphi}_0^\beta \rangle. \tag{9.17}$$

Then we define the projection operator P_0 onto the kernel of A, which we call the P_0 space, and the projection operator Q_0 onto the Q_0 space as the complement to the P_0 space by

$$P_0 \boldsymbol{\psi} := \sum_{\alpha,\beta=1}^{M_0} \boldsymbol{\varphi}_0^\alpha \, \eta_{0\alpha\beta}^{-1} \, \langle \tilde{\boldsymbol{\varphi}}_0^\beta | \boldsymbol{\psi} \rangle, \tag{9.18}$$

$$Q_0 := 1 - P_0, \tag{9.19}$$

where $\eta_{0\alpha\beta}^{-1}$ is the inverse matrix of the P_0-space metric matrix $\eta_0^{\alpha\beta}$:

$$\sum_{\gamma=1}^{M_0} \eta_{0\alpha\gamma}^{-1} \eta_0^{\gamma\beta} = \delta_\alpha^\beta. \tag{9.20}$$

We can verify the idempotency of P_0 as follows

$$\begin{aligned} P_0 P_0 \boldsymbol{\psi} &= P_0 \sum_{\alpha,\beta=1}^{M_0} \boldsymbol{\varphi}_0^\alpha \, \eta_{0\alpha\beta}^{-1} \, \langle \tilde{\boldsymbol{\varphi}}_0^\beta | \boldsymbol{\psi} \rangle \\ &= \sum_{\alpha,\beta=1}^{M_0} \sum_{\gamma_1,\gamma_2=1}^{M_0} \boldsymbol{\varphi}_0^{\gamma_1} \, \eta_{0\gamma_1\gamma_2}^{-1} \langle \tilde{\boldsymbol{\varphi}}_0^{\gamma_2} | \boldsymbol{\varphi}_0^\alpha \rangle \, \eta_{0\alpha\beta}^{-1} \, \langle \tilde{\boldsymbol{\varphi}}_0^\beta | \boldsymbol{\psi} \rangle \\ &= \sum_{\alpha,\beta=1}^{M_0} \sum_{\gamma_1,\gamma_2=1}^{M_0} \boldsymbol{\varphi}_0^{\gamma_1} \, \eta_{0\gamma_1\gamma_2}^{-1} \eta^{\gamma_2\alpha} \, \eta_{0\alpha\beta}^{-1} \, \langle \tilde{\boldsymbol{\varphi}}_0^\beta | \boldsymbol{\psi} \rangle \\ &= \sum_{\gamma,\beta=1}^{M_0} \boldsymbol{\varphi}_0^\gamma \, \eta_{0\gamma_1\beta}^{-1} \langle \tilde{\boldsymbol{\varphi}}_0^\beta | \boldsymbol{\psi} \rangle = P_0 \boldsymbol{\psi}. \end{aligned} \tag{9.21}$$

As is shown in Sect. 3.6.2, the right and left eigenvectors belonging to different eigenvalues are orthogonal to each other:

$$a \neq b \implies \langle \tilde{\boldsymbol{\varphi}}_{\lambda_a} | \boldsymbol{\varphi}_{\lambda_b} \rangle = 0. \tag{9.22}$$

Since $\lambda_a \neq 0$ is not degenerate, the eigenvectors are normalized as follows, $\langle \tilde{\boldsymbol{\varphi}}_{\lambda_a} | \boldsymbol{\varphi}_{\lambda_a} \rangle = 1$. Thus,

$$\langle \tilde{\boldsymbol{\varphi}}_{\lambda_a} | \boldsymbol{\varphi}_{\lambda_b} \rangle = \delta_{ab}. \tag{9.23}$$

The projection operator Q_0 is now expressed as

$$Q_0 = \sum_{a=M_0+1}^{N} |\tilde{\boldsymbol{\varphi}}_{\lambda_a}\rangle\langle \boldsymbol{\varphi}_{\lambda_a}|. \tag{9.24}$$

Then the idempotency of Q_0 follows directly from (9.23) as,

$$\begin{aligned} Q_0^2 &= \sum_{a=M_0+1}^{N} |\tilde{\boldsymbol{\varphi}}_{\lambda_a}\rangle\langle \boldsymbol{\varphi}_{\lambda_a}| \left(\sum_{b=M_0+1}^{N} |\tilde{\boldsymbol{\varphi}}_{\lambda_b}\rangle\langle \boldsymbol{\varphi}_{\lambda_b}| \right) \\ &= \sum_{a,b=M_0+1}^{N} |\tilde{\boldsymbol{\varphi}}_{\lambda_a}\rangle\langle \boldsymbol{\varphi}_{\lambda_a}|\tilde{\boldsymbol{\varphi}}_{\lambda_b}\rangle\langle \boldsymbol{\varphi}_{\lambda_b}| = \sum_{a,b=M_0+1}^{N} |\tilde{\boldsymbol{\varphi}}_{\lambda_a}\rangle\delta_{ab}\langle \boldsymbol{\varphi}_{\lambda_b}| \\ &= \sum_{a=M_0+1}^{N} |\tilde{\boldsymbol{\varphi}}_{\lambda_a}\rangle\langle \boldsymbol{\varphi}_{\lambda_a}| = Q_0. \end{aligned} \tag{9.25}$$

Following the general argument given in Sect. 3.6.2, we introduce a modified inner product for arbitrary two vectors $\boldsymbol{\varphi}$ and $\boldsymbol{\psi}$ with the metric as

$$\langle\, \boldsymbol{\varphi}\,,\, \boldsymbol{\psi}\,\rangle := \langle \boldsymbol{\varphi} | \boldsymbol{g} \boldsymbol{\psi}\,\rangle, \tag{9.26}$$

where

$$\boldsymbol{g} := \sum_{\alpha=1}^{M_0} |\tilde{\boldsymbol{\varphi}}_0^{\alpha}\rangle\langle \boldsymbol{\varphi}_0^{\alpha}| + \sum_{a=M_0+1}^{N} |\tilde{\boldsymbol{\varphi}}_{\lambda_a}\rangle\langle \boldsymbol{\varphi}_{\lambda_a}|. \tag{9.27}$$

Then, as is shown in Sect. 3.6.2, the following relation holds for arbitrary two vectors $\boldsymbol{\psi}$ and $\boldsymbol{\chi}$

$$\begin{aligned} \langle\, \boldsymbol{\varphi}\,,\, A\boldsymbol{\psi}\,\rangle &= \langle \boldsymbol{\varphi} | \boldsymbol{g} A \boldsymbol{\psi}\,\rangle \\ &= \sum_{a=M_0+1}^{N} \lambda_a \langle \boldsymbol{\varphi} | \tilde{\boldsymbol{\varphi}}_{\lambda_a}\rangle\langle \boldsymbol{\varphi}_{\lambda_a} | \boldsymbol{\psi}\rangle, \end{aligned} \tag{9.28}$$

$$\begin{aligned} \langle\, A\boldsymbol{\varphi}\,,\, \boldsymbol{\psi}\,\rangle &= \langle A\boldsymbol{\varphi} | \boldsymbol{g} \boldsymbol{\psi}\,\rangle \\ &= \sum_{a=M_0+1}^{N} \lambda_a \langle \boldsymbol{\varphi} | \tilde{\boldsymbol{\varphi}}_{\lambda_a}\rangle\langle \boldsymbol{\varphi}_{\lambda_a} | \boldsymbol{\psi}\rangle = \langle\, \boldsymbol{\varphi}\,,\, A\boldsymbol{\psi}\,\rangle, \end{aligned} \tag{9.29}$$

which tells us that A is self-adjoint. Here we have used the reality of λ_a;

$$\lambda_a^* = \lambda_a. \tag{9.30}$$

We will see that this self-adjoint nature of A plays an essential role in making the form of the resultant equation simpler. Similarly, the following identities for an arbitrary vector $\boldsymbol{\psi}$ follow from the self-adjointness of the projection operators;

$$\langle \boldsymbol{\varphi}_0^\alpha , P_0 \boldsymbol{\psi} \rangle = \langle P_0 \boldsymbol{\varphi}_0^\alpha , \boldsymbol{\psi} \rangle = \langle \boldsymbol{\varphi}_0^\alpha , \boldsymbol{\psi} \rangle, \quad \langle \boldsymbol{\varphi}_0^\alpha , Q_0 \boldsymbol{\psi} \rangle = \langle Q_0 \boldsymbol{\varphi}_0^\alpha , \boldsymbol{\psi} \rangle = 0. \tag{9.31}$$

9.2.2 Construction of the Approximate Solution Around Arbitrary Time

To extract the mesoscopic slow dynamics of Eq. (9.1) in the asymptotic region supposing that an initial condition for the exact solution $\boldsymbol{X}(t)$ is given, say, at $t = -\infty$, we first construct an approximate solution composed of the zero modes and the appropriate excited modes around arbitrary time by the perturbation theory.

In accordance with the general formulation of the RG method [3, 5, 6, 46] presented in Chaps. 4 and 5, we try to construct a perturbative solution $\tilde{\boldsymbol{X}}(t\,;\,t_0)$ with the 'initial' condition at $t = t_0$:

$$\tilde{\boldsymbol{X}}(t = t_0\,;\,t_0) = \boldsymbol{X}(t_0), \tag{9.32}$$

where it has been made explicit that $\tilde{\boldsymbol{X}}(t = t_0\,;\,t_0)$ may depend on t_0.

We expand the 'initial' value as well as the perturbative solution as follows:

$$\tilde{\boldsymbol{X}}(t\,;\,t_0) = \tilde{\boldsymbol{X}}_0(t\,;\,t_0) + \epsilon\, \tilde{\boldsymbol{X}}_1(t\,;\,t_0) + \epsilon^2\, \tilde{\boldsymbol{X}}_2(t\,;\,t_0) + \cdots, \tag{9.33}$$

$$\boldsymbol{X}(t_0) = \boldsymbol{X}_0(t_0) + \epsilon\, \boldsymbol{X}_1(t_0) + \epsilon^2\, \boldsymbol{X}_2(t_0) + \cdots, \tag{9.34}$$

where the respective terms are set to obey the 'initial' conditions at $t = t_0$ as

$$\tilde{\boldsymbol{X}}_l(t = t_0\,;\,t_0) = \boldsymbol{X}_l(t_0), \quad l = 0, 1, 2, \ldots, \tag{9.35}$$

respectively. We shall construct the zeroth-order 'initial' value $\tilde{\boldsymbol{X}}_0(t_0\,;\,t_0) = \boldsymbol{X}_0(t_0)$ as close as possible to the exact value $\boldsymbol{X}(t_0)$.

Insertion of the above expansions into Eq. (9.1) leads to a series of the perturbative equations by equating the coefficients of the respective powers of ϵ. In the following, we shall carry out the perturbative analysis up to the second order, which is necessary to obtain a sensible mesoscopic dynamics.

Before entering the perturbative analysis, we give the geometrical picture, in Fig. 9.1, of the way how the invariant/attractive manifold spanned solely by the zero modes is extended and thus improved so as to capture the whole mesoscopic dynamics.

Now, the zeroth-order equation reads

$$\frac{\partial}{\partial t}\tilde{\boldsymbol{X}}_0(t\,;\,t_0) = \boldsymbol{G}(\tilde{\boldsymbol{X}}_0(t\,;\,t_0)). \tag{9.36}$$

Since we are interested in the slow motion that would be realized asymptotically as $t \to \infty$, we try to find a stationary solution, which satisfies

$$\frac{\partial}{\partial t}\tilde{\boldsymbol{X}}_0(t\,;\,t_0) = 0. \tag{9.37}$$

This equation is satisfied when $\tilde{\boldsymbol{X}}_0(t\,;\,t_0)$ is a fixed point, $\boldsymbol{G}(\tilde{\boldsymbol{X}}_0(t\,;\,t_0)) = 0$, which is nothing but Eq. (9.4). Thus, on account of the uniqueness of the fixed point which has been assumed, we can identify $\tilde{\boldsymbol{X}}_0(t\,;\,t_0)$ with $\boldsymbol{X}^{\mathrm{eq}}$:

$$\tilde{\boldsymbol{X}}_0(t\,;\,t_0) = \boldsymbol{X}^{\mathrm{eq}}(t_0), \tag{9.38}$$

which implies that

$$\boldsymbol{X}_0(t_0) = \tilde{\boldsymbol{X}}_0(t = t_0\,;\,t_0) = \boldsymbol{X}^{\mathrm{eq}}(t_0). \tag{9.39}$$

We remark that $\boldsymbol{X}^{\mathrm{eq}}(t_0)$ has a t_0-dependence through the integral constants $C_\alpha(t_0)$ $(\alpha = 1,\,\ldots,\,M_0)$ defined by Eq. (9.6);

$$\boldsymbol{X}^{\mathrm{eq}}(t_0) := \boldsymbol{X}^{\mathrm{eq}}(C_1(t_0),\,\ldots,\,C_{M_0}(t_0)). \tag{9.40}$$

From now on, we suppress the t_0-dependence of the solutions when no misunderstanding is expected.

9.2.3 First-Order Solution and Introduction of the Doublet Scheme

The first-order equation is now given by

$$\frac{\partial}{\partial t}\tilde{\boldsymbol{X}}_1(t) = A\,\tilde{\boldsymbol{X}}_1(t) + \boldsymbol{F}_0, \tag{9.41}$$

with

$$\boldsymbol{F}_0 := \boldsymbol{F}(\boldsymbol{X}^{\mathrm{eq}}). \tag{9.42}$$

The general solution to the linear inhomogeneous equation (9.41) is readily obtained with the use of the method of variation of the constants: The solution can be written as $\tilde{\boldsymbol{X}}_1 = \mathrm{e}^{At}\boldsymbol{c}(t)$, and $\boldsymbol{c}(t)$ satisfies the equation $\dot{\boldsymbol{c}} = \mathrm{e}^{-At}\boldsymbol{F}_0$, which is solved by the quadrature. Then the solution with the 'initial' condition

$$\boldsymbol{X}_1(t_0) = \tilde{\boldsymbol{X}}_1(t = t_0\,;\, t_0) =: \boldsymbol{\phi}(t_0), \tag{9.43}$$

is expressed in terms of the projection operators P_0 and Q_0 as

$$\tilde{\boldsymbol{X}}_1(t\,;\, t_0) = \mathrm{e}^{A(t-t_0)}\,\boldsymbol{\phi} + (t - t_0)\, P_0\, \boldsymbol{F}_0 + (\mathrm{e}^{A(t-t_0)} - 1)\, A_{Q_0}^{-1}\, Q_0\, \boldsymbol{F}_0, \tag{9.44}$$

where $A_{Q_0}^{-1}$ was introduced in Sect. 3.6.2 and has the following property,

$$\begin{aligned} A_{Q_0}^{-1} &:= Q_0 A^{-1} Q_0 = Q_0 \sum_{a=M_0+1}^{N} A^{-1} |\tilde{\boldsymbol{\varphi}}_{\lambda_a}\rangle\langle \boldsymbol{\varphi}_{\lambda_a}| = Q_0 \sum_{a=M_0+1}^{N} \frac{1}{\lambda_a} |\tilde{\boldsymbol{\varphi}}_{\lambda_a}\rangle\langle \boldsymbol{\varphi}_{\lambda_a}| \\ &= \sum_{a=M_0+1}^{N} \frac{1}{\lambda_a} |\tilde{\boldsymbol{\varphi}}_{\lambda_a}\rangle\langle \boldsymbol{\varphi}_{\lambda_a}| = A^{-1} Q_0, \end{aligned} \tag{9.45}$$

which imply that

$$A_{Q_0}^{-1} = A_{Q_0}^{-1} Q_0 = A^{-1} Q_0. \tag{9.46}$$

In the following, we shall use the last notation.

Without loss of generality, we can suppose that $\boldsymbol{\phi}(t_0)$ is orthogonal to the tangent space spanned by the zero modes $\{\boldsymbol{\varphi}_0^{\alpha}\}_{\alpha=1,\ldots,M_0}$ and hence belongs to the Q_0 space, because the possible zero modes contained in $\boldsymbol{\phi}$ can be eliminated by the redefinition of the zeroth-order 'initial' value $\boldsymbol{X}^{\mathrm{eq}}$ through a shift of the parameters $(C_1(t_0),\ \ldots,\ C_{M_0}(t_0))$.

We note the existence of the secular term in Eq. (9.44) apparently invalidate the perturbative solution when $|t - t_0|$ becomes large.

It is found convenient for later discussions to expand $\mathrm{e}^{A(t-t_0)}$ with respect to $t - t_0$ and retain the terms in the first order as

$$\tilde{\boldsymbol{X}}_1(t\,;\, t_0) = \boldsymbol{\phi}(t_0) + (t - t_0)\,(A\,\boldsymbol{\phi} + P_0\,\boldsymbol{F}_0 + Q_0\,\boldsymbol{F}_0). \tag{9.47}$$

The neglected terms of $O((t - t_0)^n)$ (≥ 2) are irrelevant when we apply the RG/E equation.

We are now in a position to introduce one of the central ideas of the doublet scheme. The problem we must solve is how to extend the vector space beyond that spanned by the zero modes to accommodate the appropriate excited modes for constructing the closed mesoscopic dynamics. We call the additional vector space the P_1 space, which is a subspace of the Q_0 space.

Although it is admittedly not apparent a priori whether such a vector space exists, it will be found that the form of the perturbative solution (9.47) itself naturally leads to the way how to construct the P_1 space on the basis of the basic principle of the reduction theory of dynamical systems that the closed reduced system is to be composed of a as small as possible number of variables and equations. In the present case, this principle is realized by imposing the condition that the tangent space of the perturbative solution at $t = t_0$ should be spanned by a as small as possible number of independent vectors. Then we should require that

(I) $A\,\boldsymbol{\phi}$ and $Q_0\,\boldsymbol{F}_0$ belong to a common vector space.

Next, we note that the solution (9.47) at $t \sim t_0$ is a linear combination of a vector $P_0\,\boldsymbol{F}_0$ belonging to the P_0 space and three new vectors, *i.e.*, $\boldsymbol{\phi}$, $A\,\boldsymbol{\phi}$, and $Q_0\,\boldsymbol{F}_0$. Therefore, the minimal P_1 space that is closed is readily obtained if the following condition is additionally satisfied:

(II) The P_1 space is spanned by the bases of the union of $\boldsymbol{\phi}$ and $A\,\boldsymbol{\phi}$.

The first condition (I) may be restated that both $\boldsymbol{\phi}$ and $A^{-1}\,Q_0\,\boldsymbol{F}_0$ belong to a common vector space: We note that this condition is in accordance with the fact that $\boldsymbol{\phi}$ belongs to the Q_0 space.

Thus, to get the more detailed structure of the P_1 space, one only have to calculate $A^{-1}\,Q_0\,\boldsymbol{F}_0$ and identify the basis vector fields of the vector subspace which $A^{-1}\,Q_0\boldsymbol{F}_0$ belongs to, and will also constitute the basis vectors to express $\boldsymbol{\phi}$. The vector $Q_0\,\boldsymbol{F}_0[\boldsymbol{C}]$ would span a vector space when $\boldsymbol{C} = (C_1, \ldots, C_{M_0})$ is varied in the domain where $\boldsymbol{C}$ is defined. Let the dimension of the vector space is a finite number, say, M_1. Then, there exist M_1 independent vectors $\boldsymbol{\varphi}_1^{\mu}$ with $\mu = 1, \ldots, M_1$ so that $Q_0\,\boldsymbol{F}_0[\boldsymbol{C}]$ for any $\boldsymbol{C}$ can be expressed as a linear combination of them as

$$Q_0\,\boldsymbol{F}_0 = \sum_{\mu=1}^{M_1} f_{\mu}(\boldsymbol{C})\,\boldsymbol{\varphi}_1^{\mu}, \tag{9.48}$$

where $f_{\mu}(\boldsymbol{C})$ are numerical coefficients solely dependent on $\boldsymbol{F}_0$ and hence functions of $\boldsymbol{C} = (C_1, \ldots, C_{M_0})$. Note that using the freedom in the choice of the set of the basis vectors $\boldsymbol{\varphi}_1^{\mu}$ ($\mu = 1, \ldots, M_1$), we can make a suitable choice of them in actual problems on the basis of physical meaning and/or conditions such as symmetry properties.

Now, the fact $\boldsymbol{\varphi}_1^{\mu}$'s belong to the Q_0 space implies that

$$Q_0\boldsymbol{\varphi}_1^{\mu} = \boldsymbol{\varphi}_1^{\mu}, \tag{9.49}$$

then we can write as

$$A^{-1}\,\boldsymbol{\varphi}_1^{\mu} = A^{-1}\,Q_0\,\boldsymbol{\varphi}_1^{\mu}, \tag{9.50}$$

without any mathematical ambiguity. Thus, we can also write as

$$A^{-1}\,Q_0\,\boldsymbol{F}_0 = \sum_{\mu=1}^{M_1} (A^{-1}\,\varphi_1^{\mu})\,f_{\mu}. \tag{9.51}$$

Note that since $\det A_{Q_0} \neq 0$, *i.e.*, A_{Q_0} is non-singular, provided that φ_1^{μ} with $\mu = 1, \ldots, M_1$ are mutually independent, which we have assumed.

Our task for finding the adequate variables for describing the reduced (mesoscopic) dynamics is tantamount to requiring $A^{-1}\,Q_0\,\boldsymbol{F}_0$ and $\boldsymbol{\phi}$ belong to the same vector subspace. This requirement is readily met by adopting the M_1 vectors $A^{-1}\,\varphi_1^{\mu}$ as the bases of the vector space to which $A^{-1}\,Q_0\,\boldsymbol{F}_0$ and $\boldsymbol{\phi}$ belong.

Here we note that it may happen that some of the coefficients f_{μ} are identically equal to 0, and the corresponding subset of the M_1 independent vectors $\{A^{-1}\,\varphi_1^{\mu}\}_{\mu=1,2,\ldots,M_1}$ do not appear. However, to develop a general theory that is applicable to the generic case, we shall keep all the coefficients f_{μ} and the M_1 vectors $\{A^{-1}\,\varphi_1^{\mu}\}_{\mu=1,2,\ldots,M_1}$.

Now in the generic case in which we are interested, $\boldsymbol{\phi}$ can be expressed as

$$\boldsymbol{\phi}(t_0) = \sum_{\mu=1}^{M_1} (A^{-1}\,\varphi_1^{\mu})\,C'_{\mu}(t_0). \tag{9.52}$$

Here we have introduced new M_1 coefficients $C'_{\mu}(t_0)$, which have the meaning of the integral constants in the context of the solution of the present differential equation and have the t_0 dependence $C'_{\mu}(t_0)$ as $C_{\alpha}(t_0)$'s do.

It is worth emphasizing that the form of $\boldsymbol{\phi}$ given in Eq. (9.52) is the most general one that makes $A\,\boldsymbol{\phi}$ and $Q_0\,\boldsymbol{F}_0$ (as given by (9.48)) belong to a common vector space provided that $A^{-1}\,Q_0\,\boldsymbol{F}_0$ is expressed by Eq. (9.51).

Now we see that the P_1 space is the vector space spanned by $A^{-1}\,\varphi_1^{\mu}$ and φ_1^{μ} ($\mu = 1, \ldots, M_1$). We call the pair of

$$A^{-1}\,\varphi_1^{\mu} \quad \text{and} \quad \varphi_1^{\mu}, \tag{9.53}$$

the **doublet modes**. One also sees that the Q_0 space is a sum of the P_1 space spanned by the doublet modes and the Q_1 space which is the complement to the sum of the P_0 and P_1 spaces. The projection operators to the P_1 and Q_1 spaces are denoted by P_1 and Q_1, respectively: Their explicit expressions are given by

$$
\begin{aligned}
P_1 \boldsymbol{\psi} &:= \sum_{\mu,\nu=1}^{M_1} \boldsymbol{\varphi}_1^{\mu} \left(\eta_{10\mu,0\nu}^{-1} \langle \boldsymbol{\varphi}_1^{\nu}, \boldsymbol{\psi} \rangle + \eta_{10\mu,1\nu}^{-1} \langle A^{-1} \boldsymbol{\varphi}_1^{\nu}, \boldsymbol{\psi} \rangle \right) \\
&\quad + A^{-1} \sum_{\mu,\nu=1}^{M_1} \boldsymbol{\varphi}_1^{\mu} \left(\eta_{11\mu,0\nu}^{-1} \langle \boldsymbol{\varphi}_1^{\nu}, \boldsymbol{\psi} \rangle + \eta_{11\mu,1\nu}^{-1} \langle A^{-1} \boldsymbol{\varphi}_1^{\nu}, \boldsymbol{\psi} \rangle \right) \\
&= \sum_{m,n=0,1} A^{-m} \sum_{\mu,\nu=1}^{M_1} \boldsymbol{\varphi}_1^{\mu}\, \eta_{1m\mu,n\nu}^{-1} \langle A^{-n} \boldsymbol{\varphi}_1^{\nu}, \boldsymbol{\psi} \rangle, \qquad (9.54) \\
Q_1 &:= Q_0 - P_1, \qquad (9.55)
\end{aligned}
$$

respectively, where $\boldsymbol{\psi}$ is an arbitrary vector and $\eta_{1m\mu,n\nu}^{-1}$ $(m,\, n = 0,\, 1)$ denotes the inverse of the metric matrix $\eta_1^{m\mu,n\nu}$ of the P$_1$ space:

$$
\eta_1^{m\mu,n\nu} := \langle A^{-m} \boldsymbol{\varphi}_1^{\mu},\, A^{-n} \boldsymbol{\varphi}_1^{\nu} \rangle, \qquad (m,\, n = 0,\, 1;\, \mu,\, \nu = 1,\, \ldots,\, M_1). \quad (9.56)
$$

As a check, we show that P_1 satisfies the properties of the projection operator of the P$_1$ space:

$$
\begin{aligned}
P_1^2 \boldsymbol{\psi} &= \sum_{l,p=0,1} A^{-l} \sum_{\rho,\sigma=1}^{M_1} \boldsymbol{\varphi}_1^{\rho}\, \eta_{1l\rho,p\sigma}^{-1} \sum_{m,n=0,1} \sum_{\mu,\nu=1}^{M_1} \langle A^{-p} \boldsymbol{\varphi}_1^{\sigma},\, A^{-m} \boldsymbol{\varphi}_1^{\mu} \rangle\, \eta_{1m\mu,n\nu}^{-1} \langle A^{-n} \boldsymbol{\varphi}_1^{\nu}, \boldsymbol{\psi} \rangle \\
&= \sum_{l,p=0,1} A^{-l} \sum_{\rho,\sigma=1}^{M_1} \boldsymbol{\varphi}_1^{\rho}\, \eta_{1l\rho,p\sigma}^{-1} \sum_{m,n=0,1} \sum_{\mu,\nu=1}^{M_1} \eta_1^{p\sigma,m\mu}\, \eta_{1m\mu,n\nu}^{-1} \langle A^{-n} \boldsymbol{\varphi}_1^{\nu}, \boldsymbol{\psi} \rangle \\
&= \sum_{l,p=0,1} A^{-l} \sum_{\rho,\sigma=1}^{M_1} \boldsymbol{\varphi}_1^{\rho}\, \eta_{1l\rho,p\sigma}^{-1} \sum_{n=0,1} \sum_{\nu=1}^{M_1} \delta_n^p \delta_\nu^\sigma \langle A^{-n} \boldsymbol{\varphi}_1^{\nu}, \boldsymbol{\psi} \rangle \\
&= \sum_{l,p=0,1} A^{-l} \sum_{\rho,\sigma=1}^{M_1} \boldsymbol{\varphi}_1^{\rho}\, \eta_{1l\rho,p\sigma}^{-1} \langle A^{-p} \boldsymbol{\varphi}_1^{\sigma}, \boldsymbol{\psi} \rangle \\
&= P_1
\end{aligned}
$$

for any vector ψ, which shows the idempotency of P_1;

$$
P_1^2 = P_1. \qquad (9.57)
$$

Similarly,

$$
\begin{aligned}
P_1\varphi_1^\rho &= \sum_{m,n=0,1} A^{-m} \sum_{\mu,\nu=1}^{M_1} \varphi_1^\mu\, \eta_{1m\mu,n\nu}^{-1} \langle A^{-n}\, \varphi_1^\nu\,,\, \varphi_1^\rho \rangle \\
&= \sum_{m,n=0,1} A^{-m} \sum_{\mu,\nu=1}^{M_1} \varphi_1^\mu\, \eta_{1m\mu,n\nu}^{-1}\, \eta_1^{n\nu,0\rho} \\
&= \sum_{m=0,1} A^{-m} \sum_{\mu=1}^{M_1} \varphi_1^\mu\, \delta_m^0 \delta_\mu^\rho \\
&= \varphi_1^\rho,
\end{aligned} \tag{9.58}
$$

$$
\begin{aligned}
P_1\left(A^{-1}\varphi_1^\rho\right) &= \sum_{m,n=0,1} A^{-m} \sum_{\mu,\nu=1}^{M_1} \varphi_1^\mu\, \eta_{1m\mu,n\nu}^{-1} \langle A^{-n}\, \varphi_1^\nu\,,\, A^{-1}\,\varphi_1^\rho \rangle \\
&= \sum_{m,n=0,1} A^{-m} \sum_{\mu,\nu=1}^{M_1} \varphi_1^\mu\, \eta_{1m\mu,n\nu}^{-1}\, \eta_1^{n\nu,1\rho} \\
&= \sum_{m=0,1} A^{-m} \sum_{\mu=1}^{M_1} \varphi_1^\mu\, \delta_m^1 \delta_\mu^\rho \\
&= A^{-1}\,\varphi_1^\rho.
\end{aligned} \tag{9.59}
$$

Thus, on account of the self-adjointness[4] of P_1, we have the following equalities for an arbitrary vector $\boldsymbol{\psi}$,

$$
\langle \varphi_1^\mu\,,\, P_1\,\boldsymbol{\psi} \rangle = \langle \varphi_1^\mu\,,\, \boldsymbol{\psi} \rangle, \quad \langle A^{-1}\,\varphi_1^\mu\,,\, P_1\,\boldsymbol{\psi} \rangle = \langle A^{-1}\,\varphi_1^\mu\,,\, \boldsymbol{\psi} \rangle. \tag{9.60}
$$

These formulae will play an important role in Sect. 9.2.6.

9.2.4 *Second-Order Analysis*

To write down the second-order equation, it is found convenient to introduce a third-order and second-order rank tensors B and F_1 whose components are given by

$$
B_{ijk} := \left.\frac{\partial^2}{\partial X_j\, \partial X_k} G_i(X)\right|_{X=X^{\text{eq}}}, \quad F_{1ij} := \left.\frac{\partial}{\partial X_j} F_i(X)\right|_{X=X^{\text{eq}}}, \tag{9.61}
$$

respectively. Then the second-order equation is written as

$$
\frac{\partial}{\partial t}\tilde{\boldsymbol{X}}_2(t) = A\,\tilde{\boldsymbol{X}}_2(t) + \boldsymbol{K}(t - t_0), \tag{9.62}
$$

[4] We recall that the self-adjointness of A is shown in (9.29).

with the inhomogeneous term

$$\boldsymbol{K}(t-t_0) := \frac{1}{2} B\left[\tilde{\boldsymbol{X}}_1(t\,;\,t_0)\,,\,\tilde{\boldsymbol{X}}_1(t\,;\,t_0)\right] + F_1\,\tilde{\boldsymbol{X}}_1(t\,;\,t_0), \tag{9.63}$$

where we have introduced the symbol for arbitrary two vectors $\psi_i = (\boldsymbol{\psi})_i$ and $\chi_i = (\boldsymbol{\chi})_i$ as

$$(B\left[\boldsymbol{\psi}\,,\,\boldsymbol{\chi}\right])_i = \sum_{j=1}^{N}\sum_{k=1}^{N} B_{ijk}\,\psi_j\,\chi_k. \tag{9.64}$$

We are going to solve Eq. (9.62) in a way where the 'initial' values suitable to describe the slow dynamics are obtained so that the solution would describe the motion coming from the P_0 and P_1 spaces.

The general solution to Eq. (9.62) is obtained, say, by the method of variation of constants, and reads

$$\begin{aligned}\tilde{\boldsymbol{X}}_2(t) &= \mathrm{e}^{A(t-t_0)}\,\tilde{\boldsymbol{X}}_2(t_0) + \int_{t_0}^{t}\mathrm{d}t'\,\mathrm{e}^{A(t-t')}\,\boldsymbol{K}(t'-t_0)\\ &= \mathrm{e}^{A(t-t_0)}\,\tilde{\boldsymbol{X}}_2(t_0) + \int_{t_0}^{t}\mathrm{d}t'\,P_0\,\boldsymbol{K}(t'-t_0)\\ &\quad + \int_{t_0}^{t}\mathrm{d}t'\,\mathrm{e}^{A(t-t')}\,Q_0\,\boldsymbol{K}(t'-t_0),\end{aligned} \tag{9.65}$$

where we have inserted the identity $1 = P_0 + Q_0$ in front of $\boldsymbol{K}(t'-t_0)$ in the second equality. Using the simple formula

$$\boldsymbol{K}(t'-t_0) = \mathrm{e}^{(t'-t_0)\partial/\partial s}\,\boldsymbol{K}(s)\Big|_{s=0}, \tag{9.66}$$

and then carrying out integration with respect to t', we have

$$\begin{aligned}\tilde{\boldsymbol{X}}_2(t) &= \mathrm{e}^{A(t-t_0)}\,\tilde{X}_2(t_0) + (1-\mathrm{e}^{(t-t_0)\partial/\partial s})\,(-\partial/\partial s)^{-1}\,P_0\,\boldsymbol{K}(s)\Big|_{s=0}\\ &\quad + (\mathrm{e}^{A(t-t_0)} - \mathrm{e}^{(t-t_0)\partial/\partial s})\,(A-\partial/\partial s)^{-1}\,Q_0\,\boldsymbol{K}(s)\Big|_{s=0}\\ &= \mathrm{e}^{A(t-t_0)}\left[\tilde{X}_2(t_0) + Q_1\,(A-\partial/\partial s)^{-1}\,Q_0\,\boldsymbol{K}(s)\Big|_{s=0}\right]\\ &\quad + (1-\mathrm{e}^{(t-t_0)\partial/\partial s})\,(-\partial/\partial s)^{-1}\,P_0\,\boldsymbol{K}(s)\Big|_{s=0}\\ &\quad + (\mathrm{e}^{A(t-t_0)} - \mathrm{e}^{(t-t_0)\partial/\partial s})\,P_1\,(A-\partial/\partial s)^{-1}\,Q_0\,\boldsymbol{K}(s)\Big|_{s=0}\\ &\quad - \mathrm{e}^{(t-t_0)\partial/\partial s}\,Q_1\,(A-\partial/\partial s)^{-1}\,Q_0\,\boldsymbol{K}(s)\Big|_{s=0},\end{aligned} \tag{9.67}$$

where $1 = P_0 + P_1 + Q_1$ has been inserted in front of $(A - \partial/\partial s)^{-1}\, Q_0\, \boldsymbol{K}(s)$ in the second line of Eq. (9.67). We note that the contributions from the inhomogeneous term $\boldsymbol{K}(t - t_0)$ are decomposed into two parts, whose time dependencies are given by $\mathrm{e}^{A(t-t_0)}$ and $\mathrm{e}^{(t-t_0)\partial/\partial s}$, respectively. The former gives a fast motion characterized by the eigenvalues of A acting on the Q_1 space, while the time dependence of the latter is independent of the dynamics due to the absence of A. Since we are interested in the motion coming from the P_0 and P_1 spaces, we eliminate the former associated with the Q_1 space with a choice of the 'initial' value $\tilde{X}_2(t_0)$ that has not yet been specified, as follows:

$$\tilde{X}_2(t_0) = -Q_1\, (A - \partial/\partial s)^{-1}\, Q_0\, \boldsymbol{K}(s)\Big|_{s=0}, \tag{9.68}$$

with which Eq. (9.67) is reduced to

$$\begin{aligned}\tilde{X}_2(t) &= (1 - \mathrm{e}^{(t-t_0)\partial/\partial s})\, (-\partial/\partial s)^{-1}\, P_0\, \boldsymbol{K}(s)\Big|_{s=0} \\ &\quad + (\mathrm{e}^{A(t-t_0)} - \mathrm{e}^{(t-t_0)\partial/\partial s})\, P_1\, (A - \partial/\partial s)^{-1}\, Q_0\, \boldsymbol{K}(s)\Big|_{s=0} \\ &\quad - \mathrm{e}^{(t-t_0)\partial/\partial s}\, Q_1\, (A - \partial/\partial s)^{-1}\, Q_0\, \boldsymbol{K}(s)\Big|_{s=0}.\end{aligned} \tag{9.69}$$

By introducing a "propagator"

$$\mathcal{G}(s) := (A - \partial/\partial s)^{-1}, \tag{9.70}$$

in Eqs. (9.68) and (9.69), we write the 'initial' value and solution to Eq. (9.62) as

$$X_2(t_0) = \tilde{X}_2(t = t_0\, ;\, t_0) = -Q_1\, \mathcal{G}(s)\, Q_0\, \boldsymbol{K}(s)\Big|_{s=0}, \tag{9.71}$$

and

$$\begin{aligned}\tilde{X}_2(t\, ;\, t_0) &= (1 - \mathrm{e}^{(t-t_0)\partial/\partial s})\, (-\partial/\partial s)^{-1}\, P_0\, \boldsymbol{K}(s)\Big|_{s=0} \\ &\quad + (\mathrm{e}^{A(t-t_0)} - \mathrm{e}^{(t-t_0)\partial/\partial s})\, P_1\, \mathcal{G}(s)\, Q_0\, \boldsymbol{K}(s)\Big|_{s=0} \\ &\quad - \mathrm{e}^{(t-t_0)\partial/\partial s}\, Q_1\, \mathcal{G}(s)\, Q_0\, \boldsymbol{K}(s)\Big|_{s=0},\end{aligned} \tag{9.72}$$

respectively. We note the appearance of secular terms in Eq. (9.72).

Thus, up to the second order of ϵ, we have the approximate solution that is locally valid around $t \sim t_0$ as

$$
\begin{aligned}
\tilde{X}(t\,;\,t_0) = X^{\mathrm{eq}} + \epsilon &\left[\mathrm{e}^{A(t-t_0)}\,\phi + (t-t_0)\,P_0\,F_0 + (\mathrm{e}^{A(t-t_0)} - 1)\,A^{-1}\,Q_0\,F_0 \right] \\
&+ \epsilon^2 \left[(1 - \mathrm{e}^{(t-t_0)\partial/\partial s})\,(-\partial/\partial s)^{-1}\,P_0\,K(s)\Big|_{s=0} \right. \\
&+ (\mathrm{e}^{A(t-t_0)} - \mathrm{e}^{(t-t_0)\partial/\partial s})\,P_1\,\mathcal{G}(s)\,Q_0\,K(s)\Big|_{s=0} \\
&\left. - \mathrm{e}^{(t-t_0)\partial/\partial s}\,Q_1\,\mathcal{G}(s)\,Q_0\,K(s)\Big|_{s=0} \right],
\end{aligned} \tag{9.73}
$$

with the ‘initial’ value at $t = t_0$

$$
X(t_0) = X^{\mathrm{eq}} + \epsilon\,\phi - \epsilon^2\,Q_1\,\mathcal{G}(s)\,Q_0\,K(s)\Big|_{s=0} + O(\epsilon^3). \tag{9.74}
$$

9.2.5 RG Improvement of Perturbative Expansion

As we have repeatedly noted in the procedure of the solution, the perturbative solution (9.73) contains the secular terms, which make the solution (9.73) only valid around $t = t_0$.

To construct a solution that is valid in a global domain from the perturbative solution (9.73), we adopt the RG method [3, 5, 6, 40, 58]. That is, we apply the RG/E equation to the local solution (9.73)

$$
\frac{\mathrm{d}}{\mathrm{d}t_0}\tilde{X}(t\,;\,t_0)\Big|_{t_0=t} = 0, \tag{9.75}
$$

which leads to

$$
\begin{aligned}
\frac{\partial}{\partial t}X^{\mathrm{eq}} + \epsilon &\left[-A\,\phi + \frac{\partial}{\partial t}\phi - P_0\,F_0 - Q_0\,F_0 \right] \\
&+ \epsilon^2 \left[-P_0\,K(0) - (A - \partial/\partial s)\,P_1\,\mathcal{G}(s)\,Q_0\,K(s)\Big|_{s=0} \right. \\
&\left. + (\partial/\partial s)\,Q_1\,\mathcal{G}(s)\,Q_0\,K(s)\Big|_{s=0} \right] = 0,
\end{aligned} \tag{9.76}
$$

up to the second order of ϵ. We remark that the RG/E equation (9.76) actually gives the equation of motion by lifting the would-be integral constants C_α in X^{eq} and C'_μ in ϕ to dynamical variables. Furthermore, we have a globally improved solution as the envelope trajectory [5] that is given by the ‘initial’ value (9.74) as

$$\boldsymbol{X}^{\text{global}}(t) := \boldsymbol{X}(t_0 = t)$$
$$= \boldsymbol{X}^{\text{eq}} + \epsilon\, \boldsymbol{\phi} - \epsilon^2\, Q_1\, \mathcal{G}(s)\, Q_0\, \boldsymbol{K}(s)\Big|_{s=0}\Big|_{t_0=t} + O(\epsilon^3), \quad (9.77)$$

where the exact solution to the RG/E equation (9.76) is to be inserted.

We emphasize that the solution (9.77) together with the dynamical equation (9.76) constitutes the mesoscopic dynamics of Eq. (9.1): Equation (9.77) gives the (extended) invariant/attractive manifold of Eq. (9.1), and Eq. (9.76) the dynamical equation incorporating the mesoscopic dynamics defined on the manifold.

9.2.6 Reduction of RG/E Equation to Simpler Form

In this section, we reduce the RG/E equation (9.76) to a simpler form by a kind of averaging.

We first note that Eq. (9.76) contains terms belonging to the Q_1 space that are not supposed to constitute the mesoscopic variables.[5] These variables can be eliminated by taking the inner product of Eq. (9.76) with the zero modes $\boldsymbol{\varphi}_0^{\alpha}$ and the excited modes $A^{-1}\, \boldsymbol{\varphi}_1^{\mu}$ used in $\boldsymbol{\phi}$. This is a kind of averaging.

First we note the fact that $\partial \boldsymbol{X}^{\text{eq}}/\partial t$ belongs to the P_0 space as

$$\frac{\partial}{\partial t}\boldsymbol{X}^{\text{eq}} = \sum_{\alpha=1}^{M_0} \boldsymbol{\varphi}_0^{\alpha}\, \frac{\partial}{\partial t} C_{\alpha}, \quad (9.78)$$

which follows from Eqs. (9.10) and (9.40).

Then the averaging is accomplished by multiplying the projection operators P_0 and P_1 from the left-hand side of Eq. (9.76), which are reduced to

$$P_0\, \frac{\partial}{\partial t}\boldsymbol{X}^{\text{eq}} + \epsilon \left[P_0\, \frac{\partial}{\partial t}\boldsymbol{\phi} - P_0\, \boldsymbol{F}_0 \right] - \epsilon^2\, P_0\, \boldsymbol{K}(0) = 0, \quad (9.79)$$

$$\epsilon \left[- P_1\, A\, \boldsymbol{\phi} + P_1\, \frac{\partial}{\partial t}\boldsymbol{\phi} - P_1\, Q_0\, \boldsymbol{F}_0 \right]$$
$$- \epsilon^2\, P_1\, (A - \partial/\partial s)\, P_1\, \mathcal{G}(s)\, Q_0\, \boldsymbol{K}(s)\Big|_{s=0} = 0, \quad (9.80)$$

respectively, up to $O(\epsilon^2)$.

We also note that $\boldsymbol{K}(0)$ in (9.79) takes the simple form,

[5] We note that these variables in the Q_1 space may be incorporated to the mesoscopic equation as noise terms to make a stochastic mesoscopic dynamics or a Langevin equation [100, 158, 159]. Such an attempt to obtain a stochastic mesoscopic dynamics is quite interesting but beyond the scope of the present work.

$$\boldsymbol{K}(0) = \frac{1}{2} B\left[\boldsymbol{\phi}\,,\,\boldsymbol{\phi}\right] + F_1\,\boldsymbol{\phi}, \tag{9.81}$$

owing to Eqs. (9.63) and (9.47).

To get more concrete formulae of Eqs. (9.79) and (9.80), we take the inner product of them with the basis vectors $\boldsymbol{\varphi}_0^\alpha$ in the P_0 and $A^{-1}\,\boldsymbol{\varphi}_1^\mu$ in the P_1 space, respectively, which lead to

$$\langle\,\boldsymbol{\varphi}_0^\alpha\,,\,\frac{\partial}{\partial t}(\boldsymbol{X}^{\mathrm{eq}} + \epsilon\,\boldsymbol{\phi})\,\rangle - \epsilon\,\langle\,\boldsymbol{\varphi}_0^\alpha\,,\,\boldsymbol{F}_0 + \epsilon\,F_1\,\boldsymbol{\phi}\,\rangle = \epsilon^2\,\frac{1}{2}\,\langle\,\boldsymbol{\varphi}_0^\alpha\,,\,B\left[\boldsymbol{\phi}\,,\,\boldsymbol{\phi}\right]\rangle, \tag{9.82}$$

$$\begin{aligned}\epsilon\,\langle\,A^{-1}\,\boldsymbol{\varphi}_1^\mu\,,\,\frac{\partial}{\partial t}\boldsymbol{\phi}\,\rangle - \epsilon\,\langle\,A^{-1}\,\boldsymbol{\varphi}_1^\mu\,,\,\boldsymbol{F}_0 + \epsilon\,F_1\,\boldsymbol{\phi}\,\rangle &= \epsilon\,\langle\,A^{-1}\,\boldsymbol{\varphi}_1^\mu\,,\,A\,\boldsymbol{\phi}\,\rangle \\ &\quad + \epsilon^2\,\frac{1}{2}\,\langle\,A^{-1}\,\boldsymbol{\varphi}_1^\mu\,,\,B\left[\boldsymbol{\phi}\,,\,\boldsymbol{\phi}\right]\rangle,\end{aligned} \tag{9.83}$$

respectively. Here, we have omitted the terms of $O(\epsilon^3)$.

The calculation of the inner product of the last term (9.80) with $A^{-1}\,\boldsymbol{\varphi}_1^\mu$ is rather involved. Firstly, with a repeated use of the identities (9.60) due to the self-adjointness of A and the definition of P_1, we have

$$\begin{aligned}&\langle\,A^{-1}\,\boldsymbol{\varphi}_1^\mu\,,\,(A - \partial/\partial s)\,P_1\,\mathcal{G}(s)\,Q_0\,\boldsymbol{K}(s)\Big|_{s=0}\rangle \\ &= \langle\,(A - \partial/\partial s)\,A^{-1}\,\boldsymbol{\varphi}_1^\mu\,,\,P_1\,\mathcal{G}(s)\,Q_0\,\boldsymbol{K}(s)\Big|_{s=0}\rangle \\ &= \langle\,(\boldsymbol{\varphi}_1^\mu - A^{-1}\,\boldsymbol{\varphi}_1^\mu\,\partial/\partial s)\,,\,P_1\,\mathcal{G}(s)\,Q_0\,\boldsymbol{K}(s)\Big|_{s=0}\rangle. \\ &= \langle\,P_1\,(\boldsymbol{\varphi}_1^\mu - A^{-1}\,\boldsymbol{\varphi}_1^\mu\,\partial/\partial s)\,,\,\mathcal{G}(s)\,Q_0\,\boldsymbol{K}(s)\Big|_{s=0}\rangle. \\ &= \langle\,(\boldsymbol{\varphi}_1^\mu - A^{-1}\,\boldsymbol{\varphi}_1^\mu\,\partial/\partial s)\,,\,\mathcal{G}(s)\,Q_0\,\boldsymbol{K}(s)\Big|_{s=0}\rangle \\ &= \langle\,(A - \partial/\partial s)\,A^{-1}\,\boldsymbol{\varphi}_1^\mu\,,\,\mathcal{G}(s)\,Q_0\,\boldsymbol{K}(s)\Big|_{s=0}\rangle,\end{aligned} \tag{9.84}$$

which is further reduced owing to the self-adjointness of A again to

$$\begin{aligned}\langle\,A^{-1}\,\boldsymbol{\varphi}_1^\mu\,,\,(A - \partial/\partial s)\,\mathcal{G}(s)\,Q_0\,\boldsymbol{K}(s)\Big|_{s=0}\rangle &= \langle\,A^{-1}\,\boldsymbol{\varphi}_1^\mu\,,\,Q_0\,\boldsymbol{K}(s)\Big|_{s=0}\rangle \\ &= \langle\,A^{-1}\,\boldsymbol{\varphi}_1^\mu\,,\,\boldsymbol{K}(0)\,\rangle,\end{aligned} \tag{9.85}$$

where the definition of $\mathcal{G}(s)$ given in (9.70) has been used in the first equality. Then inserting (9.81) into $K(0)$, we finally arrive at

$$\begin{aligned}&\langle\,A^{-1}\,\boldsymbol{\varphi}_1^\mu\,,\,(A - \partial/\partial s)\,P_1\,\mathcal{G}(s)\,Q_0\,\boldsymbol{K}(s)\Big|_{s=0}\rangle \\ &= \frac{1}{2}\,\langle\,A^{-1}\,\boldsymbol{\varphi}_1^\mu\,,\,B\left[\boldsymbol{\phi}\,,\,\boldsymbol{\phi}\right]\rangle + \langle\,A^{-1}\,\boldsymbol{\varphi}_1^\mu\,,\,F_1\,\boldsymbol{\phi}\,\rangle.\end{aligned} \tag{9.86}$$

We note that the pair of Eqs. (9.82) and (9.83) is also the equation of motion governing C_α in $\boldsymbol{X}^{\mathrm{eq}}$ and C'_μ in $\boldsymbol{\phi}$, which is much simpler than Eq. (9.76). We shall shortly give more words on the significance of the equality (9.86).

There are some notable points in the derivation of the mesoscopic dynamics done above:

(i) The equations for the mesoscopic dynamics, that is, Eqs. (9.82) and (9.83), are local ones composed of finite number of terms, although the nonlocal term $\mathcal{G}(s)\, Q_0\, \boldsymbol{K}(s)|_{s=0}$ in Eq. (9.80) consists of an infinite number of terms without the averaging;

$$\mathcal{G}(s)\, Q_0\, \boldsymbol{K}(s)\Big|_{s=0} = \sum_{n=0}^{\infty} A^{-1-n}\, Q_0\, \frac{\partial^n}{\partial s^n} \boldsymbol{K}(s)\Big|_{s=0}, \tag{9.87}$$

This nice feature clearly originates from the averaging procedure by $A^{-1}\, \varphi_1^\mu$ (not by φ_1^μ) and the fact that the P_1 space is spanned by the doublet modes $A^{-1}\, \varphi_1^\mu$ and φ_1^μ. It is remarkable that when φ_1^μ were used for the averaging instead of $A^{-1}\, \varphi_1^\mu$, it would lead to a complicated equation with an infinite number of terms: In fact, the reduction of infinite terms into a single term $K(0)$ as was shown in Eq. (9.86) is not obtained in this case, as is shown below;

$$\begin{aligned}
&\langle \varphi_1^\mu\,,\, (A - \partial/\partial s)\, P_1\, \mathcal{G}(s)\, Q_0\, \boldsymbol{K}(s)\Big|_{s=0} \rangle \\
&- \langle (A - \partial/\partial s)\, \varphi_1^\mu\,,\, P_1\, \mathcal{G}(s)\, Q_0\, \boldsymbol{K}(s)\Big|_{s=0} \rangle \\
&= \langle (A\, \varphi_1^\mu - \varphi_1^\mu\, \partial/\partial s)\,,\, P_1\, \mathcal{G}(s)\, Q_0\, \boldsymbol{K}(s)\Big|_{s=0} \rangle \\
&\neq \langle (A\, \varphi_1^\mu - \varphi_1^\mu\, \partial/\partial s)\,,\, \mathcal{G}(s)\, Q_0\, \boldsymbol{K}(s)\Big|_{s=0} \rangle,
\end{aligned} \tag{9.88}$$

because $A\, \varphi_1^\mu$ does not belong to the P_1 space.
Needless to say, the reduced dynamical equation composed of an infinite number of terms is quite undesirable. Thus we find that the averaging by $A^{-1}\, \varphi_1^\mu$ used in the definition of $\boldsymbol{\phi}$ is essential to obtain the correct mesoscopic dynamics.

(ii) The mesoscopic dynamics of a generic equation (9.1) consists of Eqs. (9.82) and (9.83) and the invariant/attractive manifold given in Eq. (9.77). It is noteworthy that reflecting the trade-off relation between the simplicities of them, the invariant/attractive manifold is represented by a rather complicated nonlocal equation, whereas the reduced differential equations are local equations with a finite number of terms.

(iii) The mesoscopic dynamics given by Eqs. (9.82) and (9.83) is consistent with the slow dynamics described solely by the zero modes in the asymptotic regime. A proof for this natural property of the mesoscopic dynamics will be presented in Sect. 9.2.7.

(iv) The doublet scheme in the RG method has been constructed with the requirement that the tangent space of the first-order perturbative solution (9.47) at $t = t_0$ should be spanned by a as small as possible number of independent vectors. This geometrical way of construction might sound somewhat technical. More explicit and natural way of derivation of the doublet scheme will be presented in Sect. 9.3, where the structure of the P_1 space and the functional form of $\boldsymbol{\phi}$ are determined through a faithful solution of the generic evolution equation (9.1).

(v) As is clear from the derivation, the doublet scheme developed here as a reduction theory of the mesoscopic dynamics has a universal nature and should be applicable to a wide class of evolution equations, as long as the microscopic equation is expressed as Eq. (9.1) with the linear operator A having only real eigenvalues and being self-adjoint. As a demonstration, we shall apply the doublet scheme in the RG method to the Lorenz model in Sect. 9.3. In Chaps. 12, 15, and 16, we shall apply the scheme to the Boltzmann equation to derive so called the second-order fluid dynamics. An extension of the doublet scheme to the case where A has complex eigenvalues should be possible but is left as a future project for the moment.

9.2.7 Transition of the Mesoscopic Dynamics to the Slow Dynamics in Asymptotic Regime

In this section, we shall show that the motion described by the mesoscopic dynamics Eqs. (9.82) and (9.83) derived in the previous section for (9.1) asymptotically approaches the motion described by the would-be zero modes as derived by the RG method in Chap. 5.

First we derive the slow dynamics described only by the zero modes from the generic evolution equation (9.1) with the RG method developed in Chap. 5 for later comparison.

In this method, the first task is to obtain the perturbative solution $\tilde{\boldsymbol{X}}$ to Eq. (9.1) around an arbitrary 'initial' time $t = t_0$ with the 'initial' value $\boldsymbol{X}(t_0)$; $\tilde{\boldsymbol{X}}(t = t_0\,;\, t_0) = \boldsymbol{X}(t_0)$. We expand the 'initial' value as well as the solution with respect to ϵ as shown in Eqs. (9.33) and (9.34), and obtain the series of the perturbative equations with respect to ϵ.

The zeroth-order equation is the same as Eq. (9.36). Since we are interested in the slow motion realized asymptotically for $t \to \infty$, we adopt the static solution $\boldsymbol{X}^{\mathrm{eq}}$ as the zeroth-order solution:

$$\tilde{\boldsymbol{X}}_0(t\,;\, t_0) = \boldsymbol{X}^{\mathrm{eq}}, \tag{9.89}$$

which means that the zeroth-order 'initial' value reads

$$\boldsymbol{X}_0(t_0) = \tilde{\boldsymbol{X}}_0(t_0\,;\,t_0) = \boldsymbol{X}^{\mathrm{eq}}. \tag{9.90}$$

The first-order equation reads

$$\frac{\partial}{\partial t}\tilde{\boldsymbol{X}}_1(t\,;\,t_0) = A\,\tilde{\boldsymbol{X}}_1(t\,;\,t_0) + \boldsymbol{F}_0, \tag{9.91}$$

where A and $\boldsymbol{F}_0$ have been defined in Eqs. (9.7) and (9.42), respectively. The general solution to the first-order equation is given by

$$\tilde{\boldsymbol{X}}_1(t\,;\,t_0) = \mathrm{e}^{A(t-t_0)}\left[\tilde{\boldsymbol{X}}_1(t_0\,;\,t_0) + A^{-1}\,Q_0\,\boldsymbol{F}_0\right] + (t-t_0)\,P_0\,\boldsymbol{F}_0 - A^{-1}\,Q_0\,\boldsymbol{F}_0, \tag{9.92}$$

where P_0 denotes the projection operator onto the P_0 space, *i.e.*, the kernel of A, and Q_0 the projection operator onto the Q_0 space; $P_0 + Q_0 = 1$.

To obtain the slow dynamics described solely by the variables in the P_0 space, we eliminate the fast motion coming from the Q_0 space, which is possible by utilizing the freedom of the choice of the 'initial' value $\tilde{\boldsymbol{X}}_1(t_0\,;\,t_0)$ that has not yet been specified as follows:

$$\boldsymbol{X}_1(t_0) = \tilde{\boldsymbol{X}}_1(t_0\,;\,t_0) = -A^{-1}\,Q_0\,\boldsymbol{F}_0, \tag{9.93}$$

with which Eq. (9.92) is reduced to

$$\tilde{\boldsymbol{X}}_1(t\,;\,t_0) = (t-t_0)\,P_0\,\boldsymbol{F}_0 - A^{-1}\,Q_0\,\boldsymbol{F}_0. \tag{9.94}$$

Then the second-order equation now takes the form

$$\frac{\partial}{\partial t}\tilde{\boldsymbol{X}}_2(t\,;\,t_0) = A\,\tilde{\boldsymbol{X}}_2(t\,;\,t_0) + \boldsymbol{U}(t-t_0), \tag{9.95}$$

where

$$\begin{aligned}\boldsymbol{U}(s) := &\ \frac{1}{2}\,B\left[s\,P_0\,\boldsymbol{F}_0 - A^{-1}\,Q_0\,\boldsymbol{F}_0\,,\,s\,P_0\,\boldsymbol{F}_0 - A^{-1}\,Q_0\,\boldsymbol{F}_0\right]\\ &+ F_1\,(s\,P_0\,\boldsymbol{F}_0 - A^{-1}\,Q_0\,\boldsymbol{F}_0).\end{aligned} \tag{9.96}$$

Here B and F_1 is defined in Eq. (9.61).

The solution to the second-order equation (9.95) is found to be

$$\begin{aligned}\tilde{\boldsymbol{X}}_2(t\,;\,t_0) = &\ \mathrm{e}^{A(t-t_0)}\left[\tilde{\boldsymbol{X}}_2(t_0\,;\,t_0) + (A-\partial/\partial s)^{-1}\,Q_0\,\boldsymbol{U}(s)\Big|_{s=0}\right]\\ &+ (1-\mathrm{e}^{(t-t_0)\partial/\partial s})\,(-\partial/\partial s)^{-1}\,P_0\,\boldsymbol{U}(s)\Big|_{s=0}\\ &- \mathrm{e}^{(t-t_0)\partial/\partial s}\,(A-\partial/\partial s)^{-1}\,Q_0\,\boldsymbol{U}(s)\Big|_{s=0},\end{aligned} \tag{9.97}$$

which still contains components in the Q_0 space. As in the first-order case, the possible fast motion caused by the Q_0 space components can be eliminated with the choice of the 'initial' value $\tilde{\boldsymbol{X}}_2(t_0\,;\,t_0)$ in the second order as

$$\boldsymbol{X}_2(t_0) = \tilde{\boldsymbol{X}}_2(t_0\,;\,t_0) = -(A - \partial/\partial s)^{-1}\, Q_0\, \boldsymbol{U}(s)\Big|_{s=0}, \tag{9.98}$$

and accordingly we have for the second-order solution

$$\begin{aligned}\tilde{\boldsymbol{X}}_2(t\,;\,t_0) &= (1 - \mathrm{e}^{(t-t_0)\partial/\partial s})\,(-\partial/\partial s)^{-1}\, P_0\, \boldsymbol{U}(s)\Big|_{s=0} \\ &\quad - \mathrm{e}^{(t-t_0)\partial/\partial s}\,(A - \partial/\partial s)^{-1}\, Q_0\, \boldsymbol{U}(s)\Big|_{s=0}.\end{aligned} \tag{9.99}$$

Thus we have the perturbative solution up to $O(\epsilon^2)$ as

$$\begin{aligned}\tilde{\boldsymbol{X}}(t\,;\,t_0) = \boldsymbol{X}^{\mathrm{eq}} &+ \epsilon\Big[(t - t_0)\, P_0\, \boldsymbol{F}_0 - A^{-1}\, Q_0\, \boldsymbol{F}_0\Big] \\ &+ \epsilon^2\Big[(1 - \mathrm{e}^{(t-t_0)\partial/\partial s})\,(-\partial/\partial s)^{-1}\, P_0\, \boldsymbol{U}(s)\Big|_{s=0} \\ &- \mathrm{e}^{(t-t_0)\partial/\partial s}\,(A - \partial/\partial s)^{-1}\, Q_0\, \boldsymbol{U}(s)\Big|_{s=0}\Big],\end{aligned} \tag{9.100}$$

with the 'initial' value at $t = t_0$

$$\begin{aligned}\boldsymbol{X}(t_0) = \boldsymbol{X}^{\mathrm{eq}} &- \epsilon\, A^{-1}\, Q_0\, \boldsymbol{F}_0 \\ &- \epsilon^2\,(A - \partial/\partial s)^{-1}\, Q_0\, \boldsymbol{U}(s)\Big|_{s=0}.\end{aligned} \tag{9.101}$$

Note the appearance of the secular term proportional to $t - t_0$.

As before in this monograph, we utilize the RG method as formulated in [3, 5, 6, 40, 58] to obtain a globally improved solution from this local perturbative solution (9.100); namely, applying the RG/E equation

$$\mathrm{d}\tilde{\boldsymbol{X}}_1(t\,;\,t_0)/\mathrm{d}t_0|_{t_0=t} = 0, \tag{9.102}$$

to Eq. (9.100), we have

$$\begin{aligned}&\frac{\partial}{\partial t}(\boldsymbol{X}^{\mathrm{eq}} - \epsilon\, A^{-1}\, Q_0\, \boldsymbol{F}_0) - \epsilon\, P_0\, \boldsymbol{F}_0 + \epsilon^2\Big[- P_0\, \boldsymbol{U}(0) \\ &- (-\partial/\partial s)\,(A - \partial/\partial s)^{-1}\, Q_0\, \boldsymbol{U}(s)\Big|_{s=0}\Big] = 0,\end{aligned} \tag{9.103}$$

where

$$U(0) = \frac{1}{2}\, B\left[A^{-1}\, Q_0\, \boldsymbol{F}_0\,,\; A^{-1}\, Q_0\, \boldsymbol{F}_0\right] - F_1\, A^{-1}\, Q_0\, \boldsymbol{F}_0, \tag{9.104}$$

which follows from Eq. (9.96). Note that Eq. (9.103) is the equation governing the slow variables $C_\alpha(t)$ in $\boldsymbol{X}^{\mathrm{eq}}$.

To get a more explicit form of the equation for the slow variables, it is found useful to make an 'averaging'; *i.e.*, we take the inner product of Eq. (9.103) with the zero modes φ_0^α;

$$\begin{aligned} &\langle\, \varphi_0^\alpha\,,\, \frac{\partial}{\partial t}(\boldsymbol{X}^{\mathrm{eq}} - \epsilon\, A^{-1}\, Q_0\, \boldsymbol{F}_0)\,\rangle - \epsilon\,\langle\, \varphi_0^\alpha\,,\, \boldsymbol{F}_0 - \epsilon\, F_1\, A^{-1}\, Q_0\, \boldsymbol{F}_0\,\rangle \\ &= \epsilon^2\, \frac{1}{2}\,\langle\, \varphi_0^\alpha\,,\, B\left[A^{-1}\, Q_0\, \boldsymbol{F}_0\,,\; A^{-1}\, Q_0\, \boldsymbol{F}_0\right]\rangle + O(\epsilon^3). \end{aligned} \tag{9.105}$$

We are now in a position to give a proof that the mesoscopic dynamics given in the doublet scheme is consistent with the slow dynamics given above in the asymptotic regime after a long time. First we notice that there exists a *separation of time scales* between the fast motion of C'_μ belonging to the P_1 space and the slow motion of C_α in the P_0 space. Thanks to this separation of the time scales, which tends to become more significant as time goes by, the asymptotic behavior of the solution to Eq. (9.83) can be obtained by adiabatically eliminating the variables C'_μ in the P_1 space.

Since C'_μ appears in Eq. (9.82) in the order of $O(\epsilon^2)$, it suffices to construct the solution C'_μ valid up to $O(1)$ for obtaining the closed equation for C_α valid up to $O(\epsilon^2)$. Equation (9.83) for C'_μ tells us that such an equation for C'_μ takes the form

$$\sum_{\nu=1}^{M_1} \langle\, A^{-1}\, \varphi_1^\mu\,,\, A^{-1}\, \varphi_1^\nu\,\rangle\, \frac{\partial}{\partial t} C'_\nu = \sum_{\nu=1}^{M_1} \langle\, \varphi_1^\mu\,,\, A^{-1}\, \varphi_1^\nu\,\rangle\, (C'_\nu + f_\nu) + O(\epsilon), \tag{9.106}$$

where the time dependence of C_α is ignored without a loss of generality. Here recall that the eigenvalues of A except for the zero are supposed to be real negative as mentioned in Sect. 9.2.1. Then one finds that $\langle\, A^{-1}\, \varphi_1^\mu\,,\, A^{-1}\, \varphi_1^\nu\,\rangle$ is a positive definite matrix, while $\langle\, \varphi_1^\mu\,,\, A^{-1}\, \varphi_1^\nu\,\rangle$ is a negative definite. Thus, one finds that $C'_\mu + f_\mu$ decays exponentially down to 0 in the asymptotic regime after a long time,

$$C'_\mu \to -f_\mu, \quad \text{for } t \to \infty, \tag{9.107}$$

up to terms of $O(\epsilon)$. Thus on account of (9.52),

$$\boldsymbol{\phi}(t) \to -\sum_{\mu=1}^{M_1} (A^{-1}\varphi_1^\mu)\, f_\mu = -A^{-1}\, Q_0\, \boldsymbol{F}_0, \tag{9.108}$$

where the last equality follows from Eq. (9.51). Inserting Eq. (9.108) into Eq. (9.82), one arrives at a closed equation for C_α, which turns out to be the same as Eq. (9.105). This is what we wanted to show: the mesoscopic dynamics derived by the doublet scheme has a natural property that it is consistent with the slow dynamics as described with only the zero modes in the asymptotic regime after a long time.

9.3 An Example: Mesoscopic Dynamics of the Lorenz Model

In this section, we apply the doublet scheme in the RG method developed in the previous section Sect. 9.2 to extract the mesoscopic dynamics of the Lorenz model [84]. This is an extension of the analysis done in Sect. 6.2 where the RG method was applied to extract the slow dynamics utilizing solely the would-be zero modes; the results were shown to be equivalent with those given by the multiple scale analysis done in Sect. 3.6.2.

For the sake of a self-contained presentation, we shall recapitulate and repeat some expressions presented in Sects. 3.6.2 and 6.2. The Lorenz model reads

$$\begin{cases} \dot{\xi} = \sigma(-\xi + \eta), \\ \dot{\eta} = r\xi - \eta - \xi\zeta, \\ \dot{\zeta} = \xi\eta - b\zeta, \end{cases} \tag{9.109}$$

with model parameters $\sigma > 0, r > 0$, and $b > 0$. The linear stability analysis [34] shows that the origin (A) is stable for $0 < r < 1$ but unstable for $r > 1$, while the latter steady states (B) $(\xi, \eta, \zeta) = (+\sqrt{b(r-1)}, +\sqrt{b(r-1)}, r-1)$ and (C) $(\xi, \eta, \zeta) = (-\sqrt{b(r-1)}, -\sqrt{b(r-1)}, r-1)$ are stable for $1 < r < \sigma(\sigma + b + 3)/(\sigma - b - 1) =: r_c$ but unstable for $r > r_c$.

Here, we investigate the non-linear stability around the origin (A) for $r \sim 1$: A similar analysis was done in Sects. 3.6.2 and 6.2 in the level of the slow dynamics but not in the mesoscopic dynamics.

In the situation in which we are interested, the deviation of r from 1 is small and the absolute values of the amplitudes of (ξ, η, ζ) are small. As before, we introduce a small quantity ϵ instead of r as $r = 1 + \chi\epsilon^2$, where $\chi = \pm 1$ depending on the sign of $r - 1$, and we scale the dynamical variables as

$$\begin{pmatrix} \xi \\ \eta \\ \zeta \end{pmatrix} =: \epsilon \begin{pmatrix} X \\ Y \\ Z \end{pmatrix} =: \boldsymbol{X}. \tag{9.110}$$

Then the Lorenz model is cast into the form

$$\frac{\mathrm{d}}{\mathrm{d}t}\boldsymbol{X} = A\,\boldsymbol{X} + \epsilon \begin{pmatrix} 0 \\ -X\,Z \\ X\,Y \end{pmatrix} + \epsilon^2 \begin{pmatrix} 0 \\ \chi\,X \\ 0 \end{pmatrix}, \quad A = \begin{pmatrix} -\sigma & \sigma & 0 \\ 1 & -1 & 0 \\ 0 & 0 & -b \end{pmatrix}. \tag{9.111}$$

As is shown in Sects. 3.6.2 and 6.2, eigenvalues and their respective (right) eigenvectors of A are given as follows: $\lambda_1 = 0$, $\lambda_2 = -1 - \sigma$, and $\lambda_3 = -b$ and

$$\boldsymbol{U}_1 = \begin{pmatrix} 1 \\ 1 \\ 0 \end{pmatrix}, \quad \boldsymbol{U}_2 = \begin{pmatrix} \sigma \\ -1 \\ 0 \end{pmatrix}, \quad \boldsymbol{U}_3 = \begin{pmatrix} 0 \\ 0 \\ 1 \end{pmatrix}. \tag{9.112}$$

As was shown in Sect. 9.2.1, when all the eigenvalues are real numbers as in this case, the apparently asymmetric linear operator A is made symmetric with respect to a suitably defined inner product. Thus, we see that the Lorenz model (9.111) can be analyzed by the doublet scheme in the RG method developed in the last section where the symmetric property of the linear operator A is utilized.

Let $\boldsymbol{X}(t_0)$ be a point on an exact solution. We try to solve Eq. (9.111) in a perturbation method around $t = t_0$ with $\boldsymbol{X}(t_0)$ being set to the 'initial' value. We assume that the 'initial' value and the approximate solution $\tilde{\boldsymbol{X}}(t; t_0)$ can be expanded with respect to ϵ as follows:

$$\boldsymbol{X}(t_0) = \boldsymbol{X}_0(t_0) + \epsilon\, \boldsymbol{X}_1(t_0) + \epsilon^2\, \boldsymbol{X}_2(t_0) + \cdots, \tag{9.113}$$

and

$$\tilde{\boldsymbol{X}}(t; t_0) = \tilde{\boldsymbol{X}}_0(t; t_0) + \epsilon\, \tilde{\boldsymbol{X}}_1(t; t_0) + \epsilon^2\, \tilde{\boldsymbol{X}}_2(t; t_0) + \cdots, \tag{9.114}$$

where

$$\tilde{\boldsymbol{X}}_l(t = t_0; t_0) = \boldsymbol{X}_l(t_0), \qquad l = 0, 1, 2, \ldots. \tag{9.115}$$

Inserting these expansions into Eq. (9.111) and equating the coefficients of ϵ^l, we have a series of equations for $\tilde{\boldsymbol{X}}_l(t; t_0)$.

The zeroth-order equation reads

$$\frac{\mathrm{d}}{\mathrm{d}t}\tilde{\boldsymbol{X}}_0(t; t_0) = A\tilde{\boldsymbol{X}}_0(t; t_0). \tag{9.116}$$

As the asymptotic solution as $t \to \infty$, we take the zero mode solution as before;

$$\tilde{\boldsymbol{X}}_0(t; t_0) = C(t_0)\boldsymbol{U}_1, \tag{9.117}$$

where $C(t_0)$ is an integral constant and we have made it explicit that the solution may depend on the 'initial' time t_0. $C(t_0)$ corresponds to $C_\alpha(t_0)$ in the general case discussed in Sect. 9.2 with $M_0 = 1$. The solution (9.117) leads to the 'initial' value

$$\boldsymbol{X}_0(t_0) = \tilde{\boldsymbol{X}}_0(t = t_0; t_0) = C(t_0)\boldsymbol{U}_1. \tag{9.118}$$

In this case, the P_0 space is spanned solely by the zero mode $\boldsymbol{U}_1$.

The first-order equation now takes the form

$$\frac{\mathrm{d}}{\mathrm{d}t}\tilde{\boldsymbol{X}}_1(t;t_0) = A\tilde{\boldsymbol{X}}_1(t;t_0) + C^2(t_0)\boldsymbol{U}_3. \tag{9.119}$$

The general solution to (9.119) is written as

$$\begin{aligned}\tilde{\boldsymbol{X}}_1(t;t_0) &= \mathrm{e}^{A(t-t_0)}\boldsymbol{X}_1(t_0) + (\mathrm{e}^{A(t-t_0)} - 1)A^{-1}C^2(t_0)\boldsymbol{U}_3 \\ &= \boldsymbol{X}_1(t_0) + (t-t_0)(A\boldsymbol{X}_1(t_0) + C^2(t_0)\boldsymbol{U}_3) + O((t-t_0)^2).\end{aligned} \tag{9.120}$$

As was done in the general formulation in Sect. 9.2, we specify the ‘initial’ value $\boldsymbol{X}_1(t_0)$ so that the dimension of the tangent space given by the term proportional to $t - t_0$ of the solution (9.120) is as small as possible. This requirement is satisfied if $A\boldsymbol{X}_1(t_0)$ belongs to the vector space spanned by $\boldsymbol{U}_3$. Thus with the use of a constant $C'(t_0)$, we write

$$\boldsymbol{X}_1(t_0) = A^{-1}\boldsymbol{U}_3C'(t_0) = -\frac{1}{b}\boldsymbol{U}_3C'(t_0). \tag{9.121}$$

Note that $C'(t_0)$ is interpreted as another integral constant corresponding to $C'_\mu(t_0)$ in Sect. 9.2 with $M_1 = 1$.

According to the general scheme developed in the last section, the P_1 space is spanned by the doublet modes

$$\left(\boldsymbol{U}_3,\ A^{-1}\boldsymbol{U}_3\right). \tag{9.122}$$

Here a warning is necessary: the doublet modes in the present simple model with three degrees of freedom happen to belong to a common space since $\boldsymbol{U}_3$ is an eigenvector of A. Accordingly, the left vector $\boldsymbol{U}_2$ belongs to the Q_1 space that is complement to the P_0 and P_1 spaces. Thus the vector space accommodating the local solution is decomposed as follows:

$$\mathrm{P}_0 : \boldsymbol{U}_1, \quad \mathrm{P}_1 : \boldsymbol{U}_3, \quad \mathrm{Q}_1 : \boldsymbol{U}_2, \tag{9.123}$$

where the vectors in the right-hand side of the colon are the respective basis vectors.

Using the time-dependent function defined by

$$K(t-t_0) := \chi C(t_0) - \frac{C^3(t_0)}{b}(1 - \mathrm{e}^{-b(t-t_0)}) + \frac{C(t_0)\,C'(t_0)}{b}\,\mathrm{e}^{-b(t-t_0)}, \tag{9.124}$$

the second-order equation is written as

$$\frac{\mathrm{d}}{\mathrm{d}t}\tilde{\boldsymbol{X}}_2(t;t_0) = A\tilde{\boldsymbol{X}}_2(t;t_0) + K(t-t_0)\frac{1}{1+\sigma}(\sigma\boldsymbol{U}_1 - \boldsymbol{U}_2). \tag{9.125}$$

The solution to Eq. (9.125) is given by

$$\begin{aligned}\tilde{X}_2(t;t_0) &= e^{A(t-t_0)}\Big[X_2(t_0) + Q_1(A-\partial/\partial s)^{-1}Q_0K(s)\frac{1}{1+\sigma}(\sigma U_1 - U_2)\Big|_{s=0}\Big] \\ &+ (1-e^{(t-t_0)\partial/\partial s})(-\partial/\partial s)^{-1}P_0K(s)\frac{1}{1+\sigma}(\sigma U_1 - U_2)\Big|_{s=0} \\ &+ (e^{A(t-t_0)} - e^{(t-t_0)\partial/\partial s})P_1(A-\partial/\partial s)^{-1}Q_0K(s)\frac{1}{1+\sigma}(\sigma U_1 - U_2)\Big|_{s=0} \\ &- e^{(t-t_0)\partial/\partial s}Q_1(A-\partial/\partial s)^{-1}Q_0K(s)\frac{1}{1+\sigma}(\sigma U_1 - U_2)\Big|_{s=0}. \qquad (9.126)\end{aligned}$$

The unwanted fast mode belonging to the Q_1 space can be eliminated by a choice of the 'initial' value $X_2(t_0)$ as

$$\begin{aligned}X_2(t_0) = \tilde{X}_2(t=t_0;t_0) &= -Q_1(A-\partial/\partial s)^{-1}Q_0K(s)\frac{1}{1+\sigma}(\sigma U_1 - U_2)\Big|_{s=0} \\ &= -((-1-\sigma)-\partial/\partial s)^{-1}K(s)\Big|_{s=0}\frac{1}{1+\sigma}(-U_2), \qquad (9.127)\end{aligned}$$

and accordingly

$$\begin{aligned}\tilde{X}_2(t;t_0) &= (1-e^{(t-t_0)\partial/\partial s})(-\partial/\partial s)^{-1}K(s)\Big|_{s=0}\frac{1}{1+\sigma}\sigma U_1 \\ &- e^{(t-t_0)\partial/\partial s}((-1-\sigma)-\partial/\partial s)^{-1}K(s)\Big|_{s=0}\frac{1}{1+\sigma}(-U_2). \qquad (9.128)\end{aligned}$$

Thus the perturbative solution up to $O(\epsilon^2)$ is given by

$$\begin{aligned}\tilde{X}(t;t_0) &= C(t_0)U_1 + \epsilon\Big[e^{A(t-t_0)}A^{-1}U_3C'(t_0) + (e^{A(t-t_0)}-1)A^{-1}C^2(t_0)U_3\Big] \\ &+ \epsilon^2\Big[(1-e^{(t-t_0)\partial/\partial s})(-\partial/\partial s)^{-1}K(s)\Big|_{s=0}\frac{1}{1+\sigma}\sigma U_1 \\ &- e^{(t-t_0)\partial/\partial s}((-1-\sigma)-\partial/\partial s)^{-1}K(s)\Big|_{s=0}\frac{1}{1+\sigma}(-U_2)\Big], \qquad (9.129)\end{aligned}$$

with the 'initial' value at $t = t_0$

$$\begin{aligned}X(t_0) &= C(t_0)U_1 + \epsilon A^{-1}U_3C'(t_0) \\ &+ \epsilon^2\Big[-((-1-\sigma)-\partial/\partial s)^{-1}K(s)\Big|_{s=0}\frac{1}{1+\sigma}(-U_2)\Big]. \qquad (9.130)\end{aligned}$$

The solution (9.129) is valid only locally around $t = t_0$ owing to the secular terms, which diverge when $|t - t_0|$ goes infinity. A solution to Eq. (9.111) that is valid in a global domain of t can be constructed by the RG method. Application of the RG/E equation

$$\frac{\mathrm{d}}{\mathrm{d}t_0}\tilde{\boldsymbol{X}}(t;t_0)\bigg|_{t_0=t} = 0, \tag{9.131}$$

to Eq. (9.129) leads to

$$\begin{aligned}&\dot{C}\boldsymbol{U}_1 + \epsilon\Big[-\boldsymbol{U}_3C' + A^{-1}\boldsymbol{U}_3\dot{C}' - C^2\boldsymbol{U}_3\Big] + \epsilon^2\Big[-K(0)\frac{1}{1+\sigma}\sigma\boldsymbol{U}_1\\ &-(-\partial/\partial s)((-1-\sigma)-\partial/\partial s)^{-1}K(s)\Big|_{s=0}\frac{1}{1+\sigma}(-\boldsymbol{U}_2)\Big] = 0,\end{aligned} \tag{9.132}$$

which is further reduced to the dynamical equations to the would-be integral constants C and C' as

$$\dot{C} = \epsilon^2\frac{\sigma}{1+\sigma}\left(\chi C(t) + b^{-1}C(t)C'(t)\right), \tag{9.133}$$

$$\dot{C}' = -bC'(t) - bC^2(t). \tag{9.134}$$

Here we have used the following formula derived from (9.124),

$$K(0) = \chi C(t) + b^{-1}C(t)C'(t). \tag{9.135}$$

Then the globally improved solution defined on the invariant/attractive manifold is given in terms of the slow variables $C(t)$ and $C'(t)$ as

$$\begin{aligned}\boldsymbol{X}^{\mathrm{global}}(t) &:= \boldsymbol{X}(t_0 = t)\\ &= C(t)\boldsymbol{U}_1 + \epsilon A^{-1}\boldsymbol{U}_3C'(t)\\ &\quad + \epsilon^2\Big[((-1-\sigma)-\partial/\partial s)^{-1}K(s)\Big|_{s=0}\frac{1}{1+\sigma}\boldsymbol{U}_2\Big]\Big|_{t_0=t}.\end{aligned} \tag{9.136}$$

With the use of the identity

$$((-1-\sigma)-\partial/\partial s)^{-1}K(s) = -\frac{\chi C - C^3/b}{1+\sigma} + \mathrm{e}^{-bs}\frac{C(C^2+C')}{b^2 - b(1+\sigma)}, \tag{9.137}$$

one finds that the respective components ${}^t(\xi,\ \eta,\ \zeta) = \epsilon\,\boldsymbol{X}$ are given by

$$\xi = \epsilon C + \epsilon^3 \frac{\sigma}{1+\sigma}\left[-\frac{\chi C - C^3/b}{1+\sigma} + \frac{C(C^2 + C')}{b^2 - b(1+\sigma)} \right], \tag{9.138}$$

$$\eta = \epsilon C - \epsilon^3 \frac{1}{1+\sigma}\left[-\frac{\chi C - C^3/b}{1+\sigma} + \frac{C(C^2 + C')}{b^2 - b(1+\sigma)} \right], \tag{9.139}$$

$$\zeta = -\epsilon^2 \frac{C'}{b}. \tag{9.140}$$

It is to be noted that the invariant/attractive manifold (9.136) is a two-dimensional manifold parametrized by the two parameters C and C', in contrast to that obtained in Sects. 3.6.2 and 6.2 where the zero mode was only employed to construct the slow dynamics.

To see what has been obtained, let us see an asymptotic behavior after a long time, where the time dependence of the fast variable $C'(t)$ is negligible while that of the slow one $C(t)$ is not, which implies that the former becomes a slaving variable of the latter. Inserting $\dot{C}' \simeq 0$ into Eq. (9.134), we have

$$C' \simeq -C^2, \tag{9.141}$$

substitution of which into Eqs. (9.133), (9.138)–(9.140) leads to a closed equation with respect to C as

$$\dot{C} = \epsilon^2 \frac{\sigma}{1+\sigma}(\chi C(t) \; - \; b^{-1} C^3(t)), \tag{9.142}$$

and the one-dimensional invariant manifold parametrized only by C. We note that these equations written by C are the same as the reduced equations derived by employing the would-be zero mode from the outset done in Sects. 3.6.2 and 6.2.

It is worth emphasizing that the set of Eqs. (9.133) and (9.134) governing the dynamics of C and C' describes the *mesoscopic dynamics* of the Lorenz model, and the corresponding two-dimensional invariant/attractive manifold is given by Eqs. (9.138)–(9.140). It would be interesting to compare the present result with the previous ones obtained by various reduction theories, e.g., the center manifold theory [33, 34]. This is, however, beyond the scope of this monograph.

In the rest of this section, we shall show the numerical results with the parameter set

$$b = 8/3 \quad \text{and} \quad \sigma = 10. \tag{9.143}$$

First, we shall analyze the unstable motion around the origin (A) by setting

$$\chi = +1 \quad \text{and} \quad \epsilon = 0.5, \tag{9.144}$$

accordingly,

$$r = 1 + \chi\epsilon^2 = 1.25. \tag{9.145}$$

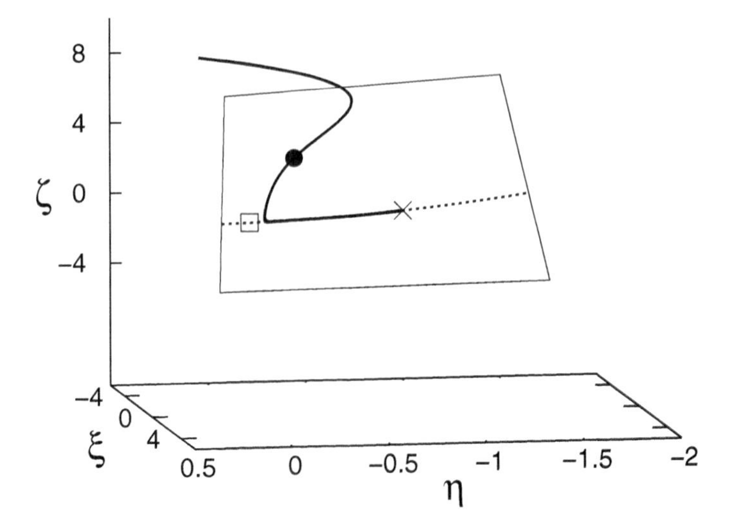

Fig. 9.2 The solid line shows the numerical solution of the Lorenz model, while the dashed line the one-dimensional manifold given by imposing the constraint $C' = -C^2$, for the parameter set $b = 8/3$, $\sigma = 10$, $\epsilon = 0.5$, and $\chi = +1$. One sees that after some time, both the two curves are confined in a surface, which is the invariant/attractive manifold (surface) given by Eqs. (9.138)–(9.140) that is obtained in the doublet scheme. The square and cross are the fixed points (A) and (C), respectively, while the big dot with the coordinate $(\xi,\ \eta,\ \zeta) = (-0.297,\ -0.212,\ 3.437)$ denotes the point from which the solution becomes to move on the invariant/attractive manifold. (The figure is slightly modified, with permission, from Fig. 3 in [69]. Copyright Elsevier (2016))

In this setting, while the origin (A) is unstable, the steady states (B) and (C) are stable because $1 < r < r_c \simeq 24.7$.

Figure 9.2 shows the numerical result for the trajectory of the solution to Eq. (9.111) with the initial values $(\xi,\ \eta,\ \zeta) = (1,\ 5,\ 15)$: The figure also includes the two-dimensional attractive manifold described by Eqs. (9.138)–(9.140) for the mesoscopic dynamics and the one-dimensional manifold for the slow dynamics obtained by imposing the constraint on Eqs. (9.138)–(9.140). We see that the trajectory tends to be attracted to the two-dimensional manifold at a rather early time, after which the trajectory remains on it. Then it tends to be confined in the one-dimensional manifold of the slow dynamics, and finally approaches the steady state (C) asymptotically.

In Fig. 9.3, we show separately the time dependence of each coordinate $\xi(t)$, $\eta(t)$, and $\zeta(t)$ of the solution to the three-dimensional original equation (9.111) of the Lorenz model together with those given by (9.138)–(9.140) with the solution to the reduced equation (9.133) and (9.134). We find that the solution given by the reduced mesoscopic dynamics are almost the same as those of the original Lorenz equation (9.111). Thus, we can conclude the validity of the doublet scheme in the RG method as a general theory for extracting the mesoscopic dynamics in an asymptotic regime. The material so far is essentially contained in [69].

Now, although the reduced equations (9.133) and (9.134) together with the representation of the attractive/invariant manifold (9.138)–(9.140) have been extracted

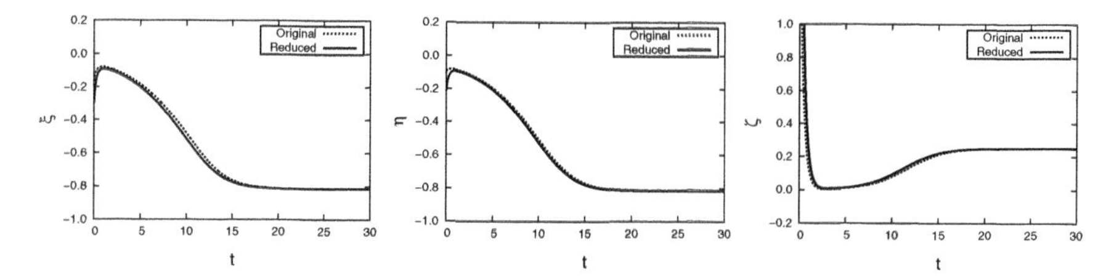

Fig. 9.3 The time dependence of ξ, η, and ζ with the initial value set at $(\xi, \eta, \zeta) = (-0.297, -0.212, 3.437)$, *i.e.*, the point denoted by the big dot in Fig. 9.2: The parameter set is the same as that for Fig. 9.2. The solid lines are the solution to the reduced equations (9.133), (9.134), and (9.138)–(9.140) of the Lorenz equation (9.111), while the dashed lines the numerical solution to (9.111). One sees that the original equation is approximated by the set of reduced equations quite well in the asymptotic region after some time. (The figure is slightly modified, with permission, from Fig. 4 in [69]. Copyright Elsevier (2016))

in the perturbation theory formally assuming that ϵ is small, the reduced equations are to be solved exactly in the RG method. It is worth emphasizing that through this process of solution terms in the infinite orders of ϵ are summed up, as stressed in [1, 3, 5, 6, 40, 46, 57, 58], and hence it may turn out that the resummed solution gives a rather good approximation to the exact solutions even with large values of ϵ, say, as large as $\epsilon > 1$. We shall now show that it is indeed the case.

Under the same initial conditions with the parameter values for b and σ being kept the previous ones, we shall examine the case of

$$\epsilon = 2.0, \tag{9.146}$$

and accordingly, $r = 5.0 < r_c \simeq 24.7$, for which the fixed point (A) is unstable while (B) and (C) are stable. Figure 9.4 shows the numerical results for $b = 8/3$, $\sigma = 10$, $\epsilon = 2.0$, and $\chi = +1$: The solid line shows the trajectory of the original Lorenz equation (9.111) while the surface is the two-dimensional manifold described by Eqs. (9.138)–(9.140) given by the doublet scheme in the RG method. We see that the trajectory is rapidly attracted to the two-dimensional attractive manifold, and is eventually confined on it, where the trajectory forms a spiral and approaches the steady state (B) asymptotically. This result tells us that the mesoscopic dynamics derived by the doublet scheme is powerful enough to reveal the essential properties of the exact solution in a wide range of time beyond the asymptotic regime even for $\epsilon = 2.0$.

This powerfulness of the doublet scheme is further confirmed in Fig. 9.5 where the separate behavior of $\xi(t)$, $\eta(t)$, and $\zeta(t)$ given by (9.138)–(9.140) with the solution of Eqs. (9.133) and (9.134) being inserted well reproduce the time dependence of the solution to the original Lorenz equation (9.111); notice that the exact solutions are denoted by dashed lines.

Thus, we can conclude that the doublet scheme in the RG method based on the perturbation theory provides us with a powerful construction method of the mesoscopic

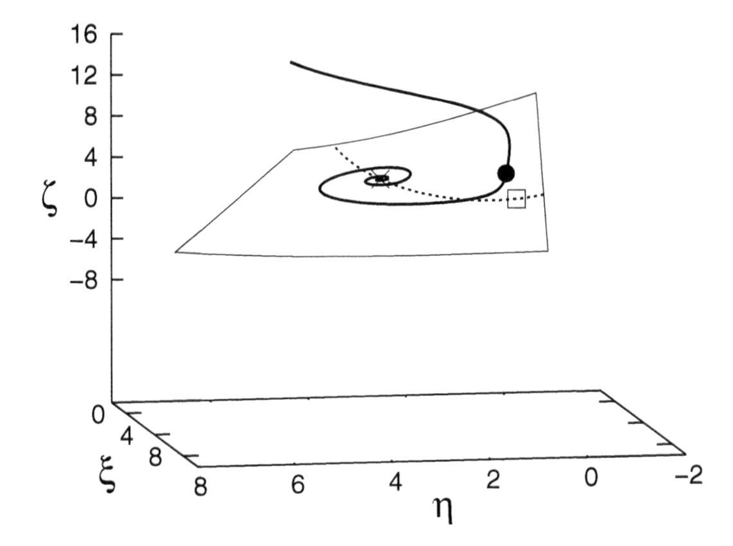

Fig. 9.4 The numerical solution to the Lorenz model (9.111) with $b = 8/3$, $\sigma = 10$, $\epsilon = 2.0$, and $\chi = +1$. The solid line represents the solution, while the surface denotes the two-dimensional manifold described by Eqs. (9.138)–(9.140). The dashed line shows the one-dimensional manifold that we obtain by substituting $C' = -C^2$ into Eqs. (9.138)–(9.140). The square and cross denote (A) and (B), respectively, while the circle a point from which the solution begins to be confined on the two-dimensional manifold

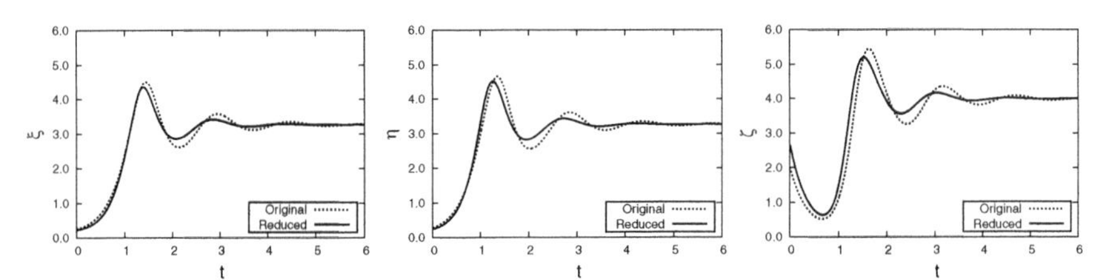

Fig. 9.5 The time dependence of ξ, η, and ζ, where the parameters are set as $b = 8/3$, $\sigma = 10$, $\epsilon = 2.0$, and $\chi = +1$. The dashed lines denote the solution to the original equation (9.111), while the solid lines the solution to the reduced equations (9.133), (9.134), and (9.138)–(9.140), whose initial values are $(\xi,\ \eta,\ \zeta) = (0.244,\ 0.240,\ 2.701)$ corresponding to the circle in Fig. 9.4

dynamics that includes non-perturbative effects caused by the perturbation terms even with $\epsilon > 1$, which would not to be taken into account in a naive perturbation theory.

Finally, we give a theoretical analysis on the qualitative difference of the behaviors between $\epsilon = 0.5$ and $\epsilon = 2.0$ cases: As shown in Figs. 9.3 and 9.5, the solution with $\epsilon = 0.5$ shows an over damping around (C), while the solution with $\epsilon = 2.0$ a damped oscillation around (B).

Let us investigate the dynamical properties of Eqs. (9.133) and (9.134) around the respective stationary solution to them. For $\chi = +1$, Eqs. (9.133) and (9.134) have two stationary solutions

$$(\pm\sqrt{b},\ -b), \tag{9.147}$$

which correspond to the steady states (B) and (C), respectively. Expanding C and C' as

$$(C, C') = (\pm\sqrt{b} + \delta C, -b + \delta C'), \tag{9.148}$$

we linearize Eqs. (9.133) and (9.134) as

$$\frac{\mathrm{d}}{\mathrm{d}t}\begin{pmatrix}\delta C \\ \delta C'\end{pmatrix} = \begin{pmatrix} 0 & \pm\epsilon^2 \frac{\sigma}{1+\sigma}\frac{1}{\sqrt{b}} \\ \mp 2b\sqrt{b} & -b \end{pmatrix}\begin{pmatrix}\delta C \\ \delta C'\end{pmatrix}. \tag{9.149}$$

Since the dynamical property of Eqs. (9.133) and (9.134) around (B) or (C) are determined by eigenvalues of the matrix in the above linearized equation, we analyze an eigenvalue equation given by

$$\lambda^2 + b\lambda + \epsilon^2 \frac{\sigma}{1+\sigma} 2b = 0, \tag{9.150}$$

with λ being the eigenvalue. We find that there exists a critical value of ϵ, which is written as

$$\epsilon^* := \sqrt{\frac{b(1+\sigma)}{8\sigma}}. \tag{9.151}$$

In fact, for $0 < \epsilon < \epsilon^*$, λ is real negative and $(\delta C, \delta C')$ shows an over damping, while for $\epsilon > \epsilon^*$, λ is complex whose real part is negative and $(\delta C, \delta C')$ shows a damped oscillation. Substituting $b = 8/3$ and $\sigma = 10$ into Eq. (9.151), we find that $\epsilon^* \sim 0.61$. It is noteworthy that this value of ϵ^* nicely explains the reason why the solution with $\epsilon = 0.5$ shows the over damping around (C), while the solution with $\epsilon = 2.0$ the damped oscillation around (B).

Appendix: Constructive Formulation of the Doublet Scheme in the RG Method

In this Appendix, we show that a faithful solution of the generic equation (9.1) aiming at obtaining a simplest closed equation without fast modes naturally leads to the doublet scheme in the RG method introduced in Sect. 9.2. The detailed strategy is as follows:

(1) We start from the zero-th order solution that is a stationary state as in Sect. 9.2.
(2) With the first-order ‘initial’ value $\boldsymbol{\phi}$, we obtain the perturbative solution $\tilde{\boldsymbol{X}}(t\,;\,t_0)$ by formally introducing the projection operators P_1 and Q_1 without the explicit forms yet specified.
(3) We eliminate the undesirable fast modes belonging to the Q_1-space from $\tilde{\boldsymbol{X}}(t\,;\,t_0)$.
(4) Then the RG/E equation applied to thus constructed perturbative solution $\tilde{\boldsymbol{X}}(t\,;\,t_0)$ lead to the reduced evolution equation solely composed of the P_0 and P_1 modes including $\boldsymbol{\phi}$.
(5) Finally, we determine the explicit forms of $\boldsymbol{\phi}$ and P_1 modes so that the reduced equation is closed and has a fewer number of terms and degrees of freedom.

As declared above, we start with $\tilde{\boldsymbol{X}}_0(t; t_0) = \boldsymbol{X}^{\mathrm{eq}}(t_0)$ as adopted in Sect. 9.2.2. Then the first-order perturbative solution is given by Eq. (9.44), which reads

$$\tilde{\boldsymbol{X}}_1(t\,;\,t_0) = \mathrm{e}^{A(t-t_0)}\left[A^{-1}\,Q_0\,\boldsymbol{F}_0 + \boldsymbol{\phi}\right] + (t-t_0)\,P_0\,\boldsymbol{F}_0 - A^{-1}\,Q_0\,\boldsymbol{F}_0. \tag{9.152}$$

The insertion of the identity $1 = P_0 + P_1 + Q_1$ behind $\mathrm{e}^{A(t-t_0)}$ converts this solution into

$$\begin{aligned}\tilde{\boldsymbol{X}}_1(t\,;\,t_0) &= \mathrm{e}^{A(t-t_0)}\,(P_0+P_1+Q_1)\left[A^{-1}\,Q_0\,\boldsymbol{F}_0+\boldsymbol{\phi}\right] + (t-t_0)\,P_0\,\boldsymbol{F}_0 - A^{-1}\,Q_0\,\boldsymbol{F}_0\\ &= \mathrm{e}^{A(t-t_0)}\left[Q_1\,A^{-1}\,Q_0\,\boldsymbol{F}_0 + Q_1\,\boldsymbol{\phi}\right]\\ &\quad + \mathrm{e}^{A(t-t_0)}\left[P_1\,A^{-1}\,Q_0\,\boldsymbol{F}_0 + P_1\,\boldsymbol{\phi}\right]\\ &\quad + (t-t_0)\,P_0\,\boldsymbol{F}_0 + P_0\,\boldsymbol{\phi} - A^{-1}\,Q_0\,\boldsymbol{F}_0,\end{aligned} \tag{9.153}$$

where the identity $P_0 A^{-1} Q_0 = 0$ has been used.

We suppose that the modes belonging to the Q_1 space are a fast and undesired modes, which are to be eliminated for constructing the mesoscopic dynamics. This elimination is simply accomplished by choosing not-yet specified ‘initial’ value as

$$Q_1\,\boldsymbol{\phi} = -Q_1\,A^{-1}\,Q_0\,\boldsymbol{F}_0, \tag{9.154}$$

which is a simple but important procedure adopted in the RG method. Thus Eq. (9.153) now takes the form consisting of the P_0 and P_1 modes only as

$$\begin{aligned}\tilde{\boldsymbol{X}}_1(t\,;\,t_0) = \mathrm{e}^{A(t-t_0)}&\left[P_1\,A^{-1}\,Q_0\,\boldsymbol{F}_0 + P_1\,\boldsymbol{\phi}\right]\\ &+ (t-t_0)\,P_0\,\boldsymbol{F}_0 + P_0\,\boldsymbol{\phi} - A^{-1}\,Q_0\,\boldsymbol{F}_0.\end{aligned} \tag{9.155}$$

The second-order perturbative analysis is carried out much the similar way as that in Sect. 9.2.4, and we arrive at the perturbative solution in the second-order approximation as

$$
\begin{aligned}
\tilde{\boldsymbol{X}}(t\,;\,t_0) = \boldsymbol{X}^{\mathrm{eq}} + \epsilon \Big[& \mathrm{e}^{A(t-t_0)}\,(P_1\,\boldsymbol{\phi} + P_1\,A^{-1}\,Q_0\,\boldsymbol{F}_0) + (t-t_0)\,P_0\,\boldsymbol{F}_0 \\
& + P_0\,\boldsymbol{\phi} - A^{-1}\,Q_0\,\boldsymbol{F}_0 \Big] \\
& + \epsilon^2 \Big[(1 - \mathrm{e}^{(t-t_0)\partial/\partial s})\,(-\partial/\partial s)^{-1}\,P_0\,\boldsymbol{K}(s)\Big|_{s=0} \\
& + (\mathrm{e}^{A(t-t_0)} - \mathrm{e}^{(t-t_0)\partial/\partial s})\,P_1\,\mathcal{G}(s)\,Q_0\,\boldsymbol{K}(s)\Big|_{s=0} \\
& - \mathrm{e}^{(t-t_0)\partial/\partial s}\,Q_1\,\mathcal{G}(s)\,Q_0\,\boldsymbol{K}(s)\Big|_{s=0} \Big].
\end{aligned} \tag{9.156}
$$

Here $\boldsymbol{K}(s)$ are defined by

$$
\boldsymbol{K}(s) = \frac{1}{2}\,B\left[\tilde{\boldsymbol{X}}_1(s+t_0\,;\,t_0)\,,\,\tilde{\boldsymbol{X}}_1(s+t_0\,;\,t_0)\right] + F_1\,\tilde{\boldsymbol{X}}_1(s+t_0\,;\,t_0). \tag{9.157}
$$

Applying the RG/E equation

$$
\mathrm{d}\tilde{\boldsymbol{X}}(t\,;\,t_0)/\mathrm{d}t_0|_{t_0=t} = 0, \tag{9.158}
$$

to $\tilde{\boldsymbol{X}}(t\,;\,t_0)$ thus obtained, we have the following reduced equation,

$$
\begin{aligned}
& \frac{\partial}{\partial t}\boldsymbol{X}^{\mathrm{eq}} + \epsilon \Big[- A\,P_1\,\boldsymbol{\phi} - A\,P_1\,A^{-1}\,Q_0\,\boldsymbol{F}_0 \\
& + \frac{\partial}{\partial t}(P_0\,\boldsymbol{\phi} + P_1\,\boldsymbol{\phi} - Q_1\,A^{-1}\,Q_0\,\boldsymbol{F}_0) - P_0\,\boldsymbol{F}_0 \Big] \\
& + \epsilon^2 \Big[- P_0\,\boldsymbol{K}(0) - (A - \partial/\partial s)\,P_1\,\mathcal{G}(s)\,Q_0\,\boldsymbol{K}(s)\Big|_{s=0} \\
& + (\partial/\partial s)\,Q_1\,\mathcal{G}(s)\,Q_0\,\boldsymbol{K}(s)\Big|_{s=0} \Big] = 0.
\end{aligned} \tag{9.159}
$$

This is a kind of the master equation for the reduced dynamics. Operating the projection operators P_0 and P_1 onto Eq. (9.159), we obtain the reduced equations

$$
\begin{aligned}
& P_0\,\frac{\partial}{\partial t}\boldsymbol{X}^{\mathrm{eq}} + \epsilon \Big[P_0\,\frac{\partial}{\partial t}(P_0\,\boldsymbol{\phi} + P_1\,\boldsymbol{\phi} - Q_1\,A^{-1}\,Q_0\,\boldsymbol{F}_0) \\
& - P_0\,\boldsymbol{F}_0 \Big] - \epsilon^2\,P_0\,\boldsymbol{K}(0) = 0,
\end{aligned} \tag{9.160}
$$

$$\epsilon \Big[- P_1\, A\, P_1\, \boldsymbol{\phi} - P_1\, A\, P_1\, A^{-1}\, Q_0\, \boldsymbol{F}_0 + P_1\, \frac{\partial}{\partial t}(P_0\, \boldsymbol{\phi} + P_1\, \boldsymbol{\phi} - Q_1\, A^{-1}\, Q_0\, \boldsymbol{F}_0) \Big] - \epsilon^2\, P_1\, (A - \partial/\partial s)\, P_1\, \mathcal{G}(s)\, Q_0\, \boldsymbol{K}(s)\Big|_{s=0} = 0, \tag{9.161}$$

respectively. The inner product between $\boldsymbol{\varphi}_0^{\alpha}$ and Eq. (9.160) leads to

$$\langle\, \boldsymbol{\varphi}_0^{\alpha}\,,\, \frac{\partial}{\partial t}(\boldsymbol{X}^{\mathrm{eq}} + \epsilon\, (P_0\, \boldsymbol{\phi} + P_1\, \boldsymbol{\phi} - Q_1\, A^{-1}\, Q_0\, \boldsymbol{F}_0))\,\rangle = \epsilon\, \langle\, \boldsymbol{\varphi}_0^{\alpha}\,,\, \boldsymbol{F}_0\,\rangle + \epsilon^2\, \langle\, \boldsymbol{\varphi}_0^{\alpha}\,,\, \boldsymbol{K}(0)\,\rangle, \tag{9.162}$$

with

$$\boldsymbol{K}(0) = \frac{1}{2}\, B\big[P_0\, \boldsymbol{\phi} + P_1\, \boldsymbol{\phi} - Q_1\, A^{-1}\, Q_0\, \boldsymbol{F}_0\,,\, P_0\, \boldsymbol{\phi} + P_1\, \boldsymbol{\phi} - Q_1\, A^{-1}\, Q_0\, \boldsymbol{F}_0\big] + F_1\, (P_0\, \boldsymbol{\phi} + P_1\, \boldsymbol{\phi} - Q_1\, A^{-1}\, Q_0\, \boldsymbol{F}_0). \tag{9.163}$$

We are now in the position to determine the structure of the P_1 space and the functional form of $\boldsymbol{\phi}$. We shall perform this task step by step.

First, from the observation of Eq. (9.162) we find that if the equality

$$Q_1\, A^{-1}\, Q_0\, \boldsymbol{F}_0 = 0, \tag{9.164}$$

is satisfied, Eq. (9.162) is reduced into a simper form as

$$\langle\, \boldsymbol{\varphi}_0^{\alpha}\,,\, \frac{\partial}{\partial t}(\boldsymbol{X}^{\mathrm{eq}} + \epsilon\, (P_0\, \boldsymbol{\phi} + P_1\, \boldsymbol{\phi}))\,\rangle = \epsilon\, \langle\, \boldsymbol{\varphi}_0^{\alpha}\,,\, \boldsymbol{F}_0\,\rangle + \epsilon^2\, \langle\, \boldsymbol{\varphi}_0^{\alpha}\,,\, \boldsymbol{K}(0)\,\rangle, \tag{9.165}$$

with

$$\boldsymbol{K}(0) = \frac{1}{2}\, B\big[P_0\, \boldsymbol{\phi} + P_1\, \boldsymbol{\phi}\,,\, P_0\, \boldsymbol{\phi} + P_1\, \boldsymbol{\phi}\big] + F_1\, (P_0\, \boldsymbol{\phi} + P_1\, \boldsymbol{\phi}). \tag{9.166}$$

In fact, the imposed condition (9.164) can be utilized to determine the P_1 space, as follows. We start from the generic representation

$$A^{-1}\, Q_0\, \boldsymbol{F}_0 = \sum_{\mu=1}^{M_1} (A^{-1}\, \boldsymbol{\varphi}_1^{\mu})\, f_{\mu}. \tag{9.167}$$

As mentioned in Sect. 9.2.3, $A^{-1}\, \boldsymbol{\varphi}_1^{\mu}$ with $\mu = 1,\, 2,\, \ldots,\, M_1$ are linearly independent, and supposed to make the basis vectors of the P_1 space.

Owing to Eq. (9.154), Eq. (9.164) is equivalent to

$$Q_1\,\boldsymbol{\phi} = 0. \tag{9.168}$$

Without loss of generality, we assume that $\boldsymbol{\phi}$ contains no zero modes

$$P_0\,\boldsymbol{\phi} = 0, \tag{9.169}$$

as discussed in Sect. 9.2.3. Thus we find that $\boldsymbol{\phi}$ belongs to the P_1 space as

$$P_1\,\boldsymbol{\phi} = \boldsymbol{\phi}, \tag{9.170}$$

and is represented as

$$\boldsymbol{\phi} = \sum_{\mu=1}^{M_1} (A^{-1}\,\boldsymbol{\varphi}_1^\mu)\,C'_\mu(t_0), \tag{9.171}$$

where the coefficients $C'_\mu(t_0)$ denotes a would-be integral constant.

By taking the inner product between $A^{-1}\,\boldsymbol{\varphi}_1^\mu$ and Eq. (9.161), we have

$$\begin{aligned}
&\epsilon\,\langle\,A^{-1}\,\boldsymbol{\varphi}_1^\mu\,,\,\frac{\partial}{\partial t}\boldsymbol{\phi}\,\rangle\\
&= \epsilon\,\langle\,A^{-1}\,\boldsymbol{\varphi}_1^\mu\,,\,A\,P_1\,A^{-1}\,Q_0\,\boldsymbol{F}_0\,\rangle + \epsilon\,\langle\,A^{-1}\,\boldsymbol{\varphi}_1^\mu\,,\,A\,P_1\,\boldsymbol{\phi}\,\rangle\\
&\quad + \epsilon^2\,\langle\,A^{-1}\,\boldsymbol{\varphi}_1^\mu\,,\,(A - \partial/\partial s)\,P_1\,\mathcal{G}(s)\,Q_0\,\boldsymbol{K}(s)\Big|_{s=0}\,\rangle,
\end{aligned} \tag{9.172}$$

the third term of which apparently produces infinite numbers of terms because of $\mathcal{G}(s)$, which is undesirable for the reduced dynamics. To consider whether it could be avoided, let us calculate it as

$$\begin{aligned}
&\langle\,A^{-1}\,\boldsymbol{\varphi}_1^\mu\,,\,(A - \partial/\partial s)\,P_1\,\mathcal{G}(s)\,Q_0\,\boldsymbol{K}(s)\Big|_{s=0}\,\rangle\\
&= \langle\,(A - \partial/\partial s)\,A^{-1}\,\boldsymbol{\varphi}_1^\mu\,,\,P_1\,\mathcal{G}(s)\,Q_0\,\boldsymbol{K}(s)\Big|_{s=0}\,\rangle\\
&= \langle\,(\boldsymbol{\varphi}_1^\mu - A^{-1}\,\boldsymbol{\varphi}_1^\mu\,\partial/\partial s)\,,\,P_1\,\mathcal{G}(s)\,Q_0\,\boldsymbol{K}(s)\Big|_{s=0}\,\rangle,
\end{aligned} \tag{9.173}$$

where we have used the self-adjointness of A. If $\boldsymbol{\varphi}_1^\mu$ are also the vectors belonging to the P_1 space, we can further reduce Eq. (9.173) as follows:

$$
\begin{aligned}
&\langle\,(\boldsymbol{\varphi}_1^{\mu} - A^{-1}\,\boldsymbol{\varphi}_1^{\mu}\,\partial/\partial s)\,,\; P_1\,\mathcal{G}(s)\,Q_0\,\boldsymbol{K}(s)\Big|_{s=0}\,\rangle \\
&= \langle\, P_1(\boldsymbol{\varphi}_1^{\mu} - A^{-1}\,\boldsymbol{\varphi}_1^{\mu}\,\partial/\partial s)\,,\; \mathcal{G}(s)\,Q_0\,\boldsymbol{K}(s)\Big|_{s=0}\,\rangle \\
&= \langle\,(A - \partial/\partial s)\,A^{-1}\,\boldsymbol{\varphi}_1^{\mu}\,,\; \mathcal{G}(s)\,Q_0\,\boldsymbol{K}(s)\Big|_{s=0}\,\rangle \\
&= \langle\, A^{-1}\,\boldsymbol{\varphi}_1^{\mu}\,,\; (A - \partial/\partial s)\,\mathcal{G}(s)\,Q_0\,\boldsymbol{K}(s)\Big|_{s=0}\,\rangle \\
&= \langle\, A^{-1}\,\boldsymbol{\varphi}_1^{\mu}\,,\; Q_0\,\boldsymbol{K}(s)\Big|_{s=0}\,\rangle \\
&= \langle\, A^{-1}\,\boldsymbol{\varphi}_1^{\mu}\,,\; \boldsymbol{K}(0)\,\rangle,
\end{aligned} \tag{9.174}
$$

where the identity $(A - \partial/\partial s)\,\mathcal{G}(s) = 1$ has been used in the third equality. Thus we are lead to identify $\boldsymbol{\varphi}_1^{\mu}$ as additional components of the P_1 space and redefine the vector space spanned by both of $A^{-1}\,\boldsymbol{\varphi}_1^{\mu}$ and $\boldsymbol{\varphi}_1^{\mu}$ as the P_1 space. These pairs of $A^{-1}\,\boldsymbol{\varphi}_1^{\mu}$ and $\boldsymbol{\varphi}_1^{\mu}$ are nothing but the doublet modes introduced in Sect. 9.2.3.

The structure of the P_1 space simplifies the other terms in the right-hand side of Eq. (9.172) as follows:

$$
\langle\, A^{-1}\,\boldsymbol{\varphi}_1^{\mu}\,,\; A\,P_1\,A^{-1}\,Q_0\,\boldsymbol{F}_0\,\rangle = \langle\,\boldsymbol{\varphi}_1^{\mu}\,,\; A^{-1}\,Q_0\,\boldsymbol{F}_0\,\rangle, \tag{9.175}
$$

$$
\langle\, A^{-1}\,\boldsymbol{\varphi}_1^{\mu}\,,\; A\,P_1\,\boldsymbol{\phi}\,\rangle = \langle\,\boldsymbol{\varphi}_1^{\mu}\,,\; \boldsymbol{\phi}\,\rangle. \tag{9.176}
$$

Substituting Eqs. (9.169), (9.170), and (9.174)–(9.176) into Eqs. (9.165) and (9.172), respectively, we arrive at the same equations as Eqs. (9.82) and (9.83). This implies that the constructive introduction of the doublet scheme in the RG method has been successfully completed.

Part II
RG/E Derivation of Second-Order Relativistic and Non-relativistic Dissipative Fluid Dynamics

Chapter 10
Introduction to Relativistic Dissipative Fluid Dynamics and Its Derivation from Relativistic Boltzmann Equation by Chapman-Enskog and Fourteen-Moment Methods

10.1 Basics of Relativistic Dissipative Fluid Dynamics

Relativistic fluid dynamics is the powerful means to describe low-frequency and long-wavelength dynamics of an interacting many-body system, when the flow velocity and/or the velocities of the constituent particles of the fluid are large and comparable to the light velocity [62]. In fact, relativistic fluid dynamic equations have been used to analyze the dynamics of a hot matter composed of quarks and gluons, *i.e.*, the quark-gluon plasma (QGP) [160, 161], which should have been being created in the experiments of relativistic heavy ion collision at the Relativistic Heavy Ion Collider (RHIC) at Brookhaven National Laboratory and the Large Hadron Collider (LHC) at CERN; see Refs. [60, 162] as recent summaries. The relativistic dissipative fluid dynamics is also relevant to the soft-mode dynamics [163–165] around the possible critical point(s) in QCD phase diagram [166, 167]; see [168] for the latest up date. Moreover, the relativistic dissipative fluid dynamic equation has been also applied to various high-energy astrophysical phenomena [59, 169–171].

As is clearly demonstrated in Sect. 8.5 of Chap. 8 for the non-relativistic case, the fluid dynamic equation can be reorganized to a set of balance equations of the energy and momenta as well as the particle number, which was assumed to be the sole conserved quantity (or 'charge'). As will be shown later, the balance equation for the energy and momenta are combined into the conservation law for the energy-momentum tensor $T_0^{\mu\nu}$ in the covariant notation as

$$\partial_\mu T_0^{\mu\nu} = 0, \tag{10.1}$$

while the particle number conservation is expressed in terms of the particle current or flow N_0^μ as

$$\partial_\mu N_0^\mu = 0. \tag{10.2}$$

T. Kunihiro et al., *Geometrical Formulation of Renormalization-Group Method as an Asymptotic Analysis*, Fundamental Theories of Physics 206,
https://doi.org/10.1007/978-981-16-8189-9_10

As is clearly shown in Sect. 133 of Ref. [62], the energy-momentum tensor of the ideal system is written as

$$T_0^{\mu\nu} = eu^\mu u^\nu - P\Delta^{\mu\nu}, \tag{10.3}$$

where e and P denote the proper internal energy and pressure, respectively, while $\Delta^{\mu\nu}$ is the projection tensor onto the space like vector given by

$$\Delta^{\mu\nu} := g^{\mu\nu} - u^\mu u^\nu, \tag{10.4}$$

with $u^\mu(x)$ being the flow velocity, which is normalized as

$$u_\mu u^\mu = g^{\mu\nu} u_\mu u_\nu = 1. \tag{10.5}$$

In the present and the following chapters, we use the Minkowski metric $g^{\mu\nu} = \mathrm{diag}(+1, -1, -1, -1)$ and the natural unit, *i.e.*,

$$\hbar = c = k_\mathrm{B} = 1. \tag{10.6}$$

The particle current of the ideal fluid is given by

$$N_0^\mu = nu^\mu, \tag{10.7}$$

in terms of the mass density n.

It is, however, to be noted that we have not necessarily reached a full understanding of the theory of relativistic fluid dynamics for *viscous* fluids, although there have been many important studies [62, 64–68, 172] since Eckart's pioneering work [61]. For instance, we can indicate following three fundamental problems (a)$\sim$(c) with the relativistic fluid dynamic equations for viscous fluids, one of which is inherent in any relativistic theory and the other two may or may not be related to the first one.

(a) Ambiguities and ad-hoc ansatz in the definition of the flow velocity [61, 62, 64, 173, 174]. The very equivalence of the mass and the energy in the relativity makes ambiguous the definition of the flow velocity or the rest frame of the fluid. Thus the form of the relativistic dissipative fluid dynamic equation depends on the definition of the flow velocity $u^\mu(x)$. One of the typical rest frames is the particle frame in which the fluid velocity is proportional to that of the particle current $N^\mu(x)$, *i.e.*,

$$u^\mu(x) := N^\mu/\sqrt{N^\mu N_\mu}, \tag{10.8}$$

The other typical one is the energy frame in which the fluid velocity is proportional to that of the energy flow $T^{\mu\nu}u_\nu$, *i.e.*,

$$u^\mu := T^{\mu\nu}u_\nu/\sqrt{u_\sigma T^{\sigma\tau}T_{\tau\rho}u^\rho}. \tag{10.9}$$

Indeed the dissipative relativistic fluid dynamic equation was first constructed on the particle frame in a *phenomenological way* by Eckart [61], whereas the famous Landau-Lifshitz equation is on the energy frame [62].
The standard way of the phenomenological construction of the relativistic dissipative fluid dynamic equation is based on the following two plausible principles and an ansatz on the rest frame:

(1) the particle-number and energy-momentum conservation laws,
(2) the law of the increase in entropy

and

(3) the choice of the rest frame of the fluid or definition of the flow velocity.

The items (1) and (2) are equally employed in the non-relativistic case, *i.e.*, the construction of the Navier–Stokes equation, while the third item (3) is specific for the relativistic case owing to the mass-energy equivalence.
To be more explicit, let $\delta T^{\mu\nu}$ and δN^{μ} be the dissipative part of the energy-momentum tensor and the particle current, respectively: The total energy-momentum tensor and the particle current are given by a sum of the ideal and dissipative parts as

$$T^{\mu\nu} = T_0^{\mu\nu} + \delta T^{\mu\nu}, \quad N^{\mu} = N_0^{\mu} + \delta N^{\mu}, \tag{10.10}$$

respectively. The point is that the energy-momentum and particle-number conservation laws together with the law of the increase in entropy do not supply sufficient conditions for a unique determination of the forms of $\delta T^{\mu\nu}$ and δN^{μ}, and hence some physical ansatz involving the flow velocity u^{μ} are necessary.
In the phenomenological derivations of the fluid dynamics by Eckart [61] and Landau-Lifshitz [62], it is commonly taken for granted that there are no contribution from the dissipation to the internal energy nor to the particle-number density, which implies that the dissipative parts $\delta T^{\mu\nu}$ and δN^{μ} should obey the following constraints,

$$\text{(i)} \quad \delta e := u_{\mu}\, \delta T^{\mu\nu}\, u_{\nu} = 0, \tag{10.11}$$

$$\text{(ii)} \quad \delta n := u_{\mu}\, \delta N^{\mu} = 0, \tag{10.12}$$

respectively.
Note that Eqs. (10.11) and (10.12) are constraints only on the longitudinal parts of the dissipative energy-momentum tensor and particle current.
It is found that if the particle frame (Eckart frame) (10.8) is adopted for the definition of the flow velocity, the transverse part of the dissipative particle current satisfies the constraint [172] as

$$\text{(iii)} \quad \nu_{\mu} := \Delta_{\mu\nu}\, \delta N^{\nu} = 0, \tag{10.13}$$

which physically means that there is no dissipative particle current ν^μ.
On the other hand, if the energy frame (10.9) is adopted for the definition of the flow velocity [62], we have a condition on the transverse part of the energy-momentum flow as

$$\text{(iv)} \quad Q_\mu := \Delta_{\mu\nu}\, \delta T^{\nu\rho}\, u_\rho = 0, \tag{10.14}$$

which means that there is no heat current Q^μ of dissipative origin.
We note that the conditions (iii) and (iv) and hence the two of frames specified by these conditions can not be connected with each other by a Lorentz transformation.
Here we remark that there is a proposal by Stewart [64] for the condition for the particle frame, as given by (ii), (iii), and

$$\text{(v)} \quad \delta T^\mu_{\ \mu} = \delta e - 3\,\delta P = 0, \tag{10.15}$$

where

$$\delta P := -\Delta_{\mu\nu}\, \delta T^{\mu\nu}/3, \tag{10.16}$$

is the dissipative pressure to be identified with the standard bulk pressure. In the present case, the condition (i) of Eckart is replaced by a different one (v), which imposes a constraint between the dissipative internal energy δe and the dissipative pressure δP. One may ask if both the Eckart and Stewart ansatz make sense or not. It is noteworthy that the most general derivation [175] of the fluid dynamic equation on the basis of the phenomenological argument gives a class of equations which can allow the existence of the dissipative internal energy δe and the dissipative particle-number density δn as well as the standard dissipative pressure δP.

(b) Unphysical instabilities of the equilibrium state [63, 176]. There arises an unphysical instability of the equilibrium state caused by a special form of the constitutive equation involving the heat flux conventionally adopted in the Eckart (particle) frame [63, 176]. The unphysical instability might be attributed to the lack of causality, and Israel-Stewart method is presently being examined in connection to this problem [177–190]. Although their equation may get rid of the instability problem with a choice of the relaxation times as shown in Ref. [63], we emphasize that there exists no connection between the unphysical instabilities and the lack of causality. In fact, the Landau-Lifshitz equation is free from the instabilities of the equilibrium state in contrast of the Eckart equation. Furthermore, one should notice that the causal equation by Israel and Stewart is an extended version of the Eckart equation and hence it can naturally exhibit unphysical instabilities depending on the values of transport coefficients and relaxation times contained in the equation [191].

(c) A lack of causality inherent in the first-order dissipative equations [65–68]. The relativistic dissipative fluid dynamic equations proposed by Eckart, Landau and

Lifshitz, and Stewart, unfortunately, suffer from instantaneous propagation of information, which ought to be completely prohibited in a consistent relativistic theory.
The origin of this deficiency is the parabolic character of the equations containing the time derivative only in the first order [64, 145]. Thus, in order to solve instantaneous propagation, we should modify the form of the equations to the hyperbolic equation by additional terms which contains the second-order time-derivative, which are called relaxation terms with relaxation times and lengths.
In 1967, Müller [146, 147] examined the origin of the causality problem due to an instantaneous propagation of the information present even in the non-relativistic case and proposed a method to introduce relaxation terms: He traced back the undesired instantaneous propagation in the conventional theory to the neglects of some higher-order contributions of dissipative effects to the entropy through the heat flow and viscous stress. Restoring these terms, Müller derived a modified Navier–Stokes equation with relaxation terms.
Some ten years later, Müller's theory was rediscovered and extended to the relativistic fluid equations by Israel [65]. The resultant causal equation is now called the *second-order* fluid dynamic equation, while the Eckart, Landau and Lifshitz, and Stewart equations the *first-order* fluid dynamic equations.
In 1996, Jou and his collaborators [148, 150] called the description by the second-order equation *mesoscopic* since it occupies an intermediate level between the descriptions by fluid dynamics and kinetic theory. In fact, in the fields of the non-relativistic fluid, Müller's second-order equation has been applied to various kinetic problems, e.g., in plasma and in photon transport, whose dynamics often cannot be described by the Navier–Stokes equation because the systems are not close to equilibrium state. Since the advent of Israel's second-order fluid dynamic equations, a number of new proposals or elaborations of the second-order equations have appeared [192–198], on the basis of the Müller-Israel method in principle.

The lesson that we can derive from the above description (a)$\sim$(c) may be that it is necessary to have recourse to a microscopic theory underlying the fluid dynamics beyond the phenomenological ones to get a hint on a unique choice of the frame or the flow velocity of the dissipative relativistic fluid, thereby derive the correct second-order as well as the first-order relativistic fluid dynamic equations.

In this chapter, after giving a brief account of some of basic properties of the relativistic Boltzmann equation with quantum statistics, we shall describe in detail two standard reduction methods for deriving fluid dynamic equations from the relativistic Boltzmann equation in a way where the mathematical requirements and physical but possibly ad-hoc assumptions become as clear as possible. The first is the Chapman-Enskog method based on the perturbation theory; the resultant fluid dynamic equation remains, however, of parabola type, and accordingly is not free from causality problem. Then we proceed to an account of the Israel-Stewart fourteen-moment method based on the Grad's moment method [144], which leads to the fluid dynamic equation of a hyperbolic nature.

Although we shall not discuss the (in)stability properties of the resultant equations, one will recognize that both the methods make specific ansatz on the distribution functions to choose but not derive the rest frames of the fluids.

10.2 Basics of Relativistic Boltzmann Equation with Quantum Statistics

For obtaining the proper relativistic fluid dynamic equation, it is legitimate and natural to employ the relativistic Boltzmann equation which is Lorentz invariant and expected to be free from causality problem [67, 68]; moreover, numerical simulations show that the relativistic Boltzmann equation is free from an apparent instability, as far as we are aware of, and the stability is proved at least for the linearized version of it [199, 200].

In this section, we summarize basic properties of the relativistic Boltzmann equation with quantum statistics [67, 68].

The relativistic Boltzmann equation reads

$$p^{\mu}\partial_{\mu} f_p(x) = C[f]_p(x), \tag{10.17}$$

where $f_p(x)$ is the one-particle distribution function with p^{μ} being the four-momentum of the on-shell particle with mass m, *i.e.*,

$$p^{\mu} p_{\mu} = m^2 \quad \text{and} \quad p^0 = \sqrt{m^2 + \boldsymbol{p}^2} > 0. \tag{10.18}$$

The collision integral $C[f]_p(x)$ is given by

$$\begin{aligned} C[f]_p(x) := \frac{1}{2!} \int \mathrm{d}p_1 \int \mathrm{d}p_2 \int \mathrm{d}p_3 \, \omega(p,\, p_1|p_2,\, p_3) \\ \times \Big((1 + a\, f_p(x))\, (1 + a\, f_{p_1}(x))\, f_{p_2}(x)\, f_{p_3}(x) \\ - f_p(x)\, f_{p_1}(x)\, (1 + a\, f_{p_2}(x))\, (1 + a\, f_{p_3}(x)) \Big), \end{aligned} \tag{10.19}$$

where a takes the values $a = +1,\ -1$ and 0 for a boson, fermion, and Boltzmann gas, respectively. Here, $\omega(p,\, p_1|p_2,\, p_3)$ is the transition probability due to the microscopic two-particle interaction, which has the symmetry property

$$\omega(p,\, p_1|p_2,\, p_3) = \omega(p_2,\, p_3|p,\, p_1) = \omega(p_1,\, p|p_3,\, p_2) = \omega(p_3,\, p_2|p_1,\, p), \tag{10.20}$$

due to the interchangeability of particles and the time-reversal invariance. We note that it includes the constraint by the energy-momentum conservation law

$$\omega(p,\ p_1|p_2,\ p_3) \propto \delta^4(p + p_1 - p_2 - p_3). \tag{10.21}$$

In Eq. (10.19), we have abbreviated an integration measure as

$$\mathrm{d}p := \frac{\mathrm{d}^3\boldsymbol{p}}{(2\pi)^3\, p^0}, \tag{10.22}$$

with $\boldsymbol{p}$ being the spatial components of the four momentum p^μ. In the following, we suppress the arguments x when no misunderstanding is expected.

For an arbitrary vector φ_p, the collision integral satisfies the following identity thanks to the above-mentioned symmetry property (10.20),

$$\begin{aligned}\int \mathrm{d}p\, \varphi_p\, C[f]_p = \frac{1}{2!}\frac{1}{4} &\int \mathrm{d}p \int \mathrm{d}p_1 \int \mathrm{d}p_2 \int \mathrm{d}p_3\, \omega(p,\ p_1|p_2,\ p_3) \\ &\times (\varphi_p + \varphi_{p_1} - \varphi_{p_2} - \varphi_{p_3}) \\ &\times \Big((1 + a\, f_p)\,(1 + a\, f_{p_1})\, f_{p_2}\, f_{p_3} \\ &- f_p\, f_{p_1}\,(1 + a\, f_{p_2})\,(1 + a\, f_{p_3}) \Big).\end{aligned} \tag{10.23}$$

Substituting $(1,\ p^\mu)$ into $\varphi_p(x)$ in Eq. (10.23), we find that $(1,\ p^\mu)$ are collision invariants

$$\int \mathrm{d}p\, C[f]_p = 0, \tag{10.24}$$

$$\int \mathrm{d}p\, p^\mu\, C[f]_p = 0, \tag{10.25}$$

due to the particle-number and energy-momentum conservation in the collision process, respectively. We note that the function

$$\varphi_{0p}(x) := \alpha(x) + p^\mu\, \beta_\mu(x), \tag{10.26}$$

is also a collision invariant where $\alpha(x)$ and $\beta_\mu(x)$ are arbitrary functions of x.

Because of the conservation laws as given by Eqs. (10.24) and (10.25), the relativistic Boltzmann equation (10.17) is nicely reduced to the balance equations

$$\partial_\mu N^\mu = 0, \quad \partial_\mu T^{\mu\nu} = 0, \tag{10.27}$$

where the particle current N^μ and the energy-momentum tensor $T^{\mu\nu}$ are defined by

$$N^\mu := \int \mathrm{d}p\, p^\mu\, f_p, \tag{10.28}$$

$$T^{\mu\nu} := \int \mathrm{d}p\, p^\mu\, p^\nu\, f_p, \tag{10.29}$$

respectively.

It should be noted, however, that any dynamical properties are not contained in these equations unless the evolution of f_p has been obtained as a solution to Eq. (10.17).

In the Boltzmann theory, the entropy current may be defined [67] by

$$S^\mu := -\int \mathrm{d}p\, p^\mu \left[f_p \ln f_p - \frac{(1 + a\, f_p) \ln(1 + a\, f_p)}{a} \right], \tag{10.30}$$

which satisfies

$$\partial_\mu S^\mu = -\int \mathrm{d}p\, C[f]_p \ln\left[\frac{f_p}{1 + a\, f_p} \right], \tag{10.31}$$

due to Eq. (10.17).

One sees that S^μ is conserved only if $\ln(f_p/(1 + a\, f_p))$ is a collision invariant, *i.e.*,

$$\ln(f_p/(1 + a\, f_p)) = \varphi_{0p} = \alpha(x) + p^\mu\, \beta_\mu(x). \tag{10.32}$$

Now it is of an essential importance [68] that the functions $\alpha(x)$ and $\beta^\mu(x)$ can be represented by the local temperature $T(x)$, chemical potential $\mu(x)$, and flow velocity $u^\mu(x)$ with the normalization condition (10.5), with the use of the Gibbs relation [55], $dS = \frac{1}{T}\,(dE\ -\ PdV)$, where S, E, and V denote the entropy, internal energy and the volume of the system at (local) thermal equilibrium. Thus it is found that the entropy-conserving distribution function can be expressed as [68]

$$f_p = \frac{1}{\mathrm{e}^{(p^\mu\, u_\mu(x) - \mu(x))/T(x)} - a} =: f_p^{\mathrm{eq}}, \tag{10.33}$$

which is identified with the local equilibrium distribution function.[1]

The collision integral identically vanishes for the local equilibrium distribution f_p^{eq} as

$$C[f^{\mathrm{eq}}]_p = 0, \tag{10.34}$$

owning to the detailed balance

[1] The quantum relativistic equilibrium distribution function given by (10.33) was first derived by Juüttner [201] in 1928. The five independent variables $T(x)$, $\mu(x)$, and $u^\mu(x)$ are called fluid dynamic variables.

We remark that an explicit derivation of (10.33) from the entropy-conserving distribution function for quantum statistics is given in Sect. 2.7 of [68] with the use of the Gibbs relation, whereas the classical statistics is discussed in [67] where the classical distribution function (Jüttner function [202]) is derived with the use of the Gibbs-Duhem relation [55].

$$\omega(p,\, p_1|p_2,\, p_3)\Big((1+a\, f_p^{\rm eq})\,(1+a\, f_{p_1}^{\rm eq})\, f_{p_2}^{\rm eq}\, f_{p_3}^{\rm eq} \\ -\, f_p^{\rm eq}\, f_{p_1}^{\rm eq}\,(1+a\, f_{p_2}^{\rm eq})\,(1+a\, f_{p_3}^{\rm eq})\Big) = 0, \tag{10.35}$$

which are guaranteed by the energy-momentum conservation (10.21).

Substituting $f_p = f_p^{\rm eq}$ into Eqs. (10.28) and (10.29), we have

$$N^\mu = n\, u^\mu = N_0^\mu, \tag{10.36}$$

$$T^{\mu\nu} = e\, u^\mu\, u^\nu - P\, \Delta^{\mu\nu} = T_0^{\mu\nu}. \tag{10.37}$$

Here, n, e, and P denote the particle-number density, internal energy, and pressure, respectively, whose microscopic representations are given by

$$n := \int {\rm d}p\, f_p^{\rm eq}\,(p\cdot u) \\ = (2\pi)^{-3}\, 4\pi\, m^3 \sum_{k=1}^{\infty} a^{k-1}\, {\rm e}^{k\mu/T}\, (km/T)^{-1}\, K_2(km/T), \tag{10.38}$$

$$e := \int {\rm d}p\, f_p^{\rm eq}\,(p\cdot u)^2 \\ = m\, n\left[\frac{\sum_{k=1}^{\infty} a^{k-1}\, {\rm e}^{k\mu/T}\, (km/T)^{-1}\, K_3(km/T)}{\sum_{l=1}^{\infty} a^{l-1}\, {\rm e}^{l\mu/T}\, (lm/T)^{-1}\, K_2(lm/T)} - \frac{\sum_{k=1}^{\infty} a^{k-1}\, {\rm e}^{k\mu/T}\, (km/T)^{-2}\, K_2(km/T)}{\sum_{l=1}^{\infty} a^{l-1}\, {\rm e}^{l\mu/T}\, (lm/T)^{-1}\, K_2(lm/T)}\right], \tag{10.39}$$

$$P := \int {\rm d}p\, f_p^{\rm eq}\,(-p^\mu\, p^\nu\, \Delta_{\mu\nu}/3) \\ = m\, n\, \frac{\sum_{k=1}^{\infty} a^{k-1}\, {\rm e}^{k\mu/T}\, (km/T)^{-2}\, K_2(km/T)}{\sum_{l=1}^{\infty} a^{l-1}\, {\rm e}^{l\mu/T}\, (lm/T)^{-1}\, K_2(lm/T)}. \tag{10.40}$$

Here, $K_2(z)$ and $K_3(z)$ are the second- and third-order modified Bessel functions, whose explicit form is given by

$$K_\ell(z) = \frac{2^\ell\, \ell!}{(2\,\ell)!}\, z^{-\ell} \int_z^\infty {\rm d}\tau\, (\tau^2 - z^2)^{\ell-1/2}\, {\rm e}^{-\tau}, \quad \ell = 2,\, 3. \tag{10.41}$$

Setting $a = 0$ in the above expressions, we can check that the classical expressions for n, e, and P [67] are reproduced. We note that N_0^μ and $T_0^{\mu\nu}$ in Eqs. (10.36) and (10.37) are identical to those in the relativistic Euler equation, which describes the fluid dynamics without dissipative effects, and n, e, and P defined by Eqs. (10.38)–(10.40) are the equations of state of the dilute gas. Since the entropy-conserving distribution function $f_p^{\rm eq}$ reproduces the relativistic Euler equation, we find that the dissipative effects are attributable to the deviation of f_p from $f_p^{\rm eq}$.

10.3 Review of Conventional Methods to Derive Relativistic Dissipative Fluid Dynamics from Relativistic Boltzmann Equation

In this section, we give a review of the Chapman-Enskog method [67, 68, 113] based on the perturbation theory and the Israel-Stewart fourteen-moment method [66–68, 113] based on the Grad' moment method [144]; they are popular methods to determine an explicit form of the deviation δf_p and derive the relativistic dissipative fluid dynamic equation.

This review is in principle based on the monograph by de Groot et al. [67]. We treat, however, the relativistic Boltzmann equation with *quantum statistics* as the underlying kinetic equation. Note that the relativistic Boltzmann equation with classical statistics is treated in Ref. [67], so please refer to that.

10.3.1 Chapman-Enskog Method

Before entering the Chapman-Enskog method [67, 68, 113] to be applied to the relativistic Boltzmann equation (10.17), we note that on account of the identity (see (10.4)),

$$g^{\mu\nu} = u^{\mu}u^{\nu} + \Delta^{\mu\nu}, \tag{10.42}$$

the Lorenz-invariant derivative in the left hand side of Eq. (10.17) is rewritten as

$$p^{\mu}\partial_{\mu} = g^{\mu\nu}p_{\nu}\partial_{\mu} = (u^{\mu}u^{\nu} + \Delta^{\mu\nu})p_{\nu}\partial_{\mu} = p \cdot u\frac{\partial}{\partial\tau} + p \cdot \nabla, \tag{10.43}$$

with $p \cdot u = p_{\nu}u^{\nu}$ and

$$\frac{\partial}{\partial\tau} := u^{\mu}\,\partial_{\mu}, \quad \nabla^{\mu} := \Delta^{\mu\nu}\,\partial_{\mu}, \tag{10.44}$$

are the covariant temporal and spatial derivatives, respectively.

10.3.1.1 Preliminaries

Now the Chapman-Enskog method starts from the physical assumption that the deformation of the distribution function from that for the local equilibrium state is solely caused by the small *inhomogeneity of the system*, and one rewrites the relativistic Boltzmann equation (10.17) into the following form with the use of (10.43),

$$p \cdot u \frac{\partial}{\partial \tau} f_p + \epsilon\, p \cdot \nabla f_p = C[f]_p. \tag{10.45}$$

Here one should notice that the parameter ϵ is attached to the partial derivative ∇^μ in Eq. (10.45) to make it explicit that the spatial inhomogeneity is small in accordance with the basic assumption of the method mentioned above. This modification can be formally achieved by introducing a scaled spatial coordinate

$$\bar{\boldsymbol{r}} = \epsilon\, \boldsymbol{r} \to \boldsymbol{r}, \tag{10.46}$$

as was done in the non-relativistic case; see Eq. (8.41).

To extract the slow dynamics from the Boltzmann equation (10.45) in the Chapman-Enskog method, one adopts the multiple-scale method [20–22] among traditional reduction methods, which are described in Chap. 3 in this monograph; see Sect. 3.6 for a self-contained account of the multiple-scale method.

According to this method, which is based on an observation of the possible existence of a hierarchy of various time scales in the dynamics, one introduces new temporal variables as

$$\tau^{(1)} := \epsilon\, \tau, \quad \tau^{(2)} := \epsilon^2\, \tau, \quad \ldots, \tag{10.47}$$

with an ad hoc assumption that they are independent variables. In accordance with this assumption, f_p is treated as a function of $\tau^{(i)}$ with $i = 1, 2, \ldots$:

$$f_p = f_p(\tau) = f_p(\tau^{(1)}, \tau^{(2)}, \ldots). \tag{10.48}$$

We note that $\tau^{(i+1)}$ represents slower scale than $\tau^{(i)}$ for $i = 1, 2, \ldots$. Then with use of the chain rule,

$$\frac{\partial}{\partial \tau} = \sum_{i=1}^{\infty} \frac{\partial \tau^{(i)}}{\partial \tau} \frac{\partial}{\partial \tau^{(i)}} = \sum_{i=1}^{\infty} \epsilon^i \frac{\partial}{\partial \tau^{(i)}}, \tag{10.49}$$

Equation (10.45) can be cast into the following form,

$$p \cdot u \left[\epsilon \frac{\partial}{\partial \tau^{(1)}} + \epsilon^2 \frac{\partial}{\partial \tau^{(2)}} + \cdots \right] f_p + \epsilon\, p \cdot \nabla f_p = C[f]_p. \tag{10.50}$$

To solve Eq. (10.50), we apply the perturbation theory by expanding f_p as

$$f_p = f_p^{(0)} + \epsilon\, f_p^{(1)} + \epsilon^2\, f_p^{(2)} + \cdots := f_p^{(0)} + \delta f_p. \tag{10.51}$$

Substituting Eq. (10.51) into Eq. (10.50) and equating the coefficient of ϵ^i, one obtains a series of equations for $f_p^{(i)}$ $(i = 0, 1, 2, \ldots)$.

10.3.1.2 Zero-th Order Solution

The zeroth-order equation is found to take the form

$$C[f^{(0)}]_p = 0, \tag{10.52}$$

which implies, on account of Eq. (10.34), that the zeroth-order solution is given by the local equilibrium distribution function (10.33) as:

$$f_p^{(0)} = f_p^{\rm eq}. \tag{10.53}$$

Here T, μ, and u^μ that parametrize $f_p^{\rm eq}$ are functions of $\tau^{(1)}$, $\tau^{(2)}$, and so on;

$$T = T(\tau^{(1)},\ \tau^{(2)},\ \ldots), \quad \mu = \mu(\tau^{(1)},\ \tau^{(2)},\ \ldots), \quad u^\mu = u^\mu(\tau^{(1)},\ \tau^{(2)},\ \ldots). \tag{10.54}$$

We remark here that the Gibbs relation is utilized to make the quantum local equilibrium distribution function expressed in terms of the fluid dynamic variables including the flow velocity [68] as given by (10.33).

10.3.1.3 Higher Orders

Let us proceed to the higher orders by solving the perturbative equations order by order. In this section, we shall carry out the perturbative analysis up to the second order of ϵ.

A couple of remarks are in order here:

(1) Generic special solutions of the higher-order equations as inhomogeneous equations may also contain secular terms proportional to the zero-th order solution $f_p^{\rm eq}$ as well as $f_p^{\rm eq}$ itself in addition to independent functions. In the following analysis, we construct the perturbative solutions so that they do not contain such secular terms and the zero-th order solution and whence δf_p defined in Eq. (10.51) will gives a net deviation of f_p from $f_p^{\rm eq}$ and produces the dissipative effects only. One will see that *solvability condition* for inhomogeneous linear differential equations [80–82] plays an important role there; see Chap.3 for a detailed account of this issue.

(2) It will, however, turn out that the unique perturbative solutions can not be obtained solely by the solvability condition in contrast to the non-relativistic case. In the traditional methods of the reduction of the relativistic Boltzmann equation, it is customary to impose a condition to fix the frame of the flow velocity, which is called 'matching conditions' or 'conditions of fit'.

On account of Eq. (10.34), the collision integral is expanded as follows,

$$
\begin{aligned}
C[f]_p &= C[f^{\text{eq}}]_p + \int \mathrm{d}q \, \frac{\delta}{\delta f_q} C[f]_p \bigg|_{f=f^{\text{eq}}} \delta f_q + O(\delta f^2) \\
&= f_p^{\text{eq}} \, \bar{f}_p^{\text{eq}} \int \mathrm{d}q \, L_{pq} \, (f_q^{\text{eq}} \, \bar{f}_q^{\text{eq}})^{-1} \, \delta f_q + O(\delta f^2),
\end{aligned} \tag{10.55}
$$

where L_{pq} denotes the kernel of the linearized collision integral defined by

$$
\begin{aligned}
L_{pq} &:= (f_p^{\text{eq}} \, \bar{f}_p^{\text{eq}})^{-1} \, \frac{\delta}{\delta f_q} C[f]_p \bigg|_{f=f^{\text{eq}}} (f_q^{\text{eq}} \, \bar{f}_q^{\text{eq}}) \\
&= -\frac{1}{2!} \int \mathrm{d}p_1 \int \mathrm{d}p_2 \int \mathrm{d}p_3 \, \omega(p \, , \, p_1 | p_2 \, , \, p_3) \\
&\quad \times \frac{f_{p_1}^{\text{eq}} \, \bar{f}_{p_2}^{\text{eq}} \, \bar{f}_{p_3}^{\text{eq}}}{\bar{f}_p^{\text{eq}}} \, (\delta_{pq} + \delta_{p_1 q} - \delta_{p_2 q} - \delta_{p_3 q}),
\end{aligned} \tag{10.56}
$$

and $\bar{f}_p^{\text{eq}}$ is given by

$$
\bar{f}_p^{\text{eq}} := 1 + a \, f_p^{\text{eq}} = \frac{\mathrm{e}^{(p \cdot u - \mu)/T}}{\mathrm{e}^{(p \cdot u - \mu)/T} - a}. \tag{10.57}
$$

In Eq. (10.56), we have introduced an abbreviated delta function as

$$
\delta_{pq} := (2\,\pi)^3 \, p^0 \, \delta^3(\boldsymbol{p} - \boldsymbol{q}), \tag{10.58}
$$

which has the following nice property when integrated with an arbitrary function F_p as

$$
\int \mathrm{d}q \, \delta_{pq} \, F_q = F_p, \tag{10.59}
$$

with the integration measure given by (10.22). In the following, we shall call the kernel L_{pq} the linearized collision operator when any misunderstanding is not expected.

10.3.1.4 First-Order Solution: Spectral Properties of the Linearized Collision Operator

Then we see that the first-order equation takes the form

$$
\int \mathrm{d}q \, L_{pq} \, (f_q^{\text{eq}} \, \bar{f}_q^{\text{eq}})^{-1} \, f_q^{(1)} = (f_p^{\text{eq}} \, \bar{f}_p^{\text{eq}})^{-1} \left[p \cdot u \, \frac{\partial}{\partial \tau^{(1)}} + p \cdot \nabla \right] f_p^{\text{eq}}. \tag{10.60}
$$

As in the non-relativistic case in Chap. 8, the spectral properties of the linearized collision operator L_{pq} has a basic importance for the subsequent analyses.

First of all, owing to the conservation laws of the energy-momentum and the particle number, L_{pq} has the zero eigenvalue, and the eigenvectors belonging to the zero eigenvalue are nothing but the collision invariants $(1,\ p^{\mu})$ $(\mu = 0,\ 1,\ 2,\ 3)$. In fact, by differentiating Eq. (10.34) with respect to the five independent variables μ/T and u^{μ}/T, we can show that the following equalities hold, respectively,

$$\int \mathrm{d}q\, L_{pq} \cdot 1 = 0 \quad \text{and} \quad \int \mathrm{d}q\, L_{pq}\, q^{\mu} = 0, \tag{10.61}$$

with $\mu = 0,\ 1,\ 2,\ 3$.

It is to be noted that L_{pq} is an asymmetric matrix and hence the above $(1,\ p^{\mu})$ are right eigenvectors of L_{pq}; see Sect. 3.6.2 for asymmetric matrices and left and right eigenvectors. For later convenience, we shall here give the left eigenvectors belonging to the zero eigenvalue of L_{pq}. By differentiating Eqs. (10.24) and (10.25) with respect to f_p and then setting $f_p = f_p^{\mathrm{eq}}$, we have

$$\int \mathrm{d}p\, 1 \cdot f_p^{\mathrm{eq}}\, \bar{f}_p^{\mathrm{eq}}\, L_{pq} = 0 \quad \text{and} \quad \int \mathrm{d}p\, p^{\mu}\, f_p^{\mathrm{eq}}\, \bar{f}_p^{\mathrm{eq}}\, L_{pq} = 0, \tag{10.62}$$

with $\mu = 0,\ 1,\ 2,\ 3$, which tells us that

$$f_p^{\mathrm{eq}}\, \bar{f}_p^{\mathrm{eq}} \quad \text{and} \quad f_p^{\mathrm{eq}}\, \bar{f}_p^{\mathrm{eq}}\, p^{\mu}, \tag{10.63}$$

are the left eigenvectors belonging to the zero eigenvalue of L_{pq}.

We note that Eqs. (10.61) and (10.62) can be derived by the explicit calculation based on the definition of L_{pq} given in Eq. (10.56).

With these spectral properties of L_{pq} taken for granted, it is still necessary to impose the *solvability condition* and the *conditions of fit* to obtain a solution to the linear inhomogeneous equation (10.60) in an unambiguous way, as noted before.

10.3.1.5 Solvability Condition for the First-Order Equation

First let us recall the notion of the solvability condition of inhomogeneous linear equations, an account of which is given in Chap. 3, in particular in Sect. 3.6.2. If we apply this notion to the linear inhomogeneous equation (10.60), we find that the equation is solvable only if the right hand side of Eq. (10.60) dose not contain any collision invariants, which requires the following two equations hold,

$$\int \mathrm{d}p\, 1 \cdot \left[p \cdot u\, \frac{\partial}{\partial \tau^{(1)}} + p \cdot \nabla \right] f_p^{\mathrm{eq}} = 0, \quad \int \mathrm{d}p\, p^{\mu} \left[p \cdot u\, \frac{\partial}{\partial \tau^{(1)}} + p \cdot \nabla \right] f_p^{\mathrm{eq}} = 0. \tag{10.64}$$

In fact, if $f_p^{\mathrm{eq}}\, \bar{f}_p^{\mathrm{eq}}$ and $f_p^{\mathrm{eq}}\, \bar{f}_p^{\mathrm{eq}}\, p^{\mu}$ are multiplied by Eq. (10.60) and then the integration of the resultant expressions is carried out with respect to p, we see that both the left- and right-hand sides of (10.60) vanish consistently, owing to Eqs. (10.62) and (10.64), respectively.

After some computation presented in a supplemental Chap. 13, Eq. (10.64) is reduced into three equations for which the physical significance is apparent, as follows:

$$\frac{\partial}{\partial \tau^{(1)}} T = -T \left.\frac{\partial P}{\partial e}\right|_n \nabla \cdot u, \tag{10.65}$$

$$\frac{\partial}{\partial \tau^{(1)}} \frac{\mu}{T} = -\frac{1}{T} \left.\frac{\partial P}{\partial n}\right|_e \nabla \cdot u, \tag{10.66}$$

$$\frac{\partial}{\partial \tau^{(1)}} u^\mu = \frac{1}{T} \nabla^\mu T + \frac{T\, n}{e + P} \nabla^\mu \frac{\mu}{T}. \tag{10.67}$$

On account of Eqs. (10.65)–(10.67), we see that the right-hand side of Eq. (10.60) now takes the following form

$$\begin{aligned} &(f_p^{\text{eq}} \bar{f}_p^{\text{eq}})^{-1} \left[p \cdot u \frac{\partial}{\partial \tau^{(1)}} + p \cdot \nabla \right] f_p^{\text{eq}} \\ &= -\Pi_p \frac{-\nabla \cdot u}{T} + J_p^\mu \frac{n}{e + P} \nabla_\mu \frac{\mu}{T} - \pi_p^{\mu\nu} \frac{\sigma_{\mu\nu}}{T}, \end{aligned} \tag{10.68}$$

where

$$\Pi_p := (p \cdot u)^2 \left[\frac{1}{3} - \left.\frac{\partial P}{\partial e}\right|_n \right] - (p \cdot u) \left.\frac{\partial P}{\partial n}\right|_e - \frac{1}{3} m^2, \tag{10.69}$$

$$J_p^\mu := -\Delta^{\mu\nu}\, p_\nu \left[(p \cdot u) - \frac{e + P}{n} \right], \tag{10.70}$$

$$\pi_p^{\mu\nu} := \Delta^{\mu\nu\rho\sigma}\, p_\rho\, p_\sigma, \tag{10.71}$$

with

$$\Delta^{\mu\nu\rho\sigma} := \frac{1}{2} \left[\Delta^{\mu\rho} \Delta^{\nu\sigma} + \Delta^{\mu\sigma} \Delta^{\nu\rho} - \frac{2}{3} \Delta^{\mu\nu} \Delta^{\rho\sigma} \right], \tag{10.72}$$

$$\sigma_{\mu\nu} := \Delta_{\mu\nu\rho\sigma} \nabla^\rho u^\sigma. \tag{10.73}$$

We note that the inhomogeneous term (10.68) contains no collision invariants thanks to the solvability condition, and that Eqs. (10.65)–(10.67) are the same as the relativistic Euler equation if we finish the perturbative analysis up to the first-order and set $\partial/\partial\tau = \partial/\partial\tau^{(1)}$. Thus we sees that the solvability condition play the essential role in giving the dynamic equations governing the slow motion of T, μ, and u^μ which were originally introduced merely to parametrize the zeroth-order solution f_p^{eq}. We note that it has been already shown in Chap. 3 that the solvability condition play a decisive role in various reduction theories [20, 30, 34, 83], not specific to the Chapman-Enskog method based on the multiple-scale analysis utilizing the perturbation theory.

10.3.1.6 Conditions of Fit for the First-Order Solution

Now that the right-hand side of Eq. (10.60) has been made free from the zero modes, we can apply the inverse operator of L_{pq} to the r.h.s. to obtain a particular solution without giving rise to a singularity. However, the solution to the inhomogeneous equation (10.60) still suffers from an ambiguity owing to the existence of the zero modes of the linear operator L_{pq}, which are nothing but the five collision invariants. In fact, the general solution to (10.60) is given by a sum of the special solution and a linear combination of the zero modes with five arbitrary coefficients, as

$$f_p^{(1)} = f_p^{\text{eq}}\,\bar{f}_p^{\text{eq}}\left\{\int \mathrm{d}q\, L_{pq}^{-1}\left[-\Pi_p\,\frac{-\nabla\cdot u}{T} + J_p^{\mu}\,\frac{n}{e+P}\,\nabla_{\mu}\frac{\mu}{T} - \pi_p^{\mu\nu}\,\frac{\sigma_{\mu\nu}}{T}\right]\right.$$
$$\left. + C(x) + C_{\mu}(x)\,p^{\mu}\right\}, \tag{10.74}$$

where $C(x)$ and $C^{\mu}(x)$ are arbitrary but dependent on x, say, through T, μ, and u^{μ}, for instance. Here L_{pq}^{-1} denotes the formal inverse matrix of the linearized collision integral L_{pq}. One can check that Eq. (10.74) solves Eq. (10.60) by substituting Eq. (10.74) into Eq. (10.60) and then using Eq. (10.61).

Thus we have ended up with a solution with arbitrary functions $C(x)$ and $C_{\mu}(x)$, and it is clear that it would be desirable if one has some appropriate constraints to determine the arbitrary functions $C(x)$ and $C_{\mu}(x)$ in a unique way. This is the place where the *matching conditions* or the *conditions of fit* play some role.

An account of the conditions of fit goes as follows [67]. First, one requires that the particle-number density and internal energy in the non-equilibrium state is the same as those in the local equilibrium state, and accordingly set

$$n = \int \mathrm{d}p\,(u\cdot p)\,f_p = \int \mathrm{d}p\,(u\cdot p)\,f_p^{\text{eq}}, \tag{10.75}$$

$$e = \int \mathrm{d}p\,(u\cdot p)^2\,f_p = \int \mathrm{d}p\,(u\cdot p)^2\,f_p^{\text{eq}}. \tag{10.76}$$

For consistency, one also imposes the constraints to the higher-order terms

$$\int \mathrm{d}p\,(u\cdot p)\,\delta f_p = 0, \tag{10.77}$$

$$\int \mathrm{d}p\,(u\cdot p)^2\,\delta f_p = 0. \tag{10.78}$$

We note that the requirement (10.75) and (10.76) exactly correspond to the ansatz (10.12) and (10.11), respectively, mentioned in Sect. 10.1.

Still we need three constraints. The remaining three conditions can be obtained by a choice of the rest frame of the flow velocity $u^\mu(x)$ with three degrees of freedom. Popular choices of them are the energy (Landau-Lifshitz) and the particle (Eckart) frames.

[Energy frame]: The fluid dynamic equation in the energy frame [62] is obtained by imposing the condition

$$\int \mathrm{d}p\,(\Delta_{\mu\nu}\,p^\nu)\,(u\cdot p)\,\delta f_p = 0, \tag{10.79}$$

which correspond to the ansatz (10.14) for the heat current mentioned in Sect. 10.1.

[Particle frame]: On the other hand, the fluid dynamic equation in the particle frame [61] is obtained by

$$\int \mathrm{d}p\,(\Delta_{\mu\nu}\,p^\nu)\,\delta f_p = 0, \tag{10.80}$$

which correspond to the ansatz (10.13) for the dissipative particle current mentioned in Sect. 10.1.

These conditions imposed to the distribution function in the higher orders are called the *conditions of fit*.

In the case of $\delta f_p = f_p^{(1)}$, a straightforward calculation gives for **the energy frame**,

$$C(x) = C^\mu(x) = 0, \tag{10.81}$$

while for **the particle frame**,

$$C(x) = 0, \tag{10.82}$$

$$C^\mu(x) = \lambda^{\mathrm{CE}}\,\frac{n\,T}{(e+P)^2}\,\nabla^\mu\frac{\mu}{T}. \tag{10.83}$$

Here λ^{CE} denotes the heat conductivity, that is, one of the transport coefficients, whose microscopic representation will be given later by Eq. (10.95).

10.3.1.7 Second-Order Solution

The second-order equation reads

$$\int \mathrm{d}q\, L_{pq}\, (f_q^{\mathrm{eq}}\, \bar{f}_q^{\mathrm{eq}})^{-1}\, f_q^{(2)} = (f_p^{\mathrm{eq}}\, \bar{f}_p^{\mathrm{eq}})^{-1} \left[p \cdot u\, \frac{\partial}{\partial \tau^{(1)}} + p \cdot \nabla \right] f_p^{(1)} + (f_p^{\mathrm{eq}}\, \bar{f}_p^{\mathrm{eq}})^{-1}\, p \cdot u\, \frac{\partial}{\partial \tau^{(2)}}\, f_p^{\mathrm{eq}}, \tag{10.84}$$

where the second-derivative term of the collision integral $C[f]_p$ is suppressed because it is irrelevant in this order of approximation.

Equation (10.84) is an inhomogeneous equation as was the case with the first-order equation. Thus the solvability condition to be imposed on Eq. (10.84) is given by

$$\int \mathrm{d}p\, (1,\, p^\mu) \left\{ \left[p \cdot u\, \frac{\partial}{\partial \tau^{(1)}} + p \cdot \nabla \right] f_p^{(1)} + p \cdot u\, \frac{\partial}{\partial \tau^{(2)}}\, f_p^{\mathrm{eq}} \right\} = 0, \tag{10.85}$$

which actually leads to the equation governing the slow dynamics of T, μ, and u^μ.

Summing up the first-order and second-order solvability conditions (10.64) and (10.85). we have

$$\begin{aligned} 0 &= \int \mathrm{d}p\, (1,\, p^\mu) \left\{ p \cdot u \left[\epsilon\, \frac{\partial}{\partial \tau^{(1)}} + \epsilon^2\, \frac{\partial}{\partial \tau^{(2)}} \right] + \epsilon\, p \cdot \nabla \right\} (f_p^{\mathrm{eq}} + \epsilon\, f_p^{(1)}) \\ &= \int \mathrm{d}p\, (1,\, p^\mu) \left[p \cdot u\, \frac{\partial}{\partial \tau} + \epsilon\, p \cdot \nabla \right] (f_p^{\mathrm{eq}} + \epsilon\, f_p^{(1)}), \end{aligned} \tag{10.86}$$

up to the second order of ϵ. Here we have used the identity,

$$\partial/\partial \tau = \epsilon\, \partial/\partial \tau^{(1)} + \epsilon^2\, \partial/\partial \tau^{(2)} + O(\epsilon^3). \tag{10.87}$$

Equation (10.86) constitutes one of the evolution equations for $T(x)$, $\mu(x)$, and $u^\mu(x)$.

Equation (10.86) can be cast into more familiar forms written in terms of the divergence of the currents. In fact, omitting $O(\epsilon^3)$, putting back $\epsilon = 1$, and using the equality $p \cdot u\, \partial/\partial \tau + p \cdot \nabla = p^\mu \partial_\mu$, we obtain

$$\partial_\nu \left[\int \mathrm{d}p\, p^\nu\, (f_p^{\mathrm{eq}} + f_p^{(1)}) \right] = 0 \quad \text{and} \quad \partial_\nu \left[\int \mathrm{d}p\, p^\mu\, p^\nu\, (f_p^{\mathrm{eq}} + f_p^{(1)}) \right] = 0, \tag{10.88}$$

which is equivalent to the set of equations

$$\partial_\mu N^\mu = 0 \quad \text{and} \quad \partial_\mu T^{\mu\nu} = 0, \tag{10.89}$$

where

$$N^\mu = \int \mathrm{d}p \, p^\mu \, (f_p^{\mathrm{eq}} + f_p^{(1)}) =: N_0^\mu + \delta N^\mu, \tag{10.90}$$

$$T^{\mu\nu} = \int \mathrm{d}p \, p^\mu \, p^\nu \, (f_p^{\mathrm{eq}} + f_p^{(1)}) =: T_0^{\mu\nu} + \delta T^{\mu\nu}. \tag{10.91}$$

Here δN^μ and $\delta T^{\mu\nu}$ denote the dissipative components of the particle-number current and energy-momentum tensor, respectively, while N_0^μ and $T_0^{\mu\nu}$ the non-dissipative (ideal) components whose definitions are given by Eqs. (10.36) and (10.37).

After the straightforward manipulations, we find that

$$\begin{aligned} \delta N^\mu &= \lambda^{\mathrm{CE}} \left(\frac{n\,T}{e+P} \right)^2 \nabla^\mu \frac{\mu}{T} \\ &\quad + C\,T \, \frac{\partial n}{\partial \mu} \, u^\mu + C_\nu \, T \left[\frac{\partial e}{\partial \mu} \, u^\mu \, u^\nu - n \, \Delta^{\mu\nu} \right], \end{aligned} \tag{10.92}$$

$$\begin{aligned} \delta T^{\mu\nu} &= \zeta^{\mathrm{CE}} \, \Delta^{\mu\nu} \, \nabla \cdot u + 2\,\eta^{\mathrm{CE}} \, \sigma^{\mu\nu} \\ &\quad + C\,T \left[\frac{\partial e}{\partial \mu} \, u^\mu \, u^\nu - n \, \Delta^{\mu\nu} \right] + C_\rho \, T \left\{ \left[T \, \frac{\partial e}{\partial T} + \mu \, \frac{\partial e}{\partial \mu} \right] u^\mu \, u^\nu \, u^\rho \right. \\ &\quad \left. - (e+P) \, (u^\mu \, \Delta^{\nu\rho} + u^\nu \, \Delta^{\rho\mu} + u^\rho \, \Delta^{\mu\nu}) \right\}, \end{aligned} \tag{10.93}$$

where ζ^{CE} and η^{CE} are kinds of transport coefficients as the thermal conductivity λ^{CE}, and called the bulk and shear viscosity, respectively: Their microscopic expressions are given by

$$\zeta^{\mathrm{CE}} = -\frac{1}{T} \int \mathrm{d}p \int \mathrm{d}q \; f_p^{\mathrm{eq}} \, \bar{f}_p^{\mathrm{eq}} \, \Pi_p \, L_{pq}^{-1} \, \Pi_q, \tag{10.94}$$

$$\lambda^{\mathrm{CE}} = \frac{1}{3\,T^2} \int \mathrm{d}p \int \mathrm{d}q \; f_p^{\mathrm{eq}} \, \bar{f}_p^{\mathrm{eq}} \, J_p^\mu \, L_{pq}^{-1} \, J_{q\mu}, \tag{10.95}$$

$$\eta^{\mathrm{CE}} = -\frac{1}{10\,T} \int \mathrm{d}p \int \mathrm{d}q \; f_p^{\mathrm{eq}} \, \bar{f}_p^{\mathrm{eq}} \, \pi_p^{\mu\nu} \, L_{pq}^{-1} \, \pi_{q\mu\nu}, \tag{10.96}$$

respectively.

Finally, we shall present the explicit forms of δN^μ and $\delta T^{\mu\nu}$ in the energy frame and the particle frame.

Substituting C and C^μ in Eqs. (10.81)–(10.83) into Eqs. (10.92) and (10.93), we find that those in the **energy frame** read

$$\delta N^{\mu} = \lambda^{\text{CE}} \left(\frac{n\,T}{e+P} \right)^2 \nabla^{\mu} \frac{\mu}{T}, \tag{10.97}$$

$$\delta T^{\mu\nu} = \zeta^{\text{CE}}\, \Delta^{\mu\nu}\, \nabla \cdot u + 2\,\eta^{\text{CE}}\, \sigma^{\mu\nu}, \tag{10.98}$$

which satisfy the ansatz (10.11), (10.12), and (10.14).

On the other hand, those in the **particle frame** read

$$\delta N^{\mu} = 0, \tag{10.99}$$

$$\delta T^{\mu\nu} = \zeta^{\text{CE}}\, \Delta^{\mu\nu}\, \nabla \cdot u + 2\,\eta^{\text{CE}}\, \sigma^{\mu\nu} - \lambda^{\text{CE}} \frac{n\,T^2}{e+P} \left[u^{\mu}\, \nabla^{\nu} \frac{\mu}{T} + u^{\nu}\, \nabla^{\mu} \frac{\mu}{T} \right], \tag{10.100}$$

which satisfy the ansatz (10.11)–(10.13) that characterize the particle frame.

We note that $\delta T^{\mu\nu}$ given by (10.100) is often rewritten in terms of a time derivative by using Eq. (10.67) as

$$\begin{aligned} \delta T^{\mu\nu} &= \zeta^{\text{CE}}\, \Delta^{\mu\nu}\, \nabla \cdot u + 2\,\eta^{\text{CE}}\, \sigma^{\mu\nu} \\ &\quad + T\,\lambda^{\text{CE}} \left\{ u^{\mu} \left[\frac{1}{T}\, \nabla^{\nu} T - \frac{\partial}{\partial \tau} u^{\nu} \right] + u^{\nu} \left[\frac{1}{T}\, \nabla^{\mu} T - \frac{\partial}{\partial \tau} u^{\mu} \right] \right\}. \end{aligned} \tag{10.101}$$

10.3.2 Israel-Stewart Fourteen-Moment Method

In the Israel-Stewart fourteen-moment method for deriving the relativistic fluid dynamics [66–68, 113], one starts from the following ansatz for the distribution function

$$f_p = f_p^{\text{eq}}(x)\,(1 + \bar{f}_p^{\text{eq}}(x)\,\Phi_p), \tag{10.102}$$

with

$$\Phi_p = a(x) + b_{\mu}(x)\, p^{\mu} + c_{\mu\nu}(x) p^{\mu}\, p^{\nu}. \tag{10.103}$$

Here $\bar{f}_p^{\text{eq}}(x)$ is the local equilibrium distribution function that may depend on space-time x through the fluid variables $T(x)$, $\mu(x)$, and $u^{\mu}(x)$, and the deviation Φ_p is supposed to be so small that the second and higher order of it can be neglected. Note that the prefactor of $p^{\mu}\, p^{\nu}$ in Eq. (10.103) obeys the constraints

$$c^{\mu\nu} = c^{\nu\mu} \quad \text{and} \quad c^{\mu}{}_{\mu} = 0, \tag{10.104}$$

without loss of generality because $p^\mu\, p^\nu = p^\nu\, p^\mu$ and $p^\mu\, p_\mu = m^2$. The coefficient functions $a(x)$, $b^\mu(x)$, and $c^{\mu\nu}(x)$ in Eq. (10.103) are yet to be determined as well as the fluid variables contained in f_p^{eq}.

We note that if the *conditions of fit* (10.77)–(10.79) is imposed on the distribution function (10.102), then $T(x)$, $\mu(x)$, $u^\mu(x)$, $a(x)$, $b^\mu(x)$, and $c^{\mu\nu}(x)$ can not be independent, and thus, the total number of independent functions within T, μ, u^μ, a, b^μ, and $c^{\mu\nu}$ is fourteen.

10.3.2.1 Constraints and Moment Equations

To determine the fourteen functions, we utilize the fourteen-moment equations which will be derived by multiplying the relativistic Boltzmann equation (10.17) by appropriate fourteen quantities dependent on the momentum p, and integrating them with respect to p.

In the Israel-Stewart method, the five collision invariants $(1,\ p^\mu)$ and the second moments $p^\mu\, p^\nu$ are adopted as the fourteen quantities: In fact, the number of the independent components of

$$1,\ \ p^\mu,\ \ p^\mu\, p^\nu, \tag{10.105}$$

is fourteen because the number of independent components of $p^\mu\, p^\nu = p^\nu\, p^\mu$ is nine due to $p^\mu\, p_\mu = m^2$.

Thus the Israel-Stewart fourteen-moment equations consist of the five constraints imposed by the collision invariant properties Eqs. (10.24) and (10.25) for 1 and p^μ,

$$\int \mathrm{d}p\, p^\nu\, \partial_\nu\Big(f_p^{\mathrm{eq}}\,(1+\bar{f}_p^{\mathrm{eq}}\,\Phi_p)\Big)=0,\quad \int \mathrm{d}p\, p^\mu\, p^\nu\, \partial_\nu\Big(f_p^{\mathrm{eq}}\,(1+\bar{f}_p^{\mathrm{eq}}\,\Phi_p)\Big)=0, \tag{10.106}$$

and the nine equations given by the second moments

$$\int \mathrm{d}p\, p^\mu\, p^\nu\, p^\rho\, \partial_\rho\Big(f_p^{\mathrm{eq}}\,(1+\bar{f}_p^{\mathrm{eq}}\,\Phi_p)\Big)=\int \mathrm{d}p \int \mathrm{d}q\, p^\mu\, p^\nu\, f_p^{\mathrm{eq}}\,\bar{f}_p^{\mathrm{eq}}\,L_{pq}\,\Phi_q. \tag{10.107}$$

Here we have made the following expansion for the collision operator in the right-hand side, with the use of Eqs. (10.34) and (10.56),

$$\begin{aligned} C[f]_p &= C[f^{\mathrm{eq}}]_p + \int \mathrm{d}q\, \frac{\delta}{\delta f_q} C[f]_p\bigg|_{f=f^{\mathrm{eq}}} f_q^{\mathrm{eq}}\,\bar{f}_q^{\mathrm{eq}}\,\Phi_q + O(\Phi^2) \\ &= f_p^{\mathrm{eq}}\,\bar{f}_p^{\mathrm{eq}} \int \mathrm{d}q\, L_{pq}\,\Phi_q + O(\Phi^2), \end{aligned} \tag{10.108}$$

where we have suppressed the terms of $O(\Phi^2)$ which are to be neglected as promised.

The five equations (10.106) are identical to the balance equations. Indeed they are written as

$$\partial_\mu N^\mu = 0 \quad \text{and} \quad \partial_\mu T^{\mu\nu} = 0, \tag{10.109}$$

where

$$N^\mu = n\, u^\mu + \delta N^\mu \quad \text{and} \quad T^{\mu\nu} = e\, u^\mu\, u^\nu - P\, \Delta^{\mu\nu} + \delta T^{\mu\nu}, \tag{10.110}$$

with

$$\delta N^\mu = \int \mathrm{d}p\, p^\mu\, f_p^{\mathrm{eq}}\, \bar{f}_p^{\mathrm{eq}}\, \Phi_p, \tag{10.111}$$

$$\delta T^{\mu\nu} = \int \mathrm{d}p\, p^\mu\, p^\nu\, f_p^{\mathrm{eq}}\, \bar{f}_p^{\mathrm{eq}}\, \Phi_p. \tag{10.112}$$

The nine equations (10.107) are called the *relaxation equations*, which are equations characteristic in the moment method, not seen in the Chapman-Enskog method. Indeed, the relaxation equations (10.107) contain novel coefficients other than the transport coefficients, which are called the *relaxation times* and so on.

10.3.2.2 Calculational Set Up

We are in a position to compute the needed integrals and derive explicit forms of the balance and relaxation equations, from which one can read off the microscopic expressions of the transport coefficients and relaxation times.

Before entering the detailed calculation, we first note that Φ_p in Eq. (10.103) is at most bilinear of the momentum. Then, with use of

$$\pi_p^\mu := \Delta^{\mu\nu}\, p_\nu \quad \text{and} \quad \pi_p^{\mu\nu} = \Delta^{\mu\nu\rho\sigma}\, p_\rho\, p_\sigma, \tag{10.113}$$

we convert (10.103) into a form with which the physical significance of respective terms are obvious as

$$\Phi_p = A_p\, \Pi + B_p\, \pi_p^\mu\, J_\mu + C_p\, \pi_p^{\mu\nu}\, \pi_{\mu\nu}, \tag{10.114}$$

where

$$A_p = A_2\, (p \cdot u)^2 + A_1\, (p \cdot u) + A_0, \tag{10.115}$$

$$B_p = B_1\, (p \cdot u) + B_0, \tag{10.116}$$

$$C_p = C_0, \tag{10.117}$$

and Π, J_μ, and $\pi_{\mu\nu}$ are introduced in place of a, b_μ, and $c_{\mu\nu}$; note that the number of independent components of the set of variables is not altered by this replacement and remains 9. We remark that J^μ and $\pi^{\mu\nu}$ are of space-like and their covariant components are obtained as

$$J_\mu = \Delta_{\mu\nu}\, J^\nu, \quad \pi_{\mu\nu} = \Delta_{\mu\nu\rho\sigma}\, \pi^{\rho\sigma}, \tag{10.118}$$

respectively. We stress that the coefficients A_2, A_1, A_0, B_1, B_0, and C_0 are yet to be determined, say, by the conditions of fits together with some additional conditions.

Substituting Eq. (10.114) into Eqs. (10.111) and (10.112), δN^μ and $\delta T^{\mu\nu}$ are calculated to be

$$\begin{aligned}
\delta N^\mu &= \left[\int \mathrm{d}p\, f_p^{\mathrm{eq}}\, \bar{f}_p^{\mathrm{eq}}\, (p\cdot u)\, A_p\right] u^\mu\, \Pi + \left[\int \mathrm{d}p\, f_p^{\mathrm{eq}}\, \bar{f}_p^{\mathrm{eq}}\, \frac{1}{3}\, \pi_p^\nu\, \pi_{p\nu}\, B_p\right] J^\mu \\
&= \left[A_2\, a_3 + A_1\, a_2 + A_0\, a_1\right] u^\mu\, \Pi \\
&\quad + \left[B_1\, \frac{1}{3}\, (m^2\, a_1 - a_3) + B_0\, \frac{1}{3}\, (m^2\, a_0 - a_2)\right] J^\mu,
\end{aligned} \tag{10.119}$$

and

$$\begin{aligned}
\delta T^{\mu\nu} &= \left[\int \mathrm{d}p\, f_p^{\mathrm{eq}}\, \bar{f}_p^{\mathrm{eq}}\, (p\cdot u)^2\, A_p\right] u^\mu\, u^\nu\, \Pi + \left[\int \mathrm{d}p\, f_p^{\mathrm{eq}}\, \bar{f}_p^{\mathrm{eq}}\, \frac{1}{3}\, \pi_p^\rho\, \pi_{p\rho}\, A_p\right] \Delta^{\mu\nu}\, \Pi \\
&\quad + \left[\int \mathrm{d}p\, f_p^{\mathrm{eq}}\, \bar{f}_p^{\mathrm{eq}}\, \frac{1}{3}\, \pi_p^\rho\, \pi_{p\rho}\, (p\cdot u)\, B_p\right] (u^\mu\, J^\nu + u^\nu\, J^\mu) \\
&\quad + \left[\int \mathrm{d}p\, f_p^{\mathrm{eq}}\, \bar{f}_p^{\mathrm{eq}}\, \frac{1}{5}\, \pi_p^{\rho\sigma}\, \pi_{p\rho\sigma}\, C_p\right] \pi^{\mu\nu} \\
&= \left[A_2\, a_4 + A_1\, a_3 + A_0\, a_2\right] u^\mu\, u^\nu\, \Pi \\
&\quad + \left[A_2\, \frac{1}{3}\, (m^2\, a_2 - a_4) + A_1\, \frac{1}{3}\, (m^2\, a_1 - a_3) + A_0\, \frac{1}{3}\, (m^2\, a_0 - a_2)\right] \Delta^{\mu\nu}\, \Pi \\
&\quad + \left[B_1\, \frac{1}{3}\, (m^2\, a_2 - a_4) + B_0\, \frac{1}{3}\, (m^2\, a_1 - a_3)\right] (u^\mu\, J^\nu + u^\nu\, J^\mu) \\
&\quad + \left[C_0\, \frac{2}{15}\, (m^4\, a_0 - 2\, m^2\, a_2 + a_4)\right] \pi^{\mu\nu},
\end{aligned} \tag{10.120}$$

respectively. Here a_0, a_1, a_2, a_3, and a_4 are defined in Eq. (13.36), which reads $a_\ell := \int \mathrm{d}p\, f_p^{\mathrm{eq}}\, \bar{f}_p^{\mathrm{eq}}\, (p\cdot u)^\ell$, $(\ell = 0,\ 1,\ 2,\ \ldots)$; we note that some of a_ℓ are expressed in terms of the thermodynamic quantities $T,\ \mu,\ n,\ e,\ P$.

Now we have come to the place where one needs to impose the conditions of fit for a further reduction of the formulae.

10.3.2.3 Energy Frame

First we consider the energy frame. Substituting $\delta f_p = f_p^{\mathrm{eq}} \bar{f}_p^{\mathrm{eq}} \Phi_p$ into Eqs. (10.77), (10.78) and (10.79), we can confirm that these conditions of fit exactly correspond to the ansatz (10.12), (10.11), and (10.14), respectively, which we recapitulate for convenience

$$\delta n = u^\mu \, \delta N_\mu = 0, \tag{10.121}$$

$$\delta e = u^\mu \, u^\nu \, \delta T_{\mu\nu} = 0, \tag{10.122}$$

$$Q^\mu = \Delta^{\mu\nu} \, u^\rho \, \delta T_{\nu\rho} = 0. \tag{10.123}$$

Applying these conditions of fit to δN^μ and $\delta T^{\mu\nu}$ in Eqs. (10.119) and (10.120), we have

$$A_2 \, a_3 + A_1 \, a_2 + A_0 \, a_1 = 0, \tag{10.124}$$

$$A_2 \, a_4 + A_1 \, a_3 + A_0 \, a_2 = 0, \tag{10.125}$$

$$B_1 \, \frac{1}{3} \, (m^2 \, a_2 - a_4) + B_0 \, \frac{1}{3} \, (m^2 \, a_1 - a_3) = 0. \tag{10.126}$$

Under these constraints, δN^μ and $\delta T^{\mu\nu}$ take the following forms

$$\delta N^\mu = \left[B_1 \, \frac{1}{3} \, (m^2 \, a_1 - a_3) + B_0 \, \frac{1}{3} \, (m^2 \, a_0 - a_2) \right] J^\mu, \tag{10.127}$$

$$\begin{aligned} \delta T^{\mu\nu} = -\Delta^{\mu\nu} \Big[& - A_2 \, \frac{1}{3} \, (m^2 \, a_2 - a_4) - A_1 \, \frac{1}{3} \, (m^2 \, a_1 - a_3) \\ & - A_0 \, \frac{1}{3} \, (m^2 \, a_0 - a_2) \Big] \Pi + C_0 \, \frac{2}{15} \, (m^4 \, a_0 - 2 \, m^2 \, a_2 + a_4) \, \pi^{\mu\nu}, \end{aligned} \tag{10.128}$$

respectively. We now impose the conditions that the prefactors of Π, J^μ, and $\pi^{\mu\nu}$ are all set to 1, which implies another three conditions

$$B_1 \, \frac{1}{3} \, (m^2 \, a_1 - a_3) + B_0 \, \frac{1}{3} \, (m^2 \, a_0 - a_2) = 1, \tag{10.129}$$

$$-A_2 \, \frac{1}{3} \, (m^2 \, a_2 - a_4) - A_1 \, \frac{1}{3} \, (m^2 \, a_1 - a_3) - A_0 \, \frac{1}{3} \, (m^2 \, a_0 - a_2) = 1, \tag{10.130}$$

$$C_0 \, \frac{2}{15} \, (m^4 \, a_0 - 2 \, m^2 \, a_2 + a_4) = 1. \tag{10.131}$$

Solving the six equations (10.124)–(10.126) and (10.129)–(10.131) for A_2, A_1, A_0, B_1, B_0, and C_0, we have

$$A_2 = \frac{1}{-a_4 \frac{m^2 (a_1^2 - a_0 a_2)}{3 (a_2^2 - a_1 a_3)} - a_3 \frac{m^2 (a_0 a_3 - a_1 a_2)}{3 (a_2^2 - a_1 a_3)} - a_2 \frac{m^2}{3}}, \tag{10.132}$$

$$A_1 = -A_2 \frac{a_2 a_3 - a_1 a_4}{a_2^2 - a_1 a_3}, \tag{10.133}$$

$$A_0 = A_2 \frac{a_3^2 - a_2 a_4}{a_2^2 - a_1 a_3}, \tag{10.134}$$

$$B_1 = \frac{3 \frac{m^2 a_1 - a_3}{m^2 a_0 - a_2}}{-(m^2 a_2 - a_4) + \frac{m^2 a_1 - a_3}{m^2 a_0 - a_2} (m^2 a_1 - a_3)}, \tag{10.135}$$

$$B_0 = -B_1 \frac{m^2 a_2 - a_4}{m^2 a_1 - a_3}, \tag{10.136}$$

$$C_0 = \frac{5}{\frac{2}{3} (m^4 a_0 - 2 m^2 a_2 + a_4)}. \tag{10.137}$$

These quantities are complicated functions of T and μ, but A_2, B_1 and C_0 can be reduced into compact forms as follows:

$$A_2 = \frac{1}{\int \mathrm{d}p\, f_p^{\mathrm{eq}} \bar{f}_p^{\mathrm{eq}} (p \cdot u)^2 \Pi_p}, \tag{10.138}$$

$$B_1 = \frac{\frac{e+P}{n}}{\frac{1}{3} \int \mathrm{d}p\, f_p^{\mathrm{eq}} \bar{f}_p^{\mathrm{eq}} (p \cdot u) \pi_{p\mu} J_p^{\mu}}, \tag{10.139}$$

$$C_0 = \frac{1}{\frac{1}{5} \int \mathrm{d}p\, f_p^{\mathrm{eq}} \bar{f}_p^{\mathrm{eq}} \pi_{p\mu\nu} \pi_p^{\mu\nu}}, \tag{10.140}$$

where Eqs. (13.41)–(13.43) have been used together with the definitions (10.69)–(10.71), that is,

$$\Pi_p = \left[\frac{1}{3} - \left.\frac{\partial P}{\partial e}\right|_n\right] \left[(p \cdot u)^2 - (p \cdot u) \frac{a_0 a_3 - a_1 a_2}{a_1^2 - a_0 a_2} + \frac{a_2^2 - a_1 a_3}{a_1^2 - a_0 a_2}\right], \tag{10.141}$$

$$J_p^{\mu} = -\pi_p^{\mu} \left[(p \cdot u) - \frac{m^2 a_1 - a_3}{m^2 a_0 - a_2}\right]. \tag{10.142}$$

Substituting the above A_2, A_1, A_0, B_1, B_0, and C_0 into Φ_p, we have

$$\Phi_p = \frac{\mathcal{P}_p \Pi}{\int \mathrm{d}q\, f_q^{\mathrm{eq}} \bar{f}_q^{\mathrm{eq}} \mathcal{P}_q \Pi_q} + \frac{e+P}{n} \frac{\mathcal{J}_p^{\mu} J_{\mu}}{\frac{1}{3} \int \mathrm{d}q\, f_q^{\mathrm{eq}} \bar{f}_q^{\mathrm{eq}} \mathcal{J}_{q\nu} J_q^{\nu}} + \frac{\pi_p^{\mu\nu} \pi_{\mu\nu}}{\frac{1}{5} \int \mathrm{d}q\, f_q^{\mathrm{eq}} \bar{f}_q^{\mathrm{eq}} \pi_{q\rho\sigma} \pi_q^{\rho\sigma}}, \tag{10.143}$$

where $\mathcal{P}_p$ and $\mathcal{J}_p^{\mu}$ are defined by

$$\mathcal{P}_p := \left[\frac{1}{3} - \left.\frac{\partial P}{\partial e}\right|_n\right]\left[(p \cdot u)^2 - (p \cdot u)\,\frac{a_2\,a_3 - a_1\,a_4}{a_2^2 - a_1\,a_3} + \frac{a_3^2 - a_2\,a_4}{a_2^2 - a_1\,a_3}\right], \tag{10.144}$$

$$\mathcal{J}_p^\mu := -\pi_p^\mu\left[(p \cdot u) - \frac{m^2\,a_2 - a_4}{m^2\,a_1 - a_3}\right]. \tag{10.145}$$

Here we have used the equalities

$$\int\!\mathrm{d}p\, f_p^{\mathrm{eq}}\,\bar{f}_p^{\mathrm{eq}}\,(p \cdot u)^2\,\Pi_p = \frac{1}{\frac{1}{3} - \left.\frac{\partial P}{\partial e}\right|_n}\int\!\mathrm{d}p\, f_p^{\mathrm{eq}}\,\bar{f}_p^{\mathrm{eq}}\,\mathcal{P}_p\,\Pi_p, \tag{10.146}$$

$$\int\!\mathrm{d}p\, f_p^{\mathrm{eq}}\,\bar{f}_p^{\mathrm{eq}}\,(p \cdot u)\,\pi_{p\mu}\,J_p^\mu = -\int\!\mathrm{d}p\, f_p^{\mathrm{eq}}\,\bar{f}_p^{\mathrm{eq}}\,\mathcal{J}_{p\mu}\,J_p^\mu, \tag{10.147}$$

which are attributed to the identities

$$\int\!\mathrm{d}p\, f_p^{\mathrm{eq}}\,\bar{f}_p^{\mathrm{eq}}\,(1,\ (p \cdot u))\,\Pi_p = 0, \tag{10.148}$$

$$\int\!\mathrm{d}p\, f_p^{\mathrm{eq}}\,\bar{f}_p^{\mathrm{eq}}\,\pi_{p\mu}\,J_p^\mu = 0. \tag{10.149}$$

By comparing Eqs. (10.141) and (10.142) with Eqs. (10.144) and (10.145), respectively, we find that the coefficients of $(p \cdot u)^2$ ($\pi_p^\mu\,(p \cdot u)$) are common between $\mathcal{P}_p$ ($\mathcal{J}_p^\mu$) and Π_p (J_p^μ), but those of the others are not the case. It is noted that the differences between $\mathcal{P}_p$ and Π_p ($\mathcal{J}_p^\mu$ and J_p^μ) come only from the collision invariants $(1,\ p^\mu)$, and hence the following identities hold,

$$\int\!\mathrm{d}q\, L_{pq}\,(\mathcal{P}_q,\ \mathcal{J}_q^\mu) = \int\!\mathrm{d}q\, L_{pq}\,(\Pi_q,\ J_p^\mu), \tag{10.150}$$

$$\int\!\mathrm{d}p\, f_p^{\mathrm{eq}}\,\bar{f}_p^{\mathrm{eq}}\,(\mathcal{P}_p,\ \mathcal{J}_p^\mu)\,L_{pq} = \int\!\mathrm{d}p\, f_p^{\mathrm{eq}}\,\bar{f}_p^{\mathrm{eq}}\,(\Pi_p,\ J_p^\mu)\,L_{pq}. \tag{10.151}$$

Let us substitute Φ_p in Eq. (10.143) into Eqs. (10.106) and (10.107) to derive the explicit forms of the balance equations and relaxation equations together with the microscopic representations of the transport coefficients and relaxation times. The balance equations (10.106) are easily found to take the forms $\partial_\mu N^\mu = 0$ and $\partial_\mu T^{\mu\nu} = 0$ with

$$N^\mu = n\,u^\mu + J^\mu, \tag{10.152}$$

$$T^{\mu\nu} = e\,u^\mu\,u^\nu - (P + \Pi)\,\Delta^{\mu\nu} + \pi^{\mu\nu}. \tag{10.153}$$

We see that the resultant N^μ and $T^{\mu\nu}$ surely satisfy the conditions of fit (10.121), (10.122), and (10.123) for the energy frame because of the space-like nature of J^μ, $\Delta^{\mu\nu}$, and $\pi^{\mu\nu}$.

Owing to the identity

$$
\begin{aligned}
p^\mu\,p^\nu &= [(p\cdot u)\,u^\mu + \pi_p^\mu]\,[(p\cdot u)\,u^\nu + \pi_p^\nu] \\
&= (u^\mu\,u^\nu - \Delta^{\mu\nu}/3)\,(p\cdot u)^2 + (u^\mu\,\pi_p^\nu + u^\nu\,\pi_p^\mu)\,(p\cdot u) + \pi_p^{\mu\nu} + m^2\,\Delta^{\mu\nu}/3 \\
&= \left[u^\mu\,u^\nu - \frac{1}{3}\,\Delta^{\mu\nu}\right]\left[\frac{1}{\frac{1}{3} - \left.\frac{\partial P}{\partial e}\right|_n}\mathcal{P}_p + (p\cdot u)\,\frac{a_2\,a_3 - a_1\,a_4}{a_2^2 - a_1\,a_3} - \frac{a_3^2 - a_2\,a_4}{a_2^2 - a_1\,a_3}\right] \\
&\quad - u^\mu\left[\mathcal{J}_p^\nu - \frac{m^2\,a_2 - a_4}{m^2\,a_1 - a_3}\,\pi_p^\nu\right] - u^\nu\left[\mathcal{J}_p^\mu - \frac{m^2\,a_2 - a_4}{m^2\,a_1 - a_3}\,\pi_p^\mu\right] \\
&\quad + \pi_p^{\mu\nu} + \frac{m^2}{3}\,\Delta^{\mu\nu}, \qquad (10.154)
\end{aligned}
$$

derived from

$$
\pi_p^{\mu\nu} = \pi_p^\mu\,\pi_p^\nu - \Delta^{\mu\nu}\,\pi_p^\rho\,\pi_{p\rho}/3 = \pi_p^\mu\,\pi_p^\nu - \Delta^{\mu\nu}\,[m^2 - (p\cdot u)^2]/3 \qquad (10.155)
$$

and the definitions of $\mathcal{P}_p$ and $\mathcal{J}_p$ in Eqs. (10.144) and (10.145), the relaxation equations (10.107) can be reduced to the the following three equations with different ranks,

$$
\begin{aligned}
&\int\mathrm{d}p\,(\mathcal{P}_p,\,\mathcal{J}_p^\mu,\,\pi_p^{\mu\nu})\left[(p\cdot u)\,\frac{\partial}{\partial\tau} + p\cdot\nabla\right](f_p^{\rm eq} + f_p^{\rm eq}\,\bar{f}_p^{\rm eq}\,\Phi_p) \\
&= \int\mathrm{d}p\int\mathrm{d}q\,(\mathcal{P}_p,\,\mathcal{J}_p^\mu,\,\pi_p^{\mu\nu})\,f_p^{\rm eq}\,\bar{f}_p^{\rm eq}\,L_{pq}\,\Phi_q. \qquad (10.156)
\end{aligned}
$$

Here we have used the fact that the components proportional to the collision invariants 1 and p^μ vanish owing to the balance equations (10.106) and Eq. (10.62).

The right-hand side of the relaxation equations (10.156) is reduced into

$$
\int\mathrm{d}p\int\mathrm{d}q\,\mathcal{P}_p\,f_p^{\rm eq}\,\bar{f}_p^{\rm eq}\,L_{pq}\,\Phi_q = \left[\frac{\int\mathrm{d}p\int\mathrm{d}q\,f_p^{\rm eq}\,\bar{f}_p^{\rm eq}\,\mathcal{P}_p\,L_{pq}\,\mathcal{P}_q}{\int\mathrm{d}p\,f_p^{\rm eq}\,\bar{f}_p^{\rm eq}\,\mathcal{P}_p\,\Pi_p}\right]\Pi, \qquad (10.157)
$$

$$
\int\mathrm{d}p\int\mathrm{d}q\,\mathcal{J}_p^\mu\,f_p^{\rm eq}\,\bar{f}_p^{\rm eq}\,L_{pq}\,\Phi_q = \frac{e+P}{n}\left[\frac{\int\mathrm{d}p\int\mathrm{d}q\,f_p^{\rm eq}\,\bar{f}_p^{\rm eq}\,\mathcal{J}_{p\nu}\,L_{pq}\,\mathcal{J}_q^\nu}{\int\mathrm{d}p\,f_p^{\rm eq}\,\bar{f}_p^{\rm eq}\,\mathcal{J}_{p\nu}\,J_p^\nu}\right]J^\mu, \qquad (10.158)
$$

$$
\int\mathrm{d}p\int\mathrm{d}q\,\pi_p^{\mu\nu}\,f_p^{\rm eq}\,\bar{f}_p^{\rm eq}\,L_{pq}\,\Phi_q = \left[\frac{\int\mathrm{d}p\int\mathrm{d}q\,f_p^{\rm eq}\,\bar{f}_p^{\rm eq}\,\pi_{p\rho\sigma}\,L_{pq}\,\pi_q^{\rho\sigma}}{\int\mathrm{d}p\,f_p^{\rm eq}\,\bar{f}_p^{\rm eq}\,\pi_{p\rho\sigma}\,\pi_p^{\rho\sigma}}\right]\pi^{\mu\nu}. \qquad (10.159)
$$

Then, we write the left-hand side of the relaxation equations (10.156) as

$$\int \mathrm{d}p\,(\mathcal{P}_p,\,\mathcal{J}_p^{\mu},\,\pi_p^{\mu\nu})\,f_p^{\mathrm{eq}}\,\bar{f}_p^{\mathrm{eq}}\left[(p\cdot u)\frac{\partial}{\partial\tau}+p\cdot\nabla\right]\frac{\mu-p\cdot u}{T}$$
$$+\int \mathrm{d}p\,(\mathcal{P}_p,\,\mathcal{J}_p^{\mu},\,\pi_p^{\mu\nu})\left[(p\cdot u)\frac{\partial}{\partial\tau}+p\cdot\nabla\right]f_p^{\mathrm{eq}}\,\bar{f}_p^{\mathrm{eq}}\,\Phi_p. \tag{10.160}$$

To convert the first terms into familiar forms, we replace the temporal derivative in the integrand by the spatial derivative on the basis of the relativistic Euler equations, which are obtained by setting $\partial/\partial\tau^{(1)}=\partial/\partial\tau$ in Eqs. (10.65)–(10.67). As a result, we have

$$\left[(p\cdot u)\frac{\partial}{\partial\tau}+p\cdot\nabla\right]\frac{\mu-p\cdot u}{T}=-\Pi_p\,\frac{-\nabla\cdot u}{T}+J_p^{\mu}\,\frac{n}{e+P}\,\nabla_{\mu}\frac{\mu}{T}-\pi_p^{\mu\nu}\,\frac{\sigma_{\mu\nu}}{T}. \tag{10.161}$$

This equation is the same as Eq. (10.68) derived in the context of the Chapman-Enskog method. Using the expression (10.161), we calculate the first terms in Eq. (10.160) to be

$$\int \mathrm{d}p\,\mathcal{P}_p\,f_p^{\mathrm{eq}}\,\bar{f}_p^{\mathrm{eq}}\left[(p\cdot u)\frac{\partial}{\partial\tau}+p\cdot\nabla\right]\frac{\mu-p\cdot u}{T}$$
$$=\left[-\frac{1}{T}\int \mathrm{d}p\,f_p^{\mathrm{eq}}\,\bar{f}_p^{\mathrm{eq}}\,\mathcal{P}_p\,\Pi_p\right](-\nabla\cdot u), \tag{10.162}$$

$$\int \mathrm{d}p\,\mathcal{J}_p^{\mu}\,f_p^{\mathrm{eq}}\,\bar{f}_p^{\mathrm{eq}}\left[(p\cdot u)\frac{\partial}{\partial\tau}+p\cdot\nabla\right]\frac{\mu-p\cdot u}{T}$$
$$=\left[\frac{1}{3T^2}\int \mathrm{d}p\,f_p^{\mathrm{eq}}\,\bar{f}_p^{\mathrm{eq}}\,\mathcal{J}_{p\nu}\,J_p^{\nu}\right]\frac{n\,T^2}{e+P}\,\nabla^{\mu}\frac{\mu}{T}, \tag{10.163}$$

$$\int \mathrm{d}p\,\pi_p^{\mu\nu}\,f_p^{\mathrm{eq}}\,\bar{f}_p^{\mathrm{eq}}\left[(p\cdot u)\frac{\partial}{\partial\tau}+p\cdot\nabla\right]\frac{\mu-p\cdot u}{T}$$
$$=\left[-\frac{1}{10T}\int \mathrm{d}p\,f_p^{\mathrm{eq}}\,\bar{f}_p^{\mathrm{eq}}\,\pi_{p\rho\sigma}\,\pi_p^{\rho\sigma}\right]2\,\sigma^{\mu\nu}. \tag{10.164}$$

Next let us calculate the second terms in Eq. (10.160). The temporal derivative of the variables Π, J^{μ}, and $\pi^{\mu\nu}$ in Φ will give the relaxation equations which we wanted. The derivative, however, hits not only Π, J^{μ}, and $\pi^{\mu\nu}$ but also $f_p^{\mathrm{eq}}\,\bar{f}_p^{\mathrm{eq}}$ and the prefactors in Φ. Thus, the second terms include several terms other than the relaxation terms given by the temporal derivative of Π, J^{μ}, and $\pi^{\mu\nu}$. In this section, we focus only the relaxation terms.

Then, the second terms in Eq. (10.160) is reduced to

$$\int \mathrm{d}p \mathcal{P}_p \left[(p\cdot u)\frac{\partial}{\partial\tau} + p\cdot\nabla \right] (f_p^{\mathrm{eq}} \bar{f}_p^{\mathrm{eq}} \Phi_p)$$
$$= \left[\frac{\int \mathrm{d}p\, f_p^{\mathrm{eq}} \bar{f}_p^{\mathrm{eq}} \mathcal{P}_p\, (p\cdot u) \mathcal{P}_p}{\int \mathrm{d}p\, f_p^{\mathrm{eq}} \bar{f}_p^{\mathrm{eq}} \mathcal{P}_p\, \Pi_p} \right] \frac{\partial}{\partial\tau}\Pi + \cdots, \tag{10.165}$$

$$\int \mathrm{d}p \mathcal{J}_p^{\mu} \left[(p\cdot u)\frac{\partial}{\partial\tau} + p\cdot\nabla \right] (f_p^{\mathrm{eq}} \bar{f}_p^{\mathrm{eq}} \Phi_p)$$
$$= \frac{e+P}{n} \left[\frac{\int \mathrm{d}p\, f_p^{\mathrm{eq}} \bar{f}_p^{\mathrm{eq}} \mathcal{J}_{p\nu}\, (p\cdot u) \mathcal{J}_p^{\nu}}{\int \mathrm{d}p\, f_p^{\mathrm{eq}} \bar{f}_p^{\mathrm{eq}} \mathcal{J}_{p\nu}\, J_p^{\nu}} \right] \Delta^{\mu\nu} \frac{\partial}{\partial\tau} J_\nu + \cdots, \tag{10.166}$$

$$\int \mathrm{d}p\, \pi_p^{\mu\nu} \left[(p\cdot u)\frac{\partial}{\partial\tau} + p\cdot\nabla \right] (f_p^{\mathrm{eq}} \bar{f}_p^{\mathrm{eq}} \Phi_p)$$
$$= \left[\frac{\int \mathrm{d}p\, f_p^{\mathrm{eq}} \bar{f}_p^{\mathrm{eq}} \pi_{p\rho\sigma}\, (p\cdot u) \pi_p^{\rho\sigma}}{\int \mathrm{d}p\, f_p^{\mathrm{eq}} \bar{f}_p^{\mathrm{eq}} \pi_{p\rho\sigma}\, \pi_p^{\rho\sigma}} \right] \Delta^{\mu\nu\rho\sigma} \frac{\partial}{\partial\tau} \pi_{\rho\sigma} + \cdots. \tag{10.167}$$

Collecting Eqs. (10.157)–(10.159), (10.162)–(10.164), and (10.165)–(10.167), we construct the relaxation equations as follows:

$$\Pi = -\zeta^{\mathrm{IS}}\, \nabla\cdot u - \tau_\Pi^{\mathrm{IS}} \frac{\partial}{\partial\tau}\Pi + \cdots, \tag{10.168}$$

$$J^\mu = \lambda^{\mathrm{IS}} \left(\frac{n\,T}{e+P} \right)^2 \nabla^\mu \frac{\mu}{T} - \tau_J^{\mathrm{IS}}\, \Delta^{\mu\nu} \frac{\partial}{\partial\tau} J_\nu + \cdots, \tag{10.169}$$

$$\pi^{\mu\nu} = 2\,\eta^{\mathrm{IS}}\, \sigma^{\mu\nu} - \tau_\pi^{\mathrm{IS}}\, \Delta^{\mu\nu\rho\sigma} \frac{\partial}{\partial\tau}\pi_{\rho\sigma} + \cdots. \tag{10.170}$$

Here ζ^{IS}, λ^{IS}, and η^{IS} denote the transport coefficients, that is, the bulk viscosity, heat conductivity, and shear viscosity given by

$$\zeta^{\mathrm{IS}} := -\frac{1}{T} \frac{\left[\int \mathrm{d}p\, f_p^{\mathrm{eq}} \bar{f}_p^{\mathrm{eq}} \mathcal{P}_p\, \Pi_p \right]^2}{\int \mathrm{d}p \int \mathrm{d}q\, f_p^{\mathrm{eq}} \bar{f}_p^{\mathrm{eq}} \mathcal{P}_p\, L_{pq}\, \mathcal{P}_q}, \tag{10.171}$$

$$\lambda^{\mathrm{IS}} := \frac{1}{3\,T^2} \frac{\left[\int \mathrm{d}p\, f_p^{\mathrm{eq}} \bar{f}_p^{\mathrm{eq}} \mathcal{J}_{p\mu}\, J_p^{\mu} \right]^2}{\int \mathrm{d}p \int \mathrm{d}q\, f_p^{\mathrm{eq}} \bar{f}_p^{\mathrm{eq}} \mathcal{J}_{p\mu}\, L_{pq}\, \mathcal{J}_q^{\mu}}, \tag{10.172}$$

$$\eta^{\mathrm{IS}} := -\frac{1}{10\,T} \frac{\left[\int \mathrm{d}p\, f_p^{\mathrm{eq}} \bar{f}_p^{\mathrm{eq}} \pi_{p\mu\nu}\, \pi_p^{\mu\nu} \right]^2}{\int \mathrm{d}p \int \mathrm{d}q\, f_p^{\mathrm{eq}} \bar{f}_p^{\mathrm{eq}} \pi_{p\mu\nu}\, L_{pq}\, \pi_q^{\mu\nu}}, \tag{10.173}$$

respectively, while τ_Π^{IS}, τ_J^{IS}, and τ_π^{IS} the relaxation times for the channels of the bulk pressure Π, thermal flux J^μ, and stress pressure $\pi^{\mu\nu}$ given by

$$\tau_{\Pi}^{\rm IS} := -\frac{\int dp\, f_p^{\rm eq}\, \bar{f}_p^{\rm eq}\, \mathcal{P}_p\, (p\cdot u)\, \mathcal{P}_p}{\int dp \int dq\, f_p^{\rm eq}\, \bar{f}_p^{\rm eq}\, \mathcal{P}_p\, L_{pq}\, \mathcal{P}_q}, \tag{10.174}$$

$$\tau_{J}^{\rm IS} := -\frac{\int dp\, f_p^{\rm eq}\, \bar{f}_p^{\rm eq}\, \mathcal{J}_{p\mu}\, (p\cdot u)\, \mathcal{J}_p^{\mu}}{\int dp \int dq\, f_p^{\rm eq}\, \bar{f}_p^{\rm eq}\, \mathcal{J}_{p\mu}\, L_{pq}\, \mathcal{J}_q^{\mu}}, \tag{10.175}$$

$$\tau_{\pi}^{\rm IS} := -\frac{\int dp\, f_p^{\rm eq}\, \bar{f}_p^{\rm eq}\, \pi_{p\mu\nu}\, (p\cdot u)\, \pi_p^{\mu\nu}}{\int dp \int dq\, f_p^{\rm eq}\, \bar{f}_p^{\rm eq}\, \pi_{p\mu\nu}\, L_{pq}\, \pi_q^{\mu\nu}}, \tag{10.176}$$

respectively. It is to be noted that all the microscopic expressions of the transport coefficients (10.171)–(10.173) are clearly different from those (10.94)–(10.96) given by the Chapman-Enskog method.

The relaxation equations (10.168)–(10.170) tell us that Π, J^{μ}, and $\pi^{\mu\nu}$ are to be relaxed to the following forms

$$\Pi = -\zeta^{\rm IS}\, \nabla\cdot u, \tag{10.177}$$

$$J^{\mu} = \lambda^{\rm IS} \left(\frac{n\,T}{e+P}\right)^2 \nabla^{\mu} \frac{\mu}{T}, \tag{10.178}$$

$$\pi^{\mu\nu} = 2\,\eta^{\rm IS}\, \sigma^{\mu\nu}, \tag{10.179}$$

after a long time, longer than $\tau_{\Pi}^{\rm IS}$, $\tau_{J}^{\rm IS}$, and $\tau_{\pi}^{\rm IS}$, respectively.

Inserting the above Π, J^{μ}, and $\pi^{\mu\nu}$ into N^{μ} and $T^{\mu\nu}$ in Eqs. (10.152) and (10.153), we find that the resultant forms of N^{μ} and $T^{\mu\nu}$ have the same forms as those by the Chapman-Enskog method, although the microscopic expressions of the transport coefficients are mutually different.

10.3.2.4 Particle Frame

Next we analyze the case of the particle frame, whose conditions of fit are expressed as Eqs. (10.121), (10.122), and (10.11), which we recapitulate here for convenience,

$$\upsilon^{\mu} = \Delta^{\mu\nu}\, \delta N_{\nu} = 0. \tag{10.180}$$

Much the same way as in the case of the energy frame, the functional form of Φ_p can be determined, as follows. The parameters A_2, A_1, A_0, and C_0 are given by Eqs. (10.132)–(10.134), (10.137), while B_1 and B_0 reads

$$B_1 = \frac{-3}{-(m^2\, a_2 - a_4) + \frac{m^2\, a_1 - a_3}{m^2\, a_0 - a_2}\, (m^2\, a_1 - a_3)}, \tag{10.181}$$

$$B_0 = -B_1\, \frac{m^2\, a_1 - a_3}{m^2\, a_0 - a_2}. \tag{10.182}$$

Furthermore, B_1 is found to be nicely reduced into a compact form as

$$B_1 = -\frac{1}{\frac{1}{3}\int\mathrm{d}p\, f_p^{\mathrm{eq}}\,\bar{f}_p^{\mathrm{eq}}\,(p\cdot u)\,\pi_{p\mu}\,J_p^{\mu}} = \frac{1}{\frac{1}{3}\int\mathrm{d}p\, f_p^{\mathrm{eq}}\,\bar{f}_p^{\mathrm{eq}}\,J_{p\mu}\,J_p^{\mu}}. \tag{10.183}$$

Thus we have

$$\Phi_p = \frac{\mathcal{P}_p\,\Pi}{\int\mathrm{d}q\, f_q^{\mathrm{eq}}\,\bar{f}_q^{\mathrm{eq}}\,\Pi_q\,\Pi_q} - \frac{J_p^{\mu}\,J_{\mu}}{\frac{1}{3}\int\mathrm{d}q\, f_q^{\mathrm{eq}}\,\bar{f}_q^{\mathrm{eq}}\,J_{q\nu}\,J_q^{\nu}} + \frac{\pi_p^{\mu\nu}\,\pi_{\mu\nu}}{\frac{1}{5}\int\mathrm{d}q\, f_q^{\mathrm{eq}}\,\bar{f}_q^{\mathrm{eq}}\,\pi_{q\rho\sigma}\,\pi_q^{\rho\sigma}}. \tag{10.184}$$

Now that Φ_p has been obtained in the form given in Eq. (10.184), one can derive the balance and relaxation equations from Eqs. (10.106) and (10.107), respectively, as was done in the case of the energy frame. The particle-number current and the energy-momentum tensor constituting the balance equations $\partial_\mu N^\mu = 0$ and $\partial_\mu T^{\mu\nu} = 0$ thus read

$$N^{\mu} = n\,u^{\mu}, \tag{10.185}$$

$$T^{\mu\nu} = e\,u^{\mu}\,u^{\nu} - (P + \Pi)\,\Delta^{\mu\nu} + u^{\mu}\,J^{\nu} + J^{\mu}\,u^{\nu} + \pi^{\mu\nu}, \tag{10.186}$$

respectively. These energy-momentum tensor and particle-number current are in accord with the conditions of fit (10.121), (10.122), and (10.180), or equivalently the ansatz (10.11)–(10.13) that characterize **particle frame**.

Although it is found that the relaxation equations for Π and $\pi^{\mu\nu}$ are also of the same forms as those in the **energy frame**, the relaxation equation for J^μ to be obtained from (10.107) takes a form different from that in the energy frame, as we shall show below. An explicit calculation of the right-hand side of (10.107) gives

$$\int\mathrm{d}p\int\mathrm{d}q\, J_p^{\mu}\, f_p^{\mathrm{eq}}\,\bar{f}_p^{\mathrm{eq}}\, L_{pq}\,\Phi_q = -\left[\frac{\int\mathrm{d}p\int\mathrm{d}q\, f_p^{\mathrm{eq}}\,\bar{f}_p^{\mathrm{eq}}\,J_{p\nu}\,L_{pq}\,J_q^{\nu}}{\int\mathrm{d}p\, f_p^{\mathrm{eq}}\,\bar{f}_p^{\mathrm{eq}}\,J_{p\nu}\,J_p^{\nu}}\right] J^{\mu}, \tag{10.187}$$

whereas the left-hand side can be expanded as

$$\int\mathrm{d}p\, J_p^{\mu}\left[(p\cdot u)\frac{\partial}{\partial\tau} + p\cdot\nabla\right](f_p^{\mathrm{eq}}\,\bar{f}_p^{\mathrm{eq}}\,\Phi_p) = -\left[\frac{\int\mathrm{d}p\, f_p^{\mathrm{eq}}\,\bar{f}_p^{\mathrm{eq}}\,J_{p\nu}\,(p\cdot u)\,J_p^{\nu}}{\int\mathrm{d}p\, f_p^{\mathrm{eq}}\,\bar{f}_p^{\mathrm{eq}}\,J_{q\nu}\,J_q^{\nu}}\right]\Delta^{\mu\nu}\frac{\partial}{\partial\tau}J_{\nu} + \cdots. \tag{10.188}$$

Thus we have the relaxation equation for J^μ as

$$J^{\mu} = -\lambda'^{\mathrm{IS}}\,\frac{n\,T^2}{e + P}\,\nabla^{\mu}\frac{\mu}{T} - \tau_J'^{\mathrm{IS}}\,\Delta^{\mu\nu}\,\frac{\partial}{\partial\tau}J_{\nu}. \tag{10.189}$$

Here $\lambda'^{\rm IS}$ and $\tau_J'^{\rm IS}$ denote the heat conductivity and relaxation time for the channel of the thermal flux, respectively, which are defined by

$$\lambda'^{\rm IS} := \frac{1}{3\,T^2}\,\frac{\left[\int \mathrm{d}p\, f_p^{\rm eq}\, \bar{f}_p^{\rm eq}\, J_{p\mu}\, J_p^{\mu}\right]^2}{\int \mathrm{d}p \int \mathrm{d}q\, f_p^{\rm eq}\, \bar{f}_p^{\rm eq}\, J_{p\mu}\, L_{pq}\, J_q^{\mu}}, \tag{10.190}$$

$$\tau_J'^{\rm IS} := -\frac{\int \mathrm{d}p\, f_p^{\rm eq}\, \bar{f}_p^{\rm eq}\, J_{p\mu}\, (p\cdot u)\, J_p^{\mu}}{\int \mathrm{d}p \int \mathrm{d}q\, f_p^{\rm eq}\, \bar{f}_p^{\rm eq}\, J_{p\mu}\, L_{pq}\, J_q^{\mu}}. \tag{10.191}$$

We note that $\lambda'^{\rm IS}$ is equal to $\lambda^{\rm IS}$ in Eq. (10.169) owing to the equalities

$$\int \mathrm{d}p\, f_p^{\rm eq}\, \bar{f}_p^{\rm eq}\, J_{p\mu}\, J_p^{\mu} = \int \mathrm{d}p\, f_p^{\rm eq}\, \bar{f}_p^{\rm eq}\, \mathcal{J}_{p\mu}\, J_p^{\mu}, \tag{10.192}$$

$$\int \mathrm{d}p \int \mathrm{d}q\, f_p^{\rm eq}\, \bar{f}_p^{\rm eq}\, J_{p\mu}\, L_{pq}\, J_q^{\mu} = \int \mathrm{d}p \int \mathrm{d}q\, f_p^{\rm eq}\, \bar{f}_p^{\rm eq}\, \mathcal{J}_{p\mu}\, L_{pq}\, \mathcal{J}_q^{\mu}, \tag{10.193}$$

which are derived from Eqs. (10.61), (10.62), and (10.149), while $\tau_J'^{\rm IS}$ is newly defined, that is, $\tau_J'^{\rm IS} \neq \tau_J^{\rm IS}$, because of

$$\int \mathrm{d}p\, f_p^{\rm eq}\, \bar{f}_p^{\rm eq}\, J_{p\mu}\, (p\cdot u)\, J_p^{\mu} \neq \int \mathrm{d}p\, f_p^{\rm eq}\, \bar{f}_p^{\rm eq}\, \mathcal{J}_{p\mu}\, (p\cdot u)\, \mathcal{J}_p^{\mu}. \tag{10.194}$$

Thus the relaxation time for the channel of the thermal flux is different between the energy frame and the particle frame.

A small remark is in order: Since the set of the balance equations and the relaxation equations thus obtained is based on the fourteen-moment method that incorporates the second-order of the space-time derivatives, the Israel-Stewart fourteen-moment method may be interpreted to provide us with the so-called mesoscopic dynamics [148, 150], which lies in the intermediate level between the fluid dynamics derived by the Chapman-Enskog method and the microscopic dynamics described by the underlying Boltzmann equation, as was indicated in Sect. 10.1.

10.3.3 Concluding Remarks

In this chapter, after giving an argument on the need of a derivation of relativistic dissipative fluid dynamic equation from the underlying microscopic theory to settle down the fundamental problems in the theory of relativistic fluid dynamics, we have given a detailed account of the Chapman-Enskog method based on the perturbation theory and the Israel-Stewart fourteen-moment method for deriving fluid dynamic equations from the relativistic Boltzmann equation. Although the two methods are different in whether having recourse to a perturbation or truncated-function method, they equally rely on the two principles, namely, the solvability condition related to the conservation laws embodied in the collision integral and the conditions of fit (or matching conditions). One might have, however, recognized the somewhat ad hoc nature of the latter, which can lead to the local rest frame of the fluid at one's disposal, which may be even different from the popular energy and particle frames.

In the subsequent chapters, we shall show that the application of the RG method, which starts from a faithful solution of the Boltzmann equation based on the perturbation theory, is free from such an ad hoc nature in principle, and can lead to the fluid dynamic equations uniquely. Although we have not discussed the stability problem of the relativistic fluid dynamic equations in this chapter, we shall see that the fluid dynamic equations derived by the RG method gets rid of the stability problem.

Chapter 11
RG/E Derivation of Relativistic First-Order Fluid Dynamics

11.1 Introduction

There have been various attempts [66–68, 113, 144, 172] to derive relativistic dissipative fluid dynamic equations from the relativistic Boltzmann equation [67, 68], and some of past works certainly succeeded in deriving the known equations in the various local rest frames by identifying the assumptions and/or approximations to reproduce the equations. In fact, we showed in the previous Chap. 10 that the standard derivation of relativistic fluid dynamic equations based on the Chapman-Enskog expansion or Israel-Stewart fourteen moment methods [67, 68] utilizes the ansatz (10.11)–(10.15) as the crucial constraints on the distribution function.

One may say, however, that it is curious that the physical meanings of these assumptions/approximations or possible uniqueness of a local rest frame have been hardly questioned nor elucidated. Our point is that this unsatisfactory situation stems from an incomplete application of the reduction theory of the dynamics or rather the incompleteness of the theories adopted. For instance, the Chapman-Enskog method is formulated as an application of the multiple-scale method [20, 21], an account of which as a reduction theory is presented in Sect. 3.6, where the reader might have recognized that the multiple-scale method does not necessarily have a solid foundation, in particular, in treating higher order terms because of the lack of the exact independence of the multiple times $t,\ t_1 = \epsilon t,\ t_2 = \epsilon^2 t,\ \ldots$.

In this respect, let us recall that the RG method [1, 38, 39, 48–51, 65, 92–96] as presented in Chaps. 4 and 5 on the basis of [3–6, 46, 203] also provides us with a powerful and systematic tool for the reduction of the dynamics even in the context of the asymptotic analysis as shown in Chap. 5, where it was shown that the RG method provides us with an elementary way to construct the invariant/attractive manifold of the dynamics in the asymptotic region as well as the reduced equation on the manifold. The fourteen moment method of Israel-Stewart can be interpreted to be an approximate theory of the construction of the invariant manifold of the asymptotic solution of the relativistic Boltzmann equation within a restricted functional space.

T. Kunihiro et al., *Geometrical Formulation of Renormalization-Group Method as an Asymptotic Analysis*, Fundamental Theories of Physics 206,
https://doi.org/10.1007/978-981-16-8189-9_11

Indeed it is noteworthy that the RG method applied to the non-relativistic Boltzmann equation successfully applied to obtain the Navier-Stokes equation together with the microscopic expressions of the transport coefficients [57, 58], as is shown in detail for the classical statistics in Chap. 8. It is also encouraging that the RG method proves to be an elementary but systematic method for obtaining the asymptotic slow dynamics of some other kinetic equations like Langevin and/or Fokker-Planck equations, where an essential role of the coarse graining of the time is elucidated, as shown in Chap. 7.

Thus, it is quite natural to apply the RG method to derive the relativistic dissipative fluid dynamic equations from the relativistic Boltzmann equation, as an extension of the non-relativistic case. It will, however, turn out that the necessity of the coarse graining of the space and time have to be made apparent in deriving the fluid dynamics as the asymptotic solution of the relativistic Boltzmann equation. As was emphasized in the non-relativistic case in Chap. 8, the RG method is based on a faithful solution of the equation, involving no ad hoc ansatz, and thus it is expected that the physical meanings and the validity of the ansatz posed in the phenomenological derivation will be clarified in the process of the reduction in the RG method.

In fact, we shall show that the resultant relativistic dissipative fluid dynamic equation is uniquely defined on the energy frame and admits the stable equilibrium state.

As for the choice of the local rest frame, it is to be noted that the fluid velocity $u^\mu(x)$ constitutes the very fluid dynamic variables together with the temperature $T(x)$ and chemical potential $\mu(x)$, which are the averaged slow variables to be determined by the reduction theory employed, in principle. In the usual derivation of the relativistic fluid dynamics from the Boltzmann equation, however, the existence of the fluid velocity $u^\mu(x)$ is taken for granted and used to define the local rest frame to rewrite the kinetic equation from the out set, which implies that the fluid velocity is treated differently from the other fluid dynamic variables. In the formulation below, we shall first *derive* the fluid velocity as an averaged variable. Then we show that it turns out to be identified with the macroscopic vector that defines the local rest frame. Moreover, this identification is found to be essential to lead the uniqueness of the energy frame.

Although the Boltzmann equation that we adopt is admittedly valid only for a dilute gas, it is expected that the form of the derived fluid dynamic equation itself can have a universal nature even for dense systems; this is found plausible if one recalls the wide applicability of the Navier-Stokes equation beyond dilute systems although it can be also derived from the non-relativistic Boltzmann equation, as is shown in [57, 58, 113] and in Chap. 8.

Although the RG method extended in the frame work of the doublet scheme developed in Chap. 9 is capable to lead to the *causal* stable relativistic fluid dynamic equation [69, 155], *i.e.*, the so called second-order equation, this task is rather involved and is postponed in the next chapter. This chapter is devoted to the derivation of the first-order relativistic dissipative fluid dynamic equation based on the RG method developed in Chap. 5.

11.2 Preliminaries

We start from recapitulating the basic facts presented in Sect. 10.2 on the relativistic Boltzmann equation, which reads

$$p^\mu \partial_\mu f_p(x) = C[f]_p(x), \qquad (10.17)$$

where $f_p(x)$ denotes the one-particle distribution function with p^μ being the four-momentum of the on-shell particle with mass m, and $C[f]_p(x)$ is the collision integral given by

$$\begin{aligned} C[f]_p(x) := \frac{1}{2!} \int \mathrm{d}p_1 \int \mathrm{d}p_2 \int \mathrm{d}p_3 \, \omega(p,\, p_1 | p_2,\, p_3) \\ \times \Big((1 + a\, f_p(x))\, (1 + a\, f_{p_1}(x))\, f_{p_2}(x)\, f_{p_3}(x) \\ - f_p(x)\, f_{p_1}(x)\, (1 + a\, f_{p_2}(x))\, (1 + a\, f_{p_3}(x)) \Big), \qquad (10.19) \end{aligned}$$

where a takes a fixed value $+1$, -1 or 0 depending whether a boson, fermion, and Boltzmann gas is treated. Here the integration measure is expressed in a compact form as

$$\mathrm{d}p := \frac{\mathrm{d}^3 \boldsymbol{p}}{(2\,\pi)^3\, p^0}, \qquad (10.22)$$

with $\boldsymbol{p}$ being the spatial components of the four momentum p^μ. The transition probability $\omega(p,\, p_1 | p_2,\, p_3)$ has the important symmetry property (10.20) due to the interchangeability of particles and the time-reversal invariance.

11.3 Introduction and Properties of Macroscopic Frame Vector

Since we are interested in the fluid dynamic regime where the time and space dependence of the physical quantities are small, we try to solve the relativistic Boltzmann equation (10.17) in the situation where the space-time variation of $f_p(x)$ is small and extract a reduced dynamics with a coarse-grained space-time scales from it.

To make a coarse graining with the Lorentz covariance retained, we introduce a time-like Lorentz covariant vector a_{F}^μ to define the local frame [204, 205],

$$a_{\mathrm{F}}^\mu(x), \qquad (11.1)$$

which is directed to the positive time direction

$$a_{\rm F}^0(x) > 0, \tag{11.2}$$

with the normalization $a_{\rm F}^\mu(x)a_{{\rm F}\mu}(x) = 1$. Notice that $a_{\rm F}^\mu(x)$ may depend on x^μ but not on the momentum p^μ.

Although the flow velocity $u^\mu(x)$ is simply adopted as the macroscopic flow vector in the standard method (see Sect. 10.3.1) without any physical reasoning, we shall give an argument [204, 206] on why $a_{\rm F}^\mu(x)$ with the above properties must coincide with the flow velocity $u^\mu(x)$ in the next section.

With the use of the projection operator on the space-like vector defined by

$$\Delta_{\rm F}^{\mu\nu}(x) := g^{\mu\nu} - a_{\rm F}^\mu(x)a_{\rm F}^\nu(x), \tag{11.3}$$

we decompose the derivative $\frac{\partial}{\partial x_\mu} =: \partial^\mu$ into time-like and space-like ones as

$$\partial^\mu = a_{\rm F}^\mu(x)a_{\rm F}^\nu(x)\,\partial_\nu + \Delta_{\rm F}^{\mu\nu}(x)\partial_\nu = a_{\rm F}^\mu(x)\frac{\partial}{\partial\tau} + \frac{\partial}{\partial\sigma_\mu}, \tag{11.4}$$

where

$$\frac{\partial}{\partial\tau} := a_{\rm F}^\mu\,\partial_\mu, \quad \frac{\partial}{\partial\sigma_\mu} := \Delta_{\rm F}^{\mu\nu}\,\partial_\nu. \tag{11.5}$$

Thus, $a_{\rm F}^\mu(x)$ specifies the covariant but macroscopic coordinate system where the local rest frame of the flow velocity and/or the flow velocity itself are defined: Since such a coordinate system is called *frame*, we call $a_{\rm F}^\mu$ the *macroscopic frame vector*.

Then, the relativistic Boltzmann equation (10.17) in the new coordinate system $(\tau\,,\,\sigma^\mu)$ is written as

$$p\cdot a_{\rm F}(\tau\,,\,\sigma)\frac{\partial}{\partial\tau}f_p(\tau\,,\,\sigma) + p\cdot\frac{\partial}{\partial\sigma}f_p(\tau\,,\,\sigma) = C[f]_p(\tau\,,\,\sigma), \tag{11.6}$$

where $a_{\rm F}^\mu(\tau\,,\,\sigma) \equiv a_{\rm F}^\mu(x)$ and $f_p(\tau\,,\,\sigma) \equiv f_p(x)$. We remark the prefactor of the time derivative is a Lorentz scalar and positive definite;

$$p\cdot a_{\rm F}(\tau\,,\,\sigma) > 0, \tag{11.7}$$

which is easily verified by taking the rest frame of p^0.

Now, as was done in the standard method in Sect. 10.3.1 and also in the non-relativistic case (see Eq. (8.41)), we introduce a scaled spatial coordinate

$$\bar{\sigma}^\mu := \epsilon\,\sigma^\mu \;\to\; \sigma^\mu. \tag{11.8}$$

Then Eq. (11.6) is converted into

$$\frac{\partial}{\partial \tau} f_p(\tau\,,\,\sigma) = \frac{1}{p \cdot a_{\mathrm{F}}(\tau\,,\,\sigma)} C[f]_p(\tau\,,\,\sigma) - \epsilon \frac{1}{p \cdot a_{\mathrm{F}}(\tau\,,\,\sigma)} p \cdot \frac{\partial}{\partial \sigma} f_p(\tau\,,\,\sigma). \tag{11.9}$$

Here, the parameter ϵ is a measure of the non-uniformity of the system and may be identified with the ratio of the mean free path to the representative macroscopic length scale of the system, *i.e.*, the Knudsen number:

$$\mathcal{K} = \frac{\ell}{L}, \quad \ell : \text{mean free path}, \quad L : \text{macroscopic length scale}. \tag{11.10}$$

Conversely speaking, the present analysis will not be applicable to the Knudsen regime where the mean free path is much larger than the macroscopic inhomogeneities of the system [67] nor Knudsen layer with a thickness as tiny as a few mean free path near the wall [68].

11.4 Perturbative Solution to Relativistic Boltzmann Equation and RG/E Equation as Macroscopic Dynamics

Because of the presence of a small parameter, Eq. (11.9) has a form suitable to apply the perturbation theory. In the physical terms, the fact that the expansion parameter ϵ is attached to the spatial derivative implies that only the spatial inhomogeneity causes the dissipative effects, as was the case in the non-relativistic case in Chap. 8.

We shall now apply the RG method to derive the fluid dynamic equation from the Boltzmann equation (11.9) in the asymptotic region on the basis of the perturbation theory. It should be noted that the parameter ϵ will be set back to unity eventually after the end of calculation.

11.4.1 Construction of Approximate Solution Around Arbitrary Time in the Asymptotic Region

Let $f_p(\tau, \sigma)$ be an exact solution (with some initial condition posed at a far past time) in the asymptotic region *i.e.*, the fluid dynamic region, where the time dependence of the system is slow and the length scales of the inhomogeneities of the system is large. We assume that the exact solution can be expanded with respect to ϵ as

$$f_p(\tau\,,\,\sigma) = f_p^{(0)}(\tau\,,\,\sigma) + \epsilon\, f_p^{(1)}(\tau\,,\,\sigma) + \epsilon^2\, f_p^{(2)}(\tau\,,\,\sigma) + \cdots. \tag{11.11}$$

In accordance with the general formulation of the RG method [3, 6, 46], we first try to obtain the perturbative solution $\tilde{f}_p$ to Eq. (11.9) around an arbitrary time $\tau = \tau_0$ in the asymptotic region with the 'initial' value at $\tau = \tau_0$ set to the exact value as

$$\tilde{f}_p(\tau = \tau_0\,,\,\sigma\,;\,\tau_0) = f_p(\tau_0\,,\,\sigma). \tag{11.12}$$

Here we have made explicit that the solution may depend on the 'initial' time τ_0.

An important remark is in order here: We suppose that the time variation of $a_{\mathrm{F}}^{\mu}(\tau\,,\,\sigma)$ is as small as, say, that of the flow velocity, and hence much smaller than the typical time variation of the microscopic processes as described by the Boltzmann equation. We are, however, seeking for a local solution to Eq. (11.9) that suffices to be valid only around $\tau \simeq \tau_0$ in the RG method. Therefore the time dependence of $a_{\mathrm{F}}^{\mu}(\tau\,,\,\sigma)$ in Eq. (11.9) may be neglected and set to the value $a_{\mathrm{F}}^{\mu}(\tau_0\,,\,\sigma)$ as

$$a_{\mathrm{F}}^{\mu}(\tau\,,\,\sigma) \to a_{\mathrm{F}}^{\mu}(\tau_0\,,\,\sigma) =: a_{\mathrm{F}}^{\mu}(\sigma;\,\tau_0), \tag{11.13}$$

for the purpose of obtaining a local solution around $\tau \simeq \tau_0$ in the RG method. Then we only have to solve the modified equation

$$\frac{\partial}{\partial\tau} f_p(\tau\,,\,\sigma) = \frac{1}{p\cdot a_{\mathrm{F}}(\sigma;\,\tau_0)}\,C[f]_p(\tau\,,\,\sigma) - \epsilon\,\frac{1}{p\cdot a_{\mathrm{F}}(\sigma;\,\tau_0)}\,p\cdot\frac{\partial}{\partial\sigma} f_p(\tau\,,\,\sigma). \tag{11.14}$$

The solution to (11.14) is expanded with respect to ϵ as follows;

$$\tilde{f}_p(\tau\,,\,\sigma\,;\,\tau_0) = \tilde{f}_p^{(0)}(\tau\,,\,\sigma\,;\,\tau_0) + \epsilon\,\tilde{f}_p^{(1)}(\tau\,,\,\sigma\,;\,\tau_0) + \epsilon^2\,\tilde{f}_p^{(2)}(\tau\,,\,\sigma\,;\,\tau_0) + \cdots \tag{11.15}$$

with the 'initial' condition posed on each term to coincide with the corresponding exact value at $\tau = \tau_0$ as

$$\tilde{f}_p^{(l)}(\tau_0\,,\,\sigma\,;\,\tau_0) = f_p^{(l)}(\tau_0\,,\,\sigma), \quad l = 0,\,1,\,2,\,\ldots. \tag{11.16}$$

Inserting the above expansions into Eq. (11.14) and equating the respective terms proportional to ϵ^n ($n = 0,\,1,\,2,\,\ldots$), we obtain the equation for $\tilde{f}_p^{(n)}(\tau_0\,,\,\sigma\,;\,\tau_0)$.

The zeroth-order equation reads

$$\frac{\partial}{\partial\tau}\tilde{f}_p^{(0)}(\tau\,,\,\sigma\,;\,\tau_0) = \frac{1}{p\cdot a_{\mathrm{F}}(\sigma\,;\,\tau_0)}\,C[\tilde{f}^{(0)}]_p(\tau\,,\,\sigma\,;\,\tau_0). \tag{11.17}$$

Since we are interested in the slow motion in the asymptotic region with large τ_0. we take the stationary solution or the fixed point,

$$\frac{\partial}{\partial\tau}\tilde{f}_p^{(0)}(\tau\,,\,\sigma\,;\,\tau_0) = 0, \tag{11.18}$$

or

$$\frac{1}{p\cdot a_{\mathrm{F}}(\sigma\,;\,\tau_0)}\,C[\tilde{f}^{(0)}]_p(\tau\,,\,\sigma\,;\,\tau_0) = 0, \qquad \forall\sigma, \tag{11.19}$$

which implies that $\ln\tilde{f}_p^{(0)}(\tau\,,\,\sigma\,;\,\tau_0)$ can be represented as a linear combination of the five collision invariants $(1\,,\,p^\mu)$ as shown in Sect. 10.2, and hence the zero-th order solution is nothing but a local equilibrium distribution function (10.33) as expressed by

$$\tilde{f}_p^{(0)}(\tau\,,\,\sigma\,;\,\tau_0) = \frac{1}{\exp\{[p\cdot u(\sigma\,;\,\tau_0) - \mu(\sigma\,;\,\tau_0)]/T(\sigma\,;\,\tau_0)\} - a} =: f_p^{\mathrm{eq}}(\sigma\,;\,\tau_0). \tag{11.20}$$

Accordingly, the 'initial' value at $\tau = \tau_0$ reads

$$f_p^{(0)}(\tau_0\,,\,\sigma) = \tilde{f}_p^{(0)}(\tau_0\,,\,\sigma\,;\,\tau_0) = f_p^{\mathrm{eq}}(\sigma\,;\,\tau_0).$$

We emphasize that the Gibbs relation plays an essential role [68] in making the quantum local equilibrium distribution function expressed in terms of the fluid dynamic variables including the flow velocity as given by (10.33) and accordingly (11.20).

Now we demonstrate that the equality

$$a_{\mathrm{F}}^\mu(\sigma\,;\,\tau_0) = u^\mu(\sigma\,;\,\tau_0). \tag{11.21}$$

Since u^μ and ∂^μ are the only available Lorentz vectors at hand, the generic Lorentz-covariant vector is expressed as

$$a_{\mathrm{F}}^\mu(x) = A_1(x)\,u^\mu + A_2(x)\,\partial^\mu T + A_3(x)\,\partial^\mu\mu + A_4(x)\,u^\nu\,\partial_\nu u^\mu, \tag{11.22}$$

where $A_i(x)$ with $i = 1,\,2,\,3,\,4$ are arbitrary Lorentz-scalar functions. We assume that the possible space-time dependence of them only come through the temperature and the chemical potential;

$$A_i(x) = A_i(T(x),\,\mu(x)). \tag{11.23}$$

Now utilizing the decomposition of the derivative ∂^μ

$$\partial^\mu = u^\mu\,u^\nu\,\partial_\nu + \Delta^{\mu\nu}\,\partial_\nu = u^\mu\,D + \nabla^\mu, \quad (D := u^\mu\,\partial_\mu;\ \nabla^\mu := (g^{\mu\nu} - u^\mu\,u^\nu)\partial_\nu), \tag{11.24}$$

we can rewrite Eq. (11.22) as

$$\begin{aligned} a_{\mathrm{F}}^{\mu} &= (A_1 + A_2\, DT + A_3\, D\mu)\, u^{\mu} + A_2\, \nabla^{\mu} T + A_3\, \nabla^{\mu}\mu + A_4\, Du^{\mu} \\ &\equiv C_t(T,\ \mu)\, u^{\mu} + \delta w^{\mu}, \end{aligned} \tag{11.25}$$

where

$$C_t(T,\ \mu) := A_1 + A_2\, DT + A_3\, D\mu, \quad \delta w^{\mu} := A_2\, \nabla^{\mu} T + A_3\, \nabla^{\mu}\mu + A_4\, Du^{\mu}. \tag{11.26}$$

Now the relative magnitudes of $C_t(T,\ \mu)$ and δw^{μ} matters in the present argument, although the equality $a_{\mathrm{F}}^2 = 1$ gives a minor constraint on them at the present stage. An important observation here is that the terms with coefficients A_2, A_3, and A_4 in δw^{μ} are all derivative terms, which are supposed to be small in the fluid dynamic regime even in the dissipative regime if the dynamics is governed by the fluid dynamics at all. In fact, as for the space-like derivative terms with the coefficients A_2 and A_3 in δw^{μ} are of higher order with respect to the dissipative effect and should be ignored in the perturbative approach which we adopt, where dissipative effects are treated as a perturbation. Then as for the time-like derivative term with A_4 in δw^{μ} should be also ignored in the outset, because the unperturbed solution in the perturbative approach adopted is stationary with no time dependence. Thus we conclude that the equality (11.21) holds.[1]

We remark that the five integral constants $T(\sigma\ ;\ \tau_0)$, $\mu(\sigma\ ;\ \tau_0)$, and $u_{\mu}(\sigma\ ;\ \tau_0)$ may equally depend on τ_0 as well as σ. For making the expressions simple, we introduce the following quantity with the equality (11.21) being taken for granted,

$$\begin{aligned} \bar{f}_p^{\mathrm{eq}}(\sigma\ ;\ \tau_0) &:= 1 + a\, f_p^{\mathrm{eq}}(\sigma\ ;\ \tau_0) \\ &= \frac{\mathrm{e}^{[p\cdot u(\sigma\ ;\ \tau_0) - \mu(\sigma\ ;\ \tau_0)]/T(\sigma\ ;\ \tau_0)}}{\mathrm{e}^{[p\cdot u(\sigma\ ;\ \tau_0) - \mu(\sigma\ ;\ \tau_0)]/T(\sigma\ ;\ \tau_0)} - a}, \end{aligned} \tag{11.27}$$

and suppress the coordinate arguments $(\sigma\ ;\ \tau_0)$ and the momentum subscript p, when no misunderstanding is expected

11.4.1.1 Properties of the Linear Evolution Operator and the Adequate Inner Product

The first-order equation reads

$$\frac{\partial}{\partial \tau}\tilde{f}_p^{(1)}(\tau) = \int \mathrm{d}q\, (f_p^{\mathrm{eq}}\, \bar{f}_p^{\mathrm{eq}})\, \hat{L}_{pq}\, (f_q^{\mathrm{eq}}\, \bar{f}_q^{\mathrm{eq}})^{-1}\, \tilde{f}_q^{(1)}(\tau) + (f_p^{\mathrm{eq}}\, \bar{f}_p^{\mathrm{eq}})\, F_p, \tag{11.28}$$

[1] We shall show in Sect. 11.6.1 that even if we were to include δw^{μ}, the leading term of δw^{μ} in the resultant fluid dynamic equation is of third order with respect to temporal and spatial derivatives of the fluid variables, which are of higher order than the dissipative terms in the fluid dynamic equations and negligible.

where the linear evolution operator $\hat{L}$ and the inhomogeneous term F_p are defined by

$$\begin{aligned}\hat{L}_{pq} &:= \frac{1}{p\cdot u} L_{pq}\\ &= -\frac{1}{p\cdot u}\frac{1}{2!}\int \mathrm{d}p_1 \int \mathrm{d}p_2 \int \mathrm{d}p_3\, \omega(p\,,\, p_1|p_2\,,\, p_3)\\ &\quad\times \frac{f^{\mathrm{eq}}_{p_1}\,\bar{f}^{\mathrm{eq}}_{p_2}\,\bar{f}^{\mathrm{eq}}_{p_3}}{\bar{f}^{\mathrm{eq}}_{p}}\,(\delta_{pq}+\delta_{p_1 q}-\delta_{p_2 q}-\delta_{p_3 q}), \end{aligned} \tag{11.29}$$

$$F_p := -(f^{\mathrm{eq}}_p\,\bar{f}^{\mathrm{eq}}_p)^{-1}\,\frac{1}{p\cdot u}\,p\cdot\nabla f^{\mathrm{eq}}_p, \tag{11.30}$$

respectively. We remark that now that the equality (11.21) holds, we have the identity $\frac{\partial}{\partial\sigma_\mu} = \nabla^\mu$. We also note that L ($\neq \hat{L}$) in the first line of Eq. (11.29) is nothing but the linearized collision integral introduced in Eq. (10.56), and the linear evolution operator $\hat{L}$ is proportional to but not the same as L.

To discuss the spectral properties of the linear operator $\hat{L}$, we first introduce an inner product

$$\langle\,\varphi\,,\,\psi\,\rangle := \int \mathrm{d}p\,(p\cdot u)\, f^{\mathrm{eq}}_p\,\bar{f}^{\mathrm{eq}}_p\,\varphi_p\,\psi_p, \tag{11.31}$$

for arbitrary two vectors φ_p and ψ_p. It should be noted that this inner product is a unique one that respects the self-adjoint nature of $\hat{L}$ shown in (11.33).[2]

We remark that the norm defined through this inner product is positive definite

$$\langle\,\varphi\,,\,\varphi\,\rangle = \int \mathrm{d}p\,(p\cdot u)\, f^{\mathrm{eq}}_p\,\bar{f}^{\mathrm{eq}}_p\,(\varphi_p)^2 > 0, \qquad \text{for } \varphi_p \neq 0, \tag{11.32}$$

since $(p\cdot u) > 0$ on account of Eqs. (11.7) and (11.21). We shall see that this positive definiteness (11.32) of the inner product plays an essential role in making the resultant fluid dynamic equation assure the stability of the thermal equilibrium state, as it should be, in contrast to some phenomenological equations.

Now one finds the following remarkable properties of the linear operator $\hat{L}$:

(1) $\hat{L}$ is self-adjoint with respect to this inner product:

$$\begin{aligned}\langle\,\varphi\,,\,\hat{L}\,\psi\,\rangle &= -\frac{1}{2!}\frac{1}{4}\int \mathrm{d}p \int \mathrm{d}p_1 \int \mathrm{d}p_2 \int \mathrm{d}p_3\, \omega(p\,,\, p_1|p_2\,,\, p_3)\, f^{\mathrm{eq}}_p\, f^{\mathrm{eq}}_{p_1}\,\bar{f}^{\mathrm{eq}}_{p_2}\,\bar{f}^{\mathrm{eq}}_{p_3}\\ &\quad\times(\varphi_p+\varphi_{p_1}-\varphi_{p_2}-\varphi_{p_3})\,(\psi_p+\psi_{p_1}-\psi_{p_2}-\psi_{p_3})\\ &= \langle\,\hat{L}\,\varphi\,,\,\psi\,\rangle. \end{aligned} \tag{11.33}$$

[2] A proof of the uniqueness of it is presented in Appendix 13.1.

(2) $\hat{L}$ is semi-negative definite:

$$\langle \varphi , \hat{L} \varphi \rangle = -\frac{1}{2!}\frac{1}{4}\int \mathrm{d}p \int \mathrm{d}p_1 \int \mathrm{d}p_2 \int \mathrm{d}p_3 \omega(p , p_1|p_2 , p_3)\, f_p^{\mathrm{eq}}\, f_{p_1}^{\mathrm{eq}}\, \bar{f}_{p_2}^{\mathrm{eq}}\, \bar{f}_{p_3}^{\mathrm{eq}} \\ \times (\varphi_p + \varphi_{p_1} - \varphi_{p_2} - \varphi_{p_3})^2 \\ \leq 0. \tag{11.34}$$

This inequality implies that the eigenvalues of $\hat{L}$ are negative or zero. Indeed, $\hat{L}$ has multiple zero eigenvalues, and the eigenvectors belonging to the zero eigenvalues are given by

$$\varphi_{0p}^{\alpha} := \begin{cases} p^{\mu}, \ \alpha = \mu, \\ 1, \quad \alpha = 4, \end{cases} , \tag{11.35}$$

which accordingly satisfy

$$\left[\hat{L}\, \varphi_0^{\alpha}\right]_p \equiv \int \mathrm{d}q\, \hat{L}_{pq}\, \varphi_{0q}^{\alpha} = 0. \tag{11.36}$$

These five eigenvectors span the kernel of $\hat{L}$, and are called the zero modes of $\hat{L}$. It is to be noted that these zero modes are collision invariants, which satisfy

$$\int \mathrm{d}p\, \varphi_{0p}^{\alpha}\, C[f]_p = 0, \tag{11.37}$$

as seen from Eqs. (10.24) and (10.25) in Chap. 10.

The kernel of $\hat{L}$ is also called the P space, and its complement space the Q space; the projection operators onto the respective spaces are denoted by P and Q, respectively. Of course,

$$Q = 1 - P. \tag{11.38}$$

To express the projection operators, it is found convenient to first define the P-space metric matrix $\eta_0^{\alpha\beta}$;

$$\eta_0^{\alpha\beta} := \langle \varphi_0^{\alpha} , \varphi_0^{\beta} \rangle. \tag{11.39}$$

Then the projection operator P is expressed as [6]

$$\left[P\, \psi\right]_p := \sum_{\alpha,\beta} \varphi_{0p}^{\alpha}\, \eta_{0\alpha\beta}^{-1}\, \langle \varphi_0^{\beta} , \psi \rangle, \tag{11.40}$$

where $\eta_{0\alpha\beta}^{-1}$ is the inverse of the metric matrix $\eta_0^{\alpha\beta}$; $\sum_{\gamma} \eta_{0\alpha\gamma}^{-1}\eta_0^{\gamma\beta} = \delta_{\alpha}^{\beta}$.

Let us show that the operator P has certainly the properties of the projection operator onto the P space spanned by the zero modes. First, the idempotency of P is shown as follows:

$$
\begin{aligned}
\left[P(P\,\psi)\right]_p &= P\left(\sum_{\gamma,\sigma} \varphi_{0p}^{\gamma}\,\eta_{0\gamma\sigma}^{-1}\,\langle\,\varphi_0^{\sigma}\,,\,\psi\,\rangle\right)\\
&= \sum_{\alpha,\beta}\sum_{\gamma,\sigma} \varphi_{0p}^{\alpha}\,\eta_{0\alpha\beta}^{-1}\,\langle\,\varphi_0^{\beta}\,,\,\varphi_0^{\gamma}\,\rangle\,\eta_{0\gamma\sigma}^{-1}\,\langle\,\varphi_0^{\sigma}\,,\,\psi\,\rangle\\
&= \sum_{\alpha,\beta}\sum_{\gamma,\sigma} \varphi_{0p}^{\alpha}\,\eta_{0\alpha\beta}^{-1}\,\eta_0^{\beta\gamma}\,\eta_{0\gamma\sigma}^{-1}\,\langle\,\varphi_0^{\sigma}\,,\,\psi\,\rangle\\
&= \sum_{\alpha,\beta}\sum_{\sigma} \varphi_{0p}^{\alpha}\,\eta_{0\alpha\beta}^{-1}\,\delta_{\sigma}^{\beta}\,\langle\,\varphi_0^{\sigma}\,,\,\psi\,\rangle\\
&= \sum_{\alpha,\beta} \varphi_{0p}^{\alpha}\,\eta_{0\alpha\beta}^{-1}\,\langle\,\varphi_0^{\beta}\,,\,\psi\,\rangle\\
&= \left[P\,\psi\right]_p, \qquad \forall\psi,
\end{aligned}
\tag{11.41}
$$

which implies that the idempotency P,

$$
P^2 = P. \tag{11.42}
$$

It follows also that $Q^2 = (1-P)^2 = 1 - 2P + P^2 = 1 - P = Q$. Accordingly, their eigenvalues are real numbers, 0 and 1. Furthermore,

$$
\begin{aligned}
\left[P\varphi_0^{\gamma}\right]_p &= \sum_{\alpha,\beta} \varphi_{0p}^{\alpha}\,\eta_{0\alpha\beta}^{-1}\,\langle\,\varphi_0^{\beta}\,,\,\varphi_0^{\gamma}\,\rangle\\
&= \sum_{\alpha,\beta} \varphi_{0p}^{\alpha}\,\eta_{0\alpha\beta}^{-1}\,\eta_0^{\beta\gamma}\\
&= \sum_{\alpha} \varphi_{0p}^{\alpha}\,\delta_{\alpha}^{\gamma}\\
&= \varphi_{0p}^{\gamma},
\end{aligned}
\tag{11.43}
$$

and accordingly,

$$
\left[Q\varphi_0^{\gamma}\right]_p = \left[(1-P)\varphi_0^{\gamma}\right]_p = 0. \tag{11.44}
$$

It is then readily shown that P and hence Q is self-adjoint with this inner product; to show that this is the case, it suffices to demonstrate that $\langle\,\varphi_0^{\alpha}\,,\,P\,\varphi_0^{\beta}\rangle = \eta_0^{\alpha\beta} = \langle\,P\,\varphi_0^{\alpha}\,,\,\varphi_0^{\beta}\rangle$, which is self-evident.

11.4.1.2 First-Order Analysis

The solution to Eq. (11.28) with the 'initial' condition at $\tau = \tau_0$

$$\tilde{f}^{(1)}(\tau = \tau_0) := \tilde{f}^{(1)}_p(\tau = \tau_0\,,\,\sigma\,;\,\tau_0) = f^{(1)} := f^{(1)}_p(\sigma\,;\,\tau_0), \qquad (11.45)$$

is given by

$$\tilde{f}^{(1)}(\tau) = f^{\mathrm{eq}}\,\bar{f}^{\mathrm{eq}}\left\{\mathrm{e}^{(\tau-\tau_0)\hat{L}}\left[(f^{\mathrm{eq}}\,\bar{f}^{\mathrm{eq}})^{-1}\,f^{(1)} + \hat{L}^{-1}\,Q\,F\right]\right.$$
$$\left. + (\tau - \tau_0)\,P\,F - L^{-1}\,Q\,F\right\}, \qquad (11.46)$$

with

$$\left[f^{\mathrm{eq}}\right]_{pq} := f^{\mathrm{eq}}_p\,\delta_{pq} \quad \text{and} \quad \left[\bar{f}^{\mathrm{eq}}\right]_{pq} := \bar{f}^{\mathrm{eq}}_p\,\delta_{pq}. \qquad (11.47)$$

The first term in Eq. (11.46) belongs to the Q space, and would create fast motion that is not included in the fluid dynamics. This undesirable term can be, however, simply eliminated by choosing the 'initial' value $f^{(1)}$, which has not yet been specified, as

$$f^{(1)} = \tilde{f}^{(1)}(\tau_0) = -f^{\mathrm{eq}}\,\bar{f}^{\mathrm{eq}}\,\hat{L}^{-1}\,Q\,F. \qquad (11.48)$$

Then, the first-order solution is now given by

$$\tilde{f}^{(1)}(\tau) = f^{\mathrm{eq}}\,\bar{f}^{\mathrm{eq}}\left[(\tau - \tau_0)\,P\,F - \hat{L}^{-1}\,Q\,F\right], \qquad (11.49)$$

with the 'initial' value (11.48). The secular term proportional to $\tau - \tau_0$ in Eq. (11.49) apparently invalidates the perturbative solution when $|\tau - \tau_0|$ becomes large.

It is worth mentioning that the standard Chapman-Enskog expansion method [67] presented in Sect. 10.3.1 includes a set of conditions for making secular terms disappear, which are nothing but the solvability condition of the balance Eqs. (10.64). In the present RG method, secular terms are allowed to appear and no constraints are imposed on the distribution function in an apparent way. In fact, the effects of the secular terms will be renormalized away into the unperturbed distribution function, which is in turn to acquire the slow time dependence.

We remark here that the application of the RG/E equation to $\tilde{f}(\tau) = \tilde{f}^{(0)}(\tau) + \epsilon\,\tilde{f}^{(1)}(\tau)$ leads to the relativistic Euler equation $\partial_\mu T^{\mu\nu}_0 = \partial_\mu N^\mu_0 = 0$ with

$$T^{\mu\nu}_0 = e\,u^\mu\,u^\nu - P\,\Delta^{\mu\nu}, \quad N^\mu_0 = n\,u^\mu, \qquad (11.50)$$

where e, P, and n denote the internal energy, pressure, particle-number density defined in Eqs. (10.38)–(10.40), respectively. We note that the explicit formulae of

$T_0^{\mu\nu}$ and N_0^{μ} calculated with the distribution function coincide with those given in Eqs. (10.36) and (10.37), respectively. A detailed derivation of them will be given in Sect. 11.4.1.4.

One thus finds that one must proceed to the second order to get a dissipative fluid dynamic equation in the RG method,

11.4.1.3 Second-Order Analysis

The second-order equation reads

$$\frac{\partial}{\partial\tau}\Big[(f^{\mathrm{eq}}\,\bar{f}^{\mathrm{eq}})^{-1}\,\tilde{f}^{(2)}(\tau)\Big] = \hat{L}\Big[(f^{\mathrm{eq}}\,\bar{f}^{\mathrm{eq}})^{-1}\,\tilde{f}^{(2)}(\tau)\Big] + (\tau-\tau_0)^2\,G + (\tau-\tau_0)\,H + I, \quad (11.51)$$

with

$$G_p := \frac{1}{2}\big[B[P\,F\,,\,P\,F]\big]_p, \quad (11.52)$$

$$H_p := -\big[B[P\,F\,,\,\hat{L}^{-1}\,Q\,F]\big]_p - (f_p^{\mathrm{eq}}\,\bar{f}_p^{\mathrm{eq}})^{-1}\,\frac{1}{p\cdot u}\,p\cdot\nabla\big[f^{\mathrm{eq}}\,\bar{f}^{\mathrm{eq}}\,P\,F\big]_p, \quad (11.53)$$

$$I_p := \frac{1}{2}\big[B[\hat{L}^{-1}\,Q\,F\,,\,\hat{L}^{-1}\,Q\,F]\big]_p + (f_p^{\mathrm{eq}}\,\bar{f}_p^{\mathrm{eq}})^{-1}\,\frac{1}{p\cdot u}\,p\cdot\nabla\big[f^{\mathrm{eq}}\,\bar{f}^{\mathrm{eq}}\,\hat{L}^{-1}\,Q\,F\big]_p. \quad (11.54)$$

Here, we have defined $\big[B[\varphi\,,\,\psi]\big]_p$ for arbitrary functions φ and ψ of momentum by

$$\big[B[\varphi\,,\,\psi]\big]_p := \int\mathrm{d}q\int\mathrm{d}r\;B_{pqr}\,\varphi_q\,\psi_r, \quad (11.55)$$

with

$$\begin{aligned} B_{pqr} &:= (f_p^{\mathrm{eq}}\,\bar{f}_p^{\mathrm{eq}})^{-1}\,\frac{1}{p\cdot u}\,\frac{\delta^2}{\delta f_q\delta f_r}C[f]_p\bigg|_{f=f^{\mathrm{eq}}}(f_q^{\mathrm{eq}}\,\bar{f}_q^{\mathrm{eq}})\,(f_r^{\mathrm{eq}}\,\bar{f}_r^{\mathrm{eq}}) \\ &= \frac{1}{p\cdot u}\,\frac{1}{2!}\int\mathrm{d}p_1\int\mathrm{d}p_2\int\mathrm{d}p_3\,\omega(p\,,\,p_1|p_2\,,\,p_3)\,\frac{f_{p_1}^{\mathrm{eq}}\,\bar{f}_{p_2}^{\mathrm{eq}}\,\bar{f}_{p_3}^{\mathrm{eq}}}{\bar{f}_p^{\mathrm{eq}}} \\ &\quad\times\bigg\{(\delta_{p_2q}\,\delta_{p_3r}+\delta_{p_3q}\,\delta_{p_2r}-\delta_{pq}\,\delta_{p_1r}-\delta_{p_1q}\,\delta_{pr})\,(1-\bar{f}_q^{\mathrm{eq}}-\bar{f}_r^{\mathrm{eq}}) \\ &\quad+\Big[(\delta_{p_2q}+\delta_{p_3q})\,(\delta_{pr}+\delta_{p_1r}) \\ &\quad-(\delta_{pq}+\delta_{p_1q})\,(\delta_{p_2r}+\delta_{p_3r})\Big]\,(\bar{f}_q^{\mathrm{eq}}-\bar{f}_r^{\mathrm{eq}})\bigg\}. \end{aligned} \quad (11.56)$$

The solution to Eq. (11.51) is found to be

$$\begin{aligned}\tilde{f}^{(2)}(\tau) = f^{\mathrm{eq}}\,\bar{f}^{\mathrm{eq}}\Big\{&\mathrm{e}^{(\tau-\tau_0)\hat{L}}\Big[(f^{\mathrm{eq}}\,\bar{f}^{\mathrm{eq}})^{-1}\,f^{(2)} + 2\,\hat{L}^{-3}\,Q\,G + \hat{L}^{-2}\,Q\,H + \hat{L}^{-1}\,Q\,I\Big]\\ &+ \frac{1}{3}\,(\tau-\tau_0)^3\,P\,G + \frac{1}{2}\,(\tau-\tau_0)^2\Big[P\,H - 2\,\hat{L}^{-1}\,Q\,G\Big]\\ &+ (\tau-\tau_0)\Big[P\,I - 2\,\hat{L}^{-2}\,Q\,G - \hat{L}^{-1}\,Q\,H\Big]\\ &- 2\,\hat{L}^{-3}\,Q\,G - \hat{L}^{-2}\,Q\,H - \hat{L}^{-1}\,Q\,I\Big\}. \end{aligned} \tag{11.57}$$

The would-be fast motion can be eliminated again by a choice of the 'initial' value $f^{(2)}$ as

$$f^{(2)} = \tilde{f}^{(2)}(\tau_0) = -f^{\mathrm{eq}}\,\bar{f}^{\mathrm{eq}}\Big[2\,\hat{L}^{-3}\,Q\,G + \hat{L}^{-2}\,Q\,H + \hat{L}^{-1}\,Q\,I\Big]. \tag{11.58}$$

Then the second-order solution is now expressed as

$$\begin{aligned}\tilde{f}^{(2)}(\tau) = f^{\mathrm{eq}}\,\bar{f}^{\mathrm{eq}}\Big\{&\frac{1}{3}\,(\tau-\tau_0)^3\,P\,G + \frac{1}{2}\,(\tau-\tau_0)^2\Big[P\,H - 2\,\hat{L}^{-1}\,Q\,G\Big]\\ &+ (\tau-\tau_0)\Big[P\,I - 2\,\hat{L}^{-2}\,Q\,G - \hat{L}^{-1}\,Q\,H\Big]\\ &- 2\,\hat{L}^{-3}\,Q\,G - \hat{L}^{-2}\,Q\,H - \hat{L}^{-1}\,Q\,I\Big\}. \end{aligned} \tag{11.59}$$

In contrast to the standard Chapman-Enskog expansion method [67], secular terms are present and that no constraints on the solution are imposed for defining the rest frame of the flow.

Thus the approximate solution up to the second order is given by

$$\begin{aligned}\tilde{f}_p(\tau\,,\,\sigma\,;\,\tau_0) &= \tilde{f}_p^{(0)}(\tau\,,\,\sigma\,;\,\tau_0) + \epsilon\,\tilde{f}_p^{(1)}(\tau\,,\,\sigma\,;\,\tau_0) + \epsilon^2\,\tilde{f}_p^{(2)}(\tau\,,\,\sigma\,;\,\tau_0) + O(\epsilon^3)\\ &= f^{\mathrm{eq}} + \epsilon\,f^{\mathrm{eq}}\,\bar{f}^{\mathrm{eq}}\Big[(\tau-\tau_0)\,P\,F - \hat{L}^{-1}\,Q\,F\Big]\\ &\quad + \epsilon^2\,f^{\mathrm{eq}}\,\bar{f}^{\mathrm{eq}}\Big\{\frac{1}{3}\,(\tau-\tau_0)^3\,P\,G + \frac{1}{2}\,(\tau-\tau_0)^2\Big[P\,H - 2\,\hat{L}^{-1}\,Q\,G\Big]\\ &\quad + (\tau-\tau_0)\Big[P\,I - 2\,\hat{L}^{-2}\,Q\,G - \hat{L}^{-1}\,Q\,H\Big]\\ &\quad - 2\,\hat{L}^{-3}\,Q\,G - \hat{L}^{-2}\,Q\,H - \hat{L}^{-1}\,Q\,I\Big\} + O(\epsilon^3), \end{aligned} \tag{11.60}$$

which is valid only locally around $\tau = \tau_0$ due the secular terms.

11.4.1.4 Extraction of the Fluid Dynamics Through RG/E Equation

Following the general argument of the RG method presented in Chap. 4, we can take a geometrical point of view that we have now a family of curves $\tilde{f}_p(\tau\,,\,\sigma\,;\,\tau_0)$ given by (11.60) which are parameterized by τ_0: They are all on the exact solution $f_p(\sigma\,;\,\tau)$ at $\tau = \tau_0$ up to $O(\epsilon^2)$, but only valid locally for τ near τ_0. So it is conceivable and actually the case that the envelope of the family of curves which contacts with each local solution at $\tau = \tau_0$ will give a global solution in the asymptotic region.

The envelope which contact with any curve in the family at $\tau = \tau_0$ is obtained by the RG/E equation,

$$\frac{\mathrm{d}}{\mathrm{d}\tau_0}\tilde{f}_p(\tau\,,\,\sigma\,;\,\tau_0)\bigg|_{\tau_0=\tau} = 0, \tag{11.61}$$

or explicitly

$$\frac{\partial}{\partial\tau}\Big[f^{\mathrm{eq}} - \epsilon\, f^{\mathrm{eq}}\,\bar{f}^{\mathrm{eq}}\,\hat{L}^{-1}\,Q\,F\Big] - \epsilon\, f^{\mathrm{eq}}\,\bar{f}^{\mathrm{eq}}\,P\,F$$
$$- \epsilon^2\, f^{\mathrm{eq}}\,\bar{f}^{\mathrm{eq}}\Big[P\,I - 2\,\hat{L}^{-2}\,Q\,G - \hat{L}^{-1}\,Q\,H\Big] + O(\epsilon^3) = 0, \tag{11.62}$$

which gives the equation of motion governing the dynamics of the five slow variables $T(\sigma\,;\,\tau)$, $\mu(\sigma\,;\,\tau)$, and $u^\mu(\sigma\,;\,\tau)$ in $f_p^{\mathrm{eq}}(\sigma\,;\,\tau)$. The global solution in the asymptotic region is given as an envelope function,

$$\begin{aligned} f_p^{\mathrm{E}}(\tau\,,\,\sigma) &= f_p(\sigma\,;\,\tau_0 = \tau) = \tilde{f}_p(\tau\,,\,\sigma\,;\,\tau_0 = \tau) \\ &= f^{\mathrm{eq}} - \epsilon\, f^{\mathrm{eq}}\,\bar{f}^{\mathrm{eq}}\,\hat{L}^{-1}\,Q\,F - \epsilon^2\, f^{\mathrm{eq}}\,\bar{f}^{\mathrm{eq}}\Big[2\,\hat{L}^{-3}\,Q\,G \\ &\quad + \hat{L}^{-2}\,Q\,H + \hat{L}^{-1}\,Q\,I\Big]\bigg|_{\tau_0=\tau} + O(\epsilon^3), \end{aligned} \tag{11.63}$$

where the solution of Eq. (11.62) is to be inserted.

As was proved in Chap. 4 in general, the envelope function $f_p^{\mathrm{E}}(\tau\,,\,\sigma)$ satisfies Eq. (11.14) in a global domain up to $O(\epsilon^2)$ owing to the condition (11.61), although $\tilde{f}_p(\tau\,,\,\sigma\,;\,\tau_0)$ itself was constructed as a local solution around $\tau \sim \tau_0$. Furthermore, one finds that $f_p^{\mathrm{E}}(\tau\,,\,\sigma)$ describes a slow motion of the one-particle distribution function since the time-derivatives of the quantities in $f_p^{\mathrm{E}}(\tau\,,\,\sigma)$ are all in the order of ϵ or higher.

In summary, we have obtained an approximate solution to Eq. (11.9) that is valid in a global domain in the asymptotic region in the form of the pair of Eqs. (11.62) and (11.63); the latter provides gives the functional form of the distribution function written solely in terms of the fluid dynamical variables while the former the time-evolution equations of the fluid dynamical variables, which is to be reduced to the fluid dynamic equation.

11.4.1.5 Reduction of RG/E Equation to the Fluid Dynamic Equation

We observe that the RG/E equation (11.62) includes fast modes that should not be identified as the fluid dynamic modes. While these modes could be incorporated to make a Langevinized fluid dynamic equation, we average out them to have the genuine fluid dynamic equation. This averaging can be made by taking the inner product of Eq. (11.62) with the zero modes φ_p^α.

To calculate the inner product explicitly, we first note that the following equalities hold;

$$\langle \varphi_0^\alpha \,,\, P\,\psi \rangle = \langle P\,\varphi_0^\alpha \,,\, \psi \rangle = \langle \varphi_0^\alpha \,,\, \psi \rangle, \tag{11.64}$$

$$\langle \varphi_0^\alpha \,,\, Q\,\psi \rangle = \langle Q\,\varphi_0^\alpha \,,\, \psi \rangle = 0. \tag{11.65}$$

for arbitrary ψ. Indeed Eqs. (11.64) and (11.65) follow from the self-adjointness of the projection operators and (11.43) and (11.44). Then one sees that following equality holds because of $\hat{L}^{-1}Q = Q\hat{L}^{-1}Q$,

$$\begin{aligned}
&\epsilon\,\langle \varphi_0^\alpha \,,\, P\,F\rangle + \epsilon^2\,\langle \varphi_0^\alpha \,,\, \Big[P\,I - 2\,\hat{L}^{-2}\,Q\,G - \hat{L}^{-1}\,Q\,H\Big]\rangle \\
&= \epsilon\,\langle \varphi_0^\alpha \,,\, F + \epsilon\,I\rangle \\
&= \epsilon\,\int \mathrm{d}p\,\varphi_{0p}^\alpha\,p\cdot\nabla\Big[f_p^{\mathrm{eq}} - \epsilon\,f_p^{\mathrm{eq}}\,\bar{f}_p^{\mathrm{eq}}\,\big[\hat{L}^{-1}\,Q\,F\big]_p\Big] \\
&\quad + \epsilon^2\,\frac{1}{2}\,\langle \varphi_0^\alpha \,,\, B[\hat{L}^{-1}\,Q\,F \,,\, L^{-1}\,Q\,F]\rangle.
\end{aligned} \tag{11.66}$$

However, owing to the fact that φ_{0p}^α are the collision invariants as shown in Eq. (11.37), one can show the equality[3]

$$\langle \varphi_0^\alpha \,,\, B[\psi \,,\, \chi]\rangle = 0. \tag{11.67}$$

Thus, Eq. (11.66) takes the form

$$\begin{aligned}
&\epsilon\,\langle \varphi_0^\alpha \,,\, P\,F\rangle + \epsilon^2\,\langle \varphi_0^\alpha \,,\, \Big[P\,I - 2\,\hat{L}^{-2}\,Q\,G - \hat{L}^{-1}\,Q\,H\Big]\rangle \\
&= \epsilon\,\int \mathrm{d}p\,\varphi_{0p}^\alpha\,p\cdot\nabla\Big[f_p^{\mathrm{eq}} - \epsilon\,f_p^{\mathrm{eq}}\,\bar{f}_p^{\mathrm{eq}}\,\big[\hat{L}^{-1}\,Q\,F\big]_p\Big],
\end{aligned} \tag{11.68}$$

and accordingly the averaging by the inner product leads to

$$\int \mathrm{d}p\,\varphi_{0p}^\alpha\Big[(p\cdot u)\,\frac{\partial}{\partial\tau} + \epsilon\,p\cdot\nabla\Big]\Big[f_p^{\mathrm{eq}} - \epsilon\,f_p^{\mathrm{eq}}\,\bar{f}_p^{\mathrm{eq}}\,\big[\hat{L}^{-1}\,Q\,F\big]_p\Big] = 0, \tag{11.69}$$

---

[3] A detailed proof of Eq. (11.67) is presented in Appendix 13.5.

up to the second order.

Putting back $\epsilon = 1$ in Eq. (11.69), and using the identity

$$(p \cdot u) \frac{\partial}{\partial \tau} + p \cdot \nabla = p^\mu \, \partial_\mu, \tag{11.70}$$

which follows from Eq. (11.5) together with $a_{\mathrm{F}}^\mu = u^\mu$, we arrive at

$$\partial_\mu J^{\mu\alpha} = 0, \tag{11.71}$$

with

$$J^{\mu\alpha} := \int \mathrm{d}p \, p^\mu \, \varphi_{0p}^\alpha \left[f_p^{\mathrm{eq}} - f_p^{\mathrm{eq}} \, \bar{f}_p^{\mathrm{eq}} \left[\hat{L}^{-1} \, Q \, F \right]_p \right]. \tag{11.72}$$

This is the fluid dynamic equation that we wanted. Indeed it is worth noting that $J^{\mu\alpha}$ perfectly agrees with the one obtained by inserting the solution $f_p^{\mathrm{E}}(\tau \, , \, \sigma)$ in Eq. (11.63) into N^μ and $T^{\mu\nu}$ in Eq. (10.27):

$$N^\mu = J^{\mu 4}, \quad T^{\mu\nu} = J^{\mu\nu}. \tag{11.73}$$

Thus we confirm that Eq. (11.71) is the relativistic dissipative fluid dynamic equation.

The current $J^{\mu\alpha}$ is decomposed into two parts as $J^{\mu\alpha} = J_0^{\mu\alpha} + \delta J^{\mu\alpha}$, where

$$J_0^{\mu\alpha} := \int \mathrm{d}p \, p^\mu \, \varphi_{0p}^\alpha \, f_p^{\mathrm{eq}}, \tag{11.74}$$

$$\delta J^{\mu\alpha} := - \int \mathrm{d}p \, p^\mu \, \varphi_{0p}^\alpha \, f_p^{\mathrm{eq}} \, \bar{f}_p^{\mathrm{eq}} \left[\hat{L}^{-1} \, Q \, F \right]_p = -\langle \, \tilde{\varphi}_1^{\mu\alpha} \, , \, \hat{L}^{-1} \, Q \, F \, \rangle, \tag{11.75}$$

with

$$\tilde{\varphi}_{1p}^{\mu\alpha} := p^\mu \, \varphi_{0p}^\alpha \, \frac{1}{p \cdot u}. \tag{11.76}$$

As one can readily recognize, $J_0^{\mu\alpha}$ and $\delta J^{\mu\alpha}$ represent the ideal and dissipative parts of the current, respectively. Correspondingly, N^μ and $T^{\mu\nu}$ in Eqs. (11.73) are also decomposed into the ideal and dissipative parts as

$$N^\mu = N_0^\mu + \delta N^\mu, \quad T^{\mu\nu} = T_0^{\mu\nu} + \delta T^{\mu\nu}, \tag{11.77}$$

respectively, where

$$N_0^\mu := J_0^{\mu 4}, \quad \delta N^\mu := \delta J^{\mu 4}, \tag{11.78}$$

$$T_0^{\mu\nu} := J_0^{\mu\nu}, \quad \delta T^{\mu\nu} := \delta J^{\mu\nu}. \tag{11.79}$$

Let us reduce the expression of the dissipative part $\delta J^{\mu\alpha}$ to a simpler form. First, we note that F in Eq. (11.30) is written as

$$F_p = -\tilde{\varphi}_{1p}^{\mu 4}\,\nabla_\mu \frac{\mu}{T} + \tilde{\varphi}_{1p}^{\mu\nu}\,\nabla_\mu \frac{u_\nu}{T} = -\tilde{\varphi}_{1p}^{\mu\alpha}\,\nabla_\mu X_\alpha, \tag{11.80}$$

where

$$X_\alpha := \begin{cases} -u_\nu/T, & \alpha = \nu, \\ \mu/T, & \alpha = 4. \end{cases} \tag{11.81}$$

We also note the identity

$$\langle\, \varphi\,,\, L^{-1}\, Q\, \psi\,\rangle = \langle\, Q\, \varphi\,,\, \hat{L}^{-1}\, Q\, \psi\,\rangle. \tag{11.82}$$

Then one finds that the dissipative part $\delta J^{\mu\alpha}$ takes the following form,

$$\delta J^{\mu\alpha} = \eta_1^{\mu\alpha\nu\beta}\,\nabla_\nu X_\beta, \tag{11.83}$$

with

$$\eta_1^{\mu\alpha\nu\beta} := \langle\, \varphi_1^{\mu\alpha}\,,\, \hat{L}^{-1}\, \varphi_1^{\nu\beta}\,\rangle, \tag{11.84}$$

where

$$\varphi_{1p}^{\mu\alpha} := \big[Q\, \tilde{\varphi}_1^{\mu\alpha}\big]_p. \tag{11.85}$$

Note that $\varphi_{1p}^{\mu\alpha} \neq \tilde{\varphi}_1^{\mu\alpha}$. Equation (11.83) expresses that the dissipative current is given by a product of quantities of different physical significances. Indeed $\eta_1^{\mu\alpha\nu\beta}$ has some information about the transport coefficients, while $\nabla_\nu X_\beta$ is identical to the corresponding thermodynamic forces.

Using Eqs. (11.69) and (11.80), we can write down the relativistic dissipative fluid dynamic equation defined in the covariant coordinate system $(\sigma\,;\,\tau)$ as

$$\int \mathrm{d}p\, \varphi_{0p}^\alpha \left[(p\cdot u)\,\frac{\partial}{\partial\tau} + p\cdot\nabla\right]\left[f_p^{\mathrm{eq}} + f_p^{\mathrm{eq}}\,\bar{f}_p^{\mathrm{eq}}\,\big[\hat{L}^{-1}\,\varphi_1^{\nu\beta}\big]_p\,\nabla_\nu X_\beta\right] = 0. \tag{11.86}$$

It will be found that this form of the relativistic dissipative fluid dynamic equation plays an essential role in the stability analysis of the fluid dynamics presented in Sect. 11.6.2.

11.5 First-Order Fluid Dynamic Equation and Microscopic Expressions of Transport Coefficients

In this section, we give the explicit form of the currents $J^{\mu\alpha} = J_0^{\mu\alpha} + \delta J^{\mu\alpha}$ and accordingly have the relativistic dissipative fluid dynamic equations with the microscopic expressions of the transport coefficients.

A simple calculation of the integral in Eq. (11.74) for $J_0^{\mu\alpha}$ leads to the following form of the ideal part,

$$J_0^{\mu\alpha} = \begin{cases} e\,u^\mu\,u^\nu - P\,\Delta^{\mu\nu}, & \alpha = \nu, \\ n\,u^\mu, & \alpha = 4, \end{cases} \tag{11.87}$$

where e, P, and n denote the internal energy, pressure, and particle-number density, respectively, the explicit forms of which are given in Eqs. (10.38)–(10.40).

The dissipative part $\delta J^{\mu\alpha}$ is given by (11.83). Since X_β in (11.83) is already given by (11.81), let us calculate the remaining factor $\eta_1^{\mu\alpha\nu\beta}$ in Eq. (11.83), which task is tantamount to that of $\varphi_{1p}^{\mu\alpha}$ as seen from Eq. (11.84).

By the straightforward calculation presented in Chap. 13. we see that $\varphi_{1p}^{\mu\alpha}$ takes the form

$$\varphi_{1p}^{\mu\alpha} = \begin{cases} -\Delta^{\mu\nu}\,\hat{\Pi}_p + \hat{\pi}_p^{\mu\nu}, & \alpha = \nu, \\ \dfrac{n}{e+P}\,\hat{J}_p^\mu, & \alpha = 4. \end{cases} \tag{11.88}$$

Here, $\hat{\Pi}_p$, $\hat{J}_p^\mu$, and $\hat{\pi}_p^{\mu\nu}$ are the microscopic representations of the dissipative currents whose definitions are given by

$$(\hat{\Pi}_p,\ \hat{J}_p^\mu,\ \hat{\pi}_p^{\mu\nu}) := \frac{1}{p\cdot u}\,(\Pi_p,\ J_p^\mu,\ \pi_p^{\mu\nu}). \tag{11.89}$$

with

$$\Pi_p = (p\cdot u)^2\left[\frac{1}{3} - \left.\frac{\partial P}{\partial e}\right|_n\right] - (p\cdot u)\left.\frac{\partial P}{\partial n}\right|_e - \frac{1}{3}m^2, \tag{11.90}$$

$$J_p^\mu = -\Delta^{\mu\nu}\,p_\nu\left[(p\cdot u) - \frac{e+P}{n}\right], \tag{11.91}$$

$$\pi_p^{\mu\nu} = \Delta^{\mu\nu\rho\sigma}\,p_\rho\,p_\sigma. \tag{11.92}$$

It is to be noted that $\varphi_{1p}^{\mu\alpha}$ is a linear combination of $\hat{\Pi}_p$, $\hat{J}_p^\mu$, and $\hat{\pi}_p^{\mu\nu}$, but not that of Π_p, J_p^μ, and $\pi_p^{\mu\nu}$ which have been treated as the microscopic representations of the dissipative currents in the Chapman-Enskog expansion method and the Israel-Stewart fourteen moment method, which are reviewed in Chap. 10.

Now that the explicit form of $\varphi_{1p}^{\mu\alpha}$ has been obtained as in Eq. (11.88), we can write down $\eta_1^{\mu\alpha\nu\beta}$ given by (11.84). Thus, noting following identities

$$\langle \hat{J}^{\mu}, \hat{L}^{-1} \hat{J}^{\nu} \rangle = \frac{1}{3} \Delta^{\mu\nu} \langle \hat{J}^{a}, \hat{L}^{-1} \hat{J}_{a} \rangle, \tag{11.93}$$

$$\langle \hat{\pi}^{\mu\nu}, \hat{L}^{-1} \hat{\pi}^{\rho\sigma} \rangle = \frac{1}{5} \Delta^{\mu\nu\rho\sigma} \langle \hat{\pi}^{ab}, \hat{L}^{-1} \hat{\pi}_{ab} \rangle, \tag{11.94}$$

we have

$$\eta_1^{\mu\rho\nu\sigma} = -T\, \zeta\, \Delta^{\mu\rho}\, \Delta^{\nu\sigma} - 2\, T\, \eta\, \Delta^{\mu\rho\nu\sigma}, \tag{11.95}$$

$$\eta_1^{\mu\rho\nu 4} = \eta_1^{\nu 4\mu\rho} = 0, \tag{11.96}$$

$$\eta_1^{\mu 4\nu 4} = \lambda \left(\frac{n\, T}{e + P} \right)^2 \Delta^{\mu\nu}, \tag{11.97}$$

where

$$\zeta := -\frac{1}{T} \langle \hat{\Pi}, \hat{L}^{-1} \hat{\Pi} \rangle, \tag{11.98}$$

$$\lambda := \frac{1}{3\, T^2} \langle \hat{J}^{\mu}, \hat{L}^{-1} \hat{J}_{\mu} \rangle, \tag{11.99}$$

$$\eta := -\frac{1}{10\, T} \langle \hat{\pi}^{\mu\nu}, \hat{L}^{-1} \hat{\pi}_{\mu\nu} \rangle, \tag{11.100}$$

which are nothing but the transport coefficients, *i.e.*, the bulk viscosity ζ, the heat conductivity λ, and the shear viscosity η.

The expressions of the transport coefficients Eqs. (11.98)–(11.100) can be cast into the forms

$$\zeta = -\frac{1}{T} \int \mathrm{d}p \int \mathrm{d}q\, f_p^{\mathrm{eq}}\, \bar{f}_p^{\mathrm{eq}}\, \Pi_p\, L_{pq}^{-1}\, \Pi_q, \tag{11.101}$$

$$\lambda = \frac{1}{3\, T^2} \int \mathrm{d}p \int \mathrm{d}q\, f_p^{\mathrm{eq}}\, \bar{f}_p^{\mathrm{eq}}\, J_p^{\mu}\, L_{pq}^{-1}\, J_{q\mu}, \tag{11.102}$$

$$\eta = -\frac{1}{10\, T} \int \mathrm{d}p \int \mathrm{d}q\, f_p^{\mathrm{eq}}\, \bar{f}_p^{\mathrm{eq}}\, \pi_p^{\mu\nu}\, L_{pq}^{-1}\, \pi_{q\mu\nu}, \tag{11.103}$$

respectively. Here, Π_p, J_p^{μ}, and $\pi_p^{\mu\nu}$ are the microscopic representations of the dissipative currents given by Eqs. (11.90)–(11.92), and $L_{pq}^{-1} = \hat{L}_{pq}^{-1}\, (q \cdot u)^{-1}$ denotes the inverse matrix of $L_{pq} = (p \cdot u)\, \hat{L}_{pq}$. It turns out that these expressions are in agreement with those obtained by the Chapman-Enskog expansion method [67]; see Eqs. (10.94), (10.95), and (10.96).

Now, let us rewrite the expressions of ζ, λ, and η in a more familiar form, *i.e.*, the Green-Kubo formula [142, 207, 208] in the linear response theory [54, 55, 115, 138, 209]. Frist we define the 'time-evolution' operator $\mathrm{e}^{s\hat{L}}$ that acts on an arbitray vector ψ_p as

$$\psi_p(s) := \left[\mathrm{e}^{s\hat{L}} \psi \right]_p = \int dq \left[\mathrm{e}^{s\hat{L}} \right]_{pq} \psi_q. \tag{11.104}$$

Then we itroduce the 'time-evolved' dissipative currents by

$$\begin{pmatrix} \hat{\Pi}_p(s) \\ \hat{J}^{\mu}_p(s) \\ \hat{\pi}^{\mu\nu}_p(s) \end{pmatrix} := \left[\mathrm{e}^{s\hat{L}} \begin{pmatrix} \hat{\Pi} \\ \hat{J}^{\mu} \\ \hat{\pi}^{\mu\nu} \end{pmatrix} \right]_p = \left[\mathrm{e}^{s\hat{L}} Q \begin{pmatrix} \hat{\Pi} \\ \hat{J}^{\mu} \\ \hat{\pi}^{\mu\nu} \end{pmatrix} \right]_p, \tag{11.105}$$

where the last equality holds since all the dissipative currents $\hat{\Pi}_p$, $\hat{J}^{\mu}_p$, and $\hat{\pi}^{\mu\nu}_p$ belong to the Q space. Integrating out (11.105) over the whole range of the 'time' s, we have

$$\int_0^{\infty} \mathrm{d}s \begin{pmatrix} \hat{\Pi}_p(s) \\ \hat{J}^{\mu}_p(s) \\ \hat{\pi}^{\mu\nu}_p(s) \end{pmatrix} = - \left[\hat{L}^{-1} \begin{pmatrix} \hat{\Pi} \\ \hat{J}^{\mu} \\ \hat{\pi}^{\mu\nu} \end{pmatrix} \right]_p. \tag{11.106}$$

where $\hat{L}^{-1}$ may be understood as $\hat{L}^{-1} Q$. Then we can rewrite Eqs. (11.98)–(11.100) as

$$\zeta = \int_0^{\infty} \mathrm{d}s \, R_{\Pi}(s), \quad \lambda = \int_0^{\infty} \mathrm{d}s \, R_J(s), \quad \eta = \int_0^{\infty} \mathrm{d}s \, R_{\pi}(s), \tag{11.107}$$

with

$$R_{\Pi}(s) := \frac{1}{T} \langle \hat{\Pi}(0) \, , \, \hat{\Pi}(s) \rangle, \tag{11.108}$$

$$R_J(s) := -\frac{1}{3\,T^2} \langle \hat{J}^{\mu}(0) \, , \, \hat{J}_{\mu}(s) \rangle, \tag{11.109}$$

$$R_{\pi}(s) := \frac{1}{10\,T} \langle \hat{\pi}^{\mu\nu}(0) \, , \, \hat{\pi}_{\mu\nu}(s) \rangle, \tag{11.110}$$

which are called the relaxation functions in the linear response theory [54, 55, 115, 138, 209].

Finally, the dissipative currents $\delta J^{\mu\alpha}$ are obtained from $\eta_1^{\mu\alpha\nu\beta}$ in Eqs. (11.95)–(11.97) and $\nabla_{\nu} X_{\beta}$ in Eq. (11.81), as follows:

$$\delta J^{\mu\alpha} = \begin{cases} \zeta \, \Delta^{\mu\nu} \, \nabla \cdot u + 2 \, \eta \, \Delta^{\mu\nu\rho\sigma} \, \nabla_{\rho} u_{\sigma} \, , \; \alpha = \nu, \\ \lambda \left(\dfrac{n\,T}{e+P} \right)^2 \nabla^{\mu} \dfrac{\mu}{T}, \qquad \alpha = 4. \end{cases} \tag{11.111}$$

Here, the following relations have been used: $u^{\mu} \, \nabla_{\mu} = 0$ and $u^{\nu} \, \nabla_{\mu} u_{\nu} = 0$.

Thus, combining Eqs. (11.87) and (11.111), we arrive at the explicit form of the energy-momentum tensor $T^{\mu\nu}$ and particle-number current N^{μ} as

$$T^{\mu\nu} = e\,u^\mu\,u^\nu - (P - \zeta\,\nabla\cdot u)\,\Delta^{\mu\nu} + 2\,\eta\,\Delta^{\mu\nu\rho\sigma}\,\nabla_\rho u_\sigma, \tag{11.112}$$

$$N^\mu = n\,u^\mu + \lambda\left(\frac{n\,T}{e+P}\right)^2 \nabla^\mu \frac{\mu}{T}, \tag{11.113}$$

which are in accordance with the identification as

$$T_0^{\mu\nu} = e\,u^\mu\,u^\nu - P\,\Delta^{\mu\nu}, \tag{11.114}$$

$$\delta T^{\mu\nu} = \zeta\,\Delta^{\mu\nu}\,\nabla\cdot u + 2\,\eta\,\Delta^{\mu\nu\rho\sigma}\,\nabla_\rho u_\sigma, \tag{11.115}$$

$$N_0^\mu = n\,u^\mu, \tag{11.116}$$

$$\delta N^\mu = \lambda\left(\frac{n\,T}{e+P}\right)^2 \nabla^\mu \frac{\mu}{T}. \tag{11.117}$$

A few comments are in order here:

1. As will be discussed in the next section, the energy-momentum tensor (11.112) and the particle current (11.113) are those for the first-order relativistic fluid dynamic equation in the energy frame proposed by Landau and Lifshitz [62].
2. We have obtained the microscopic expressions of the thermodynamic quantities and transport coefficients together with the energy-momentum tensor and the particle number current using single microscopic equation, *i.e.*, the relativistic Boltzmann equation. Then our resultant fluid dynamic equation in the energy frame has an inherent restriction and also a merit absent in the phenomenological theory.
3. The explicit forms of them are for the relativistic rarefied gas, inherently. We would like to remind the reader, however, that the main purpose of the present work is to determine the form of the relativistic fluid dynamic equations for a viscous fluid, and expect that the forms of the macroscopic fluid dynamic equations which contain the thermodynamic quantities and transport coefficients only parametrically, and hence the forms are independent of the microscopic expressions of these quantities.

11.6 Properties of First-Order Fluid Dynamic Equation

In this section, we examine some properties of the resultant first-order fluid dynamic equation, concerning the uniqueness of the local rest frame and the stability of the equilibrium state.

11.6.1 Uniqueness of Landau-Lifshitz Energy Frame

We discuss the form of $\delta T^{\mu\nu}$ and δN^{μ} in Eqs. (11.115) and (11.117).

First, as has been mentioned in the last section, these formulae completely agree with those proposed by Landau and Lifshitz [62]. Indeed, the respective dissipative parts $\delta T^{\mu\nu}$ and δN^{μ} in Eqs. (11.115) and (11.117) meet Landau and Lifshitz's ansatz

$$u_\mu\,\delta T^{\mu\nu}\,u_\nu = 0, \tag{11.118}$$

$$u_\mu\,\delta N^{\mu} = 0, \tag{11.119}$$

$$\Delta_{\mu\nu}\,\delta T^{\nu\rho}\,u_\rho = 0, \tag{11.120}$$

which are nothing but the constrains imposed in a heuristic way by Landau and Lifshitz in their phenomenological derivation [62, 67].

Next, we shall discus the underlying meaning of Eqs. (11.118)–(11.120) in the level of the kinetic equation using the distribution function on the basis of the previous results [204, 205]. As was mentioned above, these equations are usually just imposed [67] to the higher-order terms of the distribution function as the conditions of fit without any foundation to select the fluid dynamic equation in the energy frame. We shall clarify that these conditions are equivalent to the orthogonality condition for the excited modes expressed in terms of the inner product [204, 205] and hence an inevitable consequence for the relativistic fluid dynamics in our analysis which is free from any ansatz.

A manipulation shows [204, 205] that Eqs. (11.79) and (11.78) can be rewritten as

$$\delta T^{\mu\nu} = -\int \mathrm{d}p\, p^\mu\, p^\nu\, f^{\mathrm{eq}}_p\, \bar{f}^{\mathrm{eq}}_p \left[\hat{L}^{-1}\, Q\, F\right]_p, \tag{11.121}$$

$$\delta N^{\mu} = -\int \mathrm{d}p\, p^\mu\, f^{\mathrm{eq}}_p\, \bar{f}^{\mathrm{eq}}_p \left[\hat{L}^{-1}\, Q\, F\right]_p. \tag{11.122}$$

It should be emphasized that $\left[\hat{L}^{-1}\, Q\, F\right]_p$ belongs to the Q space and thus orthogonal to the zero modes,

$$\langle\, \varphi^{\alpha}_0\,,\, \hat{L}^{-1}\, Q\, F\,\rangle = 0, \quad \alpha = 0,\ 1,\ 2,\ 3,\ 4. \tag{11.123}$$

Recalling the definition Eq. (11.31) of the inner product, we see that Eq. (11.123) with $\alpha = \mu = 0,\ 1,\ 2,\ 3$ can be recast into the following form,

$$0 = \int \mathrm{d}p\,(p\cdot u)\, f^{\mathrm{eq}}_p\, \bar{f}^{\mathrm{eq}}_p\, p^\mu \left[\hat{L}^{-1}\, Q\, F\right]_p = -u_\nu\, \delta T^{\mu\nu}, \tag{11.124}$$

which readily leads to Eqs. (11.118) and (11.120). Quite similarly, Eq. (11.123) with $\alpha = 4$ is reduced to

$$0 = \int \mathrm{d}p \, (p \cdot u) \, f_p^{\mathrm{eq}} \, \bar{f}_p^{\mathrm{eq}} \left[\hat{L}^{-1} \, Q \, F \right]_p = -u_\nu \, \delta N^\nu, \tag{11.125}$$

which is nothing but Eq. (11.119).

A few remarks are in order here:

(i) Our proof of the uniqueness of the energy frame [210] can be traced back to the identification of the macroscopic frame vector a_{F}^{μ} with the flow velocity u^{μ} and the physical assumption that the dissipative effects can be solely attributed to the spatial inhomogeneity, apart from the Gibbs relation in the local equilibrium state which leads to the local equilibrium distribution function (11.20), as was mentioned in Sect. 10.2; see Sect. 2.7 of [68]. Conversely, if one of these conditions were not satisfied, the uniqueness of the energy frame could be violated, as is argued in [174]; see also [173]. Nevertheless, we shall further show below that if the macroscopic frame vector a_{F}^{μ} in Eq. (11.25) retains the space-like term δw^{μ}, the resulting terms turn out to be of third-order in the derivatives, which is of higher order than that in the Landau-Lifshitz energy frame.

(ii) In [211], the authors showed, on the basis of the projection operator method [100, 158, 212], that only the energy frame is natural one for the relativistic fluid dynamics, at least in the linear regime, if the fluid dynamics is an effective dynamics described solely by the genuine slow variables of microscopic Hamiltonian dynamics at all. In a relatively recent paper [213], Hayata et al. derived relativistic dissipative fluid dynamic equations in the energy and particle-frame from quantum field theory on the basis of the nonequilibrium statistical operator method [209, 214–216], and they showed that if the frame vector (in our terminology) is in the lowest order of the derivative expansion, then the resultant fluid dynamics becomes uniquely that in the energy frame given by Landau and Lifshitz, which is in accordance with the present analysis.

(iii) There might be fundamental principles which lead to the unique choice of the energy frame for the dissipative relativistic fluid dynamic equation as the infrared effective theory of many body systems. The exploration of such principles should be interesting.

In the rest of this section, we shall show that although we start with the generic form of the macroscopic frame vector $a_{\mathrm{F}}^{\mu} = C_t(T,\ \mu)\, u^{\mu} + \delta w^{\mu}$ given in Eq. (11.25), the leading term of the fluid dynamic equation with respect to δw^{μ} is found to become a third-order in derivatives, and accordingly of higher order than that in the Landau-Lifshitz energy frame derived by setting $a_{\mathrm{F}}^{\mu} = u^{\mu}$ from the outset.

By replacing u^{μ} by a_{F}^{μ} in all of the quantities except for that originated from f_p^{eq}, we can derive straightforwardly the fluid dynamic equation with a_{F}^{μ} not being specified. For instance, by this replacement, $\hat{L}_{pq}$, $\tilde{\varphi}_{1p}^{\mu\alpha}$, and the inner product are converted into

$$\hat{L}_{pq} = \frac{1}{p\cdot u} L_{pq} \to \frac{1}{p\cdot a_{\mathrm{F}}} L_{pq} =: \hat{L}_{pq}(a_{\mathrm{F}}), \tag{11.126}$$

$$\tilde{\varphi}^{\mu\alpha}_{1p} = \frac{1}{p\cdot u} p^{\mu} \varphi^{\alpha}_{0p} \to \frac{1}{p\cdot a_{\mathrm{F}}} p^{\mu} \varphi^{\alpha}_{0p} =: \tilde{\varphi}^{\mu\alpha}_{1p}(a_{\mathrm{F}}), \tag{11.127}$$

$$\langle\, \psi\,,\, \chi\,\rangle = \int \mathrm{d}p\, f^{\mathrm{eq}}_p\, \bar{f}^{\mathrm{eq}}_p\, (p\cdot u)\, \psi_p\, \chi_p \to \int \mathrm{d}p\, f^{\mathrm{eq}}_p\, \bar{f}^{\mathrm{eq}}_p\, (p\cdot a_{\mathrm{F}})\, \psi_p\, \chi_p, \tag{11.128}$$

for arbitrary vectors ψ_p and χ_p. We note that owing the modification of the inner product the projection operators P and Q are also modified into

$$\big[P(a_{\mathrm{F}})\,\psi\big]_p = \varphi^{\alpha}_{0p}\, \eta^{-1}_{0\alpha\beta}(a_{\mathrm{F}})\,\langle\, \varphi^{\beta}_{0}\,,\, \psi\,\rangle, \tag{11.129}$$

$$Q(a_{\mathrm{F}}) = 1 - P(a_{\mathrm{F}}), \tag{11.130}$$

respectively, where $\eta^{-1}_{0\alpha\beta}(a_{\mathrm{F}})$ is the inverse matrix of

$$\eta^{\alpha\beta}_{0}(a_{\mathrm{F}}) := \langle\, \varphi^{\alpha}_{0}\,,\, \varphi^{\beta}_{0}\,\rangle = \int \mathrm{d}p\, f^{\mathrm{eq}}_p\, \bar{f}^{\mathrm{eq}}_p\, (p\cdot a_{\mathrm{F}})\, \varphi^{\alpha}_{0p}\, \varphi^{\beta}_{0p}. \tag{11.131}$$

Using these modifications, we obtain the fluid dynamic equation as $\partial_{\mu} N^{\mu}(a_{\mathrm{F}}) = \partial_{\mu} T^{\mu\nu}(a_{\mathrm{F}}) = 0$ with

$$N^{\mu}(a_{\mathrm{F}}) = N^{\mu}_{0}(a_{\mathrm{F}}) + \delta N^{\mu}(a_{\mathrm{F}}), \tag{11.132}$$

$$T^{\mu\nu}(a_{\mathrm{F}}) = T^{\mu\nu}_{0}(a_{\mathrm{F}}) + \delta T^{\mu\nu}(a_{\mathrm{F}}), \tag{11.133}$$

where

$$N^{\mu}_{0}(a_{\mathrm{F}}) = \int \mathrm{d}p\, p^{\mu}\, f^{\mathrm{eq}}_p, \tag{11.134}$$

$$\delta N^{\mu}(a_{\mathrm{F}}) = -\int \mathrm{d}p\, p^{\mu}\, f^{\mathrm{eq}}_p\, \bar{f}^{\mathrm{eq}}_p\, (p\cdot a_{\mathrm{F}})\, \big[\hat{L}^{-1}(a_{\mathrm{F}})\, \varphi^{\nu\beta}_{1}(a_{\mathrm{F}})\big]_p \times \Delta_{\mathrm{F}\nu\rho}\, \partial^{\rho} X_{\beta}, \tag{11.135}$$

$$T^{\mu\nu}_{0}(a_{\mathrm{F}}) = \int \mathrm{d}p\, p^{\mu}\, p^{\nu}\, f^{\mathrm{eq}}_p, \tag{11.136}$$

$$\delta T^{\mu\nu}(a_{\mathrm{F}}) = -\int \mathrm{d}p\, p^{\mu}\, p^{\nu}\, f^{\mathrm{eq}}_p\, \bar{f}^{\mathrm{eq}}_p\, (p\cdot a_{\mathrm{F}})\, \big[\hat{L}^{-1}(a_{\mathrm{F}})\, \varphi^{\nu\beta}_{1}(a_{\mathrm{F}})\big]_p \times \Delta_{\mathrm{F}\nu\rho}\, \partial^{\rho} X_{\beta}, \tag{11.137}$$

where X_{α} has been defined in Eq. (11.81), $\Delta^{\mu\nu}_{\mathrm{F}}$ denotes the projection matrix introduced in Eq. (11.3), $\hat{L}^{-1}(a_{\mathrm{F}})$ is the inverse matrix of $\hat{L}(a_{\mathrm{F}})$, and $\varphi^{\mu\alpha}_{1}(a_{\mathrm{F}})$ is given by

$$\varphi^{\mu\alpha}_{1p}(a_{\mathrm{F}}) =: \left[Q(a_{\mathrm{F}})\,\tilde{\varphi}^{\mu\alpha}_{1}(a_{\mathrm{F}})\right]_p$$
$$= \tilde{\varphi}^{\mu\alpha}_{1p}(a_{\mathrm{F}}) - \varphi^{\beta}_{0p}\,\eta^{-1}_{0\beta\gamma}(a_{\mathrm{F}})\,\langle\,\varphi^{\gamma}_{0}\,,\,\tilde{\varphi}^{\mu\alpha}_{1}(a_{\mathrm{F}})\,\rangle. \tag{11.138}$$

It is important to note that the zeroth terms $N^{\mu}_{0}(a_{\mathrm{F}})$ and $T^{\mu\nu}_{0}(a_{\mathrm{F}})$ are independent of a^{μ}_{F}, and agree with the particle current and energy-momentum tensor in the relativistic Euler equation as

$$N^{\mu}_{0}(a_{\mathrm{F}}) = n\,u^{\mu} = N^{\mu}_{0}, \tag{11.139}$$
$$T^{\mu\nu}_{0}(a_{\mathrm{F}}) = e\,u^{\mu}\,u^{\nu} - P\,\Delta^{\mu\nu} = T^{\mu\nu}_{0}, \tag{11.140}$$

respectively. It is apparent that $N^{\mu}_{0}(a_{\mathrm{F}})$ and $T^{\mu\nu}_{0}(a_{\mathrm{F}})$ contain no differential operator ∂_{μ}. On the other hand, $\delta N^{\mu}(a_{\mathrm{F}})$ and $\delta T^{\mu\nu}(a_{\mathrm{F}})$ have a dependence on a^{μ}_{F}, and manifestly contain terms of the first order of ∂_{μ}.

Here, we shall count the number of ∂_{μ} in $\delta N^{\mu}(a_{\mathrm{F}})$ and $\delta T^{\mu\nu}(a_{\mathrm{F}})$. For this purpose, we first expand $\delta N^{\mu}(a_{\mathrm{F}})$ and $\delta T^{\mu\nu}(a_{\mathrm{F}})$ with respect to δw^{μ} as

$$\delta N^{\mu}(a_{\mathrm{F}}) = \delta N^{\mu}(u) + \frac{\partial}{\partial a^{\nu}_{\mathrm{F}}}\delta N^{\mu}(a_{\mathrm{F}})\bigg|_{a_{\mathrm{F}}=u}\delta w^{\nu} + \cdots, \tag{11.141}$$

$$\delta T^{\mu\nu}(a_{\mathrm{F}}) = \delta T^{\mu\nu}(u) + \frac{\partial}{\partial a^{\rho}_{\mathrm{F}}}\delta T^{\mu\nu}(a_{\mathrm{F}})\bigg|_{a_{\mathrm{F}}=u}\delta w^{\rho} + \cdots, \tag{11.142}$$

where $\delta N^{\mu}(u)$ and $\delta T^{\mu\nu}(u)$ are nothing but those in the energy frame, *i.e.*, $\delta N^{\mu}(u) = \delta N^{\mu}$ and $\delta T^{\mu\nu}(u) = \delta T^{\mu\nu}$.

Then, let us count the number of ∂_{μ} in each terms of $N^{\mu}(a_{\mathrm{F}})$. We find that

$$N^{\mu}_{0} \sim O(\partial^{0}), \tag{11.143}$$
$$\delta N^{\mu}(u) \sim O(\partial^{1}), \tag{11.144}$$
$$\frac{\partial}{\partial a^{\nu}_{\mathrm{F}}}\delta N^{\mu}(a_{\mathrm{F}})\bigg|_{a_{\mathrm{F}}=u} \sim O(\partial^{1}), \tag{11.145}$$
$$\delta w^{\nu} \sim O(\partial^{1}), \tag{11.146}$$

where we note that the counting in Eq. (11.145) has been derived from the fact that the partial derivative with respect to a^{μ}_{F} does not change the number of ∂_{μ} that manifestly exists in $\delta N^{\mu}(a_{\mathrm{F}})$. Thus, we have

$$\frac{\partial}{\partial a^{\nu}_{\mathrm{F}}}\delta N^{\mu}(a_{\mathrm{F}})\bigg|_{a_{\mathrm{F}}=u}\delta w^{\nu} \sim O(\partial^{2}). \tag{11.147}$$

Finally, by combining Eqs. (11.143), (11.144), and (11.147), we have

$$N^{\mu}(a_{\mathrm{F}}) = N^{\mu}(u) + O(\partial^{2}). \tag{11.148}$$

The same argument can be given for $T^{\mu\nu}(a_{\rm F})$, and we obtain

$$T^{\mu\nu}(a_{\rm F}) = T^{\mu\nu}(u) + O(\partial^2). \tag{11.149}$$

Thus, the fluid dynamic equation with $a^{\mu}_{\rm F} = C_t u^{\mu} + \delta w^{\mu}$ reads

$$0 = \partial_\mu N^\mu(a_{\rm F}) = \partial_\mu N^\mu(u) + O(\partial^3), \tag{11.150}$$
$$0 = \partial_\mu T^{\mu\nu}(a_{\rm F}) = \partial_\mu T^{\mu\nu}(u) + O(\partial^3). \tag{11.151}$$

The above equations tell us that even if δw^{μ} is left, the resultant fluid dynamic equation is consistent with one in the energy frame in the fluid dynamic regime.

11.6.2 Generic Stability

In this Appendix, we prove that when the fluid dynamic equation in the energy frame thus derived is applied with the microscopic representations of the transport coefficients (11.98)–(11.100) and the thermodynamic quantities, the generic constant solutions is stable against a small perturbation on account of the positive definiteness of the inner product Eq. (11.32) [205, 217]. Here the generic constant solution means a finite homogeneous flow with a constant temperature and a constant chemical potential, as given by

$$T(\sigma\,;\,\tau) = T_0, \quad \mu(\sigma\,;\,\tau) = \mu_0, \quad u_\mu(\sigma\,;\,\tau) = u_{0\mu}, \tag{11.152}$$

where T_0, μ_0, and $u_{0\mu}$ are constant. Note that these states include the thermal equilibrium state, given by $u_{0\mu} = (1,\,0,\,0,\,0)$, as a special case.

We shall make a linear stability analysis for the situation where $T(\sigma\,;\,\tau)$, $\mu(\sigma\,;\,\tau)$, and $u_\mu(\sigma\,;\,\tau)$ are close to the respective constant solutions, and represent them as

$$T(\sigma\,;\,\tau) = T_0 + \delta T(\sigma\,;\,\tau), \tag{11.153}$$
$$\mu(\sigma\,;\,\tau) = \mu_0 + \delta\mu(\sigma\,;\,\tau), \tag{11.154}$$
$$u_\mu(\sigma\,;\,\tau) = u_{0\mu} + \delta u_\mu(\sigma\,;\,\tau), \tag{11.155}$$

where the deviations δT, $\delta\mu$, and δu_μ are assumed so small that we only have to retain the first-order terms of them.

Since these six variables which are not independent of each other because of the constraint $\delta u_\mu\, u^\mu_0 = 0$, we use instead the following five independent variables in accordance with Eq. (11.81),

$$\delta X_\alpha =: \begin{cases} -\delta(u_\mu/T) = -\delta u_\mu/T_0 + \delta T\, u_{0\mu}/T_0^2, & \alpha = \mu, \\ \delta(\mu/T) = \delta\mu/T_0 - \delta T\, \mu_0/T_0^2, & \alpha = 4. \end{cases} \tag{11.156}$$

Inserting Eq. (11.156) into Eq. (11.86), we obtain the linearized equation governing δX_α, after some manipulations, as

$$\Big(\langle\, \varphi_0^\alpha\,,\, \varphi_0^\beta\,\rangle + \langle\, \varphi_0^\alpha\,,\, \hat{L}^{-1}\,\varphi_1^{\nu\beta}\,\rangle\, \nabla_\nu\Big)\, \frac{\partial}{\partial\tau}\delta X_\beta$$
$$+ \Big(\langle\, \tilde{\varphi}_1^{\mu\alpha}\,,\, \varphi_0^\beta\,\rangle\, \nabla_\mu + \langle\, \tilde{\varphi}_1^{\mu\alpha}\,,\, \hat{L}^{-1}\,\varphi_1^{\nu\beta}\,\rangle\, \nabla_\mu\, \nabla_\nu\Big)\, \delta X_\beta = 0. \quad (11.157)$$

Here, we have used the following simple relation

$$\delta(f_p^{\mathrm{eq}}) = f_p^{\mathrm{eq}}\, \bar{f}_p^{\mathrm{eq}}\, \varphi_{0p}^\alpha\, \delta X_\alpha. \quad (11.158)$$

We note that all the coefficients in Eq. (11.157) take constant values because they are solely given by the constant solution $(T,\ \mu,\ u_\mu) = (T_0,\ \mu_0,\ u_{0\mu})$. Owing to the orthogonality between the P and Q spaces, Eq. (11.157) is reduced to

$$\eta_0^{\alpha\beta}\, \frac{\partial}{\partial\tau}\delta X_\beta + D^{\alpha\beta}\, \delta X_\beta = 0. \quad (11.159)$$

Here $\eta_0^{\alpha\beta}$ is the metric tensor defined in (11.39) and $D^{\alpha\beta}$ is defined by

$$D^{\alpha\beta} =: \langle\, \tilde{\varphi}_1^{\mu\alpha}\,,\, \varphi_0^\beta\,\rangle\, \nabla_\mu + \eta_1^{\mu\alpha\nu\beta}\, \nabla_\mu\, \nabla_\nu, \quad (11.160)$$

with $\eta_1^{\mu\alpha\nu\beta} = \langle\, \varphi_1^{\mu\alpha}\,,\, \hat{L}^{-1}\,\varphi_1^{\nu\beta}\,\rangle$ given by Eqs. (11.95)–(11.97). Both of $\eta_0^{\alpha\beta}$ and $D^{\alpha\beta}$ are symmetric tensors.

Inserting the ansatz

$$\delta X_\alpha(\sigma\,;\,\tau) = \delta\tilde{X}_\alpha(k\,;\,\Lambda)\, \mathrm{e}^{ik\cdot\sigma - \Lambda\tau}, \quad (11.161)$$

into Eq. (11.159), we have the following linear equation,

$$(\Lambda\, \eta_0^{\alpha\beta} - \tilde{D}^{\alpha\beta})\, \delta\tilde{X}_\beta(k\,;\,\Lambda) = 0, \quad (11.162)$$

with

$$\tilde{D}^{\alpha\beta} := i\, \langle\, \tilde{\varphi}_1^{\mu\alpha}\,,\, \varphi_0^\beta\,\rangle\, k_\mu - \eta_1^{\mu\alpha\nu\beta}\, k_\mu\, k_\nu. \quad (11.163)$$

Since $\delta\tilde{X}_\beta(k\,;\,\Lambda) \not\equiv 0$, we have the eigenvalue equation

$$\det(\Lambda\, \eta_0 - \tilde{D}) = 0, \quad (11.164)$$

which leads to the dispersion relation

$$\Lambda = \Lambda(k). \quad (11.165)$$

Now the stability of the generic constant solution (11.152) against a small perturbation is assured when the real part of $\Lambda(k)$ is non-negative for any k^μ, *i.e.*,

$$\mathrm{Re}\,\Lambda(k) \ge 0, \quad \forall k^\mu, \tag{11.166}$$

which is shown to be the case below.

First of all, we note that the matrix η_0 is a real symmetric and positive-definite matrix:

$$\begin{aligned} w_\alpha\, \eta_0^{\alpha\beta}\, w_\beta &= \langle\, w_\alpha\, \varphi_0^\alpha\,,\, w_\beta\, \varphi_0^\beta\, \rangle \\ &= \langle\, \varphi\,,\, \varphi\, \rangle > 0, \quad w_\alpha \neq 0, \end{aligned} \tag{11.167}$$

where w_α is a component of an arbitrary vector $\boldsymbol{w}$ and $\varphi_p := w_\alpha\, \varphi_{0p}^\alpha$. Here, we have used the positive definiteness of the inner product (11.32). Equation (11.167) means that the inverse matrix η_0^{-1} exists, and η_0^{-1} is also a real symmetric positive-definite matrix. Therefore, using the Cholesky decomposition [218], we can represent η_0^{-1} as

$$\eta_0^{-1} = {}^t U\, U, \tag{11.168}$$

where U denotes a real matrix and ${}^t U$ a transposed matrix of U. Then, Eq. (11.164) is converted to

$$\det(\Lambda\, I - U\, \tilde{D}\, {}^t U) = 0, \tag{11.169}$$

where I denotes the unit matrix. Equation (11.169) tells us that $\Lambda(k)$ is an eigenvalue of $U\, \tilde{D}\, {}^t U$.

Next, we prove the following theorem for a complex matrix C:

If $\mathrm{Re}(C) := (C + C^\dagger)/2$, which is hermitian, is semi-positive definite, then the real part of the eigenvalue of C is nonnegative.

[Proof] Let $\boldsymbol{\chi}$ be a normalized eigenvector of C belonging to its eigenvalue λ;

$$C\, \boldsymbol{\chi} = \lambda\, \boldsymbol{\chi}, \tag{11.170}$$

with $\boldsymbol{\chi}^\dagger\, \boldsymbol{\chi} = 1$. If $\mathrm{Re}(C)$ is semi-positive definite, we can suppose that

$$\boldsymbol{\psi}^\dagger\, \mathrm{Re}(C)\, \boldsymbol{\psi} \ge 0, \quad \forall\, \boldsymbol{\psi}. \tag{11.171}$$

By setting $\boldsymbol{\psi} = \boldsymbol{\chi}$ in Eq. (11.171) and using Eq. (11.170) and its hermitian conjugate, we have

$$(\lambda + \lambda^*)/2 \ge 0, \tag{11.172}$$

which means that the real part of the eigenvalue of C is nonnegative. This completes the proof.

Applying this theorem to the present case, we find that the real part of $\Lambda(k)$ becomes nonnegative for any k^μ when $\mathrm{Re}(U\,\tilde{D}^t U)$ is a semi-positive definite matrix, which is shown to be the case, as follows;

$$\begin{aligned} w_\alpha[\mathrm{Re}(U\tilde{D}^t U)]^{\alpha\beta} w_\beta &= w_\alpha[U\mathrm{Re}(\tilde{D})^t U]^{\alpha\beta} w_\beta \\ &= [wU]_\alpha[\mathrm{Re}(\tilde{D})]^{\alpha\beta}[wU]_\beta \\ &= -[w\,U]_\alpha\,\eta_1^{\mu\alpha\nu\beta}\,k_\mu\,k_\nu\,[w\,U]_\beta \\ &= -\langle\,k_\mu\,[w\,U]_\alpha\,\varphi_1^{\mu\alpha}\,,\,\hat{L}^{-1}\,k_\nu\,[w\,U]_\beta\,\varphi_1^{\nu\beta}\,\rangle \\ &= -\langle\,\psi\,,\,\hat{L}^{-1}\,\psi\,\rangle \geq 0, \quad w_\alpha \neq 0, \end{aligned} \tag{11.173}$$

where

$$\psi_p := k_\mu\,[w\,U]_\alpha\,\varphi_{1p}^{\mu\alpha}. \tag{11.174}$$

This completes the proof that the generic constant solution in Eq. (11.152) is stable against a small perturbation.

Chapter 12
RG/E Derivation of Relativistic Second-Order Fluid Dynamics

12.1 Introduction

In Chap. 11, we have derived the relativistic *first-order* fluid dynamic equation to describe macroscopic dynamics of the underlying microscopic kinetic theory given by the Boltzmann equation on the basis of the RG method [1, 38, 39, 48–52, 92–96] as formulated in [3, 5, 6, 46, 57, 58] and presented in Chap. 5; in the derivation, the zero modes of the unperturbed operator are utilized.[1] The derived equation has a desirable properties that the microscopic expressions of the transport coefficients are of the same forms as those given in the established Chapman-Enskog expansion method [67, 68, 113], and it admits stable constant equilibrium states.[2] However, this equation has a serious short comings as a relativistic theory that it formally violates the causality, which is actually a common drawbacks inherent in the first-order fluid dynamic equations.

The causal dissipative fluid dynamic equation needs to be a *second-order* one in which some additional, say, excited modes beyond the zero modes are incorporated to make an extended closed 'slow' dynamics. It means that the second-order fluid dynamic equation can be characterized as the *mesoscopic* dynamics [64, 144–154], *i.e.*, the dynamics in the intermediate scales between the macroscopic and microscopic ones as described by the first-order fluid dynamic equation and the Boltzmann equation or a kinetic equation, respectively.

As is fully accounted for in Chap. 9, the *doublet scheme* [69, 217] is a general framework for the construction of the mesoscopic dynamics from the microscopic dynamics based on the RG method. In fact, the doublet scheme has been applied to the non-relativistic Boltzmann equation [69, 217], and has successfully led to the second-order non-relativistic fluid dynamic equation which happens to be a natural

---

[1] As for some historical development of the RG method, see Sect. 1.3.

[2] As for the stability problem of dissipative relativistic fluids, see the pioneering work by Hiscock and Lindblom [63, 219] and subsequent extensive analyses [174, 191, 220–225], and the references therein.

T. Kunihiro et al., *Geometrical Formulation of Renormalization-Group Method as an Asymptotic Analysis*, Fundamental Theories of Physics 206,
https://doi.org/10.1007/978-981-16-8189-9_12

extension of the first-order equation, *i.e.*, the Navier-Stokes equation with the microscopic expressions of the transport coefficients as given by the Chapman-Enskog method [67, 68, 113].

Thus, it is intriguing to apply the doublet scheme in the RG method to the derivation of relativistic second-order fluid dynamic equations from the relativistic Boltzmann equation [67, 68], as the relativistic extension of the non-relativistic case [69, 217]. In this chapter, we derive the relativistic second-order fluid dynamic equation and the microscopic formulae of the relaxation times as well as the transport coefficients in a systematic way on the basis of the doublet scheme from the relativistic Boltzmann equation with quantum statistics.

The task to be performed in this chapter is a kind of combination of those in Chaps. 11 and 9. Thus the next Sect. 12.2 starts from a recapitulation of Chap. 11 and then the doublet scheme is applied in Sect. 12.3 to construct the mesoscopic dynamics of the relativistic Boltzmann equation and thereby derive the second-order relativistic fluid dynamic equations with the microscopic expressions of the relaxation times as well as the transport coefficients. In reproducing some of equations that appeared in Chap. 11, we shall dare to reproduce the equations in this chapter for convenience and self-containedness.

In accordance with the first-order one presented in Chap. 11, the resultant fluid dynamic equation is defined uniquely in the energy frame proposed by Landau and Lifshitz, and the microscopic representations of the transport coefficients take the same forms as those given by the Chapman-Enskog expansion method. However, the resultant microscopic expressions of the relaxation times are different from those by the Israel-Stewart fourteen moment method, which fails in giving the transport coefficients given by the Chapman-Enskog method.

It is proved in Sect. 12.7, that the derived second-order fluid dynamic equation satisfies the causality in the sense that the propagating velocities of the fluctuations of the fluid dynamic variables do not exceed the speed of light. It is also confirmed that the equilibrium state is stable for any perturbation described by our equation.

12.2 Preliminaries

As preliminaries, we first recapitulate the first part of Chap. 11 for notational convenience.

To extract the fluid dynamic equation from the relativistic Boltzmann equation

$$p^{\mu}\frac{\partial}{\partial x^{\mu}} f_p(x) = C[f]_p(x), \tag{12.1}$$

in the asymptotic limit, it is best to use a coarse-grained coordinates with the Lorentz covariance retained, and we introduce the macroscopic frame vector

$$a_{\mathrm{F}}^{\mu}, \qquad (a_{\mathrm{F}\mu}a_{\mathrm{F}}^{\mu} = 1,\ a_{\mathrm{F}}^{0} > 0). \tag{12.2}$$

to define the local frame [204, 205]. Then the derivative $\frac{\partial}{\partial x_\mu} =: \partial^\mu$ is decomposed into time-like and space-like ones as

$$\partial^\mu = a_{\rm F}^\mu(x)\, a_{\rm F}^\nu(x)\, \partial_\nu + \Delta_{\rm F}^{\mu\nu}(x)\, \partial_\nu = a_{\rm F}^\mu(x)\, \frac{\partial}{\partial \tau} + \frac{\partial}{\partial \sigma_\mu}, \tag{12.3}$$

where $\Delta_{\rm F}^{\mu\nu}(x) = g^{\mu\nu} - a_{\rm F}^\mu(x)\, a_{\rm F}^\nu(x)$, $\frac{\partial}{\partial\tau} = a_{\rm F}^\mu\, \partial_\mu$, and $\frac{\partial}{\partial\sigma_\mu} = \Delta_{\rm F}^{\mu\nu}\, \partial_\nu$.

Wit the novel coordinate system $(\tau\,,\,\sigma^\mu)$, the Boltzmann equation (12.1) is expressed as

$$p \cdot a_{\rm F}(\tau\,,\,\sigma)\, \frac{\partial}{\partial \tau} f_p(\tau\,,\,\sigma) + p^\mu \frac{\partial}{\partial \sigma^\mu} f_p(\tau\,,\,\sigma) = C[f]_p(\tau\,,\,\sigma), \tag{12.4}$$

with $a_{\rm F}^\mu(\tau\,,\,\sigma) =: a_{\rm F}^\mu(x)$, $f_p(\tau\,,\,\sigma) =: f_p(x)$ and $p \cdot a_{\rm F}(\tau\,,\,\sigma) > 0$.

Using a small parameter ϵ that may be identified with the Knudsen number $\mathcal{K}$, *i.e.*, the ratio of the mean free path to the representative macroscopic length scale of the system, a coarse-grained spatial coordinate is introduced $\bar{\sigma}^\mu = \epsilon\, \sigma^\mu \to \sigma^\mu$. Then, Eq. (12.4) is converted into the form given by

$$\begin{aligned} \frac{\partial}{\partial \tau} f_p(\tau\,,\,\sigma) = & \frac{1}{p \cdot a_{\rm F}(\tau\,,\,\sigma)}\, C[f]_p(\tau\,,\,\sigma) \\ & - \epsilon\, \frac{1}{p \cdot a_{\rm F}(\tau\,,\,\sigma)}\, p^\mu \frac{\partial}{\partial \sigma^\mu} f_p(\tau\,,\,\sigma). \end{aligned} \tag{12.5}$$

We apply the doublet scheme in the RG method to the Boltzmann equation (12.5) on the basis of the perturbation theory to obtain the second-order fluid dynamic equation in the asymptotic region; The parameter ϵ will be eventually set back to unity in the end of the calculation.

Let $f_p(\tau, \sigma)$ be an exact solution to (12.5), which is assumed to be expanded with respect to ϵ as $f_p(\tau\,,\,\sigma) = f_p^{(0)}(\tau\,,\,\sigma) + \epsilon\, f_p^{(1)}(\tau\,,\,\sigma) + \epsilon^2\, f_p^{(2)}(\tau\,,\,\sigma) + \cdots$. We write the perturbative solution to Eq. (12.5) around an arbitrary time $\tau = \tau_0$ in the asymptotic region as $\tilde{f}_p$, with the 'initial' value at $\tau = \tau_0$ set to the exact value as

$$\tilde{f}_p(\tau = \tau_0\,,\,\sigma\,;\,\tau_0) = f_p(\tau_0\,,\,\sigma). \tag{12.6}$$

Now, we are interested in a local solution to Eq. (12.5) that only has to be valid around $\tau \simeq \tau_0$ in the RG method, and the time scales of the change of $a_{\rm F}^\mu(\tau\,,\,\sigma)$ is supposed to be much larger than the typical time scales of the microscopic processes as described by the Boltzmann equation. Thus, it is legitimate to neglect the time dependence of $a_{\rm F}^\mu(\tau\,,\,\sigma)$ in Eq. (12.5) and set to the value $a_{\rm F}^\mu(\tau_0\,,\,\sigma)$ as $a_{\rm F}^\mu(\tau\,,\,\sigma) \to a_{\rm F}^\mu(\tau_0\,,\,\sigma) =: a_{\rm F}^\mu(\sigma;\,\tau_0)$ for the purpose of obtaining a local solution around $\tau \simeq \tau_0$, and hence we only have to solve the following equation

$$\frac{\partial}{\partial \tau} f_p(\tau\,,\,\sigma) = \frac{1}{p \cdot a_{\mathrm{F}}(\sigma;\,\tau_0)}\,C[f]_p(\tau\,,\,\sigma) - \epsilon\,\frac{1}{p \cdot a_{\mathrm{F}}(\sigma;\,\tau_0)}\,p^{\mu}\frac{\partial}{\partial \sigma^{\mu}} f_p(\tau\,,\,\sigma), \tag{12.7}$$

the solution to which is expanded with respect to ϵ as follows; $\tilde{f}_p(\tau\,,\,\sigma\,;\,\tau_0) = \tilde{f}_p^{(0)}(\tau\,,\,\sigma\,;\,\tau_0) + \epsilon\,\tilde{f}_p^{(1)}(\tau\,,\,\sigma\,;\,\tau_0) + \epsilon^2\,\tilde{f}_p^{(2)}(\tau\,,\,\sigma\,;\,\tau_0) + \cdots$ with the 'initial' condition posed on each term to coincide with the corresponding exact value at $\tau = \tau_0$ as

$$\tilde{f}_p^{(l)}(\tau_0\,,\,\sigma\,;\,\tau_0) = f_p^{(l)}(\tau_0\,,\,\sigma), \quad l = 0,\,1,\,2,\cdots. \tag{12.8}$$

Inserting the above expansions into Eq. (12.7), we obtain the equation for $\tilde{f}_p^{(n)}(\tau_0\,,\,\sigma\,;\,\tau_0)$. The zeroth-order equation reads

$$\frac{\partial}{\partial \tau}\tilde{f}_p^{(0)}(\tau\,,\,\sigma\,;\,\tau_0) = \frac{1}{p \cdot a_{\mathrm{F}}(\sigma\,;\,\tau_0)}\,C[\tilde{f}^{(0)}]_p(\tau\,,\,\sigma\,;\,\tau_0). \tag{12.9}$$

As the solution to describe the slow motion in the asymptotic region we can take the stationary solution, $\frac{\partial}{\partial \tau}\tilde{f}_p^{(0)}(\tau\,,\,\sigma\,;\,\tau_0) = 0$, or

$$\frac{1}{p \cdot a_{\mathrm{F}}(\sigma\,;\,\tau_0)}\,C[\tilde{f}^{(0)}]_p(\tau\,,\,\sigma\,;\,\tau_0) = 0, \qquad \forall\sigma, \tag{12.10}$$

which implies that the zero-th order solution is given by a local equilibrium distribution function as

$$\tilde{f}_p^{(0)}(\tau\,,\,\sigma\,;\,\tau_0) = \frac{1}{\mathrm{e}^{\{p\cdot u(\sigma\,;\,\tau_0) - \mu(\sigma\,;\,\tau_0)\}/T(\sigma\,;\,\tau_0)} - a} =: f_p^{\mathrm{eq}}(\sigma\,;\,\tau_0), \tag{12.11}$$

where u^{μ} can be identified with the flow velocity on the basis of the Gibbs relation [68], as is mentioned in Sect. 11.4.1, where the following equality is also shown,

$$a_{\mathrm{F}}^{\mu}(\sigma\,;\,\tau_0) = u^{\mu}(\tau_0\,,\,\sigma), \tag{12.12}$$

which leads to the following identities

$$\frac{\partial}{\partial \tau} = u^{\mu}\partial_{\mu}, \quad \frac{\partial}{\partial \sigma_{\mu}} = \Delta^{\mu\nu}\partial_{\mu} =: \nabla^{\mu}. \tag{12.13}$$

Note that $\partial/\partial\tau$ and ∇^{μ} are the Lorentz-covariant temporal and spacial derivatives, respectively.

It is to be noted that the five integral constants $T(\sigma\,;\,\tau_0)$, $\mu(\sigma\,;\,\tau_0)$, and $u_{\mu}(\sigma\,;\,\tau_0)$ equally depend on the 'initial' time τ_0 as well as the space coordinate σ. To simplify the expressions, we introduce the following quantity with the equality (12.12) being taken for granted,

$$\bar{f}_p^{\mathrm{eq}}(\sigma\,;\,\tau_0) := 1 + a\, f_p^{\mathrm{eq}}(\sigma\,;\,\tau_0) = \frac{\mathrm{e}^{[p\cdot u(\sigma\,;\,\tau_0) - \mu(\sigma\,;\,\tau_0)]/T(\sigma\,;\,\tau_0)}}{\mathrm{e}^{[p\cdot u(\sigma\,;\,\tau_0) - \mu(\sigma\,;\,\tau_0)]/T(\sigma\,;\,\tau_0)} - a}. \tag{12.14}$$

In the following, we shall suppress the coordinate arguments $(\sigma\,;\,\tau_0)$ and the momentum subscript p, when no misunderstanding is expected.

Now, the first-order equation reads

$$\frac{\partial}{\partial \tau}\tilde{f}_p^{(1)}(\tau) = \int \mathrm{d}q\,(f_p^{\mathrm{eq}}\,\bar{f}_p^{\mathrm{eq}})\,\hat{L}_{pq}\,(f_q^{\mathrm{eq}}\,\bar{f}_q^{\mathrm{eq}})^{-1}\,\tilde{f}_q^{(1)}(\tau) + (f_p^{\mathrm{eq}}\,\bar{f}_p^{\mathrm{eq}})\,F_{0p}, \tag{12.15}$$

where the linear evolution operator $\hat{L} = (1/p\cdot u)L_{pq}$ and the inhomogeneous term F_{0p} are given by[3]

$$\begin{aligned} \hat{L}_{pq} &= -\frac{1}{p\cdot u}\frac{1}{2!}\int \mathrm{d}p_1 \int \mathrm{d}p_2 \int \mathrm{d}p_3\, \omega(p\,,\,p_1|p_2\,,\,p_3) \\ &\quad\times \frac{f_{p_1}^{\mathrm{eq}}\,\bar{f}_{p_2}^{\mathrm{eq}}\,\bar{f}_{p_3}^{\mathrm{eq}}}{\bar{f}_p^{\mathrm{eq}}}\,(\delta_{pq} + \delta_{p_1q} - \delta_{p_2q} - \delta_{p_3q}), \end{aligned} \tag{12.16}$$

$$F_{0p} = -(f_p^{\mathrm{eq}}\,\bar{f}_p^{\mathrm{eq}})^{-1}\,\frac{1}{p\cdot u}\,p\cdot\nabla f_p^{\mathrm{eq}}, \tag{12.17}$$

respectively. Here the operator L is the linearized collision integral introduced in Eq. (10.56).

In order to present the spectral properties of the linear operator $\hat{L}$, it is found convenient to define the inner product for arbitrary two vectors φ_p and ψ_p by

$$\langle\,\varphi\,,\,\psi\,\rangle := \int \mathrm{d}p\,(p\cdot u)\,f_p^{\mathrm{eq}}\,\bar{f}_p^{\mathrm{eq}}\,\varphi_p\,\psi_p. \tag{12.18}$$

We first remark that the norm defined through this inner product is positive definite

$$\langle\,\varphi\,,\,\varphi\,\rangle = \int \mathrm{d}p\,(p\cdot u)\,f_p^{\mathrm{eq}}\,\bar{f}_p^{\mathrm{eq}}\,(\varphi_p)^2 \geq 0, \qquad \text{for } \varphi_p \neq 0. \tag{12.19}$$

Then, as is shown in Sect. 11.4.1.1, we finds the following remarkable properties of $\hat{L}$:

(1) $\hat{L}$ is self-adjoint with respect to this inner product:

$$\langle\,\varphi\,,\,\hat{L}\,\psi\,\rangle = \langle\,\hat{L}\,\varphi\,,\,\psi\,\rangle. \tag{12.20}$$

(2) $\hat{L}$ is semi-negative definite:

$$\langle\,\varphi\,,\,L\,\varphi\,\rangle \leq 0, \tag{12.21}$$

[3] F_{0p} here is the same as F_p defined by (11.30) in Chap. 11. We have added the subscript 0 to distinguish it from a similar quantity to appear which is denoted by F_{1p}.

which inequality implies that the eigenvalues of $\hat{L}$ are zero or negative. In fact, the eigenvectors belonging to the zero eigenvalue are given by

$$\varphi^{\mu}_{0p} = p^{\mu} \quad (\mu = 0,\, 1,\, 2,\, 3) \quad \text{and} \quad \varphi^{4}_{0p} = 1, \tag{12.22}$$

which span the kernel of $\hat{L}$ and are the zero modes of $\hat{L}$ as

$$\big[\hat{L}\,\varphi^{\alpha}_{0}\big]_p := \int \mathrm{d}q\, \hat{L}_{pq}\, \varphi^{\alpha}_{0q} = 0. \qquad (\alpha = 0,\, 1 \ldots 4). \tag{12.23}$$

12.3 First-Order Solution in the Doublet Scheme

To present the solution to the first-order equation (12.15) in a comprehensive way, we introduce the projection operator P_0 onto the kernel of $\hat{L}$ which is called the P_0 space and the projection operator Q_0 onto the Q_0 space complement to the P_0 space[4]:

$$Q_0 = 1 - P_0. \tag{12.24}$$

For that, we first define the metric matrix $\eta^{\alpha\beta}_0$ for the P_0 space as

$$\eta^{\alpha\beta}_0 := \langle\, \varphi^{\alpha}_0 \,,\, \varphi^{\beta}_0 \,\rangle. \tag{12.25}$$

Then the projection operator P_0 is defined by

$$\big[P_0\, \psi\big]_p := \varphi^{\alpha}_{0p}\, \eta^{-1}_{0\alpha\beta}\, \langle\, \varphi^{\beta}_0 \,,\, \psi \,\rangle. \tag{12.26}$$

where $\eta^{-1}_{0\alpha\beta}$ is the inverse matrix of $\eta^{\alpha\beta}_0$. As is shown in Chap. 11, we have

$$P_0^2 = P_0, \quad Q_0^2 = Q_0, \quad \big[P_0\varphi^{\alpha}_0\big]_p = \varphi^{\alpha}_{0p}, \quad \big[Q_0\varphi^{\alpha}_0\big]_p = 0. \tag{12.27}$$

It is also found that P_0 and Q_0 are self-adjoint;

$$\langle\psi_1 \,,\, P_0\, \psi_2\rangle = \langle P_0\, \psi_1 \,,\, \psi_2\rangle, \quad \langle\psi_1 \,,\, Q_0\, \psi_2\rangle = \langle Q_0\, \psi_1 \,,\, \psi_2\rangle, \tag{12.28}$$

for arbitrary ψ_{1p} and ψ_{2p}.

Now the solution to (12.15) is found to be given in terms of the projection operators P_0 and Q_0 as

---

[4] The projection operator P_0 (Q_0) onto the kernel of $\hat{L}$ was denoted simply by P (Q) in Chap. 11, so do the subspace P_0 (Q_0). We have changed the notations for notational convenience in the subsequent discussions in the present chapter.

$$\tilde{f}^{(1)}(\tau) = f^{\mathrm{eq}}\,\bar{f}^{\mathrm{eq}}\Big[\mathrm{e}^{\hat{L}(\tau-\tau_0)}\,\Psi + (\tau-\tau_0)\,P_0\,F_0 + (\mathrm{e}^{\hat{L}(\tau-\tau_0)} - 1)\,\hat{L}^{-1}\,Q_0\,F_0\Big], \tag{12.29}$$

where

$$f^{(1)}(\tau_0,\sigma) = \tilde{f}^{(1)}(\tau=\tau_0,\sigma;\tau_0) = f^{\mathrm{eq}}\,\bar{f}^{\mathrm{eq}}\,\Psi, \tag{12.30}$$

with $\Psi = \Psi(\sigma;\tau_0)$ being an integral constant.

We can impose the following condition to the 'initial' value at $\tau = \tau_0$ without a loss of generality,

$$P_0\,\Psi = 0, \tag{12.31}$$

implying that the 'initial' value at $\tau = \tau_0$ contains no zero modes, in accordance with the basic prescription of the RG method. Indeed, if Ψ were to contain zero modes

$$p^\mu\,\alpha_\mu + \beta =: \delta\Psi, \tag{12.32}$$

the 'initial' value (12.30) contains in part

$$f_p^{\mathrm{eq}}\,\bar{f}_p^{\mathrm{eq}}\,\delta\Psi = f_p^{\mathrm{eq}}\,\bar{f}_p^{\mathrm{eq}}\,(p^\mu\,\alpha_\mu + \beta), \tag{12.33}$$

which can be nicely written as the derivative of f_p^{eq}

$$\delta f_p^{\mathrm{eq}} = -f_p^{\mathrm{eq}}\,\bar{f}_p^{\mathrm{eq}}\,(p^\mu\,\delta(u_\mu/T) - \delta(\mu/T)). \tag{12.34}$$

with the identification

$$-\alpha_\mu = \delta(u_\mu/T) = \delta u_\mu/T + u_\mu\,\delta(1/T), \tag{12.35}$$

$$-\beta = -\delta(\mu/T) = -\delta\mu/T - \mu\,\delta(1/T). \tag{12.36}$$

Here we remark that the transverse component of α_μ is proportional to δu_μ. Thus, we see that the possible zero modes in Ψ can be renormalized into the local temperature $T(\tau_0,\sigma)$, chemical potential $\mu(\tau_0,\sigma)$, and flow velocity $u_\mu(\tau_0,\sigma)$, which is tantamount to the redefinition of the 'initial' distribution function that is yet to be determined eventually.

We note the appearance of the secular term proportional to $\tau - \tau_0$ in Eq. (12.29), which apparently invalidate the perturbative solution as $|\tau - \tau_0|$ becomes large, although we only utilize the local properties of the perturbative solutions in the RG method.

For making the discussions simple and transparent, let us expand $\mathrm{e}^{(\tau-\tau_0)\hat{L}}$ with respect to $\tau - \tau_0$ and retain the terms up to the first order,

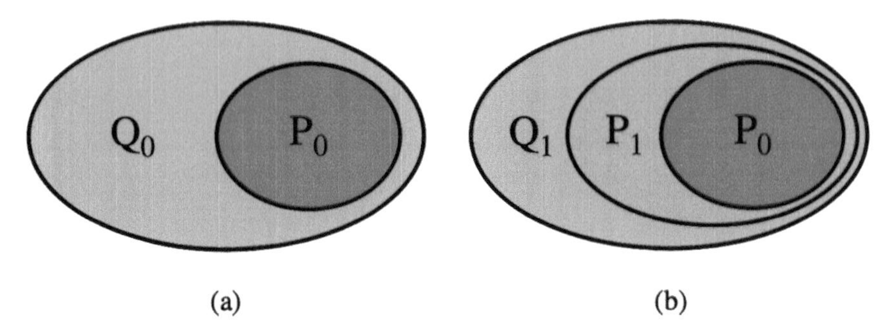

Fig. 12.1 Decomposition of the solution space of the Boltzmann equation. (a) The P_0 space is the kernel of the linearized collision operator, while the Q_0 space, spanned by excited modes, is the complement of the P_0 space. (b) The Q_0 space is further decomposed into the P_1 and Q_1 spaces: $Q_0 = P_1 + Q_1$

$$\tilde{f}^{(1)}(\tau) \simeq f^{\text{eq}}\,\bar{f}^{\text{eq}}\Big[\Psi + (\tau - \tau_0)\,P_0\,F_0 + (\tau - \tau_0)\,(\hat{L}\,\Psi + Q_0\,F_0)\Big]. \quad (12.37)$$

Here the neglected terms of $O((\tau - \tau_0)^2)$ play no role in the RG method.

We are now in a position to discuss one of the central issues in deriving the second-order fluid dynamic equation in the doublet scheme of the RG method. The problem is how to extend the vector space beyond that spanned by the zero modes to accommodate the excited modes that are responsible for the mesoscopic dynamics.

Let us call the vector space to which the excited modes belong the P_1 space. Needless to say, the P_1 space is a subspace of the Q_0 space; see Fig. 12.1. According to the general prescription of the doublet scheme given in Chap. 9, the only requirement we have to make is the following:

[R] The tangent space of the perturbative solution at $\tau = \tau_0$ is as small as possible to simplify the obtained equation.[5]

Here, it is to be noted that $\hat{L}\Psi$ and Q_0F_0 are included in the tangent space because they are multiplied by $\tau - \tau_0$ in Eq. (12.37). Thus we should impose the condition that

(I) The vectors Ψ and $\hat{L}^{-1}\,Q_0\,F_0$ belong to a common vector space,

to meet the requirement [R]. Furthermore the P_1 space should accommodate all the terms except for the zero modes in Eq. (12.37). Thus, we should also impose the additional requirement:

(II) The P_1 space is spanned by $\hat{L}\Psi$ and Ψ.

One sees now that the key ingredient is $\hat{L}^{-1}\,Q_0\,F_0$ for revealing the structure of the vector space to which it belongs. In fact the following faithful explicit calculation

[5] Incidentally one may recall that simplicity of the obtained equation is one of the basic principles in the reduction theory of dynamical systems [32].

of the deformation of the distribution function constitutes one of the essential ingredients in the derivation of the second-order fluid dynamic equation in the RG method, which is in contrast to the moment method in which some seemingly plausible ansatz is adopted without faithfully solving the Boltzmann equation.

The quantity $\hat{L}^{-1} Q_0 F_0$ is computed [6] to be

$$\left[\hat{L}^{-1} Q_0 F_0\right]_p = \left[\hat{L}^{-1} \hat{\Pi}\right]_p \frac{-\nabla \cdot u}{T} - \left[\hat{L}^{-1} \hat{J}^\mu\right]_p \frac{1}{h} \nabla_\mu \frac{\mu}{T} + \left[\hat{L}^{-1} \hat{\pi}^{\mu\nu}\right]_p \frac{\Delta_{\mu\nu\rho\sigma} \nabla^\rho u^\sigma}{T}, \tag{12.38}$$

where

$$\hat{\Pi}_p := \frac{1}{p \cdot u} \left\{ (p \cdot u)^2 \left[\frac{1}{3} - \left.\frac{\partial P}{\partial e}\right|_n \right] - (p \cdot u) \left.\frac{\partial P}{\partial n}\right|_e - \frac{m^2}{3} \right\}, \tag{12.39}$$

$$\hat{J}^\mu_p := \frac{-\Delta^{\mu\nu} p_\nu}{p \cdot u} ((p \cdot u) - h), \tag{12.40}$$

$$\hat{\pi}^{\mu\nu}_p := \frac{\Delta^{\mu\nu\rho\sigma} p_\rho p_\sigma}{p \cdot u}, \tag{12.41}$$

which are found to be the microscopic representations of the dissipative currents. Here, h denotes the enthalpy per particle

$$h := (e + P)/n, \tag{12.42}$$

and $\Delta^{\mu\nu\rho\sigma}$ is the projection matrix given by

$$\Delta^{\mu\nu\rho\sigma} := \frac{1}{2} \left[\Delta^{\mu\rho} \Delta^{\nu\sigma} + \Delta^{\mu\sigma} \Delta^{\nu\rho} - \frac{2}{3} \Delta^{\mu\nu} \Delta^{\rho\sigma} \right]. \tag{12.43}$$

From Eq. (12.38), one can read off the following nine vectors

$$\left[\hat{L}^{-1} \hat{\Pi}\right]_p, \quad \left[\hat{L}^{-1} \hat{J}^\mu\right]_p, \quad \left[\hat{L}^{-1} \hat{\pi}^{\mu\nu}\right]_p, \tag{12.44}$$

as a natural set of the bases of the vector space that $\left[\hat{L}^{-1} Q_0 F_0\right]_p$ and hence Ψ belong to. Here, we note that the nine Lorentz vectors and the tensor are transverse to the frame vector;

$$\left[\hat{L}^{-1} \hat{J}^\mu\right]_p = \Delta^{\mu\nu} \left[\hat{L}^{-1} \hat{J}_\nu\right]_p, \tag{12.45}$$

$$\left[\hat{L}^{-1} \hat{\pi}^{\mu\nu}\right]_p = \Delta^{\mu\nu\rho\sigma} \left[\hat{L}^{-1} \hat{\pi}_{\rho\sigma}\right]_p. \tag{12.46}$$

---

[6] See Sect. 13.4 for the details of the computation.

We now see that Ψ can be written as a linear combination of these basis vectors as,

$$\Psi_p = \frac{\left[\hat{L}^{-1}\,\hat{\Pi}\right]_p}{\langle\,\hat{\Pi}\,,\,\hat{L}^{-1}\,\hat{\Pi}\,\rangle}\,\Pi + \frac{3\,h\left[\hat{L}^{-1}\,\hat{J}^{\mu}\right]_p}{\langle\,\hat{J}^{\nu}\,,\,\hat{L}^{-1}\,\hat{J}_{\nu}\,\rangle}\,J_{\mu} + \frac{5\left[\hat{L}^{-1}\,\hat{\pi}^{\mu\nu}\right]_p}{\langle\,\hat{\pi}^{\rho\sigma}\,,\,\hat{L}^{-1}\,\hat{\pi}_{\rho\sigma}\,\rangle}\,\pi_{\mu\nu}. \quad (12.47)$$

Here, we have introduced the following nine integral constants as coefficients of the basis vectors:

$$\Pi(\sigma\,;\,\tau_0),\quad J^{\mu}(\sigma\,;\,\tau_0),\quad \pi^{\mu\nu}(\sigma\,;\,\tau_0). \quad (12.48)$$

We note that the factors $1/\langle\,\hat{\Pi}\,,\,\hat{L}^{-1}\,\hat{\Pi}\,\rangle, 3\,h/\langle\,\hat{J}^{\nu}\,,\,\hat{L}^{-1}\,\hat{J}_{\nu}\,\rangle$, and $5/\langle\,\hat{\pi}^{\rho\sigma}\,,\,\hat{L}^{-1}\,\hat{\pi}_{\rho\sigma}\,\rangle$ in Ψ are mere coefficients which can be scaled by the redefinitions of Π, J^{μ}, and $\pi^{\mu\nu}$, respectively. It is worth emphasizing that the form of Ψ given in Eq. (12.47) is the most general expression that makes $\hat{L}\,\Psi$ and $Q_0\,F_0$ belong to the common space.

As is clear now, we see that the P_1 space is identified with the vector space spanned by

$$\hat{\Pi}_p,\quad \hat{J}^{\mu}_p,\quad \hat{\pi}^{\mu\nu}_p,\quad \left[\hat{L}^{-1}\,\hat{\Pi}\right]_p,\quad \left[\hat{L}^{-1}\,\hat{J}^{\mu}\right]_p,\quad \text{and}\quad \left[\hat{L}^{-1}\,\hat{\pi}^{\mu\nu}\right]_p. \quad (12.49)$$

The three pairs $(\hat{\Pi},\ \hat{L}^{-1}\,\hat{\Pi})$, $(\hat{J}^{\mu},\ \hat{L}^{-1}\,\hat{J}^{\mu})$, and $(\hat{\pi}^{\mu\nu},\ \hat{L}^{-1}\,\hat{\pi}^{\mu\nu})$ are called the *doublet modes* [72]; see Chap. 9. The Q_0 space is now decomposed into the P_1 space spanned by the doublet modes and the Q_1 space which is the complement to the P_0 and P_1 spaces:

$$1 = \mathrm{P}_0 + \mathrm{Q}_0,\quad \mathrm{Q}_0 = \mathrm{P}_1 + \mathrm{Q}_1. \quad (12.50)$$

The corresponding projection operators are denoted as P_1 and Q_1, respectively.

It is to be noted that because of Eqs. (12.45) and (12.46), the coefficients J^{μ} and $\pi^{\mu\nu}$ in Eq. (12.47) are taken to be transverse without loss of generality, *i.e.*,

$$J^{\mu} = \Delta^{\mu\nu}\,J_{\nu},\quad \pi^{\mu\nu} = \Delta^{\mu\nu\rho\sigma}\,\pi_{\rho\sigma}, \quad (12.51)$$

which lead to the following identities:

$$u_{\mu}\,J^{\mu} = u_{\mu}\,\pi^{\mu\nu} = \Delta_{\mu\nu}\,\pi^{\mu\nu} = 0,\quad \pi^{\mu\nu} = \pi^{\nu\mu}. \quad (12.52)$$

It will be shown later that Π, J^{μ}, and $\pi^{\mu\nu}$ can be identified with the bulk pressure, thermal flux, and stress pressure, respectively.

12.4 Second-Order Solution in the Doublet Scheme

Let us proceed to the second-order equation, for which it is found convenient to introduce the symbol $\left[B[\varphi\,,\,\psi]\right]_p$ for arbitrary two vector φ_p and ψ_p by

$$\left[B[\varphi\,,\,\psi]\right]_p = \int \mathrm{d}q \int \mathrm{d}r\; B_{pqr}\,\varphi_q\,\psi_r. \tag{12.53}$$

Then the second-order equation reads

$$\frac{\partial}{\partial\tau}\tilde{f}^{(2)}(\tau) = f^{\mathrm{eq}}\,\bar{f}^{\mathrm{eq}}\,\hat{L}(f^{\mathrm{eq}}\,\bar{f}^{\mathrm{eq}})^{-1}\,\tilde{f}^{(2)}(\tau) + f^{\mathrm{eq}}\,\bar{f}^{\mathrm{eq}}\,K(\tau-\tau_0), \tag{12.54}$$

where

$$\begin{aligned} K(\tau-\tau_0) :=&\; F_1\,\tilde{f}^{(1)}(\tau) + \frac{1}{2}\,B[(f^{\mathrm{eq}}\,\bar{f}^{\mathrm{eq}})^{-1}\,\tilde{f}^{(1)}(\tau))\,,\,(f^{\mathrm{eq}}\,\bar{f}^{\mathrm{eq}})^{-1}\,\tilde{f}^{(1)}(\tau))] \\ =&\; F_1\,f^{\mathrm{eq}}\,\bar{f}^{\mathrm{eq}}\left[\mathrm{e}^{\hat{L}(\tau-\tau_0)}\,\Psi + (\tau-\tau_0)\,P_0\,F_0 + (\mathrm{e}^{\hat{L}(\tau-\tau_0)}-1)\,\hat{L}^{-1}\,Q_0\,F_0\right] \\ &+ \frac{1}{2}\,B[\mathrm{e}^{\hat{L}(\tau-\tau_0)}\,\Psi + (\tau-\tau_0)\,P_0\,F_0 + (\mathrm{e}^{\hat{L}(\tau-\tau_0)}-1)\,\hat{L}^{-1}\,Q_0\,F_0\,, \\ &\mathrm{e}^{\hat{L}(\tau-\tau_0)}\,\Psi + (\tau-\tau_0)\,P_0\,F_0 + (\mathrm{e}^{\hat{L}(\tau-\tau_0)}-1)\,\hat{L}^{-1}\,Q_0\,F_0], \end{aligned} \tag{12.55}$$

with F_1 and B being matrices the components of which are given by

$$F_{1\,pq} := -(f^{\mathrm{eq}}_p\,\bar{f}^{\mathrm{eq}}_p)^{-1}\,\frac{1}{p\cdot u}\,p\cdot\nabla\,\delta_{pq}, \tag{12.56}$$

$$B_{pqr} := (f^{\mathrm{eq}}_p\,\bar{f}^{\mathrm{eq}}_p)^{-1}\,\frac{1}{p\cdot u}\,\frac{\delta^2}{\delta f_q\,\delta f_r}C[f]_p\bigg|_{f=f^{\mathrm{eq}}}\,f^{\mathrm{eq}}_q\,\bar{f}^{\mathrm{eq}}_q\,f^{\mathrm{eq}}_r\,\bar{f}^{\mathrm{eq}}_r, \tag{12.57}$$

respectively.

By multiplying $(f^{\mathrm{eq}}\bar{f}^{\mathrm{eq}})^{-1}$ on the both side, Eq. (12.54) is converted to

$$\frac{\partial}{\partial\tau}X(\tau) = \hat{L}X(\tau) + K(\tau-\tau_0), \tag{12.58}$$

with

$$X(\tau) := (f^{\mathrm{eq}}\bar{f}^{\mathrm{eq}})^{-1}\,\tilde{f}^{(2)}(\tau). \tag{12.59}$$

Then, the solution is readily given by

$$
\begin{aligned}
X(\tau) &= \mathrm{e}^{\hat{L}(\tau-\tau_0)}\, X(\tau_0) + \int_{\tau_0}^{\tau} \mathrm{d}\tau'\, \mathrm{e}^{\hat{L}(\tau-\tau')}\, K(\tau'-\tau_0) \\
&= \mathrm{e}^{\hat{L}(\tau-\tau_0)}\, X(\tau_0) + \int_{\tau_0}^{\tau} \mathrm{d}\tau'\, P_0\, K(\tau'-\tau_0) \\
&\quad + \int_{\tau_0}^{\tau} \mathrm{d}\tau'\, \mathrm{e}^{\hat{L}(\tau-\tau')}\, Q_0\, K(\tau'-\tau_0),
\end{aligned} \tag{12.60}
$$

where we have inserted $1 = P_0 + Q_0$ in front of $K(\tau'-\tau_0)$ in the first line.

Now using the compact form of the Taylor expansion as

$$
K(\tau'-\tau_0) = \mathrm{e}^{(\tau'-\tau_0)\partial/\partial s}\, K(s)\Big|_{s=0}, \tag{12.61}
$$

and carrying out integration with respect to τ', Eq. (12.60) can be reduced to

$$
\begin{aligned}
X(\tau) &= \mathrm{e}^{\hat{L}(\tau-\tau_0)}\, X(\tau_0) + \Big[(1-\mathrm{e}^{(\tau-\tau_0)\partial/\partial s})\,(-\partial/\partial s)^{-1}\, P_0 \\
&\quad + (\mathrm{e}^{\hat{L}(\tau-\tau_0)} - \mathrm{e}^{(\tau-\tau_0)\partial/\partial s})\,(\hat{L}-\partial/\partial s)^{-1}\, Q_0\Big]\, K(s)\Big|_{s=0} \\
&= \mathrm{e}^{\hat{L}(\tau-\tau_0)}\Big[X(\tau_0) + Q_1\,(\hat{L}-\partial/\partial s)^{-1}\, Q_0\, K(s)\Big|_{s=0}\Big] \\
&\quad + \Big[(1-\mathrm{e}^{(\tau-\tau_0)\partial/\partial s})\,(-\partial/\partial s)^{-1}\, P_0 \\
&\quad + (\mathrm{e}^{\hat{L}(\tau-\tau_0)} - \mathrm{e}^{(\tau-\tau_0)\partial/\partial s})\, P_1\,(\hat{L}-\partial/\partial s)^{-1}\, Q_0 \\
&\quad - \mathrm{e}^{(\tau-\tau_0)\partial/\partial s}\, Q_1\,(\hat{L}-\partial/\partial s)^{-1}\, Q_0\Big]\, K(s)\Big|_{s=0},
\end{aligned} \tag{12.62}
$$

where $1 = P_0 + P_1 + Q_1$ has been inserted in front of $(\hat{L}-\partial/\partial s)^{-1}\, Q_0\, K(s)$ in the second equality of Eq. (12.62).

We note that Eq. (12.62) has a quite similar structure with Eq. (9.67) derived for a generic equation, and one sees that the inhomogeneous term $K(\tau-\tau_0)$ is composed of the modes belonging to the Q_1 space described by $\mathrm{e}^{\hat{L}(\tau-\tau_0)}$ and the mesoscopic modes contained in $\mathrm{e}^{(\tau-\tau_0)\partial/\partial s}$. As was done for (9.67), the former mode that should not be included in the mesococpic dynamics is nicely eliminated with the following choice of the 'initial' value $X(\tau_0)$

$$
X(\tau_0) = -Q_1\,(\hat{L}-\partial/\partial s)^{-1}\, Q_0\, K(s)\Big|_{s=0}, \tag{12.63}
$$

with which Eq. (12.62) is reduced to

$$
\begin{aligned}
X(\tau) &= \Big[(1-\mathrm{e}^{(\tau-\tau_0)\partial/\partial s})\,(-\partial/\partial s)^{-1}\, P_0 \\
&\quad + (\mathrm{e}^{\hat{L}(\tau-\tau_0)} - \mathrm{e}^{(\tau-\tau_0)\partial/\partial s})\, P_1\,(\hat{L}-\partial/\partial s)^{-1}\, Q_0 \\
&\quad - \mathrm{e}^{(\tau-\tau_0)\partial/\partial s}\, Q_1\,(\hat{L}-\partial/\partial s)^{-1}\, Q_0\Big]\, K(s)\Big|_{s=0}.
\end{aligned} \tag{12.64}
$$

With the use of the formula (12.59) for $X(\tau)$ and $X(\tau_0) = (f^{\rm eq}\,\bar{f}^{\rm eq})^{-1}\, f^{(2)}$, we see that Eqs. (12.64) and (12.63) imply

$$\tilde{f}^{(2)}(\tau) = f^{\rm eq}\,\bar{f}^{\rm eq}\Big[(1 - {\rm e}^{(\tau-\tau_0)\partial/\partial s})\,\mathcal{G}(s)\,P_0 + ({\rm e}^{\hat{L}(\tau-\tau_0)} - {\rm e}^{(\tau-\tau_0)\partial/\partial s})\,P_1\,\mathcal{G}(s)\,Q_0 - {\rm e}^{(\tau-\tau_0)\partial/\partial s}\,Q_1\,\mathcal{G}(s)\,Q_0\Big]\,K(s)\Big|_{s=0}, \tag{12.65}$$

$$f^{(2)} = \tilde{f}^{(2)}(\tau = \tau_0) = -f^{\rm eq}\,\bar{f}^{\rm eq}\,Q_1\,\mathcal{G}(s)\,Q_0\,K(s)\Big|_{s=0}, \tag{12.66}$$

respectively. In Eqs. (12.65) and (12.66), we have introduced a *propagator* defined by

$$\mathcal{G}(s) := (\hat{L} - \partial/\partial s)^{-1}. \tag{12.67}$$

We remark again that Eq. (12.65) contains secular terms.

Thus we have the approximate solution up to the second order as

$$\begin{aligned}\tilde{f}(\tau) = f^{\rm eq} &+ \epsilon\, f^{\rm eq}\,\bar{f}^{\rm eq}\Big[{\rm e}^{\hat{L}(\tau-\tau_0)}\,\Psi + (\tau-\tau_0)\,P_0\,F_0 + ({\rm e}^{\hat{L}(\tau-\tau_0)} - 1)\,\hat{L}^{-1}\,Q_0\,F_0\Big] \\ &+ \epsilon^2\, f^{\rm eq}\,\bar{f}^{\rm eq}\Big[(1 - {\rm e}^{(\tau-\tau_0)\partial/\partial s})\,\mathcal{G}(s)\,P_0 + ({\rm e}^{\hat{L}(\tau-\tau_0)} - {\rm e}^{(\tau-\tau_0)\partial/\partial s})\,P_1\,\mathcal{G}(s)\,Q_0 \\ &- {\rm e}^{(\tau-\tau_0)\partial/\partial s}\,Q_1\,\mathcal{G}(s)\,Q_0\Big]\,K(s)\Big|_{s=0},\end{aligned} \tag{12.68}$$

with the 'initial' value at $\tau = \tau_0$

$$f = f^{\rm eq} + \epsilon\, f^{\rm eq}\,\bar{f}^{\rm eq}\,\Psi - \epsilon^2\, f^{\rm eq}\,\bar{f}^{\rm eq}\,Q_1\,\mathcal{G}(s)\,Q_0\,K(s)\Big|_{s=0}. \tag{12.69}$$

Two remarks are in order here:

(i) The distribution function (12.69) contains terms of order higher than p^2. which are not included in the ansatz (10.103) adopted in the fourteen-moment method [66] (see Sect. 10.3.2 for the details). In view that (12.69) is obtained by faithfully solving the Boltzmann equation (12.1) without restricting the functional space though within the perturbation theory, the fourteen moment is not likely to cover the proper solution space of the Boltzmann equation.

(ii) Expanding $\mathcal{G}(s)\,Q_0$ in terms of $\hat{L}^{-1}\partial/\partial s$, the term $\mathcal{G}(s)\,Q_0\,K(s)|_{s=0}$ in Eqs. (12.68) and (12.69) is reduced to the form of infinite series as

$$\mathcal{G}(s)\, Q_0\, K(s)\Big|_{s=0} = \sum_{n=0}^{\infty} \hat{L}^{-1-n}\, Q_0\, \frac{\partial^n}{\partial s^n} K(s)\Big|_{s=0}, \tag{12.70}$$

because $\partial^n K(s)/\partial s^n|_{s=0}$ does not vanish for any n; see Eq. (12.55). The existence of such an infinite number of terms would be undesirable for the construction of the (closed) mesoscopic dynamics. It will be found, however, that an averaging procedure for obtaining the mesoscopic dynamics nicely leads to a cancellation of all the terms but single term in the resultant equation of motion thanks to the self-adjoint nature of $\hat{L}$ and the structure of the P_1 space spanned by the doublet modes; see Eq. (12.78) below.

12.5 Construction of the Distribution Function Valid in a Global Domain in the Asymptotic Regime By the RG Method

In this section, we shall construct a distribution function valid in a global domain by applying the RG method to the perturbative solution (12.68) of the relativistic Boltzmann equation (12.1).

12.5.1 RG/E Equation

Although the solution (12.68) apparently loose its validity for τ away from the 'initial' time τ_0 owing to the secular terms, we can take the following geometrical point of view and construct a distribution function that is valid in a global domain in the asymptotic regime, as has been shown in this monograph: The perturbative solution $\tilde{f}_p(\tau\,,\,\sigma\,;\,\tau_0)$ in Eq. (12.68) provides a family of curves parameterized with the 'initial' time τ_0, each of which is on the exact solution $f_p(\tau\,,\,\sigma)$ given by Eq. (12.69) at $\tau = \tau_0$ up to $O(\epsilon^2)$, but only valid locally for τ close to τ_0. Thus, it is natural that the envelope of the family of curves, which is in contact with each local solution at $\tau = \tau_0$, certainly give a global solution in our asymptotic situation [3, 5, 6]; see Chaps. 4 and 5.

As is shown in Chaps. 4 and 5, the envelope that is in contact with any curve in the family at $\tau = \tau_0$ is obtained by the envelope equation given by [3, 5]

$$\frac{\mathrm{d}}{\mathrm{d}\tau_0} \tilde{f}_p(\tau\,,\,\sigma\,;\,\tau_0)\Big|_{\tau_0=\tau} = 0, \tag{12.71}$$

where the subscript p and $(\sigma\,;\,\tau_0)$ have been restored for later convenience. We call Eq. (12.71) the RG/E equation following [5], because a similar equation was first introduced in the name of the renormalization group equation [1].

The RG/E equation (12.71) is reduced to

$$\frac{\partial}{\partial \tau}\Big[f^{\text{eq}}\,(1+\epsilon\,\bar{f}^{\text{eq}}\,\Psi)\Big] - \epsilon\, f^{\text{eq}}\,\bar{f}^{\text{eq}}\Big[\hat{L}\,\Psi + P_0\,F_0 + Q_0\,F_0\Big]$$
$$- \epsilon^2\, f^{\text{eq}}\,\bar{f}^{\text{eq}}\Big[P_0 + (\hat{L} - \partial/\partial s)\,P_1\,\mathcal{G}(s)\,Q_0$$
$$- (\partial/\partial s)\,Q_1\,\mathcal{G}(s)\,Q_0\Big]\,K(s)\Big|_{s=0} = 0, \tag{12.72}$$

up to $O(\epsilon^2)$. Note that Eq. (12.72) is actually the equation of motion governing the dynamics of the fourteen integral constants $T(\sigma\,;\,\tau)$, $\mu(\sigma\,;\,\tau)$, $u^{\mu}(\sigma\,;\,\tau)$, $\Pi(\sigma\,;\,\tau)$, $J^{\mu}(\sigma\,;\,\tau)$, and $\pi^{\mu\nu}(\sigma\,;\,\tau)$.

The envelope function as the approximate but globally valid solution is given by the 'initial' value (12.69) with $\tau_0 = \tau$:

$$f_p^{\text{E}}(\tau\,,\,\sigma) := f_p(\tau_0 = \tau\,,\,\sigma) = \tilde{f}_p(\tau\,,\,\sigma\,;\,\tau_0 = \tau)$$
$$= f^{\text{eq}}\,(1+\epsilon\,\bar{f}^{\text{eq}}\,\Psi) - \epsilon^2\, f^{\text{eq}}\,\bar{f}^{\text{eq}}\,Q_1\,\mathcal{G}(s)\,Q_0\,K(s)\Big|_{s=0}\Big|_{\tau_0=\tau} + \mathrm{O}(\epsilon^3), \tag{12.73}$$

where the solution to the RG/E equation (12.72) is to be substituted.

We show that the envelope function $f_p^{\text{E}}(\tau\,,\,\sigma)$ solves the Boltzmann equation (12.7) in a global domain up to $O(\epsilon^2)$ in the asymptotic regime: Indeed, for an arbitrary $\tau(=\tau_0)$ in the asymptotic regime, we have

$$\frac{\partial}{\partial \tau} f_p^{\text{E}}(\tau\,,\,\sigma) = \frac{\partial}{\partial \tau}\tilde{f}_p(\tau\,,\,\sigma\,;\,\tau_0)\Big|_{\tau_0=\tau} + \frac{\mathrm{d}}{\mathrm{d}\tau_0}\tilde{f}_p(\tau\,,\,\sigma\,;\,\tau_0)\Big|_{\tau_0=\tau}$$
$$= \frac{\partial}{\partial \tau}\tilde{f}_p(\tau\,,\,\sigma\,;\,\tau_0)\Big|_{\tau_0=\tau}, \tag{12.74}$$

where the fact that $\tilde{f}_p(\tau\,,\,\sigma\,;\,\tau_0)$ satisfies the RG/E equation (12.71) has been used. Furthermore, since $\tilde{f}_p(\tau\,,\,\sigma\,;\,\tau_0)$ solves Eq. (12.7) with $a_{\text{F}}^{\mu}(\sigma) = u^{\mu}(\sigma\,;\,\tau_0)$ up to $\mathrm{O}(\epsilon^2)$, the right-hand side of Eq. (12.74) is written as

$$\frac{\partial}{\partial \tau}\tilde{f}_p(\tau\,,\,\sigma\,;\,\tau_0)\Big|_{\tau_0=\tau} = \frac{1}{p\cdot u(\sigma\,;\,\tau)}\,C[\tilde{f}]_p(\tau\,,\,\sigma\,;\,\tau)$$
$$- \epsilon\,\frac{1}{p\cdot u(\sigma\,;\,\tau)}\,p^{\mu}\,\frac{\partial}{\partial \sigma^{\mu}}\tilde{f}_p(\tau\,,\,\sigma\,;\,\tau), \tag{12.75}$$

up to $O(\epsilon^2)$. Then inserting the definition $\tilde{f}_p(\tau\,,\,\sigma\,;\,\tau) = f_p^{\text{E}}(\tau\,,\,\sigma)$ of the envelope function, we see that Eq. (12.74) becomes

$$
\begin{aligned}
\frac{\partial}{\partial \tau} f_p^{\rm E}(\tau\,,\,\sigma) &= \frac{1}{p\cdot u(\sigma\,;\,\tau)}\, C[f^{\rm E}]_p(\tau\,,\,\sigma) \\
&\quad - \epsilon\,\frac{1}{p\cdot u(\sigma\,;\,\tau)}\, p^\mu\,\frac{\partial}{\partial\sigma^\mu} f_p^{\rm E}(\tau\,,\,\sigma),
\end{aligned}
\tag{12.76}
$$

up to $O(\epsilon^2)$. This shows that the envelope function $f_p^{\rm E}(\tau\,,\,\sigma)$ satisfies the Boltzmann equation (12.7) up to $O(\epsilon^2)$ in a global domain in the asymptotic region, because τ_0 is an arbitrary time in the asymptotic region.

What we have done is a derivation of the mesoscopic dynamics of the relativistic Boltzmann equation (12.7) as the pair of equations, (12.72) and (12.73). It is also noteworthy that an infinite number of terms produced by $\mathcal{G}(s)$ are included both in the dynamical equation governing the mesoscopic variables as given by the RG/E equation and the functional from of the solution as given by the envelope function.

12.5.2 Reduction of RG/E Equation to a Simpler Form

It should be noted that the RG/E equation (12.72) still contains fast modes belonging to the Q_1 space that is complement to the space spanned by the fluid dynamic modes even in the second order. Although these modes may be incorporated to make a Langevinized fluid dynamic equation, we average them out to have the genuine fluid dynamic equation in the second order. This averaging can be achieved by taking the inner product of Eq. (12.72) both with the zero modes φ_{0p}^{α} and the excited modes $\big[\hat{L}^{-1}\,(\hat{\Pi},\ \hat{J}^\mu,\ \hat{\pi}^{\mu\nu})\big]_p$; we remark that they are all orthogonal to the Q_1 space.

The averaging of Eq. (12.72) with the zero modes leads to

$$
\int {\rm d}p\,\varphi_{0p}^{\alpha}\Big[(p\cdot u)\,\frac{\partial}{\partial\tau} + \epsilon\, p\cdot\nabla\Big]\Big[f_p^{\rm eq}\,(1+\epsilon\,\bar{f}_p^{\rm eq}\,\Psi_p)\Big] = 0,
\tag{12.77}
$$

up to $O(\epsilon^2$. The reduction of the averaging with the excited modes in the P_1 space is quite involved: We first note the following equality,

$$
\begin{aligned}
&\langle\,\hat{L}^{-1}\,(\hat{\Pi},\ \hat{J}^\mu,\ \hat{\pi}^{\mu\nu})\,,\ (\hat{L}-\partial/\partial s)\,P_1\mathcal{G}(s)\,Q_0\,K(s)\Big|_{s=0}\,\rangle \\
&= \langle\,(\hat{L}-\partial/\partial s)\,\hat{L}^{-1}\,(\hat{\Pi},\ \hat{J}^\mu,\ \hat{\pi}^{\mu\nu})\,,\ P_1\,\mathcal{G}(s)\,Q_0\,K(s)\Big|_{s=0}\,\rangle \\
&= \langle\,(\hat{L}-\partial/\partial s)\,\hat{L}^{-1}\,(\hat{\Pi},\ \hat{J}^\mu,\ \hat{\pi}^{\mu\nu})\,,\ \mathcal{G}(s)\,Q_0\,K(s)\Big|_{s=0}\,\rangle \\
&= \langle\,\hat{L}^{-1}\,(\hat{\Pi},\ \hat{J}^\mu,\ \hat{\pi}^{\mu\nu})\,,\ (\hat{L}-\partial/\partial s)\,\mathcal{G}(s)\,Q_0\,K(s)\Big|_{s=0}\,\rangle \\
&= \langle\,\hat{L}^{-1}\,(\hat{\Pi},\ \hat{J}^\mu,\ \hat{\pi}^{\mu\nu})\,,\ Q_0\,K(s)\Big|_{s=0}\,\rangle \\
&= \langle\,\hat{L}^{-1}\,(\hat{\Pi},\ \hat{J}^\mu,\ \hat{\pi}^{\mu\nu})\,,\ K(0)\,\rangle,
\end{aligned}
\tag{12.78}
$$

where we have utilized the self-adjoint nature of $\hat{L}_{pq}$ shown in Eq. (12.20) and the structure of the P_1 space spanned by the doublet modes, *i.e.*, the three pairs of $(\hat{\Pi}_p,\ [\hat{L}^{-1}\,\hat{\Pi}]_p)$, $(\hat{J}^{\mu}_p,\ [\hat{L}^{-1}\,\hat{J}^{\mu}]_p)$, and $(\hat{\pi}^{\mu\nu}_p,\ [\hat{L}^{-1}\,\hat{\pi}^{\mu\nu}]_p)$. Then using the equality

$$K(0) = F_1 f^{\mathrm{eq}}\,\bar{f}^{\mathrm{eq}}\,\Psi + B[\Psi\,,\ \Psi]/2, \tag{12.79}$$

derived from Eq. (12.55), we have

$$\begin{aligned}(12.78) = &\ \langle\,\hat{L}^{-1}\,(\hat{\Pi},\ \hat{J}^{\mu},\ \hat{\pi}^{\mu\nu})\,,\ F_1\,f^{\mathrm{eq}}\,\bar{f}^{\mathrm{eq}}\,\Psi\,\rangle\\ &+\frac{1}{2}\langle\,\hat{L}^{-1}\,(\hat{\Pi},\ \hat{J}^{\mu},\ \hat{\pi}^{\mu\nu})\,,\ B[\Psi\,,\ \Psi]\,\rangle.\end{aligned} \tag{12.80}$$

Then we find that the averaging of Eq. (12.72) with the excited modes in the P_1 space is reduced to

$$\begin{aligned}&\int\mathrm{d}p\,[\hat{L}^{-1}\,(\hat{\Pi},\ \hat{J}^{\mu},\ \hat{\pi}^{\mu\nu})]_p\Big[(p\cdot u)\,\frac{\partial}{\partial\tau}+\epsilon\,p\cdot\nabla\Big]\Big[f^{\mathrm{eq}}_p\,(1+\epsilon\,\bar{f}^{\mathrm{eq}}_p\,\Psi_p)\Big]\\ &=\epsilon\,\langle\,\hat{L}^{-1}(\hat{\Pi},\ \hat{J}^{\mu},\ \hat{\pi}^{\mu\nu})\,,\ \hat{L}\,\Psi\,\rangle+\epsilon^2\,\frac{1}{2}\,\langle\,\hat{L}^{-1}\,(\hat{\Pi},\ \hat{J}^{\mu},\ \hat{\pi}^{\mu\nu})\,,\ B[\Psi\,,\ \Psi]\,\rangle,\end{aligned} \tag{12.81}$$

up to $O(\epsilon^2)$. The pair of Eqs. (12.77) and (12.81) constitutes the fluid dynamic equation in the second order, which gives the equation of motion governing T, μ, u^{μ}, Π, J^{μ}, and $\pi^{\mu\nu}$. We emphasize that this pair of equations is composed of a finite number of terms in contrast to the RG/E equation (12.72) and much simpler than it: Note that this simplification through the averaging by $\hat{L}^{-1}\,(\hat{\Pi},\ \hat{J}^{\mu},\ \hat{\pi}^{\mu\nu})$ is due to the self-adjoint nature of $\hat{L}$ and the structure of the P_1 space spanned by the doublet modes $(\hat{\Pi},\ \hat{J}^{\mu},\ \hat{\pi}^{\mu\nu})$ and $\hat{L}^{-1}(\hat{\Pi},\ \hat{J}^{\mu},\ \hat{\pi}^{\mu\nu})$.

12.6 Derivation of the Second-Order Fluid Dynamic Equation

In this section, we reduce the RG equation given by Eqs. (12.77) and (12.81) to the second-order fluid dynamic equation as the set of balance equations and relaxation equations.

12.6.1 Balance Equations and Local Rest Frame of Flow Velocity

From now on, we put back $\epsilon = 1$. Then since $(p\cdot u)\,\frac{\partial}{\partial\tau} + p\cdot\nabla = p^{\mu}\,\partial_{\mu}$, Eq. (12.77) now takes the form

$$\partial_\mu J^{\mu\alpha} = 0, \tag{12.82}$$

with

$$J^{\mu\alpha} := \int \mathrm{d}p\, p^\mu\, \varphi^\alpha_{0p}\, f^{\mathrm{eq}}_p\, (1 + \bar{f}^{\mathrm{eq}}_p\, \Psi_p), \tag{12.83}$$

We remark that Eq. (12.82) is nothing but the balance equations; $J^{\mu\nu}$ and $J^{\mu 4}$ are identified with the energy-momentum tensor $T^{\mu\nu}$ and particle current N^μ, respectively. As a consistency check, we note that one can derive the same expression as $J^{\mu\alpha}$ by inserting the distribution function $f^{\mathrm{E}}_p(\tau\,,\,\sigma)$ in Eq. (12.73) into the formulae

$$T^{\mu\nu} = \int dp\, p^\mu\, p^\nu\, f_p, \quad N^\mu = \int dp\, p^\mu\, f_p, \tag{12.84}$$

given in Eqs. (10.29) and (10.28), respectively.

We decompose $J^{\mu\alpha}$ into two parts as

$$J^{\mu\alpha} = J^{\mu\alpha}_0 + \delta J^{\mu\alpha}, \tag{12.85}$$

where

$$J^{\mu\alpha}_0 := \int \mathrm{d}p\, p^\mu\, \varphi^\alpha_{0p}\, f^{\mathrm{eq}}_p, \tag{12.86}$$

$$\delta J^{\mu\alpha} := \int \mathrm{d}p\, p^\mu\, \varphi^\alpha_{0p}\, f^{\mathrm{eq}}_p\, \bar{f}^{\mathrm{eq}}_p\, \Psi_p = \langle\, \tilde{\varphi}^{\mu\alpha}_1\,,\, \Psi\,\rangle. \tag{12.87}$$

Here we have used the definition of the inner product (12.18) and the formula (11.76), which reads

$$\tilde{\varphi}^{\mu\alpha}_{1p} = p^\mu\, \varphi^\alpha_{0p}\, \frac{1}{p\cdot u}. \tag{12.88}$$

One should readily find that $J^{\mu\alpha}_0$ are identical to the energy-momentum tensor and particle current for an ideal fluid, whose explicit forms are given by

$$J^{\mu\alpha}_0 = \begin{cases} e\, u^\mu\, u^\nu - P\, \Delta^{\mu\nu}, & \alpha = \nu, \\ n\, u^\mu, & \alpha = 4. \end{cases} \tag{12.89}$$

Here n, e, and P denote the particle-number density, internal energy, and pressure, respectively;

$$n = \int \mathrm{d}p\, f_p^{\mathrm{eq}}\,(p \cdot u), \quad e = \int \mathrm{d}p\, f_p^{\mathrm{eq}}\,(p \cdot u)^2, \tag{12.90}$$

$$P = \int \mathrm{d}p\, f_p^{\mathrm{eq}}\,(-p^\mu\, p^\nu\, \Delta_{\mu\nu}/3). \tag{12.91}$$

We note that these equations are the same as Eqs. (10.38)–(10.40).

Next, we calculate the dissipative part $\delta J^{\mu\alpha}$. First, we recall the following expression derived in Sect. 13.3,

$$\left[Q_0\, \tilde{\varphi}_1^{\mu\alpha}\right]_p = \begin{cases} -\Delta^{\mu\nu}\,\hat{\Pi}_p + \hat{\pi}_p^{\mu\nu}, & \alpha = \mu, \\ \dfrac{1}{h}\,\hat{J}_p^\mu, & \alpha = 4, \end{cases} \tag{12.92}$$

and the identity

$$\langle\, \tilde{\varphi}_1^{\mu\alpha}\,,\, \Psi\,\rangle = \langle\, Q_0\, \tilde{\varphi}_1^{\mu\alpha}\,,\, \Psi\,\rangle. \tag{12.93}$$

Then with the use of Eq. (12.47), we have

$$\delta J^{\mu\alpha} = \begin{cases} -\Pi\,\Delta^{\mu\nu} + \pi^{\mu\nu} =: \delta T^{\mu\nu}, & \alpha = \nu, \\ J^\mu =: \delta N^\mu, & \alpha = 4. \end{cases}, \tag{12.94}$$

where we have used the following symmetry properties:

$$\langle\, \hat{\Pi}\,,\, \hat{L}^{-1}\,\hat{J}^\mu\,\rangle = 0, \quad \langle\, \hat{\Pi}\,,\, \hat{L}^{-1}\,\hat{\pi}^{\mu\nu}\,\rangle = 0, \quad \langle\, \hat{J}^\mu\,,\, \hat{L}^{-1}\,\hat{\pi}^{\nu\rho}\,\rangle = 0. \tag{12.95}$$

Here we remark that Π, $\pi^{\mu\nu}$, and J^μ were the would-be integral constants introduced in Eq. (12.47), and have been now lifted to dynamical variables through the RG/E equation.

It is noteworthy that $\delta J^{\mu\alpha} = \{\delta T^{\mu\nu},\ \delta N^\mu\}$ automatically satisfy the relations that the dissipative parts of the Landau-Lifshitz equation do[7] [62]:

$$u_\mu\, \delta T^{\mu\nu}\, u_\nu = 0, \quad u_\mu\, \delta N^{\mu\nu} = 0, \quad \Delta_{\mu\nu}\, \delta T^{\nu\rho} = 0. \tag{12.96}$$

We stress that the local rest frame of the flow velocity described by the balance equations is actually the energy frame by Landau and Lifshitz [62], as expected when the macroscopic frame vector a_{F}^μ has been set equal to u^μ in Eq. (12.12).

Collecting Eqs. (12.89) and (12.94), we find that the currents $J^{\mu\alpha}$ finally takes the form

$$J^{\mu\alpha} = \begin{cases} e\,u^\mu\,u^\nu - (P + \Pi)\,\Delta^{\mu\nu} + \pi^{\mu\nu}, & \alpha = \nu, \\ n\,u^\mu + J^\mu, & \alpha = 4. \end{cases} \tag{12.97}$$

[7] See also Sect. 10.1.

12.6.2 Relaxation Equations and Microscopic Representations of Transport Coefficients and Relaxation Times

The relaxation equations are derived from Eq. (12.81). Since its derivation is straightforward but rather lengthy, we leave an account of the detailed derivation to the following subsection, and here we just write down the explicit expressions of them:

$$\begin{aligned}\Pi = &-\zeta\,\theta\\ &-\tau_\Pi\,\frac{\partial}{\partial\tau}\Pi - h\,\ell_{\Pi J}\,\nabla\cdot J\\ &+\kappa_{\Pi\Pi}\,\Pi\theta\\ &+\kappa_{\Pi J}^{(1)}\,J_\rho\,\nabla^\rho T+\kappa_{\Pi J}^{(2)}\,J_\rho\,\nabla^\rho\frac{\mu}{T}\\ &+\kappa_{\Pi\pi}\,\pi_{\rho\sigma}\,\sigma^{\rho\sigma}\\ &+b_{\Pi\Pi\Pi}\,\Pi^2+b_{\Pi JJ}\,J^\rho\,J_\rho+b_{\Pi\pi\pi}\,\pi^{\rho\sigma}\,\pi_{\rho\sigma},\end{aligned} \tag{12.98}$$

$$\begin{aligned}J^\mu = &\,\lambda\,\frac{T^2}{h^2}\,\nabla^\mu\frac{\mu}{T}\\ &-\tau_J\,\Delta^{\mu\rho}\,\frac{\partial}{\partial\tau}J_\rho - h^{-1}\,\ell_{J\Pi}\,\nabla^\mu\Pi - h^{-1}\,\ell_{J\pi}\,\Delta^{\mu\rho}\,\nabla^\sigma\pi_{\rho\sigma}\\ &+\kappa_{J\Pi}^{(1)}\,\Pi\,\nabla^\mu T+\kappa_{J\Pi}^{(2)}\,\Pi\,\nabla^\mu\frac{\mu}{T}\\ &+\kappa_{JJ}^{(1)}\,J^\mu\,\theta+\kappa_{JJ}^{(2)}\,J_\rho\,\sigma^{\mu\rho}+\kappa_{JJ}^{(3)}\,J_\rho\,\omega^{\mu\rho}\\ &+\kappa_{J\pi}^{(1)}\,\pi^{\mu\rho}\,\nabla_\rho T+\kappa_{J\pi}^{(2)}\,\pi^{\mu\rho}\,\nabla_\rho\frac{\mu}{T}\\ &+b_{J\Pi J}\,\Pi\,J^\mu+b_{JJ\pi}\,J_\rho\,\pi^{\rho\mu},\end{aligned} \tag{12.99}$$

$$\begin{aligned}\pi^{\mu\nu} = &\,2\,\eta\,\sigma^{\mu\nu}\\ &-\tau_\pi\,\Delta^{\mu\nu\rho\sigma}\,\frac{\partial}{\partial\tau}\pi_{\rho\sigma} - h\,\ell_{\pi J}\,\nabla^{\langle\mu}\,J^{\nu\rangle}\\ &+\kappa_{\pi\Pi}\,\Pi\,\sigma^{\mu\nu}\\ &+\kappa_{\pi J}^{(1)}\,J^{\langle\mu}\,\nabla^{\nu\rangle}T+\kappa_{\pi J}^{(2)}\,J^{\langle\mu}\,\nabla^{\nu\rangle}\frac{\mu}{T}\\ &+\kappa_{\pi\pi}^{(1)}\,\pi^{\mu\nu}\,\theta+\kappa_{\pi\pi}^{(2)}\,\pi_\rho{}^{\langle\mu}\,\sigma^{\nu\rangle\rho}+\kappa_{\pi\pi}^{(3)}\,\pi_\rho{}^{\langle\mu}\,\omega^{\nu\rangle\rho}\\ &+b_{\pi\Pi\pi}\,\Pi\,\pi^{\mu\nu}+b_{\pi JJ}\,J^{\langle\mu}\,J^{\nu\rangle}+b_{\pi\pi\pi}\,\pi_\rho{}^{\langle\mu}\,\pi^{\nu\rangle\rho},\end{aligned} \tag{12.100}$$

where the scalar expansion, the shear tensor, and the vorticity are denoted by

$$\theta := \nabla \cdot u, \tag{12.101}$$

$$\sigma^{\mu\nu} := \Delta^{\mu\nu\rho\sigma} \nabla_\rho u_\sigma, \tag{12.102}$$

$$\omega^{\mu\nu} := \frac{1}{2}(\nabla^\mu u^\nu - \nabla^\nu u^\mu), \tag{12.103}$$

respectively. We have also used the notation

$$A^{\langle\mu\nu\rangle} := \Delta^{\mu\nu\rho\sigma} A_{\rho\sigma}, \tag{12.104}$$

for an arbitrary space-like tensor $A^{\mu\nu}$. Here the transport coefficients, *i.e.*, the bulk viscosity, the heat conductivity λ, and the shear viscosity η have the following microscopic expressions:

$$\zeta := -\frac{1}{T}\langle \hat{\Pi}, \hat{L}^{-1}\hat{\Pi}\rangle, \tag{12.105}$$

$$\lambda := \frac{1}{3T^2}\langle \hat{J}^\mu, \hat{L}^{-1}\hat{J}_\mu\rangle, \tag{12.106}$$

$$\eta := -\frac{1}{10T}\langle \hat{\pi}^{\mu\nu}, \hat{L}^{-1}\hat{\pi}_{\mu\nu}\rangle, \tag{12.107}$$

while the corresponding relaxation times:

$$\tau_\Pi := \frac{\langle \hat{L}^{-1}\hat{\Pi}, \hat{L}^{-1}\hat{\Pi}\rangle}{T\zeta} = -\frac{\langle \hat{\Pi}, \hat{L}^{-2}\hat{\Pi}\rangle}{\langle \hat{\Pi}, \hat{L}^{-1}\hat{\Pi}\rangle}, \tag{12.108}$$

$$\tau_J := -\frac{\langle \hat{L}^{-1}\hat{J}^\mu, \hat{L}^{-1}\hat{J}_\mu\rangle}{3T^2\lambda} = -\frac{\langle \hat{J}^\mu, \hat{L}^{-2}\hat{J}_\mu\rangle}{\langle \hat{J}^\rho, \hat{L}^{-1}\hat{J}_\rho\rangle}, \tag{12.109}$$

$$\tau_\pi := \frac{\langle \hat{L}^{-1}\hat{\pi}^{\mu\nu}, \hat{L}^{-1}\hat{\pi}_{\mu\nu}\rangle}{10T\eta} = -\frac{\langle \hat{\pi}^{\mu\nu}, \hat{L}^{-2}\hat{\pi}_{\mu\nu}\rangle}{\langle \hat{\pi}^{\rho\sigma}, \hat{L}^{-1}\hat{\pi}_{\rho\sigma}\rangle}. \tag{12.110}$$

Here we have used the self-adjointness of $\hat{L}$ and the formula of the transport coefficient to obtain the respective second equality.

Our approach is based on a kind of statistical physics, and thus has successfully given the microscopic expressions of all the coefficients of the relaxation Eqs. (12.98)–(12.100), as seen above. Here we focus on the transport coefficients and relaxation times given by Eqs. (12.105)–(12.107) and (12.108)–(12.110), which are also written as ζ^{RG}, λ^{RG}, η^{RG}, τ_Π^{RG}, τ_J^{RG}, and τ_π^{RG}, respectively, for convenience in the comparison with those by other methods.

As a check of the reliability of our derivation, we first compare the microscopic formulae of the transoport coefficients and the relaxation times with those given by other typical methods. First of all, as is shown for the first-order equation in Chap. 11, ζ^{RG}, λ^{RG}, and η^{RG} are perfectly in agreement with those of the Chapman-Enskog (CE) expansion method [67], which are calculated and denoted as ζ^{CE}, λ^{CE}, and η^{CE} in Sect. 10.3.1. Furthermore, our expressions of the transport coefficients can be nicely rewritten in the form of Green-Kubo formula [226–228] in the linear response theory, as is also shown for the first-order equation in Sect. 11.5 in Chap. 11. For self-containedness, we recapitulate the derivation here.

First, the "time-evolved" dissipative currents are defined in terms of the linearized collision operator $\hat{L}$ interpreted as the time-evolution operator by;

$$\begin{pmatrix} \hat{\Pi}_p(s) \\ \hat{J}^\mu_p(s) \\ \hat{\pi}^{\mu\nu}_p(s) \end{pmatrix} := \left[\mathrm{e}^{s\hat{L}} \begin{pmatrix} \hat{\Pi} \\ \hat{J}^\mu \\ \hat{\pi}^{\mu\nu} \end{pmatrix} \right]_p = \left[\mathrm{e}^{s\hat{L}} Q_0 \begin{pmatrix} \hat{\Pi} \\ \hat{J}^\mu \\ \hat{\pi}^{\mu\nu} \end{pmatrix} \right]_p , \tag{12.111}$$

where the projection operator Q_0 can be inserted without loss of generality because all the dissipative currents belong to the Q_0 space. Then, we have

$$\zeta^{\mathrm{RG}} = \frac{1}{T} \int_0^\infty \mathrm{d}s \, \langle \, \hat{\Pi}(0) \, , \, \hat{\Pi}(s) \, \rangle, \tag{12.112}$$

$$\lambda^{\mathrm{RG}} = -\frac{1}{3\,T^2} \int_0^\infty \mathrm{d}s \, \langle \, \hat{J}^\mu(0) \, , \, \hat{J}_\mu(s) \, \rangle, \tag{12.113}$$

$$\eta^{\mathrm{RG}} = \frac{1}{10\,T} \int_0^\infty \mathrm{d}s \, \langle \, \hat{\pi}^{\mu\nu}(0) \, , \, \hat{\pi}_{\mu\nu}(s) \, \rangle. \tag{12.114}$$

We note that the integrands in the formulae have the meanings of the relaxation functions or time-correlation functions of the dissipative currents:

$$R_\Pi(s) := \frac{1}{T} \langle \, \hat{\Pi}(0) \, , \, \hat{\Pi}(s) \, \rangle, \tag{12.115}$$

$$R_J(s) := -\frac{1}{3\,T^2} \langle \, \hat{J}^\mu(0) \, , \, \hat{J}_\mu(s) \, \rangle, \tag{12.116}$$

$$R_\pi(s) := \frac{1}{10\,T} \langle \, \hat{\pi}^{\mu\nu}(0) \, , \, \hat{\pi}_{\mu\nu}(s) \, \rangle. \tag{12.117}$$

We stress that the results of the transport coefficients all show the reliability of our approach based on the doublet scheme in the RG method. Here we recall that the naïve version of the moment method by Israel and Stewart does not give the Chapman-Enskog formulae [66], as is also shown in Sect. 10.3.

Then it is rather natural that the microscopic expressions of the relaxation times (12.108)–(12.110) given above also differ from those (10.174)–(10.176) given by the simplest Israel-Stewart method [66].

We shall now argue that our formulae (12.108)–(12.110) are quite in accordance with the physical meaning of the relaxation times. To see this, we first note the expressions (12.108)–(12.110) can be rewritten as integrals of a product of 'time' and the 'time-correlation' functions of the dissipative currents (12.115)–(12.117):

$$\tau_{\Pi}^{\mathrm{RG}} = \frac{\int_0^\infty \mathrm{d}s\, s\, R_{\Pi}(s)}{\int_0^\infty \mathrm{d}s\, R_{\Pi}(s)}, \tag{12.118}$$

$$\tau_{J}^{\mathrm{RG}} = \frac{\int_0^\infty \mathrm{d}s\, s\, R_{J}(s)}{\int_0^\infty \mathrm{d}s\, R_{J}(s)}, \tag{12.119}$$

$$\tau_{\pi}^{\mathrm{RG}} = \frac{\int_0^\infty \mathrm{d}s\, s\, R_{\pi}(s)}{\int_0^\infty \mathrm{d}s\, R_{\pi}(s)}. \tag{12.120}$$

This is for the first time that the relaxation times are expressed in terms of the relaxation functions in the context of the derivation of the relativistic second-order fluid dynamic equation from the relativistic Boltzmann equation [155]; see also [69]. Furthermore the expressions (12.118)–(12.120) have nice forms that allow the natural interpretation that the respective relaxation times are effective or averaged correlation times in the respective relaxation functions.

Next let us discuss the physical meaning of each term in the relaxation Eqs. (12.98)–(12.100). The first lines in Eqs. (12.98)–(12.100) are nothing but the so-called constitutive equations, which give the relations between the dissipative variables Π, J^μ, and $\pi^{\mu\nu}$ and the thermodynamic forces given by the gradients of T, μ, and u^μ. If we were to insert only these constitutive equations, *i.e.*, the expressions in the first lines, into the conserved currents $J^{\mu\alpha}$ in Eq. (12.97), we have the first-order fluid dynamic equation in the Landau-Lifshitz frame, which was discussed in Chap. 11.

The terms in the other lines are the new ones appearing in the second-order fluid dynamic equation. The second lines represent the relaxation terms given by the temporal and spatial derivatives of the dissipative variables, which describe the relaxation processses of the dissipative variables in response to the thermodynamic forces. The third, fourth, and fifth lines are composed of the products of the thermodynamic forces and dissipative variables, among which we emphasize that the **vorticity term** $\omega^{\mu\nu}$ appears. The final lines give the non-linear terms of the dissipative variables.

A couple of comments are in order here:

(i) Temporal and spatial derivatives in the relaxation equations (12.98)–(12.100) are at most first order and hence our fluid dynamic equation is hyperbolic not parabolic, as expected. Accordingly our fluid dynamic equation satisfies the necessary condition for the causality. In Sect. 12.7, we will prove that the set of the balance equations for the currents (12.97) and the relaxation Eqs. (12.98)–(12.100) is actually causal.

(ii) The relaxation Eqs. (12.98)–(12.100) do not contain the nonlinear vortex term $\omega^{\lambda\langle\mu}\omega^{\nu\rangle}{}_{\lambda}$, which has been seen in the literatures [198]. In fact, our relaxation equations contain the nonlinear vortex terms in a implicit way, which could be made explicit by an iterative solution of the relaxation equations but with a serious problem encountered. To make the discussion clearer, let us set $\Pi = J^\mu = 0$ in Eq. (12.100) to treat the relaxation equation for only $\pi^{\mu\nu}$ and consider the leading terms of ϵ:

$$\pi^{\mu\nu} = 2\,\eta\,\sigma^{\mu\nu} - \tau_\pi\,\Delta^{\mu\nu\rho\sigma}\frac{\partial}{\partial\tau}\pi_{\rho\sigma}. \tag{12.121}$$

First, we solve this equation with respect to $\pi^{\mu\nu}$ formally as

$$\begin{aligned}\pi^{\mu\nu} &= \Delta^{\mu\nu\rho\sigma}\left[1+\tau_\pi\,\frac{\partial}{\partial\tau}\right]^{-1} 2\,\eta\,\sigma_{\rho\sigma}\\ &= \Delta^{\mu\nu\rho\sigma}\left[1-\tau_\pi\,\frac{\partial}{\partial\tau}+\tau_\pi\,\frac{\partial}{\partial\tau}\tau_\pi\,\frac{\partial}{\partial\tau}-\cdots\right]2\,\eta\,\sigma_{\rho\sigma}\\ &= 2\,\eta\,\sigma^{\mu\nu} - 2\,\eta\,\tau_\pi\,\Delta^{\mu\nu\rho\sigma}\frac{\partial}{\partial\tau}\sigma_{\rho\sigma}+\cdots.\end{aligned} \tag{12.122}$$

Then, using the identity

$$\begin{aligned}\Delta^{\mu\nu\rho\sigma}\frac{\partial}{\partial\tau}\sigma_{\rho\sigma} &= \Delta^{\mu\nu\rho\sigma}\left\{-\left[\frac{\partial}{\partial\tau}u_\rho\right]\left[\frac{\partial}{\partial\tau}u_\sigma\right]+\nabla_\rho\left[\frac{\partial}{\partial\tau}u_\sigma\right]\right\}-\frac{2}{3}\theta\,\sigma^{\mu\nu}\\ &\quad -\sigma^{\lambda\langle\mu}\,\sigma^{\nu\rangle}{}_\lambda - \omega^{\lambda\langle\mu}\,\omega^{\nu\rangle}{}_\lambda - 2\,\sigma^{\lambda\langle\mu}\,\omega^{\nu\rangle}{}_\lambda,\end{aligned} \tag{12.123}$$

and the balance Eq. (12.178), we convert Eq. (12.122) into

$$\begin{aligned}\pi^{\mu\nu} &= 2\,\eta\,\sigma^{\mu\nu}\\ &\quad + 2\,\eta\,\tau_\pi\left\{\sigma^{\lambda\langle\mu}\,\sigma^{\nu\rangle}{}_\lambda + 2\,\sigma^{\lambda\langle\mu}\,\omega^{\nu\rangle}{}_\lambda + \omega^{\lambda\langle\mu}\,\omega^{\nu\rangle}{}_\lambda + \frac{2}{3}\theta\,\sigma^{\mu\nu}\right.\\ &\quad + \Delta^{\mu\nu\rho\sigma}\left[\frac{1}{T^2}(\nabla_\rho T)(\nabla_\sigma T) + \frac{2}{h}(\nabla_\rho T)(\nabla_\sigma\frac{\mu}{T}) + \frac{T^2}{h^2}(\nabla_\rho\frac{\mu}{T})(\nabla_\sigma\frac{\mu}{T})\right]\\ &\quad \left. - \Delta^{\mu\nu\rho\sigma}\,\nabla_\rho\left[\frac{1}{T}\nabla_\sigma T\right] - \Delta^{\mu\nu\rho\sigma}\,\nabla_\rho\left[\frac{T}{h}\nabla_\sigma\frac{\mu}{T}\right]\right\}+\cdots.\end{aligned} \tag{12.124}$$

We note that the above equation apparently has the term of $\omega^{\lambda\langle\mu}\omega^{\nu\rangle}{}_\lambda$. Notice, however, that the last two terms of Eq. (12.124) have a form of the second-order spatial derivatives of fluid dynamic variables, which make the fluid dynamic equation **parabolic** and accordingly acausal. Hence we have an important observation that the iterative construction of the solution may spoil the causal property of the original fluid dynamic equation, and thus we must use the original form of the relaxation Eqs. (12.121) or (12.98)–(12.100). Furthermore, since the appearance of $\omega^{\lambda\langle\mu}\omega^{\nu\rangle}{}_\lambda$ seems to be inevitably associated with that of the second-order spatial derivative terms, the explicit appearance of such a nonlinear vortex term should be avoided in the relaxation equations although its effect should be included in Eq. (12.121) implicitly.

12.6.3 Derivation of Relaxation Equations

In this subsection, we shall show how the relaxation equations (12.98)-(12.100) are derived from Eq. (12.81). If we define

$$v_{pq}^{\alpha} := \begin{cases} v_{pq}^{\mu} := \dfrac{1}{p \cdot u} \Delta^{\mu\nu} p_{\nu} \delta_{pq}, & \alpha = \mu, \\ \delta_{pq}, & \alpha = 4, \end{cases} \tag{12.125}$$

$$D_{\alpha} := \begin{cases} \epsilon \nabla_{\mu}, \ \alpha = \mu, \\ \dfrac{\partial}{\partial \tau}, \quad \alpha = 4, \end{cases} \tag{12.126}$$

the differential operator in (12.81) are expressed in a compact form as

$$\left[(p \cdot u) \frac{\partial}{\partial \tau} + \epsilon \, p \cdot \nabla \right] \delta_{pq} = (p \cdot u) \, v_{pq}^{\alpha} \, D_{\alpha}. \tag{12.127}$$

Then Eq. (12.81) is converted into the following form with the use of the inner products:

$$\begin{aligned} &\langle \hat{L}^{-1} \hat{\psi}^i , (f^{\mathrm{eq}} \bar{f}^{\mathrm{eq}})^{-1} v^{\alpha} D_{\alpha} \Big[f^{\mathrm{eq}} (1 + \epsilon \, \bar{f}^{\mathrm{eq}} \hat{L}^{-1} \hat{\chi}^j \psi_j) \Big] \rangle \\ -\, &\epsilon \langle \hat{L}^{-1} \hat{\psi}^i , \hat{\chi}^i \psi_j \rangle \\ +\, &\epsilon^2 \frac{1}{2} \langle \hat{L}^{-1} \hat{\psi}^i , B \Big[\hat{L}^{-1} \hat{\chi}^j \psi_j , \hat{L}^{-1} \hat{\chi}^k \psi_k \Big] \rangle, \end{aligned} \tag{12.128}$$

up to $O(\epsilon^2)$, where we have introduced the following three "vectors":

$$\hat{\boldsymbol{\psi}}_p := \begin{pmatrix} \hat{\Pi}_p \\ \hat{J}_p^{\mu}/h \\ \hat{\pi}_p^{\mu\nu} \end{pmatrix}, \ \boldsymbol{\psi} := \begin{pmatrix} \Pi \\ J_{\mu} \\ \pi_{\mu\nu} \end{pmatrix}, \ \hat{\boldsymbol{\chi}}_p := \begin{pmatrix} \hat{\Pi}_p / \langle \hat{\Pi} , \hat{L}^{-1} \hat{\Pi} \rangle \\ 3 \, h \, \hat{J}_p^{\mu} / \langle \hat{J}^{\rho} , \hat{L}^{-1} \hat{J}_{\rho} \rangle \\ 5 \, \hat{\pi}_p^{\mu\nu} / \langle \hat{\pi}^{\rho\sigma} , \hat{L}^{-1} \hat{\pi}_{\rho\sigma} \rangle \end{pmatrix}. \tag{12.129}$$

For instance, $\hat{\psi}_p^1 = \hat{\Pi}_p$, $\psi_2 = J^{\mu}$ and so on.

We expand the left-hand sides of Eq. (12.128) as

$$\begin{aligned} &\langle \hat{L}^{-1} \hat{\psi}^i , (f^{\mathrm{eq}} \bar{f}^{\mathrm{eq}})^{-1} v^{\alpha} D_{\alpha} \Big[f^{\mathrm{eq}} (1 + \epsilon \, \bar{f}^{\mathrm{eq}} \hat{L}^{-1} \hat{\chi}^j \psi_j) \Big] \rangle \\ = &\langle \hat{L}^{-1} \hat{\psi}^i , (f^{\mathrm{eq}} \bar{f}^{\mathrm{eq}})^{-1} v^{\alpha} D_{\alpha} f^{\mathrm{eq}} \rangle \\ +\, &\epsilon \langle \hat{L}^{-1} \hat{\psi}^i , (f^{\mathrm{eq}} \bar{f}^{\mathrm{eq}})^{-1} v^{\alpha} D_{\alpha} \Big[f^{\mathrm{eq}} \bar{f}^{\mathrm{eq}} \hat{L}^{-1} \hat{\chi}^j \Big] \rangle \psi_j \\ +\, &c \langle \hat{L}^{-1} \hat{\psi}^i , v^{\alpha} \hat{L}^{-1} \hat{\chi}^j \rangle D_{\alpha} \psi_j. \end{aligned} \tag{12.130}$$

The first term of Eq. (12.130) is calculated to be

$$
\begin{aligned}
\langle \hat{L}^{-1} \hat{\psi}^i, (f^{\mathrm{eq}} \bar{f}^{\mathrm{eq}})^{-1} v^\alpha D_\alpha f^{\mathrm{eq}} \rangle &= \epsilon \langle \hat{L}^{-1} \hat{\psi}^i, (f^{\mathrm{eq}} \bar{f}^{\mathrm{eq}})^{-1} v^a \nabla_a f^{\mathrm{eq}} \rangle \\
&= -\epsilon \langle \hat{L}^{-1} \hat{\psi}^i, F_0 \rangle \\
&= -\epsilon \langle \hat{L}^{-1} \hat{\psi}^i, Q_0 F_0 \rangle \\
&= -\epsilon \langle \hat{\psi}^i, \hat{L}^{-1} Q_0 F_0 \rangle. \qquad (12.131)
\end{aligned}
$$

In the first equality, we have used the fact that

$$
(f^{\mathrm{eq}} \bar{f}^{\mathrm{eq}})^{-1} \frac{\partial}{\partial \tau} f^{\mathrm{eq}} = \frac{\partial}{\partial \tau}(\mu/T) - p^\mu \frac{\partial}{\partial \tau}(u_\mu/T), \qquad (12.132)
$$

belongs to the P_0 space, and hence the inner product with the vector $\hat{L}^{-1} \hat{\psi}^i$ in the Q_0-space vanishes, whereas in the second equality, we have used the expression of F_0 given by Eq. (12.17), and the self-adjointness of $\hat{L}$ in the last equality. Next, substituting $\hat{L}^{-1} Q_0 F_0$ in Eq. (12.38) into Eq. (12.131), we have

$$
\langle \hat{L}^{-1} \hat{\psi}^i, (f^{\mathrm{eq}} \bar{f}^{\mathrm{eq}})^{-1} v^\alpha D_\alpha f^{\mathrm{eq}} \rangle = \epsilon \langle \hat{L}^{-1} \hat{\psi}^i, \hat{\chi}^j \rangle X'_j, \qquad (12.133)
$$

with

$$
X'_i := (-\zeta\, \theta, \lambda\, T^2 h^{-2} \nabla_\mu(\mu/T), 2\, \eta\, \sigma_{\mu\nu}). \qquad (12.134)
$$

Here, ζ, λ, and η are the bulk viscosity, heat conductivity, and shear viscosity.

The third term of Eq. (12.130) is rewritten as

$$
\begin{aligned}
\epsilon \langle \hat{L}^{-1} \hat{\psi}^i, v^\alpha \hat{L}^{-1} \hat{\chi}^j \rangle D_\alpha \psi_j &= \epsilon \langle \hat{L}^{-1} \hat{\psi}^i, \hat{L}^{-1} \hat{\chi}^j \rangle \frac{\partial}{\partial \tau} \psi_j \\
&\quad + \epsilon^2 \langle \hat{L}^{-1} \hat{\psi}^i, v^a \hat{L}^{-1} \hat{\chi}^j \rangle \nabla_a \psi_j. \qquad (12.135)
\end{aligned}
$$

Inserting Eq. (12.130) into Eq. (12.128) with the replacement of Eqs. (12.133) and (12.135), we have the relaxation equation as follows,

$$
\begin{aligned}
\epsilon \langle \hat{L}^{-1} \hat{\psi}^i, \hat{\chi}^j \rangle \psi_j &= \epsilon \langle \hat{L}^{-1} \hat{\psi}^i, \hat{\chi}^j \rangle X'_j + \epsilon \langle \hat{L}^{-1} \hat{\psi}^i, \hat{L}^{-1} \hat{\chi}^j \rangle \frac{\partial}{\partial \tau} \psi_j \\
&\quad + \epsilon^2 \langle \hat{L}^{-1} \hat{\psi}^i, v^a \hat{L}^{-1} \hat{\chi}^j \rangle \nabla_a \psi_j \\
&\quad + \epsilon^2 \frac{1}{2} M^{i,j,k} \psi_j \psi_k + \epsilon N^{i,j} \psi_j, \qquad (12.136)
\end{aligned}
$$

up to $O(\epsilon^2)$, where

$$
M^{i,j,k} := -\langle \hat{L}^{-1} \hat{\psi}^i, B[\hat{L}^{-1} \hat{\chi}^j, \hat{L}^{-1} \hat{\chi}^k] \rangle, \qquad (12.137)
$$

$$
N^{i,j} := \langle \hat{L}^{-1} \hat{\psi}^i, (f^{\mathrm{eq}} \bar{f}^{\mathrm{eq}})^{-1} v^\alpha D_\alpha \Big[f^{\mathrm{eq}} \bar{f}^{\mathrm{eq}} \hat{L}^{-1} \hat{\chi}^j \Big] \rangle. \qquad (12.138)
$$

Equation (12.136) can be made further simpler, if we note the following useful formulae for space-like tensors A:

$$\langle A^{\mu\nu} \rangle = \frac{1}{3} \Delta^{\mu\nu} \langle A^{\rho}{}_{\rho} \rangle, \tag{12.139}$$

$$\langle A^{\langle\mu\nu\rangle\rho\sigma} \rangle = \frac{1}{5} \Delta^{\mu\nu\rho\sigma} \langle A^{\langle\alpha\beta\rangle}{}_{\langle\alpha\beta\rangle} \rangle, \tag{12.140}$$

$$\begin{aligned} \langle A^{\mu\nu\rho\sigma} \rangle &= \frac{1}{3} \Delta^{\mu\nu} \langle A^{\alpha}{}_{\alpha}{}^{\rho\sigma} \rangle + \langle A^{\langle\mu\nu\rangle\rho\sigma} \rangle + \langle A^{(\mu\nu)\rho\sigma} \rangle \\ &= \frac{1}{9} \Delta^{\mu\nu} \Delta^{\rho\sigma} \langle A^{\alpha}{}_{\alpha}{}^{\beta}{}_{\beta} \rangle + \frac{1}{5} \Delta^{\mu\nu\rho\sigma} \langle A^{\langle\alpha\beta\rangle}{}_{\langle\alpha\beta\rangle} \rangle \\ &\quad + \frac{1}{3} \Omega^{\mu\nu\rho\sigma} \langle A^{(\alpha\beta)}{}_{(\alpha\beta)} \rangle, \end{aligned} \tag{12.141}$$

$$\langle A^{\langle\mu\nu\rangle\langle\rho\sigma\rangle\langle\alpha\beta\rangle} \rangle = \frac{12}{35} \Delta^{\mu\nu\gamma\delta} \Delta^{\rho\sigma\lambda}{}_{\gamma} \Delta^{\alpha\beta}{}_{\lambda\delta} \langle A^{\langle\tau\eta\rangle}{}_{\langle\tau}{}^{\kappa\rangle}{}_{\langle\kappa\eta\rangle} \rangle, \tag{12.142}$$

$$\begin{aligned} \langle A^{\langle\mu\nu\rangle\langle\rho\sigma\rangle\alpha\beta} \rangle &= \frac{1}{3} \Delta^{\alpha\beta} \langle A^{\langle\mu\nu\rangle\langle\rho\sigma\rangle\lambda}{}_{\lambda} \rangle + \langle A^{\langle\mu\nu\rangle\langle\rho\sigma\rangle\langle\alpha\beta\rangle} \rangle + \langle A^{\langle\mu\nu\rangle\langle\rho\sigma\rangle(\alpha\beta)} \rangle \\ &= \frac{1}{15} \Delta^{\mu\nu\rho\sigma} \Delta^{\alpha\beta} \langle A^{\langle\gamma\delta\rangle}{}_{\langle\gamma\delta\rangle}{}^{\lambda}{}_{\lambda} \rangle \\ &\quad + \frac{12}{35} \Delta^{\mu\nu\gamma\delta} \Delta^{\rho\sigma\lambda}{}_{\gamma} \Delta^{\alpha\beta}{}_{\lambda\delta} \langle A^{\langle\tau\eta\rangle}{}_{\langle\tau}{}^{\kappa\rangle}{}_{\langle\kappa\eta\rangle} \rangle \\ &\quad + \frac{4}{15} \Delta^{\mu\nu\gamma\delta} \Delta^{\rho\sigma\lambda}{}_{\gamma} \Omega^{\alpha\beta}{}_{\lambda\delta} \langle A^{\langle\tau\eta\rangle}{}_{\langle\tau}{}^{\kappa\rangle}{}_{(\kappa\eta)} \rangle, \end{aligned} \tag{12.143}$$

where we have defined the notations,

$$\langle A \rangle := \langle 1 , A \rangle = \int \mathrm{d}p \, (p \cdot u) \, f^{\mathrm{eq}} \, \bar{f}^{\mathrm{eq}} \, A_p. \tag{12.144}$$

and

$$A^{(\mu\nu)} := \Omega^{\mu\nu\rho\sigma} A_{\rho\sigma}, \tag{12.145}$$

with

$$\Omega^{\mu\nu\rho\sigma} := \frac{1}{2} (\Delta^{\mu\rho} \Delta^{\nu\sigma} - \Delta^{\mu\sigma} \Delta^{\nu\rho}). \tag{12.146}$$

In the first equality of Eq. (12.141) and the first equality of Eq. (12.143), we have used the fact that a space-like rank-two tensor $B^{\mu\nu}$, which satisfies both $u_\mu B^{\mu\nu} = 0$ and $u_\nu B^{\mu\nu} = 0$, is decomposed to be

$$B^{\mu\nu} = \frac{1}{3} \Delta^{\mu\nu} B^{\rho}{}_{\rho} + B^{\langle\mu\nu\rangle} + B^{(\mu\nu)}. \tag{12.147}$$

The numerical factors may be verified by contracting both sides of equations. To see how to use these formulae, let us consider $\langle\, A^{\langle\rho\sigma\rangle\alpha\beta}\,\rangle\, \psi_{\langle\rho\sigma\rangle}\, \chi_{\alpha\beta}$, which is found in the fifth line after the first equality of Eq. (12.179),

$$\begin{aligned}\langle\, A^{\langle\rho\sigma\rangle\alpha\beta}\,\rangle\, \psi_{\langle\rho\sigma\rangle}\, \chi_{\alpha\beta} &= \frac{1}{5}\,\Delta^{\rho\sigma\alpha\beta}\,\langle\, A^{\langle\gamma\delta\rangle}{}_{\langle\gamma\delta\rangle}\,\rangle\, \psi_{\langle\rho\sigma\rangle}\, \chi_{\alpha\beta} \\ &= \frac{1}{5}\,\langle\, A^{\langle\gamma\delta\rangle}{}_{\langle\gamma\delta\rangle}\,\rangle\, \psi^{\langle\rho\sigma\rangle}\, \chi_{\langle\rho\sigma\rangle},\end{aligned} \tag{12.148}$$

where we have used Eq. (12.140) in the second equality. With the use of the formulae (12.139) and (12.140), some coefficients in Eq. (12.136) are reduced to be

$$\begin{aligned}
&\langle \hat{L}^{-1}\,\hat{\psi}^i\,,\,\hat{\chi}^j\,\rangle \\
&= \begin{pmatrix} \dfrac{\langle\,\hat{L}^{-1}\,\hat{\Pi}\,,\,\hat{\Pi}\,\rangle}{-T\,\zeta} & \dfrac{h\,\langle\,\hat{L}^{-1}\,\hat{\Pi}\,,\,\hat{J}^{\rho}\,\rangle}{T^2\,\lambda} & \dfrac{\langle\,\hat{L}^{-1}\,\hat{\Pi}\,,\,\hat{\pi}^{\rho\sigma}\,\rangle}{-2\,T\,\eta} \\ \dfrac{\langle\,\hat{L}^{-1}\,\hat{J}^{\mu}\,,\,\hat{\Pi}\,\rangle}{-T\,\zeta\,h} & \dfrac{\langle\,\hat{L}^{-1}\,\hat{J}^{\mu}\,,\,\hat{J}^{\rho}\,\rangle}{T^2\,\lambda} & \dfrac{\langle\,\hat{L}^{-1}\,\hat{J}^{\mu}\,,\,\hat{\pi}^{\rho\sigma}\,\rangle}{-2\,T\,\eta\,h} \\ \dfrac{\langle\,\hat{L}^{-1}\,\hat{\pi}^{\mu\nu}\,,\,\hat{\Pi}\,\rangle}{-T\,\zeta} & \dfrac{h\,\langle\,\hat{L}^{-1}\,\hat{\pi}^{\mu\nu}\,,\,\hat{J}^{\rho}\,\rangle}{T^2\,\lambda} & \dfrac{\langle\,\hat{L}^{-1}\,\hat{\pi}^{\mu\nu}\,,\,\hat{\pi}^{\rho\sigma}\,\rangle}{-2\,T\,\eta} \end{pmatrix} \\
&= \begin{pmatrix} 1 & 0 & 0 \\ 0 & \Delta^{\mu\rho} & 0 \\ 0 & 0 & \Delta^{\mu\nu\rho\sigma} \end{pmatrix},
\end{aligned} \tag{12.149}$$

$$\begin{aligned}
&\langle\,\hat{L}^{-1}\,\hat{\psi}^i\,,\,\hat{L}^{-1}\,\hat{\chi}^j\,\rangle \\
&= \begin{pmatrix} \dfrac{\langle\,\hat{L}^{-1}\,\hat{\Pi}\,,\,\hat{L}^{-1}\,\hat{\Pi}\,\rangle}{-T\,\zeta} & \dfrac{h\,\langle\,\hat{L}^{-1}\,\hat{\Pi}\,,\,\hat{L}^{-1}\,\hat{J}^{\rho}\,\rangle}{T^2\,\lambda} & \dfrac{\langle\,\hat{L}^{-1}\,\hat{\Pi}\,,\,\hat{L}^{-1}\,\hat{\pi}^{\rho\sigma}\,\rangle}{-2\,T\,\eta} \\ \dfrac{\langle\,\hat{L}^{-1}\,\hat{J}^{\mu}\,,\,\hat{L}^{-1}\,\hat{\Pi}\,\rangle}{-T\,\zeta\,h} & \dfrac{\langle\,\hat{L}^{-1}\,\hat{J}^{\mu}\,,\,\hat{L}^{-1}\,\hat{J}^{\rho}\,\rangle}{T^2\,\lambda} & \dfrac{\langle\,\hat{L}^{-1}\,\hat{J}^{\mu}\,,\,\hat{L}^{-1}\,\hat{\pi}^{\rho\sigma}\,\rangle}{-2\,T\,\eta\,h} \\ \dfrac{\langle\,\hat{L}^{-1}\,\hat{\pi}^{\mu\nu}\,,\,\hat{L}^{-1}\,\hat{\Pi}\,\rangle}{-T\,\zeta} & \dfrac{h\,\langle\,\hat{L}^{-1}\,\hat{\pi}^{\mu\nu}\,,\,\hat{L}^{-1}\,\hat{J}^{\rho}\,\rangle}{T^2\,\lambda} & \dfrac{\langle\,\hat{L}^{-1}\,\hat{\pi}^{\mu\nu}\,,\,\hat{L}^{-1}\,\hat{\pi}^{\rho\sigma}\,\rangle}{-2\,T\,\eta} \end{pmatrix} \\
&= \begin{pmatrix} -\tau_{\Pi} & 0 & 0 \\ 0 & -\tau_J\,\Delta^{\mu\rho} & 0 \\ 0 & 0 & -\tau_{\pi}\,\Delta^{\mu\nu\rho\sigma} \end{pmatrix},
\end{aligned} \tag{12.150}$$

$$\begin{aligned}
&\langle\,\hat{L}^{-1}\,\hat{\psi}^i\,,\,v^a\,\hat{L}^{-1}\,\hat{\chi}^j\,\rangle \\
&= \begin{pmatrix} \dfrac{\langle\,\hat{L}^{-1}\,\hat{\Pi}\,,\,v^a\,\hat{L}^{-1}\,\hat{\Pi}\,\rangle}{-T\,\zeta} & \dfrac{h\,\langle\,\hat{L}^{-1}\,\hat{\Pi}\,,\,v^a\,\hat{L}^{-1}\,\hat{J}^{\rho}\,\rangle}{T^2\,\lambda} & \dfrac{\langle\,\hat{L}^{-1}\,\hat{\Pi}\,,\,v^a\,\hat{L}^{-1}\,\hat{\pi}^{\rho\sigma}\,\rangle}{-2\,T\,\eta} \\ \dfrac{\langle\,\hat{L}^{-1}\,\hat{J}^{\mu}\,,\,v^a\,\hat{L}^{-1}\,\hat{\Pi}\,\rangle}{-T\,\zeta\,h} & \dfrac{\langle\,\hat{L}^{-1}\,\hat{J}^{\mu}\,,\,v^a\,\hat{L}^{-1}\,\hat{J}^{\rho}\,\rangle}{T^2\,\lambda} & \dfrac{\langle\,\hat{L}^{-1}\,\hat{J}^{\mu}\,,\,v^a\,\hat{L}^{-1}\,\hat{\pi}^{\rho\sigma}\,\rangle}{-2\,T\,\eta\,h} \\ \dfrac{\langle\,\hat{L}^{-1}\,\hat{\pi}^{\mu\nu}\,,\,v^a\,\hat{L}^{-1}\,\hat{\Pi}\,\rangle}{-T\,\zeta} & \dfrac{h\,\langle\,\hat{L}^{-1}\,\hat{\pi}^{\mu\nu}\,,\,v^a\,\hat{L}^{-1}\,\hat{J}^{\rho}\,\rangle}{T^2\,\lambda} & \dfrac{\langle\,\hat{L}^{-1}\,\hat{\pi}^{\mu\nu}\,,\,v^a\,\hat{L}^{-1}\,\hat{\pi}^{\rho\sigma}\,\rangle}{-2\,T\,\eta} \end{pmatrix} \\
&= \begin{pmatrix} 0 & -h\,\ell_{\Pi J}\,\Delta^{a\rho} & 0 \\ -h^{-1}\,\ell_{J\Pi}\,\Delta^{\mu a} & 0 & -h^{-1}\,\ell_{J\pi}\,\Delta^{\mu a\rho\sigma} \\ 0 & -h\,\ell_{\pi J}\,\Delta^{\mu\nu a\rho} & 0 \end{pmatrix},
\end{aligned} \tag{12.151}$$

Here, we have introduced the relaxation times τ_{Π}, τ_J, and τ_{π} and the relaxation lengths $\ell_{\Pi J}$, $\ell_{J\Pi}$, $\ell_{J\pi}$, and $\ell_{\pi J}$, whose definitions are given by

$$\tau_\Pi := \frac{\langle \hat{L}^{-1}\,\hat{\Pi}\,,\,\hat{L}^{-1}\,\hat{\Pi}\,\rangle}{T\,\zeta}, \tag{12.152}$$

$$\tau_J := -\frac{\langle \hat{L}^{-1}\,\hat{J}^{\mu}\,,\,\hat{L}^{-1}\,\hat{J}_{\mu}\,\rangle}{3\,T^2\,\lambda}, \tag{12.153}$$

$$\tau_\pi := \frac{\langle \hat{L}^{-1}\,\hat{\pi}^{\mu\nu}\,,\,\hat{L}^{-1}\,\hat{\pi}_{\mu\nu}\,\rangle}{10\,T\,\eta}, \tag{12.154}$$

$$\ell_{\Pi J} := -\frac{\langle \hat{L}^{-1}\,\hat{\Pi}\,,\,v^{\mu}\,\hat{L}^{-1}\,\hat{J}_{\mu}\,\rangle}{3\,T^2\,\lambda}, \tag{12.155}$$

$$\ell_{J\Pi} := \frac{\langle \hat{L}^{-1}\,\hat{J}^{\mu}\,,\,v_{\mu}\,\hat{L}^{-1}\,\hat{\Pi}\,\rangle}{3\,T\,\zeta}, \tag{12.156}$$

$$\ell_{J\pi} := \frac{\langle \hat{L}^{-1}\,\hat{J}^{\mu}\,,\,v^{\nu}\,\hat{L}^{-1}\,\hat{\pi}_{\mu\nu}\,\rangle}{10\,T\,\eta}, \tag{12.157}$$

$$\ell_{\pi J} := -\frac{\langle \hat{L}^{-1}\,\hat{\pi}^{\mu\nu}\,,\,v_{\mu}\,\hat{L}^{-1}\,\hat{J}_{\nu}\,\rangle}{5\,T^2\,\lambda}. \tag{12.158}$$

Thus, we find that some terms in Eq. (12.136) are obtained as follows:

$$\epsilon\,\langle\,\hat{L}^{-1}\,\hat{\psi}^i\,,\,\hat{\chi}^j\,\rangle\,\psi_j = \epsilon\,(\Pi,\ J^{\mu},\ \pi^{\mu\nu}), \tag{12.159}$$

$$\epsilon\,\langle\,\hat{L}^{-1}\,\hat{\psi}^i\,,\,\hat{\chi}^j\,\rangle\,X'_j = \epsilon\,(-\zeta\,\theta,\ \lambda\,T^2\,h^{-2}\,\nabla^{\mu}(\mu/T),\,2\,\eta\,\sigma^{\mu\nu}), \tag{12.160}$$

$$\begin{aligned}\epsilon\,\langle\,\hat{L}^{-1}\,\hat{\psi}^i\,,\,\hat{L}^{-1}\,\hat{\chi}^j\,\rangle\,\frac{\partial}{\partial\tau}\psi_j &= \epsilon\,(-\tau_\Pi\,\frac{\partial}{\partial\tau}\Pi,\ -\tau_J\,\Delta^{\mu\rho}\,\frac{\partial}{\partial\tau}J_\rho, \\ &\quad -\tau_\pi\,\Delta^{\mu\nu\rho\sigma}\,\frac{\partial}{\partial\tau}\pi_{\rho\sigma}),\end{aligned} \tag{12.161}$$

$$\begin{aligned}\epsilon^2\,\langle\,\hat{L}^{-1}\,\hat{\psi}^i\,,\,v^a\,\hat{L}^{-1}\,\hat{\chi}^j\,\rangle\,\nabla_a\psi_j &= \epsilon^2\,(-h\,\ell_{\Pi J}\,\nabla\cdot J, \\ &\quad -h^{-1}\,\ell_{J\Pi}\,\nabla^{\mu}\Pi - h^{-1}\,\ell_{J\pi}\,\Delta^{\mu a\rho\sigma}\,\nabla_a\pi_{\rho\sigma}, \\ &\quad -h\,\ell_{\pi J}\,\Delta^{\mu\nu a\rho}\,\nabla_a J_\rho).\end{aligned} \tag{12.162}$$

From now on, we examine the rest of the terms in Eq. (12.136). Using the formulae (12.139), (12.140), and (12.142), we can reduce $\epsilon^2\,\frac{1}{2}\,M^{i,j,k}\,\psi_j\,\psi_k$ for $\hat{\psi}^i = \hat{\Pi},\ \hat{J}^{\mu}/h,\ \hat{\pi}^{\mu\nu}$ to the following forms:

For $\hat{\psi}^i = \hat{\Pi}$,

$$\begin{aligned}&-\frac{\epsilon^2}{2}\,\langle L^{-1}\,\hat{\Pi},\,B[L^{-1}\,\hat{\chi}^j\,,\,L^{-1}\,\hat{\chi}^k]\rangle\,\psi_j\,\psi_k \\ &= -\epsilon^2\Bigg[\frac{\langle L^{-1}\,\hat{\Pi},\,B[L^{-1}\,\hat{\Pi},\,L^{-1}\,\hat{\Pi}]\rangle}{2\,(T\,\zeta)^2}\,\Pi^2 + \frac{\langle L^{-1}\,\hat{\Pi},\,B[L^{-1}\,\hat{J}^{\rho}\,,\,L^{-1}\,\hat{J}^{\sigma}]\rangle}{2\,(T^2\,\lambda/h)^2}\,J_\rho\,J_\sigma \\ &\quad + \frac{\langle L^{-1}\,\hat{\Pi},\,B[L^{-1}\,\hat{\pi}^{\rho\sigma}\,,\,L^{-1}\,\hat{\pi}^{\alpha\beta}]\rangle}{2\,(2\,T\,\eta)^2}\,\pi_{\rho\sigma}\,\pi_{\alpha\beta}\Bigg]\end{aligned}$$

$$= -\epsilon^2 \left[\frac{\langle L^{-1}\,\hat{\Pi}\,,\,B[L^{-1}\,\hat{\Pi}\,,\,L^{-1}\,\hat{\Pi}]\rangle}{2\,(T\,\zeta)^2}\,\Pi^2 + \frac{\langle L^{-1}\,\hat{\Pi}\,,\,B[L^{-1}\,\hat{J}^{\mu}\,,\,L^{-1}\,\hat{J}_{\mu}]\rangle}{6\,(T^2\,\lambda/h)^2}\,J^{\rho}\,J_{\rho} \right.$$
$$\left. + \frac{\langle L^{-1}\,\hat{\Pi}\,,\,B[L^{-1}\,\hat{\pi}^{\mu\nu}\,,\,L^{-1}\,\hat{\pi}_{\mu\nu}]\rangle}{10\,(2\,T\,\eta)^2}\,\pi^{\rho\sigma}\,\pi_{\rho\sigma} \right], \tag{12.163}$$

for $\hat{\psi}^i = \hat{J}^{\mu}/h$,

$$-\frac{\epsilon^2}{2}\,\langle L^{-1}\,\hat{J}^{\mu}/h\,,\,B[L^{-1}\,\hat{\psi}^j\,,\,L^{-1}\,\hat{\psi}^k]\rangle\,\chi_j\,\chi_k$$
$$= \epsilon^2 \left[\frac{\langle L^{-1}\,\hat{J}^{\mu}\,,\,B[L^{-1}\,\hat{\Pi}\,,\,L^{-1}\,\hat{J}^{\rho}]\rangle}{(T\,\zeta)(T^2\,\lambda/h)\,h}\,\Pi\,J_{\rho} \right.$$
$$\left. + \frac{\langle L^{-1}\,\hat{J}^{\mu}\,,\,B[L^{-1}\,\hat{J}^{\rho}\,,\,L^{-1}\,\hat{\pi}^{\alpha\beta}]\rangle}{(T^2\,\lambda/h)(2\,T\,\eta)\,h}\,J_{\rho}\,\pi_{\alpha\beta} \right.$$
$$= \epsilon^2 \left[\frac{\langle L^{-1}\,\hat{J}^{\mu}\,,\,B[L^{-1}\,\hat{\Pi}\,,\,L^{-1}\,\hat{J}_{\mu}]\rangle}{3\,(T\,\zeta)\,(T^2\,\lambda/h)\,h}\,\Pi\,J^{\mu} \right.$$
$$\left. + \frac{\langle L^{-1}\,\hat{J}^{\mu}\,,\,B[L^{-1}\,\hat{J}^{\nu}\,,\,L^{-1}\,\hat{\pi}_{\mu\nu}]\rangle}{5\,(T^2\,\lambda/h)\,(2\,T\,\eta)\,h}\,J^{\rho}\,\pi_{\rho}{}^{\mu} \right], \tag{12.164}$$

and, for $\hat{\psi}^i = \hat{\pi}^{\mu\nu}$,

$$-\frac{\epsilon^2}{2}\,\langle L^{-1}\,\hat{\pi}^{\mu\nu}\,,\,B[L^{-1}\,\hat{\psi}^j\,,\,L^{-1}\,\hat{\psi}^k]\rangle\,\chi_j\,\chi_k$$
$$= -\epsilon^2 \left[\frac{\langle L^{-1}\,\hat{\pi}^{\mu\nu}\,,\,B[L^{-1}\,\hat{\Pi}\,,\,L^{-1}\,\hat{\pi}^{\rho\sigma}]\rangle}{(T\,\zeta)\,(2\,T\,\eta)}\,\Pi\,\pi_{\rho\sigma} \right.$$
$$+ \frac{\langle L^{-1}\,\hat{\pi}^{\mu\nu}\,,\,B[L^{-1}\,\hat{J}^{\rho}\,,\,L^{-1}\,\hat{J}^{\sigma}]\rangle}{2\,(T^2\,\lambda/h)^2}\,J_{\rho}\,J_{\sigma}$$
$$\left. + \frac{\langle L^{-1}\,\hat{\pi}^{\mu\nu}\,,\,B[L^{-1}\,\hat{\pi}^{\rho\sigma}\,,\,L^{-1}\,\hat{\pi}^{\alpha\beta}]\rangle}{2\,(2\,T\,\eta)^2}\,\pi_{\rho\sigma}\,\pi_{\alpha\beta} \right.$$
$$= -\epsilon^2 \left[\frac{\langle L^{-1}\,\hat{\pi}^{\mu\nu}\,,\,B[L^{-1}\,\hat{\Pi}\,,\,L^{-1}\,\hat{\pi}_{\mu\nu}]\rangle}{5\,(T\,\zeta)(2\,T\,\eta)}\,\Pi\,\pi^{\mu\nu} \right.$$
$$+ \frac{\langle L^{-1}\,\hat{\pi}^{\mu\nu}\,,\,B[L^{-1}\,\hat{J}_{\mu}\,,\,L^{-1}\,\hat{J}_{\nu}]\rangle}{10\,(T^2\,\lambda/h)^2}\,J^{\langle\mu}\,J^{\nu\rangle}$$
$$\left. + \frac{\langle L^{-1}\,\hat{\pi}^{\mu\nu}\,,\,B[L^{-1}\,\hat{\pi}^{\lambda}_{\mu}\,,\,L^{-1}\,\hat{\pi}_{\lambda\nu}]\rangle}{(35/6)\,(2\,T\,\eta)^2}\,\pi^{\rho\langle\mu}\,\pi^{\nu\rangle}{}_{\rho} \right], \tag{12.165}$$

which are summarized as

$$\epsilon^2\,\frac{1}{2}\,M^{i,j,k}\,\psi_j\,\psi_k = \epsilon^2\,(b_{\Pi\Pi\Pi}\,\Pi^2 + b_{\Pi JJ}\,J^{\rho}\,J_{\rho} + b_{\Pi\pi\pi}\,\pi^{\rho\sigma}\,\pi_{\rho\sigma},$$
$$b_{J\Pi J}\,\Pi\,J^{\mu} + b_{JJ\pi}\,J^{\rho}\,\pi_{\rho}{}^{\mu},$$
$$b_{\pi\Pi\pi}\,\Pi\,\pi^{\mu\nu} + b_{\pi JJ}\,J^{\langle\mu}\,J^{\nu\rangle} + b_{\pi\pi\pi}\,\pi^{\rho\langle\mu}\,\pi^{\nu\rangle}{}_{\rho}). \tag{12.166}$$

Here, the coefficients $b_{\Pi\Pi\Pi}$, $b_{\Pi JJ}$, $b_{\Pi\pi\pi}$, $b_{J\Pi J}$, $b_{JJ\pi}$, $b_{\pi\Pi\pi}$, $b_{\pi JJ}$, and $b_{\pi\pi\pi}$ are given by

$$b_{\Pi\Pi\Pi} := -\frac{\langle \hat{L}^{-1}\,\hat{\Pi}\,,\, B[\hat{L}^{-1}\,\hat{\Pi}\,,\,\hat{L}^{-1}\,\hat{\Pi}]\rangle}{2\,T^2\,(\zeta)^2}, \tag{12.167}$$

$$b_{\Pi JJ} := -\frac{h^2\,\langle \hat{L}^{-1}\,\hat{\Pi}\,,\, B[\hat{L}^{-1}\,\hat{J}^{\mu}\,,\,\hat{L}^{-1}\,\hat{J}_{\mu}]\rangle}{6\,T^4\,(\lambda)^2}, \tag{12.168}$$

$$b_{\Pi\pi\pi} := -\frac{\langle \hat{L}^{-1}\,\hat{\Pi}\,,\, B[\hat{L}^{-1}\,\hat{\pi}^{\mu\nu}\,,\,\hat{L}^{-1}\,\hat{\pi}_{\mu\nu}]\rangle}{40\,T^2\,(\eta)^2}, \tag{12.169}$$

$$b_{J\Pi J} := \frac{\langle \hat{L}^{-1}\,\hat{J}^{\mu}\,,\, B[\hat{L}^{-1}\,\hat{\Pi}\,,\,\hat{L}^{-1}\,\hat{J}_{\mu}]\rangle}{3\,T^3\,\zeta\,\lambda}, \tag{12.170}$$

$$b_{JJ\pi} := \frac{\langle \hat{L}^{-1}\,\hat{J}^{\mu}\,,\, B[\hat{L}^{-1}\,\hat{J}^{\nu}\,,\,\hat{L}^{-1}\,\hat{\pi}_{\mu\nu}]\rangle}{10\,T^3\,\lambda\,\eta}, \tag{12.171}$$

$$b_{\pi\Pi\pi} := -\frac{\langle \hat{L}^{-1}\,\hat{\pi}^{\mu\nu}\,,\, B[\hat{L}^{-1}\,\hat{\Pi}\,,\,\hat{L}^{-1}\,\hat{\pi}_{\mu\nu}]\rangle}{10\,T^2\,\zeta\,\eta}, \tag{12.172}$$

$$b_{\pi JJ} := -\frac{h^2\,\langle \hat{L}^{-1}\,\hat{\pi}^{\mu\nu}\,,\, B[\hat{L}^{-1}\,\hat{J}_{\mu}\,,\,\hat{L}^{-1}\,\hat{J}_{\nu}]\rangle}{10\,T^4\,(\lambda)^2}, \tag{12.173}$$

$$b_{\pi\pi\pi} := -\frac{3\,\langle \hat{L}^{-1}\,\hat{\pi}^{\mu\nu}\,,\, B[\hat{L}^{-1}\,\hat{\pi}_{\mu}{}^{\lambda}\,,\,\hat{L}^{-1}\,\hat{\pi}_{\lambda\nu}]\rangle}{70\,T^2\,(\eta)^2}. \tag{12.174}$$

Next, we rewrite $\epsilon\,N^{i,j}\,\psi_j$ as follows:

$$\begin{aligned}
&\epsilon\,\langle L^{-1}\,\hat{\psi}^i\,,\,(f^{\rm eq}\,\bar{f}^{\rm eq})^{-1}\left[\frac{\partial}{\partial\tau}+\epsilon\,v\cdot\nabla\right]f^{\rm eq}\,\bar{f}^{\rm eq}\,L^{-1}\,\hat{\chi}^j\,\rangle\,\psi_j\\
&=\epsilon\,\langle L^{-1}\,\hat{\psi}^i\,,\,(f^{\rm eq}\,\bar{f}^{\rm eq})^{-1}\,\frac{\partial}{\partial T}[f^{\rm eq}\,\bar{f}^{\rm eq}\,L^{-1}\,\hat{\chi}^j]\rangle\,\psi_j\,\frac{\partial}{\partial\tau}T\\
&\quad+\epsilon^2\,\langle L^{-1}\,\hat{\psi}^i\,,\,(f^{\rm eq}\,\bar{f}^{\rm eq})^{-1}\,v^{\beta}\,\frac{\partial}{\partial T}[f^{\rm eq}\,\bar{f}^{\rm eq}\,L^{-1}\,\hat{\chi}^j]\rangle\,\psi_j\,\nabla_{\beta}T\\
&\quad+\epsilon\,\langle L^{-1}\,\hat{\psi}^i\,,\,(f^{\rm eq}\,\bar{f}^{\rm eq})^{-1}\,\frac{\partial}{\partial\frac{\mu}{T}}[f^{\rm eq}\,\bar{f}^{\rm eq}\,L^{-1}\,\hat{\chi}^j]\rangle\,\psi_j\,\frac{\partial}{\partial\tau}\frac{\mu}{T}\\
&\quad+\epsilon^2\,\langle L^{-1}\,\hat{\psi}^i\,,\,(f^{\rm eq}\,\bar{f}^{\rm eq})^{-1}\,v^{\beta}\,\frac{\partial}{\partial\frac{\mu}{T}}[f^{\rm eq}\,\bar{f}^{\rm eq}\,L^{-1}\,\hat{\chi}^j]\rangle\,\psi_j\,\nabla_{\beta}\frac{\mu}{T}\\
&\quad+\epsilon\,\langle L^{-1}\,\hat{\psi}^i\,,\,(f^{\rm eq}\,\bar{f}^{\rm eq})^{-1}\,\frac{\partial}{\partial u^{\beta}}[f^{\rm eq}\,\bar{f}^{\rm eq}\,L^{-1}\,\hat{\chi}^j]\rangle\,\psi_j\,\frac{\partial}{\partial\tau}u^{\beta}\\
&\quad+\epsilon^2\,\langle L^{-1}\,\hat{\psi}^i\,,\,(f^{\rm eq}\,\bar{f}^{\rm eq})^{-1}\,v^{\beta}\,\frac{\partial}{\partial u^{\alpha}}[f^{\rm eq}\,\bar{f}^{\rm eq}\,L^{-1}\,\hat{\chi}^j]\rangle\,\psi_j\,\nabla_{\beta}u^{\alpha}.
\end{aligned} \tag{12.175}$$

The temporal derivative of T, μ/T, and u^{μ} are rewritten by using the balance equations up to the first order with respect to ϵ, which correspond to the relativistic Euler equation:

$$\frac{\partial}{\partial \tau} T = -T \left. \frac{\partial P}{\partial e} \right|_n \epsilon\, \theta + O(\epsilon^2), \tag{12.176}$$

$$\frac{\partial}{\partial \tau} \frac{\mu}{T} = -\frac{1}{T} \left. \frac{\partial P}{\partial n} \right|_e \epsilon\, \theta + O(\epsilon^2), \tag{12.177}$$

$$\frac{\partial}{\partial \tau} u^\mu = \frac{1}{T} \epsilon\, \nabla^\mu T + \frac{T}{h} \epsilon\, \nabla^\mu \frac{\mu}{T} + O(\epsilon^2). \tag{12.178}$$

Using the formulae (12.139)–(12.143) and the Euler equations (12.176)–(12.178), we convert Eq. (12.175) into the following forms: For $\hat{\psi}^i = \hat{\Pi}$,

$$\begin{aligned}
&\epsilon \langle L^{-1} \hat{\Pi}, (f^{\rm eq} \bar{f}^{\rm eq})^{-1} \left[\frac{\partial}{\partial \tau} + \epsilon\, v \cdot \nabla \right] f^{\rm eq} \bar{f}^{\rm eq} L^{-1} \hat{\chi}^j \rangle \psi_j \\
&= \epsilon^2 \Bigg\{ \langle L^{-1} \hat{\Pi}, (f^{\rm eq} \bar{f}^{\rm eq})^{-1} \left[-T \left. \frac{\partial P}{\partial e} \right|_n \frac{\partial}{\partial T} - \frac{1}{T} \left. \frac{\partial P}{\partial n} \right|_e \frac{\partial}{\partial \frac{\mu}{T}} \right] \frac{f^{\rm eq} \bar{f}^{\rm eq} L^{-1} \hat{\Pi}}{-T\, \zeta} \rangle\, \Pi\, \theta \\
&\quad + \langle L^{-1} \hat{\Pi}, (f^{\rm eq} \bar{f}^{\rm eq})^{-1} \left[v^\beta \frac{\partial}{\partial T} + \frac{1}{T} \frac{\partial}{\partial u_\beta} \right] \frac{f^{\rm eq} \bar{f}^{\rm eq} L^{-1} \hat{J}^\rho}{T^2 \lambda / h} \rangle\, J_\rho\, \nabla_\beta T \\
&\quad + \langle L^{-1} \hat{\Pi}, (f^{\rm eq} \bar{f}^{\rm eq})^{-1} \left[v^\beta \frac{\partial}{\partial \frac{\mu}{T}} + \frac{T}{h} \frac{\partial}{\partial u_\beta} \right] \frac{f^{\rm eq} \bar{f}^{\rm eq} L^{-1} \hat{J}^\rho}{T^2 \lambda / h} \rangle\, J_\rho\, \nabla_\beta \frac{\mu}{T} \\
&\quad + \langle L^{-1} \hat{\Pi}, (f^{\rm eq} \bar{f}^{\rm eq})^{-1} v^\beta \frac{\partial}{\partial u^\alpha} \frac{f^{\rm eq} \bar{f}^{\rm eq} L^{-1} \hat{\Pi}}{-T\, \zeta} \rangle\, \Pi\, \nabla_\beta u^\alpha \\
&\quad + \langle L^{-1} \hat{\Pi}, (f^{\rm eq} \bar{f}^{\rm eq})^{-1} v^\beta \frac{\partial}{\partial u^\alpha} \frac{f^{\rm eq} \bar{f}^{\rm eq} L^{-1} \hat{\pi}^{\rho\sigma}}{-2\, T\, \eta} \rangle\, \pi_{\rho\sigma}\, \nabla_\beta u^\alpha \Bigg\} + O(\epsilon^3) \\
&= \epsilon^2 \Bigg\{ \langle L^{-1} \hat{\Pi}, (f^{\rm eq} \bar{f}^{\rm eq})^{-1} \left[-T \left. \frac{\partial P}{\partial e} \right|_n \frac{\partial}{\partial T} - \frac{1}{T} \left. \frac{\partial P}{\partial n} \right|_e \frac{\partial}{\partial \frac{\mu}{T}} \right] \frac{f^{\rm eq} \bar{f}^{\rm eq} L^{-1} \hat{\Pi}}{-T\, \zeta} \rangle\, \Pi\, \theta \\
&\quad + \frac{\Delta^{\alpha\beta}}{3} \langle L^{-1} \hat{\Pi}, (f^{\rm eq} \bar{f}^{\rm eq})^{-1} \left[v_\alpha \frac{\partial}{\partial T} + \frac{1}{T} \frac{\partial}{\partial u^\alpha} \right] \frac{f^{\rm eq} \bar{f}^{\rm eq} L^{-1} \hat{J}_\beta}{T^2 \lambda / h} \rangle\, J^\rho\, \nabla_\rho T \\
&\quad + \frac{\Delta^{\alpha\beta}}{3} \langle L^{-1} \hat{\Pi}, (f^{\rm eq} \bar{f}^{\rm eq})^{-1} \left[v_\alpha \frac{\partial}{\partial \frac{\mu}{T}} + \frac{T}{h} \frac{\partial}{\partial u^\alpha} \right] \frac{f^{\rm eq} \bar{f}^{\rm eq} L^{-1} \hat{J}_\beta}{T^2 \lambda / h} \rangle\, J^\rho\, \nabla_\rho \frac{\mu}{T} \\
&\quad + \frac{\Delta^{\alpha\beta}}{3} \langle L^{-1} \hat{\Pi}, (f^{\rm eq} \bar{f}^{\rm eq})^{-1} v_\alpha \frac{\partial}{\partial u^\beta} \frac{f^{\rm eq} \bar{f}^{\rm eq} L^{-1} \hat{\Pi}}{-T\, \zeta} \rangle\, \Pi\, \theta \\
&\quad + \frac{\Delta^{\alpha\beta\gamma\delta}}{5} \langle L^{-1} \hat{\Pi}, (f^{\rm eq} \bar{f}^{\rm eq})^{-1} v_\alpha \frac{\partial}{\partial u^\beta} \frac{f^{\rm eq} \bar{f}^{\rm eq} L^{-1} \hat{\pi}_{\gamma\delta}}{-2\, T\, \eta} \rangle\, \pi^{\rho\sigma}\, \sigma_{\rho\sigma} \Bigg\} \\
&\quad + O(\epsilon^3),
\end{aligned} \tag{12.179}$$

for $\hat{\psi}^i = \hat{J}^\mu / h$,

$$
\begin{aligned}
&\epsilon \langle L^{-1} \hat{J}^{\mu}/h, (f^{\rm eq} \bar{f}^{\rm eq})^{-1} \left[\frac{\partial}{\partial \tau} + \epsilon\, v \cdot \nabla \right] f^{\rm eq} \bar{f}^{\rm eq} L^{-1} \hat{\psi}^{j} \rangle \chi_j \\
&= \epsilon^2 \Bigg\{ \langle L^{-1} \hat{J}^{\mu}/h, (f^{\rm eq} \bar{f}^{\rm eq})^{-1} \\
&\quad \times \left[-T \left.\frac{\partial P}{\partial e}\right|_n \frac{\partial}{\partial T} - \frac{1}{T} \left.\frac{\partial P}{\partial n}\right|_e \frac{\partial}{\partial \frac{\mu}{T}} \right] \frac{f^{\rm eq} \bar{f}^{\rm eq} L^{-1} \hat{J}^{\nu}}{T^2 \lambda/h} \rangle J_\nu \theta \\
&\quad + \langle L^{-1} \hat{J}^{\mu}/h, (f^{\rm eq} \bar{f}^{\rm eq})^{-1} \left[v^{\beta} \frac{\partial}{\partial T} + \frac{1}{T} \frac{\partial}{\partial u_\beta} \right] \frac{f^{\rm eq} \bar{f}^{\rm eq} L^{-1} \hat{\Pi}}{-T \zeta} \rangle \Pi \nabla_\beta T \\
&\quad + \langle L^{-1} \hat{J}^{\mu}/h, (f^{\rm eq} \bar{f}^{\rm eq})^{-1} \left[v^{\beta} \frac{\partial}{\partial T} + \frac{1}{T} \frac{\partial}{\partial u_\beta} \right] \frac{f^{\rm eq} \bar{f}^{\rm eq} L^{-1} \hat{\pi}^{\rho\sigma}}{-2 T \eta} \rangle \pi_{\rho\sigma} \nabla_\beta T \\
&\quad + \langle L^{-1} \hat{J}^{\mu}/h, (f^{\rm eq} \bar{f}^{\rm eq})^{-1} \left[v^{\beta} \frac{\partial}{\partial \frac{\mu}{T}} + \frac{T}{h} \frac{\partial}{\partial u_\beta} \right] \frac{f^{\rm eq} \bar{f}^{\rm eq} L^{-1} \hat{\Pi}}{-T \zeta} \rangle \Pi \nabla_\beta \frac{\mu}{T} \\
&\quad + \langle L^{-1} \hat{J}^{\mu}/h, (f^{\rm eq} \bar{f}^{\rm eq})^{-1} \left[v^{\beta} \frac{\partial}{\partial \frac{\mu}{T}} + \frac{T}{h} \frac{\partial}{\partial u_\beta} \right] \frac{f^{\rm eq} \bar{f}^{\rm eq} L^{-1} \hat{\pi}^{\rho\sigma}}{-2 T \eta} \rangle \pi_{\rho\sigma} \nabla_\beta \frac{\mu}{T} \\
&\quad + \langle L^{-1} \hat{J}^{\mu}/h, (f^{\rm eq} \bar{f}^{\rm eq})^{-1} v^{\beta} \frac{\partial}{\partial u^\alpha} \frac{f^{\rm eq} \bar{f}^{\rm eq} L^{-1} \hat{J}^{\rho}}{T^2 \lambda/h} \rangle J_\rho \nabla_\beta u^\alpha \Bigg\} + O(\epsilon^3) \\
&= \epsilon^2 \Bigg\{ \frac{\Delta^{\rho\sigma}}{3h} \langle L^{-1} \hat{J}_{\rho}, (f^{\rm eq} \bar{f}^{\rm eq})^{-1} \\
&\quad \times \left[-T \left.\frac{\partial P}{\partial e}\right|_n \frac{\partial}{\partial T} - \frac{1}{T} \left.\frac{\partial P}{\partial n}\right|_e \frac{\partial}{\partial \frac{\mu}{T}} \right] \frac{f^{\rm eq} \bar{f}^{\rm eq} L^{-1} \hat{J}_{\sigma}}{T^2 \lambda/h} \rangle J^\mu \theta \\
&\quad + \frac{\Delta^{\rho\sigma}}{3h} \langle L^{-1} \hat{J}_{\rho}, (f^{\rm eq} \bar{f}^{\rm eq})^{-1} \left[v_{\sigma} \frac{\partial}{\partial T} + \frac{1}{T} \frac{\partial}{\partial u^\sigma} \right] \frac{f^{\rm eq} \bar{f}^{\rm eq} L^{-1} \hat{\Pi}}{-T \zeta} \rangle \Pi \nabla^\mu T \\
&\quad + \frac{\Delta^{\alpha\beta\gamma\delta}}{5h} \langle L^{-1} \hat{J}_{\alpha}, (f^{\rm eq} \bar{f}^{\rm eq})^{-1} \left[v_{\beta} \frac{\partial}{\partial T} + \frac{1}{T} \frac{\partial}{\partial u^\beta} \right] \frac{f^{\rm eq} \bar{f}^{\rm eq} L^{-1} \hat{\pi}_{\gamma\delta}}{-2 T \eta} \rangle \pi^{\mu\rho} \nabla_\rho T \\
&\quad + \frac{\Delta^{\rho\sigma}}{3h} \langle L^{-1} \hat{J}_{\rho}, (f^{\rm eq} \bar{f}^{\rm eq})^{-1} \left[v_{\sigma} \frac{\partial}{\partial \frac{\mu}{T}} + \frac{T}{h} \frac{\partial}{\partial u^\sigma} \right] \frac{f^{\rm eq} \bar{f}^{\rm eq} L^{-1} \hat{\Pi}}{-T \zeta} \rangle \Pi \nabla^\mu \frac{\mu}{T} \\
&\quad + \frac{\Delta^{\alpha\beta\gamma\delta}}{5h} \langle L^{-1} \hat{J}_{\alpha}, (f^{\rm eq} \bar{f}^{\rm eq})^{-1} \left[v_{\beta} \frac{\partial}{\partial \frac{\mu}{T}} + \frac{T}{h} \frac{\partial}{\partial u^\beta} \right] \frac{f^{\rm eq} \bar{f}^{\rm eq} L^{-1} \hat{\pi}_{\gamma\delta}}{-2 T \eta} \rangle \pi^{\mu\rho} \nabla_\rho \frac{\mu}{T} \\
&\quad + \frac{\Delta^{\rho\sigma} \Delta^{\alpha\beta}}{9h} \langle L^{-1} \hat{J}_{\rho}, (f^{\rm eq} \bar{f}^{\rm eq})^{-1} v_{\alpha} \frac{\partial}{\partial u^\beta} \frac{f^{\rm eq} \bar{f}^{\rm eq} L^{-1} \hat{J}_{\sigma}}{T^2 \lambda/h} \rangle J^\mu \theta \\
&\quad + \frac{\Delta^{\alpha\beta\gamma\delta}}{5h} \langle L^{-1} \hat{J}_{\alpha}, (f^{\rm eq} \bar{f}^{\rm eq})^{-1} v_{\gamma} \frac{\partial}{\partial u^\delta} \frac{f^{\rm eq} \bar{f}^{\rm eq} L^{-1} \hat{J}_{\beta}}{T^2 \lambda/h} \rangle J^\rho \sigma^{\mu}{}_{\rho} \\
&\quad + \frac{\Omega^{\alpha\beta\gamma\delta}}{3h} \langle L^{-1} \hat{J}_{\alpha}, (f^{\rm eq} \bar{f}^{\rm eq})^{-1} v_{\gamma} \frac{\partial}{\partial u^\delta} \frac{f^{\rm eq} \bar{f}^{\rm eq} L^{-1} \hat{J}_{\beta}}{T^2 \lambda/h} \rangle J^\rho \omega^{\mu}{}_{\rho} \Bigg\} \\
&\quad + O(\epsilon^3),
\end{aligned}
\tag{12.180}
$$

and, for $\hat{\psi}^i = \hat{\pi}^{\mu\nu}$,

$$
\begin{aligned}
&\epsilon \langle L^{-1} \hat{\pi}^{\mu\nu} , (f^{\rm eq} \bar{f}^{\rm eq})^{-1} \left[\frac{\partial}{\partial \tau} + \epsilon\, v \cdot \nabla \right] f^{\rm eq} \bar{f}^{\rm eq} L^{-1} \hat{\psi}^{j} \rangle \chi_j \\
&= \epsilon^2 \Bigg\{ \langle L^{-1} \hat{\pi}^{\mu\nu} , (f^{\rm eq} \bar{f}^{\rm eq})^{-1} \\
&\times \left[- T \left.\frac{\partial P}{\partial e}\right|_n \frac{\partial}{\partial T} - \frac{1}{T} \left.\frac{\partial P}{\partial n}\right|_e \frac{\partial}{\partial \frac{\mu}{T}} \right] \frac{f^{\rm eq} \bar{f}^{\rm eq} L^{-1} \hat{\pi}^{\rho\sigma}}{-2 T \eta}] \rangle \pi_{\rho\sigma} \theta \\
&+ \langle L^{-1} \hat{\pi}^{\mu\nu} , (f^{\rm eq} \bar{f}^{\rm eq})^{-1} \left[v^{\beta} \frac{\partial}{\partial T} + \frac{1}{T} \frac{\partial}{\partial u_{\beta}} \right] \frac{f^{\rm eq} \bar{f}^{\rm eq} L^{-1} \hat{J}^{\rho}}{T^2 \lambda / h} \rangle J_{\rho} \nabla_{\beta} T \\
&+ \langle L^{-1} \hat{\pi}^{\mu\nu} , (f^{\rm eq} \bar{f}^{\rm eq})^{-1} \left[v^{\beta} \frac{\partial}{\partial \frac{\mu}{T}} + \frac{T}{h} \frac{\partial}{\partial u_{\beta}} \right] \frac{f^{\rm eq} \bar{f}^{\rm eq} L^{-1} \hat{J}^{\rho}}{T^2 \lambda / h} \rangle J_{\rho} \nabla_{\beta} \frac{\mu}{T} \\
&+ \langle L^{-1} \hat{\pi}^{\mu\nu} , (f^{\rm eq} \bar{f}^{\rm eq})^{-1} v^{\beta} \frac{\partial}{\partial u^{\alpha}} \frac{f^{\rm eq} \bar{f}^{\rm eq} L^{-1} \hat{\Pi}}{-T \zeta} \rangle \Pi \nabla_{\beta} u^{\alpha} \\
&+ \langle L^{-1} \hat{\pi}^{\mu\nu} , (f^{\rm eq} \bar{f}^{\rm eq})^{-1} v^{\beta} \frac{\partial}{\partial u^{\alpha}} \frac{f^{\rm eq} \bar{f}^{\rm eq} L^{-1} \hat{\pi}^{\rho\sigma}}{-2 T \eta} \rangle \pi_{\rho\sigma} \nabla_{\beta} u^{\alpha} \Bigg\} + O(\epsilon^3) \\
&= \epsilon^2 \Bigg\{ \frac{\Delta^{\rho\sigma\alpha\beta}}{5} \langle L^{-1} \hat{\pi}_{\rho\sigma} , (f^{\rm eq} \bar{f}^{\rm eq})^{-1} \\
&\times \left[- T \left.\frac{\partial P}{\partial e}\right|_n \frac{\partial}{\partial T} - \frac{1}{T} \left.\frac{\partial P}{\partial n}\right|_e \frac{\partial}{\partial \frac{\mu}{T}} \right] \frac{f^{\rm eq} \bar{f}^{\rm eq} L^{-1} \hat{\pi}_{\alpha\beta}}{-2 T \eta} \rangle \pi^{\mu\nu} \theta \\
&+ \frac{\Delta^{\rho\sigma\alpha\beta}}{5} \langle L^{-1} \hat{\pi}_{\rho\sigma} , (f^{\rm eq} \bar{f}^{\rm eq})^{-1} v_{\alpha} \frac{\partial}{\partial u^{\beta}} \frac{f^{\rm eq} \bar{f}^{\rm eq} L^{-1} \hat{\Pi}}{-T \zeta} \rangle \Pi \sigma^{\mu\nu} \\
&+ \frac{\Delta^{\rho\sigma\alpha\beta}}{5} \langle L^{-1} \hat{\pi}_{\rho\sigma} , (f^{\rm eq} \bar{f}^{\rm eq})^{-1} \left[v_{\alpha} \frac{\partial}{\partial T} + \frac{1}{T} \frac{\partial}{\partial u^{\alpha}} \right] \frac{f^{\rm eq} \bar{f}^{\rm eq} L^{-1} \hat{J}_{\beta}}{T^2 \lambda / h} \rangle J^{\langle\mu} \nabla^{\nu\rangle} T \\
&+ \frac{\Delta^{\rho\sigma\alpha\beta}}{5} \langle L^{-1} \hat{\pi}_{\rho\sigma} , (f^{\rm eq} \bar{f}^{\rm eq})^{-1} \left[v_{\alpha} \frac{\partial}{\partial \frac{\mu}{T}} + \frac{T}{h} \frac{\partial}{\partial u^{\alpha}} \right] \frac{f^{\rm eq} \bar{f}^{\rm eq} L^{-1} \hat{J}_{\beta}}{T^2 \lambda / h} \rangle J^{\langle\mu} \nabla^{\nu\rangle} \frac{\mu}{T} \\
&+ \frac{\Delta^{\rho\sigma\alpha\beta}}{5} \langle L^{-1} \hat{\pi}_{\rho\sigma} , (f^{\rm eq} \bar{f}^{\rm eq})^{-1} v_{\alpha} \frac{\partial}{\partial u^{\beta}} \frac{f^{\rm eq} \bar{f}^{\rm eq} L^{-1} \hat{\Pi}}{-T \zeta} \rangle \Pi \sigma^{\mu\nu} \\
&+ \frac{\Delta^{\rho\sigma\alpha\beta} \Delta^{\gamma\delta}}{15} \langle L^{-1} \hat{\pi}_{\rho\sigma} , (f^{\rm eq} \bar{f}^{\rm eq})^{-1} v_{\gamma} \frac{\partial}{\partial u^{\delta}} \frac{f^{\rm eq} \bar{f}^{\rm eq} L^{-1} \hat{\pi}_{\alpha\beta}}{-2 T \eta} \rangle \pi^{\mu\nu} \theta \\
&+ \frac{12 \Delta^{\tau\eta\gamma\delta} \Delta^{\kappa\sigma\lambda}{}_{\gamma} \Delta^{\alpha\beta}{}_{\lambda\delta}}{35} \langle L^{-1} \hat{\pi}_{\tau\eta} , (f^{\rm eq} \bar{f}^{\rm eq})^{-1} \\
&\times v_{\alpha} \frac{\partial}{\partial u^{\beta}} \frac{f^{\rm eq} \bar{f}^{\rm eq} L^{-1} \hat{\pi}_{\kappa\sigma}}{-2 T \eta} \rangle \pi^{\rho\langle\mu} \sigma^{\nu\rangle}{}_{\rho} \\
&+ \frac{4 \Delta^{\tau\eta\gamma\delta} \Delta^{\kappa\sigma\lambda}{}_{\gamma} \Omega^{\alpha\beta}{}_{\lambda\delta}}{15} \langle L^{-1} \hat{\pi}_{\tau\eta} , (f^{\rm eq} \bar{f}^{\rm eq})^{-1} \\
&\times v_{\alpha} \frac{\partial}{\partial u^{\beta}} \frac{f^{\rm eq} \bar{f}^{\rm eq} L^{-1} \hat{\pi}_{\kappa\sigma}}{-2 T \eta} \rangle \pi^{\rho\langle\mu} \omega^{\nu\rangle}{}_{\rho} \Bigg\} + O(\epsilon^3),
\end{aligned}
\tag{12.181}
$$

with the vorticity $\omega^{\mu\nu} = \Omega^{\mu\nu\rho\sigma} \nabla_{\rho} u_{\sigma}$. Thus, $\epsilon\, N^{i,j} \psi_j$ finally takes the form

$$
\begin{aligned}
\epsilon\, N^{i,j}\, \psi_j = \epsilon^2\, (&\kappa_{\Pi\Pi}\, \Pi\, \theta \\
&+ \kappa_{\Pi J}^{(1)}\, J^{\rho}\, \nabla_{\rho} T + \kappa_{\Pi J}^{(2)}\, J^{\rho}\, \nabla_{\rho} \frac{\mu}{T} \\
&+ \kappa_{\Pi\pi}\, \pi^{\rho\sigma}\, \sigma_{\rho\sigma}, \\
&\kappa_{J\Pi}^{(1)}\, \Pi\, \nabla^{\mu} T + \kappa_{J\Pi}^{(2)}\, \Pi\, \nabla^{\mu} \frac{\mu}{T} \\
&+ \kappa_{JJ}^{(1)}\, J^{\mu}\, \theta + \kappa_{JJ}^{(2)}\, J^{\rho}\, \sigma^{\mu}{}_{\rho} + \kappa_{JJ}^{(3)}\, J^{\rho}\, \omega^{\mu}{}_{\rho} \\
&+ \kappa_{J\pi}^{(1)}\, \pi^{\mu\rho}\, \nabla_{\rho} T + \kappa_{J\pi}^{(2)}\, \pi^{\mu\rho}\, \nabla_{\rho} \frac{\mu}{T}, \\
&\kappa_{\pi\Pi}\, \Pi\, \sigma^{\mu\nu} \\
&+ \kappa_{\pi J}^{(1)}\, J^{\langle\mu}\, \nabla^{\nu\rangle} T + \kappa_{\pi J}^{(2)}\, J^{\langle\mu}\, \nabla^{\nu\rangle} \frac{\mu}{T} \\
&+ \kappa_{\pi\pi}^{(1)}\, \pi^{\mu\nu}\, \theta + \kappa_{\pi\pi}^{(2)}\, \pi^{\rho\langle\mu}\, \sigma^{\nu\rangle}{}_{\rho} + \kappa_{\pi\pi}^{(3)}\, \pi^{\rho\langle\mu}\, \omega^{\nu\rangle}{}_{\rho}),
\end{aligned}
\tag{12.182}
$$

where we have omitted terms of $O(\epsilon^3)$.

The coefficients $\kappa_{\Pi\Pi}$, $\kappa_{\Pi J}^{(1)}$, $\kappa_{\Pi J}^{(2)}$, $\kappa_{\Pi\pi}$, $\kappa_{J\Pi}^{(1)}$, $\kappa_{J\Pi}^{(2)}$, $\kappa_{JJ}^{(1)}$, $\kappa_{JJ}^{(2)}$, $\kappa_{JJ}^{(3)}$, $\kappa_{J\pi}^{(1)}$, $\kappa_{J\pi}^{(2)}$, $\kappa_{\pi\Pi}$, $\kappa_{\pi J}^{(1)}$, $\kappa_{\pi J}^{(2)}$, $\kappa_{\pi\pi}^{(1)}$, $\kappa_{\pi\pi}^{(2)}$, and $\kappa_{\pi\pi}^{(3)}$ are defined by

$$
\begin{aligned}
\kappa_{\Pi\Pi} := \langle &L^{-1}\, \hat{\Pi},\, (f^{\mathrm{eq}}\, \bar{f}^{\mathrm{eq}})^{-1} \\
&\times \left[-T\, \frac{\partial P}{\partial e}\bigg|_{n} \frac{\partial}{\partial T} - \frac{1}{T}\, \frac{\partial P}{\partial n}\bigg|_{e} \frac{\partial}{\partial \frac{\mu}{T}} + \frac{1}{3}\, v^{\mu}\, \frac{\partial}{\partial u^{\mu}} \right] \frac{f^{\mathrm{eq}}\, \bar{f}^{\mathrm{eq}}\, L^{-1}\, \hat{\Pi}}{-T\, \zeta} \rangle,
\end{aligned}
\tag{12.183}
$$

$$
\kappa_{\Pi J}^{(1)} := \frac{\Delta^{\mu\nu}}{3}\, \langle L^{-1}\, \hat{\Pi},\, (f^{\mathrm{eq}}\, \bar{f}^{\mathrm{eq}})^{-1} \left[v_{\mu}\, \frac{\partial}{\partial T} + \frac{1}{T}\, \frac{\partial}{\partial u^{\mu}} \right] \frac{f^{\mathrm{eq}}\, \bar{f}^{\mathrm{eq}}\, L^{-1}\, \hat{J}_{\nu}}{T^2\, \lambda / h} \rangle,
\tag{12.184}
$$

$$
\kappa_{\Pi J}^{(2)} := \frac{\Delta^{\mu\nu}}{3}\, \langle L^{-1}\, \hat{\Pi},\, (f^{\mathrm{eq}}\, \bar{f}^{\mathrm{eq}})^{-1} \left[v_{\mu}\, \frac{\partial}{\partial \frac{\mu}{T}} + \frac{T}{h}\, \frac{\partial}{\partial u^{\mu}} \right] \frac{f^{\mathrm{eq}}\, \bar{f}^{\mathrm{eq}}\, L^{-1}\, \hat{J}_{\nu}}{T^2\, \lambda / h} \rangle,
\tag{12.185}
$$

$$
\kappa_{\Pi\pi} := \frac{\Delta^{\mu\nu\rho\sigma}}{5}\, \langle L^{-1}\, \hat{\Pi},\, (f^{\mathrm{eq}}\, \bar{f}^{\mathrm{eq}})^{-1}\, v_{\mu}\, \frac{\partial}{\partial u^{\nu}}\, \frac{f^{\mathrm{eq}}\, \bar{f}^{\mathrm{eq}}\, L^{-1}\, \hat{\pi}_{\rho\sigma}}{-2\, T\, \eta} \rangle,
\tag{12.186}
$$

$$
\kappa_{J\Pi}^{(1)} := \frac{1}{3\, h}\, \langle L^{-1}\, \hat{J}^{\mu},\, (f^{\mathrm{eq}}\, \bar{f}^{\mathrm{eq}})^{-1} \left[v_{\mu}\, \frac{\partial}{\partial T} + \frac{1}{T}\, \frac{\partial}{\partial u^{\mu}} \right] \frac{f^{\mathrm{eq}}\, \bar{f}^{\mathrm{eq}}\, L^{-1}\, \hat{\Pi}}{-T\, \zeta} \rangle,
\tag{12.187}
$$

$$
\kappa_{J\Pi}^{(2)} := \frac{1}{3\, h}\, \langle L^{-1}\, \hat{J}^{\mu},\, (f^{\mathrm{eq}}\, \bar{f}^{\mathrm{eq}})^{-1} \left[v_{\mu}\, \frac{\partial}{\partial \frac{\mu}{T}} + \frac{T}{h}\, \frac{\partial}{\partial u^{\mu}} \right] \frac{f^{\mathrm{eq}}\, \bar{f}^{\mathrm{eq}}\, L^{-1}\, \hat{\Pi}}{-T\, \zeta} \rangle,
\tag{12.188}
$$

$$
\begin{aligned}
\kappa_{JJ}^{(1)} := \frac{\Delta^{\mu\nu}}{3\, h}\, \langle &L^{-1}\, \hat{J}_{\mu},\, (f^{\mathrm{eq}}\, \bar{f}^{\mathrm{eq}})^{-1} \\
&\times \left[-T\, \frac{\partial P}{\partial e}\bigg|_{n} \frac{\partial}{\partial T} - \frac{1}{T}\, \frac{\partial P}{\partial n}\bigg|_{e} \frac{\partial}{\partial \frac{\mu}{T}} + \frac{1}{3}\, v^{\rho}\, \frac{\partial}{\partial u^{\rho}} \right] \frac{f^{\mathrm{eq}}\, \bar{f}^{\mathrm{eq}}\, L^{-1}\, \hat{J}_{\nu}}{T^2\, \lambda / h} \rangle,
\end{aligned}
\tag{12.189}
$$

$$
\kappa_{JJ}^{(2)} := \frac{\Delta^{\mu\nu\rho\sigma}}{5\, h}\, \langle L^{-1}\, \hat{J}_{\mu},\, (f^{\mathrm{eq}}\, \bar{f}^{\mathrm{eq}})^{-1}\, v_{\rho}\, \frac{\partial}{\partial u^{\sigma}}\, \frac{f^{\mathrm{eq}}\, \bar{f}^{\mathrm{eq}}\, L^{-1}\, \hat{J}_{\nu}}{T^2\, \lambda / h} \rangle,
\tag{12.190}
$$

$$
\kappa_{JJ}^{(3)} := \frac{\Omega^{\mu\nu\rho\sigma}}{3\, h}\, \langle L^{-1}\, \hat{J}_{\mu},\, (f^{\mathrm{eq}}\, \bar{f}^{\mathrm{eq}})^{-1}\, v_{\rho}\, \frac{\partial}{\partial u^{\sigma}}\, \frac{f^{\mathrm{eq}}\, \bar{f}^{\mathrm{eq}}\, L^{-1}\, \hat{J}_{\nu}}{T^2\, \lambda / h} \rangle,
\tag{12.191}
$$

$$
\kappa_{J\pi}^{(1)} := \frac{\Delta^{\mu\nu\rho\sigma}}{5\, h}\, \langle L^{-1}\, \hat{J}_{\mu},\, (f^{\mathrm{eq}}\, \bar{f}^{\mathrm{eq}})^{-1} \left[v_{\nu}\, \frac{\partial}{\partial T} + \frac{1}{T}\, \frac{\partial}{\partial u^{\nu}} \right] \frac{f^{\mathrm{eq}}\, \bar{f}^{\mathrm{eq}}\, L^{-1}\, \hat{\pi}_{\rho\sigma}}{-2\, T\, \eta} \rangle,
\tag{12.192}
$$

$$
\kappa_{J\pi}^{(2)} := \frac{\Delta^{\mu\nu\rho\sigma}}{5\, h}\, \langle L^{-1}\, \hat{J}_{\mu},\, (f^{\mathrm{eq}}\, \bar{f}^{\mathrm{eq}})^{-1} \left[v_{\nu}\, \frac{\partial}{\partial \frac{\mu}{T}} + \frac{T}{h}\, \frac{\partial}{\partial u^{\nu}} \right] \frac{f^{\mathrm{eq}}\, \bar{f}^{\mathrm{eq}}\, L^{-1}\, \hat{\pi}_{\rho\sigma}}{-2\, T\, \eta} \rangle,
\tag{12.193}
$$

$$
\kappa_{\pi\Pi} := \frac{\Delta^{\mu\nu\rho\sigma}}{5}\, \langle L^{-1}\, \hat{\pi}_{\mu\nu},\, (f^{\mathrm{eq}}\, \bar{f}^{\mathrm{eq}})^{-1}\, v_{\rho}\, \frac{\partial}{\partial u^{\sigma}}\, \frac{f^{\mathrm{eq}}\, \bar{f}^{\mathrm{eq}}\, L^{-1}\, \hat{\Pi}}{-T\, \zeta} \rangle,
\tag{12.194}
$$

$$\kappa_{\pi J}^{(1)} := \frac{\Delta^{\mu\nu\rho\sigma}}{5} \langle L^{-1}\hat{\pi}_{\mu\nu}, (f^{\mathrm{eq}}\bar{f}^{\mathrm{eq}})^{-1}\left[v_\rho \frac{\partial}{\partial T} + \frac{1}{T}\frac{\partial}{\partial u^\rho}\right]\frac{f^{\mathrm{eq}}\bar{f}^{\mathrm{eq}}L^{-1}\hat{J}_\sigma}{T^2\lambda/h}\rangle, \tag{12.195}$$

$$\kappa_{\pi J}^{(2)} := \frac{\Delta^{\mu\nu\rho\sigma}}{5} \langle L^{-1}\hat{\pi}_{\mu\nu}, (f^{\mathrm{eq}}\bar{f}^{\mathrm{eq}})^{-1}\left[v_\rho \frac{\partial}{\partial \frac{\mu}{T}} + \frac{T}{h}\frac{\partial}{\partial u^\rho}\right]\frac{f^{\mathrm{eq}}\bar{f}^{\mathrm{eq}}L^{-1}\hat{J}_\sigma}{T^2\lambda/h}\rangle, \tag{12.196}$$

$$\kappa_{\pi\pi}^{(1)} := \frac{\Delta^{\mu\nu\rho\sigma}}{5} \langle L^{-1}\hat{\pi}_{\mu\nu}, (f^{\mathrm{eq}}\bar{f}^{\mathrm{eq}})^{-1}$$
$$\times\left[-T\left.\frac{\partial P}{\partial e}\right|_n \frac{\partial}{\partial T} - \frac{1}{T}\left.\frac{\partial P}{\partial n}\right|_e \frac{\partial}{\partial \frac{\mu}{T}} + \frac{1}{3}v^\mu \frac{\partial}{\partial u^\mu}\right]\frac{f^{\mathrm{eq}}\bar{f}^{\mathrm{eq}}L^{-1}\hat{\pi}_{\rho\sigma}}{-2T\eta}\rangle, \tag{12.197}$$

$$\kappa_{\pi\pi}^{(2)} := \frac{12\,\Delta^{\mu\nu\gamma\delta}\,\Delta^{\rho\sigma\lambda}{}_\gamma\,\Delta^{\alpha\beta}{}_{\lambda\delta}}{35} \langle L^{-1}\hat{\pi}_{\mu\nu}, (f^{\mathrm{eq}}\bar{f}^{\mathrm{eq}})^{-1} v_\alpha \frac{\partial}{\partial u^\beta}\frac{f^{\mathrm{eq}}\bar{f}^{\mathrm{eq}}L^{-1}\hat{\pi}_{\rho\sigma}}{-2T\eta}\rangle, \tag{12.198}$$

$$\kappa_{\pi\pi}^{(3)} := \frac{4\,\Delta^{\mu\nu\gamma\delta}\,\Delta^{\rho\sigma\lambda}{}_\gamma\,\Omega^{\alpha\beta}{}_{\lambda\delta}}{15} \langle L^{-1}\hat{\pi}_{\mu\nu}, (f^{\mathrm{eq}}\bar{f}^{\mathrm{eq}})^{-1} v_\alpha \frac{\partial}{\partial u^\beta}\frac{f^{\mathrm{eq}}\bar{f}^{\mathrm{eq}}L^{-1}\hat{\pi}_{\rho\sigma}}{-2T\eta}\rangle. \tag{12.199}$$

Inserting the above equations into Eqs. (12.136) and setting ϵ equal to 1, we finally get the explicit form of the relaxation equations as presented in (12.98)–(12.100).

12.7 Properties of Second-Order Fluid Dynamic Equation

In this section, we examine the basic properties of the resultant second-order fluid dynamic equation. First, we show that our equation respects the stability of the static solution containing the equilibrium. Then, we prove that our equation is really causal in the sense that the speed of any fluctuations around the equilibrium is less than the speed of light.

12.7.1 *Stability*

We first prove that the static solution to the relativistic second-order fluid dynamic equation, *i.e.*, the pair of Eqs. (12.77) and (12.81), is stable against a small perturbation. The strategy of the proof in this subsection is the same as the one adopted in the case of the first-order equation in Sect. 11.6.2.

A generic constant solution is written as

$$T(\sigma\,;\,\tau) = T_0, \quad \mu(\sigma\,;\,\tau) = \mu_0, \quad u^\mu(\sigma\,;\,\tau) = u_0^\mu, \tag{12.200}$$

$$\Pi(\sigma\,;\,\tau) = 0, \quad J^\mu(\sigma\,;\,\tau) = 0, \quad \pi^{\mu\nu}(\sigma\,;\,\tau) = 0, \tag{12.201}$$

where T_0, μ_0, and u_0^μ are constant. We remark that the equilibrium state is a constant solution in the special case of $u_0^\mu = (1, 0, 0, 0)$.

To show the stability of the constant solution, we apply the linear stability analysis to the relativistic second-order fluid dynamic equation (12.77) and (12.81). Thus we consider the case where T, μ, u^μ, Π, J^μ, and $\pi^{\mu\nu}$ are all expressed as a sum of the constant solution and a small correction as

$$T(\sigma\,;\,\tau) = T_0 + \delta T(\sigma\,;\,\tau), \quad \mu(\sigma\,;\,\tau) = \mu_0 + \delta\mu(\sigma\,;\,\tau), \tag{12.202}$$
$$u^\mu(\sigma\,;\,\tau) = u_0^\mu + \delta u^\mu(\sigma\,;\,\tau), \quad \Pi(\sigma\,;\,\tau) = \delta\Pi(\sigma\,;\,\tau), \tag{12.203}$$
$$J^\mu(\sigma\,;\,\tau) = \delta J^\mu(\sigma\,;\,\tau), \quad \pi^{\mu\nu}(\sigma\,;\,\tau) = \delta\pi^{\mu\nu}(\sigma\,;\,\tau), \tag{12.204}$$

where δT, $\delta\mu$, δu^μ, $\delta\Pi$, δJ^μ, and $\delta\pi^{\mu\nu}$ are assumed to be so small that the second or higher-orders can be neglected.

Since δT, $\delta\mu$, and δu^μ are not independent variables because of the constraint $\delta u_\mu\, u_0^\mu = 0$, we define the following variables composed of linear combinations of δT, $\delta\mu$, and δu^μ as mutually independent ones,

$$\delta X_{4\mu} := -\delta(u_\mu/T) = -\frac{1}{T_0}\,\delta u_\mu + \frac{u_{0\mu}}{T_0^2}\,\delta T, \tag{12.205}$$

$$\delta X_{44} := \delta(\mu/T) = \frac{1}{T_0}\,\delta\mu - \frac{\mu_0}{T_0^2}\,\delta T, \tag{12.206}$$

From now on, we suppress the subscript "0" in T_0, μ_0, and u_0^μ.

To describe fluctuations of fluid dynamic variables governed by the second-order fluid dynamics, we need to incorporate the fluctuations of the dissipative currents; $\delta\Pi$, δJ^μ, and $\delta\pi^{\mu\nu}$. Instead of them, however, we define the following variables composed of linear combinations of them as the mutually independent ones,

$$\delta X_{\mu\nu} := \frac{-\frac{1}{3}\,\Delta_{\mu\nu}\,\delta\Pi}{\langle\, \hat{\Pi}\,,\, \hat{L}^{-1}\,\hat{\Pi}\,\rangle} + \frac{\delta\pi_{\mu\nu}}{\frac{1}{5}\,\langle\, \hat{\pi}^{\rho\sigma}\,,\, \hat{L}^{-1}\,\hat{\pi}_{\rho\sigma}\,\rangle}, \tag{12.207}$$

$$\delta X_{\mu 4} := \frac{h^2\,\delta J_\mu}{\frac{1}{3}\,\langle\, \hat{J}^\rho\,,\, \hat{L}^{-1}\,\hat{J}_\rho\,\rangle}. \tag{12.208}$$

We treat

$$\delta X_{\alpha\beta} = (\delta X_{\mu\nu},\, \delta X_{\mu 4},\, \delta X_{4\nu},\, \delta X_{44}), \tag{12.209}$$

as the fundamental variables.

Inserting Eqs. (12.202)–(12.204) into the relativistic second-order fluid dynamic equation (12.77) and (12.81), we obtain the equation governing the time evolution of $\delta X_{\alpha\beta}$ as

$$\begin{aligned} &\langle\, \varphi_0^\alpha\,,\, \varphi_0^\beta\,\rangle\,\frac{\partial}{\partial\tau}\delta X_{4\beta} + \langle\, \varphi_0^\alpha\,,\, \hat{L}^{-1}\,\varphi_1^{\nu\beta}\,\rangle\,\frac{\partial}{\partial\tau}\delta X_{\nu\beta} \\ &= -\langle\, \varphi_0^\alpha\,,\, v^\rho\,\varphi_0^\beta\,\rangle\,\nabla_\rho\delta X_{4\beta} - \langle\, \varphi_0^\alpha\,,\, v^\rho\,\hat{L}^{-1}\,\varphi_1^{\nu\beta}\,\rangle\,\nabla_\rho\delta X_{\nu\beta}, \end{aligned} \tag{12.210}$$

$$\begin{aligned} &\langle\, \hat{L}^{-1}\,\varphi_1^{\mu\alpha}\,,\, \varphi_0^\beta\,\rangle\,\frac{\partial}{\partial\tau}\delta X_{4\beta} + \langle\, \hat{L}^{-1}\,\varphi_1^{\mu\alpha}\,,\, \hat{L}^{-1}\,\varphi_1^{\nu\beta}\,\rangle\,\frac{\partial}{\partial\tau}\delta X_{\nu\beta} \\ &= -\langle\, \hat{L}^{-1}\,\varphi_1^{\mu\alpha}\,,\, v^\rho\,\varphi_0^\beta\,\rangle\,\nabla_\rho\delta X_{4\beta} \quad \langle\, \hat{L}^{-1}\,\varphi_1^{\mu\alpha}\,,\, v^\rho\,\hat{L}^{-1}\,\varphi_1^{\nu\beta}\,\rangle\,\nabla_\rho\delta X_{\nu\beta} \\ &\quad - \langle\, \hat{L}^{-1}\,\varphi_1^{\mu\alpha}\,,\, \hat{L}\,\hat{L}^{-1}\,\varphi_1^{\nu\beta}\,\rangle\,\delta X_{\nu\beta}. \end{aligned} \tag{12.211}$$

In deriving Eqs. (12.210) and (12.211), we have used the following formulae

$$\delta(f_p^{\mathrm{eq}}) = f_p^{\mathrm{eq}}\,\bar{f}_p^{\mathrm{eq}}\,\varphi_{0p}^{\alpha}\,\delta X_{4\alpha}, \quad \delta(\Psi_p) = \big[\hat{L}^{-1}\,\varphi_1^{\mu\alpha}\big]_p\,\delta X_{\mu\alpha}, \tag{12.212}$$

with the relation $\varphi_{1p}^{\mu\alpha} = \big[Q_0\,\tilde{\varphi}_1^{\mu\alpha}\big]_p$, the explicit formula of which is given in Eq. (12.92).

Equations (12.210) and (12.211) are combined and expressed in single formula as

$$A^{\alpha\beta,\gamma\delta}\,\frac{\partial}{\partial\tau}\delta X_{\gamma\delta} + B^{\alpha\beta,\gamma\delta}\,\delta X_{\gamma\delta} = 0, \tag{12.213}$$

where $A^{\alpha\beta,\gamma\delta}$ and $B^{\alpha\beta,\gamma\delta}$ are given by

$$A^{\mu\beta,\nu\delta} := \langle\,\hat{L}^{-1}\,\varphi_1^{\mu\beta}\,,\,\hat{L}^{-1}\,\varphi_1^{\nu\delta}\,\rangle, \tag{12.214}$$
$$A^{\mu\beta,4\delta} := \langle\,\hat{L}^{-1}\,\varphi_1^{\mu\beta}\,,\,\varphi_0^{\delta}\,\rangle, \tag{12.215}$$
$$A^{4\beta,\nu\delta} := \langle\,\varphi_0^{\beta}\,,\,\hat{L}^{-1}\,\varphi_1^{\nu\delta}\,\rangle, \tag{12.216}$$
$$A^{4\beta,4\delta} := \langle\,\varphi_0^{\beta}\,,\,\varphi_0^{\delta}\,\rangle, \tag{12.217}$$

$$B^{\mu\beta,\nu\delta} := -\langle\,\varphi_1^{\mu\beta}\,,\,\hat{L}^{-1}\,\varphi_1^{\nu\delta}\,\rangle + \langle\,\hat{L}^{-1}\,\varphi_1^{\mu\beta}\,,\,v^{\rho}\,\hat{L}^{-1}\,\varphi_1^{\nu\delta}\,\rangle\,\nabla_{\rho}, \tag{12.218}$$
$$B^{\mu\beta,4\delta} := \langle\,\hat{L}^{-1}\,\varphi_1^{\mu\beta}\,,\,v^{\rho}\,\varphi_0^{\delta}\,\rangle\,\nabla_{\rho}, \tag{12.219}$$
$$B^{4\beta,\nu\delta} := \langle\,\varphi_0^{\beta}\,,\,v^{\rho}\,\hat{L}^{-1}\,\varphi_1^{\nu\delta}\,\rangle\,\nabla_{\rho}, \tag{12.220}$$
$$B^{4\beta,4\delta} := \langle\,\varphi_0^{\beta}\,,\,v^{\rho}\,\varphi_0^{\delta}\,\rangle\,\nabla_{\rho}. \tag{12.221}$$

Let us obtain the normal modes given by the linear equation (12.213). Inserting

$$\delta X_{\alpha\beta}(\sigma\,;\,\tau) = \delta\tilde{X}_{\alpha\beta}(k\,;\,\Lambda)\,\mathrm{e}^{ik\cdot\sigma-\Lambda\tau}, \tag{12.222}$$

with k^{μ} being a space-like vector ($k^{\mu} = \Delta^{\mu\nu}\,k_{\nu}$) into Eq. (12.213), we have

$$(\Lambda\,A^{\alpha\beta,\gamma\delta} - \tilde{B}^{\alpha\beta,\gamma\delta})\,\delta\tilde{X}_{\gamma\delta} = 0, \tag{12.223}$$

where $\tilde{B}^{\alpha\beta,\gamma\delta}$ are given by

$$\tilde{B}^{\mu\beta,\nu\delta} := -\langle\,\varphi_1^{\mu\beta}\,,\,\hat{L}^{-1}\,\varphi_1^{\nu\delta}\,\rangle + \langle\,\hat{L}^{-1}\,\varphi_1^{\mu\beta}\,,\,v^{\rho}\,\hat{L}^{-1}\,\varphi_1^{\nu\delta}\,\rangle\,i\,k_{\rho}, \tag{12.224}$$
$$\tilde{B}^{\mu\beta,4\delta} := \langle\,\hat{L}^{-1}\,\varphi_1^{\mu\beta}\,,\,v^{\rho}\,\varphi_0^{\delta}\,\rangle\,i\,k_{\rho}, \tag{12.225}$$
$$\tilde{B}^{4\beta,\nu\delta} := \langle\,\varphi_0^{\beta}\,,\,v^{\rho}\,\hat{L}^{-1}\,\varphi_1^{\nu\delta}\,\rangle\,i\,k_{\rho}, \tag{12.226}$$
$$\tilde{B}^{4\beta,4\delta} := \langle\,\varphi_0^{\beta}\,,\,v^{\rho}\,\varphi_0^{\delta}\,\rangle\,i\,k_{\rho}. \tag{12.227}$$

In the rest of this section, we use the matrix representation when no misunderstanding is expected. In order to get a nontrivial solution, we impose

$$\det(\Lambda\, A - \tilde{B}) = 0, \tag{12.228}$$

which leads to the dispersion relation

$$\Lambda = \Lambda(k). \tag{12.229}$$

The stability of the constant solution given by Eqs. (12.200)–(12.201) against a small perturbation is assured if the real part of $\Lambda(k)$ is positive for any k^μ, which implies that δX tends to decay along with the time evolution.

We first show that A is a real symmetric positive-definite matrix. In fact, for an arbitrary non-vanishing 'vector' $(\boldsymbol{w})_{\alpha\beta} := w_{\alpha\beta}$

$$\begin{aligned} w_{\alpha\beta}\, A^{\alpha\beta,\gamma\delta}\, w_{\gamma\delta} &= \langle\, w_{\mu\beta}\, \hat{L}^{-1}\, \varphi_1^{\mu\beta} + w_{4\beta}\, \varphi_0^{\beta}\,,\ w_{\nu\delta}\, \hat{L}^{-1}\, \varphi_1^{\nu\delta} + w_{4\delta}\, \varphi_0^{\delta} \rangle \\ &= \langle\, \chi\,,\ \chi \rangle > 0, \quad w_{\alpha\beta} \neq 0, \end{aligned} \tag{12.230}$$

with

$$\chi_p := w_{\mu\alpha} \left[\hat{L}^{-1}\, \varphi_1^{\mu\alpha}\right]_p + w_{4\alpha}\, \varphi_{0p}^{\alpha}. \tag{12.231}$$

In Eq. (12.230), we have used the positive-definite property (12.19) of the inner product.

Equation (12.230) means that the inverse matrix A^{-1} exists, and A^{-1} is also a real symmetric positive-definite matrix. Then with the use of the Cholesky decomposition [218], A^{-1} is factorized into a real upper triangular matrix U and its transpose ${}^tU$ as

$$A^{-1} = {}^tU\, U. \tag{12.232}$$

Now substituting Eq. (12.232) into Eq. (12.228), we have

$$\det(\Lambda\, I - U\, \tilde{B}\, {}^tU) = 0, \tag{12.233}$$

where I stands for the unit matrix, which shows that $\Lambda(k)$ is an eigenvalue of $U\, \tilde{B}\, {}^tU$. One can see that the real part of $\Lambda(k)$ is positive for any k^μ when

$$\mathrm{Re}(U\, \tilde{B}\, {}^tU) := \frac{1}{2}\left[(U\, \tilde{B}\, {}^tU) + (U\, \tilde{B}\, {}^tU)^\dagger\right], \tag{12.234}$$

is a positive definite matrix, which we now show is the case. In fact, for a real-number 'vector' $\boldsymbol{w}$ ($(\boldsymbol{w})_{\alpha\beta} =: w_{\alpha\beta}$)

$$\begin{aligned} w_{\alpha\beta}\,[\mathrm{Re}(U\,\tilde{B}\,{}^tU)]^{\alpha\beta,\gamma\delta}\,w_{\gamma\delta} &= w_{\alpha\beta}\,[U\,\mathrm{Re}(\tilde{B})\,{}^tU]^{\alpha\beta,\gamma\delta}\,w_{\gamma\delta} \\ &= [w\,U]_{\alpha\beta}\,[\mathrm{Re}(\tilde{B})]^{\alpha\beta,\gamma\delta}\,[w\,U]_{\gamma\delta} \\ &= -[w\,U]_{\mu\beta}\,\langle\,\varphi_1^{\mu\beta}\,,\,\hat{L}^{-1}\,\varphi_1^{\nu\delta}\rangle\,[w\,U]_{\nu\delta} \\ &= -\langle\,\psi\,,\,\hat{L}^{-1}\,\psi\rangle, \end{aligned} \tag{12.235}$$

with $\psi_p := [w\,U]_{\mu\alpha}\,\varphi_{1p}^{\mu\alpha}$. However, since the vector ψ_p belongs to the Q_0 space spanned by the eigenvectors with the negative eigenvalues of $\hat{L}_{pq}$, we have $-\langle\,\psi\,,\,\hat{L}^{-1}\,\psi\rangle, > 0$. Thus,

$$w_{\alpha\beta}\,[\mathrm{Re}(U\,\tilde{B}\,{}^tU)]^{\alpha\beta,\gamma\delta}\,w_{\gamma\delta} > 0, \tag{12.236}$$

which proves that the constant solution given by Eqs. (12.200)–(12.201) is stable against a small perturbation around the general constant solution.

12.7.2 Causality

We show that the speed at which the fluctuation $\delta X_{\alpha\beta}$ propagates does not exceed that of light, *i.e.*, unity in the present unit system.

We suppose that the propagation speed of $\delta X_{\alpha\beta}$ is given by the group velocity. Here we call the maximum value of the group velocity the characteristic speed. To define the characteristic speed, let us introduce a space-like vector v_{ch}^{μ}, which is defined in terms of the eigenvalue $\Lambda(k)$ given by (12.229) as

$$v_{\mathrm{ch}}^{\mu} := \lim_{-k^2\to\infty}\left[-i\,\frac{\partial}{\partial k_\mu}\Lambda(k)\right]. \tag{12.237}$$

Then the Lorentz-invariant characteristic speed is given by

$$v_{\mathrm{ch}} := \sqrt{-\Delta_{\mu\nu}\,v_{\mathrm{ch}}^{\mu}\,v_{\mathrm{ch}}^{\nu}}. \tag{12.238}$$

By a differentiation of Eq. (12.233) with respect to ik_μ, v_{ch}^{μ} is found to be given as an eigenvalue of $U\,C^{\mu}\,{}^tU$,

$$\det\left[v_{\mathrm{ch}}^{\mu}\,I - U\,C^{\mu}\,{}^tU\right] = 0, \tag{12.239}$$

where

$$\left[C^{\rho}\right]^{\alpha\beta,\gamma\delta} := \lim_{-k^2\to\infty}\left[-i\,\frac{\partial}{\partial k_\rho}\tilde{B}^{\alpha\beta,\gamma\delta}\right]. \tag{12.240}$$

With the use of Eqs. (12.224)–(12.227) for $\tilde{B}^{\alpha\beta,\gamma\delta}$, the components of $\left[C^{\rho}\right]^{\alpha\beta,\gamma\delta}$ are found to be

$$\left[C^{\rho}\right]^{\mu\beta,\nu\delta} = \langle\, \hat{L}^{-1}\, \varphi_1^{\mu\beta}\,,\, v^{\rho}\, \hat{L}^{-1}\, \varphi_1^{\nu\delta} \rangle, \tag{12.241}$$

$$\left[C^{\rho}\right]^{\mu\beta,4\delta} = \langle\, \hat{L}^{-1}\, \varphi_1^{\mu\beta}\,,\, v^{\rho}\, \varphi_0^{\delta} \rangle, \tag{12.242}$$

$$\left[C^{\rho}\right]^{4\beta,\nu\delta} = \langle\, \varphi_0^{\beta}\,,\, v^{\rho}\, \hat{L}^{-1}\, \varphi_1^{\nu\delta} \rangle, \tag{12.243}$$

$$\left[C^{\rho}\right]^{4\beta,4\delta} = \langle\, \varphi_0^{\beta}\,,\, v^{\rho}\, \varphi_0^{\delta} \rangle. \tag{12.244}$$

Then the expectation value of $U\, C^{\mu}\, {}^{t}U$ with respect to an arbitrary vector $w' := {}^{t}(U^{-1})\, w$ can be written as

$$\frac{\left[w\, U^{-1}\right]_{\alpha\beta} \left[U\, C^{\mu}\, {}^{t}U\right]^{\alpha\beta,\gamma\delta} \left[{}^{t}(U^{-1})\, w\right]_{\gamma\delta}}{w_{\alpha'\beta'} \left[U^{-1}\, {}^{t}(U^{-1})\right]^{\alpha'\beta',\gamma'\delta'} w_{\gamma'\delta'}} = \frac{w_{\alpha\beta} \left[C^{\mu}\right]^{\alpha\beta,\gamma\delta} w_{\gamma\delta}}{w_{\alpha'\beta'}\, A^{\alpha'\beta',\gamma'\delta'}\, w_{\gamma'\delta'}} = \frac{\langle\, \chi\,,\, v^{\mu}\, \chi \rangle}{\langle\, \chi\,,\, \chi \rangle} = \langle\, v^{\mu}\, \rangle_{\chi}, \tag{12.245}$$

with

$$\chi_p := w_{\mu\alpha} \left[\hat{L}^{-1}\, \varphi_1^{\mu\alpha}\right]_p + w_{4\alpha}\, \varphi_{0p}^{\alpha}, \tag{12.246}$$

where we have introduced the notation

$$\langle\, O\, \rangle_{\chi} := \frac{\langle\, \chi\,,\, O\, \chi \rangle}{\langle\, \chi\,,\, \chi \rangle}, \tag{12.247}$$

for an arbitrary operator O.

We will show the inequality

$$\sqrt{-\Delta_{\mu\nu}\, \langle\, v^{\mu}\, \rangle_{\chi}\, \langle\, v^{\nu}\, \rangle_{\chi}} \le 1, \tag{12.248}$$

holds for any χ_p, which proves the inequality

$$v_{\mathrm{ch}} = \sqrt{-\Delta_{\mu\nu}\, v_{\mathrm{ch}}^{\mu}\, v_{\mathrm{ch}}^{\nu}} \le 1. \tag{12.249}$$

The proof of (12.248) is given as follows: First, with the use of the identities

$$-\,\Delta_{\mu\nu}\, v_p^{\mu}\, v_p^{\nu} = \frac{(p\cdot u)^2 - m^2}{(p\cdot u)^2} \le 1, \tag{12.250}$$

$$\langle\, 1\, \rangle_{\chi} = 1, \tag{12.251}$$

we obtain

$$\langle -\Delta_{\mu\nu}\, v^{\mu}\, v^{\nu} \rangle_{\chi} \leq 1. \tag{12.252}$$

We also find

$$\begin{aligned} \langle -\Delta_{\mu\nu}\, v^{\mu}\, v^{\nu} \rangle_{\chi} &= -\Delta_{\mu\nu}\, \langle v^{\mu} \rangle_{\chi}\, \langle v^{\nu} \rangle_{\chi} + \langle -\Delta_{\mu\nu}\, \delta v^{\mu}\, \delta v^{\nu} \rangle_{\chi} \\ &\geq -\Delta_{\mu\nu}\, \langle v^{\mu} \rangle_{\chi}\, \langle v^{\nu} \rangle_{\chi}, \end{aligned} \tag{12.253}$$

where $\delta v^{\mu}_{pq} := \delta v^{\mu}_{p}\, \delta_{pq}$ with $\delta v^{\mu}_{p} := v^{\mu}_{p} - \langle v^{\mu} \rangle_{\chi}$. We have used

$$-\,\Delta_{\mu\nu}\, \delta v^{\mu}_{p}\, \delta v^{\nu}_{p} \geq 0, \tag{12.254}$$

due to the fact that δv^{μ}_{p} is also a space-like vector. By combing Eq. (12.253) with Eq. (12.252), the proof is completed.

In short, the relativistic second-order fluid dynamic equation given by Eqs. (12.77) and (12.81) respects the causality in the linear analysis around the homogeneous steady state (12.200)–(12.201), in addition to the stability around the static solution.

Chapter 13
Appendices for Chaps. 10, 11, and 12

13.1 Foundation of the Symmetrized Inner Product defined by Eqs. (11.31) and (12.18)

In this section, by applying the general theory of asymmetric linear operators given in Sect. 3.6.2, we shall prove that the inner product defined by Eqs. (11.31) and (12.18) is the unique one which satisfies manifestly the self-adjointness of the linearized evolution operator $\hat{L}$.

To make the discussion clear, we suppose that the dimension of the vector space operated by $\hat{L}$ is n. For this $n \times n$ matrix $\hat{L}$, we define right eigenvectors U_i and left eigenvectors $\tilde{U}_i^\dagger$ with $i = 1, \ldots, n$ as

$$\hat{L}\, U_i = \lambda_i\, U_i, \tag{13.1}$$

$$\tilde{U}_i^\dagger\, \hat{L} = \lambda_i\, \tilde{U}_i^\dagger, \tag{13.2}$$

respectively. Here, λ_i with $i = 1, \ldots, n$ are corresponding eigenvalues. A Hermitian conjugate of Eq. (13.2) is

$$A^\dagger\, \tilde{U}_i = \lambda_i^*\, \tilde{U}_i, \tag{13.3}$$

with λ_i^* being a complex conjugate of λ_i. Without loss of generality, we can impose the orthogonality and completeness as follows:

$$\tilde{U}_i^\dagger\, U_j = \delta_{ij}, \tag{13.4}$$

$$\sum_{i=1}^{n} U_i\, \tilde{U}_i^\dagger = 1, \tag{13.5}$$

respectively.

T. Kunihiro et al., *Geometrical Formulation of Renormalization-Group Method as an Asymptotic Analysis*, Fundamental Theories of Physics 206,
https://doi.org/10.1007/978-981-16-8189-9_13

We define an inner product for φ and ψ in terms of the right eigenvectors U_i with $i = 1, \cdots, n$ as

$$\langle \varphi , \psi \rangle := \varphi^\dagger g \psi, \tag{13.6}$$

where g is a metric tensor

$$g = \sum_{i=1}^{n} \tilde{U}_i \tilde{U}_i^\dagger . \tag{13.7}$$

We note that the following symmetry property with respect to U_i with $i = 1, \cdots, n$ holds:

$$\langle U_i , U_j \rangle = \delta_{ij}, \tag{13.8}$$

which can be derived from Eq. (13.4).

Using this symmetrized inner product, we obtain

$$\begin{aligned}\langle \varphi , \hat{L} \psi \rangle = \varphi^\dagger g \hat{L} \psi &= \sum_{i=1}^{n} \varphi^\dagger \tilde{U}_i \tilde{U}_i^\dagger \hat{L} \psi \\ &= \sum_{i=1}^{n} \lambda_i \varphi^\dagger \tilde{U}_i \tilde{U}_i^\dagger \psi,\end{aligned} \tag{13.9}$$

and

$$\begin{aligned}\langle \hat{L} \varphi , \psi \rangle = (\hat{L} \varphi)^\dagger g \psi = \varphi^\dagger \hat{L}^\dagger g \psi &= \sum_{i=1}^{n} \varphi^\dagger \hat{L}^\dagger \tilde{U}_i \tilde{U}_i^\dagger \psi \\ &= \sum_{i=1}^{n} \lambda_i^* \varphi^\dagger \tilde{U}_i \tilde{U}_i^\dagger \psi.\end{aligned} \tag{13.10}$$

We find that, when all the eigenvalues are real number as $\lambda_i^* = \lambda_i$ with $i = 1, \cdots, n$, the self-adjoint nature of $\hat{L}$ is apparent,

$$\langle \varphi , \hat{L} \psi \rangle = \langle \hat{L} \varphi , \psi \rangle. \tag{13.11}$$

Let us construct an explicit form of the metric tensor g. First we shall show that there exisits a hermitian matrix G so that $\hat{L}$ and $\hat{L}^\dagger$ satisfy

$$\hat{L}^\dagger = G \hat{L} G^{-1}, \tag{13.12}$$

with

$$G^{\dagger} = G. \tag{13.13}$$

In fact, using the definition of $\hat{L}$ in Eq. (11.29), we have

$$\begin{aligned}(p \cdot u)\, f_p^{\rm eq}\, \bar{f}_p^{\rm eq}\, \hat{L}_{pq} &= -\frac{1}{2!} \int \mathrm{d}p_1 \int \mathrm{d}p_2 \int \mathrm{d}p_3\, \omega(p\,,\, p_1|p_2\,,\, p_3) \\ &\quad \times f_p^{\rm eq}\, f_{p_1}^{\rm eq}\, \bar{f}_{p_2}^{\rm eq}\, \bar{f}_{p_3}^{\rm eq}\, (\delta_{pq} + \delta_{p_1q} - \delta_{p_2q} - \delta_{p_3q}) \\ &= -\frac{1}{2!} (\delta_{pq}\, a_p + b_{pq} - c_{pq} - d_{pq}),\end{aligned} \tag{13.14}$$

with

$$a_p := \int \mathrm{d}p_1 \int \mathrm{d}p_2 \int \mathrm{d}p_3\, \omega(p\,,\, p_1|p_2\,,\, p_3)\, f_p^{\rm eq}\, f_{p_1}^{\rm eq}\, \bar{f}_{p_2}^{\rm eq}\, \bar{f}_{p_3}^{\rm eq}, \tag{13.15}$$

$$b_{pq} := \int \mathrm{d}p_2 \int \mathrm{d}p_3\, \omega(p\,,\, q|p_2\,,\, p_3)\, f_p^{\rm eq}\, f_q^{\rm eq}\, \bar{f}_{p_2}^{\rm eq}\, \bar{f}_{p_3}^{\rm eq}, \tag{13.16}$$

$$c_{pq} := \int \mathrm{d}p_1 \int \mathrm{d}p_3\, \omega(p\,,\, p_1|q\,,\, p_3)\, f_p^{\rm eq}\, f_{p_1}^{\rm eq}\, \bar{f}_q^{\rm eq}\, \bar{f}_{p_3}^{\rm eq}, \tag{13.17}$$

$$d_{pq} := \int \mathrm{d}p_1 \int \mathrm{d}p_2\, \omega(p\,,\, p_1|p_2\,,\, q)\, f_p^{\rm eq}\, f_{p_1}^{\rm eq}\, \bar{f}_{p_2}^{\rm eq}\, \bar{f}_q^{\rm eq}. \tag{13.18}$$

Here, it can be readily shown that

$$b_{pq} = b_{qp}, \quad c_{pq} = c_{qp}, \quad d_{pq} = d_{qp}. \tag{13.19}$$

For instance,

$$\begin{aligned}d_{pq} &= \int \mathrm{d}p_1 \int \mathrm{d}p_2\, \omega(p\,,\, p_1|p_2\,,\, q)\, f_p^{\rm eq}\, f_{p_1}^{\rm eq}\, \bar{f}_{p_2}^{\rm eq}\, \bar{f}_q^{\rm eq} \\ &= \int \mathrm{d}p_1 \int \mathrm{d}p_2\, \omega(q\,,\, p_2|p_1\,,\, p)\, f_p^{\rm eq}\, f_{p_1}^{\rm eq}\, \bar{f}_{p_2}^{\rm eq}\, \bar{f}_q^{\rm eq} \\ &= \int \mathrm{d}p_1 \int \mathrm{d}p_2\, \omega(q\,,\, p_2|p_1\,,\, p)\, \bar{f}_p^{\rm eq}\, \bar{f}_{p_1}^{\rm eq}\, f_{p_2}^{\rm eq}\, f_q^{\rm eq} \\ &= \int \mathrm{d}p_1 \int \mathrm{d}p_2\, \omega(q\,,\, p_1|p_2\,,\, p)\, f_q^{\rm eq}\, f_{p_1}^{\rm eq}\, \bar{f}_{p_2}^{\rm eq}\, \bar{f}_p^{\rm eq} = d_{qp},\end{aligned} \tag{13.20}$$

where the symmetry property of the transition probability (10.20) given by

$$\omega(p\,,\, p_1|p_2\,,\, q) = \omega(p_2\,,\, q|p\,,\, p_1) = \omega(q\,,\, p_2|p_1\,,\, p), \tag{13.21}$$

and the detailed balance (10.35) given by

$$\omega(q\,,\, p_2|p_1\,,\, p)\, (f_p^{\rm eq}\, f_{p_1}^{\rm eq}\, \bar{f}_{p_2}^{\rm eq}\, \bar{f}_q^{\rm eq} - \bar{f}_p^{\rm eq}\, \bar{f}_{p_1}^{\rm eq}\, f_{p_2}^{\rm eq}\, f_q^{\rm eq}) = 0, \tag{13.22}$$

have been used in the second and third lines, respectively, and the dummy variables p_1 and p_2 have been interchanged in the final line. Thus we find that $(p \cdot u)\, f_p^{\mathrm{eq}}\, \bar{f}_p^{\mathrm{eq}}\, \hat{L}_{pq}$ is a symmetric matrix, that is,

$$(p \cdot u)\, f_p^{\mathrm{eq}}\, \bar{f}_p^{\mathrm{eq}}\, \hat{L}_{pq} = (q \cdot u)\, f_q^{\mathrm{eq}}\, \bar{f}_q^{\mathrm{eq}}\, \hat{L}_{qp}. \tag{13.23}$$

Noticing that $\hat{L}_{qp}$ is real as

$$\hat{L}_{qp} = \hat{L}^*_{qp} = \hat{L}^\dagger_{pq}, \tag{13.24}$$

we rewrite Eq. (13.23) as

$$\hat{L}^\dagger_{pq} = (p \cdot u)\, f_p^{\mathrm{eq}}\, \bar{f}_p^{\mathrm{eq}}\, \hat{L}_{pq}[(q \cdot u)\, f_q^{\mathrm{eq}}\, \bar{f}_q^{\mathrm{eq}}]^{-1}. \tag{13.25}$$

Since $(p \cdot u)\, f_p^{\mathrm{eq}}\, \bar{f}_p^{\mathrm{eq}}$ is real, we can identify the Hermite matrix G with

$$G_{pq} = (p \cdot u)\, f_p^{\mathrm{eq}}\, \bar{f}_p^{\mathrm{eq}}\, \delta_{pq}. \tag{13.26}$$

Substituting Eq. (13.12) into Eq. (13.3), we have

$$G\, \hat{L}\, G^{-1}\, \tilde{U}_i = \lambda_i^*\, \tilde{U}_i, \tag{13.27}$$

which can be converted to

$$\hat{L}\, G^{-1}\, \tilde{U}_i = \lambda_i^*\, G^{-1}\, \tilde{U}_i. \tag{13.28}$$

By comparing this equation with Eq. (13.1), one finds

$$\lambda_i^* = \lambda_i, \tag{13.29}$$

$$G^{-1}\, \tilde{U}_i = U_i. \tag{13.30}$$

Using Eq. (13.30) and the completeness (13.5), we find that the metric tensor g is identical to G:

$$g = \sum_{i=1}^{n} G\, U_i\, \tilde{U}_i^\dagger = G. \tag{13.31}$$

It is noted that, as shown in Eq. (13.29), all eigenvalues of $\hat{L}$ are real and hence the inner product constructed by G shows in an apparent way the self-adjoint nature of $\hat{L}$.

Substituting Eq. (13.31) combined with Eq. (13.26) into Eq. (13.6), we have the inner product

$$
\begin{aligned}
\langle \varphi , \psi \rangle &= \int \mathrm{d}p \int \mathrm{d}q \, \varphi_p \, G_{pq} \, \psi_q \\
&= \int \mathrm{d}p \int \mathrm{d}q \, \varphi_p \, (p \cdot u) \, f_p^{\mathrm{eq}} \, \bar{f}_p^{\mathrm{eq}} \, \delta_{pq} \, \psi_q \\
&= \int \mathrm{d}p \, (p \cdot u) \, f_p^{\mathrm{eq}} \, \bar{f}_p^{\mathrm{eq}} \, \varphi_p \, \psi_p .
\end{aligned}
\tag{13.32}
$$

We note that the symmetrized inner product constructed in Eq. (13.32), which is a unique one that ensures the self-adjoint nature of $\hat{L}$, is nothing but the inner product introduced by Eqs. (11.31) and (12.18).

13.2 Derivation of Eqs. (10.65)–(10.67)

In this section, we present a detailed derivation of Eqs. (10.65)–(10.67).

Some manipulations of Eq. (10.64) with the use of the orthogonality relations between the zero modes give

$$
\frac{\partial}{\partial \tau^{(1)}} T = -T \left[\frac{1}{3} + \frac{m^2 (a_2 a_0 - a_1^2)}{3 (a_3 a_1 - a_2^2)} \right] \nabla \cdot u , \tag{13.33}
$$

$$
\frac{\partial}{\partial \iota^{(1)}} \frac{\mu}{T} = \frac{1}{T} \frac{m^2 (a_3 a_0 - a_2 a_1)}{3 (a_3 a_1 \;\; a_2^2)} \nabla \cdot u , \tag{13.34}
$$

$$
\frac{\partial}{\partial \tau^{(1)}} u^\mu = \frac{1}{T} \nabla^\mu T + T \frac{m^2 a_0 - a_2}{m^2 a_1 - a_3} \nabla^\mu \frac{\mu}{T} , \tag{13.35}
$$

where

$$
a_\ell := \int \mathrm{d}p \, f_p^{\mathrm{eq}} \, \bar{f}_p^{\mathrm{eq}} \, (p \cdot u)^\ell , \quad \ell = 0, \, 1, \, 2, \, \cdots , \tag{13.36}
$$

some of which are expressed in terms of the thermodynamic quantities T, μ, n, e, P, and their derivatives as

$$
\frac{\partial n}{\partial T} = \frac{1}{T^2} a_2 - \frac{\mu}{T^2} a_1 , \quad \frac{\partial n}{\partial \mu} = \frac{1}{T} a_1 , \tag{13.37}
$$

$$
\frac{\partial e}{\partial T} = \frac{1}{T^2} a_3 - \frac{\mu}{T^2} a_2 , \quad \frac{\partial e}{\partial \mu} = \frac{1}{T} a_2 , \tag{13.38}
$$

$$
\frac{\partial P}{\partial T} = \frac{1}{3T^2} (a_3 - \mu a_2 - m^2 a_1 + m^2 \mu a_0) , \tag{13.39}
$$

$$
\frac{\partial P}{\partial \mu} = \frac{1}{3T} (a_2 - m^2 a_0) . \tag{13.40}
$$

We note that the explicit forms of n, e, and P given by Eqs. (10.38)–(10.40) have been used.

Combinations of these relations further lead to the following equalities:

$$-\frac{m^2(a_2a_0-a_1^2)}{3(a_3a_1-a_2^2)}=\frac{1}{3}-\frac{\frac{\partial P}{\partial T}\frac{\partial n}{\partial \mu}-\frac{\partial P}{\partial \mu}\frac{\partial n}{\partial T}}{\frac{\partial e}{\partial T}\frac{\partial n}{\partial \mu}-\frac{\partial e}{\partial \mu}\frac{\partial n}{\partial T}}=\frac{1}{3}-\left.\frac{\partial P}{\partial e}\right|_n, \tag{13.41}$$

$$\frac{m^2(a_3a_0-a_2a_1)}{3(a_3a_1-a_2^2)}=-\frac{\frac{\partial P}{\partial T}\frac{\partial e}{\partial \mu}-\frac{\partial P}{\partial \mu}\frac{\partial e}{\partial T}}{\frac{\partial n}{\partial T}\frac{\partial e}{\partial \mu}-\frac{\partial n}{\partial \mu}\frac{\partial e}{\partial T}}=-\left.\frac{\partial P}{\partial n}\right|_e, \tag{13.42}$$

$$\frac{m^2a_1-a_3}{m^2a_0-a_2}=T\frac{\frac{\partial P}{\partial T}}{\frac{\partial P}{\partial \mu}}+\mu=\frac{e+P}{n}. \tag{13.43}$$

In Eq. (13.43), we have used the following relations derived from the Gibbs-Duhem equation $\mathrm{d}P=s\mathrm{d}T+n\mathrm{d}\mu$,

$$\frac{\partial P}{\partial T}=s=\frac{e+P-\mu n}{T}, \tag{13.44}$$

$$\frac{\partial P}{\partial \mu}=n, \tag{13.45}$$

with s being the entropy density. We note that the relations (13.44) and (13.45) can be shown not only by the Gibbs-Duhem equation but also by a straightforward manipulation based on the explicit forms of n, e, and P.

With use of Eqs. (13.41)–(13.43), Eqs. (13.33)–(13.35) are nicely simplified as Eqs. (10.65)–(10.67).

13.3 Detailed Derivation of Explicit Form of $\varphi_1^{\mu\alpha}$

We derive the expression of $\varphi_{1p}^{\mu\alpha}$ in Eq. (11.88) whose calculation can be done as follows,

$$\begin{aligned}\varphi_{1p}^{\mu\alpha}&=\left[Q_0\,\tilde{\varphi}_1^{\mu\alpha}\right]_p=\tilde{\varphi}_{1p}^{\mu\alpha}-\left[P_0\,\tilde{\varphi}_1^{\mu\alpha}\right]_p\\&=\frac{1}{p\cdot u}\Big(p^\mu\,\varphi_{0p}^{\alpha}-(p\cdot u)\,\varphi_{0p}^{\beta}\,\eta_{0\beta\gamma}^{-1}\,\langle\varphi_0^{\gamma}\,,\,\tilde{\varphi}_1^{\mu\alpha}\rangle\Big),\end{aligned} \tag{13.46}$$

Using a_ℓ with $\ell=0,\,1,\,\cdots$ defined in Eq. (13.36), then we express the metric $\eta_0^{\alpha\beta}=\langle\varphi_0^\alpha\,,\,\varphi_0^\beta\rangle$ as

$$\eta_0^{\mu\nu} = a_3 u^\mu u^\nu + (m^2 a_1 - a_3)\frac{1}{3}\Delta^{\mu\nu}, \tag{13.47}$$

$$\eta_0^{\mu 4} = \eta_0^{4\mu} = a_2 u^\mu, \tag{13.48}$$

$$\eta_0^{44} = a_1, \tag{13.49}$$

while the inverse metric $\eta_{0\alpha\beta}^{-1}$ read

$$\eta_{0\mu\nu}^{-1} = \frac{a_1 u^\mu u^\nu}{a_3 a_1 - a_2^2} + \frac{3\Delta^{\mu\nu}}{m^2 a_1 - a_3}, \tag{13.50}$$

$$\eta_{0\mu 4}^{-1} = \eta_{04\mu}^{-1} = \frac{-a_2 u^\mu}{a_3 a_1 - a_2^2}, \tag{13.51}$$

$$\eta_{044}^{-1} = \frac{a_3}{a_3 a_1 - a_2^2}. \tag{13.52}$$

The inner products $\langle\, \varphi_0^\alpha\,,\, \tilde{\varphi}_1^{\mu\beta}\,\rangle$ are evaluated as follows:

$$\langle\, \varphi_0^a\,,\, \tilde{\varphi}_1^{\mu b}\,\rangle = a_3\, u^a\, u^\mu\, u^b + (m^2\, a_1 - a_3)\, \frac{1}{3}\, (u^a\, \Delta^{\mu b} + u^\mu\, \Delta^{ba} + u^b\, \Delta^{a\mu}), \tag{13.53}$$

$$\langle\, \varphi_0^a\,,\, \tilde{\varphi}_1^{\mu 4}\,\rangle = a_2\, u^a\, u^\mu + (m^2\, a_0 - a_2)\, \frac{1}{3}\, \Delta^{a\mu}, \tag{13.54}$$

$$\langle\, \varphi_0^4\,,\, \tilde{\varphi}_1^{\mu b}\,\rangle = a_2\, u^\mu\, u^b + (m^2\, a_0 - a_2)\, \frac{1}{3}\, \Delta^{\mu b}, \tag{13.55}$$

$$\langle\, \varphi_0^4\,,\, \tilde{\varphi}_1^{\mu 4}\,\rangle = a_1\, u^\mu. \tag{13.56}$$

Inserting the inverse metric $\eta_{0\alpha\beta}^{-1}$ in Eqs. (13.50)–(13.52) and the inner products $\langle\, \varphi_0^\alpha\,,\, \tilde{\varphi}_1^{\mu\beta}\,\rangle$ in Eqs. (13.53)–(13.56) into Eq. (13.46), we have

$$\varphi_{1p}^{\mu\alpha} = \begin{cases} \dfrac{1}{p\cdot u}\,(-\Delta^{\mu\nu}\,\Pi_p + \pi_p^{\mu\nu}),\ \alpha = \mu, \\[2mm] \dfrac{1}{p\cdot u}\,\dfrac{m^2\, a_0 - a_2}{m^2\, a_1 - a_3}\, J_p^\mu, \qquad \alpha = 4. \end{cases} \tag{13.57}$$

Here, we have introduced the following quantities

$$\Pi_p := -\frac{m^2(a_2 a_0 - a_1^2)}{3(a_3 a_1 - a_2^2)}(p\cdot u)^2 + \frac{m^2(a_3 a_0 - a_2 a_1)}{3(a_3 a_1 - a_2^2)}(p\cdot u) - \frac{m^2}{3}, \tag{13.58}$$

$$J_p^\mu := -\Delta^{\mu\nu}\, p_\nu\left[(p\cdot u) - \frac{m^2 a_1 - a_3}{m^2 a_0 - a_2}\right], \tag{13.59}$$

$$\pi_p^{\mu\nu} := \Delta^{\mu\nu\rho\sigma}\, p_\rho\, p_\sigma, \tag{13.60}$$

with

$$\Delta^{\mu\nu\rho\sigma} = \frac{1}{2}\left[\Delta^{\mu\rho}\,\Delta^{\nu\sigma} + \Delta^{\mu\sigma}\,\Delta^{\nu\rho} - \frac{2}{3}\,\Delta^{\mu\nu}\,\Delta^{\rho\sigma}\right]. \tag{13.61}$$

Using Eqs. (13.41)–(13.43), which are derived from the explicit forms of a_ℓ with $\ell = 0, 1, 2, 3$, we can convert Eqs. (13.57)–(13.60) into

$$\varphi_{1p}^{\mu\alpha} = \begin{cases} \frac{1}{p\cdot u}\,(-\Delta^{\mu\nu}\,\Pi_p + \pi_p^{\mu\nu}) = -\Delta^{\mu\nu}\,\hat{\Pi}_p + \hat{\pi}_p^{\mu\nu}, & \alpha = \mu, \\ \frac{1}{p\cdot u}\,\frac{n}{e+P}\,J_p^{\mu} = \frac{n}{e+P}\,\hat{J}_p^{\mu}, & \alpha = 4, \end{cases} \tag{13.62}$$

where $(\hat{\Pi}_p,\ \hat{J}_p^{\mu},\ \hat{\pi}_p^{\mu\nu}) := \frac{1}{p\cdot u}\,(\Pi_p,\ J_p^{\mu},\ \pi_p^{\mu\nu})$ as given in (11.89) with

$$\Pi_p = (p\cdot u)^2\left[\frac{1}{3} - \left.\frac{\partial P}{\partial e}\right|_n\right] - (p\cdot u)\left.\frac{\partial P}{\partial n}\right|_e - \frac{1}{3}\,m^2, \tag{13.63}$$

$$J_p^{\mu} = -\Delta^{\mu\nu}\,p_\nu\left[(p\cdot u) - \frac{e+P}{n}\right]. \tag{13.64}$$

One sees that the expression of $\varphi_1^{\mu\alpha}$ in Eq. (13.62) agrees with that given in Eq. (11.88), *i.e.*, we have derived Eq. (11.88), as promised.

13.4 Computation of $\hat{L}\ Q_0\ F_0$ in Eq. (12.38)

We derive the expression of $\hat{L}^{-1}\,Q_0\,F_0$ given by Eq. (12.38).

Noticing that F_0 in Eq. (12.17) is expressed as

$$F_{0p} = \tilde{\varphi}_{1p}^{\mu\nu}\,\nabla_\mu\frac{u_\nu}{T} - \tilde{\varphi}_{1p}^{\mu 4}\,\nabla_\mu\frac{\mu}{T}, \tag{13.65}$$

with $\tilde{\varphi}_{1p}^{\mu\alpha} = p^\mu\,\varphi_{0p}^{\alpha}\,\frac{1}{p\cdot u}$, we have

$$\big[Q_0\,F_0\big]_p = \varphi_{1p}^{\mu\nu}\,\nabla_\mu\frac{u_\nu}{T} - \varphi_{1p}^{\mu 4}\,\nabla_\mu\frac{\mu}{T}, \tag{13.66}$$

with $\varphi_{1p}^{\mu\alpha} := \big[Q_0\,\tilde{\varphi}_1^{\mu\alpha}\big]_p$. Substituting the expressions of $\varphi_{1p}^{\mu\alpha}$ given by Eq. (13.62) into Eq. (13.66), we have

$$\big[Q_0\,F_0\big]_p = \hat{\Pi}_p\,\frac{-\nabla\cdot u}{T} - \hat{J}_p^{\mu}\,\frac{1}{h}\,\nabla_\mu\frac{\mu}{T} + \hat{\pi}_p^{\mu\nu}\,\frac{\Delta_{\mu\nu\rho\sigma}\,\nabla^\rho u^\sigma}{T}. \tag{13.67}$$

In the above equation, we have introduced the enthalpy per particle h given by

$$h := \frac{e+P}{n}. \tag{13.68}$$

By operating $\hat{L}^{-1}$ on Eq. (13.67) from the left, we obtain Eq. (12.38).

13.5 Proof of Vanishing of Inner Product Between Collision Invariants and *B*

In this section, we present a proof for the identity given by Eq. (11.67). This proof is a relativistic extension of the proof in the non-relativistic dissipative fluid dynamics, which is presented in Sect. 8.5.

We begin from the definition of the collision invariants,

$$\int \mathrm{d}p\, \varphi^{\alpha}_{0p}\, C[f]_p = 0. \tag{13.69}$$

This identity is satisfied for an arbitrary distribution function f_p, and hence we can take the functional derivatives with respect to f_p. The second derivative reads

$$\int \mathrm{d}p\, \varphi^{\alpha}_{0p}\, \frac{\delta^2}{\delta f_q \delta f_r} C[f]_p = 0. \tag{13.70}$$

We take the value of Eq. (13.70) at $f_p = f^{\mathrm{eq}}_p$ to obtain the following relation:

$$\int \mathrm{d}p\, \varphi^{\alpha}_{0p}\, f^{\mathrm{eq}}_p\, \bar{f}^{\mathrm{eq}}_p\, B_{pqr}\, (f^{\mathrm{eq}}_q\, \bar{f}^{\mathrm{eq}}_q)^{-1}\, (f^{\mathrm{eq}}_r\, \bar{f}^{\mathrm{eq}}_r)^{-1} = 0, \tag{13.71}$$

where the definition of B given by Eq. (11.56) has been used.

Multiplying Eq. (13.71) by arbitrary two vectors $f^{\mathrm{eq}}_q\, \bar{f}^{\mathrm{eq}}_q\, \psi_q$ and $f^{\mathrm{eq}}_r\, \bar{f}^{\mathrm{eq}}_r\, \chi_r$ and then taking the integration with respect to q and r, we have

$$\begin{aligned} 0 &= \int \mathrm{d}p\, \varphi^{\alpha}_{0p}\, f^{\mathrm{eq}}_p\, \bar{f}^{\mathrm{eq}}_p \int \mathrm{d}q \int \mathrm{d}r\, B_{pqr}\, \psi_q\, \chi_r \\ &= \int \mathrm{d}p\, \varphi^{\alpha}_{0p}\, f^{\mathrm{eq}}_p\, \bar{f}^{\mathrm{eq}}_p\, B[\psi\,,\, \chi] \\ &= \langle \varphi^{\alpha}_0\,,\, B[\psi\,,\, \chi] \rangle, \end{aligned} \tag{13.72}$$

where the notation (11.55) has been used. The putting of $\psi = \chi = \hat{L}^{-1}\, Q\, F$ reduces Eq. (13.72) into Eq. (11.67), and hence the proof of Eq. (11.67) is completed.

Chapter 14
Demonstration of Numerical Calculations of Transport Coefficients and Relaxation Times: Typical Three Models

14.1 Introduction

In this chapter, as a direct continuation of Chap. 12, we demonstrate how numerically calculated are the microscopic expressions of the transport coefficients and relaxation times derived in Chap. 12 from the relativistic Boltzmann equation [67, 68], and discuss the properties of them thus obtained for three systems of some physical interest. The derivation of these expressions is based on [155] where the *doublet scheme* in the RG method [69] is applied, which is an extension of the RG method [1, 38, 39, 48–52, 92–96] as formulated in [3, 5, 6, 46, 57, 58] so as to incorporate the appropriate excited modes in addition to the zero modes to make a reduced dynamics from a microscopic theory.

According to the formulae derived in Chap. 12, a main task in the calculation of the transport coefficients and relaxation times is to evaluate the following quantities:

$$\begin{pmatrix} \phi_{\Pi p} \\ \phi^{\mu}_{Jp} \\ \phi^{\mu\nu}_{\pi p} \end{pmatrix} := \left[\hat{L}^{-1} \begin{pmatrix} \hat{\Pi} \\ \hat{J}^{\mu} \\ \hat{\pi}^{\mu\nu} \end{pmatrix} \right]_p , \tag{14.1}$$

where $\hat{L}_{pq}$ is the linear evolution operator (linearized collision operator) (12.16), and $\hat{\Pi}_p$, $\hat{J}^{\mu}_p$, and $\hat{\pi}^{\mu\nu}_p$ are the microscopic representations of dissipative currents given by (12.39)–(12.41), respectively. Indeed, with the use of $\phi_{\Pi p}$, ϕ^{μ}_{Jp}, and $\phi^{\mu\nu}_{\pi p}$, we have given the microscopic expressions of the transport coefficients and relaxation times in (12.105)–(12.107) and (12.108)–(12.110), respectively, which are reproduced below

T. Kunihiro et al., *Geometrical Formulation of Renormalization-Group Method as an Asymptotic Analysis*, Fundamental Theories of Physics 206,
https://doi.org/10.1007/978-981-16-8189-9_14

$$\zeta^{\mathrm{RG}} = -\frac{1}{T}\langle \hat{\Pi}\,,\,\phi_{\Pi}\rangle,\quad \lambda^{\mathrm{RG}} = \frac{1}{3T^2}\langle \hat{J}^{\mu}\,,\,\phi_{J\mu}\rangle, \tag{14.2}$$

$$\eta^{\mathrm{RG}} = -\frac{1}{10T}\langle \hat{\pi}^{\mu\nu}\,,\,\phi_{\pi\mu\nu}\rangle,\quad \tau_{\Pi}^{\mathrm{RG}} = -\frac{\langle \phi_{\Pi}\,,\,\phi_{\Pi}\rangle}{\langle \hat{\Pi}\,,\,\phi_{\Pi}\rangle}, \tag{14.3}$$

$$\tau_{J}^{\mathrm{RG}} = -\frac{\langle \phi_{J}^{\mu}\,,\,\phi_{J\mu}\rangle}{\langle \hat{J}^{\rho}\,,\,\phi_{J\rho}\rangle},\quad \tau_{\pi}^{\mathrm{RG}} = -\frac{\langle \phi_{\pi}^{\mu\nu}\,,\,\phi_{\pi\mu\nu}\rangle}{\langle \hat{\pi}^{\rho\sigma}\,,\,\phi_{\pi\rho\sigma}\rangle}, \tag{14.4}$$

with the inner product $\langle \varphi\,,\,\psi\rangle = \int \mathrm{d}p\,(p\cdot u)\, f_p^{\mathrm{eq}}\,\bar{f}_p^{\mathrm{eq}}\,\varphi_p\,\psi_p$.

Let us convert Eq. (14.1) into the following linear equations

$$\left[\hat{L}\begin{pmatrix}\phi_{\Pi}\\ \phi_{J}^{\mu}\\ \phi_{\pi}^{\mu\nu}\end{pmatrix}\right]_p = \begin{pmatrix}\hat{\Pi}_p\\ \hat{J}_p^{\mu}\\ \hat{\pi}_p^{\mu\nu}\end{pmatrix}. \tag{14.5}$$

These equations are linear integral equations,[1] and called *linearized transport equations* [67]. We note that a solution to the linearized transport equations gives explicit forms of $\phi_{\Pi p}$, ϕ_{Jp}^{μ}, and $\phi_{\pi p}^{\mu\nu}$. Our main task in this chapter is thus reduced to solving the linear integral equations (14.5), which will be achieved in a numerical method.

This chapter is organized as follows: In Sect. 14.2, we reduce the linearized transport equations (14.5) and also the microscopic expressions of the transport coefficients and relaxation times to simpler forms that are suitable for the numerical calculation, and then provide a method to numerically solve the integral equations. In Sect. 14.3, we compute the transport coefficients and relaxation times numerically using the linear evolution operator $\hat{L}_{pq}$ for three physical models with the cross section given, and briefly discuss the properties of the results thus obtained.

14.2 Linearized Transport Equations and Solution Method

In this section, following the calculational procedure given in [67] for the transport coefficients in the case of *classical* statistics, we start with the linearized transport equations (14.5) in the case of *quantum* statistics and convert the microscopic expressions of the transport coefficients and relaxation times to simpler forms that are tractable numerically. First we reduce the linear calculus in the left-hand side of the linearized transport equations (14.5) to that of a system of three linear integral equations with some kernel functions, each of which is given as a multiple integral of sixth order. Then, the kernel functions are obtained by carrying out multiple integrals of fourth and sixth orders. Next, the three-dimensional momentum integrals are reduced to one-dimensional integrals. The final part of this section is devoted to an account of a not only accurate but also efficient numerical method for solving the integral equations.

[1] More precisely, Fredholm integral equations of the first kind [80].

14.2.1 *Reduction of the Integrals in the Linearized Transport Equations in Terms of the Differential Cross Section*

We consider a generic form of the integral operation of the left-hand side of the linearized transport equations (14.5) as

$$[\hat{L}\,\phi]_p = -\frac{1}{p\cdot u}\frac{1}{2!}\int \mathrm{d}p_1 \int \mathrm{d}p_2 \int \mathrm{d}p_3\, \omega(p,\, p_1|p_2,\, p_3) \times \frac{f^{\mathrm{eq}}_{p_1}\,\bar{f}^{\mathrm{eq}}_{p_2}\,\bar{f}^{\mathrm{eq}}_{p_3}}{\bar{f}^{\mathrm{eq}}_{p}}\,(\phi_p + \phi_{p_1} - \phi_{p_2} - \phi_{p_3}), \tag{14.6}$$

for an arbitrary vector ϕ_p with the integral measure

$$\mathrm{d}p = \frac{\mathrm{d}^3\boldsymbol{p}}{(2\pi)^3\, p^0}. \tag{14.7}$$

For a reduction of the formula, we exploit the fact that the transition probability $\omega(p,\, p_1|p_2,\, p_3)$ is expressed in terms of the differential cross section $\sigma(q,\, \Theta)$ as [67],

$$\omega(p,\, p_1|p_2,\, p_3) = (2\pi)^6\, P^2\, \sigma(q,\, \Theta)\, \delta^4(p + p_1 - p_2 - p_3), \tag{14.8}$$

where the following Lorentz invariant quantities have been introduced:

$$P^2 := (p + p_1)^2, \tag{14.9}$$

$$q^2 := -(p - p_1)^2, \tag{14.10}$$

$$\cos\Theta := -(p - p_1)\cdot(p_2 - p_3)/q^2. \tag{14.11}$$

In fact, one can confirm that the total cross section σ_{tot} is expressed in two equivalent ways as

$$\sigma_{\mathrm{tot}} = \int \mathrm{d}p_2 \int \mathrm{d}p_3\, \frac{1}{F}\,\omega(p,\, p_1|p_2,\, p_3), \tag{14.12}$$

$$= 2\pi \int_0^{\pi} \mathrm{d}\Theta\, \sin\Theta\, \sigma(q,\, \Theta). \tag{14.13}$$

with the invariant flux $F := \sqrt{(p\cdot p_1)^2 - m^4}$.

Substituting Eq. (14.8) into Eq. (14.6), and using the integral measure (14.7), we have

$$
\begin{aligned}
\left[\hat{L}\,\phi\right]_p = -\frac{1}{p\cdot u}\int\frac{\mathrm{d}^3\boldsymbol{p}_1}{p_1^0}\int\frac{\mathrm{d}^3\boldsymbol{p}_2}{p_2^0}\int\frac{\mathrm{d}^3\boldsymbol{p}_3}{p_3^0}\,\frac{1}{(2\,\pi)^3}\,P^2\,\frac{1}{2!}\,\sigma(q,\,\Theta) \\
\times\,\delta^4(p+p_1-p_2-p_3)\,\frac{f_{p_1}^{\mathrm{eq}}\,\bar{f}_{p_2}^{\mathrm{eq}}\,\bar{f}_{p_3}^{\mathrm{eq}}}{\bar{f}_p^{\mathrm{eq}}}\,(\phi_p+\phi_{p_1}-\phi_{p_2}-\phi_{p_3}).
\end{aligned}
\tag{14.14}
$$

For convenience, we decompose this equation into four contributions as

$$
\left[\hat{L}\,\phi\right]_p = I[\phi]_p + I_1[\phi]_p + I_2[\phi]_p + I_3[\phi]_p,
\tag{14.15}
$$

where

$$
\begin{aligned}
\begin{pmatrix} I[\phi]_p \\ I_1[\phi]_p \\ I_2[\phi]_p \\ I_3[\phi]_p \end{pmatrix} := -\frac{1}{p\cdot u}\int\frac{\mathrm{d}^3\boldsymbol{p}_1}{p_1^0}\int\frac{\mathrm{d}^3\boldsymbol{p}_2}{p_2^0}\int\frac{\mathrm{d}^3\boldsymbol{p}_3}{p_3^0}\,\frac{1}{(2\,\pi)^3}\,P^2\,\frac{1}{2!}\,\sigma(q,\,\Theta) \\
\times\,\delta^4(p+p_1-p_2-p_3)\,\frac{f_{p_1}^{\mathrm{eq}}\,\bar{f}_{p_2}^{\mathrm{eq}}\,\bar{f}_{p_3}^{\mathrm{eq}}}{\bar{f}_p^{\mathrm{eq}}}\begin{pmatrix} \phi_p \\ \phi_{p_1} \\ -\phi_{p_2} \\ -\phi_{p_3} \end{pmatrix}.
\end{aligned}
\tag{14.16}
$$

Let us convert the four contributions to more compact forms. We rewrite $I[\phi]_p$ and $I_1[\phi]_p$ as

$$
I[\phi]_p = -\frac{1}{p\cdot u}\,\frac{1}{\bar{f}_p^{\mathrm{eq}}}\,\phi_p\int\frac{\mathrm{d}^3\boldsymbol{p}_1}{(2\,\pi)^3\,p_1^0}\,f_{p_1}^{\mathrm{eq}}\,K_1(p,\,p_1),
\tag{14.17}
$$

$$
I_1[\phi]_p = -\frac{1}{p\cdot u}\,\frac{1}{\bar{f}_p^{\mathrm{eq}}}\int\frac{\mathrm{d}^3\boldsymbol{p}_1}{(2\,\pi)^3\,p_1^0}\,f_{p_1}^{\mathrm{eq}}\,\phi_{p_1}\,K_1(p,\,p_1),
\tag{14.18}
$$

where we have introduced a kernel function

$$
K_1(p,\,p_1) := \int\frac{\mathrm{d}^3\boldsymbol{p}_2}{p_2^0}\int\frac{\mathrm{d}^3\boldsymbol{p}_3}{p_3^0}\,P^2\,\frac{1}{2!}\,\sigma(q,\,\Theta)\,\delta^4(p+p_1-p_2-p_3)\,\bar{f}_{p_2}^{\mathrm{eq}}\,\bar{f}_{p_3}^{\mathrm{eq}}.
\tag{14.19}
$$

With the definition

$$
k(p) := \frac{1}{T_0^2}\int\frac{\mathrm{d}^3\boldsymbol{p}_1}{(2\,\pi)^3\,p_1^0}\,f_{p_1}^{\mathrm{eq}}\,K_1(p,\,p_1),
\tag{14.20}
$$

$I[\phi]_p$ is written as

$$
I[\phi]_p = -\frac{1}{p\cdot u}\,\frac{1}{\bar{f}_p^{\mathrm{eq}}}\,\phi_p\,k(p)\,T_0^2,
\tag{14.21}
$$

where T_0 with the dimension of temperature is supposed to take a typical value of the system and has been introduced to make $k(p)$ dimensionless; T_0 will be set to T eventually, for notational convenience.

By switching the dummy variables in the definitions of $I_2[\phi]_p$ and $I_3[\phi]_p$ as $(p_1, p_2) \to (p_2, p_1)$ and $(p_1, p_2, p_3) \to (p_2, p_3, p_1)$, respectively, we obtain

$$I_2[\phi]_p = \frac{1}{p \cdot u} \frac{1}{\bar{f}_p^{\text{eq}}} \int \frac{\mathrm{d}^3 \boldsymbol{p_1}}{(2\pi)^3 p_1^0} f_{p_1}^{\text{eq}} \phi_{p_1} \frac{1}{2} K_2(p, p_1), \tag{14.22}$$

$$I_3[\phi]_p = \frac{1}{p \cdot u} \frac{1}{\bar{f}_p^{\text{eq}}} \int \frac{\mathrm{d}^3 \boldsymbol{p_1}}{(2\pi)^3 p_1^0} f_{p_1}^{\text{eq}} \phi_{p_1} \frac{1}{2} K_3(p, p_1), \tag{14.23}$$

where the following kernel functions have been introduced:

$$\begin{aligned} \frac{1}{2} K_2(p, p_1) := \frac{\bar{f}_{p_1}^{\text{eq}}}{f_{p_1}^{\text{eq}}} \int \frac{\mathrm{d}^3 \boldsymbol{p_2}}{p_2^0} \int \frac{\mathrm{d}^3 \boldsymbol{p_3}}{p_3^0} \bar{P}^2 \frac{1}{2!} \sigma(\bar{q}, \bar{\Theta}) \\ \times \delta^4(p + p_2 - p_1 - p_3) f_{p_2}^{\text{eq}} \bar{f}_{p_3}^{\text{eq}}, \end{aligned} \tag{14.24}$$

$$\begin{aligned} \frac{1}{2} K_3(p, p_1) := \frac{\bar{f}_{p_1}^{\text{eq}}}{f_{p_1}^{\text{eq}}} \int \frac{\mathrm{d}^3 \boldsymbol{p_2}}{p_2^0} \int \frac{\mathrm{d}^3 \boldsymbol{p_3}}{p_3^0} \tilde{P}^2 \frac{1}{2!} \sigma(\tilde{q}, \tilde{\Theta}) \\ \times \delta^4(p + p_2 - p_3 - p_1) f_{p_2}^{\text{eq}} \bar{f}_{p_3}^{\text{eq}}, \end{aligned} \tag{14.25}$$

with

$$\bar{P}^2 := (p + p_2)^2, \quad \bar{q}^2 := -(p - p_2)^2, \tag{14.26}$$

$$\cos\bar{\Theta} := -(p - p_2) \cdot (p_1 - p_3)/\bar{q}^2, \tag{14.27}$$

$$\tilde{P}^2 := (p + p_2)^2, \quad \tilde{q}^2 := -(p - p_2)^2, \tag{14.28}$$

$$\cos\tilde{\Theta} := -(p - p_2) \cdot (p_3 - p_1)/\tilde{q}^2. \tag{14.29}$$

Now we show that the following equality holds;

$$I_3[\phi]_p = I_2[\phi]_p. \tag{14.30}$$

Indeed, we first note the following equalities are satisfied on account of their definitions,

$$\tilde{P}^2 = \bar{P}^2, \quad \tilde{q}^2 = \bar{q}^2, \quad \tilde{\Theta} = \pi - \bar{\Theta}. \tag{14.31}$$

Then we have

$$\tilde{P}^2 \sigma(\tilde{q}, \tilde{\Theta}) = \bar{P}^2 \sigma(\bar{q}, \pi - \bar{\Theta}). \tag{14.32}$$

However, for the system composed of identical particles, the following symmetry property of the differential cross section holds

$$\sigma(q,\,\Theta) = \sigma(q,\,\pi - \Theta). \tag{14.33}$$

Thus we have

$$\tilde{P}^2\,\sigma(\tilde{q},\,\tilde{\Theta}) = \bar{P}^2\,\sigma(\bar{q},\,\bar{\Theta}). \tag{14.34}$$

Substituting the above equation into $K_3(p,\ p_1)$ in Eq. (14.25) and comparing it with $K_2(p,\ p_1)$ in Eq. (14.24), we find the equality

$$K_3(p,\ p_1) = K_2(p,\ p_1), \tag{14.35}$$

which implies the equality (14.30).

Collecting the four contributions, we obtain

$$\begin{aligned}\left[\hat{L}\,\phi\right]_p = -T_0^2\,\frac{1}{p\cdot u}\,\frac{1}{\bar{f}_p^{\mathrm{eq}}}\Bigg[&k(p)\,\phi_p \\ &+\frac{1}{T_0^2}\int\frac{\mathrm{d}^3\boldsymbol{p_1}}{(2\,\pi)^3\,p_1^0}\,f_{p_1}^{\mathrm{eq}}\Big(K_1(p,\ p_1)-K_2(p,\ p_1)\Big)\,\phi_{p_1}\Bigg].\end{aligned} \tag{14.36}$$

We remark that $k(p)$, $K_1(p,\ p_1)$, and $K_2(p,\ p_1)$ are Lorentz invariant quantities. This means that $k(p)$ is a function of

$$p\cdot u, \tag{14.37}$$

although $k(p)$ seemingly depends on p^μ and u^μ, separately. We note that $p\cdot p$ and $u\cdot u$ are constants; $p\cdot p = m^2$ and $u\cdot u = 1$. Similarly, $K_1(p,\ p_1)$ and $K_2(p,\ p_1)$ are functions of

$$p\cdot u,\quad p_1\cdot u,\quad p\cdot p_1. \tag{14.38}$$

Thus, we write the kernel functions as

$$k(p) = k(\tau),\quad K_1(p,\,p_1) = K_1(\tau,\,\tau_1,\,\chi),\quad K_2(p,\,p_1) = K_2(\tau,\,\tau_1,\,\chi), \tag{14.39}$$

where the following independent Lorentz invariant quantities have been defined:

$$\tau := \frac{p\cdot u}{T_0}, \tag{14.40}$$

$$\tau_1 := \frac{p_1\cdot u}{T_0}, \tag{14.41}$$

$$\cos\chi := \frac{(p\cdot u)\,(p_1\cdot u) - p\cdot p_1}{\sqrt{(p\cdot u)^2 - m^2}\,\sqrt{(p_1\cdot u)^2 - m^2}}. \tag{14.42}$$

We note that χ in Eq. (14.42) denotes an angle between $\Delta^{\mu\nu} p_\nu$ and $\Delta^{\mu\nu} p_{1\nu}$ and reproduces a natural definition $\cos\chi = \frac{\boldsymbol{p}\cdot\boldsymbol{p}_1}{|\boldsymbol{p}||\boldsymbol{p}_1|}$ in the rest frame by $u^\mu = (1, 0, 0, 0)$, which gives $\Delta^{\mu\nu} = \mathrm{diag}(0, -1, -1, -1)$.

The integral element based on these quantities reads

$$\frac{\mathrm{d}^3 \boldsymbol{p}_1}{(2\pi)^3 p_1^0} = \frac{1}{(2\pi)^3} T_0^2 \sqrt{\tau_1^2 - z^2}\, \mathrm{d}\tau_1 \sin\chi \,\mathrm{d}\chi \,\mathrm{d}\psi, \tag{14.43}$$

with

$$z := \frac{m}{T_0}. \tag{14.44}$$

We note that ψ denotes a polar angle around $\Delta^{\mu\nu} p_\nu$. Using this integral element, we calculate $k(p)$ in Eq. (14.20) to have

$$k(p) = \int \mathrm{d}\mu(\tau_1)\, 2\pi \int_0^\pi \mathrm{d}\chi \,\sin\chi \, K_1(\tau, \tau_1, \chi). \tag{14.45}$$

Here we have introduced another integral measure

$$\int \mathrm{d}\mu(\tau) := \frac{1}{(2\pi)^3} \int_z^\infty \mathrm{d}\tau \, f^{\mathrm{eq}}(\tau) \sqrt{\tau^2 - z^2}, \tag{14.46}$$

with

$$f^{\mathrm{eq}}(\tau) = \frac{1}{\exp\left[\frac{\tau - \mu/T_0}{T/T_0}\right] - a}. \tag{14.47}$$

With the Lorentz invariant quantities, the left-hand side of the linearized transport equations (14.36) reads

$$-\tau \frac{1}{T_0} \bar{f}^{\mathrm{eq}}(\tau) \left[\hat{L}\phi\right]_p = k(\tau)\phi_p + \int \mathrm{d}\mu(\tau_1) \int_0^\pi \mathrm{d}\chi \,\sin\chi \times (K_1(\tau, \tau_1, \chi) - K_2(\tau, \tau_1, \chi)) \int_0^{2\pi} \mathrm{d}\psi\, \phi_{p_1}. \tag{14.48}$$

For $\int_0^{2\pi} \mathrm{d}\psi\, \phi_{p_1}$, we shall here write down the results in three cases which will be used later:

$$\int_0^{2\pi} d\psi = 2\pi, \tag{14.49}$$

$$\int_0^{2\pi} d\psi\, \Delta^{\mu\nu}\, p_{1\nu} = 2\pi\, \Delta^{\mu\nu}\, p_\nu \cos\chi \sqrt{\frac{\tau_1^2 - z^2}{\tau^2 - z^2}}, \tag{14.50}$$

$$\int_0^{2\pi} d\psi\, \Delta^{\mu\nu\rho\sigma}\, p_{1\rho}\, p_{1\sigma} = 2\pi\, \Delta^{\mu\nu\rho\sigma}\, p_\rho\, p_\sigma\, \frac{3\cos^2\chi - 1}{2} \frac{\tau_1^2 - z^2}{\tau^2 - z^2}. \tag{14.51}$$

We note that Eqs. (14.50) and (14.51) can be checked by being multiplied by p_μ and $p_\mu\, p_\nu$, respectively, and using Eqs. (14.40)–(14.42).

14.2.2 Explicit Forms of Kernel Functions

Let us simplify the kernel functions $K_1(p, p_1)$ and $K_2(p, p_1)$ by carrying out the integration with respect to the momenta. To this end, we can choose a frame as follows,

$$\boldsymbol{p} = \boldsymbol{s} = (0,\, 0,\, |\boldsymbol{s}|), \tag{14.52}$$

$$\boldsymbol{p}_1 = -\boldsymbol{s} = -(0,\, 0,\, |\boldsymbol{s}|), \tag{14.53}$$

$$\boldsymbol{u} = (u^1,\, 0,\, u^3), \tag{14.54}$$

without loss of generality. This is the center-of-mass frame where two colliding particles moving along the z axis. In the present frame, $K_1(\tau,\, \tau_1,\, \chi)$ and $K_2(\tau,\, \tau_1,\, \chi)$ are functions of $(|\boldsymbol{s}|,\, u^1,\, u^3)$. To make their Lorentz invariance manifest, we express $(|\boldsymbol{s}|,\, u^1,\, u^3)$ with the Lorentz invariant variables $(\tau,\, \tau_1,\, \cos\chi)$. In this frame, the variables $(\tau,\, \tau_1,\, \cos\chi)$ are expressed as

$$\tau = \sqrt{\frac{|\boldsymbol{s}|^2}{T_0^2} + z^2}\, \sqrt{1 + (u^1)^2 + (u^3)^2} - \frac{|\boldsymbol{s}|}{T_0}\, u^3, \tag{14.55}$$

$$\tau_1 = \sqrt{\frac{|\boldsymbol{s}|^2}{T_0^2} + z^2}\, \sqrt{1 + (u^1)^2 + (u^3)^2} + \frac{|\boldsymbol{s}|}{T_0}\, u^3, \tag{14.56}$$

$$\cos\chi = \frac{\tau\,\tau_1 - \frac{2|\boldsymbol{s}|^2}{T_0^2} - z^2}{\sqrt{\tau^2 - z^2}\sqrt{\tau_1^2 - z^2}}, \tag{14.57}$$

on account of Eqs. (14.40)–(14.42). The relations in turn lead to

$$\frac{|\boldsymbol{s}|}{T_0} = \frac{1}{\sqrt{2}}\left[\tau\,\tau_1 - \sqrt{\tau^2 - z^2}\sqrt{\tau_1^2 - z^2}\,\cos\chi - z^2\right]^{1/2}, \tag{14.58}$$

$$u^1 = \sqrt{\left(\frac{\tau + \tau_1}{P/T_0}\right)^2 - \left(\frac{\tau - \tau_1}{q/T_0}\right)^2 - 1}, \tag{14.59}$$

$$u^3 = -\frac{\tau - \tau_1}{q/T_0}, \tag{14.60}$$

where P and q are given in Eqs. (14.9) and (14.10), respectively, and expressed in terms of $(\tau,\ \tau_1,\ \cos\chi)$ as

$$\frac{P}{T_0} = \sqrt{2}\left[\tau\,\tau_1 - \sqrt{\tau^2 - z^2}\sqrt{\tau_1^2 - z^2}\,\cos\chi + z^2\right]^{1/2}, \tag{14.61}$$

$$\frac{q}{T_0} = \sqrt{2}\left[\tau\,\tau_1 - \sqrt{\tau^2 - z^2}\sqrt{\tau_1^2 - z^2}\,\cos\chi - z^2\right]^{1/2}. \tag{14.62}$$

For convenience we present alternative forms of P and q written by $|\boldsymbol{s}|$,

$$\frac{P}{T_0} = 2\sqrt{\frac{|\boldsymbol{s}|^2}{T_0^2} + z^2}, \qquad \frac{q}{T_0} = 2\,\frac{|\boldsymbol{s}|}{T_0}. \tag{14.63}$$

First, we shall obtain an explicit form of $K_1(\tau,\ \tau_1,\ \chi)$. By carrying out the integration with respect to $\boldsymbol{p_3}$ in the center-of-mass system, we have

$$\begin{aligned} K_1(\tau,\ \tau_1,\ \chi) = &\int \mathrm{d}^3\boldsymbol{p_2}\,\frac{1}{|\boldsymbol{p_2}|^2 + m^2}\,(4\,|\boldsymbol{s}|^2 + 4\,m^2)\,\frac{1}{2!}\,\sigma\left(2\,|\boldsymbol{s}|,\ \cos^{-1}\frac{\boldsymbol{s}\cdot\boldsymbol{p_2}}{|\boldsymbol{s}|^2}\right) \\ &\times\ \delta(2\sqrt{|\boldsymbol{s}|^2 + m^2} - 2\sqrt{|\boldsymbol{p_2}|^2 + m^2}) \\ &\times\ \bar{f}^{\mathrm{eq}}\left(\frac{\sqrt{|\boldsymbol{p_2}|^2 + m^2}\,\sqrt{1 + (u^1)^2 + (u^3)^2} - p_2^1\,u^1 - p_2^3\,u^3}{T_0}\right) \\ &\times\ \bar{f}^{\mathrm{eq}}\left(\frac{\sqrt{|\boldsymbol{p_2}|^2 + m^2}\,\sqrt{1 + (u^1)^2 + (u^3)^2} + p_2^1\,u^1 + p_2^3\,u^3}{T_0}\right), \end{aligned} \tag{14.64}$$

with

$$\bar{f}^{\mathrm{eq}}(\tau) = \frac{\exp\left[\frac{\tau - \mu/T_0}{T/T_0}\right]}{\exp\left[\frac{\tau - \mu/T_0}{T/T_0}\right] - a}. \tag{14.65}$$

In Eq. (14.64), we have used the equality

$$\cos\Theta = \frac{\boldsymbol{s}\cdot\boldsymbol{p_2}}{|\boldsymbol{s}|^2}, \tag{14.66}$$

which is derived from Eq. (14.11).

Next, we carry out the integration with respect to $\boldsymbol{p}_2$. After the integration over $|\boldsymbol{p}_2|$ in the polar coordinates $\boldsymbol{p}_2 = |\boldsymbol{p}_2|\,(\sin\theta\,\cos\psi,\ \sin\theta\,\sin\psi,\ \cos\theta)$, we have

$$
\begin{aligned}
&K_1(\tau,\ \tau_1,\ \chi)\\
&= 2\,|\boldsymbol{s}|\,\sqrt{|\boldsymbol{s}|^2+m^2}\int_0^{\pi}\mathrm{d}\theta\int_0^{2\pi}\mathrm{d}\psi\ \sin\theta\,\frac{1}{2!}\,\sigma(2\,|\boldsymbol{s}|,\ \theta)\\
&\quad\times\ \bar{f}^{\mathrm{eq}}\Big(\frac{\sqrt{|\boldsymbol{s}|^2+m^2}\,\sqrt{1+(u^1)^2+(u^3)^2}-|\boldsymbol{s}|\,\sin\theta\,\cos\psi\,u^1-|\boldsymbol{s}|\,\cos\theta\,u^3}{T_0}\Big)\\
&\quad\times\ \bar{f}^{\mathrm{eq}}\Big(\frac{\sqrt{|\boldsymbol{s}|^2+m^2}\,\sqrt{1+(u^1)^2+(u^3)^2}+|\boldsymbol{s}|\,\sin\theta\,\cos\psi\,u^1+|\boldsymbol{s}|\,\cos\theta\,u^3}{T_0}\Big),
\end{aligned}
\tag{14.67}
$$

Using $(|\boldsymbol{s}|,\ u^1,\ u^3)$ given by Eqs. (14.58)–(14.60), we reduce the above equation to

$$
\begin{aligned}
K_1(\tau,\ \tau_1,\ \chi) &= \pi\,\frac{q\,P}{T_0^2}\int_0^{\pi}\mathrm{d}\theta\ \sin\theta\left[\frac{1}{2!}\,\sigma(q,\ \theta)\,T_0^2\right]\\
&\quad\times\int_0^{2\pi}\frac{\mathrm{d}\psi}{2\pi}\ \bar{f}^{\mathrm{eq}}\Big(\frac{1}{2}\,(\tau+\tau_1)+X\Big)\,\bar{f}^{\mathrm{eq}}\Big(\frac{1}{2}\,(\tau+\tau_1)-X\Big).
\end{aligned}
\tag{14.68}
$$

where X is defined by

$$
X := \frac{1}{2}\,(\tau-\tau_1)\,\cos\theta-\frac{1}{2}\,\frac{q}{T_0}\,\sqrt{\Big(\frac{\tau+\tau_1}{P/T_0}\Big)^2-\Big(\frac{\tau-\tau_1}{q/T_0}\Big)^2-1}\ \sin\theta\,\cos\psi.
\tag{14.69}
$$

Since X is a function of $\cos\psi$, we can simplify the ψ integration in Eq. (14.68) as

$$
\begin{aligned}
K_1(\tau,\ \tau_1,\ \chi) &= \pi\,\frac{q\,P}{T_0^2}\int_0^{\pi}\mathrm{d}\theta\ \sin\theta\left[\frac{1}{2!}\,\sigma(q,\ \theta)\,T_0^2\right]\\
&\quad\times\int_0^{\pi}\frac{\mathrm{d}\psi}{\pi}\ \bar{f}^{\mathrm{eq}}\Big(\frac{1}{2}\,(\tau+\tau_1)+X\Big)\,\bar{f}^{\mathrm{eq}}\Big(\frac{1}{2}\,(\tau+\tau_1)-X\Big).
\end{aligned}
\tag{14.70}
$$

Next, we calculate $K_2(\tau,\ \tau_1,\ \chi)$ in the center-of-mass frame. The integration over $\boldsymbol{p}_3$ results in

$$
\begin{aligned}
&\frac{1}{2}\,K_2(\tau,\ \tau_1,\ \chi)\\
&= \frac{\bar{f}^{\mathrm{eq}}_{p_1}}{f^{\mathrm{eq}}_{p_1}}\int\mathrm{d}^3\boldsymbol{p}_2\,\frac{1}{\sqrt{|\boldsymbol{p}_2|^2+m^2}}\,\frac{1}{\sqrt{|\boldsymbol{p}_2+2\,\boldsymbol{s}|^2+m^2}}\,\bar{P}^2\,\frac{1}{2!}\,\sigma(\bar{q},\ \bar{\Theta})\\
&\quad\times\ \delta\big(\sqrt{|\boldsymbol{p}_2|^2+m^2}-\sqrt{|\boldsymbol{p}_2+2\,\boldsymbol{s}|^2+m^2}\big)\\
&\quad\times\ f^{\mathrm{eq}}\Big(\frac{\sqrt{|\boldsymbol{p}_2|^2+m^2}\,\sqrt{1+(u^1)^2+(u^3)^2}-p_2^1\,u^1-p_2^3\,u^3}{T_0}\Big)\\
&\quad\times\ \bar{f}^{\mathrm{eq}}\Big(\frac{\sqrt{|\boldsymbol{p}_2+2\,\boldsymbol{s}|^2+m^2}\,\sqrt{1+(u^1)^2+(u^3)^2}-(p_2^1+2\,q^1)\,u^1-(p_2^3+2\,q^3)\,u^3}{T_0}\Big).
\end{aligned}
\tag{14.71}
$$

We change p_2 to $p_2' := p_2 + s$ to obtain

$$
\begin{aligned}
&\frac{1}{2} K_2(\tau, \tau_1, \chi) \\
&= \frac{\bar{f}^{\rm eq}_{p_1}}{f^{\rm eq}_{p_1}} \int \mathrm{d}^3 p_2' \frac{1}{|p_2'|^2 + |s|^2 + m^2} \bar{P}^2 \frac{1}{2!} \sigma(\bar{q}, \bar{\Theta}) \\
&\times \frac{1}{2} \sqrt{|p_2'|^2 + |s|^2 + m^2}\, \delta(p_2' \cdot s) \\
&\times f^{\rm eq}\Big(\frac{\sqrt{|p_2'|^2 + |s|^2 + m^2}\sqrt{1 + (u^1)^2 + (u^3)^2} - (p_2'^1 - q^1) u^1 - (p_2'^3 - q^3) u^3}{T_0}\Big) \\
&\times \bar{f}^{\rm eq}\Big(\frac{\sqrt{|p_2'|^2 + |s|^2 + m^2}\sqrt{1 + (u^1)^2 + (u^3)^2} - (p_2'^1 + q^1) u^1 - (p_2'^3 + q^3) u^3}{T_0}\Big),
\end{aligned}
\tag{14.72}
$$

where $|p_2' \pm s|^2 = |p_2'|^2 + |s|^2$ holds due to $\delta(p_2' \cdot s)$. We note that $\bar{q}$, $\bar{P}^2$, and $\bar{\Theta}$ depend on $|p_2'|$, not p_2', under the condition of $p_2' \cdot s = 0$. In fact, the following relations hold from the definitions,

$$
\bar{q} = \frac{q}{\sin\frac{\bar{\Theta}}{2}}, \tag{14.73}
$$

$$
\bar{P}^2 = 4\,m^2 + \frac{q^2}{\sin^2\frac{\bar{\Theta}}{2}}, \tag{14.74}
$$

$$
|p_2'| = \frac{q}{P} \cot\frac{\bar{\Theta}}{2} \sqrt{4\,m^2 + \frac{q^2}{\sin^2\frac{\bar{\Theta}}{2}}}, \tag{14.75}
$$

where P and q are given by Eqs. (14.61) and (14.62). Since it is difficult to determine the dependence of $\bar{\Theta}$ on $|p_2'|$ by solving Eq. (14.75), we shall treat $\bar{\Theta}$ as an integral variable instead of $|p_2'|$. Then, the integral measure is rewritten as

$$
\frac{|p_2'|\,\mathrm{d}|p_2'|}{\sqrt{|p_2'|^2 + |s|^2 + m^2}} = \frac{q^2}{2\,P} \frac{1}{\sin^4\frac{\bar{\Theta}}{2}} \sin\bar{\Theta}\,\mathrm{d}\bar{\Theta}. \tag{14.76}
$$

Introducing polar coordinates $p_2' = |p_2'|\,(\sin\theta\,\cos\psi,\ \sin\theta\,\sin\psi,\ \cos\theta)$, carrying out the integration with θ with the use of the formula $\delta(p_2' \cdot s) = \delta(\cos\theta)/|p_2'||s|$, and finally rewriting $\bar{\Theta}$ as θ, we have

$$
\begin{aligned}
K_2(\tau, \tau_1, \chi) = & \frac{\bar{f}^{\rm eq}(\tau_1)}{f^{\rm eq}(\tau_1)} 2\,\pi\, \frac{q}{P} \int_0^{\pi} \mathrm{d}\theta\, \sin\theta\, \frac{1}{\sin^4\frac{\theta}{2}} \left[4\,z^2 + \frac{(q/T_0)^2}{\sin^2\frac{\theta}{2}}\right] \\
& \times \left[\frac{1}{2!}\sigma\Big(\frac{q}{\sin\frac{\theta}{2}}, \theta\Big) T_0^2\right] \int_0^{2\pi} \frac{\mathrm{d}\psi}{2\pi} f^{\rm eq}(\tau_1 + Y)\, \bar{f}^{\rm eq}(\tau + Y),
\end{aligned}
\tag{14.77}
$$

with

$$
\begin{aligned}
Y := (\tau + \tau_1)\,\frac{q^2}{P^2}\,\cot^2\frac{\theta}{2} \\
- \sqrt{\left(\frac{\tau + \tau_1}{P/T_0}\right)^2 - \left(\frac{\tau - \tau_1}{q/T_0}\right)^2 - 1}\;\frac{q}{P}\sqrt{4\,z^2 + \frac{(q/T_0)^2}{\sin^2\frac{\theta}{2}}}\;\cot\frac{\theta}{2}\;\cos\psi.
\end{aligned}
\tag{14.78}
$$

We have used Eqs. (14.73)–(14.75) and converted $(|s|,\ u^1,\ u^3)$ into $(\tau,\ \tau_1,\ \cos\chi)$ with the use of Eqs. (14.58)–(14.60). The ψ integration in Eq. (14.77) can be simplified as

$$
\begin{aligned}
K_2(\tau,\ \tau_1,\ \chi) = \frac{\bar{f}^{\mathrm{eq}}(\tau_1)}{f^{\mathrm{eq}}(\tau_1)}\,2\pi\,\frac{q}{P}\int_0^{\pi}\mathrm{d}\theta\,\sin\theta\,\frac{1}{\sin^4\frac{\theta}{2}}\left[4\,z^2 + \frac{(q/T_0)^2}{\sin^2\frac{\theta}{2}}\right] \\
\times\left[\frac{1}{2!}\,\sigma\!\left(\frac{q}{\sin\frac{\theta}{2}},\ \theta\right)T_0^2\right]\int_0^{\pi}\frac{\mathrm{d}\psi}{\pi}\,f^{\mathrm{eq}}(\tau_1 + Y)\,\bar{f}^{\mathrm{eq}}(\tau + Y).
\end{aligned}
\tag{14.79}
$$

14.2.3 Linearized Transport Equations as Integral Equations

Next we substitute ϕ in the left-hand side of (14.48) with $\phi_{\Pi p}$, ϕ^{μ}_{Jp}, and $\phi^{\mu\nu}_{\pi p}$ to obtain the linearized transport equations given in Eq. (14.1). Without loss of generality, we can parametrize them as [67]

$$
\phi_{\Pi p} = \frac{1}{\sqrt{k(\tau)}}\,A(\tau), \tag{14.80}
$$

$$
\phi^{\mu}_{Jp} = \frac{1}{\sqrt{k(\tau)}}\,B(\tau)\,\frac{1}{\sqrt{\tau^2 - z^2}}\,\Delta^{\mu\nu}\,p_\nu/T_0, \tag{14.81}
$$

$$
\phi^{\mu\nu}_{\pi p} = \frac{1}{\sqrt{k(\tau)}}\,C(\tau)\,\frac{1}{\tau^2 - z^2}\,\Delta^{\mu\nu\rho\sigma}\,p_\rho\,p_\sigma/T_0^2. \tag{14.82}
$$

We note that $A(\tau)$, $B(\tau)$, and $C(\tau)$ are the variables to be determined and their coefficients have been introduced to avoid possible unessential computations .

Substituting the above expressions of $\phi_{\Pi p}$, ϕ^{μ}_{Jp}, and $\phi^{\mu\nu}_{\pi p}$ into ϕ_p in Eq. (14.48) and using the equalities (14.49)–(14.51), we have

$$
-\,\bar{f}^{\mathrm{eq}}(\tau)\,\frac{\tau}{T_0\sqrt{k(\tau)}}\left[\hat{L}\,\phi_\Pi\right]_p = A(\tau) + \int\mathrm{d}\mu(\tau_1)\,L(\tau,\ \tau_1)\,A(\tau_1), \tag{14.83}
$$

$$
-\,\bar{f}^{\mathrm{eq}}(\tau)\,\frac{\tau\sqrt{\tau^2 - z^2}}{T_0\sqrt{k(\tau)}}\left[\hat{L}\,\phi^{\mu}_J\right]_p = \left[B(\tau) + \int\mathrm{d}\mu(\tau_1)\,M(\tau,\ \tau_1)\,B(\tau_1)\right]\Delta^{\mu\nu}\,p_\nu/T_0, \tag{14.84}
$$

$$- \bar{f}^{\text{eq}}(\tau) \frac{\tau\,(\tau^2 - z^2)}{T_0 \sqrt{k(\tau)}} \left[\hat{L}\, \phi_\pi^{\mu\nu}\right]_p = \left[C(\tau) + \int \mathrm{d}\mu(\tau_1)\, N(\tau, \tau_1)\, C(\tau_1) \right] \times \Delta^{\mu\nu\rho\sigma}\, p_\rho\, p_\sigma / T_0^2, \tag{14.85}$$

with

$$L(\tau, \tau_1) := \frac{2\pi}{\sqrt{k(\tau)\,k(\tau_1)}} \int_0^\pi \mathrm{d}\chi\ \sin\chi\, (K_1(\tau, \tau_1, \chi) - K_2(\tau, \tau_1, \chi)), \tag{14.86}$$

$$M(\tau, \tau_1) := \frac{2\pi}{\sqrt{k(\tau)\,k(\tau_1)}} \int_0^\pi \mathrm{d}\chi\ \sin\chi\ \cos\chi\, (K_1(\tau, \tau_1, \chi) - K_2(\tau, \tau_1, \chi)), \tag{14.87}$$

$$N(\tau, \tau_1) := \frac{2\pi}{\sqrt{k(\tau)\,k(\tau_1)}} \int_0^\pi \mathrm{d}\chi\ \sin\chi\ \frac{3\cos^2\chi - 1}{2}\, (K_1(\tau, \tau_1, \chi) - K_2(\tau, \tau_1, \chi)). \tag{14.88}$$

We prepare the right-hand side of the linearized transport equations (14.5) as follows:

$$\hat{\Pi}_p = -\frac{T_0\sqrt{k(\tau)}}{\tau} a(\tau), \tag{14.89}$$

$$\hat{J}_p^\mu = -\frac{T_0\sqrt{k(\tau)}}{\tau} b(\tau)\, \frac{1}{\sqrt{\tau^2 - z^2}}\, \Delta^{\mu\nu}\, p_\nu / T_0, \tag{14.90}$$

$$\hat{\pi}_p^{\mu\nu} = -\frac{T_0\sqrt{k(\tau)}}{\tau} c(\tau)\, \frac{1}{\tau^2 - z^2}\, \Delta^{\mu\nu\rho\sigma}\, p_\rho\, p_\sigma / T_0^2, \tag{14.91}$$

with

$$a(\tau) := -\frac{1}{\sqrt{k(\tau)}} \frac{z^2}{3} \left[-\frac{a_2 a_0 - a_1^2}{a_3 a_1 - a_2^2} \tau^2 + \frac{a_3 a_0 - a_2 a_1}{a_3 a_1 - a_2^2} \tau - 1 \right], \tag{14.92}$$

$$b(\tau) := -\frac{1}{\sqrt{k(\tau)}} \sqrt{\tau^2 - z^2} \left[-\tau + \frac{z^2 a_1 - a_3}{z^2 a_0 - a_2} \right], \tag{14.93}$$

$$c(\tau) := -\frac{1}{\sqrt{k(\tau)}} (\tau^2 - z^2). \tag{14.94}$$

Here, we have defined the parameters $a_{0,1,2,3}$ as

$$a_\ell = 4\pi \int \mathrm{d}\mu(\tau)\, \bar{f}^{\mathrm{eq}}(\tau)\, \tau^\ell, \quad (\ell = 0,\, 1,\, 2,\, 3). \tag{14.95}$$

Combining Eqs. (14.83)–(14.85) and (14.89)–(14.91), we arrive at

$$A(\tau) + \int \mathrm{d}\mu(\tau_1)\, L(\tau,\, \tau_1)\, A(\tau_1) = \bar{f}^{\mathrm{eq}}(\tau)\, a(\tau), \tag{14.96}$$

$$B(\tau) + \int \mathrm{d}\mu(\tau_1)\, M(\tau,\, \tau_1)\, B(\tau_1) = \bar{f}^{\mathrm{eq}}(\tau)\, b(\tau), \tag{14.97}$$

$$C(\tau) + \int \mathrm{d}\mu(\tau_1)\, N(\tau,\, \tau_1)\, C(\tau_1) = \bar{f}^{\mathrm{eq}}(\tau)\, c(\tau). \tag{14.98}$$

It is worth emphasizing that the task to obtain $A(\tau)$, $B(\tau)$, and $C(\tau)$ is nicely reduced to separately solving these simple one-dimensional integral equations. It is also to be noted that the integral equations (14.96)–(14.98) reproduce those of Ref. [67] in the classical limit with $a \to 0$ in f_p^{eq} and the setting of $T_0 = T$.

We now show that the transport coefficients and relaxation times can be expressed in compact forms in terms of $a(\tau)$, $b(\tau)$, $c(\tau)$, $A(\tau)$, $B(\tau)$, and $C(\tau)$. Noticing that the inner product $\langle \varphi\,,\, \psi \rangle = \int \mathrm{d}p\, (p \cdot u)\, f_p^{\mathrm{eq}}\, \bar{f}_p^{\mathrm{eq}}\, \varphi_p\, \psi_p$ is written as

$$\langle \varphi\,,\, \psi \rangle = 4\pi\, T_0^3 \int \mathrm{d}\mu(\tau)\, \tau\, \bar{f}^{\mathrm{eq}}(\tau)\, \varphi(\tau)\, \psi(\tau), \tag{14.99}$$

for arbitrary Lorentz invariant vectors $\varphi_p = \varphi(\tau)$ and $\psi_p = \psi(\tau)$, we can convert Eqs. (14.2)–(14.4) into

$$T\, \zeta^{\mathrm{RG}} / T_0^4 = 4\pi \int \mathrm{d}\mu(\tau)\, \bar{f}^{\mathrm{eq}}(\tau)\, a(\tau)\, A(\tau), \tag{14.100}$$

$$3\, T^2\, \lambda^{\mathrm{RG}} / T_0^4 = 4\pi \int \mathrm{d}\mu(\tau)\, \bar{f}^{\mathrm{eq}}(\tau)\, b(\tau)\, B(\tau), \tag{14.101}$$

$$15\, T\, \eta^{\mathrm{RG}} / T_0^4 = 4\pi \int \mathrm{d}\mu(\tau)\, \bar{f}^{\mathrm{eq}}(\tau)\, c(\tau)\, C(\tau), \tag{14.102}$$

$$\tau_\Pi^{\mathrm{RG}}\, T_0 = \frac{\int \mathrm{d}\mu(\tau)\, \tau\, \bar{f}^{\mathrm{eq}}(\tau)\, A(\tau)\, A(\tau)/k(\tau)}{\int \mathrm{d}\mu(\tau)\, \bar{f}^{\mathrm{eq}}(\tau)\, a(\tau)\, A(\tau)}, \tag{14.103}$$

$$\tau_J^{\mathrm{RG}}\, T_0 = \frac{\int \mathrm{d}\mu(\tau)\, \tau\, \bar{f}^{\mathrm{eq}}(\tau)\, B(\tau)\, B(\tau)/k(\tau)}{\int \mathrm{d}\mu(\tau)\, \bar{f}^{\mathrm{eq}}(\tau)\, b(\tau)\, B(\tau)}, \tag{14.104}$$

$$\tau_\pi^{\mathrm{RG}}\, T_0 = \frac{\int \mathrm{d}\mu(\tau)\, \tau\, \bar{f}^{\mathrm{eq}}(\tau) C(\tau)\, C(\tau)/k(\tau)}{\int \mathrm{d}\mu(\tau)\, \bar{f}^{\mathrm{eq}}(\tau)\, c(\tau)\, C(\tau)}. \tag{14.105}$$

We note that all the transport coefficients and relaxation times are represented by the τ integrations of bilinear forms with $a(\tau)$, $b(\tau)$, $c(\tau)$, $A(\tau)$, $B(\tau)$, and $C(\tau)$.

14.2.4 Direct Matrix-Inversion Method Based on Discretization

Since Eqs. (14.96)–(14.98) are difficult to solve in an analytical way for an arbitrary interaction, one has often recourse to approximations [67]. However, a natural and direct way to numerically solve the integral equations (14.96)–(14.98) is the discretization of Eqs. (14.96)–(14.98), which reduces $A(\tau)$, $B(\tau)$, and $C(\tau)$ into finite degrees of freedom. This discretization approach is free from any ansatz, but might demand possibly heavy numerical work because the numerical convergence of results for the discretization width must be carefully checked.

Since the integral measure $\mathrm{d}\mu(\tau)$ contains $f^{\mathrm{eq}}(\tau)$ which takes the asymptotic form $\mathrm{e}^{-\tau/(T/T_0)}$ at $\tau \to \infty$, we can improve the convergence of the discretization by adopting the **double exponential formula** [229]. In this formula, we first change the integration valuable from τ to t by

$$\tau = z + \frac{T}{T_0}\,\mathrm{e}^{t-\mathrm{e}^{-t}} =: \tau(t), \tag{14.106}$$

which leads to

$$\int \mathrm{d}\mu(\tau)\, F(\tau) = \frac{1}{(2\pi)^3}\frac{T}{T_0}\int_{-\infty}^{\infty} \mathrm{d}t\, \mathrm{e}^{t-\mathrm{e}^{-t}}(1+\mathrm{e}^{-t})\, f^{\mathrm{eq}}(\tau(t))\sqrt{\tau^2(t)-z^2}\, F(\tau(t)), \tag{14.107}$$

with $F(\tau)$ being an arbitrary function. One finds that a factor of the integrand in Eq. (14.107) has the following asymptotic forms,

$$\mathrm{e}^{t-\mathrm{e}^{-t}}(1+\mathrm{e}^{-t})\, f^{\mathrm{eq}}(\tau(t)) \propto \begin{cases} \mathrm{e}^{-\mathrm{e}^{t}}, & t \to \infty, \\ \mathrm{e}^{-\mathrm{e}^{-t}}, & t \to -\infty, \end{cases} \tag{14.108}$$

which converges to zero as $|t| \to \infty$ with *double-exponential forms.* It implies that the upper and lower bounds of the integration may be replaced with finite cutoff parameters t^+ and $-t^-$ with t^+, $t^- > 0$, respectively, without serious numerical errors.

With the use of the rectangular formulae, we divide t between t^+ and $-t^-$ into $N+1$ pieces as

$$t_n = -t^- + \frac{t^+ + t^-}{N}\, n = -t^- + \Delta t\, n, \tag{14.109}$$

where

$$\Delta t := \frac{t^+ + t^-}{N} \tag{14.110}$$

is supposed to be small enough and n is a non-negative integer,

$$n = 0,\ 1,\ \ldots,\ N. \tag{14.111}$$

Accordingly, the integration given by Eq. (14.107) is expressed as

$$\int d\mu(\tau)\, F(\tau) = \sum_{n=0}^{N} \Delta\mu_n\, F(\tau(t_n)), \tag{14.112}$$

where

$$\Delta\mu_n := \frac{\Delta t}{(2\,\pi)^3}\,\frac{T}{T_0}\,\mathrm{e}^{t_n - \mathrm{e}^{-t_n}}\,(1 + \mathrm{e}^{-t_n})\, f_n^{\mathrm{eq}}\,\sqrt{\tau_n^2 - z^2}, \tag{14.113}$$

with

$$\tau_n := \tau(t_n), \quad f_n^{\mathrm{eq}} := f^{\mathrm{eq}}(\tau_n). \tag{14.114}$$

Similarly, the other functions introduced in the previous sections are also discretized as

$$\bar{f}_n^{\mathrm{eq}} := \bar{f}^{\mathrm{eq}}(\tau_n), \quad k_n := k(\tau_n), \tag{14.115}$$

$$\begin{aligned} L_{mn} &:= \sqrt{\Delta\mu_m}\, L(\tau_m,\, \tau_n)\, \sqrt{\Delta\mu_n}, \quad M_{mn} := \sqrt{\Delta\mu_m}\, M(\tau_m,\, \tau_n)\, \sqrt{\Delta\mu_n}, \\ N_{mn} &:= \sqrt{\Delta\mu_m}\, N(\tau_m,\, \tau_n)\, \sqrt{\Delta\mu_n}, \end{aligned} \tag{14.116}$$

$$a_n := \sqrt{\Delta\mu_n}\, a(\tau_n), \quad b_n := \sqrt{\Delta\mu_n}\, b(\tau_n), \quad c_n := \sqrt{\Delta\mu_n}\, c(\tau_n), \tag{14.117}$$

$$A_n := \sqrt{\Delta\mu_n}\, A(\tau_n), \quad B_n := \sqrt{\Delta\mu_n}\, B(\tau_n), \quad C_n := \sqrt{\Delta\mu_n}\, C(\tau_n). \tag{14.118}$$

Then the one-dimensional integral equations (14.96)–(14.98) are converted to linear equations with finite dimensions as

$$\sum_{n=0}^{N} (\delta_{mn} + L_{mn})\, A_n = f_m^{\mathrm{eq}}\, a_m, \tag{14.119}$$

$$\sum_{n=0}^{N} (\delta_{mn} + M_{mn})\, B_n = f_m^{\mathrm{eq}}\, b_m, \tag{14.120}$$

$$\sum_{n=0}^{N} (\delta_{mn} + N_{mn})\, C_n = f_m^{\mathrm{eq}}\, c_m, \tag{14.121}$$

respectively. We note that these equations can be solved by numerical methods, e.g., the LU decomposition method [218].

A remark is in order here: In constructing the linear equations (14.119)–(14.121), the integrations with respect to the angles χ, θ, and ψ for k_n, L_{mn}, M_{mn}, and N_{mn} are performed numerically by applying the double exponential formula [229] as well as the τ and τ_1 integrations. For that, we convert χ, θ, and ψ into t_χ, t_θ, and t_ψ, respectively, as follows:

$$\cos\chi = \tanh\left(\frac{\pi}{2}\sinh t_\chi\right), \tag{14.122}$$

$$\theta = \frac{\pi}{2}\left[\tanh\left(\frac{\pi}{2}\sinh t_\theta\right) + 1\right], \quad \psi = \frac{\pi}{2}\left[\tanh\left(\frac{\pi}{2}\sinh t_\psi\right) + 1\right]. \tag{14.123}$$

Then we use the rectangular formulae to divide t_χ, t_θ, and t_ψ into N_χ, N_θ, and N_ψ pieces, respectively, as

$$t_{\chi n} = -t_\chi^- + \frac{t_\chi^+ + t_\chi^-}{N_\chi}\, n, \quad n = 0, 1, \ldots, N_\chi, \tag{14.124}$$

$$t_{\theta n} = -t_\theta^- + \frac{t_\theta^+ + t_\theta^-}{N_\theta}\, n, \quad n = 0, 1, \ldots, N_\theta, \tag{14.125}$$

$$t_{\psi n} = -t_\psi^- + \frac{t_\psi^+ + t_\psi^-}{N_\psi}\, n, \quad n = 0, 1, \ldots, N_\psi, \tag{14.126}$$

where $t_\chi^\pm$, $t_\theta^\pm$, and $t_\psi^\pm$ denote cutoff parameters for the integrations.

We shall see in Sect. 14.3.1 that the following parameter values of N, N_χ, N_θ, N_ψ, $t^\pm$, $t_\chi^\pm$, $t_\theta^\pm$, and $t_\psi^\pm$ are already sufficient for the convergence of the numerical results:

$$t^+ = 6,\; t^- = t_\chi^\pm = t_\theta^\pm = t_\psi^\pm = 3,\; N = 200, \text{ and } N_\chi = N_\theta = N_\psi = 100. \tag{14.127}$$

Using the results for A_n, B_n, and C_n thus obtained, the discretized versions of the transport coefficients (12.105)–(12.107) and relaxation times (12.108)–(12.110) derived with the use of the RG method in Chap. 12 are expressed as follows:

$$T\,\zeta^{\mathrm{RG}}/T_0^4 = 4\,\pi \sum_{n=0}^{N} \bar{f}_n^{\mathrm{eq}}\, a_n\, A_n, \tag{14.128}$$

$$3\,T^2\,\lambda^{\mathrm{RG}}/T_0^4 = 4\,\pi \sum_{n=0}^{N} \bar{f}_n^{\mathrm{eq}}\, b_n\, B_n, \tag{14.129}$$

$$15\,T\,\eta^{\mathrm{RG}}/T_0^4 = 4\,\pi \sum_{n=0}^{N} \bar{f}_n^{\mathrm{eq}}\, c_n\, C_n, \tag{14.130}$$

$$\tau_\Pi^{\mathrm{RG}}\,T_0 = \frac{\sum_{n=0}^{N} \bar{f}_n^{\mathrm{eq}}\, \tau_n\, A_n\, A_n/k_n}{\sum_{n=0}^{N} \bar{f}_n^{\mathrm{eq}}\, a_n\, A_n}, \tag{14.131}$$

$$\tau_J^{\mathrm{RG}}\,T_0 = \frac{\sum_{n=0}^{N} \bar{f}_n^{\mathrm{eq}}\, \tau_n\, B_n\, B_n/k_n}{\sum_{n=0}^{N} \bar{f}_n^{\mathrm{eq}}\, b_n\, B_n}, \tag{14.132}$$

$$\tau_\pi^{\mathrm{RG}}\,T_0 = \frac{\sum_{n=0}^{N} \bar{f}_n^{\mathrm{eq}}\, \tau_n\, C_n\, C_n/k_n}{\sum_{n=0}^{N} \bar{f}_n^{\mathrm{eq}}\, c_n\, C_n}. \tag{14.133}$$

14.3 Numerical Demonstration: Transport Coefficients and Relaxation Times of Physical Systems

In this section, after demonstrating the accuracy and efficiency of the numerical scheme developed in Sect. 14.2, we show the numerical results of the transport coefficients and relaxation times using the microscopic representations derived by the RG method as well as the Israel–Stewart fourteen-moment method [67] for a comparison. We set $T_0 = T$ in this section.

14.3.1 Accuracy and Efficiency of the Numerical Method: Discretization Errors and Convergence

We first check and demonstrate the accuracy and efficiency of our numerical method presented in the last subsection. In the present approach based on the Boltzmann equation, the model is solely characterized by the cross section. We take up the following three models;

(A) the classical Boltzmann gas ($a = 0$) with the constant differential cross section

$$\sigma(q,\,\Theta) = \frac{\sigma_{\mathrm{tot}}}{4\,\pi}. \tag{14.134}$$

(B) the Fermi gas ($a = -1$) with a differential cross section induced by the Yukawa interaction, and

(C) the bosonic gas ($a = 1$) with a differential cross section based on the chiral Lagrangian [230, 231].

For the models (B) and (C), we present the explicit forms of the cross sections calculated in the tree approximation in the quantum field theory for self-containedness.

In the quantum field theory, the cross section written as

$$\sigma(q, \Theta) = \frac{|\mathcal{M}|^2}{64\,\pi^2\,(q^2 + 4\,m^2)}, \tag{14.135}$$

where $\mathcal{M}$ denotes the Lorentz-invariant scattering amplitude.

In the model (B), the Lagrangian density is given by

$$\mathcal{L} = \bar{\psi}(i\gamma^\mu\partial_\mu - m)\psi - G\bar{\psi}\psi\phi + \frac{1}{2}\left[(\partial^\mu\phi)(\partial_\mu\phi) - M^2\phi^2\right], \tag{14.136}$$

where the fermion and boson fields are denoted by ψ and ϕ, respectively. For (14.136), the absolute square of the scattering amplitudes $|\mathcal{M}|^2$ in the lowest order of the perturbative expansion is given by

$$|\mathcal{M}|^2 = G^2\left[\frac{(4\,m^2 - t)^2}{(M^2 - t)^2} + \frac{(4\,m^2 - u)^2}{(M^2 - u)^2} + \frac{(4\,m^2 - s)^2 - (4\,m^2 - t)^2 - (4\,m^2 - u)^2}{2\,(M^2 - t)\,(M^2 - u)}\right], \tag{14.137}$$

where s, t, and u the Mandelstum variables defined by

$$s := (p + p_1)^2 = (p_2 + p_3)^2 = q^2 + 4\,m^2, \tag{14.138}$$

$$t := (p - p_2)^2 = (p_1 - p_3)^2 = -q^2\,\sin^2\frac{\Theta}{2}, \tag{14.139}$$

$$u := (p - p_3)^2 = (p_1 - p_2)^2 = -q^2\,\cos^2\frac{\Theta}{2}. \tag{14.140}$$

Here we note that the averaging of the initial spins and summing up of the final spins of the interacting two particles have been performed in Eq. (14.137).

On the other hand, the model (C) gives for the same spin and isospin averaged scattering amplitude squared

$$|\mathcal{M}|^2 = \frac{1}{9\,f_\pi^4}\left[21\,m^4 + 9\,s^2 - 24\,m^2\,s + 3\,(t - u)^2\right], \tag{14.141}$$

where f_π denotes the pion decay constant.

In the following, we exclusively discuss the case with a vanishing chemical potential ($\mu = 0$) for simplicity, for which the transport coefficients and relaxation times depend only on the ratio $z = m/T$.

The main ingredients in the numerical method developed in Sect. 14.2 are the discretization of the integral equations and the use of the double exponential formula for the numerical integration. Here we examine possible discretization errors and the convergence of the numerical integrations by the double exponential formula.

To this end, we check the numerical convergence of the results by varying the numbers of the mesh points, N, N_χ, N_θ, and N_ψ, for the integrals by τ, τ_1, χ, θ, and ψ, respectively; we call $N = 200$ and $N_\chi = N_\theta = N_\psi = 100$ the respective standard values.

In Fig. 14.1, we show the transport coefficients and relaxation times for the models (A), (B), and (C) at $z = 0.01$ and $\mu = 0$ obtained by the discretization method as a function of the total number of the mesh points,

$$N_{\text{mp}} := N^2 \times N_\chi \times N_\theta \times N_\psi. \tag{14.142}$$

From the figure, one finds that all the quantities quickly reach constants as N_{mp} becomes large, and show a good convergence already at the standard number of the mesh points, $N_{\text{mp}} = 200 \times 200 \times 100 \times 100 \times 100 = 4 \times 10^{10}$. Thus, we have confirmed the excellent convergence properties of our numerical method.

Next let us see how efficient our numerical method is by comparing the results of the transport coefficients and relaxation times with those obtained in other numerical scheme. Figure 14.2 shows the numerical results with the use of the rectangular formulae in the calculations of τ, τ_1, χ, θ, and ψ with N_{mp} being varied. One sees that the use of the rectangular formulae does not show convergent behavior at all, except for the shear viscosity, even for $N_{\text{mp}} = 4 \times 10^{10}$ that is far large enough for giving the convergent results with the use of the double exponential formulae. Conversely speaking, one finds that our numerical scheme with the use of the double exponential formulae is not only an accurate but also far more efficient numerical method than conventional numerical methods.

A clear check of the accuracy of the numerical method is provided when the analytic formula is available. It is known [67] that the Israel–Stewart fourteen-moment method gives analytic expressions of the transport coefficients ζ^{IS}, λ^{IS}, and η^{IS} for a classical system ($a = 0$ in $f^{\text{eq}}(\tau)$) with a constant differential cross section (14.134). Here, as an independent check of the numerical method, we dare to compute the transport coefficients derived with the Israel–Stewart fourteen-moment method for the constant cross section by our numerical scheme[2] and compare with the numerical values given by the analytic formula.[3]

The black dots in Fig. 14.3 shows the temperature dependence of the transport coefficients for the model (A) (14.134) calculated using the analytic formulae given by the Israel–Stewart fourteen moment method [67]. The squares attached to the bold lines in Fig. 14.3 shows the temperature dependence of the transport coefficients calculated by our numerical method using the formulae (14.161)–(14.163) with $a = 0$

[2] See Sect. 14.3.2.

[3] The numbers are taken from Table 1, 2, and 3 in CHAPTER XI of [67].

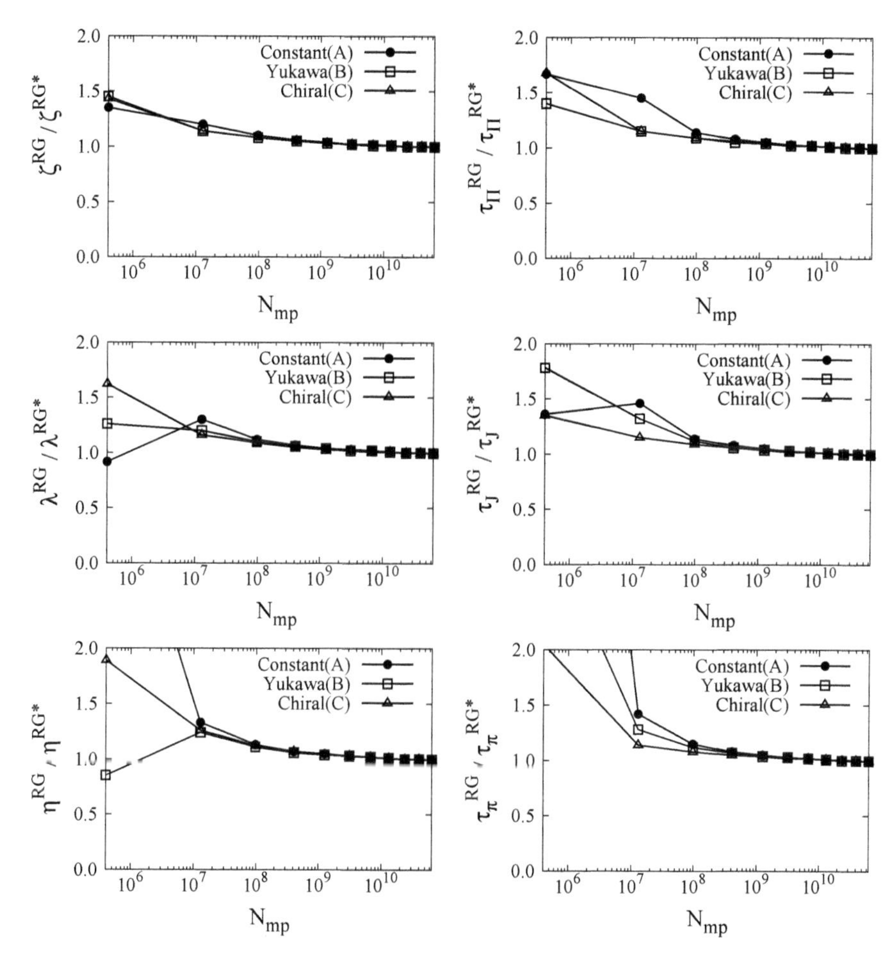

Fig. 14.1 The N_{mp} dependence of the transport coefficients and relaxation times ζ^{RG}, λ^{RG}, η^{RG}, τ_{Π}^{RG}, τ_{J}^{RG}, and τ_{π}^{RG} for the model (A), (B), and (C) with $z = 0.01$ and $\mu = 0$, which are calculated by the numerical method with the use of the double exponential method. The values are normalized by $\zeta^{\mathrm{RG}*}$, $\lambda^{\mathrm{RG}*}$, $\eta^{\mathrm{RG}*}$, $\tau_{\Pi}^{\mathrm{RG}*}$, $\tau_{J}^{\mathrm{RG}*}$, and $\tau_{\pi}^{\mathrm{RG}*}$, respectively, which are the transport coefficients and relaxation times obtained with $N_{\mathrm{mp}} = 4 \times 10^{10}$. In the model (B), we set M equal to 100 m. All the results are independent of σ_{tot}, G, and f_{π} in the models (A), (B), and (C), respectively, because the parameters can be factorized in the transport coefficients and relaxation times and hence be canceled out in the ratios

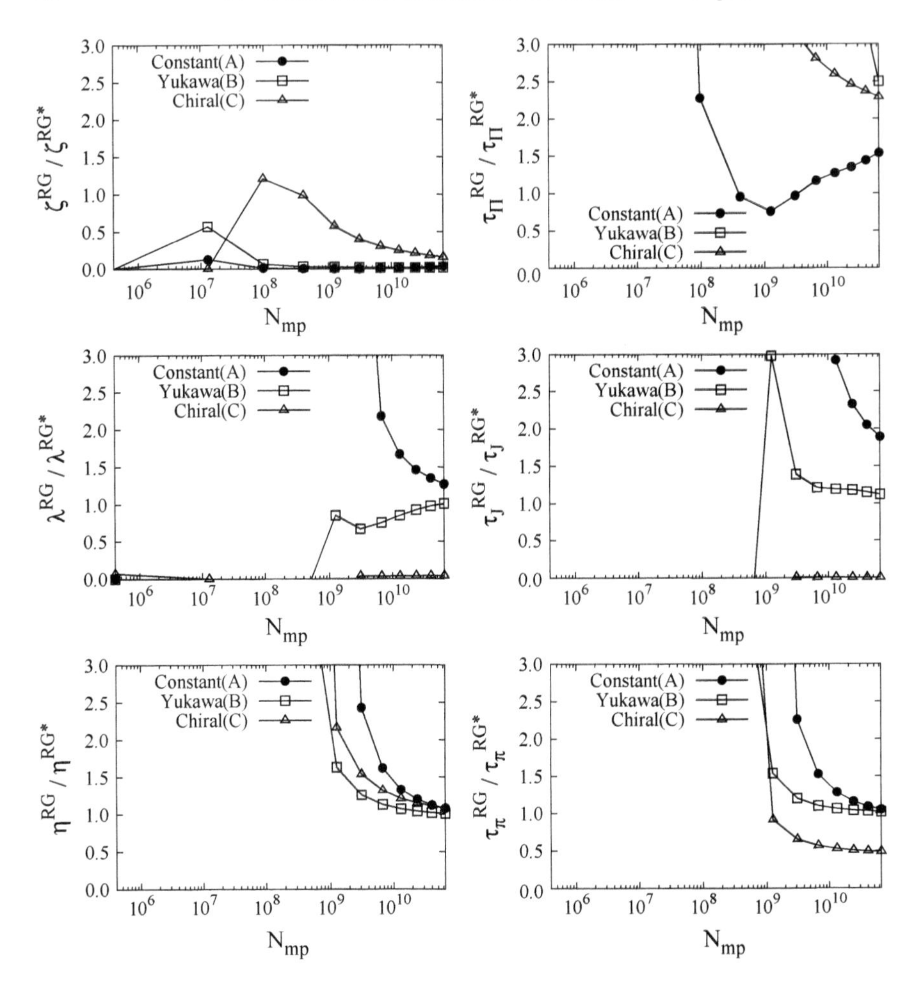

Fig. 14.2 The N_{mp} dependence of the transport coefficients and relaxation times computed with the use of the rectangular formulae for the models (A), (B), and (C) with $z = 0.01$ and $\mu = 0$. The values are normalized by those obtained by applying the double exponential formulae to the same microscopic representations with $N_{\mathrm{mp}} = 4 \times 10^{10}$, which are denoted by $\zeta^{\mathrm{RG}*}$, $\lambda^{\mathrm{RG}*}$, $\eta^{\mathrm{RG}*}$, $\tau_{\Pi}^{\mathrm{RG}*}$, $\tau_{J}^{\mathrm{RG}*}$, and $\tau_{\pi}^{\mathrm{RG}*}$, respectively, The parameters in the models are the same as those in Fig. 14.1

in $f^{\mathrm{eq}}(\tau)$ for the same model (A). One sees an excellent agreement between the two results.

In summary, we have seen that our scheme provides an accurate and efficient numerical method for evaluating not only the transport coefficients but also the relaxation times.

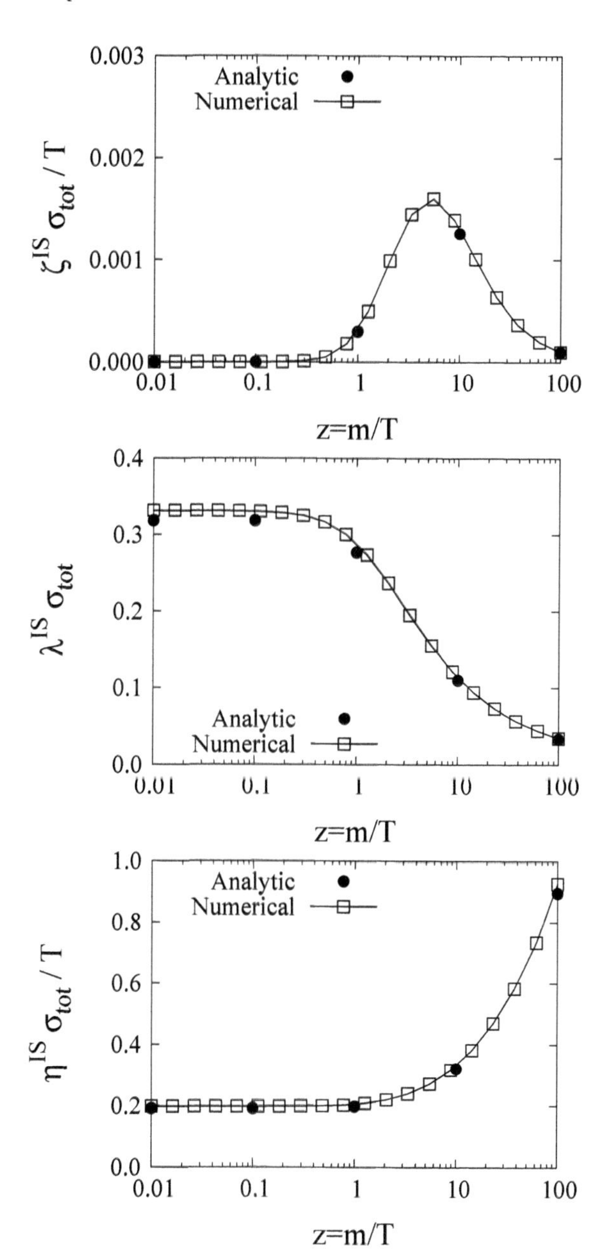

Fig. 14.3 A comparison of the bulk viscosity ζ^{IS}, heat conductivity λ^{IS}, and shear viscosity η^{IS} by the numerical integration based on the scheme developed in Sect. 14.2 with their analytic values reported in Ref. [67]

14.3.2 Numerical Results for Classical, Fermionic, and Bosonic Systems: Comparison of RG and Israel–Stewart Fourteen Moment Method

We show the numerical results of not only the transport coefficients but also the relaxation times using the formulae Eqs. (14.128)–(14.133) derived by the RG method and also the formulae (14.161)–(14.166) derived by the Israel–Stewart moment method; our focus in this subsection will be put on clarifying how different the numerical results are in the two methods, but not on discussing possible physical significance of them, which might be done elsewhere in future.

In Fig. 14.4, we show the ratios of the transport coefficients and relaxation times by the RG method to those by the Israel–Stewart fourteen moment method [67]. All the ratios in the model (A) are close to 1 for any z, while the ratios in the models (B) and (C) are not the case but significantly deviate from unity, in particular, for small z. For instance, the ratio of the bulk viscosity and its relaxation time in the model (C) increase up to 1000 when z approaches the ultra relativistic region given by $z \simeq 0.01$. Thus, one may conclude that the formulae based on the Israel–Stewart's fourteen moment method is not reliable in capturing the transport properties of the relativistic system in a quantitative way.

Appendix: Formulae for the Numerical Calculation of the Transport Coefficients and Relaxation Times by the Israel–Stewart Fourteen Moment Method

In this Appendix, we present the formulae for the numerical calculation of the transport coefficients and relaxation times with the microscopic expressions derived by the Israel–Stewart fourteen moment method [67], for self-containedness. The microscopic expressions given in Eqs. (10.171)–(10.176) are rewritten as

$$\zeta^{\rm IS} = -\frac{1}{T}\frac{\left[\langle\,\hat{\Pi}\,,\,\mathcal{P}\,\rangle\right]^2}{\langle\,\mathcal{P}\,,\,\hat{L}\,\mathcal{P}\,\rangle}, \quad \lambda^{\rm IS} = \frac{1}{3\,T^2}\frac{\left[\langle\,\hat{J}^{\mu}\,,\,\mathcal{J}_{\mu}\,\rangle\right]^2}{\langle\,\mathcal{J}^{\nu}\,,\,\hat{L}\,\mathcal{J}_{\nu}\,\rangle}, \tag{14.143}$$

$$\eta^{\rm IS} = -\frac{1}{10\,T}\frac{\left[\langle\,\hat{\pi}^{\mu\nu}\,,\,\pi_{\mu\nu}\,\rangle\right]^2}{\langle\,\pi^{\rho\sigma}\,,\,\hat{L}\,\pi_{\rho\sigma}\,\rangle}, \quad \tau_{\Pi}^{\rm IS} = -\frac{\langle\,\mathcal{P}\,,\,\mathcal{P}\,\rangle}{\langle\,\mathcal{P}\,,\,\hat{L}\,\mathcal{P}\,\rangle}, \tag{14.144}$$

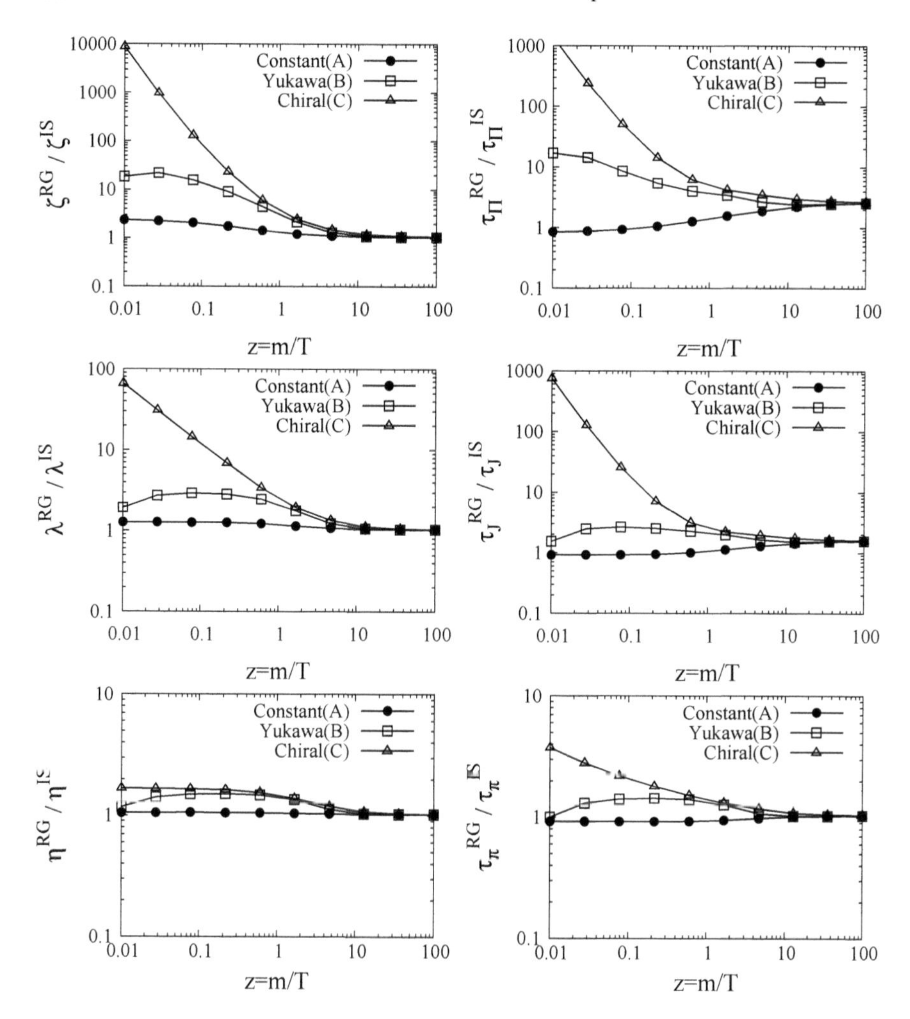

Fig. 14.4 The Ratios of the transport coefficients and relaxation times by the RG method to those by the Israel–Stewart fourteen moment method [67] for the three models: (A) the constant differential cross section and the classical limit, (B) the differential cross section by the Yukawa interaction and the Fermi statistics, and (C) the differential cross section by the chiral perturbation theory and the Bose statistics. A treatment of the parameters in the models is the same as that in Fig. 14.1

$$\tau_J^{\mathrm{IS}} = -\frac{\langle \mathcal{J}^\mu , \mathcal{J}_\mu \rangle}{\langle \mathcal{J}^\nu , \hat{L}\, \mathcal{J}_\nu \rangle}, \quad \tau_\pi^{\mathrm{IS}} = -\frac{\langle \pi^{\mu\nu} , \pi_{\mu\nu} \rangle}{\langle \pi^{\rho\sigma} , \hat{L}\, \pi_{\rho\sigma} \rangle}. \tag{14.145}$$

Here $\mathcal{P}_p$, $\mathcal{J}_p^\mu$, and $\pi_p^{\mu\nu}$ are the following vectors:

$$\mathcal{P}_p = -T_0^2 \frac{z^2 (a_2 a_0 - a_1^2)}{3 (a_3 a_1 - a_2^2)} \sqrt{k(\tau)}\, \tilde{a}(\tau), \tag{14.146}$$

$$\mathcal{J}_p^\mu = -T_0^2 \frac{1}{\sqrt{\tau^2 - z^2}} \sqrt{k(\tau)}\, \tilde{b}(\tau)\, \Delta^{\mu\nu}\, p_\nu / T_0, \tag{14.147}$$

$$\pi_p^{\mu\nu} = T_0^2 \frac{1}{\tau^2 - z^2} \sqrt{k(\tau)}\, \tilde{c}(\tau)\, \Delta^{\mu\nu\rho\sigma}\, p_\rho\, p_\sigma / T_0^2, \tag{14.148}$$

where

$$\tilde{a}(\tau) := \frac{1}{\sqrt{k(\tau)}} \left[\tau^2 - \frac{a_1 a_4 - a_2 a_3}{a_1 a_3 - a_2^2} \tau - \frac{a_3^2 - a_2 a_4}{a_1 a_3 - a_2^2} \right], \tag{14.149}$$

$$\tilde{b}(\tau) := \frac{1}{\sqrt{k(\tau)}} \sqrt{\tau^2 - z^2} \left[\tau - \frac{a_4 - a_2 z^2}{a_3 - a_1 z^2} \right], \tag{14.150}$$

$$\tilde{c}(\tau) := \frac{1}{\sqrt{k(\tau)}} (\tau^2 - z^2), \tag{14.151}$$

with $\tau = (p \cdot u)/T_0$.

Inserting the above expressions of $\mathcal{P}_p$, $\mathcal{J}_p^\mu$, and $\pi_p^{\mu\nu}$ into ϕ_p in Eq. (14.48) and using the equalities (14.49)–(14.51), we calculate $\left[\hat{L}\,\mathcal{P}\right]_p$, $\left[\hat{L}\,\mathcal{J}^\mu\right]_p$, and $\left[\hat{L}\,\pi^{\mu\nu}\right]_p$ to be

$$\begin{aligned} &-\bar{f}^{\text{eq}}(\tau) \frac{\tau}{T_0 \sqrt{k(\tau)}} \left[\hat{L}\,\mathcal{P}\right]_p \\ &= -T_0^2 \frac{z^2 (a_2 a_0 - a_1^2)}{3 (a_3 a_1 - a_2^2)} \left[k(\tau)\, \tilde{a}(\tau) + \int \mathrm{d}\mu(\tau_1)\, L(\tau,\, \tau_1)\, k(\tau_1)\, \tilde{a}(\tau_1) \right], \end{aligned} \tag{14.152}$$

$$\begin{aligned} &-\bar{f}^{\text{eq}}(\tau) \frac{\tau \sqrt{\tau^2 - z^2}}{T_0 \sqrt{k(\tau)}} \left[\hat{L}\,\mathcal{J}^\mu\right]_p \\ &= -T_0^2 \left[k(\tau)\, \tilde{b}(\tau) + \int \mathrm{d}\mu(\tau_1)\, M(\tau,\, \tau_1)\, k(\tau_1)\, \tilde{b}(\tau_1) \right] \Delta^{\mu\nu}\, p_\nu / T_0, \end{aligned} \tag{14.153}$$

$$\begin{aligned} &-\bar{f}^{\text{eq}}(\tau) \frac{\tau (\tau^2 - z^2)}{T_0 \sqrt{k(\tau)}} \left[\hat{L}\,\pi^{\mu\nu}\right]_p \\ &= T_0^2 \left[k(\tau)\, \tilde{c}(\tau) + \int \mathrm{d}\mu(\tau_1)\, N(\tau,\, \tau_1)\, k(\tau_1)\, \tilde{c}(\tau_1) \right] \Delta^{\mu\nu\rho\sigma}\, p_\rho\, p_\sigma / T_0^2, \end{aligned} \tag{14.154}$$

respectively.

Substituting these equations into Eqs. (14.143)–(14.145), we have

$$T\,\zeta^{\rm IS}/T_0^4 = \frac{4\pi\left[\int d\mu(\tau)\,k(\tau)\,\tilde{a}(\tau)\,\bar{f}^{\rm eq}(\tau)\,a(\tau)\right]^2}{\int d\mu(\tau)\,k(\tau)\,\tilde{a}(\tau)\left[k(\tau)\,\tilde{a}(\tau)+\int d\mu(\tau_1)\,L(\tau,\,\tau_1)\,k(\tau_1)\,\tilde{a}(\tau_1)\right]}, \tag{14.155}$$

$$3\,T^2\,\lambda^{\rm IS}/T_0^4 = \frac{4\pi\left[\int d\mu(\tau)\,k(\tau)\,\tilde{b}(\tau)\,\bar{f}^{\rm eq}(\tau)\,b(\tau)\right]^2}{\int d\mu(\tau)\,k(\tau)\,\tilde{b}(\tau)\left[k(\tau)\,\tilde{b}(\tau)+\int d\mu(\tau_1)\,M(\tau,\,\tau_1)\,k(\tau_1)\,\tilde{b}(\tau_1)\right]}, \tag{14.156}$$

$$15\,T\,\eta^{\rm IS}/T_0^4 = \frac{4\pi\left[\int d\mu(\tau)\,k(\tau)\,\tilde{c}(\tau)\,\bar{f}^{\rm eq}(\tau)\,c(\tau)\right]^2}{\int d\mu(\tau)\,k(\tau)\,\tilde{c}(\tau)\left[k(\tau)\,\tilde{c}(\tau)+\int d\mu(\tau_1)\,N(\tau,\,\tau_1)\,k(\tau_1)\,\tilde{c}(\tau_1)\right]}, \tag{14.157}$$

$$\tau_\Pi^{\rm IS}\,T_0 = \frac{\int d\mu(\tau)\,k(\tau)\,\tilde{a}(\tau)\,\tau\,\bar{f}^{\rm eq}(\tau)\,\tilde{a}(\tau)}{\int d\mu(\tau)\,k(\tau)\,\tilde{a}(\tau)\left[k(\tau)\,\tilde{a}(\tau)+\int d\mu(\tau_1)\,L(\tau,\,\tau_1)\,k(\tau_1)\,\tilde{a}(\tau_1)\right]}, \tag{14.158}$$

$$\tau_J^{\rm IS}\,T_0 = \frac{\int d\mu(\tau)\,k(\tau)\,\tilde{b}(\tau)\,\tau\,\bar{f}^{\rm eq}(\tau)\,\tilde{b}_2(\iota)}{\int d\mu(\tau)\,k(\tau)\,\tilde{b}(\tau)\left[k(\tau)\,\tilde{b}(\tau)+\int d\mu(\tau_1)\,M(\tau,\,\tau_1)\,k(\tau_1)\,\tilde{b}(\tau_1)\right]}, \tag{14.159}$$

$$\tau_\pi^{\rm IS}\,T_0 = \frac{\int d\mu(\tau)\,k(\tau)\,\tilde{c}(\tau)\,\tau\,\bar{f}^{\rm eq}(\tau)\,\tilde{c}(\tau)}{\int d\mu(\tau)\,k(\tau)\,\tilde{c}(\tau)\left[k(\tau)\,\tilde{c}(\tau)+\int d\mu(\tau_1)\,N(\tau,\,\tau_1)\,k(\tau_1)\,\tilde{c}(\tau_1)\right]}, \tag{14.160}$$

whose discretized forms read

$$T\,\zeta^{\rm IS}/T_0^4 = 4\pi\,\frac{\left[\sum_{n=0}^N \bar{f}_n^{\rm eq}\,a_n\,\tilde{a}_n\,k_n\right]^2}{\sum_{m=0}^N\sum_{n=0}^N \tilde{a}_m\,k_m\,(\delta_{mn}+L_{mn})\,\tilde{a}_n\,k_n}, \tag{14.161}$$

$$3\,T^2\,\lambda^{\rm IS}/T_0^4 = 4\pi\,\frac{\left[\sum_{n=0}^N \bar{f}_n^{\rm eq}\,b_n\,\tilde{b}_n\,k_n\right]^2}{\sum_{m=0}^N\sum_{n=0}^N \tilde{b}_m\,k_m\,(\delta_{mn}+M_{mn})\,\tilde{b}_n\,k_n}, \tag{14.162}$$

$$15\, T\, \eta^{\rm IS}/T_0^4 = 4\,\pi\, \frac{\left[\sum_{n=0}^{N} \bar{f}_n^{\rm eq}\, c_n\, \tilde{c}_n\, k_n\right]^2}{\sum_{m=0}^{N}\sum_{n=0}^{N} \tilde{c}_m\, k_m\, (\delta_{mn} + N_{mn})\, \tilde{c}_n\, k_n}, \tag{14.163}$$

$$\tau_\Pi^{\rm IS}\, T_0 = \frac{\sum_{n=0}^{N} \bar{f}_n^{\rm eq}\, \tau_n\, \tilde{a}_n\, \tilde{a}_n\, k_n}{\sum_{m=0}^{N}\sum_{n=0}^{N} \tilde{a}_m\, k_m\, (\delta_{mn} + L_{mn})\, \tilde{a}_n\, k_n}, \tag{14.164}$$

$$\tau_J^{\rm IS}\, T_0 = \frac{\sum_{n=0}^{N} \bar{f}_n^{\rm eq}\, \tau_n\, \tilde{b}_n\, \tilde{b}_n\, k_n}{\sum_{m=0}^{N}\sum_{n=0}^{N} \tilde{b}_m\, k_m\, (\delta_{mn} + M_{mn})\, \tilde{b}_n\, k_n}, \tag{14.165}$$

$$\tau_\pi^{\rm IS}\, T_0 = \frac{\sum_{n=0}^{N} \bar{f}_n^{\rm eq}\, \tau_n\, \tilde{c}_n\, \tilde{c}_n\, k_n}{\sum_{m=0}^{N}\sum_{n=0}^{N} \tilde{c}_m\, k_m\, (\delta_{mn} + N_{mn})\, \tilde{c}_n\, k_n}, \tag{14.166}$$

with

$$\tilde{a}_n := \sqrt{\Delta\mu_n}\, \tilde{a}(\tau_n), \quad \tilde{b}_n := \sqrt{\Delta\mu_n}\, \tilde{b}(\tau_n), \quad \tilde{c}_n := \sqrt{\Delta\mu_n}\, \tilde{c}(\tau_n). \tag{14.167}$$

These discretized forms are used in the numerical calculation that is presented in the text.

Chapter 15
RG/E Derivation of Reactive-Multi-component Relativistic Fluid Dynamics

15.1 Introduction

In this chapter, we deal with quantum relativistic gasses composed of multiple types of particles, in which the collisions between the particles can be reactive, *i.e.*, not only elastic but also inelastic with a change of particle numbers accompanied; the constituent particles are allowed to mix with each other under the constraints of conservation laws to results in additional types of diffusions that would not exist in single-component fluid. Such systems include molecular/ionic gasses in the universe at high temperature, the primordial matter created in the birth of the universe, the intermediate stages of nuclear collisions including relativistic heavy-ion collisions which may be composed of various hadrons and/or even quarks and gluons leading to a 'soup' of them, *i.e.*, the quark-gluon plasma (QGP), and so on.

In particular, since the end of the last century or the beginning of the present century, a tremendous amount of experimental and theoretical efforts have been devoted to unravel the dynamics of quarks and gluons under highly extreme conditions [160, 161]. The problem has been experimentally addressed through the ultra-relativistic heavy-ion collision experiments [162, 186, 188–190, 232–238], where high-energy collisions of heavy ions at nearly the speed of light drive the system to a far-from-equilibrium situation followed by thermalization, relaxation, hadronization, etc.

Fluid dynamics is one of the powerful tools to understand such nonequilibrium phenomena of complex underlying theory, quantum chromodynamics (QCD) in the case of the QGP, for instance. While the conventional second-order fluid dynamics have been applied to describing the QGP, it is natural to add more expressibility by exploiting the fact that the QGP is composed of distinct types of ingredients, that is, quarks and gluons with flavor and color degrees of freedom.

For the purpose of incorporating the properties specific to multi-component systems in the context of the relativistic fluid dynamics, several attempts have been made to derive multi-component fluid dynamics from microscopic theories [59, 66–68, 239–245]. The efforts in this direction were initiated by [240] in order to study

T. Kunihiro et al., *Geometrical Formulation of Renormalization-Group Method as an Asymptotic Analysis*, Fundamental Theories of Physics 206,
https://doi.org/10.1007/978-981-16-8189-9_15

the properties of hot hadronic matter. They derived the multi-component second order fluid dynamics for relativistic gas by extending the seminal work by Israel and Stewart [66]. After a while, [241] proposed an improved Israel–Stewart's 14-moment methods. They identified the functional form of the one-body distribution function by requiring the positivity of the entropy-production rate and the Onsager's reciprocal relation combined with careful order counting. However, some elaboration of the previous attempts can be possible as is the case in the single-component dynamics, as discussed in Chap. 10.

In the present Chapter, the renormalization group (RG) method [1, 3, 5, 6, 46, 57, 58] is applied to the derivation of multi-component fluid dynamic equations [156] as an extension of the relativistic fluid dynamic equations derived in Chaps. 11 and 12 (see Chaps. 5 and 9 for the respective general framework of the RG/E method). The characteristic features of multi-component fluid are naturally built in the fluid dynamics that is carefully derived from the Boltzmann theory containing multiple types of particles as microscopic ingredients. The properties of derived transport coefficients will be discussed in detail. As it turns out along the detailed study of the derived transport coefficients, the positive entropy production rate and the Onsager's reciprocal relations are readily confirmed as consequences of our derivation. These are particularly promising aspects of our methodology because, in the conventional derivations following [66], these properties are imposed in an ad hoc manner.

15.2 Boltzmann Equation in Relativistic Reactive-Multi-component Systems

Throughout this chapter, we consider a system composed of N distinct types of degrees of freedom (particles) and M conserved currents in addition to the energy and momentum conservations [67]. Since the conservation laws are our central tools to characterize and derive fluid dynamics, we first introduce conserved currents in the context of the Boltzmann's kinetic theory with quantum statistics. The kinetic theory is the semiclassical theory that describes the nonequilibrium dynamics of systems in terms of a one-body distribution function defined in phase space. The evolution of distribution function is dictated by the Boltzmann equation (15.3).

In this framework, the energy-momentum tensor is given by,

$$T^{\mu\nu}(x) = \sum_{k=1}^{N} \int \mathrm{d}p_k \, p_k^{\mu} p_k^{\nu} f_{k,p_k}(x), \tag{15.1}$$

with the Lorentz indices μ, ν. The one-particle distribution function of type k and four momentum $p_k := (\sqrt{m_k^2 + \boldsymbol{p}_k^2}, \boldsymbol{p}_k)$ is denoted by $f_{k,p_k}(x)$. The integration measure $\mathrm{d}p_k$ stands for $\mathrm{d}^3\boldsymbol{p}_k/[(2\pi)^3 p_k^0]$. By the assumption we have M conserved currents,

$$N_A^\mu(x) = \sum_{k=1}^{N} q_k^A \int \mathrm{d}p_k p_k^\mu f_{k,p_k}(x), \quad (A = 1, \ldots, M). \tag{15.2}$$

Here, we have assigned a charge q_k^A to a particle of type k ($k = 1, \ldots, N$) associated to the Ath ($A = 1, \ldots, M$) conservation law. The particle number current N_A^μ is associated to the Ath conservation laws. We shall momentarily see that these currents indeed obey conservation laws.

The relativistic Boltzmann equation governs the time evolution of the distribution functions [67, 68],

$$p_k^\mu \partial_\mu f_{k,p_k}(x) = \sum_{l=1}^{N} C_{kl}[f]_{p_k}(x), \tag{15.3}$$

with the collision integral,

$$\begin{aligned} C_{kl}[f]_{p_k} = \frac{1}{2} \sum_{i,j=1}^{N} \int \mathrm{d}p_l \mathrm{d}p_i \mathrm{d}p_j \\ \times \big[f_{i,p_i} f_{j,p_j} (1 + a_k f_{k,p_k})(1 + a_l f_{l,p_l}) \mathcal{W}_{ij|kl} \\ - (1 + a_i f_{i,p_i})(1 + a_j f_{j,p_j}) f_{k,p_k} f_{l,p_l} \mathcal{W}_{kl|ij}\big]. \end{aligned} \tag{15.4}$$

The quantum statistical effect is represented by a_k, i.e., $a_k = +1$ for bosons, $a_k = -1$ for fermions, and $a_k = 0$ for classical particles. The transition matrix $\mathcal{W}_{ij|kl}$, that is responsible for two-particle interactions, takes the form of

$$\mathcal{W}_{ij|kl} = (2\pi)^4 |\mathcal{M}|^2 \delta^4(p_i + p_j - p_k - p_l) \prod_{A=1}^{M} \delta_{q_i^A + q_j^A, q_k^A + q_l^A}, \tag{15.5}$$

where $\mathcal{M}$ is a binary scattering amplitude. It is readily confirmed that the matrix $\mathcal{W}_{ij|kl}$ satisfies the following symmetry properties,

$$0 = \mathcal{W}_{ij|kl} - \mathcal{W}_{ji|lk}, \tag{15.6}$$

$$0 = \sum_{i,j=1}^{N} \int \mathrm{d}p_i \mathrm{d}p_j \big[\mathcal{W}_{ij|kl} - \mathcal{W}_{kl|ij}\big]. \tag{15.7}$$

The second equation represents the detailed balance property of a multi-component system.

15.2.1 Collision Invariants and Conservation Laws

Let us discuss some properties of the Boltzmann equation (15.3). For an arbitrary function φ_{k,p_k} of p_k, we find that the following identity holds,

$$\sum_{k,l=1}^{N} \int \mathrm{d}p_k \varphi_{k,p_k} C_{kl}[f]_{p_k} \\ = \frac{1}{4} \sum_{i,j,k,l=1} \int \mathrm{d}p_i \mathrm{d}p_j \mathrm{d}p_k \mathrm{d}p_l (\varphi_{k,p_k} + \varphi_{l,p_l} - \varphi_{i,p_i} - \varphi_{j,p_j}) \\ \times \mathrm{W}_{ij|kl} f_{i,p_i} f_{j,p_j} (1 + f_{k,p_k})(1 + f_{l,p_l}), \tag{15.8}$$

where we have used the symmetries of $\mathcal{W}_{ij|kl}$ (15.6) and (15.7). A function φ^0_{k,p_k} is called a collision invariant if it satisfies,

$$\sum_{k,l=1}^{N} \int \mathrm{d}p_k \varphi^0_{k,p_k} C_{kl}[f]_{p_k} = 0. \tag{15.9}$$

The existence of the collision invariant results in an important consequence. An integral over the momentum p_k and a sum over the components k of the Boltzmann equation (15.3) convoluted with the collision invariant φ^0_{k,p_k} leads to

$$\partial_\mu \Big(\sum_{k=1}^{N} \int \mathrm{d}p_k \varphi^0_{k,p_k} p^\mu_k f_{k,p_k} \Big) = \sum_{k=1}^{N} \int \mathrm{d}p_k \varphi^0_{k,p_k} p^\mu_k \partial_\mu f_{k,p_k} \\ = \sum_{k,l=1}^{N} \int \mathrm{d}p_k \varphi^0_{k,p_k} C_{kl}[f]_{p_k} = 0. \tag{15.10}$$

Hence, one finds that a conserved quantity is given in the form,

$$\sum_{k=1}^{N} \int \mathrm{d}p_k \varphi^0_{k,p_k} p^\mu_k f_{k,p_k}, \tag{15.11}$$

for each collision invariant in the kinetic theory. In fact, the relation (15.8) and the form of the transition matrix (15.5) imply that q^A_k and p^μ_k are collision invariants, *i.e.*,

$$\sum_{k,l=1}^{N} q^A_k \int \mathrm{d}p_k C_{kl}[f]_{p_k} = 0, \tag{15.12}$$

$$\sum_{k,l=1}^{N} \int \mathrm{d}p_k p^\mu_k C_{kl}[f]_{p_k} = 0. \tag{15.13}$$

Thus, one find that the particle currents (15.2) and (15.1) are indeed conserved quantities,

$$\begin{aligned}\partial_\mu N_A^\mu &= \sum_{k=1}^{N} q_k^A \int \mathrm{d}p_k p_k^\mu \partial_\mu f_{k,p_k} = \sum_{k,l=1}^{N} q_k^A \int \mathrm{d}p_k C_{kl}[f]_{p_k} = 0, \\ \partial_\mu T^{\mu\nu} &= \sum_{k=1}^{N} \int \mathrm{d}p_k p_k^\mu p_k^\nu \partial_\mu f_{k,p_k} = \sum_{k,l=1}^{N} \int \mathrm{d}p_k p_k^\nu C_{kl}[f]_{p_k} = 0.\end{aligned} \tag{15.14}$$

Note that it is a pivotal step to identify the conserved quantities in the system under consideration as the fluid dynamics consists of a set of conservation laws expressed in terms of fluid dynamic variables.

15.2.2 Entropy Current

The relaxation of a non-equilibrium system is driven by dissipative processes, that is accompanied by increase of entropy in the non-relativistic case. Its relativistic counterpart is the entropy four current defined by,

$$s^\mu = -\sum_{k=1}^{N} \int \mathrm{d}p_k^\mu \left[f_{k,p_k} \ln f_{k,p_k} - \frac{(1 + a_k f_{k,p_k}) \ln (1 + a_k f_{k,p_k})}{a_k} \right]. \tag{15.15}$$

The dissipation of a relativistic system is then characterized by entropy-production rate,

$$\begin{aligned}\partial_\mu s^\mu &= -\sum_{k=1}^{N} \int \mathrm{d}p_k p_k^\mu \partial_\mu f_{k,p_k} \ln \left(\frac{f_{k,p_k}}{1 + a_k f_{k,p_k}} \right) \\ &= -\sum_{k,l=1}^{N} \int \mathrm{d}p_k C_{kl}[f]_{k,p_k} \ln \left(\frac{f_{k,p_k}}{1 + a_k f_{k,p_k}} \right),\end{aligned} \tag{15.16}$$

where we have used the Boltzmann equation (15.3) in the second equality. Vanishing entropy-production rate implies the non-dissipative dynamics, that is the case when $\ln[f_{k,p_k}/(1 + a_k f_{k,p_k})]$ is a collision invariant. It is amount to requiring the following,

$$\ln \left(\frac{f_{k,p_k}}{1 + a_k f_{k,p_k}} \right) = \sum_{A=1}^{M} q_k^A c_A(x) + p_k^\mu c_\mu(x), \tag{15.17}$$

with coefficients, $c_A(x)$ and $c_\mu(x)$, of the aforementioned collision invariants q_k^A and p_k^μ. We can solve it to obtain the entropy-conserving distribution function,

$$f_{k,p_k}(x) = \left[\exp\left(\frac{p_k \cdot u(x)}{T(x)} + \sum_{A=1}^{M} q_k^A \frac{\mu_A(x)}{T(x)} \right) - a_k \right]^{-1} =: f_{k,p_k}^{\mathrm{eq}}(x), \qquad (15.18)$$

where we have reparametrized as $c_A(x) = \mu_A(x)/T(x)$ and $c_\mu(x) = u_\mu(x)/T(x)$ with a constraint $u_\mu u^\mu = 1$. We recognize it as the local equilibrium distribution function by identifying $M + 4$ parameters $T(x)$, $\mu_A(x)$, and $u_\mu(x)$ with local temperature, chemical potentials conjugate to the M conserved charges, and flow velocities, respectively.

15.3 Reduction of Boltzmann Equation to Reactive-Multi-component Fluid Dynamics

The derivation of the fluid dynamics is accomplished by solving the Boltzmann equation. Given a small spatial gradient, we carry out the perturbative expansion of a distribution function around the local equilibrium one. After finding the perturbative solution, which contains infrared divergence in the form of secular terms, we apply the RG/Envelope method to remove the divergence and obtain the global solution [1, 3, 5, 6, 40, 46]. The fluid dynamic variables show up as a consequence of the renormalization group analysis.

15.3.1 Solving Perturbative Equations

We begin with solving the Boltzmann equation perturbatively with respect to small spatial gradients, characterized by the Knudsen number $\mathcal{K}$,

$$\mathcal{K} = \frac{\ell}{L}, \quad \ell : \text{mean free path}, \quad L : \text{macroscopic length scale}. \qquad (15.19)$$

For the purpose of performing coarse-graining, we introduce the macroscopic frame vector a_{F}^μ satisfying $a_{\mathrm{F}\mu} a_{\mathrm{F}}^\mu = 1$ and $a_{\mathrm{F}}^0 > 0$. The explicit form of a_{F}^μ is yet to be identified. Then, we employ (τ, σ)-coordinates, where the vector a_{F}^μ is aligned to the temporal (τ) direction and perpendicular to the spatial (σ) direction:

$$\frac{\partial}{\partial \tau} := a_{\mathrm{F}}^\mu \partial_\mu, \qquad \nabla^\mu := \frac{\partial}{\partial \sigma_\mu} := (g^{\mu\nu} - a_{\mathrm{F}}^\mu a_{\mathrm{F}}^\nu)\partial_\nu. \qquad (15.20)$$

Introducing a bookkeeping parameter ϵ, that keeps track of the small spatial gradient and is identified with the Knudsen number $\mathcal{K}$, we have the Boltzmann equation in the new coordinate,

$$\frac{\partial}{\partial \tau} f_{k,p_k}(x) = \frac{1}{p_k \cdot a_{\mathrm{F}}} \sum_{l=1}^{N} C_{kl}[f]_{p_k}(x) - \epsilon \frac{1}{p_k \cdot a_{\mathrm{F}}} p_k \cdot \nabla f_{k,p_k}(x), \tag{15.21}$$

where we used $\partial_\mu = a_{\mathrm{F}}^\mu \partial/\partial\tau + \nabla_\mu$. The perturbative computation proceeds by solving the Boltzmann equation (15.21) order by order with respect to ϵ,

$$\tilde{f}(\tau;\tau_0) = \tilde{f}^{(0)}(\tau;\tau_0) + \epsilon \tilde{f}^{(1)}(\tau;\tau_0) + \epsilon^2 \tilde{f}^{(2)}(\tau;\tau_0) + O(\epsilon^3), \tag{15.22}$$

along with the 'initial' distribution function,

$$\tilde{f}(\tau_0;\tau_0) = f(\tau_0) = f^{(0)}(\tau_0) + \epsilon f^{(1)}(\tau_0) + \epsilon^2 f^{(2)}(\tau_0) + O(\epsilon^3). \tag{15.23}$$

The perturbative distribution function $\tilde{f}(\tau;\tau_0)$ is formally set to the exact one $f(\tau;\tau_0)$ at 'initial' time τ_0.

We start with the zeroth order equation,

$$\frac{\partial}{\partial \tau} f_{k,p_k} = \frac{1}{p_k \cdot a_{\mathrm{F}}} \sum_{l=1}^{N} C_{kl}[f]_{p_k}. \tag{15.24}$$

The leading-order solution, representing $\tau \to \infty$ asymptotic steady solution of the near-equilibrium dynamics, is obtained by requiring

$$\frac{\partial}{\partial \tau} f_{k,p_k} = 0, \tag{15.25}$$

which, at the zeroth order, is amount to

$$\frac{1}{p_k \cdot a_{\mathrm{F}}} \sum_{l=1}^{N} C_{kl}[f]_{p_k} = 0. \tag{15.26}$$

One can readily confirm that the equilibrium distribution function f^{eq}_{k,p_k} (15.18) satisfies,

$$\begin{aligned} C_{kl}[f^{\mathrm{eq}}]_{p_k} &= \frac{1}{2} \sum_{i,j=1}^{N} \int \mathrm{d}p_l \mathrm{d}p_i \mathrm{d}p_j \mathcal{W}_{ij|kl} \\ &\times \big[f^{\mathrm{eq}}_{i,p_i} f^{\mathrm{eq}}_{j,p_j} (1 + a_k f^{\mathrm{eq}}_{k,p_k})(1 + a_l f^{\mathrm{eq}}_{l,p_l}) \\ &- (1 + a_i f^{\mathrm{eq}}_{i,p_i})(1 + a_j f^{\mathrm{eq}}_{j,p_j}) f^{\mathrm{eq}}_{k,p_k} f^{\mathrm{eq}}_{l,p_l} \big] \\ &= 0, \end{aligned} \tag{15.27}$$

and hence, is the zeroth order solution. We have used (15.6) in the first equality. In what follows, we carry out the perturbative expansion around the local equilibrium distribution function,

$$\tilde{f}^{(0)}(\tau;\tau_0) = f^{\mathrm{eq}}(\tau_0). \tag{15.28}$$

Here, the thermodynamic variables are introduced as integral constants of the zeroth order solution, *i.e.*, $u^\mu = u^\mu(\tau_0)$, $T = T(\tau_0)$, and $\mu_A = \mu_A(\tau_0)$. As discussed in detail in Sect. 11.4.1, we can now identify the macroscopic vector with the flow velocity,

$$a_{\mathrm{F}}^\mu(\tau_0) = u^\mu(\tau_0). \tag{15.29}$$

As discussed in Sect. 11.6.1, this identification of the frame vector automatically leads to the fluid dynamic equation in the Landau–Lifshitz energy frame.

Next, we consider the equation of $O(\epsilon)$,

$$\begin{aligned}
\frac{\partial}{\partial\tau}\tilde{f}^{(1)}_{k,p_k} &= \sum_m \int \mathrm{d}p_m \tilde{f}^{(1)}_{m,p_m} \frac{\delta}{\delta f_{m,p_m}}\left(\frac{1}{p_k\cdot u}\sum_{l=1}^{N} C_{kl}[f]_{p_k}\right)\Bigg|_{f=f^{\mathrm{eq}}} \\
&\quad - \frac{1}{p_k\cdot u} p_k\cdot\nabla f^{\mathrm{eq}}_{k,p_k} \\
&= f^{\mathrm{eq}}_{k,p_k}\bar{f}^{\mathrm{eq}}_{k,p_k}\sum_m\int \mathrm{d}q_m L_{k,p_k;m,p_m}\left(f^{\mathrm{eq}}_{m,p_m}\bar{f}^{\mathrm{eq}}_{m,p_m}\right)^{-1}\tilde{f}^{(1)}_{m,p_m} \\
&\quad + f^{\mathrm{eq}}_{k,p_k}\bar{f}^{\mathrm{eq}}_{k,p_k}F^{(0)}_{k,p_k},
\end{aligned} \tag{15.30}$$

where we defined

$$\bar{f}^{\mathrm{eq}}_{k,p_k} := 1 + a_k f^{\mathrm{eq}}_{k,p_k}, \tag{15.31}$$

$$F^{(i)}_{k,p_k} := -\left(f^{\mathrm{eq}}_{k,p_k}\bar{f}^{\mathrm{eq}}_{k,p_k}\right)^{-1}\frac{1}{p_k\cdot u}p_k\cdot\nabla f^{(i)}_{k,p_k}, \tag{15.32}$$

and the linearized collision operator

$$\begin{aligned}
&L_{k,p_k;m,p_m} \\
&:= (f^{\mathrm{eq}}_{k,p_k}\bar{f}^{\mathrm{eq}}_{k,p_k})^{-1}\frac{\delta}{\delta f_{m,p_m}}\left(\frac{1}{p_k\cdot u}\sum_{l=1}^{N}C_{kl}[f]_{p_k}\right)\Bigg|_{f=f^{\mathrm{eq}}} f^{\mathrm{eq}}_{m,p_m}\bar{f}^{\mathrm{eq}}_{m,p_m} \\
&= -\frac{1}{2p_k\cdot u}\sum_{i,j,l}\int \mathrm{d}p_l\mathrm{d}p_i\mathrm{d}p_j \mathcal{W}_{ij|kl}\frac{\bar{f}^{\mathrm{eq}}_{l,p_l}f^{\mathrm{eq}}_{i,p_i}f^{\mathrm{eq}}_{j,p_j}}{f^{\mathrm{eq}}_{k,p_k}} \\
&\times\left[\delta_{km}\delta(p_k-p_m)+\delta_{lm}\delta(p_l-p_m)-\delta_{im}\delta(p_i-p_m)-\delta_{jm}\delta(p_j-p_m)\right].
\end{aligned} \tag{15.33}$$

For the notational simplicity, we suppress the subscripts for momenta and types of particle in what follows by employing the matrix notation. Then, the first order equation (15.30) is concisely written as

$$\frac{\partial}{\partial \tau}\tilde{f}^{(1)} = f^{\mathrm{eq}}\bar{f}^{\mathrm{eq}}L\big(f^{\mathrm{eq}}\bar{f}^{\mathrm{eq}}\big)^{-1}\tilde{f}^{(1)} + f^{\mathrm{eq}}\bar{f}^{\mathrm{eq}}F^{(0)}, \tag{15.34}$$

with

$$\big[f^{\mathrm{eq}}\bar{f}^{\mathrm{eq}}L\big(f^{\mathrm{eq}}\bar{f}^{\mathrm{eq}}\big)^{-1}\big]_{k,p_k;m,p_m} := f^{\mathrm{eq}}_{k,p_k}\bar{f}^{\mathrm{eq}}_{k,p_k}L_{k,p_k;m,p_m}\big(f^{\mathrm{eq}}_{m,p_m}\bar{f}^{\mathrm{eq}}_{m,p_m}\big)^{-1}, \tag{15.35}$$

$$\big[f^{\mathrm{eq}}\bar{f}^{\mathrm{eq}}F^{(0)}\big]_{k,p_k} := f^{\mathrm{eq}}_{k,p_k}\bar{f}^{\mathrm{eq}}_{k,p_k}F^{(0)}_{k,p_k}. \tag{15.36}$$

Furthermore, we equip the vector space with the inner product defined by,

$$\langle \psi, \chi\rangle = \sum_{k=1}^{N}\int \mathrm{d}p_k (p_k\cdot u) f^{\mathrm{eq}}_{k,p_k}\bar{f}^{\mathrm{eq}}_{k,p_k}\psi_{k,p_k}\chi_{k,p_k}. \tag{15.37}$$

See Sect. 13.1 for the detailed discussion on the inner product.

We note three important properties of the linearized collision operator L. It is self-adjoint and semi-negative definite, and possesses zero modes:

$$\langle \psi, L\chi\rangle = \langle L\psi, \chi\rangle, \tag{15.38}$$

$$\langle \psi, L\psi\rangle \le 0, \tag{15.39}$$

$$L\varphi^{\alpha}_{k,p_k} = 0, \tag{15.40}$$

where ψ and χ are arbitrary vectors, and the zero modes are given by

$$\varphi^{\alpha}_{k,p_k} = \begin{cases} p_k^{\mu},\ \alpha = \mu = 0, 1, 2, 3, \\ q_k^{A},\ \alpha = A + 3 = 4, \ldots, M + 3. \end{cases} \tag{15.41}$$

The self-adjointness can be shown as follows,

$$\begin{aligned}
&\langle \psi, L\chi\rangle \\
&= -\frac{1}{8}\sum_{i,j,k,l}\int \mathrm{d}p_i\mathrm{d}p_j\mathrm{d}p_k\mathrm{d}p_l \mathcal{W}_{ij|kl} f^{\mathrm{eq}}_{i,p_i} f^{\mathrm{eq}}_{j,p_j}\bar{f}^{\mathrm{eq}}_{k,p_k}\bar{f}^{\mathrm{eq}}_{l,p_l} \\
&\times \big(\psi_{k,p_k} + \psi_{l,p_l} - \psi_{i,p_i} - \psi_{j,p_j}\big)\big(\chi_{k,p_k} + \chi_{l,p_l} - \chi_{i,p_i} - \chi_{j,p_j}\big) \\
&= \langle L\psi, \chi\rangle,
\end{aligned} \tag{15.42}$$

where the symmetries of $\mathcal{W}_{ij|kl}$ (15.6) and (15.7) have been used. Then, the semi-negative definiteness immediately follows,

$$
\begin{aligned}
&\langle \psi, L\psi \rangle \\
&= -\frac{1}{8} \sum_{i,j,k,l} \int \mathrm{d}p_i \mathrm{d}p_j \mathrm{d}p_k \mathrm{d}p_l \mathcal{W}_{ij|kl} f^{\mathrm{eq}}_{i,p_i} f^{\mathrm{eq}}_{j,p_j} \bar{f}^{\mathrm{eq}}_{k,p_k} \bar{f}^{\mathrm{eq}}_{l,p_l} \\
&\times \left(\psi_{k,p_k} + \psi_{l,p_l} - \psi_{i,p_i} - \psi_{j,p_j}\right)^2 \\
&\leq 0.
\end{aligned} \tag{15.43}
$$

Note that the transition probability $\mathcal{W}_{ij|kl}$ is always positive by the definition (15.5). Finally, the zero modes arise due to the conservation laws,

$$
\begin{aligned}
{}_{k,p_k} &= -\frac{1}{2p_k \cdot u} \sum_{i,j,l} \int \mathrm{d}p_l \mathrm{d}p_i \mathrm{d}p_j \mathcal{W}_{ij|kl} \frac{\bar{f}^{\mathrm{eq}}_{l,p_l} f^{\mathrm{eq}}_{i,p_i} f^{\mathrm{eq}}_{j,p_j}}{f^{\mathrm{eq}}_{k,p_k}} \\
&\times \left(\varphi^{\alpha}_{k,p_k} + \varphi^{\alpha}_{l,p_l} - \varphi^{\alpha}_{i,p_i} - \varphi^{\alpha}_{j,p_j}\right).
\end{aligned} \tag{15.44}
$$

This expression vanishes when φ^{α} is the energy-momentum or the conserved charges because they are invariant in the scattering processes.

The first-order equation is solved by introducing projection operator P_0 on the space spanned by the zero modes and its complement Q_0 such that $P_0 + Q_0 = 1$. The associated spaces are denoted by P_0 and Q_0. The projection operator P_0 is defined by,

$$
[P_0 \psi]_{k,p_k} = \varphi^{\alpha}_{k,p_k} \eta^{-1}_{\alpha\beta} \langle \varphi^{\beta}, \psi \rangle, \tag{15.45}
$$

where φ is a zero mode and $\eta^{-1}_{\alpha\beta}$ is the inverse of $\eta^{\alpha\beta} := \langle \varphi^{\alpha}, \varphi^{\beta} \rangle$, that plays a role of metric in the P_0 space. With some 'initial' condition $\tilde{f}^{(1)}(\tau_0) = f^{\mathrm{eq}} \bar{f}^{\mathrm{eq}} \Psi^{(1)}$ to be specified later, the $O(\epsilon)$ Eq. (15.34) is solved as follows:

$$
\begin{aligned}
\tilde{f}^{(1)} &= f^{\mathrm{eq}} \bar{f}^{\mathrm{eq}} \left(\mathrm{e}^{L(\tau-\tau_0)} \Psi^{(1)} + \int_{\tau_0}^{\tau} \mathrm{d}\tau' \mathrm{e}^{L(\tau-\tau')} F^{(0)} \right) \\
&= f^{\mathrm{eq}} \bar{f}^{\mathrm{eq}} \left(\mathrm{e}^{L(\tau-\tau_0)} \Psi^{(1)} + \int_{\tau_0}^{\tau} \mathrm{d}\tau' \left(P_0 + \mathrm{e}^{L(\tau-\tau')} Q_0\right) F^{(0)} \right) \\
&= f^{\mathrm{eq}} \bar{f}^{\mathrm{eq}} \left(\mathrm{e}^{L(\tau-\tau_0)} \Psi^{(1)} + (\tau - \tau_0) P_0 F^{(0)} + (\mathrm{e}^{L(\tau-\tau_0)} - 1) L^{-1} Q_0 F^{(0)} \right).
\end{aligned} \tag{15.46}
$$

The solution $\tilde{f}^{(1)}$ contains secular terms which are responsible for the divergence as $|\tau - \tau_0| \to \infty$. The divergence will be eliminated by the renormalization group analysis as we will see later.

We turn to the equation of $O(\epsilon^2)$,

$$
\frac{\partial}{\partial \tau} f^{(2)} = f^{\mathrm{eq}} \bar{f}^{\mathrm{eq}} L (f^{\mathrm{eq}} \bar{f}^{\mathrm{eq}})^{-1} \tilde{f}^{(2)} + f^{\mathrm{eq}} \bar{f}^{\mathrm{eq}} K(\tau - \tau_0), \tag{15.47}
$$

where $K(\tau - \tau_0)$ is defined by,

$$K(\tau - \tau_0) = F^{(1)}(\tau) + \frac{1}{2} B[\tilde{f}^{(1)}, \tilde{f}^{(1)}](\tau), \tag{15.48}$$

$$B[\psi, \chi]_{k,p_k;m,p_m;n,p_n} = \sum_{m,n} \int_{p_m,p_n} (f^{\mathrm{eq}}_{k,p_k} \bar{f}^{\mathrm{eq}}_{k,p_k})^{-1} \frac{\delta^2}{\delta f_{m,p_m} \delta f_{n,p_n}} \left(\frac{1}{p_k \cdot u} \sum_l C_{kl}[f]_{p_k} \right) \Bigg|_{f=f^{\mathrm{eq}}} \psi_{m,p_m} \chi_{n,p_n}. \tag{15.49}$$

The definition of $F^{(1)}(\tau)$ is already given in (15.32).

Provided some initial condition $\tilde{f}^{(2)}(\tau_0) = f^{\mathrm{eq}} \bar{f}^{\mathrm{eq}} \Psi^{(2)}$, the second-order equation is solved as,

$$\begin{aligned}
\tilde{f}^{(2)} &= f^{\mathrm{eq}} \bar{f}^{\mathrm{eq}} \left(\mathrm{e}^{L(\tau-\tau_0)} \Psi^{(2)} + \int_{\tau_0}^{\tau} \mathrm{d}\tau' \mathrm{e}^{L(\tau-\tau')} K(\tau' - \tau_0) \right) \\
&= f^{\mathrm{eq}} \bar{f}^{\mathrm{eq}} \left(\mathrm{e}^{L(\tau-\tau_0)} \Psi^{(2)} + \int_{\tau_0}^{\tau} \mathrm{d}\tau' \mathrm{e}^{L(\tau-\tau')} \mathrm{e}^{(\tau'-\tau_0)\frac{\partial}{\partial s}} K(s) \Big|_{s=0} \right) \\
&= f^{\mathrm{eq}} \bar{f}^{\mathrm{eq}} \left(\mathrm{e}^{L(\tau-\tau_0)} \Psi^{(2)} + (1 - \mathrm{e}^{(\tau-\tau_0)\frac{\partial}{\partial s}}) \left(-\frac{\partial}{\partial s} \right)^{-1} P_0 K(s) \Big|_{s=0} \right. \\
&\quad \left. + (\mathrm{e}^{L(\tau-\tau_0)} - \mathrm{e}^{(\tau-\tau_0)\frac{\partial}{\partial s}}) \left(L - \frac{\partial}{\partial s} \right)^{-1} Q_0 K(s) \Big|_{s=0} \right).
\end{aligned} \tag{15.50}$$

We have solved the Boltzmann equation perturbatively up to $O(\epsilon^2)$. We summarize the perturbative solution:

$$\tilde{f}^{(0)} = f^{\mathrm{eq}}, \tag{15.51}$$

$$\tilde{f}^{(1)} = f^{\mathrm{eq}} \bar{f}^{\mathrm{eq}} \left(\mathrm{e}^{L(\tau-\tau_0)} \Psi^{(1)} + (\tau - \tau_0) P_0 F^{(0)} + (\mathrm{e}^{L(\tau-\tau_0)} - 1) L^{-1} Q_0 F^{(0)} \right), \tag{15.52}$$

$$\begin{aligned}
\tilde{f}^{(2)} &= f^{\mathrm{eq}} \bar{f}^{\mathrm{eq}} \left(\mathrm{e}^{L(\tau-\tau_0)} \Psi^{(2)} + (1 - \mathrm{e}^{(\tau-\tau_0)\frac{\partial}{\partial s}}) \left(-\frac{\partial}{\partial s} \right)^{-1} P_0 K(s) \Big|_{s=0} \right. \\
&\quad \left. + (\mathrm{e}^{L(\tau-\tau_0)} - \mathrm{e}^{(\tau-\tau_0)\frac{\partial}{\partial s}}) \left(L - \frac{\partial}{\partial s} \right)^{-1} Q_0 K(s) \Big|_{s=0} \right),
\end{aligned} \tag{15.53}$$

under the initial condition,

$$\tilde{f}^{(0)}(\tau_0) = f^{\mathrm{eq}}, \tag{15.54}$$

$$\tilde{f}^{(1)}(\tau_0) = f^{\mathrm{eq}} \bar{f}^{\mathrm{eq}} \Psi^{(1)}, \tag{15.55}$$

$$\tilde{f}^{(2)}(\tau_0) = f^{\mathrm{eq}} \bar{f}^{\mathrm{eq}} \Psi^{(2)}. \tag{15.56}$$

Here we come to the crucial step, where the proper initial condition/integral constants are identified. At the zeroth-order perturbation we introduced $M + 4$ would-be slow variables, T, μ_A, and u_μ, which form the ordinary/first-order fluid dynamics and accompany the modes belonging to the P_0 space. To go beyond the description of ordinary dissipative fluid, we need to specify relevant fast variables, which we call quasi-slow variables. These variables are then used to identify the first-order initial value $\Psi^{(1)}$. We apply the doublet scheme [69] in order to specify the quasi-slow variables and associated vector space. See Chap. 9 for a detailed account of the doublet scheme.

To this end, we expand the first-order perturbative solution around $\tau = \tau_0$:

$$\tilde{f}^{(1)} = f^{\mathrm{eq}}\bar{f}^{\mathrm{eq}}\Psi^{(1)} + (\tau - \tau_0)f^{\mathrm{eq}}\bar{f}^{\mathrm{eq}}\big(P_0 F^{(0)} + L\Psi^{(1)} + Q_0 F^{(0)}\big) + \mathrm{O}\big((\tau - \tau_0)^2\big). \tag{15.57}$$

As will be seen later, the $O\big((\tau - \tau_0)^2\big)$ terms do not contribute to the RG/E equation. We divide the Q_0 space into P_1 and Q_1 spaces ($Q_0 = P_1 + Q_1$), the former of which consists of the quasi-slow modes. In line with the general idea of the reduction theory, the doublet scheme requires that the tangent space of the perturbative solution at $\tau = \tau_0$ should be as small as for the resultant fluid dynamic equation to be simplified. Here, we spell out the three criteria to identify $\Psi^{(1)}$ and P_1 space:

1. $\Psi^{(1)}$ and $L^{-1}Q_0F^{(0)}$ belongs to a common vector space, which is orthogonal to P_0 space.
2. Define P_1 space by a vector space spanned by $\Psi^{(1)}$ and $L\Psi^{(1)}$.

The first condition gives a prescription to minimally extend the vector space in order to incorporate the quasi-slow modes on top of the zero modes. These conditions are required in order to minimize the dimension of the tangent space of the perturbative solution $\tilde{f}^{(1)}$ at $\tau = \tau_0$. This is indeed fulfilled by choosing $\Psi^{(1)}$ so that $L^{-1}Q_0F^{(0)}$ and $\Psi^{(1)}$ belong to a common vector space as seen from (15.57). The second condition is also required so that the P_1 space accommodates all the terms except for the zero modes in (15.57).

As detailed in the next section $L^{-1}Q_0F^{(0)}$ is computed as,

$$\big[L^{-1}Q_0F^{(0)}\big]_{k,p_k} = [L^{-1}\hat{\Pi}]_{k,p_k}\left(-\frac{\theta}{T}\right) + \sum_{A=1}^{M}[L^{-1}\hat{J}^{\mu}_{A}]_{k,p_k}\left(-\frac{1}{h}\nabla_\mu\frac{\mu_A}{T}\right) + [L^{-1}\hat{\pi}]^{\mu\nu}_{k,p_k}\frac{\sigma^{\mu\nu}}{T}, \tag{15.58}$$

with $(\hat{\Pi}, \hat{J}^{\mu}, \hat{\pi}^{\mu\nu})_{k,p_k} := (\Pi, J^{\mu}, \pi^{\mu\nu})_{k,p_k}/(p_k \cdot u)$ and Π_{k,p_k}, J^{μ}_{A,k,p_k}, and $\pi^{\mu\nu}_{k,p_k}$ are given by

$$
\begin{aligned}
\Pi_{k,p_k} &:= (p_k \cdot u)^2 \Bigg(\mathcal{A} \sum_{l=1}^{N} \frac{m_l^2 M_1^l - M_3^l}{3} \\
&\quad + \sum_{l=1}^{N} \sum_{A=1}^{M} C_A q_l^A \frac{m_l^2 M_0^l - M_2^l}{3} + \frac{1}{3} \Bigg) \\
&\quad + (p_k \cdot u) \Bigg(\sum_{l=1}^{N} C_A q_k^A \frac{m_l^2 M_1^l - M_3^l}{3} \\
&\quad + \sum_{l=1}^{N} \sum_{A,B=1}^{M} \mathcal{D}_{AB} q_k^A q_l^B \frac{m_l^2 M_0^l - M_2^l}{3} \Bigg) - \frac{1}{3} m_k^2, \qquad (15.59) \\
J^{\mu}_{A,k,p_k} &:= \left(q_k^A h - \frac{n_A}{n} (p_k \cdot u) \right) \Delta^{\mu\nu} p_{k\nu} \qquad (15.60) \\
\pi^{\mu\nu}_{k,p_k} &:= \Delta^{\mu\nu\rho\sigma} p_{k\rho} p_{k\sigma}. \qquad (15.61)
\end{aligned}
$$

From (15.58), we see that $L^{-1} Q_0 F^{(0)}$ is given by a linear combination of the vectors $\left([L^{-1}\hat{\Pi}]_{k,p_k}, [L^{-1}\hat{J}^{\mu}_A]_{k,p_k}, [L^{-1}\hat{\pi}^{\mu\nu}]_{k,p_k}\right)$. Therefore, we set Ψ as a linear combination of these basis vectors,

$$
\begin{aligned}
\Psi^{(1)}_{k,p_k} &= \left[\frac{[L^{-1}\hat{\Pi}]_{k,p_k}}{\langle \hat{\Pi}, L^{-1}\hat{\Pi} \rangle} \right] \Pi \\
&\quad + \sum_{A,B=1}^{M} \left[3h [L^{-1} \hat{J}^{\mu}_A]_{k,p_k} \langle \hat{J}^{\nu}, L^{-1} \hat{J}_{\nu} \rangle^{-1}_{AB} \right] J_{B,\mu} \\
&\quad + \left[\frac{5[L^{-1}\hat{\pi}^{\mu\nu}]_{k,p_k}}{\langle \hat{\pi}^{\rho\sigma}, L^{-1}\hat{\pi}_{\rho\sigma} \rangle} \right] \pi_{\mu\nu}. \qquad (15.62)
\end{aligned}
$$

where (A, B)-component of a matrix $\langle \hat{J}^{\nu}, L^{-1}\hat{J}_{\nu} \rangle$ is given by $\langle \hat{J}^{\nu}_A, L^{-1}\hat{J}_{B,\nu} \rangle$, and $\langle \hat{J}^{\nu}, L^{-1}\hat{J}_{\nu} \rangle^{-1}_{AB}$ is (A, B)-component of the inverse of the matrix $\langle \hat{J}^{\nu}, L^{-1}\hat{J}_{\nu} \rangle$. Being a linear combination of the quasi-slow modes, $\Psi^{(1)}$ belongs to the P_1 space. We have introduced the following nine integral constants as coefficients of the basis vectors:

$$
\Pi(\sigma; \tau_0), \quad J^{\mu}_A(\sigma; \tau_0), \quad \pi^{\mu\nu}(\sigma; \tau_0), \qquad (15.63)
$$

which will be additional fluid dynamic variables responsible for describing the mesoscopic dynamics and be identified as the bulk pressure, the heat flow, and the stress tensor, respectively.

Consequently, the fluid dynamics of our interest is capable of describing dynamics in the P_1 space in addition to the P_0 space. Keeping that in mind we eliminate the fast-decaying modes in the Q_1 space from the second-order perturbative solution,

$$\tilde{f}^{(2)} = f^{\text{eq}} \bar{f}^{\text{eq}} \bigg(\mathrm{e}^{L(\tau-\tau_0)} \Psi^{(2)} + \big(1 - \mathrm{e}^{(\tau-\tau_0)\frac{\partial}{\partial s}}\big) \Big(-\frac{\partial}{\partial s} \Big)^{-1} P_0 K(s) \Big|_{s=0}$$
$$+ \big(\mathrm{e}^{L(\tau-\tau_0)} - \mathrm{e}^{(\tau-\tau_0)\frac{\partial}{\partial s}}\big)(P_1 + Q_1)\mathcal{G}(s) Q_0 K(s) \Big|_{s=0} \bigg), \tag{15.64}$$

by setting the initial value to

$$\Psi^{(2)} = -Q_1 \mathcal{G}(s) Q_0 K(s) \Big|_{s=0}, \tag{15.65}$$

with $\mathcal{G}(s) = \big(L - \partial/\partial s\big)^{-1}$. Hence, the second-order perturbative solution takes the following form,

$$\tilde{f}^{(2)} = f^{\text{eq}} \bar{f}^{\text{eq}} \bigg[(\tau - \tau_0) P_0 + (\tau - \tau_0) \mathcal{G}(s)^{-1} P_1 \mathcal{G}(s) Q_0$$
$$- \Big(1 + (\tau - \tau_0) \frac{\partial}{\partial s} \Big) Q_1 \mathcal{G}(s) Q_0 \bigg] K(s) \Big|_{s=0} + O\big((\tau - \tau_0)^2\big). \tag{15.66}$$

We expanded with respect to $(\tau - \tau_0)$ as the terms of $O\big((\tau - \tau_0)^2\big)$ disappear in the RG/E equation, and hence, do not contribute to the global solution.

Let us wrap up the discussion so far by writing down the perturbative solution $\tilde{f}(\tau; \tau_0)$ and the initial value $\tilde{f}(\tau_0)$ up to $O(\epsilon^2)$,

$$\tilde{f}(\tau; \tau_0) = f^{\text{eq}}(\tau_0) + \epsilon f^{\text{eq}} \bar{f}^{\text{eq}} \Big(\big[1 + (\tau - \tau_0) L \big] \Psi^{(1)} + (\tau - \tau_0) F^{(0)} \Big)$$
$$+ \epsilon^2 f^{\text{eq}} \bar{f}^{\text{eq}} \bigg[(\tau - \tau_0) P_0 + (\tau - \tau_0) \mathcal{G}(s)^{-1} P_1 \mathcal{G}(s) Q_0$$
$$- \Big(1 + (\tau - \tau_0) \frac{\partial}{\partial s} \Big) Q_1 \mathcal{G}(s) Q_0 \bigg] K(s) \Big|_{s=0}, \tag{15.67}$$

$$\tilde{f}(\tau_0) = f^{\text{eq}} + \epsilon f^{\text{eq}} \bar{f}^{\text{eq}} \Psi^{(1)} - \epsilon^2 f^{\text{eq}} \bar{f}^{\text{eq}} Q_1 \mathcal{G}(s) Q_0 K(s) \Big|_{s=0}. \tag{15.68}$$

15.3.2 Computation of $L^{-1} Q_0 F^{(0)}$

To explicitly write down $\Psi^{(1)}$ we need to compute $L^{-1} Q_0 F^{(0)}$. Using the definition (15.45) of the projection operator P_0 and Q_0, we find,

$$[Q_0 F^{(0)}]_{k,p_k} = [F^{(0)} - P_0 F^{(0)}]_{k,p_k} = F^{(0)}_{k,p_k} - \varphi^{\alpha}_{k,p_k} \eta^{-1}_{\alpha\beta} \langle \varphi^{\beta}, F^{(0)} \rangle, \tag{15.69}$$

with

$$F^{(0)}_{k,p_k} = \frac{1}{p_k \cdot u}\Big[p_k^\mu p_k^\nu \nabla_\nu \frac{u_\nu}{T} - p_k^\mu \sum_{A=1}^{M} q_k^A \nabla_\mu \frac{\mu_A}{T}\Big]. \tag{15.70}$$

In order to proceed the computation we introduce

$$M_\ell^k := \int \mathrm{d}p_k f^{\mathrm{eq}}_{k,p_k} \bar{f}^{\mathrm{eq}}_{k,p_k} (p_k \cdot u)^\ell, \quad \ell = 0,\ 1,\ \ldots. \tag{15.71}$$

Then, each component of the metric $\eta^{\alpha\beta} = \langle \varphi^\alpha, \varphi^\beta \rangle$ is expressed as,

$$\eta^{\mu\nu} = \sum_{k=1}^{N} \Big[a_3^k u^\mu u^\nu + \frac{m_k^2 M_1^k - M_3^k}{3} \Delta^{\mu\nu}\Big], \tag{15.72}$$

$$\eta^{\mu\ A+3} = \eta^{A+3\ \mu} = \sum_{k=1}^{N} q_k^A M_2^k u^\mu, \tag{15.73}$$

$$\eta^{A+3\ B+3} = \sum_{k=1}^{N} q_k^A q_k^B M_1^k. \tag{15.74}$$

The inverse metric $\eta^{-1}_{\alpha\beta}$ is given by,

$$\eta^{-1}_{\mu\nu} = \mathcal{A} u^\mu u^\nu + \mathcal{B} \Delta^{\mu\nu}, \tag{15.75}$$

$$\eta^{-1}_{\mu\ A+3} = \eta^{-1}_{A+3\ \mu} = C_A u^\mu, \tag{15.76}$$

$$\eta^{-1}_{A+3\ B+3} = \mathcal{D}_{AB}, \tag{15.77}$$

where each coefficient satisfies,

$$\mathcal{B} = \left(\sum_{k=1}^{N} \frac{m^2 M_1^k - M_3^k}{3}\right)^{-1}, \tag{15.78}$$

$$\sum_{k=1}^{N} M_3^k \mathcal{A} - 1 + \sum_{A=1}^{M} \sum_{k=1}^{N} q_k^A M_2^k C_A = 0, \tag{15.79}$$

$$\sum_{k=1}^{N} q_k^A M_2^k \mathcal{A} + \sum_{B=1}^{M} \sum_{k=1}^{N} q_k^A q_k^B M_1^k C_B = 0, \tag{15.80}$$

$$\sum_{k=1}^{N} q_k^A M_2^k C_B + \sum_{C=1}^{M} \sum_{k=1}^{N} q_k^A q_k^C M_1^k \mathcal{D}_{CB} = \delta_{AB}. \tag{15.81}$$

The inner products are evaluated as follows:

$$\langle \varphi^\mu, F^{(0)} \rangle$$
$$= \sum_{k=1}^{N} \left[\frac{m_k^2 M_1^k - M_3^k}{3} \left(-\frac{1}{T^2} \nabla^\mu T + u^\mu \frac{1}{T} \nabla \cdot u \right) - \frac{m^2 M_0^k - M_2^k}{3} \sum_{A=1}^{M} q_k^A \nabla^\mu \frac{\mu_A}{T} \right], \tag{15.82}$$

$$\langle \varphi^{A+3}, F^{(0)} \rangle = \sum_{k=1}^{N} q_k^A \frac{m_k^2 M_0^k - M_2^k}{3} \frac{1}{T} \nabla \cdot u. \tag{15.83}$$

Inserting the inverse metric (15.75)–(15.77) and the inner products (15.82) and (15.83) into (15.69), we find

$$\left[Q_0 F^{(0)} \right]_{k, p_k} = \hat{\Pi}_{k, p_k} \left(-\frac{\theta}{T} \right) + \sum_{A=1}^{M} \hat{J}^\mu_{A,k,p_k} \left(-\frac{1}{h} \nabla_\mu \frac{\mu_A}{T} \right) + \hat{\pi}^{\mu\nu}_{k,p_k} \frac{\sigma^{\mu\nu}}{T}. \tag{15.84}$$

Applying L^{-1} yields (15.58).

15.3.3 RG Improvement by Envelope Equation

The obtained perturbative solution (15.67) breaks down due to the secular terms exhibiting divergences as $|\tau - \tau_0|$ gets bigger. In order to circumvent the issue, we promote the integral constant $T(\tau_0)$, $\mu_A(\tau_0)$, u_μ $\Pi(\tau_0)$, $J_{B,\mu}(\tau_0)$, and $\pi_{\mu\nu}(\tau_0)$ to dynamical variables in such a way that the divergences are eliminated. The requirement is fulfilled by imposing the RG/E equation on the perturbative solution,

$$\frac{\mathrm{d}}{\mathrm{d}\tau_0} \tilde{f}(\tau; \tau_0) \Big|_{\tau_0 = \tau} = 0. \tag{15.85}$$

This equation can be viewed as a dynamical equation for $T(\tau_0)$, $\mu_A(\tau_0)$, u_μ, $\Pi(\tau_0)$, $J_{B,\mu}(\tau_0)$, and $\pi_{\mu\nu}(\tau_0)$, which is solved to specify their time evolution. Provided the solution to the RG/E equation, we obtain the globally well-defined solution,

$$f^{\text{global}} = f(\tau_0 = \tau) = f^{\text{eq}} + \epsilon f^{\text{eq}} \bar{f}^{\text{eq}} \Psi^{(1)} - \epsilon^2 f^{\text{eq}} \bar{f}^{\text{eq}} Q_1 G(s) Q_0 K(s) \Big|_{s=0}, \tag{15.86}$$

with the solution of (15.85) inserted. In what follows, we attempt to convert the RG/E equation (15.85) to the fluid dynamic equation.

We firstly project the RG/E equation onto the P_0 space by taking the inner product (15.37) with the zero modes,

$$0 = \left\langle (f^{\text{eq}}\bar{f}^{\text{eq}})^{-1}\varphi^{\alpha}, \frac{\mathrm{d}}{\mathrm{d}\tau_0}\tilde{f}(\tau;\tau_0)\Big|_{\tau_0=\tau} \right\rangle$$
$$= \sum_k \int \mathrm{d}p_k \varphi^{\alpha}_{k,p_k}\Big[(p_k\cdot u)\frac{\partial}{\partial\tau} + \epsilon p_k\cdot\nabla\Big]\Big[f^{\text{eq}}\big(1+\epsilon\bar{f}^{\text{eq}}\Psi^{(1)}\big)\Big]_{k,p_k} + O(\epsilon^3), \tag{15.87}$$

which forms the equation of continuity. It can indeed be written as follows,

$$\partial_\mu J^{\mu\alpha} = 0, \tag{15.88}$$

where, with $\epsilon = 1$, $J^{\mu\alpha}$ reads

$$J^{\mu\alpha} = \sum_{k=1}^{N}\int \mathrm{d}p_k p_k^{\mu}\varphi^{\alpha}_{k,p_k}\Big[f^{\text{eq}}\big(1+\bar{f}^{\text{eq}}\Psi^{(1)}\big)\Big]_{k,p_k}$$
$$= \begin{cases} T^{\mu\nu} = eu^{\mu}u^{\nu} - (P+\Pi)\Delta^{\mu\nu} + \pi^{\mu\nu}, & \alpha = \nu, \\ N^{\mu A} = n_A u^{\mu} + J^{\mu}_A, & \alpha = A+3. \end{cases} \tag{15.89}$$

Let us go through the second equality in more detail. Non-dissipative contributions come from the first term of $J^{\mu\alpha}$. For $\alpha = \nu$,

$$T^{\mu\nu}_{(0)} = \sum_{k=1}^{N}\int \mathrm{d}p_k p_k^{\mu} p_k^{\nu} f^{\text{eq}}_{k,p_k} = eu^{\mu}u^{\nu} - P\Delta^{\mu\nu}, \tag{15.90}$$

where the subscript "0" indicates the leading-order contribution. The internal energy e and pressure P are given by,

$$e := u_\mu u_\nu T^{\mu\nu} = \sum_{k=1}^{N}\int \mathrm{d}p_k (p_k\cdot u)^2 f_{k,p_k}, \tag{15.91}$$

$$P := -\frac{1}{3}\Delta_{\mu\nu}T^{\mu\nu} = -\frac{1}{3}\sum_{k=1}^{N}\int \mathrm{d}p_k \Delta_{\mu\nu} p_k^{\mu} p_k^{\nu} f_{k,p_k}. \tag{15.92}$$

For $\alpha = A+3$,

$$N^{\mu A}_{(0)} = \sum_{k=1}^{N} q_k^A \int \mathrm{d}p_k p_k^{\mu} f^{\text{eq}}_{k,p_k} = n_A u^{\mu}, \tag{15.93}$$

with the microscopic expressions of particle-number density n_A,

$$n_A := u_\mu N^{\mu}_A = \sum_{k=1}^{N} q_k^A \int \mathrm{d}p_k (p_k\cdot u) f_{k,p_k}. \tag{15.94}$$

Next, we compute the dissipative correction to the current $J^{\mu\alpha}$. For $\alpha = \nu$,

$$
\begin{aligned}
\delta T^{\mu\nu} &= \sum_{k=1}^{N} \int \mathrm{d}p_k p_k^\mu p_k^\nu [f^{\mathrm{eq}} \bar{f}^{\mathrm{eq}} \Psi^{(1)}]_{k,p_k} \\
&= \sum_k \int \mathrm{d}p_k (p_k \cdot u) f^{\mathrm{eq}}_{k,p_k} \bar{f}^{\mathrm{eq}}_{k,p_k} \left[Q_0 \frac{p_k^\mu p_k^\nu}{p_k \cdot u} \right] \\
&\times \left(\left[\frac{[L^{-1}\hat{\Pi}]_{k,p_k}}{-T\zeta} \right] \Pi + \sum_{A=1}^{M} \left[\frac{[L^{-1} \hat{J}_A^\rho]_{k,p_k}}{T^2/h} \right] \sum_{B=1}^{M} (\lambda^{-1})_{AB} J_{B,\rho} + \left[\frac{[L^{-1}\hat{\pi}^{\rho\sigma}]_{k,p_k}}{-2T\eta} \right] \pi_{\rho\sigma} \right) \\
&= \sum_k \int \mathrm{d}p_k (p_k \cdot u) f^{\mathrm{eq}}_{k,p_k} \bar{f}^{\mathrm{eq}}_{k,p_k} (-\Delta^{\mu\nu} \hat{\Pi}_{k,p_k} + \hat{\pi}^{\mu\nu}_{k,p_k}) \\
&\times \left(\left[\frac{[L^{-1}\hat{\Pi}]_{k,p_k}}{-T\zeta} \right] \Pi + \sum_{A=1}^{M} \left[\frac{[L^{-1} \hat{J}_A^\rho]_{k,p_k}}{T^2/h} \right] \sum_{B=1}^{M} (\lambda^{-1})_{AB} J_{B,\rho} + \left[\frac{[L^{-1}\hat{\pi}^{\rho\sigma}]_{k,p_k}}{-2T\eta} \right] \pi_{\rho\sigma} \right) \\
&= -\Delta^{\mu\nu} \left(\sum_k \int \mathrm{d}p_k (p_k \cdot u) f^{\mathrm{eq}}_{k,p_k} \bar{f}^{\mathrm{eq}}_{k,p_k} \hat{\Pi}_{k,p_k} [L^{-1}\hat{\Pi}]_{k,p_k} \right) \frac{\Pi}{-T\zeta} \\
&+ \left(\sum_k \int \mathrm{d}p_k (p_k \cdot u) f^{\mathrm{eq}}_{k,p_k} \bar{f}^{\mathrm{eq}}_{k,p_k} \hat{\pi}^{\mu\nu}_{k,p_k} [L^{-1}\hat{\pi}^{\rho\sigma}]_{k,p_k} \right) \frac{\pi_{\rho\sigma}}{-2T\eta} \\
&= -\Delta^{\mu\nu} \Pi + \pi^{\mu\nu}.
\end{aligned} \tag{15.95}
$$

We have inserted the projection operator Q_0 in the second equality by noting that $\Psi^{(1)}$ is in the Q_0 space, *i.e.*, $Q_0 \Psi^{(1)} = \Psi^{(1)}$. For $\alpha = A + 3$, we compute the dissipative correction to the Ath conserved current,

$$
\begin{aligned}
\delta N^{\mu A} &= \sum_{k=1}^{N} q_k^A \int \mathrm{d}p_k p_k^\mu [f^{\mathrm{eq}} \bar{f}^{\mathrm{eq}} \Psi^{(1)}]_{k,p_k} \\
&= \sum_k \int \mathrm{d}p_k (p_k \cdot u) f^{\mathrm{eq}}_{k,p_k} \bar{f}^{\mathrm{eq}}_{k,p_k} \left[Q_0 \frac{p_k^\mu q_k^A}{p_k \cdot u} \right] \\
&\times \left(\left[\frac{[L^{-1}\hat{\Pi}]_{k,p_k}}{-T\zeta} \right] \Pi + \sum_{B=1}^{M} \left[\frac{[L^{-1} \hat{J}_B^\rho]_{k,p_k}}{T^2/h} \right] \sum_{C=1}^{M} (\lambda^{-1})_{BC} J_{C,\rho} + \left[\frac{[L^{-1}\hat{\pi}^{\rho\sigma}]_{k,p_k}}{-2T\eta} \right] \pi_{\rho\sigma} \right) \\
&= \sum_k \int \mathrm{d}p_k (p_k \cdot u) f^{\mathrm{eq}}_{k,p_k} \bar{f}^{\mathrm{eq}}_{k,p_k} \left(\frac{1}{h} \hat{J}^\mu_{A,k,p_k} \right) \\
&\times \left(\left[\frac{[L^{-1}\hat{\Pi}]_{k,p_k}}{-T\zeta} \right] \Pi + \sum_{b=1}^{M} \left[\frac{[L^{-1} \hat{J}_B^\rho]_{k,p_k}}{T^2/h} \right] \sum_{C=1}^{M} (\lambda^{-1})_{BC} J_{C,\rho} + \left[\frac{[L^{-1}\hat{\pi}^{\rho\sigma}]_{k,p_k}}{-2T\eta} \right] \pi_{\rho\sigma} \right) \\
&= J_A^\mu.
\end{aligned} \tag{15.96}
$$

Thus, we have obtained (15.89).

Next, we project the RG/E equation onto the P_1 space by taking the inner product with the quasi-slow modes $[L^{-1}\hat{\Pi}]_{k,p_k}$, $[L^{-1}\hat{J}^{\mu}_{A}]_{k,p_k}$, and $[L^{-1}\hat{\pi}^{\mu\nu}]_{k,p_k}$,

$$
\begin{aligned}
0 &= \left\langle (f^{\mathrm{eq}}\bar{f}^{\mathrm{eq}})^{-1}L^{-1}(\hat{\Pi},\hat{J}^{\mu}_{A},\hat{\pi}^{\mu\nu}),\ \frac{\mathrm{d}}{\mathrm{d}\tau_0}\tilde{f}(\tau;\tau_0)\Big|_{\tau_0=\tau}\right\rangle \\
&= \sum_{k=1}^{N}\int \mathrm{d}p_k\Big[L^{-1}(\hat{\Pi},\hat{J}^{\mu}_{A},\hat{\pi}^{\mu\nu})\Big]_{k,p_k} \\
&\quad\times\Big[(p_k\cdot u)\frac{\partial}{\partial\tau}+\epsilon p_k\cdot\nabla\Big]\Big[f^{\mathrm{eq}}\big(1+\epsilon\bar{f}^{\mathrm{eq}}\Psi^{(1)}\big)\Big]_{k,p_k} \\
&\quad-\Big\langle L^{-1}(\hat{\Pi},\hat{J}^{\mu}_{A},\hat{\pi}^{\mu\nu}),L\Psi^{(1)}\Big\rangle-\frac{1}{2}\Big\langle L^{-1}(\hat{\Pi},\hat{J}^{\mu}_{A},\hat{\pi}^{\mu\nu}),B[\Psi^{(1)},\Psi^{(1)}]\Big\rangle+O(\epsilon^3).
\end{aligned}
\tag{15.97}
$$

A lengthy algebraic manipulation converts the above equations into the relaxation equations, which, combined with the continuity equations, form the full second-order fluid dynamic equation. The detailed derivation is delegated to the following section. The relaxation dynamics of each dissipative current is governed by,

$$
\begin{aligned}
\Pi &= -\zeta\theta-\tau_{\Pi}\frac{\partial}{\partial\tau}\Pi-\sum_{a=1}^{M}\ell^{a}_{\Pi J}\nabla_{\rho}J^{\rho}_{A} \\
&\quad+\kappa_{\Pi\Pi}\Pi\theta+\sum_{A=1}^{M}\kappa^{(1)A}_{\Pi J}J_{A,\rho}\nabla^{\rho}T+\sum_{A,B=1}^{M}\kappa^{(2)BA}_{\Pi J}J_{A,\rho}\nabla^{\rho}\frac{\mu_B}{T}+\kappa_{\Pi\pi}\pi_{\rho\sigma}\sigma^{\rho\sigma} \\
&\quad+b_{\Pi\Pi\Pi}\Pi^2+\sum_{A,B=1}^{M}b^{AB}_{\Pi JJ}J^{\rho}_{A}J_{B,\rho}+b_{\Pi\pi\pi}\pi^{\rho\sigma}\pi_{\rho\sigma},
\end{aligned}
\tag{15.98}
$$

$$
\begin{aligned}
J^{\mu}_{A} &= \sum_{B=1}^{M}\lambda_{AB}\frac{T^2}{h^2}\nabla^{\mu}\frac{\mu_B}{T}-\sum_{B=1}^{M}\tau^{AB}_{J}\Delta^{\mu\rho}\frac{\partial}{\partial\tau}J_{B,\rho}-\ell^{A}_{J\Pi}\nabla^{\mu}\Pi-\ell^{A}_{J\pi}\Delta^{\mu\rho}\nabla_{\nu}\pi^{\nu}{}_{\rho} \\
&\quad+\kappa^{(1)A}_{J\Pi}\Pi\nabla^{\mu}T+\sum_{B=1}^{M}\kappa^{(2)AB}_{J\Pi}\Pi\nabla^{\mu}\frac{\mu_B}{T} \\
&\quad+\sum_{B=1}^{M}\kappa^{(1)AB}_{JJ}J^{\mu}_{B}\theta+\sum_{B=1}^{M}\kappa^{(2)AB}_{JJ}J_{B,\rho}\sigma^{\mu\rho}+\kappa^{(3)AB}_{JJ}J_{B,\rho}\omega^{\mu\rho} \\
&\quad+\kappa^{(1)A}_{J\pi}\pi^{\mu\rho}\nabla_{\rho}T+\sum_{B=1}^{M}\kappa^{(2)AB}_{J\pi}\pi^{\mu\rho}\nabla_{\rho}\frac{\mu_B}{T} \\
&\quad+\sum_{B=1}^{M}b^{AB}_{J\Pi J}\Pi J^{\mu}_{B}+\sum_{B=1}^{M}b^{AB}_{JJ\pi}J_{B,\rho}\pi^{\rho\mu},
\end{aligned}
\tag{15.99}
$$

$$
\begin{aligned}
\pi^{\mu\nu} &= 2\eta\sigma^{\mu\nu} - \tau_\pi \Delta^{\mu\nu\rho\sigma}\frac{\partial}{\partial\tau}\pi_{\rho\sigma} - \sum_{A=1}^{M}\ell^{A}_{\pi J}\nabla^{\langle\mu}J^{\nu\rangle}_{A} \\
&+ \kappa_{\pi\Pi}\Pi\sigma^{\mu\nu} + \sum_{A=1}^{M}\kappa^{(1)A}_{\pi J}J^{\langle\mu}_{A}\nabla^{\nu\rangle}T + \sum_{A,B=1}^{M}\kappa^{(2)BA}_{\pi J}J^{\langle\mu}_{A}\nabla^{\nu\rangle}\frac{\mu_B}{T} \\
&+ \kappa^{(1)}_{\pi\pi}\pi^{\mu\nu}\theta + \kappa^{(2)}_{\pi\pi}\pi^{\lambda\langle\mu}\sigma^{\nu\rangle}{}_{\lambda} + \kappa^{(3)}_{\pi\pi}\pi^{\lambda\langle\mu}\omega^{\nu\rangle}{}_{\lambda} \\
&+ b_{\pi\Pi\pi}\Pi\pi^{\mu\nu} + \sum_{A,B=1}^{M}b^{AB}_{\pi JJ}J^{\langle\mu}_{A}J^{\nu\rangle}_{B} + b_{\pi\pi\pi}\pi^{\lambda\langle\mu}\pi^{\nu\rangle}{}_{\lambda},
\end{aligned}
\tag{15.100}
$$

where we have defined a traceless symmetric tensor $A^{\langle\mu\nu\rangle} := \Delta^{\mu\nu\rho\sigma}A_{\rho\sigma}$ for an arbitrary rank-two tensor A, the scalar expansion $\theta := \nabla \cdot u$, the shear tensor $\sigma^{\mu\nu} := \Delta^{\mu\nu\rho\sigma}\nabla_\rho u_\sigma$, and the vorticity $\omega^{\mu\nu} := \frac{1}{2}(\nabla^\mu u^\nu - \nabla^\nu u^\mu)$.

The microscopic expressions of derived transport coefficients, ζ, λ_{AB}, and η, are

$$
\zeta = -\frac{1}{T}\langle\hat{\Pi}, L^{-1}\hat{\Pi}\rangle, \tag{15.101}
$$

$$
\lambda_{AB} = \frac{1}{3T^2}\langle\hat{J}^{\mu}_{A}, L^{-1}\hat{J}_{B,\mu}\rangle, \tag{15.102}
$$

$$
\eta = -\frac{1}{10T}\langle\hat{\pi}^{\mu\nu}, L^{-1}\hat{\pi}_{\mu\nu}\rangle. \tag{15.103}
$$

We also write down the relaxation times τ_Π, τ^{AB}_{J}, and τ_π,

$$
\tau_\Pi = -\frac{\langle\hat{\Pi}, L^{-2}\hat{\Pi}\rangle}{\langle\hat{\Pi}, L^{-1}\hat{\Pi}\rangle}, \tag{15.104}
$$

$$
\tau^{AB}_{J} = -\sum_{C=1}^{M}\langle\hat{J}^{\nu}, L^{-2}\hat{J}_{\nu}\rangle_{AC}\langle\hat{J}^{\mu}, L^{-1}\hat{J}_{\mu}\rangle^{-1}_{CB}, \tag{15.105}
$$

$$
\tau_\pi = -\frac{\langle\hat{\pi}^{\mu\nu}, L^{-2}\hat{\pi}_{\mu\nu}\rangle}{\langle\hat{\pi}^{\rho\sigma}, L^{-1}\hat{\pi}_{\rho\sigma}\rangle}. \tag{15.106}
$$

These microscopic formulas allow us to derive crucial properties that the proper fluid dynamics are expected to obey. This fact, in turn, supports the validity of our derivation of the fluid dynamics. Those properties are discussed in Sect. 15.4.2.

15.3.4 Derivation of Relaxation Equations and Transport Coefficients

We present the full derivation of the relaxation equations (15.98)–(15.100) as well as microscopic expressions of all the transport coefficients that appeared in the equation.

To this end, we carry out algebraic manipulation on Eq. (15.97), that is repeated here:

$$
\begin{aligned}
0 = & \sum_{k=1}^{N} \int \mathrm{d}p_k \Big[L^{-1}\big(\hat{\Pi}, \hat{J}^{\mu}_{A}, \hat{\pi}^{\mu\nu}\big)\Big]_{k,p_k} \\
& \times \Big[(p_k \cdot u)\frac{\partial}{\partial \tau} + \epsilon p_k \cdot \nabla\Big]\Big[f^{\mathrm{eq}}\big(1 + \epsilon \bar{f}^{\mathrm{eq}}\Psi^{(1)}\big)\Big]_{k,p_k} \\
& - \Big\langle L^{-1}\big(\hat{\Pi}, \hat{J}^{\mu}_{A}, \hat{\pi}^{\mu\nu}\big), L\Psi^{(1)}\Big\rangle - \frac{1}{2}\Big\langle L^{-1}\big(\hat{\Pi}, \hat{J}^{\mu}_{A}, \hat{\pi}^{\mu\nu}\big), B[\Psi^{(1)}, \Psi^{(1)}]\Big\rangle + O(\epsilon^3).
\end{aligned}
\tag{15.107}
$$

We introduce the following quantities for the sake of notational convenience:

$$
\hat{\psi}^{i}_{k,p_k} := \big\{\hat{\Pi}_{k,p_k}, \hat{J}^{\mu}_{A,k,p_k}, \hat{\pi}^{\mu\nu}_{k,p_k}\big\}, \tag{15.108}
$$

$$
\psi_i := \big\{\Pi, J^{\mu}_{A}, \pi^{\mu\nu}\big\}, \tag{15.109}
$$

$$
\hat{\chi}^{i} := \left\{\frac{\hat{\Pi}_{k,p_k}}{-T\zeta}, \frac{\sum_B \hat{J}^{\mu}_{B,k,p_k}(\lambda^{-1})_{BA}}{T^2/h}, \frac{\hat{\pi}^{\mu\nu}_{k,p_k}}{-2T\eta}\right\}, \tag{15.110}
$$

$$
X_i := \left\{-\zeta\theta, \frac{T^2}{h^2}\sum_{B=1}^{M}\lambda_{AB}\nabla_{\mu}\frac{\mu_B}{T}, 2\eta\sigma^{\mu\nu}\right\}, \tag{15.111}
$$

$$
v^{\mu}_{k,p_k} := \frac{1}{p_k \cdot u}\Delta^{\mu\nu}p_{k,\nu}. \tag{15.112}
$$

Then, Eq. (15.107) is concisely written as

$$
\begin{aligned}
& \langle \hat{L}^{-1}\hat{\psi}^{i}, (f^{\mathrm{eq}}\bar{f}^{\mathrm{eq}})^{-1}[\partial_\tau + \epsilon v \cdot \nabla][f^{\mathrm{eq}}(1 + \epsilon \bar{f}^{\mathrm{eq}}\hat{L}^{-1}\hat{\psi}^{j}\chi_j)]\rangle \\
& = \epsilon\langle \hat{L}^{-1}\,\hat{\psi}^{i}, \hat{\psi}^{j}\chi_j\rangle \\
& + \epsilon^2\frac{1}{2}\langle \hat{L}^{-1}\hat{\psi}^{i}, B[\hat{L}^{-1}\hat{\psi}^{j}\chi_j, \hat{L}^{-1}\hat{\psi}^{k}\chi_k]\rangle + O(\epsilon^3).
\end{aligned}
\tag{15.113}
$$

Expanding the term on the left-hand side, we have

$$
\begin{aligned}
& \epsilon\big\langle L^{-1}\hat{\psi}^{i}, \hat{\chi}^{j}\big\rangle X_j \\
& + \epsilon\Big\langle L^{-1}\hat{\psi}^{i}, (f^{\mathrm{eq}}\bar{f}^{\mathrm{eq}})^{-1}\left[\frac{\partial}{\partial \tau} + \epsilon v \cdot \nabla\right] f^{\mathrm{eq}}\bar{f}^{\mathrm{eq}}L^{-1}\hat{\chi}^{j}\Big\rangle \psi_j \\
& + \epsilon\big\langle L^{-1}\hat{\psi}^{i}, L^{-1}\hat{\chi}^{j}\big\rangle\frac{\partial}{\partial \tau}\psi_j + \epsilon^2\big\langle L^{-1}\hat{\psi}^{i}, v^{\alpha}L^{-1}\hat{\psi}^{j}\big\rangle\nabla_{\alpha}\psi_j \\
& = \epsilon\big\langle L^{-1}\hat{\psi}^{i}, \hat{\chi}^{j}\big\rangle\psi_j \\
& + \epsilon^2\frac{1}{2}\big\langle L^{-1}\hat{\psi}^{i}, B[L^{-1}\hat{\chi}^{j}, L^{-1}\hat{\chi}^{k}]\big\rangle\psi_j\psi_k + O(\epsilon^3).
\end{aligned}
\tag{15.114}
$$

The coefficients of the first term on the left-hand side and the first and second terms on the right-hand side of Eq. (15.114) can be written as

$$\langle \hat{L}^{-1}\hat{\psi}^i, \hat{\chi}^j\rangle = \begin{pmatrix} \frac{\langle \hat{L}^{-1}\hat{\Pi}, \hat{\Pi}\rangle}{-T\zeta} & \sum_C \frac{\langle \hat{L}^{-1}\hat{\Pi}, \hat{J}^\rho_C\rangle}{T^2/h}(\lambda^{-1})_{CB} & \frac{\langle \hat{L}^{-1}\hat{\Pi}, \hat{\pi}^{\rho\sigma}\rangle}{-2T\eta} \\ \frac{\langle \hat{L}^{-1}\hat{J}^\mu_A, \hat{\Pi}\rangle}{-T\zeta} & \sum_C \frac{\langle \hat{L}^{-1}\hat{J}^\mu_A, \hat{J}^\rho_C\rangle}{T^2/h}(\lambda^{-1})_{CB} & \frac{\langle \hat{L}^{-1}\hat{J}^\mu_A, \hat{\pi}^{\rho\sigma}\rangle}{-2T\eta} \\ \frac{\langle \hat{L}^{-1}\hat{\pi}^{\mu\nu}, \hat{\Pi}\rangle}{-T\zeta} & \sum_C \frac{\langle \hat{L}^{-1}\hat{\pi}^{\mu\nu}, \hat{J}^\rho_B\rangle}{T^2/h}(\lambda^{-1})_{CB} & \frac{\langle \hat{L}^{-1}\hat{\pi}^{\mu\nu}, \hat{\pi}^{\rho\sigma}\rangle}{-2T\eta} \end{pmatrix}$$

$$= \begin{pmatrix} 1 & 0 & 0 \\ 0 & h\delta_{AB}\Delta^{\mu\rho} & 0 \\ 0 & 0 & \Delta^{\mu\nu\rho\sigma} \end{pmatrix}, \tag{15.115}$$

$$\langle \hat{L}^{-1}\hat{\psi}^i, \hat{L}^{-1}\hat{\chi}^j\rangle = \begin{pmatrix} \frac{\langle \hat{L}^{-1}\hat{\Pi}, \hat{L}^{-1}\hat{\Pi}\rangle}{-T\zeta} & \sum_C \frac{\langle \hat{L}^{-1}\hat{\Pi}, \hat{L}^{-1}\hat{J}^\rho_C\rangle}{T^2/h}(\lambda^{-1})_{CB} & \frac{\langle \hat{L}^{-1}\hat{\Pi}, \hat{L}^{-1}\hat{\pi}^{\rho\sigma}\rangle}{-2T\eta} \\ \frac{\langle \hat{L}^{-1}\hat{J}^\mu_A, \hat{L}^{-1}\hat{\Pi}\rangle}{-T\zeta} & \sum_C \frac{\langle \hat{L}^{-1}\hat{J}^\mu_A, \hat{L}^{-1}\hat{J}^\rho_C\rangle}{T^2/h}(\lambda^{-1})_{CB} & \frac{\langle \hat{L}^{-1}\hat{J}^\mu_A, \hat{L}^{-1}\hat{\pi}^{\rho\sigma}\rangle}{-2T\eta} \\ \frac{\langle \hat{L}^{-1}\hat{\pi}^{\mu\nu}, \hat{L}^{-1}\hat{\Pi}\rangle}{-T\zeta} & \sum_C \frac{\langle \hat{L}^{-1}\hat{\pi}^{\mu\nu}, \hat{L}^{-1}\hat{J}^\rho_C\rangle}{T^2/h}(\lambda^{-1})_{CB} & \frac{\langle \hat{L}^{-1}\hat{\pi}^{\mu\nu}, \hat{L}^{-1}\hat{\pi}^{\rho\sigma}\rangle}{-2T\eta} \end{pmatrix}$$

$$= \begin{pmatrix} -\tau_\Pi & 0 & 0 \\ 0 & -h\tau_J^{AB}\Delta^{\mu\rho} & 0 \\ 0 & 0 & -\tau_\pi\Delta^{\mu\nu\rho\sigma} \end{pmatrix}, \tag{15.116}$$

$$\langle \hat{L}^{-1}\hat{\psi}^i, v^\alpha\hat{L}^{-1}\hat{\chi}^j\rangle = \begin{pmatrix} \frac{\langle \hat{L}^{-1}\hat{\Pi}, v^\alpha\hat{L}^{-1}\hat{\Pi}\rangle}{-T\zeta} & \sum_C \frac{\langle \hat{L}^{-1}\hat{\Pi}, v^\alpha\hat{L}^{-1}\hat{J}^\rho_C\rangle}{T^2/h}(\lambda^{-1})_{CB} & \frac{\langle \hat{L}^{-1}\hat{\Pi}, v^\alpha\hat{L}^{-1}\hat{\pi}^{\rho\sigma}\rangle}{-2T\eta} \\ \frac{\langle \hat{L}^{-1}\hat{J}^\mu_A, v^\alpha\hat{L}^{-1}\hat{\Pi}\rangle}{-T\zeta} & \sum_C \frac{\langle \hat{L}^{-1}\hat{J}^\mu_A, v^\alpha\hat{L}^{-1}\hat{J}^\rho_C\rangle}{T^2/h}(\lambda^{-1})_{CB} & \frac{\langle \hat{L}^{-1}\hat{J}^\mu_A, v^\alpha\hat{L}^{-1}\hat{\pi}^{\rho\sigma}\rangle}{-2T\eta} \\ \frac{\langle \hat{L}^{-1}\hat{\pi}^{\mu\nu}, v^\alpha\hat{L}^{-1}\hat{\Pi}\rangle}{-T\zeta} & \sum_C \frac{\langle \hat{L}^{-1}\hat{\pi}^{\mu\nu}, v^\alpha\hat{L}^{-1}\hat{J}^\rho_C\rangle}{T^2/h}(\lambda^{-1})_{CB} & \frac{\langle \hat{L}^{-1}\hat{\pi}^{\mu\nu}, v^\alpha\hat{L}^{-1}\hat{\pi}^{\rho\sigma}\rangle}{-2T\eta} \end{pmatrix}$$

$$= \begin{pmatrix} 0 & -h\ell^B_{\Pi J}\Delta^{\alpha\rho} & 0 \\ -\ell^A_{J\Pi}\Delta^{\mu\alpha} & 0 & -\ell^A_{J\pi}\Delta^{\mu\alpha\rho\sigma} \\ 0 & -\ell^B_{\pi J}\Delta^{\mu\nu\rho\alpha} & 0 \end{pmatrix}, \tag{15.117}$$

respectively, where we have introduced the relaxation times

$$\tau_\Pi := \frac{1}{T\zeta}\big\langle L^{-1}\hat{\Pi}, L^{-1}\hat{\Pi}\big\rangle, \tag{15.118}$$

$$\tau_J^{AB} := -\frac{1}{3T^2}\sum_{C=1}^{M}\langle L^{-1}\hat{J}_A^{\mu}, L^{-1}\hat{J}_{C,\mu}\rangle(\lambda^{-1})_{CB}, \tag{15.119}$$

$$\tau_\pi := \frac{1}{10T\eta}\langle L^{-1}\hat{\pi}^{\mu\nu}, L^{-1}\hat{\pi}_{\mu\nu}\rangle, \tag{15.120}$$

and the relaxation lengths,

$$\ell_{\Pi J}^{A} := -\frac{1}{3T^2}\sum_{B=1}^{M}(\lambda^{-1})_{AB}\langle L^{-1}\hat{\Pi}, v^{\mu}L^{-1}\hat{J}_{B,\mu}\rangle, \tag{15.121}$$

$$\ell_{J\Pi}^{A} := \frac{1}{T\zeta}\langle L^{-1}\hat{J}_A^{\mu}, v_{\mu}L^{-1}\hat{\Pi}\rangle, \tag{15.122}$$

$$\ell_{J\pi}^{A} := \frac{1}{10T\eta}\langle L^{-1}\hat{J}_A^{\mu}, v^{\nu}L^{-1}\hat{\pi}_{\mu\nu}\rangle, \tag{15.123}$$

$$\ell_{\pi J}^{A} := -\frac{1}{5T^2}\sum_{B=1}^{M}(\lambda^{-1})_{AB}\langle L^{-1}\hat{\pi}^{\mu\nu}, v_{\mu}L^{-1}\hat{J}_{B,\nu}\rangle. \tag{15.124}$$

Then, the last term on the right-hand side of Eq. (15.114) is written as

$$-\frac{1}{2}\langle L^{-1}\hat{\Pi}, B[L^{-1}\hat{\chi}^{j}, L^{-1}\hat{\chi}^{k}]\rangle\psi_j\psi_k$$

$$= b_{\Pi\Pi\Pi}\Pi^2 + \sum_{A,B=1}^{M} b_{\Pi JJ}^{AB}J_A^{\rho}J_{B,\rho} + b_{\Pi\pi\pi}\pi^{\rho\sigma}\pi_{\rho\sigma}, \tag{15.125}$$

$$-\frac{1}{2}\langle L^{-1}\hat{J}_A^{\mu}, B[L^{-1}\hat{\chi}^{j}, L^{-1}\hat{\chi}^{k}]\rangle\psi_j\psi_k$$

$$= \sum_{B=1}^{M}(b_{J\Pi J}^{AB}\Pi J_B^{\mu} + b_{JJ\pi}^{AB}J_{B,\rho}\pi_{\rho\mu}), \tag{15.126}$$

$$-\frac{1}{2}\langle L^{-1}\hat{\pi}^{\mu\nu}, B[L^{-1}\hat{\chi}^{j}, L^{-1}\hat{\chi}^{k}]\rangle\psi_j\psi_k$$

$$= b_{\pi\Pi\pi}\Pi\pi^{\mu\nu} + \sum_{A,B=1}^{M} b_{\pi JJ}^{AB}\Delta^{\mu\nu\rho\sigma}J_{A,\rho}J_{B,\sigma}$$

$$+ b_{\pi\pi\pi}\pi_{\rho}{}^{\lambda}\pi_{\lambda\sigma}, \tag{15.127}$$

where the transport coefficients are defined by

$$b_{\Pi\Pi\Pi} := -\frac{\langle L^{-1}\hat{\Pi}, B[L^{-1}\hat{\Pi}, L^{-1}\hat{\Pi}]\rangle}{2(T\zeta)^2}, \tag{15.128}$$

$$
b^{AB}_{\Pi JJ} := -\sum_{C,D=1}^{M} \frac{\langle L^{-1}\hat{\Pi},\, B[L^{-1}\hat{J}^{\mu}_{C},\, L^{-1}\hat{J}_{D,\mu}]\rangle}{6(T^2/h)^2} \\
\times (\lambda^{-1})_{CA}(\lambda^{-1})_{DB}, \tag{15.129}
$$

$$
b_{\Pi\pi\pi} := -\frac{\langle L^{-1}\hat{\Pi},\, B[L^{-1}\hat{\pi}^{\mu\nu},\, L^{-1}\hat{\pi}_{\mu\nu}]\rangle}{10(2T\eta)^2}, \tag{15.130}
$$

$$
b^{AB}_{J\Pi J} := \sum_{C=1}^{M} \frac{\langle L^{-1}\hat{J}^{\mu}_{A},\, B[L^{-1}\hat{\Pi},\, L^{-1}\hat{J}_{C,\mu}]\rangle}{3(T\zeta)(T^2/h)} (\lambda^{-1})_{CB}, \tag{15.131}
$$

$$
b_{JJ\pi} := \sum_{C,D=1}^{M} \frac{\langle L^{-1}\hat{J}^{\mu}_{A},\, B[L^{-1}\hat{J}^{\nu}_{C},\, L^{-1}\hat{\pi}_{\mu\nu}]\rangle}{5(T^2/h)(2T\eta)} (\lambda^{-1})_{CB}, \tag{15.132}
$$

$$
b_{\pi\Pi\pi} := -\frac{\langle L^{-1}\hat{\pi}^{\mu\nu},\, B[L^{-1}\hat{\Pi},\, L^{-1}\hat{\pi}_{\mu\nu}]\rangle}{5(T\zeta)(T\eta)}, \tag{15.133}
$$

$$
b_{\pi JJ} := -\sum_{C,D=1}^{M} \frac{\langle L^{-1}\hat{\pi}^{\mu\nu},\, B[L^{-1}\hat{J}_{C,\mu},\, L^{-1}\hat{J}_{D,\nu}]\rangle}{10(T^2/h)^2} \\
\times (\lambda^{-1})_{CA}(\lambda^{-1})_{DB}, \tag{15.134}
$$

$$
b_{\pi\pi\pi} := -\frac{\langle L^{-1}\hat{\pi}^{\mu\nu},\, B[L^{-1}\hat{\pi}^{\lambda}_{\mu},\, L^{-1}\hat{\pi}_{\lambda\nu}]\rangle}{(35/6)(2T\eta)^2}. \tag{15.135}
$$

Let us rewrite the second term on the right-hand side of Eq. (15.114):

$$
\Big\langle L^{-1}\hat{\psi}^{i}, (f^{\mathrm{eq}}\bar{f}^{\mathrm{eq}})^{-1}\left[\frac{\partial}{\partial\tau} + \epsilon v\cdot\nabla\right] f^{\mathrm{eq}}\bar{f}^{\mathrm{eq}} L^{-1}\hat{\chi}^{j}\Big\rangle \psi_j \\
= \Big\langle L^{-1}\hat{\psi}^{i}, (f^{\mathrm{eq}}\bar{f}^{\mathrm{eq}})^{-1}\frac{\partial}{\partial T}[f^{\mathrm{eq}}\bar{f}^{\mathrm{eq}} L^{-1}\hat{\chi}^{j}]\Big\rangle \psi_j \frac{\partial}{\partial\tau}T \\
+ \sum_{A=1}^{M}\Big\langle L^{-1}\hat{\psi}^{i}, (f^{\mathrm{eq}}\bar{f}^{\mathrm{eq}})^{-1}\frac{\partial}{\partial\frac{\mu_A}{T}}[f^{\mathrm{eq}}\bar{f}^{\mathrm{eq}} L^{-1}\hat{\chi}^{j}]\Big\rangle \psi_j \frac{\partial}{\partial\tau}\frac{\mu_A}{T} \\
+ \Big\langle L^{-1}\hat{\psi}^{i}, (f^{\mathrm{eq}}\bar{f}^{\mathrm{eq}})^{-1}\frac{\partial}{\partial u^{\beta}}[f^{\mathrm{eq}}\bar{f}^{\mathrm{eq}} L^{-1}\hat{\chi}^{j}]\Big\rangle \psi_j \frac{\partial}{\partial\tau}u^{\beta} \\
+ \epsilon\Big\langle L^{-1}\hat{\psi}^{i}, (f^{\mathrm{eq}}\bar{f}^{\mathrm{eq}})^{-1} v^{\beta}\frac{\partial}{\partial T}[f^{\mathrm{eq}}\bar{f}^{\mathrm{eq}} L^{-1}\hat{\chi}^{j}]\Big\rangle \psi_j \nabla_{\beta}T \\
+ \epsilon\sum_{A=1}^{M}\Big\langle L^{-1}\hat{\psi}^{i}, (f^{\mathrm{eq}}\bar{f}^{\mathrm{eq}})^{-1} v^{\beta}\frac{\partial}{\partial\frac{\mu_A}{T}}[f^{\mathrm{eq}}\bar{f}^{\mathrm{eq}} L^{-1}\hat{\chi}^{j}]\Big\rangle \psi_j \nabla_{\beta}\frac{\mu_A}{T} \\
+ \epsilon\Big\langle L^{-1}\hat{\psi}^{i}, (f^{\mathrm{eq}}\bar{f}^{\mathrm{eq}})^{-1} v^{\beta}\frac{\partial}{\partial u^{\alpha}}[f^{\mathrm{eq}}\bar{f}^{\mathrm{eq}} L^{-1}\hat{\chi}^{j}]\Big\rangle \psi_j \nabla_{\beta}u^{\alpha}. \tag{15.136}
$$

The temporal derivatives of T, μ_A/T, and u^μ are rewritten by using the balance equations up to the first order with respect to ϵ, which correspond to the Euler equations

$$\frac{\partial}{\partial\tau}n_A = -\epsilon n_A \nabla\cdot u + O(\epsilon^2), \tag{15.137}$$

$$\frac{\partial}{\partial\tau}e = -\epsilon nh \nabla\cdot u + O(\epsilon^2), \tag{15.138}$$

$$\frac{\partial}{\partial\tau}u^\mu = \epsilon\frac{1}{nh}\nabla^\mu P + O(\epsilon^2). \tag{15.139}$$

These equations can be written as

$$\sum_{k=1}^{N} q_k^A a_2^k \frac{1}{T^2}\frac{\partial}{\partial\tau}T + \sum_{B=1}^{M}\sum_{k=1}^{N} q_k^A q_k^B a_1^k \frac{\partial}{\partial\tau}\frac{\mu_B}{T}$$
$$= -\epsilon n_A\theta + O(\epsilon^2), \tag{15.140}$$

$$\sum_{k=1}^{N} a_3^k \frac{1}{T^2}\frac{\partial}{\partial\tau}T + \sum_{B=1}^{M}\sum_{k=1}^{N} q_k^B a_2^k \frac{\partial}{\partial\tau}\frac{\mu_B}{T}$$
$$= -\epsilon nh\theta + O(\epsilon^2), \tag{15.141}$$

$$\frac{\partial}{\partial\tau}u^\mu = \epsilon\frac{1}{T}\nabla^\mu T + \epsilon\frac{T}{h}\sum_{A=1}^{M}\sum_{k=1}^{N} x_k q_k^A \nabla^\mu \frac{\mu_A}{T} + O(\epsilon^2), \tag{15.142}$$

where the definitions of a_1^k, a_2^k, and a_3^k are given by

$$a_\ell^k = \int \mathrm{d}p_k f_{k,p_k}^{\mathrm{eq}} \bar{f}_{k,p_k}^{\mathrm{eq}} (p_k\cdot u)^\ell. \tag{15.143}$$

We define a matrix $\mathcal{E}$ by writing Eqs. (15.140) and (15.141) as

$$\mathcal{E}_{A0}\frac{\partial}{\partial\tau}T + \sum_{B=1}^{M}\mathcal{E}_{AB}\frac{\partial}{\partial\tau}\frac{\mu_B}{T} = \epsilon\theta + O(\epsilon^2), \tag{15.144}$$

$$\mathcal{E}_{00}\frac{\partial}{\partial\tau}T + \sum_{B=1}^{M}\mathcal{E}_{0B}\frac{\partial}{\partial\tau}\frac{\mu_B}{T} = \epsilon\theta + O(\epsilon^2). \tag{15.145}$$

Therefore, Eqs. (15.140)–(15.142) can be written as

$$\frac{\partial}{\partial\tau}T = \epsilon I\theta + O(\epsilon^2), \tag{15.146}$$

$$\frac{\partial}{\partial\tau}\frac{\mu_A}{T} = \epsilon I_A\theta + O(\epsilon^2), \tag{15.147}$$

$$\frac{\partial}{\partial\tau}u^{\mu} = \epsilon\frac{1}{T}\nabla^{\mu}T + \epsilon\frac{n_A T}{nh}\sum_{A=1}^{M}\nabla^{\mu}\frac{\mu_A}{T} + O(\epsilon^2), \tag{15.148}$$

where we have defined the coefficients as $\mathcal{I} := \sum_B(\mathcal{E}^{-1})_{0B}$ and $\mathcal{I}_A := \sum_B(\mathcal{E}^{-1})_{AB}$ for notational simplicity. Then, Eq. (15.136) takes the form

$$\begin{aligned}&\left\langle L^{-1}\hat{\Pi}, (f^{\rm eq}\bar{f}^{\rm eq})^{-1}\left[\frac{\partial}{\partial\tau} + \epsilon v\cdot\nabla\right] f^{\rm eq}\bar{f}^{\rm eq}L^{-1}\hat{\chi}^j\right\rangle\psi_j \\ &= \epsilon\Bigg[\kappa_{\Pi\Pi}\Pi\theta + \sum_{A=1}^{M}\kappa_{\Pi J}^{(1)A}J_A^{\rho}\nabla_{\rho}T + \sum_{A,B=1}^{M}\kappa_{\Pi J}^{(2)AB}J_B^{\rho}\nabla_{\rho}\frac{\mu_A}{T} + \kappa_{\Pi\pi}\pi^{\rho\sigma}\sigma_{\rho\sigma}\Bigg],\end{aligned} \tag{15.149}$$

$$\begin{aligned}&\left\langle L^{-1}\hat{J}^{\mu}, (f^{\rm eq}\bar{f}^{\rm eq})^{-1}\left[\frac{\partial}{\partial\tau} + \epsilon v\cdot\nabla\right] f^{\rm eq}\bar{f}^{\rm eq}L^{-1}\hat{\psi}^j\right\rangle\chi_j \\ &= \epsilon\Bigg[\kappa_{J\Pi}^{(1)A}\Pi\nabla^{\mu}T + \sum_{B=1}^{M}\kappa_{J\Pi}^{(2)AB}\Pi\nabla^{\mu}\frac{\mu_B}{T} + \sum_{B=1}^{M}\kappa_{JJ}^{(1)AB}J_B^{\mu}\theta + \sum_{B=1}^{M}\kappa_{JJ}^{(2)AB}J_B^{\rho}\sigma^{\mu}{}_{\rho} \\ &+ \sum_{B=1}^{M}\kappa_{JJ}^{(3)AB}J_B^{\rho}\omega^{\mu}{}_{\rho} + \kappa_{J\pi}^{(1)A}\pi^{\mu\rho}\nabla_{\rho}T + \sum_{B=1}^{M}\kappa_{J\pi}^{(2)AB}\pi^{\mu\rho}\nabla_{\rho}\frac{\mu_B}{T}\Bigg],\end{aligned} \tag{15.150}$$

$$\begin{aligned}&\left\langle L^{-1}\hat{\pi}^{\mu\nu}, (f^{\rm eq}\bar{f}^{\rm eq})^{-1}\left[\frac{\partial}{\partial\tau} + \epsilon v\cdot\nabla\right] f^{\rm eq}\bar{f}^{\rm eq}L^{-1}\hat{\psi}^j\right\rangle\chi_j \\ &= \epsilon\Bigg[\kappa_{\pi\Pi}\Pi\sigma^{\mu\nu} + \sum_{A=1}^{M}\kappa_{\pi J}^{(1)A}J_A^{\langle\mu}\nabla^{\nu\rangle}T + \sum_{A,B=1}^{M}\kappa_{\pi J}^{(2)AB}J_A^{\langle\mu}\nabla^{\nu\rangle}\frac{\mu_B}{T} + \kappa_{\pi\pi}^{(1)}\pi^{\mu\nu}\theta \\ &+ \kappa_{\pi\pi}^{(2)}\pi^{\rho\langle\mu}\sigma^{\nu\rangle}{}_{\rho} + \kappa_{\pi\pi}^{(3)}\pi^{\rho\langle\mu}\omega^{\nu\rangle}{}_{\rho}\Bigg],\end{aligned} \tag{15.151}$$

where we have used $\nabla_{\mu}u_{\nu} = \sigma_{\mu\nu} + \Delta_{\mu\nu}\theta/3 + \omega_{\mu\nu}$ with the vorticity $\omega_{\mu\nu} := (\nabla_{\mu}u_{\nu} - \nabla_{\nu}u_{\mu})/2$, and the transport coefficients are defined as follows:

$$\kappa_{\Pi\Pi} = \left\langle L^{-1}\hat{\Pi}, (f^{\rm eq}\bar{f}^{\rm eq})^{-1}\left[\mathcal{I}\frac{\partial}{\partial T} + \sum_{A=1}^{M}\mathcal{I}_A\frac{\partial}{\partial\frac{\mu_A}{T}} + \frac{1}{3}v^{\mu}\frac{\partial}{\partial u^{\mu}}\right]\frac{f^{\rm eq}\bar{f}^{\rm eq}L^{-1}\hat{\Pi}}{-T\zeta}\right\rangle, \tag{15.152}$$

$$\kappa_{\Pi J}^{(1)A} = \frac{\Delta^{\mu\nu}}{3}\left\langle L^{-1}\hat{\Pi}, (f^{\rm eq}\bar{f}^{\rm eq})^{-1}\left[v_{\mu}\frac{\partial}{\partial T} + \frac{1}{T}\frac{\partial}{\partial u^{\mu}}\right]\sum_{B=1}^{M}\frac{f^{\rm eq}\bar{f}^{\rm eq}L^{-1}\hat{J}_{B,\nu}}{T^2/h}(\lambda^{-1})_{BA}\right\rangle, \tag{15.153}$$

$$\kappa_{\Pi J}^{(2)AB} = \frac{\Delta^{\mu\nu}}{3}\left\langle L^{-1}\hat{\Pi}, (f^{\rm eq}\bar{f}^{\rm eq})^{-1}\left[v_{\mu}\frac{\partial}{\partial\frac{\mu_A}{T}} + \frac{n_A T}{nh}\frac{\partial}{\partial u^{\mu}}\right]\sum_{C=1}^{M}\frac{f^{\rm eq}\bar{f}^{\rm eq}L^{-1}\hat{J}_{C,\nu}}{T^2\lambda/h}(\lambda^{-1})_{CB}\right\rangle, \tag{15.154}$$

$$\kappa_{\Pi\pi} = \frac{\Delta^{\mu\nu\rho\sigma}}{5}\left\langle L^{-1}\hat{\Pi}, (f^{\rm eq}\bar{f}^{\rm eq})^{-1}v_{\mu}\frac{\partial}{\partial u^{\nu}}\frac{f^{\rm eq}\bar{f}^{\rm eq}L^{-1}\hat{\pi}_{\rho\sigma}}{-2T\eta}\right\rangle, \tag{15.155}$$

$$\kappa_{J\Pi}^{(1)A} = \frac{1}{3}\left\langle L^{-1}\hat{J}_A^{\mu}, (f^{\rm eq}\bar{f}^{\rm eq})^{-1}\left[v_{\mu}\frac{\partial}{\partial T} + \frac{1}{T}\frac{\partial}{\partial u^{\mu}}\right]\frac{f^{\rm eq}\bar{f}^{\rm eq}L^{-1}\hat{\Pi}}{-T\zeta}\right\rangle, \tag{15.156}$$

$$\kappa_{J\Pi}^{(2)AB} = \frac{1}{3}\Big\langle L^{-1}\hat{J}_A^{\mu}, (f^{\rm eq}\bar{f}^{\rm eq})^{-1}\left[v_\mu \frac{\partial}{\partial\frac{\mu_B}{T}} + \frac{n_B T}{nh}\frac{\partial}{\partial u^\mu}\right]\frac{f^{\rm eq}\bar{f}^{\rm eq}L^{-1}\hat{\Pi}}{-T\zeta}\Big\rangle, \tag{15.157}$$

$$\kappa_{JJ}^{(1)AB} = \frac{\Delta^{\mu\nu}}{3}\Big\langle L^{-1}\hat{J}_{A,\mu}, (f^{\rm eq}\bar{f}^{\rm eq})^{-1}\left[\mathcal{I}\frac{\partial}{\partial T} + \sum_{D=1}^{M}\mathcal{I}_D\frac{\partial}{\partial\frac{\mu_D}{T}} + \frac{1}{3}v^\rho\frac{\partial}{\partial u^\rho}\right]$$
$$\times \sum_{C=1}^{M}\frac{f^{\rm eq}\bar{f}^{\rm eq}L^{-1}\hat{J}_{C,\nu}}{T^2\lambda/h}(\lambda^{-1})_{CB}\Big\rangle, \tag{15.158}$$

$$\kappa_{JJ}^{(2)AB} = \frac{\Delta^{\mu\nu\rho\sigma}}{5}\Big\langle L^{-1}\hat{J}_{A,\mu}, (f^{\rm eq}\bar{f}^{\rm eq})^{-1}v_\rho\frac{\partial}{\partial u^\sigma}\sum_{C=1}^{M}\frac{f^{\rm eq}\bar{f}^{\rm eq}L^{-1}\hat{J}_{C,\nu}}{T^2\lambda/h}(\lambda^{-1})_{CB}\Big\rangle, \tag{15.159}$$

$$\kappa_{JJ}^{(3)AB} = \frac{\Omega^{\mu\nu\rho\sigma}}{3}\Big\langle L^{-1}\hat{J}_{A,\mu}, (f^{\rm eq}\bar{f}^{\rm eq})^{-1}v_\rho\frac{\partial}{\partial u^\sigma}\sum_{C=1}^{M}\frac{f^{\rm eq}\bar{f}^{\rm eq}L^{-1}\hat{J}_{C,\nu}}{T^2\lambda/h}(\lambda^{-1})_{CB}\Big\rangle, \tag{15.160}$$

$$\kappa_{J\pi}^{(1)A} = \frac{\Delta^{\mu\nu\rho\sigma}}{5}\Big\langle L^{-1}\hat{J}_{A,\mu}, (f^{\rm eq}\bar{f}^{\rm eq})^{-1}\left[v_\nu\frac{\partial}{\partial T} + \frac{1}{T}\frac{\partial}{\partial u^\nu}\right]\frac{f^{\rm eq}\bar{f}^{\rm eq}L^{-1}\hat{\pi}_{\rho\sigma}}{-2T\eta}\Big\rangle, \tag{15.161}$$

$$\kappa_{J\pi}^{(2)AB} = \frac{\Delta^{\mu\nu\rho\sigma}}{5}\Big\langle L^{-1}\hat{J}_{A,\mu}, (f^{\rm eq}\bar{f}^{\rm eq})^{-1}\left[v_\nu\frac{\partial}{\partial\frac{\mu_B}{T}} + \frac{n_B T}{nh}\frac{\partial}{\partial u^\nu}\right]\frac{f^{\rm eq}\bar{f}^{\rm eq}L^{-1}\hat{\pi}_{\rho\sigma}}{-2T\eta}\Big\rangle, \tag{15.162}$$

$$\kappa_{\pi\Pi} = \frac{\Delta^{\mu\nu\rho\sigma}}{5}\Big\langle L^{-1}\hat{\pi}_{\mu\nu}, (f^{\rm eq}\bar{f}^{\rm eq})^{-1}v_\rho\frac{\partial}{\partial u^\sigma}\frac{f^{\rm eq}\bar{f}^{\rm eq}L^{-1}\hat{\Pi}}{-T\zeta}\Big\rangle, \tag{15.163}$$

$$\kappa_{\pi J}^{(1)A} = \frac{\Delta^{\mu\nu\rho\sigma}}{5}\Big\langle L^{-1}\hat{\pi}_{\mu\nu}, (f^{\rm eq}\bar{f}^{\rm eq})^{-1}\left[v_\rho\frac{\partial}{\partial T} + \frac{1}{T}\frac{\partial}{\partial u^\rho}\right]\sum_{B=1}^{M}\frac{f^{\rm eq}\bar{f}^{\rm eq}L^{-1}\hat{J}_{B,\sigma}}{T^2\lambda/h}(\lambda^{-1})_{BA}\Big\rangle, \tag{15.164}$$

$$\kappa_{\pi J}^{(2)} = \frac{\Delta^{\mu\nu\rho\sigma}}{5}\Big\langle L^{-1}\hat{\pi}_{\mu\nu}, (f^{\rm eq}\bar{f}^{\rm eq})^{-1}\left[v_\rho\frac{\partial}{\partial\frac{\mu}{T}} + \frac{T}{h}\frac{\partial}{\partial u^\rho}\right]\frac{f^{\rm eq}\bar{f}^{\rm eq}L^{-1}\hat{J}_\sigma}{T^2\lambda/h}\Big\rangle, \tag{15.165}$$

$$\kappa_{\pi\pi}^{(1)} = -\frac{\Delta^{\mu\nu\rho\sigma}}{5}\Big\langle L^{-1}\hat{\pi}_{\mu\nu}, (f^{\rm eq}\bar{f}^{\rm eq})^{-1}\left[\mathcal{I}\frac{\partial}{\partial T} + \sum_{A=1}^{M}\mathcal{I}_A\frac{\partial}{\partial\frac{\mu_A}{T}} + \frac{1}{3}v^\mu\frac{\partial}{\partial u^\mu}\right]\frac{f^{\rm eq}\bar{f}^{\rm eq}L^{-1}\hat{\pi}_{\rho\sigma}}{-2T\eta}\Big\rangle, \tag{15.166}$$

$$\kappa_{\pi\pi}^{(2)} = \frac{12}{35}\Delta^{\mu\nu\gamma\delta}\Delta^{\rho\sigma\lambda}{}_\gamma\Delta^{\alpha\beta}{}_{\lambda\delta}\Big\langle L^{-1}\hat{\pi}_{\mu\nu}, (f^{\rm eq}\bar{f}^{\rm eq})^{-1}v_\alpha\frac{\partial}{\partial u^\beta}\frac{f^{\rm eq}\bar{f}^{\rm eq}L^{-1}\hat{\pi}_{\rho\sigma}}{-2T\eta}\Big\rangle, \tag{15.167}$$

$$\kappa_{\pi\pi}^{(3)} = \frac{4}{15}\Delta^{\mu\nu\gamma\delta}\Delta^{\rho\sigma\lambda}{}_\gamma\Omega^{\alpha\beta}{}_{\lambda\delta}\Big\langle L^{-1}\hat{\pi}_{\mu\nu}, (f^{\rm eq}\bar{f}^{\rm eq})^{-1}v_\alpha\frac{\partial}{\partial u^\beta}\frac{f^{\rm eq}\bar{f}^{\rm eq}L^{-1}\hat{\pi}_{\rho\sigma}}{-2T\eta}\Big\rangle. \tag{15.168}$$

Substituting the above equations into Eq. (15.114), we arrive at the explicit form of the relaxation equations:

$$\epsilon\Pi = -\epsilon\zeta\theta - \epsilon\tau_\Pi\frac{\partial}{\partial\tau}\Pi + \epsilon^2\Big(-\sum_{a=1}^{M}\ell_{\Pi J}^{a}\nabla_\rho J_A^\rho$$
$$+\kappa_{\Pi\Pi}\Pi\theta + \sum_{A=1}^{M}\kappa_{\Pi J}^{(1)A}J_{A,\rho}\nabla^\rho T + \sum_{A,B=1}^{M}\kappa_{\Pi J}^{(2)BA}J_{A,\rho}\nabla^\rho\frac{\mu_B}{T} + \kappa_{\Pi\pi}\pi_{\rho\sigma}\sigma^{\rho\sigma}$$
$$+b_{\Pi\Pi\Pi}\Pi^2 + \sum_{A,B=1}^{M}b_{\Pi JJ}^{AB}J_A^\rho J_{B,\rho} + b_{\Pi\pi\pi}\pi^{\rho\sigma}\pi_{\rho\sigma}\Big), \tag{15.169}$$

$$
\begin{aligned}
\epsilon J_A^\mu &= \epsilon \sum_{B=1}^{M} \lambda_{AB} \frac{T^2}{h^2} \nabla^\mu \frac{\mu_B}{T} - \epsilon \sum_{B=1}^{M} \tau_J^{AB} \Delta^{\mu\rho} \frac{\partial}{\partial \tau} J_{B,\rho} + \epsilon^2 \Big(- \ell_{J\Pi}^A \nabla^\mu \Pi \\
&\quad - \ell_{J\pi}^A \Delta^{\mu\rho} \nabla_\nu \pi^\nu{}_\rho + \kappa_{J\Pi}^{(1)A} \Pi \nabla^\mu T + \sum_{B=1}^{M} \kappa_{J\Pi}^{(2)AB} \Pi \nabla^\mu \frac{\mu_B}{T} + \sum_{B=1}^{M} \kappa_{JJ}^{(1)AB} J_B^\mu \theta \\
&\quad + \sum_{B=1}^{M} \kappa_{JJ}^{(2)AB} J_{B,\rho} \sigma^{\mu\rho} + \kappa_{JJ}^{(3)AB} J_{B,\rho} \omega^{\mu\rho} + \kappa_{J\pi}^{(1)A} \pi^{\mu\rho} \nabla_\rho T + \sum_{B=1}^{M} \kappa_{J\pi}^{(2)AB} \pi^{\mu\rho} \nabla_\rho \frac{\mu_B}{T} \\
&\quad + \sum_{B=1}^{M} b_{J\Pi J}^{AB} \Pi J_B^\mu + \sum_{B=1}^{M} b_{JJ\pi}^{AB} J_{B,\rho} \pi^{\rho\mu} \Big),
\end{aligned}
\tag{15.170}
$$

$$
\begin{aligned}
\epsilon \pi^{\mu\nu} &= \epsilon 2 \eta \sigma^{\mu\nu} - \epsilon \tau_\pi \Delta^{\mu\nu\rho\sigma} \frac{\partial}{\partial \tau} \pi_{\rho\sigma} - \epsilon^2 \Big(\sum_{a=1}^{M} \ell_{\pi J}^a \nabla^{\langle\mu} J_a^{\nu\rangle} + \kappa_{\pi\Pi} \Pi \sigma^{\mu\nu} \\
&\quad + \sum_{A=1}^{M} \kappa_{\pi J}^{(1)A} J_A^{\langle\mu} \nabla^{\nu\rangle} T + \sum_{A,B=1}^{M} \kappa_{\pi J}^{(2)BA} J_A^{\langle\mu} \nabla^{\nu\rangle} \frac{\mu_B}{T} + \kappa_{\pi\pi}^{(1)} \pi^{\mu\nu} \theta + \kappa_{\pi\pi}^{(2)} \pi^{\lambda\langle\mu} \sigma^{\nu\rangle}{}_\lambda \\
&\quad + \kappa_{\pi\pi}^{(3)} \pi^{\lambda\langle\mu} \omega^{\nu\rangle}{}_\lambda + b_{\pi\Pi\pi} \Pi \pi^{\mu\nu} + \sum_{A,B=1}^{M} b_{\pi JJ}^{AB} J_A^{\langle\mu} J_B^{\nu\rangle} + b_{\pi\pi\pi} \pi^{\lambda\langle\mu} \pi^{\nu\rangle}{}_\lambda \Big).
\end{aligned}
\tag{15.171}
$$

After setting $\epsilon = 1$, we find that these equations become Eqs. (15.169)–(15.171).

15.4 Properties of Derived Fluid Dynamic Equations

In the last section, we have derived the second-order fluid dynamic equation as well as the microscopic expressions of all the transport coefficients. Here, we will derive pivotal properties that the fluid dynamic equation is expected to satisfy: the positivity of transport coefficients, the Onsager's reciprocal relation [246], and the positivity of the entropy production rate.

15.4.1 *Positivity of Transport Coefficients*

We first use the microscopic expression of bulk viscosity (15.101) to prove its positivity. To this end, we recall that the inverse of linearized collision operator L^{-1} is symmetric and negative definite when restricted in the Q_0 space. Note that $\hat{\Pi}$, $\hat{J}_A^\mu$, and $\hat{\pi}^{\mu\nu}$ live in the $\mathrm{P}_1 (\subset \mathrm{Q}_0)$ space. For such L^{-1}, the Cholesky decomposition[1] states the existence of a real lower triangular matrix U such that $L^{-1} = {}^t U U$. With use of the Cholesky decomposition the bulk viscosity ζ is shown to be positive,

[1] See, for example, [218].

$$\zeta = \frac{1}{T}\langle \hat{\Pi}, {}^tUU\hat{\Pi}\rangle = \frac{1}{T}\langle U\hat{\Pi}, U\hat{\Pi}\rangle \geq 0. \tag{15.172}$$

We prove the positive definiteness of λ_{AB} by showing its eigenvalues are all positive. To prove the positive definiteness of λ_{AB} we write it as follows,

$$\lambda_{AB} = \frac{1}{3T^2}\langle \hat{J}_{A,\mu}, L^{-1}\Delta^{\mu\nu}\hat{J}_{B,\nu}\rangle. \tag{15.173}$$

Note the negative definiteness of $\Delta^{\mu\nu}$ in Q_0 space, that is most easily seen by going to the local rest frame $u^\mu = (1, 0, 0, 0)^t$, where $\Delta^{\mu\nu} = g^{\mu\nu} - u^\mu u^\nu = \text{diag}(0, -1, -1, -1)$. Thus, the Cholesky decomposition can be applied to $\Delta^{\mu\nu}$, resulting in $\Delta^{\mu\nu} = -(d^t)^{\mu\rho}d^{\rho\nu}$ with a real lower triangular matrix d in the Lorentz space. Let V be a diagonalizing matrix of λ_{AB}. Then the eigenvalues are evaluated as,

$$\begin{aligned}(V^t\lambda V)_{AA} &= \frac{1}{3T^2}\langle (V\hat{J}_\mu)_A, {}^tUU(d^td)^{\mu\nu}(V\hat{J}_\nu)_A\rangle \\ &= \frac{1}{3T^2}\langle U(V(d\hat{J})_\mu)_A, U(V(d\hat{J})_\mu)_A\rangle \geq 0.\end{aligned} \tag{15.174}$$

Similarly, the shear viscosity is proved to be positive:

$$\begin{aligned}\eta &= \frac{1}{10T}\langle \hat{\pi}_{\mu\nu}, {}^tUU\Delta^{\mu\nu\rho\sigma}\hat{\pi}_{\rho\sigma}\rangle = \frac{1}{10T}\langle \hat{\pi}_{\mu\nu}, {}^tUU({}^tDD)^{\mu\nu\rho\sigma}\hat{\pi}_{\rho\sigma}\rangle \\ &= \frac{1}{10T}\langle U(D\hat{\pi})_{\mu\nu}, U(D\hat{\pi})_{\mu\nu}\rangle \geq 0.\end{aligned} \tag{15.175}$$

Noting that $\Delta^{\mu\nu\rho\sigma} = \Delta^{\mu\rho}\Delta^{\nu\sigma}/2 + \Delta^{\mu\sigma}\Delta^{\nu\rho}/2 - \Delta^{\mu\nu}\Delta^{\rho\sigma}/3$ is a semipositive-definite matrix, we again applied the Cholesky decomposition to have

$$\Delta^{\mu\nu\rho\sigma} = ({}^tD)^{\mu\nu\alpha\beta}D^{\alpha\beta\rho\sigma} \tag{15.176}$$

with a real lower triangular matrix D.

Next, we proceed to a proof of the positivity of the relaxation times (15.104)–(15.106). The positivity of those associated with the bulk pressure and shear tensor

$$\tau_\Pi = \frac{1}{T\zeta}\langle L^{-1}\hat{\Pi}, L^{-1}\hat{\Pi}\rangle, \tag{15.177}$$

$$\tau_\pi = \frac{1}{10T\eta}\langle L^{-1}\hat{\pi}^{\mu\nu}, L^{-1}\hat{\pi}_{\mu\nu}\rangle, \tag{15.178}$$

are immediate as ζ and η are positive.

We finally prove the positivity of the relaxation time τ_J^{AB}:

$$
\begin{aligned}
\tau_J^{AB} &= -\frac{1}{3T^2}\sum_{C=1}^{M}\langle L^{-1}\hat{J}_{A,\mu},\,\Delta^{\mu\nu}L^{-1}\hat{J}_{C,\nu}\rangle(\lambda^{-1})_{CB} \\
&= -\frac{1}{3T^2}\sum_{C=1}^{M}\langle L^{-1}\hat{J}_{A,\mu},\,\Delta^{\mu\nu}L^{-1}\hat{J}_{C,\nu}\rangle({}^t\Lambda\Lambda)_{CB},
\end{aligned}
\tag{15.179}
$$

where the Cholesky decomposition has been employed to have

$$
\lambda^{-1} = {}^t\Lambda\Lambda. \tag{15.180}
$$

Then, we note that the following matrix,

$$
\tilde{\tau}_J^{AB} = -\frac{1}{3T^2}\sum_{C,D=1}^{M}\Lambda_{AC}\langle L^{-1}\hat{J}_{C,\mu},\,\Delta^{\mu\nu}L^{-1}\hat{J}_{D,\nu}\rangle({}^t\Lambda)_{DB}, \tag{15.181}
$$

has the eigenvalues identical to τ_J^{AB}. Now it is readily shown that $\tilde{\tau}_J^{AB}$ is positive definite by the argument parallel to that of λ^{AB}. This in turn implies that τ_J^{AB} is a positive definite matrix.

15.4.2 Onsager's Reciprocal Relation

Let us move on to another important consequence on the transport coefficients. One of the prominent characteristics of multi-component fluid is the cross correlation between different components. For instance, the dynamics associated with Ath conserved current has an influence on the dissipation of Bth current through λ_{AB} and other coefficients for $A \neq B$.

More generally, let $\{X_i\}$ be a set of external forces of any tensor structure in a near-equilibrium system, then the induced current J_i is given, under linear approximation, by a linear combination of the external forces,

$$
J_i = \sum_j \gamma_{ij} X_j, \tag{15.182}
$$

with a set of scalar coefficients γ_{ij}. Note that X_j participating in the sum is of the same tensor structure as J_i. The Onsager's reciprocal relation asserts the symmetry of the coefficients,

$$
\gamma_{ij} = \gamma_{ji}. \tag{15.183}
$$

Identifying J_i and X_j in (15.182) with J_A^μ and $\nabla^\mu \frac{\mu_B}{T}$ in the leading order of (15.99),

$$J_A^\mu = \sum_{B=1}^{M} \lambda_{AB} \frac{T^2}{h^2} \nabla^\mu \frac{\mu_B}{T} + O(\epsilon), \tag{15.184}$$

γ_{ij} is found to be $\lambda_{AB} \frac{T^2}{h^2}$. It readily satisfies the desired relation (15.183), *i.e.*,

$$\lambda_{AB} = \lambda_{BA} \tag{15.185}$$

as seen from its microscopic expression (15.102).

15.4.3 Positivity of Entropy Production Rate

The positivity of entropy production rate is normally regarded as one of the guiding principles and imposed as an extra assumption to derive legitimate fluid dynamics [66]. In contrast, we did not impose such an assumption, and indeed, the positivity naturally follows the expression of entropy production rate (15.15) as we will see here.

Let us recall the entropy production rate,

$$\partial_\mu s^\mu = -\sum_{k=1}^{N} \int \mathrm{d}p_k p_k^\mu \partial_\mu f_{k,p_k} \ln\left(\frac{f_{k,p_k}}{1 + a_k f_{k,p_k}}\right), \tag{15.186}$$

and the distribution function that we obtained by solving the Boltzmann equation up to $O(\epsilon)$,

$$\begin{aligned} f_{k,p_k}^{\text{global}} &= f_{k,p_k}^{\text{eq}}\left(1 + \epsilon \bar{f}_{k,p_k}^{\text{eq}} \Psi_{k,p_k}^{(1)}\right) + O(\epsilon^2) \\ &= \left[\exp\left\{\frac{p_k \cdot u}{T} - \sum_{A=1}^{M} q_k^A \frac{\mu_A}{T} - \epsilon \Psi_{k,p_k}^{(1)}\right\} - a_k\right]^{-1} + O(\epsilon^2). \end{aligned} \tag{15.187}$$

The equilibrium distribution function does not contribute to the entropy production rate,[2] and thus, it starts with the term of order ϵ. Plugging it into (15.186) yields,

[2] This is a way to define the equilibrium distribution function as discussed around (15.18).

$$
\begin{aligned}
\partial_\mu s^\mu &= \epsilon \sum_{k=1}^{N} \int \mathrm{d}p_k f^{\mathrm{eq}}_{k,p_k} \bar{f}^{\mathrm{eq}}_{k,p_k} \Psi_{k,p_k} \\
&\times \left(p^\mu_k p^\nu_k \partial_\mu \frac{u_\nu}{T} - p^\mu_k \partial_\mu \sum_{A=1}^{M} q^A_k \frac{\mu_A}{T} - \epsilon p^\mu_k \partial_\mu \Psi_{k,p_k} \right) + O(\epsilon^3) \\
&= \epsilon^2 \delta T^{\mu\nu} \nabla_\mu \frac{u_\nu}{T} - \epsilon^2 \sum_{A=1}^{M} \delta N^\mu_A \nabla_\mu \frac{\mu_A}{T} - \epsilon^2 \left\langle \Psi, \frac{\partial}{\partial\tau} \Psi \right\rangle + O(\epsilon^3),
\end{aligned} \tag{15.188}
$$

where the decomposition of the derivative $\partial_\mu = u_\mu \partial/\partial\tau + \epsilon\nabla_\mu$ and the orthogonality between the fluid velocity u_μ and the dissipative contributions to the energy momentum tensor and the particle current, $\delta T^{\mu\nu}$ and δN^μ_A, have been used. Each term can be further rewritten by using the following transformations,

$$
\delta T^{\mu\nu} = \sum_{k=1}^{N} \int \mathrm{d}p\, p^\mu_k p^\nu_k f^{\mathrm{eq}}_{k,p_k} \bar{f}^{\mathrm{eq}}_{k,p_k} \Psi_{k,p_k} = -\Delta^{\mu\nu}\Pi + \pi^{\mu\nu}, \tag{15.189}
$$

$$
\delta N^\mu_A = \sum_{k=1}^{N} q^A_k \int \mathrm{d}p\, p^\mu_k f^{\mathrm{eq}}_{k,p_k} \bar{f}^{\mathrm{eq}}_{k,p_k} \Psi_{k,p_k} = J^\mu_A, \tag{15.190}
$$

and

$$
\begin{aligned}
\left\langle \Psi, \frac{\partial}{\partial\tau} \Psi \right\rangle &= \Pi \frac{1}{T\zeta} \frac{\partial}{\partial\tau} \Pi + \frac{h^2}{T^2} \sum_{A,B,C} J^\mu_A (\lambda^{-1})_{AB} \tau^{BC}_J \frac{\partial}{\partial\tau} J_{C,\mu} \\
&\quad + \pi^{\mu\nu} \frac{1}{2T\eta} \frac{\partial}{\partial\tau} \pi_{\mu\nu} + O(\epsilon).
\end{aligned} \tag{15.191}
$$

Combining these yields a compact expression,

$$
\partial_\mu s^\mu = \epsilon^2 \left(\frac{1}{T\zeta} \Pi^2 + \frac{1}{2T\eta} \pi^{\mu\nu}\pi_{\mu\nu} - \frac{h^2}{T^2} \sum_{A,B=1}^{M} (\lambda^{-1})_{AB} J^\mu_A J_{B,\mu} \right) + O(\epsilon^3), \tag{15.192}
$$

where we have also used the relaxation equations in the leading order,

$$
\Pi = -\zeta\theta - \tau_\Pi \frac{\partial}{\partial\tau} \Pi + O(\epsilon), \tag{15.193}
$$

$$
J^\mu_A = \sum_{B=1}^{M} \lambda_{AB} \frac{T^2}{h^2} \nabla^\mu \frac{\mu_B}{T} - \sum_{B=1}^{M} \tau^{AB}_J \Delta^{\mu\rho} \frac{\partial}{\partial\tau} J_{B,\rho} + O(\epsilon), \tag{15.194}
$$

$$
\pi^{\mu\nu} = 2\eta\sigma^{\mu\nu} - \tau_\pi \Delta^{\mu\nu\rho\sigma} \frac{\partial}{\partial\tau} \pi_{\rho\sigma} + O(\epsilon). \tag{15.195}
$$

It is now straightforward to see that Eq. (15.192) is positive. The viscosities ζ and η are positive quantities, as shown in the last section. Positive definiteness of the third term of Eq. (15.192) is proved as

$$\begin{aligned}
&-\frac{h^2}{T^2}\sum_{A,B=1}^{M}(\lambda^{-1})_{AB}J_A^{\mu}J_{B,\mu}\\
&=-\frac{h^2}{T^2}\sum_{A,B=1}^{M}\lambda_{AB}(\lambda^{-1}J^{\mu})_A(\lambda^{-1}J_{\mu})_B\\
&=-\frac{h^2}{T^2}\sum_{A,B=1}^{M}\langle\hat{J}_A^{\nu},L^{-1}\hat{J}_{B,\nu}\rangle(\lambda^{-1}J^{\mu})_A(\lambda^{-1}J_{\mu})_B\\
&=\frac{h^2}{T^2}\left\langle\sum_{A=1}^{M}U\hat{J}_A^{\nu}(\lambda^{-1}J^{\mu})_A,\ \sum_{B=1}^{M}U\hat{J}_{B,\nu}(\lambda^{-1}J_{\mu})_B\right\rangle\\
&\geq 0,
\end{aligned}\tag{15.196}$$

where the Cholesky decomposition $L^{-1}={}^tUU$ has been used. Consequently, the entropy production rate (15.192) is explicitly shown to exhibits the positivity.

Chapter 16
RG/E Derivation of Non-relativistic Second-Order Fluid Dynamics and Application to Fermionic Atomic Gases

16.1 Derivation of Second-Order Fluid Dynamics in Non-relativistic Systems

So far we have focused on relativistic fluid, that is for instance expected to be realized in the relativistic heavy ion collision experiments. The system is particularly interesting due to its strong coupling nature. Even in such a complicated system the fluid dynamics provides a powerful tool to investigate its long-distance and long-time behavior. A conventional wisdom is that the naïve inclusion of viscous effects in the relativistic fluid dynamic equation leads to the notorious causality issue, and thus, the second-order fluid dynamics is required to circumvent the issue.

A cold atomic system is another intriguing platform, where precise controllability of interactions allows us to simulate the formation and diffusion of strong-coupling fluid and tremendous amount of efforts have been devoted to unravel strong-coupling systems in non-relativistic context (see [247–252], for example).

From the viewpoint of fluid dynamics, one may expect that the analysis is much simpler than the relativistic case as the causality problem does not show up. Indeed, the Navier–Stokes equation suffices to deal with conventional non-relativistic fluid if the dynamics is relatively close to local equilibrium. However, the validity of the Navier–Stokes equation becomes questionable when the system is not close enough to the local equilibrium. Let us consider an experimentally created cold atomic system. Unsurprisingly, the peripheral region is dilute relative to its core. Therefore, it is not hard to imagine that atoms on the edge of the trapped system have less chances to interact with each other due to its sparsity than those in the central area. Accordingly, the equilibration process takes place slower in the peripheral region. In particular, some region is so sparse that it is not close enough to equilibrium and hence the Navier–Stokes equation is not capable of capturing the essential dynamics. That is the very situation where the mesoscopic dynamics of dissipative currents needs to be incorporated in order to describe beyond-Navier–Stokes regime. This leads us to the non-relativistic second-order fluid dynamic equations [35, 149, 152–154, 157].

T. Kunihiro et al., *Geometrical Formulation of Renormalization-Group Method as an Asymptotic Analysis*, Fundamental Theories of Physics 206,
https://doi.org/10.1007/978-981-16-8189-9_16

Besides the derivation of the fluid dynamic equations, the computations of transport coefficients have attracted a lot of attentions across wide range of disciplines in physics [253, 254] following the seminal papers proposing the universal lower bound on the ratio of shear viscosity to entropy density [255, 256]. As already mentioned, the cold atomic system provides an ideal platform to experimentally examine microscopic properties of non-relativistic fluids through the measurements of transport coefficients. Indeed, the small shear viscosity in the system implies that the strongly correlated fluid is realized [257–262]. The Boltzmann's kinetic theory offers a tool to compute transport coefficients such as shear viscosity and bulk viscosity in strongly correlated cold atomic systems [258, 263–266]. The second-order fluid dynamics requires further transport coefficients. Among others, the viscous relaxation times play prominent roles to dictate the relaxation processes of dissipative currents [258, 263, 267–270].

We apply the doublet scheme [69] in the RG method [1, 3, 5, 6, 46, 57, 58] to derive the second-order fluid dynamic equation starting from the Boltzmann equation [69, 157]. The RG method allows us to solve the Boltzmann equation faithfully by following the two steps:

1. Solve the Boltzmann equation based on the perturbative expansion with respect to small spatial inhomogeneity. The perturbative solution typically suffers from the infrared divergence.
2. Remove the infrared divergence to obtain the global solution by the renormalization-group analysis. The renormalization-group/envelope (RG/E) equation turns out to be the desired fluid dynamic equations.

The resultant fluid dynamic equations are accompanied by microscopic expressions of transport coefficients. We carry out detailed numerical studies of the transport coefficients and viscous relaxation times when the collision process is modeled by the s-wave scattering [157]. Their dependence on temperature and scattering length will be discussed in detail based on our numerical calculations. In particular, the importance of quantum statistical effects is clearly observed at low temperatures. Upon incorporating the fermionic statistical effects, the transport coefficients sharply increase at low temperatures in contrast to those of classical Boltzmann gas.

16.1.1 Non-relativistic Boltzmann Equation

Our starting point is the Boltzmann equation,

$$
\begin{aligned}
&\left(\frac{\partial}{\partial t} + \frac{\partial \boldsymbol{x}}{\partial t} \cdot \boldsymbol{\nabla} + \frac{\partial \boldsymbol{p}}{\partial t} \cdot \boldsymbol{\nabla}_p\right) f_p(t, \boldsymbol{x}) = C[f]_p(t, \boldsymbol{x}), \\
&\frac{\partial \boldsymbol{x}}{\partial t} = -\boldsymbol{\nabla}_p E_p =: \boldsymbol{v}, \\
&\frac{\partial \boldsymbol{p}}{\partial t} = -\boldsymbol{\nabla} E_p =: \boldsymbol{F},
\end{aligned} \tag{16.1}
$$

where $\boldsymbol{\nabla} = \partial/\partial \boldsymbol{x}$, $\boldsymbol{\nabla}_p = \partial/\partial \boldsymbol{p}$, and $E_p = \boldsymbol{p}^2/(2m) + V(\boldsymbol{x})$. The collision integral is given by

$$C[f]_p(t, \boldsymbol{x}) = \frac{1}{2} \int \mathrm{d}p_1 \mathrm{d}p_2 \mathrm{d}p_3 \mathcal{W}(p, p_1|p_2, p_3)(\bar{f}_p \bar{f}_{p_1} f_{p_2} f_{p_3} - f_p f_{p_1} \bar{f}_{p_2} \bar{f}_{p_3}), \tag{16.2}$$

where $\mathrm{d}p := \mathrm{d}^3 p/(2\pi)^3$ and $\bar{f}_p := 1 + a f_p$ with quantum statistics represented by $a = +1(-1)$ for boson (fermion) and $a = 0$ for the classical Boltzmann gas. The transition matrix $\mathcal{W}$ accounts for binary collisions and is given by

$$\mathcal{W}(p, p_1|p_2, p_3) = |\mathcal{M}|^2 (2\pi)^2 \delta(E + E_1 - E_2 - E_3) \delta^3(\boldsymbol{p} + \boldsymbol{p}_1 - \boldsymbol{p}_2 - \boldsymbol{p}_3), \tag{16.3}$$

which has the following symmetry,

$$\mathcal{W}(p, p_1|p_2, p_3) = \mathcal{W}(p_2, p_3|p, p_1) = \mathcal{W}(p_1, p|p_3, p_2) = \mathcal{W}(p_3, p_2|p_1, p). \tag{16.4}$$

$\mathcal{M}$ is a scattering amplitude associated with the details of collision process. For instance, we shall later use a scattering amplitude for elastic binary collisions [253],

$$\mathcal{M} = \frac{4\pi}{a_s^{-1} - i|\boldsymbol{q}|}, \tag{16.5}$$

in order to study the fluid of fermionic cold atoms. Here, a_s is the s-wave scattering length and $\boldsymbol{q}$ is the relative momentum $(\boldsymbol{p} - \boldsymbol{p}_1)/2$.

On account of the symmetry (16.4), the convolution of the collision integral and an arbitrary function of $\boldsymbol{p}$, denoted by Φ_p, is written as follows,

$$\begin{aligned} \int \mathrm{d}p\, \Phi_p C[f]_p = \frac{1}{8} \int & \mathrm{d}p \mathrm{d}p_1 \mathrm{d}p_2 \mathrm{d}p_3 \mathcal{W}(p, p_1|p_2, p_3) \\ & \times (\Phi_p + \Phi_{p_1} - \Phi_{p_2} - \Phi_{p_3})(\bar{f}_p \bar{f}_{p_1} f_{p_2} f_{p_3} - f_p f_{p_1} \bar{f}_{p_2} \bar{f}_{p_3}). \end{aligned} \tag{16.6}$$

We say a function Φ_p is a collision invariant if the convolution vanishes. From the above expression, it is clear that energy E_p and momentum $\boldsymbol{p}$ are collision invariants due to the energy and momentum conservation (16.3). Thus, the collision invariant Φ_p^{inv} is generally expressed as

$$\Phi_p^{\mathrm{inv}} = \alpha + \boldsymbol{\beta} \cdot \boldsymbol{p} + \gamma E_p, \tag{16.7}$$

for arbitrary constants α, $\boldsymbol{\beta}$, and γ.

16.1.2 Derivation of Navier–Stokes Equation

We have derived the first-order fluid dynamic equation, *i.e.*, the Navier–Stokes equation, starting from the Boltzmann equation describing the classical Boltzmann gas in Chap. 8. Here, we firstly repeat the derivation incorporating the quantum statistical effect and external field. The derivation of first-order equation has been well established by Chapman and Enskog [113]. Nevertheless, deriving the fluid dynamics and reproducing all the transport coefficients with the RG method are important consistency checks [57]. Later we shall extend it to derive the second-order equation.

16.1.2.1 Perturbative Analysis of the Boltzmann Equation

Our first task is to solve the Boltzmann equation (16.1) perturbatively with respect to small spatial gradient. In order to keep track of the order of perturbation we introduce a parameter ϵ,

$$\left(\frac{\partial}{\partial t} + \epsilon \boldsymbol{v} \cdot \nabla + \epsilon \boldsymbol{F} \cdot \nabla_p\right) f_p(t, \boldsymbol{x}) = C[f]_p(t, \boldsymbol{x}). \tag{16.8}$$

Notice that we also need ϵ for the third term in the left-hand side because $\boldsymbol{F} = -\nabla E_p$. Accordingly, we expand the distribution function in terms of ϵ,

$$\tilde{f}(t; t_0) = \tilde{f}^{(0)}(t; t_0) + \epsilon \tilde{f}^{(1)}(t; t_0) + \epsilon \tilde{f}^{(2)}(t; t_0) + O(\epsilon^3). \tag{16.9}$$

We solve the Boltzmann equation order by order provided the 'initial' condition, which is also expanded as

$$f(t = t_0; t_0) = \tilde{f}(t = t_0; t_0) = \tilde{f}^{(0)}(t_0) + \epsilon \tilde{f}^{(1)}(t_0) + \epsilon \tilde{f}^{(2)}(t_0) + O(\epsilon^3). \tag{16.10}$$

The perturbative solution $\tilde{f}(t; t_0)$ is set to the unknown exact solution $f(t; t_0)$ at the 'initial' time t_0.

The equation of $O(\epsilon^0)$ reads

$$\frac{\partial}{\partial t} \tilde{f}_p^{(0)} = C[\tilde{f}^{(0)}]_p. \tag{16.11}$$

As we are carrying out the perturbative expansion around the equilibrium state, we seek a solution satisfying

$$\frac{\partial}{\partial t} \tilde{f}_p^{(0)} = 0, \tag{16.12}$$

as a leading order solution. It implies

$$C[\tilde{f}]_p^{(0)} = 0, \tag{16.13}$$

which leads to the condition

$$\bar{\tilde{f}}_p^{(0)} \bar{\tilde{f}}_{p_1}^{(0)} \tilde{f}_{p_2}^{(0)} \tilde{f}_{p_3}^{(0)} = \tilde{f}_p^{(0)} \tilde{f}_{p_1}^{(0)} \bar{\tilde{f}}_{p_2}^{(0)} \bar{\tilde{f}}_{p_3}^{(0)}. \tag{16.14}$$

This is satisfied by requiring that $\ln(\tilde{f}_p^{(0)}/\bar{\tilde{f}}_p^{(0)})$ is a conserved quantity, *i.e.*, expressed as a linear combination of $\{1, \boldsymbol{p}, E\}$. Their coefficients are fixed by the initial condition and depend on the initial time t_0 and spatial coordinate $\boldsymbol{x}$. Then, the zeroth order solution takes the following form,

$$\tilde{f}_p^{(0)} = \left(\exp\left[\frac{(m/2)\left(\boldsymbol{v} - \boldsymbol{u}(t_0, \boldsymbol{x})\right)^2 - \tilde{\mu}(t_0, \boldsymbol{x})}{T(t_0, \boldsymbol{x})} \right] - a \right)^{-1}, \tag{16.15}$$

with $\tilde{\mu}(t_0, \boldsymbol{x}) := \mu(t_0, \boldsymbol{x}) - V(\boldsymbol{x})$. Here, $\boldsymbol{u}(t_0, \boldsymbol{x})$, $\mu(t_0, \boldsymbol{x})$, and $T(t_0, \boldsymbol{x})$ are the integral constants fixed by the initial condition, and $\tilde{f}_p^{(0)}$ is given by the t-independent equilibrium distribution function. The five integration constants will be lifted to fluid variables through the RG/E equation.

Let us draw an important observation from the fact that $\ln(\tilde{f}_p^{(0)}/\bar{\tilde{f}}_p^{(0)})$ is a conserved quantity, which implies that it is also a collision invariant (see the discussion below (16.6)). From the definition of the collision invariant, we deduce that

$$0 = \int \mathrm{d}p \ln\left(\frac{\tilde{f}_p^{(0)}}{\bar{\tilde{f}}_p^{(0)}}\right) C[f]_p = \int \mathrm{d}p \ln\left(\frac{\tilde{f}_p^{(0)}}{\bar{\tilde{f}}_p^{(0)}}\right) \left(\frac{\partial}{\partial t} + \epsilon \boldsymbol{v} \cdot \nabla + \epsilon \boldsymbol{F} \cdot \nabla_p\right) \tilde{f}_p^{(0)}. \tag{16.16}$$

Given that $\boldsymbol{F}$ is independent of $\boldsymbol{p}$, the third term disappears,

$$\begin{aligned} \int \mathrm{d}p \ln\left(\frac{\tilde{f}_p^{(0)}}{\bar{\tilde{f}}_p^{(0)}}\right) \boldsymbol{F} \cdot \nabla_p \tilde{f}_p^{(0)} &= -\boldsymbol{F} \cdot \int \mathrm{d}p \left(\frac{\nabla_p \tilde{f}_p^{(0)}}{\tilde{f}_p^{(0)}} - \frac{a \nabla_p \tilde{f}_p^{(0)}}{\bar{\tilde{f}}_p^{(0)}}\right) \tilde{f}_p^{(0)} \\ = \boldsymbol{F} \cdot \int \mathrm{d}p a \tilde{f}_p^{(0)} \frac{\nabla_p \tilde{f}_p^{(0)}}{\bar{\tilde{f}}_p^{(0)}} &= \boldsymbol{F} \cdot \int \mathrm{d}p \frac{\nabla_p \tilde{f}_p^{(0)}}{\bar{\tilde{f}}_p^{(0)}} = \frac{1}{a} \boldsymbol{F} \cdot \int \mathrm{d}p \nabla_p \ln \bar{\tilde{f}}_p^{(0)} = 0, \end{aligned} \tag{16.17}$$

where the surface term has been dropped in the second equality. We find that (16.16) amounts to,

$$0 = \frac{\partial s}{\partial t} + \nabla \cdot \boldsymbol{J}_s, \tag{16.18}$$

with use of the definition of entropy density s and entropy current density $\boldsymbol{J}_s$ defined by

$$\{s, \boldsymbol{J}_s\} := -\int \mathrm{d}p\{1, \boldsymbol{v}\}\left(f_p \ln f_p - \frac{\bar{f}_p \ln \bar{f}_p}{a}\right), \tag{16.19}$$

respectively. Thus, we have shown that $\tilde{f}_p^{(0)}$ does not produce entropy, that is an expected property as it represents the equilibrium state. The first-order solution calculated below gives a perturbative correction entailing dissipation to the leading order equilibrium solution.

Next, we solve the $O(\epsilon)$ equation,

$$\frac{\partial}{\partial t}\tilde{f}_p^{(1)}(t) = f_p^{\mathrm{eq}}\bar{f}_p^{\mathrm{eq}} L_{pq}(f_q^{\mathrm{eq}}\bar{f}_q^{\mathrm{eq}})^{-1}\tilde{f}_p^{(1)}(t) + f_p^{\mathrm{eq}}\bar{f}_p^{\mathrm{eq}} F_p^{(0)}, \tag{16.20}$$

under the initial condition

$$\tilde{f}_p^{(1)}(t_0) = f_p^{\mathrm{eq}}\bar{f}_p^{\mathrm{eq}}\Psi_p^{(1)}(t_0). \tag{16.21}$$

$\Psi_p^{(1)}(t_0)$ is to be fixed later. The linearized collision operator L_{pq} and the inhomogeneous term $F_p^{(0)}$ are defined by

$$\begin{aligned} L_{pq} &:= (f_p^{\mathrm{eq}}\bar{f}_p^{\mathrm{eq}})^{-1}\frac{\delta}{\delta f_q}C[f]_p(t)\bigg|_{f=f^{\mathrm{eq}}} f_q^{\mathrm{eq}}\bar{f}_q^{\mathrm{eq}} \\ &= -\frac{1}{2\bar{f}_p^{\mathrm{eq}}}\int \mathrm{d}p_1\mathrm{d}p_2\mathrm{d}p_3\mathcal{W}(p, p_1|p_2, p_3) f_{p_1}^{\mathrm{eq}}\bar{f}_{p_2}^{\mathrm{eq}}\bar{f}_{p_3}^{\mathrm{eq}} \\ &\quad\times\left[\delta^3(\boldsymbol{p}-\boldsymbol{q}) + \delta^3(\boldsymbol{p}_1-\boldsymbol{q}) - \delta^3(\boldsymbol{p}_2-\boldsymbol{q}) - \delta^3(\boldsymbol{p}_3-\boldsymbol{q})\right], \end{aligned} \tag{16.22}$$

$$F_p^{(i)} := F[\tilde{f}^{(i)}]_p := -(f_p^{\mathrm{eq}}\bar{f}_p^{\mathrm{eq}})^{-1}\left(\boldsymbol{v}\cdot\nabla + \boldsymbol{F}\cdot\nabla_p\right)\tilde{f}_p^{(i)}. \tag{16.23}$$

For arbitrary vectors ψ_p and χ_p, the linearized collision operator has the following three properties:

$$\langle\psi, L\chi\rangle = \langle L\psi, \chi\rangle, \quad \langle\psi, L\psi\rangle \le 0, \quad L\varphi^\alpha = 0, \tag{16.24}$$

with the definition of the inner product given by

$$\langle\psi, \chi\rangle := \int \mathrm{d}p f_p^{\mathrm{eq}}\bar{f}_p^{\mathrm{eq}}\psi_p\chi_p. \tag{16.25}$$

The linearized collision operator has five zero modes,

$$\varphi_{0p}^0 = 1, \quad \varphi_{0p}^i = \delta p^i, \quad \varphi_{0p}^4 = \frac{|\boldsymbol{\delta p}|^2}{2m} - \frac{3nT}{2c^0}. \tag{16.26}$$

The zero modes satisfy the orthogonality relation

$$\langle\varphi_0^\alpha, \varphi_0^\beta\rangle = c^\alpha\delta^{\alpha\beta}, \tag{16.27}$$

with the following normalization factors

$$c^0 = \int_p f^{\mathrm{eq}} \bar{f}^{\mathrm{eq}}, \tag{16.28}$$

$$c^i = \int \mathrm{d}p f^{\mathrm{eq}} \bar{f}^{\mathrm{eq}} \delta p_i \delta p_i = mnT, \tag{16.29}$$

$$c^4 = \int \mathrm{d}p f^{\mathrm{eq}} \bar{f}^{\mathrm{eq}} \left(\frac{|\boldsymbol{\delta p}|^2}{2m} - \frac{3nT}{2c^0} \right)^2. \tag{16.30}$$

The integral of c^4 can be reduced to,

$$\begin{aligned} c^4 &= \frac{1}{2m} \int \mathrm{d}p f^{\mathrm{eq}} \bar{f}^{\mathrm{eq}} |\boldsymbol{\delta p}|^4 - \frac{3nT}{4mc^0} \int \mathrm{d}p f^{\mathrm{eq}} \bar{f}^{\mathrm{eq}} |\boldsymbol{\delta p}|^2 \\ &= \frac{1}{2m} \int \mathrm{d}p f^{\mathrm{eq}} \bar{f}^{\mathrm{eq}} |\boldsymbol{\delta p}|^4 - \frac{(3nT)^2}{4c^0}, \end{aligned} \tag{16.31}$$

$$\begin{aligned} \int \mathrm{d}p f^{\mathrm{eq}} \bar{f}^{\mathrm{eq}} |\boldsymbol{\delta p}|^4 &= 3 \int \mathrm{d}p f^{\mathrm{eq}} \bar{f}^{\mathrm{eq}} |\boldsymbol{\delta p}|^2 \delta p_1^2 = -3mT \int \mathrm{d}p |\boldsymbol{\delta p}|^2 \delta p_1 \frac{\partial f^{\mathrm{eq}}}{\partial p_1} \\ &= 5mT \int \mathrm{d}p |\boldsymbol{\delta p}|^2 f^{\mathrm{eq}} = 15m^2 T P. \end{aligned} \tag{16.32}$$

We have used the isotropy of f^{eq} for the computations of c^i and c^4. The particle number density n, energy density e, and pressure P are given by

$$n(t, \boldsymbol{x}) := \int \mathrm{d}p f_p^{\mathrm{eq}}(t, \boldsymbol{x}), \tag{16.33}$$

$$e(t, \boldsymbol{x}) := \frac{1}{n} \int \mathrm{d}p f_p^{\mathrm{eq}}(t, \boldsymbol{x}) \left(\frac{|\boldsymbol{\delta p}|^2}{2m} + V(\boldsymbol{x}) \right), \tag{16.34}$$

$$P(t, \boldsymbol{x}) := \frac{1}{3} \int \mathrm{d}p f_p^{\mathrm{eq}}(t, \boldsymbol{x}) \boldsymbol{\delta v} \cdot \boldsymbol{\delta p}. \tag{16.35}$$

We define P_0 and Q_0 spaces as the vector spaces spanned by the zero modes given in Eq. (16.26) and its complementary space, respectively; we denote the respective associated projection operators as P_0 and Q_0, whose operations are given by

$$[P_0 \psi]_p := \sum_{\alpha=0}^{4} \frac{\varphi_{0p}^{\alpha}}{c^{\alpha}} \langle \varphi_0^{\alpha}, \psi \rangle, \tag{16.36}$$

$$Q_0 := 1 - P_0, \tag{16.37}$$

for an arbitrary vector ψ_p, respectively.

We are now ready to solve the $O(\epsilon)$ equation,

$$
\begin{aligned}
\tilde{f}^{(1)} &= f^{\mathrm{eq}}\bar{f}^{\mathrm{eq}}\Big(\mathrm{e}^{(t-t_0)L}\Psi^{(1)} + \int_{t_0}^{t}\mathrm{d}t'\mathrm{e}^{(t-t')L}F^{(0)}\Big) \\
&= f^{\mathrm{eq}}\bar{f}^{\mathrm{eq}}\Big(\mathrm{e}^{(t-t_0)L}\Psi^{(1)} + \int_{t_0}^{t}\mathrm{d}t'\big(P_0 + \mathrm{e}^{(t-t')L}Q_0\big)F^{(0)}\Big) \\
&= f^{\mathrm{eq}}\bar{f}^{\mathrm{eq}}\Big[\mathrm{e}^{(t-t_0)L}\big(\Psi^{(1)} + L^{-1}Q_0F_0\big) + (t-t_0)P_0F_0 - L^{-1}Q_0F^{(0)}\Big],
\end{aligned}
\tag{16.38}
$$

where momentum indices are suppressed. Here we come to the point to fix the initial condition $\Psi^{(1)}$ in such a way that the fast-decaying term containing $\mathrm{e}^{(t-t_0)L}$ goes away, *i.e.*,

$$
\Psi^{(1)} = -L^{-1}Q_0F_0. \tag{16.39}
$$

This expression is further converted to a suggestive form as a dissipative correction. From the definition (16.23), $F^{(0)}$ is written as follows

$$
\begin{aligned}
F_p^{(0)} &= -(f_p^{\mathrm{eq}}\bar{f}_p^{\mathrm{eq}})^{-1}(\boldsymbol{v}\cdot\boldsymbol{\nabla} + \boldsymbol{F}\cdot\boldsymbol{\nabla}_p)f_p^{\mathrm{eq}} \\
&= -v^i\frac{\nabla^i T}{T^2}\left(\frac{\boldsymbol{\delta p}^2}{2m} - \mu_{\mathrm{TF}}\right) - v^i\frac{\delta p^j\nabla^i u^j}{T} - v^i\frac{\nabla^i\mu_{\mathrm{TF}}}{T} + F^i\frac{\delta p^i}{mT}.
\end{aligned}
\tag{16.40}
$$

Its projection onto Q_0 space is

$$
\begin{aligned}
[Q_0F^{(0)}]_p = [F^{(0)} - P_0F^{(0)}]_p &= F_p^{(0)} - \sum_{\alpha=0}^{4}\frac{\varphi_p^\alpha}{c^\alpha}\langle\varphi^\alpha, F^{(0)}\rangle \\
&= -\frac{\sigma^{ij}}{T}\hat{\pi}_p^{ij} + \frac{\nabla^i T}{T^2}\hat{J}_p^i,
\end{aligned}
\tag{16.41}
$$

with

$$
\hat{\pi}_p^{ij} := \delta v^{\langle i}\delta p^{j\rangle}, \qquad \hat{J}_p^i := \left(\frac{|\boldsymbol{\delta p}|^2}{2m} - \tilde{h}\right)\delta v^i, \tag{16.42}
$$

where $\boldsymbol{\delta v} := \boldsymbol{v} - \boldsymbol{u}$, $\boldsymbol{\delta p} := m\boldsymbol{\delta v}$, and $A^{\langle ij\rangle} := \Delta^{ijkl}A^{kl}$ for an arbitrary rank-two tensor A, with

$$
\Delta^{ijkl} = \frac{1}{2}\delta^{ik}\delta^{jl} + \frac{1}{2}\delta^{il}\delta^{jk} - \frac{1}{3}\delta^{ij}\delta^{kl}. \tag{16.43}
$$

Then, the shear tensor σ^{ij} is given by

$$\sigma^{ij} = \nabla^{\langle i} u^{j\rangle}. \tag{16.44}$$

The enthalpy density $\tilde{h}$ is defined by

$$\tilde{h}(t, \boldsymbol{x}) := h(t, \boldsymbol{x}) - V(\boldsymbol{x}) := e(t, \boldsymbol{x}) + \frac{P(t, \boldsymbol{x})}{n(t, \boldsymbol{x})} - V(\boldsymbol{x}), \tag{16.45}$$

where $h(t, \boldsymbol{x})$ is the standard enthalpy density while $\tilde{h}(t, \boldsymbol{x})$ contains the contribution of the external potential. Thus, we arrive at

$$-[L^{-1} Q_0 F^{(0)}]_p = \frac{\sigma^{ij}}{T}[L^{-1}\hat{\pi}^{ij}]_p + \frac{\nabla^i T}{T^2}[L^{-1}\hat{J}^i]_p. \tag{16.46}$$

Next, we solve the $O(\epsilon^2)$ equation,

$$\frac{\partial}{\partial t}\tilde{f}^{(2)}(t; t_0) = f^{\mathrm{eq}}\bar{f}^{\mathrm{eq}} L (f^{\mathrm{eq}}\bar{f}^{\mathrm{eq}})^{-1}\tilde{f}^{(2)}(t; t_0) + f^{\mathrm{eq}}\bar{f}^{\mathrm{eq}} K(t - t_0), \tag{16.47}$$

with the definitions,

$$K(t - t_0) := F[\tilde{f}^{(1)}(t)] + \frac{1}{2} B\big[(f^{\mathrm{eq}}\bar{f}^{\mathrm{eq}})^{-1}\tilde{f}^{(1)}(t), (f^{\mathrm{eq}}\bar{f}^{\mathrm{eq}})^{-1}\tilde{f}^{(1)}(t)\big], \tag{16.48}$$

$$B[\chi, \psi]_p := -(f^{\mathrm{eq}}_p \bar{f}^{\mathrm{eq}}_p)^{-1} \int \mathrm{d}p_1 \mathrm{d}p_2 \, \frac{\delta^2}{\delta f_{p_1} \delta f_{p_2}} C[f]_p \bigg|_{f = f^{\mathrm{eq}}} f^{\mathrm{eq}}_{p_1}\bar{f}^{\mathrm{eq}}_{p_1}\chi_{p_1} f^{\mathrm{eq}}_{p_2}\bar{f}^{\mathrm{eq}}_{p_2}\psi_{p_2}. \tag{16.49}$$

The solution to this equation is calculated to be,

$$\begin{aligned}\tilde{f}^{(2)} = f^{\mathrm{eq}}\bar{f}^{\mathrm{eq}}\bigg(&\mathrm{e}^{(t-t_0)L}\Psi^{(2)} + \big(1 - \mathrm{e}^{(t-t_0)\frac{\partial}{\partial s}}\big)\Big(-\frac{\partial}{\partial s}\Big)^{-1} P_0 K(s)\bigg|_{s=0} \\ &+ \big(\mathrm{e}^{(t-t_0)L} - \mathrm{e}^{(t-t_0)\frac{\partial}{\partial s}}\big)\mathcal{G}(s) Q_0 K(s)\bigg|_{s=0}\bigg),\end{aligned} \tag{16.50}$$

with

$$\mathcal{G} := \Big(L - \frac{\partial}{\partial s}\Big)^{-1}. \tag{16.51}$$

We choose $\Psi^{(2)}$ so that the fast-decaying term in the Q_0 space is eliminated,

$$\Psi^{(2)} = -\mathcal{G}(s) Q_0 K(s)\bigg|_{s=0}. \tag{16.52}$$

Then, the $O(\epsilon^2)$ solution is reduced to

$$\tilde{f}^{(2)} = f^{\text{eq}}\bar{f}^{\text{eq}}\Big[(t-t_0)P_0 - \Big(1+(t-t_0)\frac{\partial}{\partial s}\Big)\mathcal{G}(s)Q_0\Big]K(s)\Big|_{s=0} + O\big((t-t_0)^2\big). \tag{16.53}$$

Let us collect the perturbative solution and initial condition up to $O(\epsilon^2)$,

$$\begin{aligned}\tilde{f}(t;t_0) &= f^{\text{eq}} + \epsilon f^{\text{eq}}\bar{f}^{\text{eq}}\Big[(t-t_0)P_0F^{(0)} - L^{-1}Q_0F^{(0)}\Big]\\ &\quad+\epsilon^2 f^{\text{eq}}\bar{f}^{\text{eq}}\Big[(t-t_0)P_0 - \Big(1+(t-t_0)\frac{\partial}{\partial s}\Big)\mathcal{G}(s)Q_0\Big]K(s)\Big|_{s=0} + O((t-t_0)^2),\end{aligned} \tag{16.54}$$

$$f(t_0) = f^{\text{eq}} - \epsilon f^{\text{eq}}\bar{f}^{\text{eq}}L^{-1}Q_0F^{(0)} - \epsilon^2 f^{\text{eq}}\bar{f}^{\text{eq}}\mathcal{G}(s)Q_0K(s)\Big|_{s=0}. \tag{16.55}$$

This perturbative solution is not satisfactory because the infrared divergence breaks its validity as $|t-t_0|$ grows. In other word, $f(t;t_0)$ is well approximated by $\tilde{f}(t;t_0)$ only around the 'initial' time t_0.

16.1.2.2 Renormalization Group/Envelope Equation

To obtain the globally valid solution we impose the RG condition on the integral constants via the RG/E equation,

$$0 = \frac{\mathrm{d}}{\mathrm{d}t_0}\tilde{f}(t;t_0)\Big|_{t_0=t}, \tag{16.56}$$

which promotes the integral constants $\boldsymbol{u}$, μ, and T to dynamical variables. This equation is indeed the seed of desired fluid dynamic equation. Plugging the solution of the RG/E equation back into $f(t_0)|_{t_0=t}$ gives the global solution of the Boltzmann equation.

We convert the RG/E equation (16.56) into the fluid dynamic equation. Projecting it onto the P_0 space by taking the inner product with the zero mode results in

$$\int \mathrm{d}p\,\varphi_p^{\alpha}\Big[\frac{\partial}{\partial t} + \epsilon\big(\boldsymbol{v}\cdot\nabla + \boldsymbol{F}\cdot\nabla_p\big)\Big]\Big[f^{\text{eq}} - \epsilon f^{\text{eq}}\bar{f}^{\text{eq}}L^{-1}Q_0F^{(0)}\Big]_p = O(\epsilon^3). \tag{16.57}$$

We reduce Eq. (16.57) into the equation of continuity. We find

$$\int \mathrm{d}p\,\varphi_{0p}^{\alpha}\frac{\mathrm{D}}{\mathrm{D}t}f_p^{\text{eq}} = \Big\langle \varphi_0^{\alpha}, \frac{1}{T^2}\Big(\frac{\delta p^2}{2m} - \tilde{\mu}\Big)\frac{\mathrm{D}T}{\mathrm{D}t} + \frac{1}{T}\delta p^i\frac{\mathrm{D}u^i}{\mathrm{D}t} + \frac{1}{T}\frac{\mathrm{D}\tilde{\mu}}{\mathrm{D}t}\Big\rangle$$

$$= \left\langle \varphi_0^\alpha, \frac{1}{T^2}\frac{\mathrm{D}T}{\mathrm{D}t}\varphi_0^4 + \frac{1}{c_0}\frac{\mathrm{D}n}{\mathrm{D}t}\varphi_0^0 + \frac{1}{T}\frac{\mathrm{D}u^i}{\mathrm{D}t}\varphi_0^i \right\rangle$$

$$= \begin{cases} \frac{\mathrm{D}n}{\mathrm{D}t}, & \alpha = 0, \\ mn\frac{\mathrm{D}u^i}{\mathrm{D}t}, & \alpha = i, \\ \frac{c^4}{T^2}\frac{\mathrm{D}T}{\mathrm{D}t}, & \alpha = 4, \end{cases} \tag{16.58}$$

with

$$\mathrm{D}/\mathrm{D}t = \partial/\partial t + \epsilon \boldsymbol{u} \cdot \nabla. \tag{16.59}$$

The spatial derivative and force term of the equilibrium distribution function:

$$\int \mathrm{d}p\, \varphi_{0p}^\alpha \left(\boldsymbol{\delta v} \cdot \nabla + \boldsymbol{F} \cdot \nabla_{\boldsymbol{p}}\right) f_p^{\mathrm{eq}} = -\langle \varphi_0^\alpha, F^{(0)} \rangle$$

$$= \begin{cases} n\nabla \cdot \boldsymbol{u}, & \alpha = 0, \\ \nabla_i P - nF_i, & \alpha = i, \\ c^4 \frac{2}{3T} \nabla \cdot \boldsymbol{u}, & \alpha = 4. \end{cases} \tag{16.60}$$

The dissipative part of the distribution function:

$$\int \mathrm{d}p\, \varphi_{0p}^\alpha \left[\frac{\mathrm{D}}{\mathrm{D}t} + \epsilon\left(\boldsymbol{\delta v} \cdot \nabla + \boldsymbol{F} \cdot \nabla_{\boldsymbol{p}}\right)\right] [-\epsilon f^{\mathrm{eq}} \bar{f}^{\mathrm{eq}} L^{-1} Q_0 F^{(0)}]_p$$

$$= -\epsilon \int \mathrm{d}p \varphi_{0p}^\alpha \left[\frac{\mathrm{D}}{\mathrm{D}t} + \epsilon\left(\boldsymbol{\delta v} \cdot \nabla + \boldsymbol{F} \cdot \nabla_{\boldsymbol{p}}\right)\right] \left[f^{\mathrm{eq}} \bar{f}^{\mathrm{eq}} \left(-\frac{\sigma^{ij}}{T} L^{-1}\hat{\pi}^{ij} - \frac{\nabla^i T}{T^D 2} L^{-1}\hat{j}^i\right)\right]_p$$

$\underline{\alpha = 0}$

$$= -\left[\epsilon \frac{\mathrm{D}}{\mathrm{D}t}\langle 1, L^{-1} Q_0 F^{(0)}\rangle + \epsilon^2 \nabla^j \langle \delta v^j, L^{-1} Q_0 F^{(0)}\rangle\right] = 0, \tag{16.61}$$

$\underline{\alpha = i}$

$$= -\left[\epsilon \frac{\mathrm{D}}{\mathrm{D}t}\langle \delta p^i, L^{-1} Q_0 F^{(0)}\rangle + \epsilon^2 \nabla^j \langle \delta v^j \delta p^i, L^{-1} Q_0 F^{(0)}\rangle\right.$$

$$\left. - \left\langle \left(\frac{\mathrm{D}}{\mathrm{D}t} + \epsilon \boldsymbol{\delta v} \cdot \nabla + \epsilon \boldsymbol{F} \cdot \nabla_{\boldsymbol{p}}\right) \delta v^i, \epsilon L^{-1} Q_0 F^{(0)} \right\rangle\right]$$

$$= -\epsilon^2 \nabla^j \langle \delta v^j \delta p^i, L^{-1} Q_0 F^{(0)}\rangle = -2\epsilon^2 \eta \nabla^j \sigma^{ij}, \tag{16.62}$$

$\underline{\alpha = 4}$

$$= -\left[\epsilon \frac{\mathrm{D}}{\mathrm{D}t}\left\langle \frac{|\boldsymbol{\delta p}|^2}{2m} - \frac{3nT}{2A}, L^{-1} Q_0 F^{(0)}\right\rangle + \epsilon^2 \nabla^j \left\langle \delta v^j \left(\frac{|\boldsymbol{\delta p}|^2}{2m} - \frac{3nT}{2c^0}\right), L^{-1} Q_0 F^{(0)}\right\rangle\right.$$

$$\left. - \left\langle \left(\frac{\mathrm{D}}{\mathrm{D}t} + \epsilon \boldsymbol{\delta v} \cdot \nabla + \epsilon \boldsymbol{F} \cdot \nabla_{\boldsymbol{p}}\right) \left(\frac{|\boldsymbol{\delta p}|^2}{2m} - \frac{3nT}{2c^0}\right), \epsilon L^{-1} Q_0 F^{(0)} \right\rangle\right]$$

$$
\begin{aligned}
&= -\epsilon^2 \Bigg[\nabla^j \left\langle \delta v^j \left(\frac{|\boldsymbol{\delta p}|^2}{2m} - \frac{3nT}{2c^0} \right), L^{-1} Q_0 F^{(0)} \right\rangle \\
&- \left\langle \boldsymbol{\delta v} \cdot \boldsymbol{\nabla} \left(\frac{|\boldsymbol{\delta p}|^2}{2m} - \frac{3nT}{2c^0} \right), L^{-1} Q_0 F^{(0)} \right\rangle \Bigg] \\
&= -\epsilon^2 \left[\nabla^i (\lambda \nabla^i T) + 2\eta \sigma^{ij} \sigma^{ij} \right].
\end{aligned}
\tag{16.63}
$$

In the third equality of Eq. (16.62), the following transformation has been made

$$
\begin{aligned}
\langle \delta v^j \delta p^i, L^{-1} Q_0 F^{(0)} \rangle &= -\langle \delta v^j \delta p^i, L^{-1} \hat{\pi}^{kl} \rangle \frac{\sigma^{ij}}{T} = -\langle \hat{\pi}^{ij}, L^{-1} \hat{\pi}^{kl} \rangle \frac{\sigma^{ij}}{T} \\
&= -\langle \hat{\pi}^{kl}, L^{-1} \hat{\pi}^{kl} \rangle \frac{\sigma^{ij}}{5T} = 2\eta \sigma^{ij},
\end{aligned}
\tag{16.64}
$$

and in the third equality of Eq. (16.63), we have used the following computations,

$$
\begin{aligned}
\left\langle \delta v^j \left(\frac{\delta p^2}{2m} - \frac{3nT}{2A} \right), L^{-1} Q_0 F^{(0)} \right\rangle &= -\left\langle \hat{J}^j, L^{-1} \hat{J}^k \right\rangle \frac{\nabla^k T}{T^2} \\
&= -\left\langle \hat{J}^j, L^{-1} \hat{J}^j \right\rangle \frac{\nabla^i T}{3T^2} = \lambda \nabla^i T,
\end{aligned}
\tag{16.65}
$$

$$
\begin{aligned}
\left\langle \delta v^j \nabla^j \left(\frac{\delta p^2}{2m} - \frac{3nT}{2A} \right), L^{-1} Q_0 F^{(0)} \right\rangle &= \nabla^j u^m \langle \delta v^j \delta p^m, L^{-1} \hat{\pi}^{kl} \rangle \frac{\sigma^{ij}}{T} \\
&= \langle \hat{\pi}^{kl}, L^{-1} \hat{\pi}^{kl} \rangle \frac{\sigma^{ij} \sigma^{ij}}{5T} = -2\eta \sigma^{ij} \sigma^{ij}.
\end{aligned}
\tag{16.66}
$$

We substitute Eqs. (16.58)–(16.63) into Eq. (16.57) to obtain the equations of continuity:

$$
\frac{\mathrm{D}n}{\mathrm{D}t} + \epsilon n \nabla \cdot \boldsymbol{u} = 0,
\tag{16.67}
$$

$$
mn \frac{\mathrm{D}u^i}{\mathrm{D}t} + \epsilon \nabla^i P - \epsilon n F^i - 2\epsilon^2 \nabla^j \sigma^{ij} = 0,
\tag{16.68}
$$

$$
\frac{c_4}{T^2} \frac{\mathrm{D}T}{\mathrm{D}t} + \epsilon \frac{2c_4}{3T} \boldsymbol{\nabla} \cdot \boldsymbol{u} - \epsilon^2 \left(\boldsymbol{\nabla} \cdot (\lambda \boldsymbol{\nabla} T) + 2\eta \sigma^{jk} \sigma^{jk} \right) = 0.
\tag{16.69}
$$

These equations are further reduced into the well-known forms

$$
\frac{\mathrm{D}n}{\mathrm{D}t} = -n\epsilon \boldsymbol{\nabla} \cdot \boldsymbol{u},
\tag{16.70}
$$

$$
mn \frac{\mathrm{D}u^i}{\mathrm{D}t} = -\epsilon \nabla^i P + \epsilon n F^i + 2\epsilon^2 \eta \nabla^j \sigma^{ij},
\tag{16.71}
$$

$$
n \frac{\mathrm{D}e}{\mathrm{D}t} = -\epsilon P \boldsymbol{\nabla} \cdot \boldsymbol{u} + \epsilon^2 \boldsymbol{\nabla} \cdot (\lambda \boldsymbol{\nabla} T) + 2\epsilon^2 \eta \sigma^{jk} \sigma^{jk}.
\tag{16.72}
$$

Equation (16.72) is derived as follows:

$$\begin{aligned} n\frac{\mathrm{D}e}{\mathrm{D}t} &= \frac{\mathrm{D}(ne)}{\mathrm{D}t} - e\frac{\mathrm{D}n}{\mathrm{D}t} = \frac{c^4}{T^2}\frac{\mathrm{D}T}{\mathrm{D}t} + \frac{3nT}{2c^0}\frac{\mathrm{D}n}{\mathrm{D}t} - e\frac{\mathrm{D}n}{\mathrm{D}t} \\ &= -\epsilon\left(\frac{2c^4}{3T} + \frac{3n^2T}{2c^0} + ne\right)\nabla\cdot\boldsymbol{u} + (\sigma^{jk}\sigma^{jk} + \nabla\cdot(\lambda\nabla T)) \\ &= -\epsilon P\nabla\cdot\boldsymbol{u} + \epsilon^2\sigma^{jk}\sigma^{jk} + \epsilon^2\nabla\cdot(\lambda\nabla T). \end{aligned} \tag{16.73}$$

We have used (16.31) to reach the final expression.

Microscopic expressions of the shear viscosity η and the heat conductivity λ are given by,

$$\eta = -\frac{1}{10T}\langle\hat{\pi}^{ij}, L^{-1}\hat{\pi}^{ij}\rangle, \qquad \lambda = -\frac{1}{3T^2}\langle\hat{J}^i, L^{-1}\hat{J}^i\rangle. \tag{16.74}$$

We note that the bulk viscosity is identically zero in contrast to the relativistic case. This is attributed to the fact that $L^{-1}Q_0F^{(0)}$ (16.46) does not contain the term proportional to the scalar expansion $\theta = \nabla\cdot\boldsymbol{u}$ which would be coupled to the microscopic bulk pressure. Alternatively, the vanishing bulk pressure can be seen as follows. In the kinetic theory, the bulk pressure is defined by,

$$\Pi = \frac{1}{3m}\int \mathrm{d}\boldsymbol{p}\,\delta\boldsymbol{p}^2(f_p - f_p^{\mathrm{eq}}). \tag{16.75}$$

We recall that our solution given by the RG method turns out to satisfy the matching conditions in the energy frame as:

$$n = \int \mathrm{d}\boldsymbol{p} f_p = \int \mathrm{d}\boldsymbol{p} f_p^{\mathrm{eq}}, \qquad e = \int \mathrm{d}\boldsymbol{p} E_p f_p = \int \mathrm{d}\boldsymbol{p} E_p f_p^{\mathrm{eq}}. \tag{16.76}$$

The latter condition implies that

$$0 = \int \mathrm{d}\boldsymbol{p}\frac{\delta\boldsymbol{p}^2}{2m}(f_p - f_p^{\mathrm{eq}}), \tag{16.77}$$

given $E_p = \delta\boldsymbol{p}^2/(2m)$, and thus the bulk pressure Π identically vanishes. More elaborated approaches to the computation of the bulk viscosity for non-relativistic systems have been explored [265, 266, 271, 272]. There, the finite bulk viscosity is induced by the interactions breaking scale invariance, which is not captured by the standard Boltzmann equation (16.1). Thus, it is desirable and should be possible to derive the fluid dynamic equation with such effects incorporated [265] with the RG method but it is beyond the scope of the present book.

16.1.3 Derivation of Second-Order Non-relativistic Fluid Dynamic Equation

Having derived the Navier–Stokes equations, we now turn to the second-order fluid dynamics. In case of the Navier–Stokes equations, the crucial step was to identify the P_0 space, spanned by the zero modes of linearized collision operator. For the higher-order fluid dynamic equations, we need to take into account quasi-slow modes on top of the slow/zero modes. Extraction of such modes are carried out by the doublet scheme developed in [69] and described in Chap. 9, where some low-lying modes of the linearized collision operator are properly picked up. They, in turn, govern the relaxation dynamics of dissipative currents.

Let us recall that in the Navier–Stokes equations, the fluid dynamic variables initially show up as integration constant, $\boldsymbol{u}(t_0)$, $\mu(t_0)$, and $T(t_0)$, in the $O(\epsilon^0)$-solution of Boltzmann equation (16.15). Analogously, we introduce additional integration constants in the $O(\epsilon)$-solution, that, given an 'initial' condition $\Psi^{(1)}$, takes the following form,

$$\tilde{f}^{(1)} = f^{\text{eq}}\bar{f}^{\text{eq}}\Big[\mathrm{e}^{(t-t_0)L}\big(\Psi^{(1)} + L^{-1}Q_0F_0\big) + (t-t_0)P_0F_0 - L^{-1}Q_0F_0\Big]. \quad (16.78)$$

To go beyond the description of ordinary dissipative fluid, namely, the mesoscopic regime, we need to specify the quasi-slow variables in the first-order 'initial' value $\Psi^{(1)}$. To this end, we expand the first-order perturbative solution around $t = t_0$:

$$\begin{aligned}\tilde{f}^{(1)} &= f^{\text{eq}}\bar{f}^{\text{eq}}\Psi^{(1)} + (t-t_0)f^{\text{eq}}\bar{f}^{\text{eq}}\big(P_0F^{(0)} + L\Psi^{(1)} + Q_0F^{(0)}\big)\\ &\quad + O\big((t-t_0)^2\big).\end{aligned} \quad (16.79)$$

The $O\big((t-t_0)^2\big)$ terms are not important here as they do not contribute to the RG/E equation. We divide the Q_0 space into P_1 and Q_1 spaces. The P_1 space accommodates the quasi-slow modes that play the crucial role in describing the mesoscopic dynamics. Keeping these expansion in mind, we spell out the two criteria to identify $\Psi^{(1)}$ and P_1 space:

1. $\Psi^{(1)}$ and $L^{-1}Q_0F^{(0)}$ belongs to a common vector space, which is orthogonal to P_0 space.
2. Define P_1 space by a vector space spanned by $\Psi^{(1)}$ and $L\Psi^{(1)}$.

The first criteria indeed makes the vector space to which $\mathrm{d}\tilde{f}^{(1)}/\mathrm{d}\tau|_{\tau=\tau_0}$ belongs the smallest. The second criteria defines the P_1 space so that it accommodates all the terms in (16.79) but the zero modes.

As computed in (16.46), $L^{-1}Q_0F^{(0)}$ is given by a linear combination of the vectors $\big([L^{-1}\hat{J}^{\mu}]_p, [L^{-1}\hat{\pi}^{\mu\nu}]_p\big)$. Therefore, instead of (16.39) we impose the 'initial' condition for the first-order equation,

$$\tilde{f}^{(1)}(t_0) = f^{\rm eq}\bar{f}^{\rm eq}\Psi_p^{(1)}(t_0),$$
$$\Psi_p^{(1)}(t_0) = -5\frac{[L^{-1}\hat{\pi}^{ij}]_p}{\langle\hat{\pi}^{ij}, L^{-1}\hat{\pi}^{ij}\rangle}\pi^{ij}(t_0) - 3\frac{[L^{-1}\hat{J}^i]_p}{\langle\hat{J}^i, L^{-1}\hat{J}^i\rangle}J^i(t_0), \tag{16.80}$$

where we have introduced $\pi^{ij}(t_0)$ and $J^i(t_0)$ as would-be fluid variables. Furthermore, the P_1 space are spanned by quasi-slow doublet modes $\{\hat{\pi}_p^{ij}, \hat{J}_p^i\}$ and $\{[L^{-1}\hat{\pi}^{ij}]_p, [L^{-1}\hat{J}^i]_p\}$ according to the second condition raised above.

The $\mathrm{O}(\epsilon^2)$ equation and its solution are already given by (16.47) and (16.53) under the 'initial' condition (16.52). Thus, the perturbative solution up to the second order is summarized as

$$\begin{aligned}\tilde{f} &= f^{\rm eq} + \epsilon f^{\rm eq}\bar{f}^{\rm eq}\big[(1+(t-t_0)L)\Psi^{(1)} + (t-t_0)F_0\big]\\ &+\epsilon^2 f^{\rm eq}\bar{f}^{\rm eq}\Big[(t-t_0)P_0 + (t-t_0)\mathcal{G}(s)^{-1}P_1\mathcal{G}(s)Q_0\\ &-\Big(1+(t-t_0)\frac{\partial}{\partial s}\Big)Q_1\mathcal{G}(s)Q_0\Big]K(s)\Big|_{s=0} + O\big((t-t_0)^2\big).\end{aligned} \tag{16.81}$$

Applying the projection operator P_0 onto the RG/E equation

$$\frac{\mathrm{d}}{\mathrm{d}t_0}\tilde{f}(t;t_0)\Big|_{t_0=t} = 0, \tag{16.82}$$

as we have done in the previous subsection, we have the following set of equations

$$\frac{Dn}{Dt} = -n\nabla\cdot\boldsymbol{u}, \tag{16.83}$$
$$mn\frac{Du^i}{Dt} = -\nabla^i P + nF^i + \nabla^j\pi^{ij}, \tag{16.84}$$
$$n\frac{De}{Dt} = -P\nabla\cdot\boldsymbol{u} + \nabla\cdot\boldsymbol{J} + \sigma^{jk}\pi^{jk}. \tag{16.85}$$

These are the same as the Navier–Stokes equations (16.70)–(16.72). Except that the stress tensor π^{ij} and the heat current $\boldsymbol{J}$ have their own dynamics, that are to be dictated by the relaxation equations.

In order to obtain the relaxation equation, we perform the projection of the RG/E equation onto the P_1 space by using the quasi-slow modes $\{[L^{-1}\hat{\pi}^{ij}]_p, [L^{-1}\hat{J}^i]_p\}$,

$$\begin{aligned}&\int\mathrm{d}p\Big[L^{-1}\{\hat{\pi}^{ij}, \hat{J}^i\}\Big]_p\Big[\frac{\partial}{\partial t} + \epsilon\big(\boldsymbol{v}\cdot\nabla + \boldsymbol{F}\cdot\nabla_p\big)\Big]\Big[f^{\rm eq} + \epsilon f^{\rm eq}\bar{f}^{\rm eq}\Psi\Big]_p\\ &= \epsilon\Big\langle L^{-1}\{\hat{\pi}^{ij}, \hat{J}^i\}, L\Psi^{(1)}\Big\rangle + \frac{\epsilon^2}{2}\Big\langle L^{-1}\{\hat{\pi}^{ij}, \hat{J}^i\}, B\big[\Psi^{(1)}, \Psi^{(1)}\big]\Big\rangle + O(\epsilon^3),\end{aligned} \tag{16.86}$$

where the definition of the symbol of $B[\phi, \psi]$ is given in (16.49). As we will detail in the next subsection, these equations are converted to the following set of equations,

$$
\begin{aligned}
\pi^{ij} &= 2\eta\sigma^{ij} - \tau_\pi \frac{\mathrm{D}}{\mathrm{D}t}\pi^{ij} - \ell_{\pi J}\nabla^{\langle i}J^{j\rangle} \\
&\quad + \kappa^{(1)}_{\pi\pi}\pi^{ij}\nabla\cdot\boldsymbol{u} + \kappa^{(2)}_{\pi\pi}\pi^{k\langle i}\sigma^{j\rangle k} - 2\tau_\pi\pi^{k\langle i}\omega^{j\rangle k} \\
&\quad + \kappa^{(1)}_{\pi J}J^{\langle i}\nabla^{j\rangle}n + \kappa^{(2)}_{\pi J}J^{\langle i}\nabla^{j\rangle}P + \kappa^{(3)}_{\pi J}J^{\langle i}F^{j\rangle} \\
&\quad + b_{\pi\pi\pi}\pi^{k\langle i}\pi^{j\rangle k} + b_{\pi JJ}J^{\langle i}J^{j\rangle},
\end{aligned}
\tag{16.87}
$$

$$
\begin{aligned}
J^i &= \lambda\nabla^i T - \tau_J \frac{\mathrm{D}}{\mathrm{D}t}J^i - \ell_{J\pi}\nabla^j\pi^{ij} \\
&\quad + \kappa^{(1)}_{J\pi}\pi^{ij}\nabla^j n + \kappa^{(2)}_{J\pi}\pi^{ij}\nabla^j P + \kappa^{(3)}_{J\pi}\pi^{ij}F^j \\
&\quad + \kappa^{(1)}_{JJ}J^i\nabla\cdot\boldsymbol{u} + \kappa^{(2)}_{JJ}J^j\sigma^{ij} + \tau_J J^j\omega^{ij} \\
&\quad + b_{JJ\pi}J^j\pi^{ij},
\end{aligned}
\tag{16.88}
$$

after setting $\epsilon = 1$. The vorticity ω^{ij} is defined by

$$
\omega^{ij} := (\nabla^i u^j - \nabla^j u^i)/2. \tag{16.89}
$$

They are reduced to the Navier–Stokes equations if only the first term is kept in the right-hand side of each equation. The shear viscosity η and heat conductivity λ are given by (16.74). The other terms are the second-order corrections accompanied by second-order transport coefficients. In particular, the dissipative currents $\hat{\pi}^{ij}$ and $\hat{J}^i$ have their own dynamics due to their time derivative terms, and their dynamical time scales are characterized by the relaxation times τ_π and τ_J. Their microscopic expressions are derived to be

$$
\tau_\pi = \frac{1}{10T\eta}\langle\hat{\pi}^{ij}, L^{-2}\hat{\pi}^{ij}\rangle, \qquad \tau_J = \frac{1}{3T^3\lambda}\langle\hat{J}^i, L^{-2}\hat{J}^i\rangle. \tag{16.90}
$$

16.1.3.1 Derivation of Relaxation Equations

We present the detailed derivation of relaxation equations (16.87), (16.88) starting from (16.86),

$$
\begin{aligned}
&\int \mathrm{d}p\Big[L^{-1}\{\hat{\pi}^{ij}, \hat{J}^i\}\Big]_p\Big[\frac{\partial}{\partial t} + \epsilon\big(\boldsymbol{v}\cdot\nabla + \boldsymbol{F}\cdot\nabla_p\big)\Big]\Big[f^{\mathrm{eq}} + \epsilon f^{\mathrm{eq}}\bar{f}^{\mathrm{eq}}\Psi\Big]_p \\
&= \epsilon\Big\langle L^{-1}\{\hat{\pi}^{ij}, \hat{J}^i\}, L\Psi^{(1)}\Big\rangle + \frac{\epsilon^2}{2}\Big\langle L^{-1}\{\hat{\pi}^{ij}, \hat{J}^i\}, B\big[\Psi^{(1)}, \Psi^{(1)}\big]\Big\rangle + O(\epsilon^3).
\end{aligned}
\tag{16.91}
$$

For the sake of notational simplicity, we introduce the following vectors,

$$\hat{\psi}_p^\alpha = \{\hat{\pi}_p^{ij}, \hat{J}_p^i\}, \tag{16.92}$$

$$\hat{\chi}_p^\alpha = \left\{\frac{\hat{\pi}_p^{ij}}{2T\eta}, \frac{\hat{J}_p^i}{T^2\lambda}\right\}, \tag{16.93}$$

$$\psi^\alpha = \{\pi^{ij}, J^i\}, \tag{16.94}$$

$$X^\alpha = \left\{2\eta\sigma^{ij}, \lambda\nabla^i T\right\}, \tag{16.95}$$

with which we can write as $\Psi = L^{-1}\hat{\chi}^\alpha\psi^\alpha$ and $Q_0F_0 = -\hat{\chi}_1^\alpha X^\alpha$. Equation (16.86) can be converted into the following form,

$$\begin{aligned}
&\epsilon\langle L^{-1}\hat{\psi}^\alpha, \hat{\chi}^\beta\rangle\psi^\beta \\
&= \epsilon\langle L^{-1}\hat{\psi}^\alpha, \hat{\chi}^\beta\rangle X^\beta + \langle L^{-1}\hat{\psi}^\alpha, L^{-1}\hat{\chi}^\beta\rangle\frac{\mathrm{D}}{\mathrm{D}t}\psi^\beta + \epsilon\langle L^{-1}\hat{\psi}^\alpha, \delta K^i L^{-1}\hat{\chi}^\beta\rangle\nabla^i\psi^\beta \\
&+ \epsilon\left\langle L^{-1}\hat{\psi}^\alpha, f^{\mathrm{eq}}\bar{f}^{\mathrm{eq}}\left(\frac{\mathrm{D}}{\mathrm{D}t} + \epsilon\boldsymbol{\delta K}\cdot\nabla + \epsilon\boldsymbol{F}\cdot\nabla_K\right) f^{\mathrm{eq}}\bar{f}^{\mathrm{eq}}L^{-1}\hat{\chi}^\beta\right\rangle\psi^\beta \\
&- \epsilon^2\frac{1}{2}\langle L^{-1}\hat{\psi}^\alpha, B[L^{-1}\hat{\chi}_1^\beta, L^{-1}\hat{\chi}^\gamma]\rangle\psi^\beta\psi^\gamma,
\end{aligned} \tag{16.96}$$

where we have defined $[\delta K^i]_p := \delta p^i/m = \delta v^i$ and $[\nabla_K^i]_p := \nabla_p^i$.

The coefficients of the first, third, and fourth terms in the right-hand side of Eq. (16.96) can be written as

$$\left\langle L^{-1}\hat{\psi}^\alpha, \hat{\psi}^\beta\right\rangle = \begin{pmatrix} -2T\eta\Delta^{ijkl} & 0 \\ 0 & -T^2\lambda\Delta^{ij} \end{pmatrix}, \tag{16.97}$$

$$\left\langle L^{-1}\hat{\psi}^\alpha, L^{-1}\hat{\psi}^\beta\right\rangle = \begin{pmatrix} 2T\eta\tau_\pi\Delta^{ijkl} & 0 \\ 0 & T^2\lambda\tau_J\Delta^{ij} \end{pmatrix}, \tag{16.98}$$

$$\left\langle L^{-1}\hat{\psi}^\alpha, \delta K^m L^{-1}\hat{\psi}^\beta\right\rangle = \begin{pmatrix} 0 & T^2\lambda\ell_{\pi J}\Delta^{ijkl} \\ 2T\eta\ell_{J\pi}\Delta^{ijkl} & 0 \end{pmatrix}, \tag{16.99}$$

where transport coefficients introduced here are defined as follows:

$$\tau_\pi = \frac{1}{10T\eta}\left\langle\hat{\pi}^{ij}, L^{-2}\hat{\pi}^{ij}\right\rangle, \tag{16.100}$$

$$\tau_J = \frac{1}{3T^2\lambda}\left\langle\hat{J}^i, L^{-2}\hat{J}^i\right\rangle, \tag{16.101}$$

which are viscous relaxation times for the stress tensor and heat flow, respectively, and

$$\ell_{\pi J} = \frac{1}{5T^2\lambda}\langle L^{-1}\hat{\pi}^{ij}, \delta K^i L^{-1}\hat{J}^j\rangle, \tag{16.102}$$

$$\ell_{J\pi} = \frac{1}{10T\eta}\langle L^{-1}\hat{J}^i, \delta K^j L^{-1}\hat{\pi}^{ij}\rangle, \tag{16.103}$$

which are so called viscous relaxation lengths.

Then, let us rewrite the last term in the right-hand side of Eq. (16.96) as

$$\frac{\epsilon^2}{2}\langle L^{-1}\hat{\pi}^{ij}, B[L^{-1}\hat{\chi}^\beta][L^{-1}\hat{\chi}^\gamma]\rangle\psi^\beta\psi^\gamma = b_{\pi\pi\pi}\pi^{m\langle k}\pi^{l\rangle m} + b_{\pi JJ}J^{\langle k}J^{l\rangle}, \tag{16.104}$$

$$\frac{\epsilon^2}{2}\langle L^{-1}\hat{J}^i, B[L^{-1}\hat{\chi}^\beta][L^{-1}\hat{\chi}^\gamma]\rangle\psi^\beta\psi^\gamma = b_{J\pi J}\pi^{ij}J^j, \tag{16.105}$$

where the transport coefficients are defined by

$$b_{\pi\pi\pi} = \frac{3}{70T^2\eta^2}\langle L^{-1}\hat{\pi}^{ij}, B[L^{-1}\hat{\pi}^{ik}][L^{-1}\hat{\pi}^{jk}]\rangle, \tag{16.106}$$

$$b_{\pi JJ} = \frac{1}{10T^4\lambda^2}\langle L^{-1}\hat{\pi}^{ij}, B[L^{-1}\hat{J}^i][L^{-1}\hat{J}^j]\rangle, \tag{16.107}$$

$$b_{JJ\pi} = \frac{1}{10T^3\eta\lambda}\langle L^{-1}\hat{J}^i, B[L^{-1}\hat{J}^j][L^{-1}\hat{\pi}^{ij}]\rangle. \tag{16.108}$$

We consider the forth term in the right-hand side of Eq. (16.96):

$$\begin{aligned}
&\epsilon\Big\langle L^{-1}\hat{\psi}^\alpha, (f^{\mathrm{eq}}\bar{f}^{\mathrm{eq}})^{-1}\left[\frac{\mathrm{D}}{\mathrm{D}t} + \epsilon\boldsymbol{\delta K}\cdot\nabla + \epsilon\boldsymbol{F}\cdot\nabla_K\right] f^{\mathrm{eq}}\bar{f}^{\mathrm{eq}}L^{-1}\hat{\chi}^\beta\Big\rangle\hat{\psi}^\beta \\
&= \epsilon\Big\langle L^{-1}\hat{\psi}^\alpha, (f^{\mathrm{eq}}\bar{f}^{\mathrm{eq}})^{-1}\frac{\partial}{\partial T}[f^{\mathrm{eq}}\bar{f}^{\mathrm{eq}}L^{-1}\hat{\chi}^\beta]\Big\rangle\hat{\psi}^\beta\frac{\mathrm{D}T}{\mathrm{D}t} \\
&+ \epsilon^2\Big\langle L^{-1}\hat{\psi}^\alpha, (f^{\mathrm{eq}}\bar{f}^{\mathrm{eq}})^{-1}\delta K^a\frac{\partial}{\partial T}[f^{\mathrm{eq}}\bar{f}^{\mathrm{eq}}L^{-1}\hat{\chi}^\beta]\Big\rangle\hat{\psi}^\beta\nabla^a T \\
&+ \epsilon\Big\langle L^{-1}\hat{\psi}^\alpha, (f^{\mathrm{eq}}\bar{f}^{\mathrm{eq}})^{-1}\frac{\partial}{\partial\tilde{\mu}}[f^{\mathrm{eq}}\bar{f}^{\mathrm{eq}}L^{-1}\hat{\chi}^\beta\Big\rangle\hat{\psi}^\beta\frac{\mathrm{D}\tilde{\mu}}{\mathrm{D}t} \\
&+ \epsilon^2\Big\langle L^{-1}\hat{\psi}^\alpha, (f^{\mathrm{eq}}\bar{f}^{\mathrm{eq}})^{-1}\delta K^a\frac{\partial}{\partial\tilde{\mu}}[f^{\mathrm{eq}}f^{\mathrm{eq}}L^{-1}\hat{\chi}^\beta]\Big\rangle\hat{\psi}^\beta\nabla^a\tilde{\mu} \\
&+ \epsilon\Big\langle L^{-1}\hat{\psi}^\alpha, (f^{\mathrm{eq}}\bar{f}^{\mathrm{eq}})^{-1}\frac{\partial}{\partial u^b}[f^{\mathrm{eq}}f^{\mathrm{eq}}L^{-1}\hat{\chi}^\beta]\Big\rangle\hat{\psi}^\beta\frac{\mathrm{D}u^b}{\mathrm{D}t} \\
&+ \epsilon^2\Big\langle L^{-1}\hat{\psi}^\alpha, (f^{\mathrm{eq}}\bar{f}^{\mathrm{eq}})^{-1}\delta K^a\frac{\partial}{\partial u^b}[f^{\mathrm{eq}}f^{\mathrm{eq}}L^{-1}\hat{\chi}^\beta]\Big\rangle\hat{\psi}^\beta\nabla^a u^b \\
&+ \epsilon^2\Big\langle L^{-1}\hat{\psi}^\alpha, (f^{\mathrm{eq}}\bar{f}^{\mathrm{eq}})^{-1}\nabla_K^a[f^{\mathrm{eq}}f^{\mathrm{eq}}L^{-1}\hat{\chi}^\beta]\Big\rangle\hat{\psi}^\beta\frac{F^a}{m}.
\end{aligned} \tag{16.109}$$

The Lagrange derivative of T, $\tilde{\mu}$, and u^b are rewritten by using the balance equation up to the first order with respect to ϵ, which corresponds to the Euler's equation:

$$\frac{\mathrm{D}T}{\mathrm{D}t} = -\epsilon\frac{2T}{3}\nabla\cdot\boldsymbol{u} + O(\epsilon^2), \tag{16.110}$$

$$\frac{\mathrm{D}\tilde{\mu}}{\mathrm{D}t} = -\epsilon\frac{2\tilde{\mu}}{3}\nabla\cdot\boldsymbol{u} + O(\epsilon^2), \tag{16.111}$$

$$\frac{\mathrm{D}u^b}{\mathrm{D}t} = -\epsilon\nabla^i P + \epsilon n F^i + O(\epsilon^2), \tag{16.112}$$

where we have used the relation $\mathrm{d}n = (\partial n/\partial T)\mathrm{d}T + (\partial n/\partial\tilde{\mu})\mathrm{d}\tilde{\mu}$ in the derivation of Eq. (16.111). Then, Eq. (16.109) takes the following forms

$$\begin{aligned}
&\epsilon\Big\langle L^{-1}\hat{\pi}^{ij}, (f^{\mathrm{eq}}\bar{f}^{\mathrm{eq}})^{-1}\left[\frac{\mathrm{D}}{\mathrm{D}t} + \epsilon\boldsymbol{\delta K}\cdot\nabla + \epsilon\boldsymbol{F}\cdot\nabla_K\right] f^{\mathrm{eq}}\bar{f}^{\mathrm{eq}}\hat{\chi}^\beta\Big\rangle\psi^\beta \\
&= -\epsilon^2\left(\kappa^{(1)}_{\pi\pi}\pi^{ij}\nabla\cdot\boldsymbol{u} + \kappa^{(2)}_{\pi\pi}\pi^{k\langle i}\sigma^{j\rangle k} + \kappa^{(3)}_{\pi\pi}\pi^{k\langle i}\omega^{j\rangle k}\right. \\
&\quad \left. + \kappa^{(1)}_{\pi J}J^{\langle i}\nabla^{j\rangle}n + \kappa^{(2)}_{\pi J}J^{\langle i}\nabla^{j\rangle}P + \kappa^{(3)}_{\pi J}J^{\langle i}F^{j\rangle}\right),
\end{aligned} \tag{16.113}$$

$$\begin{aligned}
&\epsilon\Big\langle L^{-1}\hat{J}^k, (f^{\mathrm{eq}}\bar{f}^{\mathrm{eq}})^{-1}\left[\frac{\mathrm{D}}{\mathrm{D}t} + \epsilon\boldsymbol{\delta K}\cdot\nabla + \epsilon\boldsymbol{F}\cdot\nabla_K\right] f^{\mathrm{eq}}\bar{f}^{\mathrm{eq}}L^{-1}\hat{\chi}^\beta\Big\rangle\psi^\beta \\
&= -\epsilon^2\left(\kappa^{(1)}_{J\pi}\pi^{ij}\nabla^j n + \kappa^{(2)}_{J\pi}\pi^{ij}\nabla^j P + \kappa^{(3)}_{J\pi}\pi^{ij}F^j\right. \\
&\quad \left. + \kappa^{(1)}_{JJ}J^i\nabla\cdot\boldsymbol{u} + \kappa^{(2)}_{J\pi}J^j\sigma^{ij} + \kappa^{(3)}_{J\pi}J^j\omega^{ij}\right),
\end{aligned} \tag{16.114}$$

where we have used the identity

$$\nabla^i u^j = \sigma^{ij} + \omega^{ij} + \delta^{ij}\nabla\cdot\boldsymbol{u}/3 \tag{16.115}$$

with the vorticity $\omega^{ij} := (\nabla^i u^j - \nabla^j u^i)/2$, and the transport coefficients are defined as follows:

$$\kappa^{(1)}_{\pi\pi} = -\frac{1}{10}\Big\langle L^{-1}\hat{\pi}^{ij}, (f^{\mathrm{eq}}\bar{f}^{\mathrm{eq}})^{-1}\left[-\frac{2T}{3}\frac{\partial}{\partial T} - \frac{2\tilde{\mu}}{3}\frac{\partial}{\partial\tilde{\mu}} + \frac{1}{3}\delta K^a\frac{\partial}{\partial u^a}\right]\frac{f^{\mathrm{eq}}\bar{f}^{\mathrm{eq}}L^{-1}\hat{\pi}^{ij}}{T\eta}\Big\rangle, \tag{16.116}$$

$$\kappa^{(2)}_{\pi\pi} = -\frac{6}{35}\Delta^{kjab}\Big\langle L^{-1}\hat{\pi}^{ij}, (f^{\mathrm{eq}}\bar{f}^{\mathrm{eq}})^{-1}\delta K^a\frac{\partial}{\partial u^b}\frac{f^{\mathrm{eq}}\bar{f}^{\mathrm{eq}}L^{-1}\hat{\pi}^{ki}}{T\eta}\Big\rangle, \tag{16.117}$$

$$\kappa^{(3)}_{\pi\pi} = -\frac{2}{15}\Omega^{kjab}\Big\langle L^{-1}\hat{\pi}^{ij}, (f^{\mathrm{eq}}\bar{f}^{\mathrm{eq}})^{-1}\delta K^a\frac{\partial}{\partial u^b}\frac{f^{\mathrm{eq}}\bar{f}^{\mathrm{eq}}L^{-1}\hat{\pi}^{ki}}{T\eta}\Big\rangle = -2\tau_\pi, \tag{16.118}$$

$$\kappa^{(1)}_{\pi J} = -\frac{1}{5}\Big\langle L^{-1}\hat{\pi}^{ij}, (f^{\mathrm{eq}}\bar{f}^{\mathrm{eq}})^{-1}\delta K^i\frac{2T^2}{3nT - 2Ah_{TF}}\left[\frac{\partial}{\partial T} - s\frac{\partial}{\partial\tilde{\mu}}\right]\frac{f^{\mathrm{eq}}\bar{f}^{\mathrm{eq}}L^{-1}\hat{J}^j}{T^2\lambda}\Big\rangle, \tag{16.119}$$

$$\kappa^{(2)}_{\pi J} = -\frac{1}{5}\Big\langle L^{-1}\hat{\pi}^{ij}, (f^{\rm eq}\bar{f}^{\rm eq})^{-1} \times \left[-\delta K^i \frac{2AT/n}{3nT - 2Ah_{TF}}\frac{\partial}{\partial T} + \delta K^i \frac{1}{n}\frac{\partial}{\partial \tilde{\mu}} - \frac{\partial}{\partial u^i}\right]\frac{f^{\rm eq}\bar{f}^{\rm eq}L^{-1}\hat{J}^j}{T^2\lambda}\Big\rangle, \tag{16.120}$$

$$\kappa^{(3)}_{\pi J} = -\frac{1}{5}\Big\langle L^{-1}\hat{\pi}^{ij}, (f^{\rm eq}\bar{f}^{\rm eq})^{-1}\left[n\frac{\partial}{\partial u^i} + \nabla^i_K\right]\frac{f^{\rm eq}\bar{f}^{\rm eq}L^{-1}\hat{J}^j}{T^2\lambda}\Big\rangle, \tag{16.121}$$

$$\kappa^{(1)}_{J\pi} = -\frac{1}{10}\Big\langle L^{-1}\hat{J}^i, (f^{\rm eq}\bar{f}^{\rm eq})^{-1}\delta K^j \frac{2T^2}{3nT - 2Ah_{TF}}\left[\frac{\partial}{\partial T} - s\frac{\partial}{\partial \tilde{\mu}}\right]\frac{f^{\rm eq}\bar{f}^{\rm eq}L^{-1}\hat{\pi}^{ij}}{T\eta}\Big\rangle, \tag{16.122}$$

$$\kappa^{(2)}_{J\pi} = -\frac{1}{10}\Big\langle L^{-1}\hat{J}^i, (f^{\rm eq}\bar{f}^{\rm eq})^{-1} \times \left[-\delta K^j \frac{2AT/n}{3nT - 2Ah_{TF}}\frac{\partial}{\partial T} + \delta K^j \frac{1}{n}\frac{\partial}{\partial \tilde{\mu}} - \frac{\partial}{\partial u^j}\right]\frac{f^{\rm eq}\bar{f}^{\rm eq}L^{-1}\hat{\pi}^{ij}}{T\eta}\Big\rangle, \tag{16.123}$$

$$\kappa^{(3)}_{J\pi} = -\frac{1}{10T\eta}\Big\langle L^{-1}\hat{J}^i, (f^{\rm eq}\bar{f}^{\rm eq})^{-1}\left[n\frac{\partial}{\partial u^j} + \nabla^j_K\right]\frac{f^{\rm eq}\bar{f}^{\rm eq}L^{-1}\hat{\pi}^{ij}}{T\eta}\Big\rangle, \tag{16.124}$$

$$\kappa^{(1)}_{JJ} = -\frac{1}{3}\Big\langle L^{-1}\hat{J}^i, (f^{\rm eq}\bar{f}^{\rm eq})^{-1}\left[-\frac{2T}{3}\frac{\partial}{\partial T} - \frac{2\tilde{\mu}}{3}\frac{\partial}{\partial \tilde{\mu}} + \frac{1}{3}\delta K^a \frac{\partial}{\partial u^a}\right]\frac{f^{\rm eq}\bar{f}^{\rm eq}L^{-1}\hat{J}^i}{T^2\lambda}\Big\rangle, \tag{16.125}$$

$$\kappa^{(2)}_{JJ} = -\tfrac{1}{5}\Delta^{ijkl}\Big\langle L^{-1}\hat{J}^i, (f^{\rm eq}\bar{f}^{\rm eq})^{-1}\delta K^k \tfrac{\partial}{\partial u^l}\tfrac{f^{\rm eq}\bar{f}^{\rm eq}L^{-1}\hat{J}^j}{T^2\lambda}\Big\rangle, \tag{16.126}$$

$$\kappa^{(3)}_{JJ} = -\tfrac{1}{3}\Omega^{ijkl}\Big\langle L^{-1}\hat{J}^i, (f^{\rm eq}\bar{f}^{\rm eq})^{-1}\delta K^k \tfrac{\partial}{\partial u^l}\tfrac{f^{\rm eq}\bar{f}^{\rm eq}L^{-1}\hat{J}^j}{T^2\lambda}\Big\rangle = \tau_J, \tag{16.127}$$

where

$$\Omega^{ijkl} := (\delta^{ik}\delta^{jl} - \delta^{il}\delta^{jk})/2 \tag{16.128}$$

is an antisymmetric projection operator, and A is defined by

$$A := T\partial n/\partial\tilde{\mu}. \tag{16.129}$$

Here, we show that

$$\kappa^{(3)}_{\pi\pi} = -2\tau_\pi, \tag{16.130}$$

by analytically evaluating the inner product of Eq. (16.117). Without loss of generality, $L^{-1}\hat{\pi}^{\mu\nu}$ may be written as

$$L^{-1}\hat{\pi}^{ij} = C(|\boldsymbol{\delta K}|)\Delta^{ijkl}\delta K^k \delta K^l. \tag{16.131}$$

We do not need the specific form of $C(|\boldsymbol{\delta K}|)$ in this computation. Then $\kappa^{(3)}_{\pi\pi}$ can be written as

$$\kappa^{(3)}_{\pi\pi} = -\frac{2}{15}\Omega^{kjab}\Big\langle C(|\boldsymbol{\delta K}|)\Delta^{ijcd}\delta K^c\delta K^d, (f^{\mathrm{eq}}\bar{f}^{\mathrm{eq}})^{-1}\delta K^a\frac{\partial}{\partial u^b}\frac{f^{\mathrm{eq}}\bar{f}^{\mathrm{eq}}C(|\boldsymbol{\delta K}|)\Delta^{kief}\delta K^e\delta K^f}{T\eta}\Big\rangle. \tag{16.132}$$

Here we write down useful formulas for further conversion:

$$\Omega^{kjab}\delta K^a\frac{\partial}{\partial u^b}|\boldsymbol{\delta K}| = \Omega^{kjab}\delta K^a\frac{-\delta K^b}{|\boldsymbol{\delta K}|} = 0, \tag{16.133}$$

$$\begin{aligned}&\Omega^{kjab}\Delta^{ijcd}\delta K^c\delta K^d\delta K^a\Delta^{kief}[-\delta^{be}\delta K^f - \delta^{bf}\delta K^e]\\ &= \frac{3}{2}\Delta^{ijkl}\delta K^i\delta K^j\delta K^k\delta K^l = \frac{3}{2}\Delta^{ijkl}\delta K^k\delta K^l\Delta^{ijab}\delta K^a\delta K^b.\end{aligned} \tag{16.134}$$

By using these formulas, Eq. (16.132) is calculated to be

$$\begin{aligned}\kappa^{(3)}_{\pi\pi} &= -\frac{2}{15}\Omega^{kjab}\Big\langle C(|\boldsymbol{\delta K}|)\Delta^{ijcd}\delta K^c\delta K^d, \delta K^a\frac{C(|\boldsymbol{\delta K}|)\Delta^{kief}(-\delta^{be}\delta K^f - \delta^{bf}\delta K^e)}{T\eta}\Big\rangle\\ &= -\frac{2}{15}\frac{3/2}{2T\eta}\langle\hat{L}^{-1}\hat{\pi}^{\mu\nu}, \hat{L}^{-1}\hat{\pi}_{\mu\nu}\rangle\\ &= -2\tau_\pi.\end{aligned} \tag{16.135}$$

Similarly we can show that

$$\kappa^{(3)}_{JJ} = \tau_J, \tag{16.136}$$

by evaluating the inner product of Eq. (16.126).

Combining the formulas derived so far, we can rewrite the relaxation equation (16.96) in the following forms:

$$\begin{aligned}\epsilon\pi^{ij} &= \epsilon 2\eta\sigma^{ij} - \epsilon^2\tau_\pi\frac{\mathrm{D}}{\mathrm{D}t}\pi^{ij} - \epsilon^2\ell_{\pi J}\nabla^{\langle i}J^{j\rangle}\\ &\quad + \epsilon^2\kappa^{(1)}_{\pi\pi}\pi^{ij}\nabla\cdot\boldsymbol{u} + \epsilon^2\kappa^{(2)}_{\pi\pi}\pi^{k\langle i}\sigma^{j\rangle k} - \epsilon^2 2\tau_\pi\pi^{k\langle}\omega^{j\rangle k}\end{aligned}$$

$$
\begin{aligned}
&+\epsilon^2\kappa^{(1)}_{\pi J} J^{\langle i}\nabla^{j\rangle} n + \epsilon^2\kappa^{(2)}_{\pi J} J^{\langle i}\nabla^{j\rangle} P + \epsilon^2\kappa^{(3)}_{\pi J} J^{\langle i} F^{j\rangle} \\
&+\epsilon^2 b_{\pi\pi\pi}\pi^{k\langle i}\pi^{j\rangle k} + \epsilon^2 b_{\pi JJ} J^{\langle i} J^{j\rangle}, \qquad (16.137)
\end{aligned}
$$

$$
\begin{aligned}
\epsilon J^i &= \epsilon\lambda\nabla^i T - \epsilon^2\tau_J \frac{\mathrm{D}}{\mathrm{D}t} J^i - \epsilon^2 \ell_{J\pi}\nabla^j\pi^{ij} \\
&+\epsilon^2\kappa^{(1)}_{J\pi}\pi^{ij}\nabla^j n + \epsilon^2\kappa^{(2)}_{J\pi}\pi^{ij}\nabla^j P + \epsilon^2\kappa^{(3)}_{J\pi}\pi^{ij} F^j \\
&+\epsilon^2\kappa^{(1)}_{JJ} J^i \nabla\cdot\boldsymbol{u} + \epsilon^2\kappa^{(2)}_{JJ} J^j\sigma^{ij} + \epsilon^2\tau_J J^j\omega^{ij} \\
&+\epsilon^2 b_{JJ\pi} J^j\pi^{ij}. \qquad (16.138)
\end{aligned}
$$

Putting back $\epsilon = 1$, we arrive at Eqs. (16.87) and (16.88).

16.2 Transport Coefficients and Relaxation Times in Non-relativistic Fluid Dynamics

We have managed to derive the second-order fluid dynamics in the non-relativistic system, and the form of equations are expected to be universal regardless of its underlying microscopic ingredients. Those details are, however, encoded in the transport coefficients such as shear viscosity, viscous relaxation times, etc. Our derivation did provide the microscopic expressions of all the transport coefficients based on the underlying Boltzmann theory. In this section, we carry out numerical analyses of these transport coefficients as well as viscous relaxation times in fermionic cold atoms [157]. As discussed around (16.75) the bulk viscosity vanishes by construction in our derivation of the fluid dynamics. The bulk viscosity can be finite when the higher order effects are taken into account [265, 266, 271, 272].

To be specific, we assume that the collision process is dominated by s-wave scattering, whose amplitude is given by [253]

$$
\mathcal{M} = \frac{1}{a_{\mathrm{s}}^{-1} - iq}, \qquad (16.139)
$$

where a_{s} is the s-wave scattering length and q is relative momentum of two incoming particles.

16.2.1 *Analytic Reduction of Transport Coefficients and Relaxation Times for Numerical Studies*

For the purpose of evaluating the following quantities,

$$
\eta = -\frac{1}{10T}\langle\hat{\pi}^{ij}, L^{-1}\hat{\pi}^{ij}\rangle, \qquad \lambda = -\frac{1}{3T^2}\langle\hat{J}^i, L^{-1}\hat{J}^i\rangle, \qquad (16.140)
$$

$$
\tau_\pi = \frac{1}{10T\eta}\langle\hat{\pi}^{ij}, L^{-2}\hat{\pi}^{ij}\rangle, \qquad \tau_J = \frac{1}{3T^3\lambda}\langle\hat{J}^i, L^{-2}\hat{J}^i\rangle, \qquad (16.141)
$$

we reduce them to the formulas suitable for numerical computation. The major part of the reduction lies in how to invert the linearized collision operator L, which in turn allows us to numerically evaluate the transport coefficients and viscous relaxation times.

For the purpose of numerical calculation we introduce the dimensionless variables,

$$\boldsymbol{p}' := \frac{\delta \boldsymbol{p}}{p_\mathrm{F}} = \frac{\delta \boldsymbol{v}}{v_\mathrm{F}}, \quad \mu' := \frac{\tilde{\mu}}{\varepsilon_\mathrm{F}}, \quad T' := \frac{T}{\varepsilon_\mathrm{F}}, \quad n' := \frac{n}{p_\mathrm{F}^3} = \frac{1}{3\pi^2}, \tag{16.142}$$

with the Fermi velocity and the Fermi energy,

$$p_\mathrm{F} = (3\pi^2 n)^{1/3}, \quad v_\mathrm{F} = \frac{p_\mathrm{F}}{m}, \quad \varepsilon_\mathrm{F} = \frac{p_\mathrm{F}^2}{2m}. \tag{16.143}$$

Here, the primed variables are defined to be dimensionless. In what follows, we suppress the primes and $\boldsymbol{p}, \mu, T, n$ are understood to be dimensionless.

Then, the dimensionless linearized collision operator L' is written as

$$\begin{aligned} L'[\phi]_p &:= \frac{1}{4\varepsilon_\mathrm{F}} L[\phi]_p \\ &= -\frac{1}{2 f_p^\mathrm{eq}} \int \frac{\mathrm{d}^3 p_1 \mathrm{d}^3 p_2 \mathrm{d}^3 p_3}{(2\pi)^3} \mathcal{W}'(p, p_1 | p_2, p_3) f_{\varepsilon_1}^\mathrm{eq} \bar{f}_{\varepsilon_2}^\mathrm{eq} \bar{f}_{\varepsilon_3}^\mathrm{eq} (\phi_p + \phi_{p_1} - \phi_{p_2} - \phi_{p_3}), \end{aligned} \tag{16.144}$$

with the equilibrium distribution function,

$$f_{\varepsilon_i}^\mathrm{eq} = \frac{1}{\mathrm{e}^{(\varepsilon_i - \mu)/T} - a}. \tag{16.145}$$

We again stress that p, μ, T are now all dimensionless and the dimensionless one-particle energy is $\varepsilon_i = p_i^2$. The dimensionless temperature and chemical potential are related by

$$n = \frac{1}{3\pi^2} = \int \frac{\mathrm{d}^3 p}{2\pi} \frac{1}{\mathrm{e}^{(\varepsilon - \mu)/T} - a}. \tag{16.146}$$

The transition matrix is written as

$$\mathcal{W}'(p, p_1 | p_2, p_3) = |\mathcal{M}'|^2 (2\pi^4) \delta(p^2 + p_1^2 - p_2^2 - p_3^2) \delta^3(\boldsymbol{p} + \boldsymbol{p}_1 - \boldsymbol{p}_2 - \boldsymbol{p}_3), \tag{16.147}$$

with the dimensionless scattering amplitude

$$\mathcal{M}' := m p_\mathrm{F} \mathcal{M}. \tag{16.148}$$

The scattering amplitude of our interest (16.139) is

$$|\mathcal{M}'|^2 = \frac{16\pi^2}{(p_{\mathrm{F}}a_{\mathrm{s}})^{-2} + q^2}. \tag{16.149}$$

With these dimensionless quantities, the linearized collision operator is converted to

$$L'[\phi]_p = -k(\boldsymbol{p})\phi_p - \int \frac{\mathrm{d}^3 p_1}{2\pi} f^{\mathrm{eq}}_{\varepsilon_1}[K_1(\boldsymbol{p}, \boldsymbol{p}_1) - K_2(\boldsymbol{p}, \boldsymbol{p}_1)]\phi_{p_1}, \tag{16.150}$$

where we have defined

$$k(\boldsymbol{p}) = \int \frac{\mathrm{d}^3 p_1}{2\pi} f^{\mathrm{eq}}_{\varepsilon_1} K_1(\boldsymbol{p}, \boldsymbol{p}_1), \tag{16.151}$$

$$K_1(\boldsymbol{p}, \boldsymbol{p}_1) = \frac{1}{2\bar{f}^{\mathrm{eq}}_{\varepsilon}} \int \frac{\mathrm{d}^3 p_2 \mathrm{d}^3 p_3}{(2\pi)^2} |\mathcal{M}'(|\boldsymbol{p} - \boldsymbol{p}_1|/2)|^2 \bar{f}^{\mathrm{eq}}_{\varepsilon_2} \bar{f}^{\mathrm{eq}}_{\varepsilon_3} \times (2\pi)^4 \delta(p^2 + p_1^2 - p_2^2 - p_3^2)\delta^3(\boldsymbol{p} + \boldsymbol{p}_1 - \boldsymbol{p}_2 - \boldsymbol{p}_3), \tag{16.152}$$

$$K_2(\boldsymbol{p}, \boldsymbol{p}_1) = \frac{\mathrm{e}^{(p_1^2 - \mu)/T}}{\bar{f}^{\mathrm{eq}}_{\varepsilon}} \int \frac{\mathrm{d}^3 p_2 \mathrm{d}^3 p_3}{(2\pi)^2} |\mathcal{M}'(|\boldsymbol{p} - \boldsymbol{p}_2|/2)|^2 f^{\mathrm{eq}}_{\varepsilon_2} \bar{f}^{\mathrm{eq}}_{\varepsilon_3} \times (2\pi)^4 \delta(p^2 + p_2^2 - p_1^2 - p_3^2)\delta^3(\boldsymbol{p} + \boldsymbol{p}_2 - \boldsymbol{p}_1 - \boldsymbol{p}_3). \tag{16.153}$$

With the help of the following momentum variables,

$$\boldsymbol{P} := \boldsymbol{p} + \boldsymbol{p}_1, \quad \boldsymbol{q} := \frac{\boldsymbol{p} - \boldsymbol{p}_1}{2}, \quad \boldsymbol{P}' := \boldsymbol{p}_2 + \boldsymbol{p}_3, \quad \boldsymbol{q}' := \frac{\boldsymbol{p}_2 - \boldsymbol{p}_3}{2}, \tag{16.154}$$

the integral (16.152) can be computed as,

$$\begin{aligned} K_1(\boldsymbol{p}, \boldsymbol{p}_1) &= \frac{1}{2\bar{f}^{\mathrm{eq}}_{\varepsilon}} \int \frac{\mathrm{d}^3 P' \mathrm{d}^3 q'}{(2\pi)^2} |\mathcal{M}'(q)|^2 \bar{f}^{\mathrm{eq}}_{|\boldsymbol{P}' + 2\boldsymbol{q}'|^2/4} \bar{f}^{\mathrm{eq}}_{|\boldsymbol{P}' - 2\boldsymbol{q}'|^2/4} \\ &\quad \times (2\pi)^4 \delta\Big(\frac{P^2}{2} + 2q^2 - \frac{P'^2}{2} - 2q'^2\Big)\delta^3(\boldsymbol{P} - \boldsymbol{P}') \\ &= \frac{q}{16\pi \bar{f}^{\mathrm{eq}}_{\varepsilon}} |\mathcal{M}'(q)|^2 \int \mathrm{d}\cos\theta \, \bar{f}^{\mathrm{eq}}_{P^2/4 + q^2 + Pq\cos\theta} \bar{f}^{\mathrm{eq}}_{P^2/4 + q^2 - Pq\cos\theta} \\ &= \frac{q}{16\pi \bar{f}^{\mathrm{eq}}_{\varepsilon}} |m p_{\mathrm{F}} \mathcal{M}(q)|^2 \frac{T}{(1 - a^2 \mathrm{e}^{-(P^2 + 4q^2 - 4\mu)/2T}) Pq} \\ &\quad \times \left[\frac{\big(1 - a\mathrm{e}^{-(P^2/4 + q^2 + 4Pq - 4\mu)/4T}\big)\big(\mathrm{e}^{(P^2/4 + q^2 + 4Pq - 4\mu)/4T} - a\big)}{\big(1 - a\mathrm{e}^{-(P^2/4 + q^2 - 4Pq - 4\mu)/4T}\big)\big(\mathrm{e}^{(P^2/4 + q^2 - 4Pq - 4\mu)/4T} - a\big)} \right], \end{aligned} \tag{16.155}$$

where we have defined

$$
\begin{aligned}
P &= \sqrt{p^2 + p_1^2 + 2pp_1 \cos\chi}, \quad \cos\theta = \frac{\boldsymbol{P}\cdot\boldsymbol{q}'}{Pq'}, \\
q &= \frac{\sqrt{p^2 + p_1^2 - 2pp_1\cos\chi}}{2}, \quad \cos\chi = \frac{\boldsymbol{p}\cdot\boldsymbol{p}_1}{pp_1}.
\end{aligned}
\tag{16.156}
$$

Accordingly, we may write $K_1(\boldsymbol{p}, \boldsymbol{p}_1) = K_1(p, p_1, \chi)$. This, in turn, allows us to write $k(\boldsymbol{p})$ (16.151) as follows,

$$
k(\boldsymbol{p}) = \frac{1}{(2\pi)^2} \int \mathrm{d}p_1 p_1^2 f_{\varepsilon_1}^{\mathrm{eq}} \int \mathrm{d}\cos\chi \, K_1(p, p_1, \chi) =: k(p). \tag{16.157}
$$

The integral (16.153) is also partially carried out analytically to give

$$
\begin{aligned}
K_2(\boldsymbol{p}, \boldsymbol{p}_1) &= \frac{\mathrm{e}^{(p_1^2-\mu)/T}}{\bar{f}_\varepsilon^{\mathrm{eq}}} \int \frac{\mathrm{d}^3P'\mathrm{d}^3q'}{(2\pi)^2} |\mathcal{M}'(|\boldsymbol{P} - \boldsymbol{P}' + 2\boldsymbol{q} - 2\boldsymbol{q}'|/4)|^2 f^{\mathrm{eq}}_{|\boldsymbol{P}'+2\boldsymbol{q}|^2/4} \bar{f}^{\mathrm{eq}}_{|\boldsymbol{P}'-2\boldsymbol{q}|^2/4} \\
&\quad \times (2\pi)^4 \delta(2\boldsymbol{P}\cdot\boldsymbol{q} + 2\boldsymbol{P}'\cdot\boldsymbol{q}')\delta^3(2\boldsymbol{q} + 2\boldsymbol{q}') \\
&= \frac{\mathrm{e}^{(p_1^2-\mu)/T}}{64\pi^2 q \bar{f}_\varepsilon^{\mathrm{eq}}} \int \mathrm{d}P''P''\mathrm{d}\phi |\mathcal{M}'(\sqrt{q^2 + P''^2/16})|^2 \\
&\quad \times f^{\mathrm{eq}}_{p_1^2 + P''^2/4 + \boldsymbol{P}''\cdot\boldsymbol{P}/2} \bar{f}^{\mathrm{eq}}_{p^2 + P''^2/4 + \boldsymbol{P}''\cdot\boldsymbol{P}/2} \\
&=: K_2(p, p_1, \chi).
\end{aligned}
\tag{16.158}
$$

The integration variable has been changed from $\boldsymbol{P}'$ to $\boldsymbol{P}'' = \boldsymbol{P}' - \boldsymbol{P}$. It is convenient to introduce the Cartesian coordinate where the vectors $\boldsymbol{P}$, $\boldsymbol{P}''$, and $\boldsymbol{q}$ are parametrized as

$$
\boldsymbol{P} = (P\sin\chi', 0, P\cos\chi')^T, \quad \boldsymbol{P}'' = (P''\cos\phi, P''\sin\phi, 0)^T, \quad \boldsymbol{q} = (0, 0, q)^T, \tag{16.159}
$$

where χ' is related to χ through

$$
\cos\chi' = \frac{\boldsymbol{P}\cdot\boldsymbol{q}}{Pq} = \frac{p^2 - p_1^2}{\sqrt{p^2 + 2pp_1\cos\chi + p_1^2}\sqrt{p^2 - 2pp_1\cos\chi + p_1^2}}. \tag{16.160}
$$

It is visualized in Fig. 16.1. In this coordinate, we have

$$
\boldsymbol{P}''\cdot\boldsymbol{P} = P''P\sin\chi'\cos\phi. \tag{16.161}
$$

Let us move on to the reduction of $[L^{-1}\hat{\pi}^{ij}]_p$ and $[L^{-1}\hat{J}^i]_p$. The stress tensor and the heat flow are given by

$$
\hat{\pi}_p^{ij} = -p_{\mathrm{F}}v_{\mathrm{F}} \frac{\sqrt{k(p)}}{p^2} c(p) p^{\langle i} p^{j\rangle}, \quad \hat{J}_p^i = -\varepsilon_{\mathrm{F}} v_{\mathrm{F}} \frac{\sqrt{k(p)}}{p} b(p) p^i, \tag{16.162}
$$

Fig. 16.1 The momentum vectors in the Cartesian coordinate (16.159)

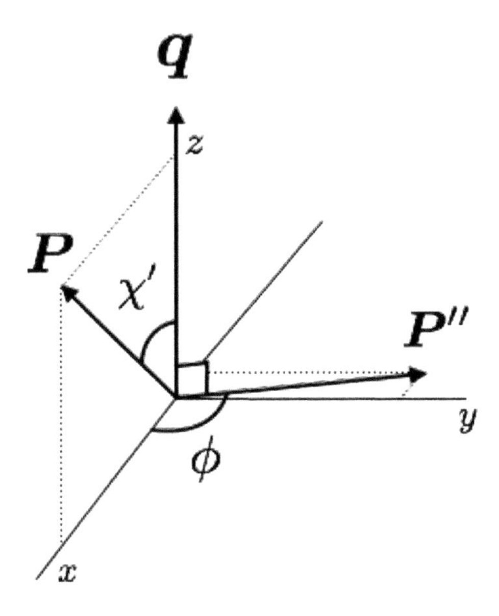

with

$$c(p) = -\frac{p^2}{\sqrt{k(p)}}, \quad b(p) = -\frac{p}{\sqrt{k(p)}}(p^2 - h_{\mathrm{TF}}), \tag{16.163}$$

and the dimensionless enthalpy

$$\tilde{h} = \frac{5}{2}\int \mathrm{d}p p^4 f_\varepsilon^{\mathrm{eq}}. \tag{16.164}$$

On account of the tensor structures, $L'^{-1}\hat{\pi}^{ij}$ and $L'^{-1}\hat{J}^i$ are expressed as

$$[L'^{-1}\hat{\pi}^{ij}]_p = \frac{p_{\mathrm{F}}v_{\mathrm{F}}}{p^2\sqrt{k(p)}}C(p)p^{\langle i}p^{j\rangle}, \quad [L'^{-1}\hat{J}^i]_p = \frac{\varepsilon_{\mathrm{F}}v_{\mathrm{F}}}{p\sqrt{k(p)}}B(p)p^i, \tag{16.165}$$

with functions $C(p)$ and $B(p)$ to be computed numerically. The equations to be solved are $\hat{\pi}_p^{ij} = [L'(L'^{-1}\hat{\pi}^{ij})]_p$ and $\hat{J}_p^i = [L'(L'^{-1}\hat{J}^i)]_p$. Upon substituting (16.150), (16.162), and (16.165) to the equations, we have

$$\begin{aligned} c(p) &= C(p) + \frac{1}{(2\pi)^2}\int \mathrm{d}p_1 p_1^2 f_{\varepsilon_1}^{\mathrm{eq}} \frac{1}{\sqrt{k(p)k(p_1)}} \\ &\quad \times \int \mathrm{d}\cos\chi \frac{3\cos^2\chi - 1}{2}[K_1(p, p_1, \chi) - K_2(p, p_1, \chi)]C(p_1), \\ b(p) &= B(p) + \frac{1}{(2\pi)^2}\int \mathrm{d}p_1 p_1^2 f_{\varepsilon_1}^{\mathrm{eq}} \frac{1}{\sqrt{k(p)k(p_1)}} \\ &\quad \times \int \mathrm{d}\cos\chi \cos\chi[K_1(p, p_1, \chi) - K_2(p, p_1, \chi)]B(p_1). \end{aligned} \tag{16.166}$$

We obtain $C(p)$ and $B(p)$ by numerically solving them as detailed in Sect. 16.2.2. Equation (16.165) then leads to $L'^{-1}\hat{\pi}^{ij}$ and $L'^{-1}\hat{J}^i$.

We are finally in position to evaluate the transport coefficients,

$$\eta = -\frac{1}{10(\varepsilon_F T)}\langle\hat{\pi}^{ij}, L^{-1}\hat{\pi}^{ij}\rangle = \frac{n}{10T}\int \mathrm{d}p p^2 f_\varepsilon^{\mathrm{eq}}\bar{f}_\varepsilon^{\mathrm{eq}} c(p) C(p), \tag{16.167}$$

$$\lambda = -\frac{1}{3(\varepsilon_F T)^2}\langle\hat{J}^i, L^{-1}\hat{J}^i\rangle = \frac{n}{m}\frac{1}{4T^2}\int \mathrm{d}p p^2 f_\varepsilon^{\mathrm{eq}}\bar{f}_\varepsilon^{\mathrm{eq}} b(p) B(p), \tag{16.168}$$

and the viscous relaxation times,

$$\tau_\pi = \frac{1}{10(\varepsilon_F T)\eta}\langle L^{-1}\hat{\pi}^{ij}, L^{-1}\hat{\pi}^{ij}\rangle = \frac{1}{\varepsilon_F}\frac{1}{40T(\eta/n)}\int \mathrm{d}p p^2 f_\varepsilon^{\mathrm{eq}}\bar{f}_\varepsilon^{\mathrm{eq}}\frac{1}{k(p)} C(p) C(p), \tag{16.169}$$

$$\tau_J = \frac{1}{3(\varepsilon_F T)^2\lambda}\langle L^{-1}\hat{J}^i, L^{-1}\hat{J}^i\rangle = \frac{1}{\varepsilon_F}\frac{1}{16T^2(m\lambda/n)}\int \mathrm{d}p p^2 f_\varepsilon^{\mathrm{eq}}\bar{f}_\varepsilon^{\mathrm{eq}}\frac{1}{k(p)} B(p) B(p). \tag{16.170}$$

16.2.2 Numerical Method

We present how to numerically solve (16.166) using the double exponential method employed in the relativistic case in Sect. 14.2.4.

With a change of the integral variable by

$$p(t) = \sqrt{T}\mathrm{e}^{t-\mathrm{e}^{-t}}, \tag{16.171}$$

we introduce the integral measure $\mathrm{d}\mu(t)$ as

$$\mathrm{d}\mu(t) := \frac{1}{(2\pi)^3} f_{\varepsilon(t)}^{\mathrm{eq}} p(t)^2 \mathrm{d}p(t) = \frac{T^{3/2}}{(2\pi)^3}(1+\mathrm{e}^{-t})\mathrm{e}^{3(t-\mathrm{e}^{-t})}\frac{1}{\exp(\mathrm{e}^{2(t-\mathrm{e}^{-t})} - \mu/T) + 1}\mathrm{d}t, \tag{16.172}$$

which has a convenient asymptotic form,

$$\mathrm{d}\mu(t) \to \begin{cases} \mathrm{e}^{-\mathrm{e}^{2t}}\mathrm{d}t & (t \to +\infty), \\ \mathrm{e}^{-3\mathrm{e}^{-t}}\mathrm{d}t & (t \to -\infty). \end{cases} \tag{16.173}$$

Since the magnitude of the measure drops rapidly at large $|t|$, we can introduce cutoffs, $(t_{\min}, t_{\max})$, to the t-integral with small systematic uncertainty. With the change of variable applied, the equations (16.166) is converted to

$$
\begin{aligned}
c(t) &= C(t) + \int_{t_{\min}}^{t_{\min}} \mathrm{d}\mu(t_1) N(t, t_1) C(t_1), \\
b(p) &= B(p) + \int_{t_{\min}}^{t_{\min}} \mathrm{d}\mu(t_1) M(t, t_1) B(t_1),
\end{aligned} \tag{16.174}
$$

where $N(t, t_1)$ and $M(t, t_1)$ are defined by

$$
\begin{aligned}
N(t, t_1) &:= \frac{2\pi}{\sqrt{k(t)k(t_1)}} \int \mathrm{d}\cos\chi \frac{3\cos^2\chi - 1}{2} [K_1(p, p_1, \chi) - K_2(p, p_1, \chi)], \\
M(t, t_1) &:= \frac{2\pi}{\sqrt{k(t)k(t_1)}} \int \mathrm{d}\cos\chi \cos\chi [K_1(p, p_1, \chi) - K_2(p, p_1, \chi)].
\end{aligned} \tag{16.175}
$$

Similarly, the transport coefficients and viscous relaxation times are transformed to

$$
\frac{\eta}{n} = \frac{1}{10T} \int_{t_{\min}}^{t_{\max}} \mathrm{d}\mu(t) \bar{f}_\varepsilon^{\mathrm{eq}} c(t) C(t), \tag{16.176}
$$

$$
\frac{m\lambda}{n} = \frac{1}{4T^2} \int_{t_{\min}}^{t_{\max}} \mathrm{d}\mu(t) \bar{f}_\varepsilon^{\mathrm{eq}} b(t) B(t), \tag{16.177}
$$

and

$$
\begin{aligned}
\varepsilon_{\mathrm{F}} \tau_\pi &= \frac{1}{40T(\eta/n)} \int_{t_{\min}}^{t_{\max}} \mathrm{d}\mu(t) \bar{f}_\varepsilon^{\mathrm{eq}} \frac{1}{k(t)} C(t) C(t) \\
&= \frac{1}{4} \frac{\int_{t_{\min}}^{t_{\max}} \mathrm{d}\mu(t) \bar{f}_\varepsilon^{\mathrm{eq}} \frac{1}{k(t)} C(t) C(t)}{\int_{t_{\min}}^{t_{\max}} \mathrm{d}\mu(t) \bar{f}_\varepsilon^{\mathrm{eq}} c(t) C(t)},
\end{aligned} \tag{16.178}
$$

$$
\begin{aligned}
\varepsilon_{\mathrm{F}} \tau_J &= \frac{1}{16T^2(m\lambda/n)} \int_{t_{\min}}^{t_{\max}} \mathrm{d}\mu(t) \bar{f}_\varepsilon^{\mathrm{eq}} \frac{1}{k(t)} B(t) B(t) \\
&= \frac{1}{4} \frac{\int_{t_{\min}}^{t_{\max}} \mathrm{d}\mu(t) \bar{f}_\varepsilon^{\mathrm{eq}} \frac{1}{k(t)} B(t) B(t)}{\int_{t_{\min}}^{t_{\max}} \mathrm{d}\mu(t) \bar{f}_\varepsilon^{\mathrm{eq}} b(t) B(t)}.
\end{aligned} \tag{16.179}
$$

Next, we adopt the rectangular discretization,

$$
t_s = t_{\min} + s\Delta t, \tag{16.180}
$$

with $\Delta t = (t_{\max} - t_{\min})/N$ and an integer $s \in [1, N]$ for some sufficiently large integer. We set $N = 150$ in the numerical simulation presented here. We apply the same discretization to the angular (χ) integrals with $N = 50$. We confirmed that the numerical integrals converge well. Accordingly, an integral of some function $F(t)$ with the measure $\mathrm{d}\mu(t)$ is replaced by a sum over s,

$$\int_{t_{\min}}^{t_{\max}} \mathrm{d}\mu(t) F(t) \approx \sum_{s=1}^{N} \Delta\mu_s F(t_s),$$
$$\Delta\mu_s := \frac{\sqrt{T}}{(2\pi)^3} \frac{(1+\mathrm{e}^{-t_s})\mathrm{e}^{t_s - \mathrm{e}^{-t_s}}}{\exp(\mathrm{e}^{2(t_s - \mathrm{e}^{-t_s})} - \mu/T) + 1} \Delta t. \tag{16.181}$$

It is found convenient to introduce the following quantities,

$$\begin{aligned} N_{s,s_1} &:= \sqrt{\Delta\mu_s} N(t_s, t_{s_1}) \sqrt{\Delta\mu_{s_1}}, \\ M_{s,s_1} &:= \sqrt{\Delta\mu_s} M(t_s, t_{s_1}) \sqrt{\Delta\mu_{s_1}}, \\ a_s &:= \sqrt{\Delta\mu_s} a(t_s), \quad A_s := \sqrt{\Delta\mu_s} A(t_s), \\ b_s &:= \sqrt{\Delta\mu_s} b(t_s), \quad B_s := \sqrt{\Delta\mu_s} B(t_s). \end{aligned} \tag{16.182}$$

Then, the discretized versions of (16.174) are concisely written as,

$$\begin{aligned} c_s &= \sum_{s=1}^{N} (\delta_{s,s_1} + N_{s,s_1}) C_{s_1}, \\ b_s &= \sum_{s=1}^{N} (\delta_{s,s_1} + M_{s,s_1}) B_{s_1}, \end{aligned} \tag{16.183}$$

which are readily solved numerically.

Once the solutions C_s and B_s are obtained, we find the transport coefficients and viscous relaxation times via,

$$\frac{\eta}{n} = \frac{1}{10T} \sum_{s=1}^{N} \bar{f}_s^{\mathrm{eq}} c_s C_s, \tag{16.184}$$

$$\frac{m\lambda}{n} = \frac{1}{4T^2} \sum_{s=1}^{N} \bar{f}_s^{\mathrm{eq}} b_s B_s, \tag{16.185}$$

$$\varepsilon_{\mathrm{F}} \tau_\pi = \frac{1}{4} \frac{\sum_{s=1}^{N} \bar{f}_s^{\mathrm{eq}} \frac{1}{k(t_s)} C_s C_s}{\sum_{s=1}^{N} \bar{f}_s^{\mathrm{eq}} c_s C_s}, \tag{16.186}$$

$$\varepsilon_{\mathrm{F}} \tau_J = \frac{1}{4} \frac{\sum_{s=1}^{N} \bar{f}_s^{\mathrm{eq}} \frac{1}{k(t_s)} B_s B_s}{\sum_{s=1}^{N} \bar{f}_s^{\mathrm{eq}} b_s B_s}. \tag{16.187}$$

16.2.3 Shear Viscosity and Heat Conductivity

We study temperature dependence of the shear viscosity and the heat conductivity for the classical Boltzmann gas and the Fermi gas, where the quantum statistical parameter a is set to 0 and -1, respectively. Accordingly, the equilibrium distribution function reads $f_p^{\rm eq} = {\rm e}^{-(|\delta \boldsymbol{p}|^2/2m-\mu_{\rm TF})/T}$ for the classical Boltzmann gas and $f_p^{\rm eq} = \left[{\rm e}^{-(|\delta \boldsymbol{p}|^2/2m-\mu_{\rm TF})/T}+1\right]^{-1}$ for the Fermi gas. We will present the result of viscous relaxation time in the following subsection along with the comparison to those under the relaxation-time approximation.

The temperature dependences of the shear viscosity η and the heat conductivity λ are shown in Fig. 16.2. The scattering length is $p_{\rm F}a_{\rm s} = 0.1$, that is close to the unitary limit $p_{\rm F}a_{\rm s} = 0$. The computations at the unitarity have been worked out in [157]. The difference between the Boltzmann gas and the Fermi gas is manifest at low temperatures, implying that the quantum statistical effect is significant there. One can understand the difference as a consequence of the Pauli-blocking effect. At low temperatures, a sharp Fermi surface is formed and most of the states are occupied inside the surface. Then, the scattering rate is significantly suppressed as the phase space for outgoing particles are limited after a scattering event. Allowed processes are mostly forward scatterings, where two particles simply exchange their momenta. Therefore, the momentum and energy transfers occur efficiently, resulting in larger value of the shear viscosity and heat conductivity relative to those expected in the Boltzmann gas.

We remark that there exists the pairing-formation temperature, below which our formulation is not applicable because the pairing formation is not taken into account in our microscopic theory. Even more importantly, the Boltzmann theory provides a good description at weak-coupling regime while the Fermi gas around unitary limit is a strong-coupling system. Hence, we do not expect that our result can be directly compared with the experimental results. Our point is that the putative agreement between the experimentally observed shear viscosity and that from the classical Boltzmann theory is not convincing considering the fact that its temperature dependence is largely modified once the Fermi statistics is incorporated.

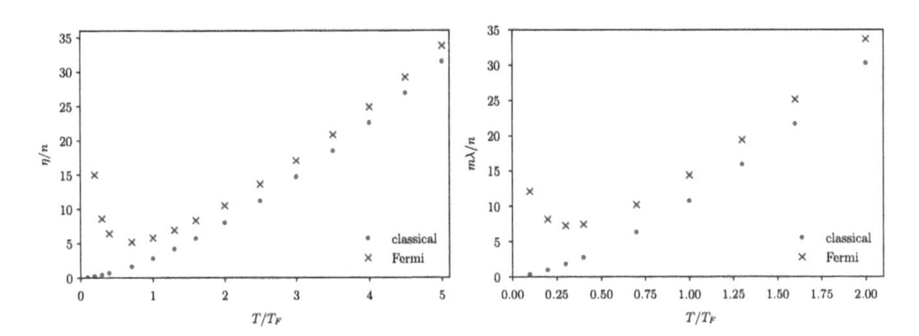

Fig. 16.2 Temperature dependence of the shear viscosity and the heat conductivity at $(p_{\rm F}a_{\rm s})^{-1} = 0.1$. The red points and blue crosses correspond to the Boltzmann gas and the Fermi gas, respectively

16.2.4 Viscous-Relaxation Time

Let us turn to the analyses of the viscous relaxation times. Their temperature dependence is shown in Fig. 16.3. Their qualitative behaviors are similar to the transport coefficients. At high temperatures, the Boltzmann gas and the Fermi gas take the same value, while the deviations are clearly seen at low temperatures.

It is intriguing to ask how close the viscous relaxation time under the relaxation-time approximation (RTA) is to our results because it is a widely used approximation in fluid dynamic simulations. In RTA, the collision integral in the Boltzmann equation (16.1) is replaced by the following simple form,

$$C[f]_p = -\frac{f_p - f^{\text{eq}}}{\tau}. \tag{16.188}$$

It is designed so that the distribution function f_p approaches the equilibrium one f^{eq} with single relaxation time τ, that is given as a free parameter.

In the derivation of fluid dynamics, the linearized collision operator L is simply replaced by $-\tau^{-1}$ under the RTA. Then, the shear viscosity and the heat conductivity (16.74) are reduced to

$$\begin{aligned}\eta^{\text{RTA}} &= -\frac{\tau}{10T}\langle\hat{\pi}^{ij}, \hat{\pi}^{ij}\rangle = -\frac{\tau}{10T}\int \mathrm{d}p f_p^{\text{eq}} \bar{f}_p^{\text{eq}} \Delta^{ijkl} \delta v^i \delta p^j \delta v^k \delta p^l \\ &= \frac{\tau}{3}\int \mathrm{d}p f_p^{\text{eq}} \delta v^i \delta p^i - \tau P,\end{aligned} \tag{16.189}$$

$$\begin{aligned}\lambda^{\text{RTA}} &= -\frac{\tau}{3T^2}\langle\hat{J}^i, \hat{J}^i\rangle = -\frac{\tau}{3T^2}\int \mathrm{d}p f_p^{\text{eq}} \bar{f}_p^{\text{eq}} \left(\frac{\delta \boldsymbol{p}^2}{2m} - h_{\text{TF}}\right) \delta \boldsymbol{v}^2 \\ &= \frac{\tau}{12mT}\left(7Q - \frac{75P^2}{n}\right),\end{aligned} \tag{16.190}$$

where the pressure P is given in (16.35) and Q is defined by

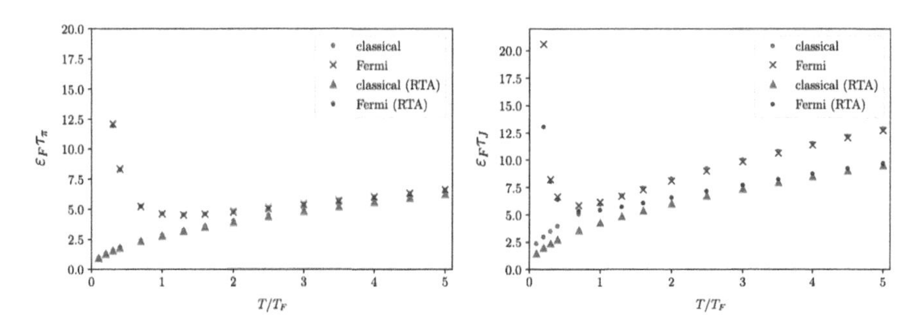

Fig. 16.3 Temperature dependence of the viscous relaxation times of stress tensor and heat flow at $(p_F a_s)^{-1} = 0.1$. The red points and blue crosses represent the relaxation times in the Boltzmann gas and the Fermi gas, respectively. The green triangles and purple points are the relaxation times in the Boltzmann gas and the Fermi gas with the RTA

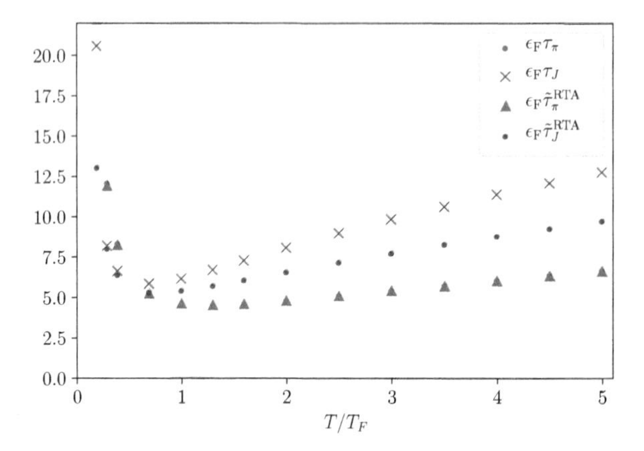

Fig. 16.4 Temperature dependence of the viscous relaxation times of stress tensor and heat flow at $(p_F a_s)^{-1} = 0.1$. The red points and blue crosses represent the relaxation times in the Boltzmann gas and the Fermi gas, respectively. The green triangles and purple points are the relaxation times in the Boltzmann gas and the Fermi gas with the RTA

$$Q = \int \mathrm{d}p f_p^{\mathrm{eq}} \delta v^2 \delta p^2. \tag{16.191}$$

Similarly, the viscous relaxation times (16.90) are simplified to,

$$\tau_\pi^{\mathrm{RTA}} = \tau_J^{\mathrm{RTA}} = \tau. \tag{16.192}$$

Although τ_π and τ_J generally take different values, they are fixed by a single parameter τ under the RTA.

For the purpose of comparison between our formulas for the viscous relaxation times (16.90) and the ones with the RTA, based on the relation between transport coefficients and relaxation time (16.189) and (16.190), we use the following expressions for the latter,

$$\tilde{\tau}_\pi^{\mathrm{RTA}} = \frac{\eta}{P}, \qquad \tilde{\tau}_J^{\mathrm{RTA}} = \frac{12mT\lambda}{7Q - 75P^2/n}, \tag{16.193}$$

where η and λ are evaluated based on our formulas (16.74).

Here, we test the reliability of the relations (16.189) and (16.190) derived under RTA, by comparing the viscous-relaxation times calculated from (16.90) with (16.193) provided η and λ via (16.74) at $p_F a_a = 0.1$. Detailed analyses at the unitarity, $p_F a_a = 0$, have been presented in [157]. Note that these formulas (16.193) yield different viscous-relaxation time in contrast to the standard RTA (16.192). This is indeed thought of as an improvement over the standard RTA. In Fig. 16.3 the viscous relaxation time from our formulas and those of the improved RTA (16.193) are shown. The resultant data of τ_π and τ_J clearly show distinct temperature dependences. The improved RTA (16.193) capture this behavior to some extent, although $\tilde{\tau}_J^{\mathrm{RTA}}$ does show the deviation from τ_J. From Fig. 16.4, we see that the temperature dependences of improved RTA $\tilde{\tau}_\pi^{\mathrm{RTA}}$ and $\tilde{\tau}_J^{\mathrm{RTA}}$ reproduce those of τ_π and τ_J qualitatively. While the agreement of $\tilde{\tau}_\pi$ is very good, $\tilde{\tau}_J$ is not well approximated by RTA.

References

1. N. Goldenfeld, O. Martin, Y. Oono , J. Sci. Comput. **4**, 4 (1989); N. Goldenfeld, O. Martin, Y. Oono, F. Liu, Phys. Rev. Lett. **64**, 1361 (1990); N.D. Goldenfeld, *Lectures on Phase Transitions and the Renormalization Group* (Addison-Wesley, Reading, MA, 1992), L.Y. Chen, N. Goldenfeld, Y. Oono, G. Paquette, Phys. A **204**, 111 (1994); G. Paquette, L.Y. Chen, N. Goldenfeld , Y. Oono, Phys. Rev. Lett. **72**, 76 (1994); L.Y. Chen, N. Goldenfeld, Y. Oono, Phys. Rev. Lett. **73**, 1311 (1994); L.Y. Chen, N. Goldenfeld, Y. Oono, Phys. Rev. E **54**, 376 (1996)
2. J. Veysey II, N. Goldenfeld, Rev. Mod. Phys. **79**(3), 883 (2007); Y. Oono, *The Nonlinear World: Conceptual Analysis and Phenomenology*, (Springer Science and Business Media, 2012)
3. T. Kunihiro, Prog. Theor. Phys. **94**(1995), 503, (E) ibid., **95**, 835 (1996)
4. T. Kunihiro, Jpn. J. Ind. Appl. Math. **14**, 51 (1997)
5. T. Kunihiro, Prog. Theor. Phys. **97**, 179 (1997)
6. S.-I. Ei, K. Fujii, T. Kunihiro, Ann. Phys. **280**, 236 (2000)
7. A. van Helden, *Galileo Italian Philosopher, Astronomer and Mathematician*, Britannica: https://www.britannica.com/biography/Galileo-Galilei
8. E.C.G. Stueckelberg, A. Petermann, Helv. Phys. Acta **26**, 499 (1953)
9. M. Gell-Mann, F.E. Low, Phys. Rev. **95**, 1300 (1953)
10. N.N. Bogolyubov, D.V. Shirkov, Introduction to the Theory of Quantized Fields (Wiley, 1957); Dokl. Akad. Nauk SSSR **103**, 203 (1955)
11. D.V. Shirkov, V.F. Kovalev, Phys. Rep. **352**, 219 (2001)
12. S. Weinberg, in *Asymptotic Realms of Physics*, ed. by A.H. Guth et al. (MIT Press, 1983)
13. K.G. Wilson, J. Kogut, Phys. Rep. **12C**, 75 (1974)
14. F. Wegner, A. Houghton, Phys. Rev. A **8**, 401 (1973)
15. J. Polchinski, Nucl. Phys. B **231**, 269 (1984); G. Keller, C. Kopper, M. Salmhofer, Helv. Phys. Acta **65**, 32 (1992)
16. C. Wetterich, Phys. Lett. B **301**, 90 (1993)
17. S.K. Ma, *Modern Theory of Critical Phenomena* (W. A. Benjamin, New York, 1976)
18. J. Zinn-Justin, *Quantum Field Theory and Critical Phenomena* (Clarendon Press, Oxford, 1989)
19. S. Weinberg, *The Quantum Theory of Fields II* (Cambridge U.P., 1996)
20. A.H. Nayfeh, *Perturbation Methods* (Wiley, 2008)

T. Kunihiro et al., *Geometrical Formulation of Renormalization-Group Method as an Asymptotic Analysis*, Fundamental Theories of Physics 206,
https://doi.org/10.1007/978-981-16-8189-9

21. C.M. Bender, S.A. Orszag, *Advanced Mathematical Methods for Scientists and Engineers I: Asymptotic Methods and Perturbation Theory* (Springer Science & Business Media, 2013)
22. A.H. Nayfeh, D.T. Mook, *Nonlinear Oscillations* (Wiley, 2008)
23. J. Bricmont, A. Kupiainen, Commun. Math. Phys. **150**, 193 (1992); J. Bricmont, A. Kupiainen, G. Lin, Cooun. Pure. Appl. Math. **47**, 893 (1994); J. Bricmont, A. Kupiainen, chao-dyn/9411015
24. See any text on mathematical analysis, for example, R. Courant, D. Hilbert, *Methods of Mathematical Physics*, vol. 2 (Interscience Publishers, New York, 1962)
25. N.N. Bogoliubov, Y.A. Mitropolski, *Asymptotic Methods in the Theory of Nonlinear Oscillations* (Gordon and Breach, 1961)
26. G.I. Barlenblatt, *Similarity, Self-Similarity, and Intermediate Asymptotics (Consultant Bureau, New York, 1979); Dimensional Analysis* (Gordon and Breach, New York, 1987)
27. J. Swift, P.C. Hohenberg, Phys. Rev. A **15**, 319 (1977)
28. Y. Kuramoto, Prog. Theor. Phys. **55**, 356 (1976)
29. G.I. Sivashinsky, Acta Astronautica **4**, 1177 (1977)
30. Y. Kuramoto, *Chemical Oscillations, Waves, and Turbulence* (Springer, 1984)
31. P. Manneville, *Dissipative Structures and Weak Turbulence* (Academic, INC, 1990)
32. Y. Kuramoto, Prog. Theor. Phys. Suppl. **99**, 244 (1989); Bussei Kenkyu **49**, 299 (1987) (in Japanese)
33. J. Carr, *Applications of Centre Manifold Theory* (Springer, 1981)
34. See for example, J. Guckenheimer, P. Holmes, *Nonlinear Oscillators, Dynamical Systems, and Bifurcations of Vector Fields* (Springer, 1983)
35. A.N. Gorban, I.V. Karlin, *Invariant Manifolds for Physical and Chemical Kinetics* (Springer Science and Business Media, 2005)
36. D.J. Benney, J. Math. Phys. **45**, 150 (1966)
37. R. Graham, Phys. Rev. Lett. **76**, 2185 (1996); see also K. Matsuba, K. Nozaki, Phys. Rev. E **56**, R4926 (1997)
38. S. Sasa, Physica D **108**, 45 (1997)
39. T. Maruo, K. Nozaki, A. Yoshimori, Prog. Theor. Phys. **101**, 243 (1999)
40. T. Kunihiro, J. Matsukidaira, Phys. Rev. E **57**, 4817 (1998)
41. H.J. de Vega, J.F.J. Salgado, Phys. Rev. D **56**, 6524 (1997)
42. A. Taruya, Y. Nambu, Phys. Lett. B **428**, 37 (1998); Y. Nambu, Y.Y. Yamaguchi, Phys. Rev. D, **60**, 104011 (1999)
43. D. Boyanovsky, H.J. de Vega, Phys. Rev. D **59**, 105019 (1999); D. Boyanovsky, H.J. de Vega, R. Holman, M. Simionato, Phys. Rev. D **60**, 65503 (1999); D. Boyanovsky, H.J. de Vega, Ann. Phys. **307**, 335 (2003)
44. M. Frasca, Phys. Rev. A **56**, 1549 (1997); see also I.L. Egusquiza, M.A. Valle Basagoiti
45. I.L. Egusquiza, M.A. Valle Basagoiti, Phys. Rev. A **57**, 1586 (1998)
46. T. Kunihiro, Phys. Rev. D **57**, R2035 (1998); Prog. Theor. Phys. Suppl. **131**, 459 (1998)
47. Yoshiyuki Y. Yamaguchi, Yasusada Nambu, Prog. Theor. Phys. **100**, 199 (1998)
48. R.L. DeVille, A. Harkin A, M. Holzer, K. Josić, T.J. Kaper, Phys. D: Nonlinear Phenom. D **237/8**, 1029 (2008)
49. H. Chiba, J. Math. Phys. **49**(10), 102703 (2008); **246**(5), 1991 (2009); **62**(6), 062703 (2021)
50. H. Chiba, SIAM, J. Appl. Dyn. Syst. **8**(3), 1066 (2009)
51. E. Kirkinis, SIAM Rev. **54**, 374 (2012)
52. M. Holzer, T.J. Kaper, Adv. Differ. Equ. **19**, 245 (2014)
53. L. Boltzmann, *Lectures on Gas Theory* (University of California Press, Berkeley, 1964)
54. N.N. Bogoliubov, in *Studies in Statistical Mechanics*, ed. by J. de Boer, G.E. Uhlenbeck, vol. 1 (North-Holland, 1960)
55. L.E. Reichl, *A Modern Course in Statistical Physics*, 2nd edn. (Wiley, New York, 1998)
56. O. Pashko, Y. Oono, Int. J. Mod. Phys. B **14**, 555 (2000)
57. Y. Hatta, T. Kunihiro, Ann. Phys. **298**, 24 (2002)
58. T. Kunihiro, K. Tsumura, J. Phys. A **39**, 8089 (2006)
59. L. Rezzolla, O. Zanotti, *Relativistic Hydrodynamics* (paperback, Oxford U.P., 2018)

60. P. Romatschke, U. Romatschke, *Relativistic Fluid Dynamics In and Out of Equilibrium And Applications to Relativistic Nuclear Collisions* (Cambridge U.P., 2019)
61. C. Eckart, Phys. Rev. **58**, 919 (1940)
62. L.D. Landau, E.M. Lifshitz, *Fluid Mechanics*, 2nd edn. (Elsevier, London, 1987)
63. W.A. Hiscock, L. Lindblom, Phys. Rev. D **31**, 725 (1985)
64. J.M. Stewart, *Non-equilibrium Relativistic Kinetic Theory*, Lecture Notes in Physics, vol. 10. (Springer, Berlin, 1971)
65. W. Israel, Ann. Phys. **100**, 310 (1976)
66. W. Israel, J.M. Stewart, Ann. Phys. **118**, 341 (1979)
67. S.R. de Groot, W.A. van Leeuwen, Ch.G. van Weert, *Relativistic Kinetic Theory* (Elsevier, North-Holland, 1980)
68. C. Cercignani, G.M. Kremer, *Relativistic Boltzmann Equation* (Springer, Berlin, 2002)
69. K. Tsumura, Y. Kikuchi, T. Kunihiro, Physica D**336**, 1 (2016); A preliminary description of the doublet scheme is presented in, K. Tsumura and T. Kunihiro, Eur. Phys. J. A **48**, 162 (2012)
70. P. Langevin, Comptes. Rendus Acad. Sci. (Paris) **146**, 530 (1908)
71. G.E. Uhlenbeck, L.S. Ornstein, Phys. Rev. **34**, 823 (1930)
72. S. Chandrasekhar, Rev. Mod. Phys. **15**, 1 (1943)
73. R.L. Stratonovich, *Topics in the Theory of Random Noise*, vols. 1 and 2 (Gordon and Breach, 1963)
74. C.W. Gardiner, *Handbook of Stochastic Methods for Physics, Chemistry and the Natural Sciences*, 2nd edn. (Springer, 1985)
75. H. Risken, *The Fokker-Planck Equation—Methods of Solution and Applications*, 2nd edn. (Springer, 1989)
76. N.G. van Kampen, *Stochastic Processes in Physics and Chemistry*, rev. and enlarged ed. (North-Holland, 1992)
77. A.D. Fokker, Ann. Phys. **43**, 810 (1914)
78. M. Planck, Sitzungsber. Preuss. Akad. Wiss. 324 (1917)
79. S.H. Strogatz, *Nonlinear Dynamics and Chaos: With Applications to Physics, Biology, Chemistry, and Engineering*, 2nd edn. (CRC Press, 2018), Chap. 7
80. R. Courant, D. Hilbert, *Methods of Mathematical Physics*, vol. 1, (Wiley-VCH, 1989), Chap. 3
81. T. Kato, *Perturbation Theory for Linear Operators*, 2nd edn. (Springer, 1976)
82. E. Zeidler, *Applied Functional Analysis: Main Principles and Their Applications* (Springer eBooks, 1995), ISBN 9781461208211
83. P. Glansdorff, I. Prigogine, *Thermodynamic Theory of Structure, Stability, and Fluctuations* (Wiley, London, 1971)
84. E.N. Lorenz, J. Atmos. Sci. **20**, 130 (1963)
85. B. Saltzman, J. Atmos. Sci. **19**, 329 (1962)
86. S.H. Strogatz, *Nonlinear Dynamics and Chaos: With Applications to Physics, Biology, Chemistry, and Engineering*, 2nd edn. (CRC Press, 2018), Chap. 9
87. K.G. Wilson, M.E. Fisher, Phys. ReV. Lett. **28**, 240 (1972); K.G. Wilson, Phys. Rev. Lett. **28**, 548 (1972), K.G. Wilson, J. Kogut, Phys. Rep. **12C**, 75 (1974); See also, as a comprehensive review with an emphasis on the conceptual aspects, M.E. Fisher, Rev. Mod. Phys. **70**, 653 (1998)
88. For instance, S. Pokorski, *Gauge Field Theories*, 2nd edn. (Cambridge University Press, 2000)
89. M. Suzuki, J. Phys. Soc. Jpn. **55**, 4205 (1986), M. Suzuki et al. *Coherent Anomaly Method, Mean Field, Fluctuations and Systematics* (World Scientific, 1995)
90. T. Kunihiro (2018), unpublished
91. L.D. Landau, E.M. Lifschic, Course of Theoretical Physics. vol. 1: *Mechanics* (Oxford, 1978)
92. R. Graham, Phys. Rev. Lett. **76**, 2185 (1996)
93. K. Matsuba, K. Nozaki, Phys. Rev. E **56**, R4926 (1997)
94. S. Goto, Y. Masutomi, K. Nozaki, Prog. Theor. Phys. **102**, 471 (1999)
95. M. Ziane, J. Math. Phys. **41**, 3290 (2000)

96. K. Nozaki, Y. Oono, Phys. Rev. E **63**, 046101 (2001)
97. A. Nayfeh, *Methods of Normal Forms* 2nd edn. (Wiley, 2010)
98. T. Taniuti, C.-C. Wei, J. Phys. Soc. Jpn. **24**, 941 (1968)
99. H. Haken, *Advanced Synergetics*, 2nd edn. (Springer, 1983)
100. H. Mori, Prog. Theor. Phys. **33**, 423 (1965)
101. F. Takens, Publ. Math. IHES **43**, 47 (1974)
102. N.J. Zabusky, M.D. Kruskal, Phys. Rev. Lett. **15**, 240 (1965)
103. E. Fermi, J. Pasta, S. Ulam, *Studies of Nonlinear Problems*, Document LA-1940. Los Alamos National Laboratory (1955), see also T. Dauxois, *Fermi, Pasta, Ulam, and a mysterious lady*. Physics Today. **6**(1), 55 (2008)
104. S.-I. Ei, T. Ohta, Phys. Rev. E **50**, 4672 (1994). ((and references cited therein))
105. T. Kunihiro (2000), unpublished
106. R. Brown, Phil. Mag. N.S., **4**, 161 (1828)
107. A. Einstein, Ann. Physik **17**, 549 (1905); **19**, 371 (1906)
108. M. von Smoluchowski, Ann. d. Physik, 4e série, t. **XXI**, 756 (1906)
109. M.J. Perrin, *Brownian Movement and Molecular Reality* (Taylor and Francis, London, 1910)
110. J.D. Crawford, Rev. Mod. Phys. **63**, 991 (1991)
111. J.L. Lebowitz, Physica A **194**, 1 (1993); *25 Years of Non-Equilibrium Statistical Mechanics, Proceedings, Sitges Conference, Barcelona, Spain, 1994*, in Lecture Notes in Physics, ed. by J.J. Brey, J. Marro, J.M. Rubí, M. San Miguel (Springer, 1995)
112. K. Kawasaki, *Non-equilibrium and Phase Transition—Statistical Physics in Meso Scales* (Asakura Shoten, 2000), Chap. 7 (in Japanese)
113. S. Chapman, T.G. Cowling, *The Mathematical Theory of Non-Uniform Gases*, 3rd edn. (Cambridge University Press, UK, 1970)
114. R. Kubo, J. Math. Phys. **4**, 174 (1963)
115. R. Kubo, M. Toda, N. Hashitsume, *Statistical Physics II Non-equilibrium Statistical Mechanics* (Springer, 1985)
116. P.A. Dirac, *The Principles of Quantum Mechanics* (Oxford University Press, 1981)
117. A. Messiah, *Quantum Mechanics*, vol. II (North-Holland, Amsterdam, 1965), Chap. XVI, sec.III
118. L.D. Landau, E. M. Lifshitz, *Quantum Mechanics: Non-relativistic Theory* (Elsevier, 2013)
119. O. Klein, Ark. Mat., Astro., och. Fys. **16**, 1 (1922); H.A. Kramers, Physica **7**, 284 (1940)
120. H.C. Brinkman, Physica **22**, 29 (1956)
121. G. Wilemski, J. Stat. Phys. **14**, 153 (1976)
122. U.M. Titulaer, Physica **91A**, 321 (1978); **100A**, 251 (1980)
123. U.M. Titulaer, Physica **100A**, 234 (1980)
124. K. Kaneko, Prog. Theor. Phys. **66**, 129 (1981)
125. N.G. van Kampen, Phys. Rep. **124**, 69 (1985)
126. T. Kato, Prog. Theor. Phys. **IV**, 514 (1949)
127. C. Bloch, Nucl. Phys. **6**, 329 (1958)
128. M. von Smoluchowski, Ann. Phys. **48**, 1103 (1915)
129. M. Matsuo, S. Sasa, Physica A **276**, 188 (2000)
130. R.P. Feynman, R.B. Leighton, M. Sands, *The Feynman Lectures on Physics*, vol. I (Addison-Wesley, Reading, 1963)
131. K. Sekimoto, J. Phys. Soc. Jpn. **66**, 1234 (1997)
132. K. Sekimoto, *Stochastic Energetics*, vol. 799 (Springer, 2010)
133. For instance, S. Weinberg, *The Quantum Theory of Fields*, vol. I (Cambridge University Press, 1995)
134. H. Mori, J. Phys. Soc. Jpn. **11**, 1029 (1956); Phys. Rev. **112**, 1829 (1958), **115**, 298 (1959)
135. J.G. Kirkwood, J. Chem. Phys. **14**, 180 (1946)
136. I. Ojima, J. Stat. Phys. **56**, 203 (1989)
137. T. Kunihiro (2009), unpublished
138. D. Zubarev, V. Morozov, G. Roepke, *Statistical Mechanics of Nonequilibrium Processes*, vols. 1 and 2 (Akademie Verlag GmbH, Berlin, 1996, 1997)

139. P. Resibois, J. Stat. Phys. **2**, 21 (1970); R. Balescu, *Equilibrium and Nonequilibrium Statistical Mechanics* (Wiley, New York, 1975)
140. B.J. Berne (ed.), *Statistical Mechanics: Part B: Time-Dependent Processes* (Springer, 1977)
141. E.M. Lifshitz, L.P. Pitaevskii, *Physical Kinetics* (Butterworth-Heinemann, 1981)
142. R. Kubo, J. Phys. Soc. **12**, 570 (1957)
143. I.V. Karlin, A.N. Gorban, G. Dukek, T.F. Nonnenmacher, Phys. Rev. E **57**, 1668 (1998)
144. H. Grad, Commun. Pure Appl. Math. **2**, 331 (1949)
145. C. Cattaneo, C.R. Acad, Sci. Paris Ser. A-B. **247**, 431 (1958)
146. I. Müller, Z. Physik **198**, 329 (1967)
147. I. Müller, T. Ruggeri, *Extended Thermodynamics* (Springer, Berlin, 1993)
148. D. Jou, J. Casas-Vazquez, G. Lebon, *Extended Irreversible Thermodynamics* (Springer, 1993)
149. C.D. Levermore, J. Stat. Phys. **83**(5–6), 1021 (1996)
150. T. Dedeurwaerdere, J. Casas-Vazquez, D. Jou, G. Lebon, Phys. Rev. E **53**, 498 (1996)
151. H.C. Öttinger, Phys. Rev. Lett. **104**, 120601 (2010)
152. H. Struchtrup, M. Torrilhon, Phys. Fluids **15**(9), 2668 (2003)
153. M. Torrilhon, Special issues on moment methods in kinetic gas theory. Contin. Mech. Thermodyn. **21**, 341 (2009). ((and references therein))
154. M. Torrilhon, Commun. Comput. Phys. **7**(4), 639 (2010)
155. K. Tsumura, Y. Kikuchi, T. Kunihiro, Phys. Rev. D **92**(8), 085048 (2015)
156. Y. Kikuchi, K. Tsumura, T. Kunihiro, Phys. Rev. C **92**(6), 064909 (2015)
157. Y. Kikuchi, K. Tsumura, T. Kunihiro, Phys. Lett. A **380**, 2075 (2016)
158. R. Zwanzig, J. Chem. Phys. **33**, 1338 (1960)
159. R. Zwanzig, J. Stat. Phys. **9**, 215 (1973); *Nonequilibrium Statistical Mechanics* (Oxford U. Press, 2001)
160. M. Gyulassy, L. McLerran, Nucl. Phys. A **750**, 30 (2005)
161. E.V. Shuryak, Prog. Part. Nucl. Phys. **53**, 273 (2004)
162. T. Hirano, P. Huovinen, K. Murase, Y. Nara, Prog. Part. Nucl. Phys. **70**, 108 (2013)
163. H. Fujii, M. Ohtani, Phys. Rev. D **70**, 014016 (2004)
164. D.T. Son, M.A. Stephanov, Phys. Rev. D **70**, 056001 (2004)
165. Y. Minami, T. Kunihiro, Prog. Theor. Phys. **122**, 881 (2010)
166. M. Asakawa, K. Yazaki, Nucl. Phys. A **504**, 668 (1989)
167. A. Barducci, R. Casalbuoni, S. De Curtis, R. Gatto, G. Pettini, Phys. Lett. B **231**, 463 (1989)
168. Z. Zhang, T. Kunihiro, Phys. Rev. D **83**, 114003 (2011)
169. As a review article, see, for example, D. Balsara, Astrophys. J. Suppl. **132**, 83 (2001)
170. J.C. Fabris, S.V.B. Goncalves, R. de Sa Ribeiro, Gen. Rel. Grav. **38**, 495 (2006)
171. R. Colistete Jr., J.C. Fabris, J. Tossa, W. Zimdahl, Phys. Rev. D **76**, 103516 (2007)
172. N.G. van Kampen, J. Stat. Phys. **46**, 709 (1987)
173. T. Osada, Phys. Rev. C **85**, 014906 (2012)
174. P. Van, T.S. Biro, Phys. Lett. B **709**, 106 (2012)
175. K. Tsumura, T. Kunihiro, Phys. Lett. B **690**, 255 (2010)
176. K. Tsumura, T. Kunihiro, Phys. Lett. B **668**, 425 (2008)
177. A. Muronga, Phys. Rev. Lett. **88**, 062302 (2002)
178. A. Muronga, Phys. Rev. C **69**, 034903 (2004)
179. A. Muronga, D.H. Rischke, arXiv:nucl-th/0407114
180. A. Muronga, Phys. Rev. C **76**, 014910 (2007)
181. A.K. Chaudhuri, U.W. Heinz, arXiv:nucl-th/0504022
182. A.K. Chaudhuri, arXiv:nucl-th/0703027
183. A.K. Chaudhuri, arXiv:nucl-th/0703029
184. A.K. Chaudhuri, arXiv:0704.0134
185. U.W. Heinz, H. Song, A.K. Chaudhuri, Phys. Rev. C **73**, 034904 (2006)
186. R. Baier, P. Romatschke, U.A. Wiedemann, Nucl. Phys. A **782**, 313 (2007)
187. P. Romatschke, U. Romatschke, Phys. Rev. Lett. **99**, 172301 (2007)
188. R. Baier, P. Romatschke, U.A. Wiedemann, Phys. Rev. C **73**, 064903 (2006)
189. R. Baier, P. Romatschke, Eur. Phys. J. C **51**, 677 (2007)

190. P. Romatschke, Eur. Phys. J. C **52**, 203 (2007)
191. P. Van, T.S. Biro, Eur. Phys. J. ST **155**, 201 (2008)
192. M. Natsuume, T. Okamura, Phys. Rev. D **77**, 066014 (2008)
193. P. Huovinen, D. Molnar, Phys. Rev. C **79**, 014906 (2009); Nucl. Phys. A **830**, 475C (2009)
194. A. El., Z. Xu, C. Greiner, Phys. Rev. C **81**, 041901 (2010)
195. I. Bouras, E. Molnar, H. Niemi, Z. Xu, A. El, O. Fochler, C. Greiner, R.H. Rischke, Phys. Rev. C **82**, 024910 (2010)
196. G.S. Denicol, X.-G. Huang, T. Koide, D.H. Rischke, Phys. Lett. B **708**, 174 (2012)
197. G.S. Denicol, T. Koide, D.H. Rischke, Phys. Rev. Lett. **105**, 162501 (2010)
198. G.S. Denicol, H. Niemi, E. Molnar, D.H. Rischke, Phys. Rev. D **85**, 114047 (2012)
199. M. Dudynski, M.L. Ekiel-Jezewska, Phys. Rev. Lett. **55**, 2831 (1985)
200. R.M. Strain, Commun. Math. Phys. **300**, 529 (2010)
201. F. Jüttner, Zeitschr. Physik **47**, 542 (1911)
202. F. Jüttner, Ann. Physik und Chemie **339**, 856 (1911)
203. T. Kunihiro, Buturi **65**, 683 (2010). ([in Japanese])
204. K. Tsumura, T. Kunihiro, K. Ohnishi, Phys. Lett. B **646**, 134 (2007)
205. K. Tsumura, T. Kunihiro, Prog. Theor. Phys. **126**, 761 (2011)
206. K. Tsumura, T. Kunihiro, Prog. Theor. Phys. Suppl. **195**, 19 (2012)
207. M.S. Green, J. Chem. Phys. **20**, 1281 (1952); ibid, **22**, 398 (1954)
208. H. Nakano, Prog. Theor. Phys. **15**, 77 (1956)
209. See, for example, D.N. Zubarev, *Nonequilibrium Statistical Thermodynamics* (Plenum Press, N. Y. 1974)
210. K. Tsumura, T. Kunihiro, Phys. Rev. E **87**, 053008 (2013)
211. Y. Minami, Y. Hidaka, Phys. Rev. E **87**, 023007 (2013)
212. R. Zwangzig, *Nonequilibrium Statistical Mechanics*, (Oxford U.P., 2001)
213. T. Hayata, Y. Hidaka, T. Noumi, M. Hongo, Phys. Rev. D **92**(6), 065008 (2015)
214. D.N. Zubarev, A.V. Prozorkevich, S.A. Smolyanskii, Theor. Math. Phys. **40**, 821 (1979)
215. S.-I. Sasa, Phys. Rev. Lett. **112**, 100602 (2014)
216. F. Becattini, L. Bucciantini, E. Grossi, L. Tinti, Eur. Phys. J. C **75**, 191 (2015)
217. K. Tsumura, T. Kunihiro, Eur. Phys. J. A **48**, 162 (2012)
218. W.H. Press, H. William, S.A. Teukolsky, W.T. Vetterling, A. Saul, B.P. Flannery, *Numerical Recipes 3rd Edition: The Art of Scientific Computing*, (Cambridge University Press, 2007)
219. W.A. Hiscock, L. Lindblom, Ann. Phys. **151**, 466 (1983)
220. A. Sandoval-Villalbazo, A.L. Garcia-Perciante, L.S. Garcia-Colin, Physica A **388**, 3765 (2009)
221. L. Stricker, H.C. Öttinger, Phys. Rev. E **99**(1), 013105 (2019)
222. P. Kovtun, JHEP **10**, 034 (2019)
223. R.E. Hoult, P. Kovtun, JHEP **06**, 067 (2020)
224. L. Gavassino, M. Antonelli, Front. Astron. Space Sci. **8**, 686344 (2021)
225. C.V. Brito, G.S. Denicol, arXiv:2107.10319 [nucl-th]
226. S. Jeon, Phys. Rev. D **52**, 3591 (1995)
227. S. Jeon, L.G. Yaffe, Phys. Rev. D **53**, 5799 (1996)
228. Y. Hidaka, T. Kunihiro, Phys. Rev. D **83**, 076004 (2011)
229. H. Takahasi, M. Mori, Publ. RIMS **9**, 721 (1974)
230. S. Weinberg, Physica A **96**, 327 (1979)
231. S. Scherer, Adv. Nucl. Phys. **27**, 277 (2003)
232. S. Bass, A. Dumitru, Phys. Rev. C **61**, 064909 (2000)
233. D. Teaney, J. Lauret, E.V. Shuryak, Phys. Rev. Lett. **86**, 4783 (2001)
234. D. Teaney, Phys. Rev. C **68**, 034913 (2003)
235. T. Hirano, M. Gyulassy, Nucl. Phys. A **769**, 71 (2006)
236. T. Hirano, U.W. Heinz, D. Kharzeev, R. Lacey, Y. Nara, Phys. Lett. B **636**, 299 (2006)
237. C. Nonaka, S.A. Bass, Phys. Rev. C **75**, 014902 (2007)
238. P. Bozek, Acta Phys. Polon. B **43**, 689 (2012)
239. J.L. Anderson, General Relativ. Gravit. **7**(1), 53 (1976)

240. M. Prakash, M. Prakash, R. Venugopalan, G. Welke, Phys. Rep. **227**, 321 (1993)
241. A. Monnai, T. Hirano, Nucl. Phys. A **847**, 283 (2010)
242. A. El, A. Muronga, Z. Xu, C. Greiner, Nucl. Phys. A **848**, 428 (2010)
243. A. El, I. Bouras, C. Wesp, Z. Xu, C. Greiner, Eur. Phys. J. A **48**, 166 (2012)
244. V. Moratto, G.M. Kremer, Phys. Rev. E **91**(5), 052139 (2015)
245. W. Florkowski, E. Maksymiuk, R. Ryblewski, L. Tinti, Phys. Rev. C **92**(5), 054912 (2015)
246. L. Onsager, I. Phys. Rev. **37**(4), 405 (1931)
247. K.M. O'hara, S.L. Hemmer, M.E. Gehm, S. Granade, J.E. Thomas, Science **298**(5601), 2179 (2002)
248. J. Kinast, A. Turlapov, J.E. Thomas, Phys. Rev. A **70**, 051401 (2004)
249. M. Bartenstein, A. Altmeyer, S. Riedl, S. Jochim, C. Chin, J.H. Denschlag, R. Grimm, Phys. Rev. Lett. **92**, 203201 (2004)
250. T. Schäfer, Phys. Rev. A **76**, 063618 (2007)
251. C. Cao, E. Elliott, J. Joseph, H. Wu, J. Petricka et al., Science **331**, 58 (2011)
252. E. Elliott, J.A. Joseph, J.E. Thomas, Phys. Rev. Lett. **113**, 020406 (2014)
253. T. Schäfer, D. Teaney, Rept. Prog. Phys. **72**, 126001 (2009)
254. A. Adams, L.D. Carr, T. Schäfer, P. Steinberg, J.E. Thomas, New J. Phys. **14**, 115009 (2012)
255. G. Policastro, D.T. Son, A.O. Starinets, Phys. Rev. Lett. **87**, 081601 (2001)
256. P. Kovtun, D.T. Son, A.O. Starinets, Phys. Rev. Lett. **94**, 111601 (2005)
257. B.A. Gelman, E.V. Shuryak, I. Zahed, Phys. Rev. A **72**, 043601 (2005)
258. G.M. Bruun, H. Smith, Phys. Rev. A **75**, 043612 (2007)
259. G. Rupak, T. Schäfer, Phys. Rev. A **76**, 053607 (2007)
260. T. Enss, R. Haussmann, W. Zwerger, Ann. Phys. **326**(3), 770 (2011)
261. H. Guo, D. Wulin, C.-C. Chien, K. Levin, New J. Phys. **13**(7), 075011 (2011)
262. T. Enss, Phys. Rev. A **86**, 013616 (2012)
263. P. Massignan, G.M. Bruun, H. Smith, Phys. Rev. A **71**, 033607 (2005)
264. T. Schäfer, Phys. Rev. A **85**, 033623 (2012)
265. K. Dusling, T. Schäfer, Phys. Rev. Lett. **111**(12), 120603 (2013)
266. C. Chafin, T. Schäfer, Phys. Rev. A **88**, 043636 (2013)
267. G.M. Bruun, H. Smith, Phys. Rev. A **72**, 043605 (2005)
268. G.M. Bruun, H. Smith, Phys. Rev. A **76**, 045602 (2007)
269. M. Braby, J. Chao, T. Schäfer, New J. Phys. **13**(3), 035014 (2011)
270. J. Chao, T. Schäfer, Ann. Phys. **327**(7), 1852–1867 (2012). (Special Issue)
271. Y. Nishida, Ann. Phys. **410**, 167949 (2019)
272. B. Frank, W. Zwerger, T. Enss, Phys. Rev. Res. **2**, 023301 (2020)

Index

T. Kunihiro et al., *Geometrical Formulation of Renormalization-Group Method as an Asymptotic Analysis*, Fundamental Theories of Physics 206,
https://doi.org/10.1007/978-981-16-8189-9

Printed by Books on Demand, Germany